THE ROUTLEDGE INTERNATIONAL HANDBOOK OF ENGINEERING ETHICS EDUCATION

Responding to the need for a timely and authoritative volume dedicated to this burgeoning and expansive area of research, this handbook will provide readers with a map of themes, topics, and arguments in the field of engineering ethics education (EEE).

Featuring critical discussion, research collaboration, and a team of international contributors of globally recognized standing, this volume comprises six key sections which elaborate on the foundations of EEE, teaching methods, accreditation and assessment, and interdisciplinary contributions. Over 100 researchers of EEE from around the globe consider the field from the perspectives of teaching, research, philosophy, and administration. The chapters cover fast-moving topics central to our current understanding of the world such as the general data protection regulation (GDPR), artificial intelligence (AI), biotechnology, and ChatGPT; and they offer new insights into best practices research to equip program leaders and instructors delivering ethics content to students.

This Open Access volume will be of interest to researchers, scholars, postgraduate students, and faculty involved with engineering education, engineering ethics, and philosophy of education. Curriculum designers, staff developers teaching pedagogical courses to faculty, and engineering professionals may also benefit from this volume.

Shannon Chance is Lecturer and Program Chair at Technological University Dublin, Ireland; Honorary Professor at University College London, UK; Deputy Editor of the *European Journal of Engineering Education* (EJEE); Registered Architect; and former Professor of Architecture at Hampton University, USA.

Tom Børsen is Associate Professor in Techno-Anthropology, Department of Sustainability and Planning, Aalborg University, Denmark, and a member of the research group in Techno-Anthropology and Participation (TAPAR) as well as of the board of the think tank Techno-Ethics.

Diana Adela Martin is a Senior Researcher with the Centre for Engineering Education, University College London, UK, holding a doctorate in engineering ethics education from Technological University Dublin (TU Dublin) in Ireland.

Roland Tormey is a sociologist and a Senior Scientist in EPFL (Swiss Federal Institute of Technology), Switzerland, where he leads the Teaching Support Centre.

Thomas Taro Lennerfors is Professor and Head of the Division of Industrial Engineering and Management, Uppsala University, Sweden.

Gunter Bombaerts is Assistant Professor in Philosophy and Ethics, Industrial Engineering and Innovation Sciences, Eindhoven University of Technology, the Netherlands.

Routledge International Handbooks of Education

The Routledge International Handbook of Student-centred Learning and Teaching in Higher Education
Edited by Sabine Hoidn and Manja Klemenčič

The Routledge Handbook of English Language Education in Bangladesh
Edited by Shaila Sultana, M. Moninoor Roshid, Md. Zulfeqar Haider, Mia Md. Naushaad Kabir and Mahmud Hasan khan

The Routledge Handbook of Language Learning and Teaching Beyond the Classroom
Edited by Hayo Reinders, Chun Lai and Pia Sundqvist

The Routledge Handbook of Dyslexia in Education
Edited by Gad Elbeheri and Siang Lee

The Routledge Handbook of International Social Work
Edited by Stephen Webb

The Routledge International Handbook of Critical Autism Studies
Edited by Damian Milton and Sara Ryan

The Routledge International Handbook of Work-Integrated Learning
Edited by Karsten E. Zegwaard and T. Judene Pretti

The Routledge International Handbook of Gender Beliefs, Stereotype Threat, and Teacher Expectations
Edited by Penelope W. St J. Watson, Christine M. Rubie-Davies and Bernhard Ertl

The Routledge International Handbook of Equity and Inclusion in Education
Edited by Paul Downes, Guofang Li, Lore van Praag and Stephen Lamb

The Routledge International Handbook of Life and Values Education in Asia
Edited by John Chi-Kin Lee and Kerry J. Kennedy

The Routledge International Handbook of Engineering Ethics Education
Edited by Shannon Chance, Tom Børsen, Diana Adela Martin, Roland Tormey, Thomas Taro Lennerfors and Gunter Bombaerts

For more information about this series, please visit: https://www.routledge.com/Routledge-International-Handbooks-of-Education/book-series/HBKSOFED

THE ROUTLEDGE INTERNATIONAL HANDBOOK OF ENGINEERING ETHICS EDUCATION

Edited by Shannon Chance and Tom Børsen

Diana Adela Martin, Roland Tormey, Thomas Taro Lennerfors and Gunter Bombaerts

LONDON AND NEW YORK

Designed cover image: Westend61 / Getty Images

First published 2025
by Routledge
4 Park Square, Milton Park, Abingdon, Oxon OX14 4RN

and by Routledge
605 Third Avenue, New York, NY 10158

Routledge is an imprint of the Taylor & Francis Group, an informa business

British Library Cataloguing-in-Publication Data
A catalogue record for this book is available from the British Library

Library of Congress Cataloging-in-Publication Data
Names: Chance, Shannon, editor.
Title: The Routledge international handbook of engineering ethics education / edited by Shannon Chance & Tom Børson, Diana Martin, Roland Tormey, Thomas Taro Lennerfors, and Gunter Bombaerts.
Description: Abingdon, Oxon; New York, NY: Routledge, 2025. | Series: Routledge international handbooks of education | Includes bibliographical references and index. | Identifiers: LCCN 2024029306 (print) | LCCN 2024029307 (ebook) | ISBN 9781032678528 (hardback) | ISBN 9781032734491 (paperback) | ISBN 9781003464259 (ebook)
Subjects: LCSH: Engineering ethics–Study and teaching.
Classification: LCC TA157 .R67 2025 (print) | LCC TA157 (ebook) | DDC 620.0071–dc23/eng/20241016
LC record available at https://lccn.loc.gov/2024029306
LC ebook record available at https://lccn.loc.gov/2024029307

ISBN: 978-1-032-67852-8 (hbk)
ISBN: 978-1-032-73449-1 (pbk)
ISBN: 978-1-003-46425-9 (ebk)

DOI: 10.4324/9781003464259

Typeset in Times New Roman
by Deanta Global Publishing Services, Chennai, India

Our engineering ethics education community dedicates this handbook to the memory of our esteemed colleague, Kenichi Natsume, Professor of Science and Technology Studies at the Kanazawa Institute of Technology. A leading voice in engineering ethics in Japan and an influential contributor globally, Kenichi enriched our field with profound insights into the ethical dimensions of science, technology, and engineering. His work, including his 2021 book, *Japan's Engineering Ethics and Western Culture*, reflects his unique perspective on how societal, political, and ideological forces shape ethical frameworks. His kindness, dedication, and wisdom will continue to inspire our community.

As we carry Kenichi's spirit forward, we also dedicate this handbook to all engineering educators committed to integrating ethics into their teaching, to students and engineers striving to act ethically, to those who pioneered engineering ethics education, and to the researchers advancing this crucial field. The editorial team extends heartfelt thanks to all contributors to this ambitious project for embracing our highly collaborative approach and to our families for their love and support, including Kenichi's wife, Misaki, and son, Yoshiaki.

Shannon Chance and Tom Børsen, the co-lead editors, hope this volume (Chance, Børsen, et al., 2025) will prove helpful to our growing global community.

CONTENTS

EDITORS AND CONTRIBUTORS

Diana Bairaktarova, Ph.D., is Associate Professor in the Department of Engineering Education and an affiliate faculty member in the Department of Mechanical Engineering and the Human-centered Design Program at Virginia Tech. She has received the W. S. 'Pete' White Innovation in Engineering Education and XCaliber Awards for exceptional contributions to technology-enriched learning.

Balamuralithara Balakrishnan, Ph.D., is Associate Professor at Sultan Idris Education University, Malaysia. Bala has received several international and national awards, fellowships, and grants in the area of educational creative studies, which amalgamate STEM education, ethics and sustainability, and creativity in education.

Jonathan Beever, Ph.D., is Associate Professor in the Department of Philosophy at the University of Central Florida and director of the university's Center for Ethics. His interdisciplinary research engages intersections of ethics, technology, and environments. You can learn more about his work at https://jonathan.beever.org.

Renato Bezerra Rodrigues was born and raised in Brazil and is now a settler in Winnipeg, Canada. He is a Ph.D. candidate in engineering education at the University of Manitoba with a background in engineering and philosophy. His general interests are in Science and Technology Studies (STS), Emancipatory Education, and Critical Pedagogy.

Angela R. Bielefeldt, Ph.D., P.E., is Professor of Civil, Environmental and Architectural Engineering at the University of Colorado Boulder (USA), where she directs the Integrated Design Engineering program and the doctoral degree program in Engineering Education. Her research interests include community engagement, social responsibility, and education for sustainability.

Gunter Bombaerts, Ph.D., is Assistant Professor at Eindhoven University of Technology in the Netherlands. His research focuses on system philosophy, attention, and ethics (energy and attention economy). He is an experienced Challenge-Based Learning teacher and coordinator of a 20 European Credit Transfer and Accumulation System (ECTS) program of non-technical courses. He also studies engineering ethics education to improve the program.

Tijn Borghuis, Ph.D., is a teacher and project manager in the Philosophy and Ethics group of Eindhoven University of Technology. In his educational research, Tijn focuses on competence-based curriculum analysis and design, as well as methods and materials for engineering ethics education.

Tom Børsen, Ph.D., is Associate Professor in Techno-Anthropology in the Department of Sustainability and Planning at Aalborg University, Denmark. He teaches students ethics and technology assessment at the Technical Faculty of IT and Design. Tom's research themes include participatory and ethical technology assessment, robust technologies through action research, and teaching responsibility in science, technology, engineering, and mathematics (STEM).

Christine Boshuijzen-van Burken, Ph.D., is Senior Research Associate at the University of New South Wales, Australia. She holds a doctorate in the ethics of technology from Eindhoven University of Technology in the Netherlands. Christine's current research focuses on the ethics of autonomous military systems and artifical intelligence. Her research interests include the ethics of technology, military ethics, refugee and humanitarian logistics, and reformational philosophy.

Corin L. Bowen, Ph.D., is Assistant Professor of Engineering Education at California State University in Los Angeles (USA). Her research focuses on structural and cultural oppression in engineering and engineering educational systems and organizing for equitable change. She teaches structural mechanics and sociotechnical topics in engineering education and practice.

Robert Braun, Ph.D., is Senior Researcher at the Institute for Advanced Studies in Vienna, Austria and an associate professor at Masaryk University in Brno, Czechia. His research focus is the ontological politics of technology transitions and Anthropocene violence. His last book, *Post-Automobility Futures* (with Richard Randell), was published in 2022.

Andrew O. Brightman, Ph.D., is Professor of Engineering Practice in the Weldon School of Biomedical Engineering at Purdue University, Indiana (USA). His research and teaching focus on ethical engineering of emerging medical technologies. He continues to be curious about how people involved in engineering think about ethics and about what influences how they act in their engineering practice.

Jeff R. Brown, Ph.D., is Professor of Civil Engineering at Embry-Riddle Aeronautical University in Daytona Beach, Florida (USA). His research interests include ethics in undergraduate engineering education and structural health monitoring of reinforced concrete structures.

Silvia Bruzzone, Ph.D., is Associate Professor at Mälardalen University in Västerås, Sweden. Her work is grounded in STS, urban studies, and posthumanist practice theory. Her interests involve sustainable transition and cross-disciplinary research work from a sociomaterial perspective.

Shannon Chance, Ph.D., is Lecturer and Program Chair at Technological University Dublin in Ireland, Honorary Professor at University College London (UCL) in the United Kingdom, Deputy Editor of the *European Journal of Engineering Education* (EJEE), and former Full Professor of Architecture at Hampton University in Virginia (USA). She is a Registered Architect who has chaired the global Research in Engineering Education Network (REEN) and guest-edited special issues on student development, diversity, and ethics. She hosts www.IrelandByChance.com.

Rockwell Clancy, Ph.D., conducts research at the intersection of technology ethics, moral psychology, and Chinese philosophy. He is Program Manager and Research Scientist in the Department of Engineering Education at Virginia Tech in Blacksburg, Virginia (USA). Rockwell holds a Ph.D. from Purdue University, an M.A. from Katholieke Universiteit in Leuven, Belgium, and a B.A. from Fordham University in New York, NY.

Cristiano Cruz, Ph.D., is a Brazilian researcher with a background in engineering and philosophy who investigates emancipating engineering interventions aiming to help decolonize (a) engineering practice and education and (b) philosophical reflection on engineering and technology. He is Assistant Professor at the Federal University of São João del-Rei, Brazil, and a visiting researcher at the State University of Campinas (Unicamp) and Aeronautics Technological Institute (ITA), Brazil.

Scott Daniel, Ph.D., is Senior Lecturer in Humanitarian Engineering at the University of Technology Sydney, Australia. He conducts qualitative research in engineering education and practice, particularly in humanitarian engineering contexts. Scott co-leads the Centre for Research into Engineering and Information Technology Education. He is Deputy Editor of the *Australasian Journal of Engineering Education.*

Inês Direito, Ph.D., is Assistant Researcher at the University of Aveiro in Portugal, and an Honorary Senior Research Fellow at the University College of London (UCL) Centre for Engineering Education (CEE) in the United Kingdom and the University of Johannesburg, South Africa. Her research interests include empathy and emotions in engineering, diversity and inclusion, skills development, and career pathways of engineering students.

Elise Dykhuis, Ph.D., researches individual pathways for character development among college students. Her work integrates the concept of character virtues with developmental theory and meta-theory. She has spent much of her professional career examining the aspects of person-centric, longitudinal statistical models of character in higher education. She received a doctorate in Child Study and Human Development from Tufts University in Massachusetts (USA) and a B.A. in Psychology and German from the University of Notre Dame in Indiana (USA).

Eva Eriksson, Ph.D., is Associate Professor at the Department of Digital Design and Information Studies at Aarhus University in Denmark. She holds a doctorate in Interaction Design from Chalmers University of Technology in Gothenburg, Sweden. Her research focuses on participatory design with children and interaction design in public knowledge institutions.

Chunping Fan, Ph.D., is Full Professor, interdisciplinary researcher, and practitioner at the Faculty of Humanities and Social Sciences, Beijing University of Technology in China. Her fields include engineering philosophy, engineering education, and ethics of science, technology, and education (STE). These experiences have allowed her to consider the technological progress and future destiny of humankind from a broader perspective.

Andrea R. Gammon, Ph.D., is Assistant Professor of Ethics and Philosophy at the Delft University of Technology (TU Delft) in the Netherlands. She researches environmental philosophy and engineering ethics education, develops curricula and approaches for ethics teaching in STEM fields, and co-ordinates ethics teaching at TU Delft.

Jie Gao is a Ph.D. candidate in learning sciences at EPFL (Swiss Federal Institute of Technology) and the Federal Institute of Technology (ETH) in Zurich, Switzerland. Her thesis examines inquiry-based learning in sustainability and ethics education. She earned a master's in philosophy from Sorbonne University and another master's in instructional design. In addition to her research, she actively contributes to humanities teaching at EPFL.

Julianna Gesun, Ph.D., is currently working at Embry-Riddle Aeronautical University in Daytona Beach, Florida (USA) and building global partnerships to improve the engineering education system at scale. Her research explores how and why some people thrive while others barely survive in the current engineering education ecosystem. Her more recent updates and adventures are documented at www.juliannagesun.com.

Silvia Gherardi is Senior Professor of Sociology of Organization at the University of Trento in Italy and a visiting professor at Mälardalen University in Sweden. Her research involves feminist new materialism, entrepreneurship, the epistemology of practice, and post-qualitative methodologies.

David Gillette, Ph.D., is Co-Director of the Liberal Arts and Engineering Studies program at California Polytechnic State University (Cal Poly) in San Luis Obispo, California (USA). David's recent research and project work includes science, technology, engineering, arts, and technology (STEAM)-based educational outreach to underserved local communities, media-driven technical documentation, and system design and testing for various forms of interactive, place-based storytelling.

Ester Gimenez-Carbo, Ph.D., is Associate Professor at the Universitat Politècnica de València in Spain – a teacher and researcher with a background in civil engineering and philosophy. One of Ester's interests is ensuring that engineers become excellent technicians and committed citizens who are aware of their environment during their training.

Alison Gwynne-Evans is Senior Lecturer at the University of Cape Town, South Africa, where she teaches professional communication and ethics in the Faculty of Engineering and the Built Environment. Alison's transdisciplinary research includes a strong practical focus on teaching and learning ethics within engineering. In her professional capacity, she served as Chairperson of the Ethics Committee of the South African Institute of Civil Engineers (SAICE).

Cécile Hardebolle, Ph.D., is a teaching advisor and learning scientist at the Center for Digital Education of EPFL (Swiss Federal Institute of Technology). Building on her computer science background, Cécile accompanies teachers in the introduction of digital-specific ethical skills into engineering courses. Her expertise in practice-oriented, experiential approaches to learning is recognized through numerous publications.

Isis Hazewindus is an ethicist and educational advisor with a Master of Arts in Philosophy and an additional Master of Arts for the Teacher Training Program in Philosophy. She is currently part of the team at Filosofie in actie (www.filosofieinactie.nl/english), where she specializes in guiding ethical conversations concerning issues on data and technology and designing and facilitating workshops and training programs for the public sector. In addition, she is an ethicist in three ethical committees and a visiting lecturer in philosophy and ethics at universities of applied sciences.

Amir Hedayati-Mehdiabadi, Ph.D., is Assistant Professor of Organization, Information and Learning Sciences at the University of New Mexico (USA). His research on talent and professional development seeks to answer how we can enhance ethical decision-making among professionals, including engineers and computer scientists, by understanding their ethical judgment processes.

Mihály Héder, Ph.D., is a philosopher and computer scientist interested in engineering design, epistemology, and ethics, especially regarding AI and other kinds of software. He teaches philosophy at Budapest University of Technology and Economics in Hungary.

Marion Hersh, Ph.D., is Senior Lecturer/Associate Professor at the University of Glasgow, Scotland. Their research covers blind people's spatial representations and (co-production) technology design to overcome barriers experienced by disabled and older people, particularly blind and autistic ones; collective and individual responsibility; and the social and environmental aspects impacts of engineering.

Christian Herzog, Ph.D., is Professor of Ethical, Legal, and Social Aspects of Artificial Intelligence at the University of Lübeck, Germany. He is a transdisciplinary researcher with expertise in algorithmic systems and applied ethics. His current focus is the ethics and epistemology of AI. He is also dedicated to advancing engineering ethics education.

Rafaela Hillerbrand, Ph.D., is Professor of Ethics of Technology and Philosophy of Science at the Institute for Technology Assessment and System Analysis (ITAS) at Karlsruhe Institute of Technology (KIT) in Germany. Rafaela leads the research group Philosophy of Engineering, Technology Assessment, and Science (PhilETAS). She is (founding) director of the KIT Academy for Responsible Research, Teaching, and Innovation (ARRTI).

Sarah Jayne Hitt, Ph.D., specializes in the integration of the arts, humanities, and social sciences within engineering education, especially focusing on ethics, sustainability, and communication. In addition to her roles at the New Model Institute for Technology and Engineering (NMITE) in Hereford and Edinburgh Napier University (both in the United Kingdom), she serves as project manager for the Ethics and Sustainability Toolkits hosted by the United Kingdom's Engineering Professors Council.

Stephanie Hladik, Ph.D., is Assistant Professor in the Centre for Engineering Professional Practice and Engineering Education at the University of Manitoba in Canada. Drawing from her interdisciplinary background in electrical engineering and educational research, Stephanie is interested in qualitative and design-based research projects that challenge inequities in STEM education.

Jette Egelund Holgaard, M.Sc., Ph.D., is Associate Professor at Aalborg University in Problem-Based Learning (PBL) in Engineering Education. Her current research focuses specifically on the relation between PBL competencies, interdisciplinarity, and societal challenges in the reshaping of engineering education.

Siara Isaac, Ph.D., is Researcher at the Centre for Learning Sciences at EPFL (Swiss Federal Institute of Technology). The book *Facilitating Experiential Learning in Higher Education* (2022) and the article 'Educating Engineering Students to Address Bias and Discrimination Within Their Project Teams' (2023) are representative of her work.

Yousef Jalali, Ph.D., is Postdoctoral Researcher at the Center for Learning Sciences (LEARN) at EPFL (Swiss Federal Institute of Technology). His current research focuses on developing and documenting the impact of activities targeting specific transversal skills. He collaborates with the EPFL Teaching Support Center (CAPE) to offer workshops linking research findings with educational practices.

Camilla Gyldendahl Jensen, Ph.D., is Lecturer in Architecture and Construction Management at University College Nordjylland (UCN) in Denmark. Camilla's research focuses on reflective practice-based learning, assessment, and learning designs.

Brent K. Jesiek, Ph.D., is Professor of Engineering Education and Electrical and Computer Engineering at Purdue University (USA). He also serves as Director of the National Institute for Engineering Ethics (NIEE). His research is focused on engineering practice, emphasizing ethics, global competency, boundary spanning, and other professional skills.

Aaron W. Johnson, Ph.D., is Assistant Professor in the Aerospace Engineering Department and a core faculty member of the Engineering Education Research Program at the University of Michigan (USA). His design-based research focuses on how to re-contextualize engineering science engineering courses to better prepare students for ill-defined, sociotechnical engineering practice.

Aditya Johri, Ph.D., is Professor of Information Sciences and Technology at the College of Engineering and Computing at George Mason University (GMU) in the USA. He is co-editor of the *Cambridge Handbook of Engineering Education Research* (CHEER) (2014) and edited the *International Handbook of Engineering Education Research* (IHEER) (Routledge/2013).

Irene Josa, Ph.D., is Lecturer at the Bartlett School of Sustainable Construction at University College London (UCL) in the United Kingdom. Her research focuses on the assessment of environmental and social impacts of construction. She holds a doctorate in Construction Engineering from the Technical University of Catalonia (BarcelonaTECH) and an M.Sc. in Human Rights from the Open University of Catalonia, both in Spain. Irene combines her expertise in these fields to promote sustainable and ethical construction practices.

Sarah Junaid, Ph.D., is Senior Lecturer and Programme Director of Mechanical Engineering and Design Engineering at Aston University in the United Kingdom. Her pedagogical research interests are in ethics education and team working with a global perspective. She has led international teams on initiatives across engineering education communities within Conceive Design Implement Operate (CDIO) and the European Society for Engineering Education (SEFI).

Dayoung Kim, Ph.D., is Assistant Professor of the Department of Engineering Education at Virginia Tech in Blacksburg, Virginia (USA). Her research interests include engineering practice (e.g., practices and experiences of, and competencies required for, engineers in various employment settings, such as business organizations and government agencies), engineering ethics, and policy.

John Bernhard Kleba, Ph.D., is Professor of Social Sciences at the Aeronautics Technological Institute (ITA) in Brazil. He leads the Laboratory of Citizenship and Social Technologies (LabCTS). John co-edited the trilingual book trilogy *Engineering and Other Engaged Technical Practices*.

Anette Kolmos, Ph.D., is Professor of Engineering Education and Problem-Based Learning (PBL) and the Founding Director (2014–2023) of the Aalborg Centre for PBL in Engineering Science and Sustainability, which is under the auspices of the United Nations Educational, Scientific and Cultural Organization (UNESCO). She is leading a research project on Interdisciplinarity and PBL (InterPBL) in engineering education.

Nihat Kotluk, Ph.D., is Researcher at the Center for Learning Sciences (LEARN) at EPFL (Swiss Federal Institute of Technology). He focuses mainly on emotions in engineering ethics education, but he also works on inequality and diversity issues in education. Nihat's academic background includes STEM and learning sciences.

Andreas Kotsios is Associate Professor of commercial law at the Department of Business Studies in Uppsala, Sweden, a researcher in the Division of Media Technology and Interaction Design at the Royal Institute of Technology (KTH) in Stockholm, and a fellow at the Information Society Law Center at the University of Milan, Italy. His current research focuses on AI regulation.

Helena Kovacs, Ph.D., is a social scientist at EPFL (Swiss Federal Institute of Technology). Her research focuses on transversal skills, teaching, and learning in engineering education, and she has served as a co-chair of the Ethics Special Interest Group within the European Society for Engineering Education (SEFI).

Jolanta Kowal, Ph.D., is Professor at the University of Wrocław and Gdańsk University of Technology in Poland, Jungian analyst, coach, and Vice-President of Research at the Association for Information Systems (AIS) SIG GlobDev for information and communications technology (ICT) and global development. She has published over 140 papers focused on economic psychology, socio-economic methodology, and the multicultural influences of psychoanalysis and has received funding from prestigious institutions.

Michael Kühler, Ph.D., is Professor of Applied Ethics of Social Responsibility at the University of Applied Sciences and Arts in Dortmund, Germany. He has taught ethics in a wide variety of study programs, including engineering, since the early 2000s. His research focuses on meta-ethics, ethics, and applied ethics, notably medical ethics and the ethics of technology.

Mikael Laaksoharju, M.S.E., Ph.D., is Associate Professor in Human–Computer Interaction at Uppsala University in Sweden. He has taught interaction design, full-stack programming, and computer ethics for over a decade. His research focuses on developing processes and methods to incorporate human values in the designing of computing systems.

Thomas Taro Lennerfors, Ph.D., is Professor and Head of the Division of Industrial Engineering and Management at Uppsala University in Sweden. He has taught ethics in engineering and management programs since the early 2000s, and recently published the book *Ethics and Sustainability in Digital Cultures* (Routledge, 2023), co-edited with Kiyoshi Murata.

Leroy Long III, Ph.D., is a Black male educator from the Midwest United States who is a community servant and mechanical engineering technology department chair. He has two mechanical engineering degrees and a doctorate in STEM (engineering) education. He has published research and designed curricula on equity and ethics in engineering.

Johanna Lönngren, Ph.D., is Associate Professor at the Department of Science and Mathematics Education at Umeå University in Sweden. Her research focuses on ethics and sustainability in engineering education, particularly the role of emotions in teaching and learning about wicked sustainability problems.

Susan M. Lord, Ph.D., is Professor and Chair of Integrated Engineering at the University of San Diego in the United States. Her degrees are all in Electrical Engineering. Her research focuses on the study and promotion of equity in engineering. She is a Fellow of the IEEE and American Society for Engineering Education (ASEE) and co-authored *The Borderlands of Education: Latinas in Engineering.*

Heather A. Love, Ph.D., is Assistant Professor of English at the University of Waterloo in Canada, where her research and teaching span topics of literature, culture, technology, health, and engineering communication. She is the author of *Cybernetic Aesthetics: Modernist Networks of Information and Data* (Cambridge University Press, 2023).

Stephanie J. Lunn, Ph.D., is Assistant Professor in the School of Universal Computing, Construction, and Engineering Education and the STEM Transformation Institute at Florida International University (USA). She also has a secondary appointment in the Knight Foundation School of Computing and Information Sciences in Florida. Her research spans computing and engineering education.

Xin Luo, Lecturer at Guangdong University of Finance, holds a master's from Beijing Institute of Technology in China and is pursuing a Ph.D. at the University of Malaya in Kuala Lumpur,

Malaysia. Her research spans Engineering Ethics Education and Education for Sustainable Development. Xin also works on cross-cultural adaptation and student career counseling.

Irene Magara is Assistant Lecturer of Science and Technology in the Department of Electrical and Electronics Engineering at Mbarara University in Uganda. Her research focus is on ethics and its application to engineering practice. She is the coordinator of student industrial internships and mentorship. She liaises with industry to ensure students receive professional skills training that emphasizes ethics since students are exposed to the real world of work.

Elena Mäkiö, Ph.D., is Professor of Software and Systems Engineering at the University of Applied Sciences Furtwangen, Germany. Her research interests include industrial digitalization, approaches in software and cyber-physical systems engineering education, engineering ethics education, and teaching approaches to promote critical thinking and other transferable skills.

Juho Mäkiö is Full Professor of computer sciences at the University of Applied Sciences Emden/Leer in Germany. He has played a leading role in international and national industrial and research projects. His research interests include the development of STEM education, teaching transferable skills, industrial digitalization, and the development of brain–computer interfaces for industrial applications.

Curwyn Mapaling, Ph.D., Senior Lecturer and clinical psychologist at the University of Johannesburg in South Africa, integrates psychological principles into engineering education. His research, focusing on well-being, combines theory with practice. He has been honored as a Mandela Rhodes Scholar and Abe Bailey Fellow, and received the 2022 Alumni Rising Star Award from Nelson Mandela University.

Lavinia Marin, Ph.D., is Assistant Professor at the Ethics and Philosophy of Technology Section, TU Delft in the Netherlands. Through Comet, a 4TU.CEE-funded project, she is designing new teaching methods for embodied ethics education in engineering. As part of the Ethics of Socially Disruptive Technologies (ESDIT) project, she conceptualizes epistemic and moral agency for social-media users.

Diana Adela Martin, Ph.D., is Senior Researcher with the Centre for Engineering Education at University College London (UCL) in the United Kingdom. Technological University Dublin (TU Dublin) in Ireland awarded her a doctorate in engineering ethics education. Diana has taught and researched various aspects of engineering ethics, societal responsibility, sustainability, and the philosophy of technology. Diana is the co-chair of the SEFI Ethics SIG, a European representative to the governing board of the Research in Engineering Education Network (REEN), and Associate Editor of both the *European Journal of Engineering Education* and *Science and Engineering Ethics*.

Esther Matemba, Ph.D., is an independent engineering education researcher, consultant, and part-time educator at Curtin University in Western Australia. Her research interests include curriculum for global engineers, problem-based learning, work-integrated learning, experiential education, and student success. She is the co-lead of the Engineering Education Research Network for Africa (EERN-Africa) and a governing board member of the Research in Engineering Education Network (REEN).

Aline Medeiros Ramos, Ph.D., is an instructor of philosophy at the Université du Québec à Trois-Rivières in Canada. She specializes in medieval philosophy and is also interested in the history of ethics, professional ethics, and virtue theory. She co-edited the book *Women's Perspectives on Ancient and Medieval Philosophy* (Springer, 2021).

Jan Mehlich, Ph.D., combines his educational background in chemistry and applied ethics in his research on science, technology, and innovation ethics. At the Center for Life Ethics at the University of Bonn in Germany, his academic focus is on options for integrating environmental- and ecological-ethical knowledge into the research and innovation process.

José Fernando Jiménez Mejía, Ph.D., is Associate Professor of Fluid Mechanics, Geosciences and Environment at the Universidad Nacional de Colombia in Colombia. Since 2019, he has been a researcher supporting the Engineering Education Institute of the Faculty of Mines in subjects focusing on Colombian engineers' ethical formation and responsible professional engagement.

Gaston Meskens, Ph.D., is Senior Researcher working with the Science, Technology and Society group of the Belgian Nuclear Research Centre and with the Centre for Ethics and Value Inquiry of Ghent University. His research focus is on the ethics of global governance and the particular role of science therein.

Taylor Joy Mitchell, Ph.D., is Associate Professor of Humanities and Composition at Embry-Riddle Aeronautical University in Daytona Beach, Florida (USA). She currently teaches interdisciplinary literature courses and honors seminars on Cold War Culture and Environmental Literature. Her publications reflect her work in cultural studies, pedagogy, and assessment.

John Mitchell, Ph.D., is Director of the Centre for Engineering Education in the UCL Faculty of Engineering and Professor of Communications Systems Engineering in the Department of Electronic and Electrical Engineering at University College London (UCL) in the United Kingdom.

L. Alexandra Morrison, Ph.D., is a Canadian who is currently Associate Professor of Philosophy in the Humanities Department at Michigan Technological University in Houghton, Michigan (USA). She specializes in twentieth-century continental philosophy (especially phenomenology and existentialism), social and political philosophy, feminist philosophy, the philosophy of technology, and ethics, including STEM ethics pedagogy.

Kenichi Natsume, Ph.D., (1974–2024) was Professor of Science and Technology Studies at Kanazawa Institute of Technology in Japan. He researched the history of science, technology, and engineering ethics. Kenichi was a committee member on engineering ethics at the Japanese Society for Engineering Education and the Japan Federation of Engineering Societies.

Christina Nick, Ph.D., is Lecturer in Applied Ethics at the University of Leeds in the United Kingdom. Her research area is moral and political philosophy, with a focus on understanding moral conflicts and responsibility. Christina has been teaching engineering ethics for many years and is interested in inclusive pedagogies and teaching ethics to non-philosophers.

Nienke Nieveen, Ph.D., is Full Professor of Curriculum Design in STEM Education and program director of the teacher education programs at the Eindhoven University of Technology (TU/e) in the Netherlands. Her orientations are in teacher professional learning concerning school curriculum (re)design, educational design research, and curriculum design approaches and tools.

Rikke Toft Nørgård, Ph.D., is Associate Professor at the Danish School of Education at Aarhus University in Denmark, where she leads the Horizon Europe project EPIC-WE (2023–2026) on values-based game-making and culture-making. She is in the steering group of Centre for Higher

Education Futures and on the board of the Danish Network for Educational Development in Higher Education.

Mauryn Chika Nweke, Ph.D., is Associate Professor at University College London (UCL) in the Department of Biochemical Engineering. Her research areas include downstream processing and engineering education, and she is particularly interested in the professional skills development of engineering students throughout their academic careers and beyond.

William (Bill) Oakes, Ph.D., is Assistant Dean for Experiential Learning, 150th Anniversary Professor, Director of the EPICS Program, Professor of Engineering Education at Purdue University (USA), and also a registered professional engineer. Bill is a fellow of both the US-based National Society for Professional Engineers (NSPE) and the American Society for Engineering Education (ASEE). He is a member of the ASEE Hall of Fame.

Robyn Mae Paul is Assistant Professor in Sustainable Systems Engineering at the University of Calgary in Canada. Her research and teaching focuses on applying frameworks from social justice, queer theories, and ecofeminism to broaden the narratives of engineering culture and foster more inclusive spaces and more socially just engineering designs.

Olga Pierrakos, Ph.D., is Program Director at the National Science Foundation in the United States. She is also Professor and served as Founding Academic Leader (for 6 years) of Wake Forest Engineering in North Carolina (USA), leading visioning and building from the ground up. Wake Forest Engineering has been nationally recognized for its diversity (faculty and students) and innovation. Olga also founded Engineering at James Madison University in Virginia (USA).

Madeline Polmear, Ph.D., is Lecturer in engineering education at King's College London (UK). Her research relates to engineering ethics education and societal responsibility. Madeline received a Ph.D. in civil engineering from the University of Colorado (USA) and previously worked as a Marie Skłodowska-Curie, EUTOPIA Cofund Fellow at Vrije Universiteit Brussel in Belgium.

Prajish Prasad, Ph.D., is Assistant Professor in the School of Computing and Data Sciences at FLAME University in Pune, India. He completed his doctorate with the Interdisciplinary Programme in Educational Technology at the Indian Institute of Technology Bombay (IIT Bombay). His current research interests are in computing education and the applications of artificial intelligence in education.

Vivek Ramachandran, Ph.D., is Lecturer at the Centre for Engineering Education at University College London (UCL), focusing on ethics, computing, sustainability, and responsible innovation within the Integrated Engineering Programme. His research focuses on the interplay between ethics, emotions, and engineering using digital tools for education.

Richard Randell, Ph.D., is a Senior Researcher in the Department of Sociology at Masaryk University in Brno, Czech Republic, and a fellow at the Institut für Höhere Studien in Vienna, Austria. He received his doctorate in sociology from the University of Wisconsin-Madison (USA). His current research focuses on the ontopolitics of the Anthropocene.

Donna Riley, Ph.D., is Jim and Ellen King Dean of Engineering and Computing at the University of New Mexico (USA). Donna's career has focused on integration of ethics, sociotechnical analysis, and other aspects of liberal education with engineering education. She is the author of *Engineering and Social Justice*.

Henrik Worm Routhe, M.Sc.EE, has 30 years of experience outside academia and is currently a Ph.D. student at Denmark's Aalborg University in Problem-Based Learning (PBL) in Engineering Education. His research is focused on management and leadership in interdisciplinary contexts related to engineering education.

Jillian Seniuk Cicek is Assistant Professor at the University of Manitoba in Winnipeg, Canada, where she grew up a settler on the original lands of the Anishinaabeg, Cree, Ojibwe-Cree, Dakota, and Dene peoples and the National Homeland of the Red River Métis. Her research explores the integration of Indigenous knowledges and worldviews in engineering curricula.

Yann Serreau, Ed.D., is Professor Emeritus at CESI LINEACT innovation and research laboratory in Paris, France. After working as an engineer, Yann joined the higher education sector and was responsible for training and projects-gathering in engineering schools. Yann focuses his research on educating engineering students in ethics and in decision-making in dynamic environments. He holds council responsibilities in ethics at various levels.

Satya Sundar Sethy, Ph.D., is Professor of Philosophy and Higher Education in the Department of Humanities and Social Sciences of Indian Institute of Technology Madras. He specializes in applied ethics (engineering ethics, academic ethics), engineering education, consciousness studies, logic, and Indian philosophy. In 2017, the Indian Council of Philosophical Research of the Ministry of Education of the Government of India named him "Young Philosopher." The United States–India Educational Foundation (USIEF) awarded him a Scholar-in-Residence Fulbright Fellowship to carry out teaching and research tasks in Utah (USA) for 2022–2023.

Ashley Shew, Ph.D., is Associate Professor in Science, Technology, and Society at Virginia Tech in Blacksburg, Virginia (USA). She is a co-editor-in-chief of *Techné*, the journal of the Society for Philosophy and Technology. Based on her research on disabled expertise and disability-led narratives about technologies, her newest book is *Against Technoableism* (W. W. Norton, 2023).

Lauren Shumaker, Ph.D., is Director of Thorson First-Year Honors and Teaching Assistant Professor at Colorado School of Mines (USA). With a background in geology and science communication, she is drawn to interdisciplinary education and is fascinated by the intersections of science and society.

Folashade Akinmolayan Taiwo, Ph.D., is Reader in Engineering Education, Director of Student Experience in the School of Engineering and Materials Science, and Deputy Director of the Centre for Academic Inclusion in Science and Engineering at Queen Mary University of London (QMUL) in the United Kingdom. Her research focuses on innovative and inclusive pedagogies embedding graduate attributes and enhancing students' sense of belonging and identity development through embedded support mechanisms within group work.

Emanuela Tilley is Full Professor of Engineering Education, Director of Education at the UCL Centre for Engineering Education, and Director of the Integrated Engineering Programme (IEP) in the Faculty of Engineering Sciences at University College London in the United Kingdom. She is internationally known for leading curriculum design and development as well as cultural change required to foster innovation in engineering education.

Fumihiko Tochinai, Ph.D., is involved in engineering ethics education, first-year education, and international collaboration programs at Kanazawa Institute of Technology in Nonoichi, Japan. His research is related to engineering ethics and the history of geology in Japan. He earned doctoral and M.A. degrees from Hokkaido University and a B.A. from International Christian University.

Roland Tormey, Ph.D., is a sociologist and Senior Scientist at EPFL (Swiss Federal Institute of Technology) where he leads the Teaching Support Centre. Roland's current research focuses on affective dimensions of pedagogical relationships and on emotions in engineering ethics education. His last book, *Facilitating Experiential Learning in Higher Education*, was published by Routledge in 2022.

Gouri Vinod is a doctoral candidate in engineering and education at University College London (UCL) in the UK. Her thesis focuses on the perceptions of AI in undergraduate engineering curricula by both faculty and students. She previously earned an M.Sc. from UCL in engineering education and a B.Sc. in electrical engineering from the University of Pittsburgh in Pennsylvania (USA).

Cristina Voinea, Ph.D., is a Marie Skłodowska-Curie Postdoctoral Fellow at the Uehiro Center for Practical Ethics at the University of Oxford (UK). Her research revolves around understanding the impact of digital technologies on both individual and societal well-being.

Natalie Wint, Ph.D., is an engineering educator at University College London (UCL) in the United Kingdom. She is interested in professional skills and the philosophy of both engineering and education. She has published in the *European Journal of Engineering Education* and is co-author of a chapter in *Engineering, Social Sciences, and the Humanities: Have Their Conversations Come of Age*? (Springer, 2022).

Richmond Y. Wong, Ph.D., is Assistant Professor of Digital Media at the Georgia Institute of Technology (USA). He directs the Creating Ethics Infrastructures Lab, where his research seeks to create tools, methods, and organizational environments that can support technologists and designers in ethical decision-making.

Adetoun Yeaman, Ph.D., is Assistant Teaching Professor in the First-Year Engineering Program at Northeastern University in Boston, Massachusetts (USA). Her research interests include empathy, design education, ethics education, and community engagement in engineering. She teaches Cornerstone of Engineering, which incorporates computer-aided design, programming, and ethics in a project-based learning environment.

Kari Zacharias, Ph.D., is Assistant Professor in the Centre for Engineering Professional Practice and Engineering Education at the University of Manitoba in Winnipeg, Canada. Her background is in engineering and science and technology studies, and her current research focuses on engineering cultures and epistemologies in Canada.

Qin Zhu, Ph.D., is Associate Professor in the Department of Engineering Education at Virginia Tech in Blacksburg, Virginia (USA). Qin leads the Laboratory and Network for the Cultural Studies of Engineering and Technology (LANCSET) research group. His research focuses on engineering ethics education, global engineering, and the ethics of AI and human–robot interaction.

Annuska Zolyomi, Ph.D., is Assistant Professor in Computing and Software Systems at the University of Washington Bothell in Washington State (USA). She leads the Interaction Design for Education and Accessibility (IDEA) lab. Employing human-centered and value-sensitive design, she researches ways to make human–computer interactions and sociotechnical collaboration more inclusive to disabled and neurodiverse communities.

MAPPING ENGINEERING ETHICS EDUCATION

Roland Tormey, Tom Børsen, Shannon Chance, Thomas Taro Lennerfors, Diana Adela Martin, and Gunter Bombaerts

Introduction

Bringing together and editing a handbook feels like a humbling task. The term 'handbook' typically refers to a reference book, a source to which one can turn to find the fundamental information needed to complete a professional task. Many engineers will be familiar with classic handbooks like Perry's (1950) *Chemical Engineers' Handbook* or *Marks' Standard Handbook for Mechanical Engineers* (Marks, 1978). Academic handbooks such as Routledge's *International Handbook of Engineering Education Research* are a little different in style, but they still convey the sense that they are a point of reference, the first place to go if one wants to understand a field. Implicitly, they convey that we know what the professional tasks are in a field and we know the fundamental information needed to complete them. Writing and editing a handbook would undoubtedly be more straightforward in a context in which the field in question was mature, the professional tasks well defined, and the required information readily identifiable. What made our editorial task daunting was that this did not feel, to us, to be the case with engineering ethics education (EEE).

We knew there were quite a few excellent engineering ethics textbooks (e.g. van de Poel & Royakkers, 2023; Fledderman, 2014; Lennerfors, 2019) that could serve as an introduction to engineering ethics for students and which could implicitly provide a course structure and pedagogical approach for new engineering ethics teachers. Yet, when we dug a little beneath these textbooks, we found differences in perspectives regarding the content to be taught, who should deliver it, optimal teaching methods, effective assessment strategies, and the subsequent actions based on assessment results. Alongside these differences, we saw unrecognized similarities: authors working on the same themes and ideas but from different disciplinary perspectives and without substantial interaction with each other. Our task, then, felt less like we were describing a clearly defined terrain and more like we were mapping a previously unmapped space, one in which many people had been working, often in collaboration with each other and sometimes in splendid isolation from those who were (intellectually) nearby to them. Our hope with this handbook was to create a map of EEE that would allow others – teachers of engineering ethics, engineering ethics education researchers, and especially graduate students who are new to the research area – to find their way more easily in the EEE terrain, to recognize the pathways that connect ideas and positions and the topography that separates them, and to allow them to better understand the range of perspectives and approaches that make up the terrain of engineering ethics education. Yet we were also con-

DOI: 10.4324/9781003464259-1

scious that we were mapping quite a diverse space, which was home to people who spoke different languages, had different epistemologies and different disciplinary methods, cared about different things, and lived in different places. In mapping the space of engineering ethics education to make it more navigable, we wanted to respect the positions of those within the space. All this fostered our sense that framing such a handbook was a profoundly humbling task.

In the pages that follow, we will explain something about our process in doing this, in the hope that our decisions make sense to those who bring engineering ethics education into being and to those who will use this handbook to do so in the future.

Identifying the 'space' of engineering ethics education

In thinking about the content of a handbook to map the space of engineering ethics education, we needed to think about what that space is. One way to reflect on the space between engineering, ethics, and education is to imagine a Venn diagram with three sets, partly overlapping, representing engineering, ethics, and education respectively. The space of EEE emerges as the common space expanded by the three sets. Thus, this handbook provides a rich overview of the common space between engineering, ethics, and education, and we will, in the following text, begin our reflections within 'engineering' and see where that leads us in terms of filling in the space of EEE.

Engineering

We are, first and foremost, educating future engineers. Engineering must be central to our discussion of ethics education – engineering and its various subfields, or disciplines, help set crucial boundaries and focal points for investigating ethics – *engineering ethics* – and how to teach and learn it. The first two sections of this handbook provide perspectives from disciplines outside engineering, seeking to explain how foundational ethics concepts and perspectives from other disciplines tie to engineering education. The third section, 'Ethical issues in different engineering disciplines,' provides a glimpse of the many ways that ethics is applied, considered, and practiced in the various subfields/disciplines of engineering.

Specifically, the third section discusses what content should be included in ethical teaching in five different engineering disciplines: civil engineering, mechanical and aerospace engineering, electrical and electronic engineering, chemical engineering, and software engineering (Chapters 14–18). The section concludes that the content of EEE needs to overlap with, or in other ways relate to, the engineering discipline in which engineering ethics is taught. If there is no connectivity between the ethical content and the engineering content surrounding it, students who are developing an engineer's identity will be unable to find meaning in the ethical issues taught and may thus reject ethical content as irrelevant and without meaning (Lönngren, 2021). Relating engineering ethics to particular engineering disciplines can be achieved in different ways. It can be done by including ethical questions within the list of questions asked in conventional engineering exercises of a technical character. The idea here is that ethical questions be asked side-by-side with technical questions. When a material's properties are discussed, ethical questions regarding its origin, the environmental costs for obtaining it, its health features, and such can be probed – opening up the discussion for ethical deliberation and reflection. Section 3 identifies several issues that can be related to specific technical content of engineering disciplines. Ethics teaching can depart from and be related to the ethical codes within specific engineering disciplines. It can present the unexpected and broader socio-techno-ecological implications of concrete historical engineering solutions, including the implications for employees, users, the environment, and public health. It can also discuss actual use, including potential military applications in engineering. It can discuss how engineering solutions relate to colonial structures and whether they promote diversity, equity, and inclusion. The crucial point is that ethical issues discussed in engineering ethics classes must be related to and exemplified with illustrations from the specific engineering discipline. Case studies, role plays, and exercises need to reflect specific disciplinary activities, issues, and dilemmas.

Section 4 (comprising Chapters 19–25) expands on different teaching approaches in EEE. Some of these overlap with teaching approaches used in other parts of engineering education, for example, project work, extra-curricular activities, internships, or design activities. When there are overlaps in the teaching approaches between engineering ethics education and engineering education in general, it is not difficult to add ethical questions or integrate examples of ethical issues in relation to, for example, humanitarian engineering or engineering project work, or to include an extra-curricular guest lecture from industry in which they speak about their ethical challenges. Engineering design is one area where this handbook includes many ethical tools to incorporate into the design process: value-sensitive design, participatory design, and empathic design, to mention just a few.

An important point is that engineering students have typically not come to study to become mini-philosophers, or to earn a bachelor's or minor in philosophy, sociology, or anthropology. Rather, the ethical content will often engage them most when it represents an integral part of their engineering discipline. Proximity of engineering and ethics in engineering education is key. Ethics is integral to engineering education and engineering practices because engineers co-create society, the constructed environment, and reality. When engineering ethics course planners and teachers incorporate content from specific engineering disciplines and practices, it facilitates students' integration of engineering and ethics concepts.

Several chapters across this book assert that EEE must include thick descriptions of the specific psychological, social, cultural, and organizational context in which engineering is conducted. Thus, engineering ethics needs to be linked to specific engineering disciplines and also to the non-technical spaces surrounding engineering work and decisions. Different disciplines from the social sciences and the humanities can contribute meaningful insights into the non-technical content that is key to engineering ethics. One can ask why engineering ethics is not called something else since it is not only engineering and ethics content that is important to teach engineering students. 'Responsible engineering' might be another way to label the type of content we are proposing: engineers must be taught to take responsibility for the implications of their work. Yet, responsibility requires knowledge of implications as well as possibilities to act (see, e.g., Doorn and Van de Poel (2011) for an exploration of the issues involved in engineering responsibility, as well as Chapter 11 in this handbook). Thus, knowledge of the concrete circumstances – including power relations (Foucault, 2001), legislation, organizational structures, and psychological factors – is imperative to include as content in engineering ethics education.

Ethics

One of the contributions to the content of EEE highlighted in the first section of this handbook, 'Foundations of engineering ethics education,' involves ethical and normative frameworks. This section (comprising Chapters 1–7) overviews a selection of normative frameworks that can help underpin ethical reflections, assessments, and decision-making in and beyond the classroom and into engineering practice. The existence of normative frameworks in ethics provides criteria for individuals to reflect upon what is right and wrong in a given situation. A central point in this handbook is that there is not only one (or even a few) normative framework(s); there are many, and they originate from different traditions. Some are rooted in engineering, such as the many ethical codes of conduct provided by engineering societies around the globe. Some are rooted in Western philosophy, for example, virtue ethics, consequentialism, deontology, contractualism, discourse ethics, and the ethics of care. Others are rooted in sustainability science, for example, circular and ecological economy, deep ecology, land and material ethics, future ethics, precaution, energy justice, energy democracy, and sustainability. Ethical frameworks are not necessarily occidental. Many normative frameworks originate from non-Western knowledge systems/traditions, such as *Ubuntu* from sub-Saharan Africa, *Buen Vivir* from indigenous Latin America, Confucianism from China, and Buddhism from different Asian countries. Normative frameworks are referred to in ethical and political discussions of engineering and technological innovation and serve as assessment criteria for what counts as 'good' engineering or technological solutions. Many people involved in the construction of this handbook seek to identify and cultivate rich, multi-dimensional understandings of these other traditions, particularly in light of the rapidly standardizing nature of engineering processes, products, and education systems – a standardization process hastened by capitalist production systems and by engineering accreditation processes. Accreditation of engineering education is the focus of Section 6 of this handbook; benefits and drawbacks of the system are discussed and critiques and criticisms raised.

A central assumption underpinning most contributions in this handbook is that applying normative frameworks is not an individual or solitary endeavor. Normative frameworks operate in or underpin the socio-techno-ecological configurations in which engineering students and engineers are located. Even when they are not openly discussed or reflected upon, they influence backstage or behind-the-scenes decisions, conventions, habits, practices, and engineering solutions. Thus, an argument put forward in this handbook is that EEE must address a broad palette of ethical frameworks so that engineering students, as future engineers, become acquainted with some existing normative frameworks *and* a broad section of engineering processes, practices, and solutions for enacting these frameworks. The frameworks can additionally inform and qualify individual and collective ethical decision-making.

The handbook's chapters reveal a lack of consensus within the EEE community regarding the status of these normative frameworks: Are they universal, objective truths? Do people construct them? Or are they somehow in between? Engineering education provides fertile in-between ground, given that some phenomena (like gravity, heat transfer, or solubility) are taken as given/objective/fact, whereas other phenomena are more readily recognized as being socially constructed (e.g., design briefs and quality standards).

An additional point where consensus is lacking regards whether insights from different normative frameworks can be combined or whether one must adhere to just one framework at a time in ethical work. In some cases, ways of weighing alternative solutions within one normative framework cannot be mixed with the rationale underpinning a different framework because the basic assumptions are at odds. These discussions emerge in various chapters of the first section; the argument is latent in many chapters and explicit in Chapter 2 on ethical theories and Chapter 8 on philosophical contributions. Moreover, the plurality is manifested in Chapter 7 on the ethics of artificial intelligence (AI) and appears explicitly in the editor's introduction to the first section of the handbook.

Education

Another way of thinking about the space of EEE is to ask what is or should be distinctive in the knowledge of those who teach engineering ethics. Throughout the twentieth century, scholars sought to identify what makes expert teachers so good at teaching. Essentially, two different approaches emerged (Darling-Hammond, 2006). The dominant tradition followed a (more or less) behaviorist path in seeking to identify the observable teaching behaviors that were associated with increases in student knowledge. Once these approaches were identified, they could be simplified, codified, built into curricula, and issued as guidance or imposed as requirements for new teachers entering the field. This approach to improving teaching owed much to engineering modes of thought and was closely allied with the emerging 'scientific management' approach pioneered by the mechanical engineer F. W. Taylor (Au, 2011). It fed through the efforts to describe 'learning outcomes' (rather than learning experiences) that emerged in psychology in the 1950s and 1960s, and it remains influential today via accreditation approaches and via the Bologna reforms across the European Higher Education region (see Gleeson, 2013; the framing of engineering curricula in terms of 'learning outcomes' is a theme that emerges in the sections on assessment and accreditation in this handbook).

A second, less dominant, tradition followed the work of John Dewey (1929), who tried to identify the (less observable) knowledge of methods, learners, and subject discipline that would empower teachers to make more flexible and adaptive decisions to respond to the different and changing needs of students. Dewey's focus on the knowledge base of flexible, adaptive educators came back into focus in the early part of the twenty-first century via the work of Linda Darling-Hammond (2006). Like Dewey, Darling-Hammond sought to map the knowledge base of the expert teacher, this time informed by a further 80 years of teaching practices and educational research. The features she identified were:

- Understanding the *subject(s) being taught*
- Understanding *learners and learning*
- Understanding *teaching*

Regarding *understanding the subject(s) being taught*, Darling-Hammond (2006) described effective educators as those who recognize that the same teaching techniques cannot simply be transported from one subject or discipline to another. In engineering ethics education, this would imply that – because the epistemologies, the modes of inquiry, and the central concepts of engineering ethics education are different from those that apply in, for example, mathematics or physics – engineering ethics education may well need to take a different format from those of other subjects being studied by engineering students. To be able to flexibly choose appropriate pedagogies, educators need to understand the epistemologies, ways of working, and core ideas of engineering ethics education. This, in turn, implies needing to clarify what these things are. Given the fragmentation and disciplinary diversity we have already identified as underpinning engineering ethics education, mapping the space of the subject(s) being taught has required an inclusive approach from us as editors (this is reflected in Sections 1 and 2 of this handbook). Since most students come to our engineering ethics classes with the intention of being engineers and not ethicists, understanding the specific disciplinary epistemologies of engineering disciplines (including mechanical, electrical, computer, environmental, or biomedical) is also important in making good decisions about what it means to teach engineering ethics within that discipline (this is reflected in Section 4 of this handbook).

Related to *understanding learners and learning*, not every technical expert grasps social dimensions well. The phenomenon of extremely gifted academics who are unable to effectively teach is perhaps so well-known as to hardly require elaboration. For Darling-Hammond (2006), the key issue is that it is not enough to be an expert in a subject domain; one also needs to understand the learners who must learn and the social and psychological processes through which they learn. Learners come to any ethics class with a store of prior knowledge and experiences, which effective teachers understand. These teachers also understand how to leverage and question students' prior knowledge in order to help them learn. Understanding our learners – in their rich diversity – is therefore central to the knowledge base of an effective educator. Our learners differ not only in their backgrounds but also in the identities and senses of self that they are in the process of constructing during their time in engineering education. These diverse learners build competence through cognitive, emotional, and embodied processes – and knowledge of these processes informs the pedagogies that are chosen in engineering ethics education. Learners also build their competencies in social contexts, in interaction with each other and with their teachers. This, too, influences the design and choice of teaching approaches of effective teachers. This focus on learning and learners shaped our editorial team's decisions regarding what to include in this handbook. Traditional engineering ethics textbooks would not usually dwell on the psychology of moral development (Chapter 10 within Section 2 and Chapter 28 within Section 5) or on the curriculum and accreditation models for engineering programs (Section 6). The relationship between emotion and rationality in ethics and in ethics learning is also a theme that is not widely considered in other texts but is included here (in Sections 1 and 2, particularly Chapter 4). This book reaches beyond the technical aspects of engineering, drawing from all these domains and more, because they all make important contributions to the expertise of engineering ethics educators.

Understanding teaching, the third component of Darling-Hammond's (2006) framework, corresponds to Dewey's focus on 'methods': the good teacher brings together knowledge of their subject and knowledge of learners and learning to incorporate diverse teaching strategies appropriate to both the content and the learners. In engineering ethics education, this is reflected in the use of a range of different teaching strategies, including case studies, art-based methods, reflection, role play, and challenge- and project-based learning, among other methods (covered in Section 3). Teaching also involves assessment, and crucial aspects include understanding how to evaluate whether or not students have developed the capabilities that were intended (Section 5). Our attention encompasses not only how teachers create assessments for their classes but also their utilization of research-based tools. These tools aid teachers and educational researchers in comprehending the impact of their work and how it might be further developed, as discussed within this handbook.

The three sets – engineering, ethics, and education – provide a framework for thinking about the choices the editorial team made in deciding what would go into a comprehensive handbook for engineering ethics educators and researchers.

The 'mapping' metaphor

The three-component description of engineering ethics education that we have used here is a useful narrative device for describing part of what is distinctive about this handbook and what distinguishes it from other excellent books and resources that exist. It is, however, a post-hoc rationalization; we did not start this project with the framework outlined above in mind. Earlier in this introduction, we described our approach as being to 'map a previously unmapped space,' and it provides a valuable way of thinking about what we intended to do when our team decided in consultation with publishers – to provide as comprehensive coverage of the overall terrain of engineering ethics education as possible. Extending the metaphor also allows us to think about some dimensions of the work that were important to us.

Mapping as a change in perspective

Because so many of us use maps so often, it is easy to forget that mapping requires the capacity to visualize a space from what seems to be an entirely unnatural angle. While we look at the world around us, we see it in a horizontal plane, but mapping requires shifting one's perspective to imagine how that world looks from above. In our world of airplane travel and satellite images, this is perhaps not such a shocking change in perspective, but the earliest known maps may date back some 27,000 years (Wolodtschenko & Forner, 2007). It is worth reflecting on what an effort it must have taken to wrench one's perspective free from the world as it was experienced and to imagine how it looked from above. Looking at maps even today can be a disorienting experience, with roads that feel parallel in experienced reality being seen to diverge in plan, while regularly traveled routes that feel short can be seen to be longer than less familiar ones.

Mapping the terrain of engineering ethics education requires a similar change in perspective. Those working in the field experience it through their interactions within the field, yet, in cultivating this handbook, we needed to present a view of the field that may feel unfamiliar to some of those who live and work in it. At the same time, we wanted to respect the different perspectives of those who 'live the world into being.' For this reason, we built on an emerging practice in engineering education to include positionality statements (see, e.g., Gani & Khan, 2024, Hampton et al., 2021; Secules et al., 2021). Since this is an emerging practice, it was not always easy for authors to know how to describe their position and perspective in engineering ethics education. Our hope is that this approach provides an appropriate counterpoint of 'horizontal' and 'vertical' views of the field and that readers will perceive the utility of seeing the field from these different perspectives.

Mapping as a colonial process

For much of the world, the process of 'being mapped' was intrinsically linked to colonialism; as Richard Phillips has written, "Imperialism went hand-in-hand with mapping, by which Europeans imaginatively and materially possessed much of the rest of the world" (1996, p. 6; see also Anderson, 1991; Gauba, 2002). Maps often served an administrative or military purpose, allowing colonial rulers to more efficiently manage populations, extract resources, and move troops. Imperial and colonial maps also shaped the way in which colonized people lived in their world: as the play *Translations* by Brian Friel (1981) explored, for example, Ordnance Surveyors anglicized and changed place names in 'their' colonies as they mapped and, in doing so, assigned places new names – often stripped of their original meaning – that are still used long after independence.

For those who, like us, want to engage in even a metaphorical mapping project, this is an important reminder. Given Europe's colonial past and postcolonial present, this reminder is perhaps even more crucial for a project like this handbook, which emerged from the European Society for Engineering Education (SEFI). We were acutely aware of the risk of a colonial or postcolonial gaze in mapping the space of engineering ethics education. Our request to all authors to include a positionality statement was one response to this danger. Another was that we tried to be as open and inclusive as possible in the perspectives and voices that contribute to this handbook. A third way in which we sought to acknowledge this risk was by asking authors to go beyond the hegemonic (Western and masculine) moral theories of consequentialism, deontology, and virtue ethics and to engage with a broader set of ethical theories that were also representative of voices previously excluded from the mainstream of engineering ethics education.

Mapping as an unfinishable process

The first attempt at modern mapping of a country is thought to be the Cassini maps of the Kingdom of France, which were developed in the period between the 1740s and the 1780s. It is perhaps interesting to note that, having spent 40 years involved in the project, Cassini died before his mapping work was completed (Brotton, 2013). Even if his work had been completed before his death, by the time it was finished, the world would certainly have changed, and his maps, once finished, would immediately have required revision.

Much changed even during the time we were working on editing this handbook. This project began in the spring of 2021, when travel was still restricted due to the Coronavirus pandemic and COVID vaccines were being distributed (and hoarded) in Western countries. The pandemic raised huge ethical questions for engineers regarding, for instance, the way the virus originated via humans' interaction with the natural world; about the way transport technology facilitated its spread; about the surveillance, monitoring, and privacy of individuals and populations as the spread of the virus was being checked; about the risks involved in rapidly developing vaccines; and about the ethics of nudging populations towards vaccination. As we were working on this book, the release of ChatGPT in November 2022 seemed a seismic moment in how people and technology interact. It also raised big ethical engineering questions about who owned the data on which it was trained, whether or not the privacy of those interacting with it was being protected, the welfare of those involved in checking toxic content, and the risks of the algorithm producing convincing but false information (see Chapter 7 on AI and Chapter 18 on software engineering).

It is not only the world that is changing; educational research also continues to give new insights into how people learn to be ethical. In the years this book was under development, we saw a notable growth in focus on de-colonizing engineering education (Cruz, 2021; Seniuk Cicek et al., 2023); on emotion in engineering education (Lönngren et al., 2021, 2023); on arts-based and drama-based methods in engineering ethics classes (e.g., Martin et al., 2019; Hitt & Lennerfors, 2022); and on challenge- and problem-based learning in engineering ethics education (e.g., Bombaerts et al., 2021; Sukackė et al., 2023), for example.

Because technology and educational science are both changing, what we know about engineering ethics education is changing, too. And so, any attempt to map the space of engineering ethics education will only ever represent a snapshot of that territory at a given moment in time.

Our team sees mapping as an unending process and this handbook as a map of sorts. It is important to remember that 'the map is not the territory.' We render a territory – in this case, the territory of engineering ethics education – on a smaller scale. The process of transposing the territory as a map does not produce a 1:1 rendition of the territory but rather an interpretation that highlights only some features. This recognition emphasizes that our endeavor is not terminated; it can and should be complemented by further attempts at mapping, via different interpretations, all incomplete and, as such, continuously open and never-ending.

Nevertheless, we also believe a good map can still last a long time. The Cassini map number 53 of central France – mapped and engraved in the mid-1700s – maps the small mountain hamlet of four houses where one of the editors of this book spends holidays each year. Of course, there are differences; the spelling is not the same, and the mountain road to the hamlet was not paved until the 1960s, two centuries after Cassini's team passed through, so roads and bridges are undoubtedly different. Yet that map from two and a half centuries ago still shows the names and locations of places that exist today. Our aim with this handbook is to provide a snapshot of the territory that is detailed enough for researchers, teachers, and even students to find it a useful first point of reference – to help them to situate themselves in the territory of engineering ethics education, even as that territory shifts over the coming years.

Our positionality as editors and the origin of this handbook

This handbook grew out of the work of the Ethics Special Interest Group (SIG) of the European Society for Engineering Education (SEFI). The SEFI Ethics SIG Spring School in March 2021 was hosted by EPFL in Switzerland but was held online due to travel restrictions still in place amid a slow rollout of COVID-19 vaccination programs across Europe. One of the themes of the 2021 Spring School was 'collaboratively writing engineering ethics' and, to address that theme, Tom Børsen from Aalborg University presented a review titled "Lessons Learned about Engineering Ethics Education at the SEFI 2020 Conference." One of Tom's conclusions was that a substantial amount of high-quality research on engineering ethics education was being presented at SEFI and that there was a basis there to make a considerable contribution to the field's growth. A participant in the event, Shannon Chance, recommended that Tom curate a handbook or special focus issue based on the expansive terrain he had identified. Coincidentally, just before that Spring School, in winter 2020, and also inspired by the annual SEFI conference, Shannon, Diana Adela Martin, and Thomas Taro Lennerfors had met and discussed the possibility of developing a handbook of EEE.

The stars were aligning, and two weeks after the Spring School, at Tom's invitation, he and Shannon met to develop a strategy, which included inviting as editors Diana, Thomas, and Roland Tormey. Tom and Shannon credit Diana with bringing Gunter Bombaerts onto the project. Tom extended invitations, and almost immediately, the constellation of editors for this handbook (Tom, Shannon, Roland, Diana, Thomas, and Gunter) gathered to discuss how we could build on the work underway via SEFI to 'collaboratively write' engineering ethics education. The composition and characteristics of the team influenced a host of decisions made during the creation of the handbook.

The team is interdisciplinary. It includes editors with diverse backgrounds in multiple disciplines with profiles such as architecture and higher education (Shannon); chemistry, philosophy, and technology studies (Tom); philosophy, liberal arts, and engineering education (Diana); sociology, mathematics, and educational sciences (Roland); industrial engineering and humanities (Thomas); and nuclear physics and philosophy (Gunter). We all have prior experience in editorial roles for journals, books, handbooks, and/or special focus volumes. Some of us also have experience developing inclusive educational projects outside academia (Shannon, Diana, Roland, and Gunter). Through this handbook project, we generated new knowledge, extended geographic scope, and helped build additional publishing and project management skills among ourselves and our larger engineering and ethics education community.

The editorial team is gender diverse (two women and four men). Its members come from or have worked in various places, including Belgium, Denmark, Ireland, Japan, the Netherlands, Romania, Sweden, Switzerland, the United States, and the United Kingdom. We share a high level of respect and advocacy for diversity, equity, and inclusion. Alongside our gender, national, and disciplinary diversity, we also have similarities that present limitations in our endeavor to represent diverse viewpoints: we are predominantly (although not entirely) white and of European descent, and we all currently work in technical universities or engineering faculties in Europe.

Most of us are research-active in engineering ethics, having published on topics that we see as linked to engineering ethics and which appear in chapters throughout this book – like energy policy, energy justice, global responsibility, corruption, AI ethics, innovation for sustainable transitions, innovation ethics, corporate social responsibility, insecticide use, and interpersonal discrimination in teams (see, e.g., Chapters 6, 7, 11, and 13). All of us are research-active in engineering ethics education, having published on pedagogical methods in ethics education (including case studies, role play, use of film, and challenge- and problem-based learning); on global and historical perspectives in engineering ethics education; on the impact of accreditation on ethics education; on emotions in learning ethics; and on the integration of socio-ecological responsibility into engineering study programs. We have drawn upon various epistemologies and research methodologies, including philosophical inquiry, quantitative and qualitative social research methods, and systematic literature review. These research interests and features of our epistemological positions certainly impacted our choices in framing this book.

Cultivating researcher capacity and building community were central concerns for the editors. Two of the editors (Roland and Shannon) have been involved in organizing SEFI annual conferences, and three of us (Gunter, Roland, and Diana) have been organizers or co-organizers of four SEFI Ethics Spring Schools. Many of us (Diana, Roland, Gunter, Thomas, and Tom) have also been involved in organizing the SEFI ethics webinar series, the SEFI ethics newsletter, the SIG annual conference workshops, and ethics working group meetings, among other things. At the start of this project, one of us (Shannon) was just finishing a term as chair of the global Research in Engineering Education Network (REEN), having led a transformation toward more geographically diverse representation and extending REEN's work to build the capacity of educators to use and conduct engineering education research. Our shared focus on community-building is evident in the choices we made in developing this handbook.

Different members of the editorial team played different roles during the handbook's development; however, we each took responsibility for one section: recruiting authors, supporting them, managing feedback workshops and peer reviews, and editing their chapters.

Tom was the conceptual lead for the project, and Shannon was the organizational and production lead and the point-of-contact with publishers. Together, Shannon and Tom collaboratively managed the overall effort. Shannon took the lead in crafting detailed proposals and negotiating with publishers; Tom managed our team meetings. The two mapped the overall process, timelines, and deadlines. In the final months, Shannon copy-edited all the chapters and introductory statements, ensuring a degree of coherence in linguistic style and approach across the 105 different authors and providing extensive editorial assistance in some cases. She contributed a significant amount of content for most of the overview statements (sections 1, 3, 4, and 6) and made substantial contributions to several of the chapters (including Chapters 27 and 33). Shannon has also taken the lead on publicizing and promoting the handbook, offering workshops, and facilitating panel discussions at conferences around the globe during the production phase of publication. The editorial team encourages all contributors to promote the work and to organize similar events – that present the content, engage and empower people to use it, and invite others to join our engineering ethics education efforts – whenever possible.

The whole team worked to identify gaps and ways to fill them, adding several new chapters along the way. Diana, Shannon, and Tom worked closely to identify and address ethical concerns related to author attribution and acknowledgments, aiming to achieve a high level of transparency and fairness. Nevertheless, our team recognizes the probability of inadvertently failing to acknowledge some of the invaluable contributions made, and we apologize to anyone whose contribution has not been rightfully named and attributed.

Our first step in bringing this project together was to organize a participatory workshop at the SEFI conference hosted by TU Berlin and held online in September 2021. Working with the participants at that workshop (titled 'Eager to Contribute to an Engineering Ethics Education Handbook?') helped us co-construct an outline of the six sections of the handbook and identify some content, structure, and potential authors for each section. Building on this, we, as editors, developed a plan for the handbook and an outline of what it should contain.

In March 2022, we launched an open call for authors, outlining the proposed structure and content and inviting educators and scholars to express their interest in participating in the project. We connected with various organizations, asking them to distribute our call for authors. We posted the call on the SEFI website. We reached out to personal contacts and as wide and globally expansive an array of engineering education organizations as possible. We asked interested people to tell us about their background and motivation, what chapter topics interested them, and whether or not they already had co-authors to propose. We were delighted to receive over 100 responses from across the globe.

Using this list of interested authors, we set about building author teams. We aimed to assemble an international author team for each chapter; ideally, each team would include four people from different institutions and countries and, wherever feasible, from different continents. We identified and invited a lead author for each chapter, outlining our collaboration aims and recommending potential co-authors for the chapter team while leaving the lead authors leeway to choose collaborators.

At the end of the process, we asked each author to let us know all the places they had lived or worked for a year or more. We developed Figure 0.1 using that dataset; the map helps us see the geographic extent of the community contributing to this resource. The map helps demonstrate our diversity, but it also makes visible where our community has not yet connected. It provides us with a valuable benchmark for the future.

Although the map communicates some core ideas about our cultural roots and our mobility, there are some things we would improve for subsequent versions: we would ask authors to pin their own locations for greater accuracy, and we would select a different base map (the one we used here was available open access on Wikipedia). We recognize that 'mapping and colonialism' are themes in this introduction, yet the equirectangular distorts space and makes equatorial regions appear smaller and, implicitly, 'less important' (Africa looks 2.5 times the size of Greenland but is, in fact, 14.5 times the size). Two-dimensional depictions of the world map projection are the subject of lively debate and are heavily contested in de-colonial studies of mapping – we support the push to evolve them.

Many of the teams that crystallized were interdisciplinary, and some were composed of people who were strangers to each other at the beginning of the process. As editors, we pondered and discussed if we were making things unnecessarily difficult for ourselves and for our authors, but the author teams approached their tasks with openness, good humor, and diligence – and they began to craft rich and coherent narratives. The pairings worked in almost every case, and the rate of success across teams was even higher than we have experienced with other book projects. We are incredibly proud of the quality, coherence, and comprehensive nature of the overall compilation – and of the community spirit and collaborative nature of the undertaking.

It was not enough to build coherence based simply on the diversity of authors within each chapter. We also needed to build coherence across chapters – within and across sections. Our first step was to cultivate coherence within themes by organizing workshops where authors presented their work to others within the same section and provided each other with feedback. This allowed authors to start to see overlaps, commonalities, and tensions within their respective sections. Once draft chapters were ready, we began to scale up – to look for coherence and tensions across sections. We did this by having each chapter reviewed by at least two other authors, typically one from within the same section and one writing in a different section. In this way, those working on an individual chapter could position their work within the broader context of the handbook. We enlisted a few additional experts to fill gaps during the review process.

Tom and Shannon dedicated themselves to fostering connectivity, coherence, and comprehensiveness across the entire collection, striving to impart a sense of unity and completeness to the book. They added chapters during the development of the book to cover emerging or

Figure 0.1 Map of authors' homelands. The circles indicate the countries where our authors have lived and/or worked for a year or more, and a line connects each contributor's countries.

under-recognized issues. Shannon edited the chapters for grammar and style, cross-referencing where possible. Tom and Shannon worked together to analyze the section overviews; they sought to ensure the overviews describe the respective themes in relation to the larger contexts of engineering, education, and the full set of manuscripts the handbook encompasses. Shannon and Tom also worked closely with Diana to ensure that individuals were adequately recognized for their contibutions.

As noted above, growing out of the work of the SEFI Ethics SIG, our starting point was the idea that we could facilitate a process to collaboratively 'write' and simultaneously 'map' engineering ethics education in a way that transgressed disciplinary borders and colonial and postcolonial (or hegemonic and oppressed) narratives; and moreover, that we could do so by involving a wide set of voices and that we could collaborate to build something that was at once rich, multi-perspectival, and coherent. We recognized the ambitious scope of the task we set ourselves and our team of authors, but with trust in each other, in the process, and in our authors, we have arrived at this handbook that charts the emergence of a coherent and vital field of study: engineering ethics education.

The structure of this handbook

This handbook is divided into six sections. The first three sections address the content and purpose of engineering ethics education, and how disciplines like engineering, environmental science, ethics, and other social sciences and humanities disciplines (and non-Western knowledge systems) feed into EEE. The subsequent three sections address processes surrounding and forming EEE: teaching methods, assessment, and accreditation.

The first section, 'Foundations of engineering ethics education,' discusses some foundational issues that underpin engineering ethics education. The issues covered in this section range from the purpose(s) of engineering ethics education to the relationship between engineering ethics education and the field of ethics, individual judgment and decision-making versus collective issues, professionalism related to engineering ethics education, how engineering ethics education relates to reason and emotion, how environmental concerns relate to EEE, and emerging issues in AI. This section contains seven chapters.

The second section, 'Interdisciplinary contributions to engineering ethics education,' makes the assumption that engineering is an interdisciplinary and applied field that draws from foundations provided by mathematics, natural sciences, computer science, management studies, social sciences, and the arts. Ethics education is equally an interdisciplinary field, drawing notably from psychology, philosophy, sociology, and social policy studies. It follows that understanding engineering ethics education requires a multidisciplinary foundation. This second section aims to provide an understanding of the foundational concepts, approaches, and problematics central to engineering ethics education. In Section 2, these concepts, approaches, and problematics are explored within the context of the disciplines from which they emerge. Six chapters comprise this section.

The third section features chapters articulating the ethical challenges of different engineering disciplines. Each of the chapters delineates which ethical issues, dilemmas, and challenges are discussed in the specific discipline and explores how the discipline's students and practitioners might address them. The ethical issues included cover both processual issues (like user involvement, codes of conduct, early warning systems, distribution of responsibility, ethical design, etc.) and the wider ambiguous implications of technology and engineering solutions (in relation to digitalization, energy, human rights and dignity, pollution/environmental impacts and climate change, colonization, big tech's influence on technological solutions, military technologies, technological accidents, etc.). Section 3 outlines how these issues, dilemmas, and challenges are approached and suggests solutions in engineering ethics education within various engineering disciplines. The section examines both the similarities and differences in topics suitable for inclusion in the ethics education of various engineering disciplines, as well as the diverse approaches to addressing ethical issues, dilemmas, and challenges within each discipline. Although this section encompasses only five chapters, the set provides a glimpse into the breadth of engineering work within subfields, showcasing a diverse range of concerns and approaches.

The role of the fourth section, 'Teaching methods in engineering ethics education,' is to delineate the established and emergent methods used to teach engineering ethics. Current research reflects a deep fragmentation of pedagogical approaches and confusion as to which approaches are most suitable in preparing socially responsible engineers. There are limited empirical findings to serve as guidance in the implementation and teaching of engineering ethics. However, there is a significant body of knowledge in relation to medicine, business studies, and other science, technology, engineering, and mathematics (STEM) fields that can serve as inspiration. Given that a coherent curriculum strategy requires alignment, Section 4 dialogues with the other handbook sections concerned with theoretical frameworks and assessment strategies. It is important to address the topics falling under these sections in conjunction, as a lack of clarity and alignment might lead to missed educational opportunities. This section includes seven chapters.

Section 5, 'Assessment in engineering ethics education,' deals with the difficult and challenging topic of assessment in engineering ethics education. It encompasses both the assessment of students and the evaluation of courses. Those assessing students in ethics education always have to balance measurability on one hand with aiming for the richness of topics (and developing competencies like moral reasoning or moral attitudes) on the other. Course evaluation poses a similar challenge. *What should be the aims of an ethics course, and how can the course be judged to be good (enough)?* Often, students' satisfaction is considered, but it can also be asked, *What can be reasonably said about a course's effectiveness in realizing moral sensitivity or moral attitude?* This section contains six chapters.

Section 6, the final section, 'Accreditation and engineering ethics education,' addresses accreditation policies and practices that have driven the adoption of ethics education within engineering courses worldwide, considering that expectations (particularly regarding student performance or 'learning outcomes' related to ethics) have been difficult to define and assess. This section on accreditation, as related to engineering ethics education (EEE), considers the background history of ethics in accreditation, maps national and international accreditation values and practices, discusses the role of accreditation in licensure, and reflects critically on whether and how accreditation promotes EEE at local and global levels. Five chapters are included in this section.

The handbook is structured so that this introduction provides an overview of some of the transversal conclusions offered within the book's pages to engineering ethics education researchers and teachers, as well as to higher education management. Each section opens with an editorial introduction that identifies what the respective section editor finds most important to highlight about the section. We recommend reading both the book introduction and the introductions to the six book sections in addition to the specific chapters that align with the reader's interests. In this way, readers can gain both gain an overview of engineering ethics education as a research and teaching discipline, and an in-depth understanding of some of its many facets. Reading section introductions will also contextualize the individual chapters of the book.

Acknowledgments

The editorial team thanks each author for their invaluable contributions and commitment to this handbook endeavor; the authors' collective efforts have resulted in a comprehensive and illuminating exploration of engineering ethics, shedding light on the evolution of how ethics is perceived, defined, and practiced in engineering and engineering education as well as challenges and implications in today's globalized world. It has been a privilege to lead this project and witness the depth of knowledge and dedication demonstrated by each chapter team throughout the process. Together, the editors and authors have navigated complex topics and discussions, striving to provide readers with fresh insights and innovative perspectives on critical aspects of engineering education. From the outset, we editors sought to provide a comprehensive overview of the state-of-the-art in engineering ethics education and to push the boundaries forward collaboratively. We are proud of the collegial spirit and intellectual rigor demonstrated, and we look forward to seeing the impact of our collective work in shaping the future of engineering ethics education.

The editorial team is indebted to the scholars who provided peer review feedback for the chapters.

Reviewers for Section 1 on the foundations of engineering ethics education were Aline Medeiros Ramos, Annuska Zolyomi, Ester Giménez Carbó, Gaston Meskens, Jie Gao, Kari Zacharias, Michael Kühler, Mihály Héder, Mikael Laaksoharju, Nienke Nieveen, Nihat Kotluk, Per Fors, Shannon Chance, and Taylor Joy Mitchell. Thomas Taro Lennerfors led the development of this section, and other editorial team members who attended discussions with the chapter authors were Roland Tormey and Tom Børsen. Shannon Chance and Tom Børsen contributed significantly to the development of the overview statement.

Reviewers for Section 2 on interdisciplinary contributions to engineering ethics education were Curwyn Mapaling, Emanuela Tilly, Irene Josa i Cullere, Jacqueline Eng, Jolanta Kowal, Julianna Gesun, Madeline Polmear, Nihat Kotluk, Richard Randell, Robert Braun, Silvia Bruzzone, and Silvia Gherardi. Roland Tormey led the development of this section. Thomas Taro Lennerfors and Tom Børsen provided feedback to Roland Tormey regarding the section introduction.

Reviewers for Section 3 on ethical issues in different engineering disciplines were Aaron Johnson, Amir Hedayati, Brent K. Jesiek, Mauryn C. Nweke, Diana Bairaktarova, Diana Martin, Erin Henslee, Irene Josa i Cullere, Jan Mehlich, Michael Kühler, Mike Murphy, Shannon Chance, and Vivek Ramachandran. Tom Børsen led the development of this section, and editorial team members Diana Martin, Gunter Bombaerts, and Roland Tormey attended discussions with the chapter authors. Shannon Chance contributed significantly to the development of the overview statement.

Reviewers for Section 4 on teaching methods in engineering ethics education were Andrew Brightman, Christina Boshuijzen van Burken, Cristiano Cordeiro Cruz, Cristina Voinea, Esther Matemba, Heather A. Love, Inês Direito, John Bernhard Kleba, Karolina Doulougeri, Kaushik Mallibhat, Natalie Wint, Olga Pierakkos, and William (Bill) Oakes. We also value various contributions to this section from Gunter Bombaerts, Irene Magara, Jorge Membrillo Hernandez, Aida Guerra, and Karolina Doulougeri. Diana Martin led the development of this section, and other editorial team members who attended discussions with the chapter authors and provided input during the process were Tom Børsen and Shannon Chance. Diana led the writing of the section overview, with extensive contributions from Shannon Chance.

Reviewers for Section 5 on the assessment in engineering ethics education were Adetoun Yeaman, Alaa Abdalla, Alison Gwynne-Evans, Andrea Gammon, Elena Mäkiö, Elise Dykhuis, Irene Magara, Jorge Membrillo Hernández, Juho Mäkiö, Helena Kovacs, Rockwell Clancy, and Sarah Junaid. Gunter Bombaerts led the development of this section, and other editorial team members who attended discussions with the chapter authors and provided input during the process were Diana Martin, Tom Børsen, and Shannon Chance.

Reviewers for Section 6 on accreditation and engineering ethics education were Corin L. Bowen, Diana Martin, Yousef Jalali, Jose Fernando Jimenez Mejia, Kenichi Natsume, Lavinia Marin, Robyn Mae Paul, Scott Daniel, Stephanie Hladik, and Yann Serreau. Shannon Chance led the development of this section; Tom Børsen and Diana Martin attended discussions with the chapter authors and provided input during the process. Shannon led the writing of this overview, with important contributions from Diana Martin.

Shannon Chance and Tom Børsen sincerely thank our editors at Routledge – AnnaMary Goodall and Kanishka Jangir – for their unwavering support. Additionally, we extend our heartfelt gratitude to Pat Gordon-Smith of UCL Press for the enthusiasm, mentorship, and extensive advice she provided during the early stages of the project.

We drew inspiration from the *Cambridge Handbook of Engineering Education Research* (Johri & Olds, 2014) and the *International Handbook of Engineering Education Research* (Johri, 2023). The second of these had similar aims regarding institutional and geographical diversity; Diana and Shannon served as Associate Editors (as well as authors and peer reviewers), supporting the handbook's Editor, Aditya Johri, in his efforts to reach farther outside North America for perspectives and contributions. Aditya helped pave the way for this new ethics-specific handbook by negotiating with Routledge the open-access purchase agreement that our present team was able to use and sharing advice and insights into the process with Shannon.

Each handbook editor facilitated a contribution to the financial costs of ensuring open access to this new *Routledge International Handbook of Engineering Ethics Education*. The open access cost was covered by Technological University Dublin (Shannon and Diana), the Department of Sustainability and Planning at Aalborg University (Tom), EPFL (Roland), the Division of Industrial Engineering and Management at Uppsala University (Thomas), Eindhoven University of Technology (Gunter and Diana), and University College London (Diana and Shannon). Providing this handbook as a free and openly available digital resource was a primary goal of Shannon and Tom from the outset; the whole editorial team embraced this vision and secured funding to support the project. We and the whole editorial team are indebted to our institutions for supporting us in our work and in purchasing open access.

We explicitly sought to ensure that newer voices could confidently contribute to the book by matching less experienced authors (i.e., those undertaking PhDs or just entering the realm of EEE research) with scholars who have more writing experience. We are thankful to the more senior researchers who welcomed newer entrants to the field onto their teams and, in many cases, provided mentorship. We recognize that those who did so provided an extra, and invaluable, service to our global EEE and engineering education research community. We hope that readers will recognize these mentoring and capacity-building characteristics in many of the projects and activities we lead, including and beyond this handbook.

We collectively thank our families and significant others who put up with us and were supportive as we spent long hours on the project. We also thank our colleagues for the conversations that pushed us to think better about the work at hand.

In conclusion, our journey in crafting this handbook on engineering ethics education has been one of joyful co-creation, mutual respect, and collective endeavor. With a diverse array of voices contributing to the discourse, our editing process aimed to maintain coherence within chapters, thematic sections, and the book overall, all while reflecting the dynamic landscape of engineering ethics education today, both as a research domain and a pedagogical subject. Engaging with 105 authors, we have cultivated a collaborative, multi-perspectival text that we hope will be seen as a significant contribution to our field. The bi-weekly meetings of our editorial team, and the many breakout sessions conducted with various clusters of editors and contributors, have proven fruitful and intellectually enriching, fostering friendships and trusted collegial bonds. The overall experience has been immensely fulfilling, yielding a resource that we hope will be key to catalyzing action to further advance engineering ethics education. We welcome readers to join our community by reaching out to the editors and authors regarding topics of mutual interest and possibilities for future collaboration. Together, we want to celebrate and extend the value of this handbook and grow the community of voices contributing to our understanding of engineering ethics and engineering ethics education.

References

Anderson, B. (1991). *Imagined communities: Reflections on the origin and spread of nationalism* (Rev. ed.). Verso.

Au, W. (2011). Teaching under the new Taylorism: High-stakes testing and the standardization of the 21st century curriculum. *Journal of Curriculum Studies, 43*(1), 25–45.

Bombaerts, G., Doulougeri, K., Tsui, S., Laes, E., Spahn, A., & Martin, D. A. (2021). Engineering students as co-creators in an ethics of technology course. *Science and Engineering Ethics, 27*(4), 48. https://doi.org/10.1007/s11948-021-00326-5

Brotton, J. (2013). *A history of the world in 12 maps*. Penguin.

Cruz, C. C. (2021). Brazilian grassroots engineering: A decolonial approach to engineering education. *European Journal of Engineering Education, 46*(5), 690–706.

Darling-Hammond, L. (2006). *Powerful teacher education: Lessons from exemplary programs*. Jossey-Bass.

Dewey, J. (1929). *The sources of a science of education*. H. Liveright.

Doorn, N., & Van de Poel, I. (2012). Editors' overview: Moral responsibility in technology and engineering. *Science and Engineering Ethics, 18*, 1–11.

Fleddermann, C. B. (2014). *Engineering ethics* (4th ed.). Pearson Education Limited.

Foucault, M. (2001). *The order of things* (2nd ed.). Routledge.

Friel, B. (1981). *Translations*. Faber & Faber.

Gani, J. K., & Khan, R. M. (2024). Positionality statements as a function of coloniality: Interrogating reflexive methodologies. *International Studies Quarterly, 68*(2), sqae038.

Garuba, H. (2002). Mapping the land/body/subject: Colonial and postcolonial geographies in African narrative. *Alternation, 9*(1), 87–116.

Gleeson, J. (2013). The European credit transfer system and curriculum design: Product before process?. *Studies in Higher Education, 38*(6), 921–938.

Hampton, C., Reeping, D., & Ozkan, D. S. (2021). Positionality statements in engineering education research: A look at the hand that guides the methodological tools. *Studies in Engineering Education, 1*(2), 126–141.

Hitt, S. J., & Lennerfors, T. T. (2022). Fictional film in engineering ethics education: With Miyazaki's the wind rises as exemplar. *Science and Engineering Ethics, 28*(5), 44.

IPCC. (2023). Summary for policymakers. In H. Lee & J. Romero (Eds.), *Climate change 2023: Synthesis report. Contribution of working groups I, II and III to the sixth assessment report of the intergovernmental panel on climate change* [Core Writing Team] (pp. 1–34). IPCC.

Johri, A. (2023). *International handbook of engineering education research*. Taylor & Francis.

Johri, A., & Olds, B. M. (Eds.). (2014). *Cambridge handbook of engineering education research*. Cambridge University Press.

Lennerfors, T. T. (2019). *Ethics in engineering*. Studentlitteratur.

Lönngren, J. (2021). Exploring the discursive construction of ethics in an introductory engineering course. *Journal of Engineering Education, 110*(1), 44–69. https://doi.org/10.1002/jee.20367

Lonngren, J., Bellocchi, A., Bogelund, P., Direito, I., Huff, J., Mohd-Yusof, K., ... Tormey, R. (2021). Emotions in engineering education: Preliminary results from a scoping review. In *REES AAEE 2021 conference: Engineering education research capability development: Engineering education research capability development* (pp. 225–233). Engineers Australia.

Lönngren, J., Direito, I., Tormey, R., & Huff, J. L. (2023). Emotions in engineering education. In A. Johri (Ed.), *International handbook of engineering education research* (pp. 156–182). Routledge.

Marks, L. S. (1978). *Marks' standard handbook for mechanical engineers. Standard handbook for mechanical engineers.*
Martin, D. A., Conlon, E., & Bowe, B. (2019). The role of role-play in student awareness of the social dimension of the engineering profession. *European Journal of Engineering Education, 44*(6), 882–905.
Perry, J. H. (1950). *Chemical engineers' handbook*. New York: McGraw4-Hill, 388.
Phillips, R. (1996). *Mapping men and empire: Geographies of adventure*. Routledge.
Secules, S., McCall, C., Mejia, J. A., Beebe, C., Masters, A. S., L. Sánchez-Peña, M., & Svyantek, M. (2021). Positionality practices and dimensions of impact on equity research: A collaborative inquiry and call to the community. *Journal of Engineering Education, 110*(1), 19–43.
Seniuk Cicek, J., Masta, S., Goldfinch, T., & Kloot, B. (2023). Decolonization in engineering education. In A. Johri (Ed.) *International handbook of engineering education research* (pp. 51–86). Routledge.
Sukackė, V., Guerra, A. O. P. D. C., Ellinger, D., Carlos, V., Petronienė, S., Gaižiūnienė, L., Blanch, S., Marbà-Tallada, A., & Brose, A. (2022). Towards active evidence-based learning in engineering education: A systematic literature review of PBL, PjBL, and CBL. *Sustainability, 14*(21), 13955.
Van de Poel, I., & Royakkers, L. (2023). *Ethics, technology, and engineering: An introduction*. John Wiley & Sons.
Wolodtschenko, A., & Forner, T. (2007). Prehistoric and early historic maps in europe: Conception of cd-atlas. *e-Perimetron, 2*(2), 114–116.

SECTION 1

Foundations of engineering ethics education

Thomas Taro Lennerfors

This opening section of our handbook delves into the foundational principles, topics, and themes that underpin the multifaceted landscape of engineering ethics education. While the term 'foundations' may suggest a definitive framework, we embrace the dynamic and evolving nature of ethical discourse within engineering practice, teaching, and research. Recognizing ethics as an ongoing dialogue, we approach this section humbly, acknowledging our vibrant research and teaching community's perpetual flux of perspectives and priorities.

Thus, contrary to implying a conclusive stance, our use of 'foundations' denotes pivotal themes and concepts permeating this handbook, including purposes of engineering ethics education, normative ethical theories, professional responsibilities, ethical decision-making, emotions, and environmental concerns. In the specialized discussions that take place in an increasingly fragmented academic discourse, we see it as being of utmost importance to confront these foundational issues directly, albeit with a recognition of the inherent complexities and inexhaustible demands they entail.

At the base of this 'foundations' section lies a fundamental question: What are the purposes of engineering ethics education? This question serves as this section's guiding beacon, illuminating subsequent explorations into normative ethical theory; individual and collective decision-making; the nuanced interplay of emotions; the role of professionalism and professional organizations; the oft-overlooked dimension of environmental, ecology, and nature; and the burgeoning domain of artificial intelligence (AI) and its intersection with engineering ethics education (EEE).

By situating these chapters within the broader field, we provide a panoramic view of foundational elements, which support many other concepts presented throughout the book: ethical theories, professional ethical codes, emotions and reason, EEE objectives, and individual and collective ethical decision-making. From the outset, we acknowledge the inherent incompleteness of our endeavor and invite ongoing discourse to further interrogate the landscape.

Chapter topics, trends, and implications

Perhaps the most foundational question about engineering ethics education regards why we have EEE in the first place. Given the substantial power and knowledge that engineers wield in shaping, constructing, implementing, and deconstructing the intricate and interconnected material frame-

DOI: 10.4324/9781003464259-2

work that constitutes society, the environment, and reality, engineers must grasp the weight of their influence. This material framework not only significantly impacts the natural environment but also becomes the milieu of human existence and affects life in all its forms on this planet. Engineers should thus be cognizant of their agency and expertise, employing them judiciously. Nevertheless, one can critique this imposition of agency on engineers as they are also part of a socio-techno-ecological system with its own directionality, placing obstacles in the way of engineers' responsibility-taking. Whether one highlights the agency of either individual or groups of engineers or rather chooses to put the structures, networks, and constraints on the agency of engineers to fulfill their professional responsibilities, our conclusion is that engineers always have agency (albeit to varying degrees) to take responsibility. It is upon this agency that much of engineering ethics education rests. Given that we have now highlighted how even this agency is contested, for engineering ethics educators and researchers, it is important to think about agency and responsibility and how it is highlighted or downplayed in research literature, textbooks, assignments, and classroom exercises. Perhaps one could even claim that much of engineering ethics education's purpose is to 'perform' this agency in an educational setting.

Given our understanding of the existing literature on the purposes of EEE, we can view EEE as serving several purposes. EEE is claimed to have the purpose of enhancing the sensitivity of future engineers to ethical issues, increasing their knowledge of relevant standards, building their ethical judgment, and increasing their ethical willpower (Herkert, 2000). Lennerfors et al. (2020) argue that EEE should instill awareness, responsibility, critical thinking, and action in students. Tormey et al. (2015) maintain that EEE should nurture moral sensitivity, judgment, motivation, and character. Our view is that although the wordings of purpose differ among scholars, their suggestions are quite similar, and there seems to be an agreement. Yet, is the agreement fictional, in the sense that there is no real discussion about foundations? A discussion about purposes is not enough. There is a need to *align* various learning activities to the purposes of EEE. Previous research literature indicates that the purposes of EEE have not always been implemented in practice. One major critique is that EEE has been focused on what Herkert would call sensitivity, knowledge, and judgment; what Lennerfors and colleagues would call awareness and critical thinking; and what Tormey et al. would call moral sensitivity and moral judgment – thus excluding the important dimensions of taking responsibility and engaging in ethical action (Lennerfors et al., 2020). If this critique is valid, it seems as if the purposes of EEE are not present in the day-to-day teaching, which is yet another reason for re-starting discussions about the purposes of EEE.

Chapter 1, 'The purposes of engineering ethics education' by Qin Zhu, Lavinia Marin, Aline Medeiros Ramos, and Satya Sundar Sethy, contributes to this discourse on the purposes of EEE by presenting a novel framework. The authors' framework highlights individual aspects such as knowledge, actions, personal habit formation, and values in artifacts and addresses more holistic considerations, emphasizing relationships, the environment, and other systems. We consider this contribution essential for reigniting and broadening discussions about the objectives of engineering ethics education. We also see that this should be regarded as part of a debate that needs to be re-ignited. And perhaps more importantly, this is a call to engineering ethics educators and researchers to bring a reflexive discussion about purposes into our daily practices.

In contemporary EEE, the role of normative ethical theories – such as consequentialism, deontology, rights-based theories, virtue ethics, relational ethics, ethics of care, existential ethics, theories of justice and fairness, and environmental ethics – is a subject of ongoing debate. The designation of the field as 'engineering ethics education' inherently connects it to normative ethical theories as discussed within philosophy. Yet, alternative labels like 'responsible engineering' may challenge the centrality of normative ethical theories, although responsible engineering, of course, would include some other normative framework. In any case, within EEE, a significant divergence exists in the treatment of such theories. Whereas some courses emphasize including normative ethical theories, others intentionally omit them, either because of knowledge or resource constraints or because the theories might not seem practical enough. There is a contentious discussion surrounding which theories should be included, the rationale for their inclusion, and how they should be applied. Consequently, the role of normative ethical theory within EEE remains a pressing concern that requires comprehensive discussion. Moreover, there have been notable criticisms regarding the perceived Eurocentric, speciesist, anthropocentric, and male-centric nature of some normative ethical theories. Over the past five decades, efforts to diversify the discourse have intensified, with increasing emphasis on incorporating a spectrum of alternative theories, such as ethics of care and ethical frameworks from diverse cultural contexts.

Chapter 2, 'Ethical theories' by Michael Kühler, Natalie Wint, Rafaela Hillerbrand, and Ester Gimenez-Carbo, tackles the role of normative ethical theories head-on. The authors argue that while such theories are undoubtedly important, careful attention must be paid to how they are integrated into EEE. They advocate for an approach that avoids oversimplification and ensures the inclusion of underrepresented theories. By adopting such a nuanced approach, EEE can effectively fulfill its mission of fostering ethical competence among students, reinforcing the broader purposes outlined in preceding chapters. The chapter takes a normative stance that we should systematically use normative ethical theories, avoid logical contradictions, and strive for internal consistency. As part of a philosophy of openness and breadth, it is worth mentioning that another strand of thinking highlights the instrumental role that normative ethical theories can have alongside other normative ethical frameworks. In other words, rather than using normative ethical theories in a consistent and logical way, some see such theories as providing helpful but not conclusive foundations on what can be considered ethical. Rather, these theories' instrumental nature and usefulness are the core concerns. Some see the theories as being possible to combine and include as components in an ethical decision-making framework, with the argument that such use of normative ethical theories is enough for "non-ethicists," that is, practical professions such as engineering (Lennerfors, 2019).

Ethical decision-making constitutes a crucial pillar of EEE, involving the process by which individuals or collectives navigate a spectrum of alternative courses of action, each with its own advantages and drawbacks. This approach resonates particularly with engineering students, offering them an understanding of ethics not as something fundamentally distinct from general decision-making processes but as a realm where similar frameworks can be applied. This can also have pedagogical implications, since ethical decision-making lends itself to practical problem-solving rather than academic seminars (although we, of course, see the latter's value). What is usually perceived as the major difference between ethical decision-making and decision-making in general is the existence of normative frameworks, which one can (or should) be informed by when making decisions concerning ethics. Numerous frameworks have been developed to support ethically sound decision-making, and a thorough review by Walter Maner (2002) identified hundreds of frameworks developed for this purpose. However, the ethical decision-making literature has faced criticism for its focus on individual decision-makers, often neglecting the dynamics of collective decision-making and the inherent power differentials. Critics argue that these frameworks prioritize rationality and agency while downplaying the influence of power dynamics and inequalities. Moreover, there's concern that an excessive emphasis on analysis may lead to decision paralysis, favoring endless argumentation over tangible action and perpetuating the status quo rather than fostering socially desirable change.

In Chapter 3, 'Individual and collective dimensions of ethical decision-making in engineering' Kari Zacharias, Marion Hersh, Andrew O. Brightman, and Jonathan Beever explicitly explore the roles of individuals and collectives across various levels in ethical decision-making (teams, organizations, professions, societies, and ecosystems). The chapter offers concrete examples of frameworks for educational practice, highlighting opportunities for enhancing ethical decision-making by bridging various levels with contrasting priorities. By connecting individuals to the diverse communities to which they belong, ethical decision-making can be significantly enriched and contextualized.

Historically, engineering has emphasized rationality, often sidelining the role of emotions in decision-making and ethics. A cognitive-rationalist approach, prioritizing critical thinking, has long dominated engineering discourse. This focus has been linked to masculinity, implying a normative focus on rationality rather than emotion. However, there has been a gradual evolution in the perception of emotions. They are increasingly recognized as important factors and potential facilitators of ethical decision-making rather than merely threats to critical thinking. Although mainstream EEE discourse still typically fails to acknowledge the significance of emotions as core elements shaping actions and outcomes, this gap is gradually subsiding as more attention is directed toward emotions in both research and pedagogical development within engineering education.

A growing body of literature is exploring the role of emotions in ethical decision-making, accompanied by efforts to integrate emotional intelligence into educational practices. Based on existing research on this topic, presented in Chapter 4, 'Reason and emotion in engineering ethics education' by Nihat Kotluk, Johanna Lönngren, and Roland Tormey, we can say that both reason and emotions are needed in EEE. They complement each other; neither reason nor emotion can stand alone. They also shape each other given that one can train one's gut feelings, and feelings can shape and even reject a rational ethical argument. Chapter 4 draws from diverse disciplines, and the authors build a comprehensive knowledge base tailored to educators and researchers in the field. Through this exploration, the chapter emphasizes the significance of addressing emotions not only within the realm of EEE but also in broader contexts, highlighting their multifaceted role in ethical decision-making and professional practice.

Another foundational issue within EEE is the engineer's professionalism and adherence to a professional code of conduct, yet another ethical framework often used in EEE. As future engineers, students are expected to be versed in the professional code of conduct and understand its role in providing guidance when facing ethical dilemmas. This raises crucial questions about the purpose of the professional code, its origins, and how engineers should engage with it. Should engineers blindly adhere to codes, view them as one among many tools for ethical engineering practice, or adopt a critical and reflexive stance toward them, recognizing potential contradictions and limitations?

Chapter 5, 'Professional organizations and codes of ethics' by Jeff R. Brown, Leroy Long III, Taylor Mitchell, and Renato Bezerra Rodrigues, delves into the role of professions in engineering ethics education, applying Freidson's (2001) framework of professionalism to examine the historical and contemporary significance of professionalism within engineering. The chapter draws on empirical examples from the United States and Canada to illustrate this discussion. Interestingly, upon reading this contribution, it becomes evident how distinct the North American context is from other national contexts, such as that in Sweden, where the section's lead editor lives and works. In Sweden, an ethos of professionalism is prevalent among engineers, but this does not always align with membership in professional organizations or a deep understanding of the ethical codes governing engineering work. The status of the professional societies in varying cultural contexts has important ramifications for EEE. In some contexts, it can be a valid educational element to frame codes of conduct as something that future engineers must subscribe to and must relate to. In other contexts, such as the Swedish one, students are often surprised to hear about the existence of an engineering ethics code of conduct, and there is usually plenty of discussion about why the code exists in the first place and a range of critical remarks about the code. In contexts where the code has less of a definitive normative character, it might be easier to think critically about the codes and the power relations that have gone into shaping them than in contexts where future engineers need to abide by the code.

In this handbook, Chapters 1 and 3, among others, touch on the relationship between engineering, EEE, and the natural environment. The natural environment deserves dedicated focus, especially given that we live in an era defined by humanity's significant impact on non-human life and the planet itself.

Chapter 6, 'A post-normal environment-centered approach to engineering ethics education' by Tom Børsen, Shannon Chance, and Gaston Meskens, contends that ethics is intricately intertwined with socio-ecological sustainability. Thus, the chapter describes several normative frameworks from sustainability science and environmental ethics. This intertwining goes beyond mere issues of knowledge and extends into politics (within political contexts, specific individuals or groups – stakeholders – possess interests, situated knowledge, and values that are considered crucial to address). The chapter foregrounds the relationship between EEE and the environment by calling for a paradigm shift in engineering education and introducing the concept of post-normal engineering (PNE), which advocates reflexivity as a way forward.

This chapter connects to a debate that should be more present within EEE, namely the connection between 'ethics' and 'sustainability.' Here, it is interesting to see how different scholarly communities create an understanding of themselves and the foundational topics concerning both their own domain and that of the other domain. There might thus be a tendency among sustainability scholars to see ethics as one limited but important strand within their field. In contrast, ethics scholars might see sustainability (depending on how it is defined) as one of the many normative values that need to be promoted as part of ethical judgment and action. The handbook editors see many potential connections between ethics and sustainability and perceive that academic and pedagogical discussion would benefit from being more open to fruitful interchange. In a text that the section lead has previously written (Lennerfors & Murata, 2023), a preliminary investigation into commonalities and differences between the two domains was conducted, focusing on the issues of concern (where both fields promote a range of normative values related to the society and environment, among which there are inevitably value conflicts), methodologies (where some impact assessment techniques are shared, but where there are also discrepancies, e.g., the use of normative ethical theories in ethics and life-cycle assessments in sustainability). Finally, there is an argument that neither ethics nor sustainability (within engineering and beyond) are harmonious bodies of knowledge. Rather, they are fragmented and contested, which implies that the domains can make connections between the domains more straightforward. The underlying argument is that because there is no fundamentally distinct essence separating the fields, bridging them should be feasible.

As researchers and educators in EEE, we are acutely aware of the far-reaching implications of new technologies on our educational practices. As mentioned in the introduction to this handbook, artificial intelligence (AI) has recently gained significant momentum. The advent of transformative technologies like ChatGPT and other powerful applications has underscored the need for ongoing adaptation and innovation in our educational approaches.

Chapter 7, 'Engineering ethics education and artificial intelligence' by Cécile Hardebolle, Mihály Héder, and Vivek Ramachandran, provides a comprehensive discussion on the implications of AI for engineering and engineering education and points to related ethical issues. Beyond addressing issues like cheating, the chapter delves into essential considerations regarding what aspects of AI ethics should be incorporated into engineering curricula and how best to do so. The authors explore the potential for AI to be leveraged positively within the engineering ethics classroom, envisioning scenarios where AI is utilized to enhance and support students in their pursuit of fulfilling the objectives of engineering ethics education.

Conclusions from the section editor

In this section, a range of foundational issues have been presented. We (editors) believe all those issues are crucial for engineering ethics educators and researchers to consider. Two themes that come through in the section are openness and breadth.

We want to maintain an open approach, which is why we have provided different perspectives on how all these foundational issues can be approached. For example, regarding the purposes of EEE, we not only presented a new set of purposes but also argued that these purposes should be the start of a new debate and that what matters eventually is how these purposes (or indeed other purposes) make their way practically into EEE. We recognize that some organizations identify their values primarily for external purposes, while their everyday work diverges from the values. We believe that EEE researchers and practitioners should often, or at least occasionally, scrutinize their work concerning the purposes, to help ensue validity and ongoing alignment. Similarly, concerning normative ethical theories, we have left the situation unresolved. We argue that EEE researchers and practitioners need to consider normative ethical theories; we do not provide a conclusive position but merely guidance regarding how they could be used in teaching.

We also have promoted the broadening of the scope of EEE. Rather than only covering human-to-human or human-to-technology relationships, we have included the natural environment in a range of chapters, which we think has been downplayed in earlier scholarship. Yet again, we have found that previous research and teaching have focused on rationality while downplaying emotions. To highlight the role of emotions in EEE is another way of broadening the scope. We have also broadened ethical decision-making from an individualist focus to more collective forms, as well as intending to eschew the critique of normative ethical theories neglecting a range of stakeholders and concerns.

Still, despite the openness and breadth that we have tried to promote, the work is always incomplete, and the section is an invitation to continue the discussion.

Positionality

As elaborated in the introductory chapter, the genesis of foundational topics emerged organically through collaborative, bottom-up deliberations within our editorial team and the broader EEE community.

However, the section has also been influenced by the position of the section editor, Thomas Taro Lennerfors. He is an industrial engineering and management graduate, which means that from the beginning he was taught that he was expected to bridge the ‘management’ domain and the ‘engineering’ domain of companies. Although he never worked in industry but embarked upon an academic career, he has always been working in the interstices, where there is contestation and where people do not agree with each other. This has also led to a career which is quite open and broad, which connects to the concluding remarks of the section. He has also, with the support of academic mentors and benefitting from being part of an open academic environment, integrated a range of theories and methods into his work. For Thomas, living in a range of countries has contributed to being open to cross-cultural differences. Furthermore, the chapters concerning purposes, normative ethical theory, ethical decision-making, professions, and emotions connect directly to how engineering ethics is taught at Uppsala University, where Thomas works. In editing this section, Thomas shared his perspectives with the author teams, but was also careful not to overly influence the content of the chapters with his own views.

References

Freidson, E. (2001). *Professionalism, the third logic: On the practice of knowledge*. University of Chicago Press.

Herkert, J. R. (2000). Engineering ethics education in the USA: Content, pedagogy and curriculum. *European Journal of Engineering Education, 25*(4), 303–313.

Lennerfors, T.T. (2019). *Ethics in Engineering*. Studentlitteratur.

Lennerfors, T. T., Laaksoharju, M., Davis, M., Birch, P., & Fors, P. (2020). A pragmatic approach for teaching ethics to engineers and computer scientists. In *2020 IEEE frontiers in education conference (FIE)* (pp. 1–9).

Lennerfors, T.T. & Murata, K. (2023). Ethics and sustainability in digital cultures: A prolegomena. In T. Lennerfors & K. Murata (Eds.), *Ethics and sustainability in digital cultures* (pp. 1–16). Routledge.

Maner, W. (2002). Heuristic methods for computer ethics. *Metaphilosophy, 33*(3), 339–365.

Tormey, R., LeDuc, I., Isaac, S., Hardebolle, C., & Cardia, I. V. (2015). *The formal and hidden curricula of ethics in engineering education*. Paper presented at the 43rd Annual SEFI Conference, Orléans, France.

1
THE PURPOSES OF ENGINEERING ETHICS EDUCATION

Qin Zhu, Lavinia Marin, Aline Medeiros Ramos, and Satya Sundar Sethy

Introduction

Engineering ethics, as a field of applied and professional ethics, assumes an essential role in forming professional identity and the everyday decision-making for engineers. Clearly articulating and rigorously assessing the purposes of engineering ethics education (EEE) constitutes the foundation of successful engineering ethics programs. Furthermore, understanding the purposes of EEE is critical for all stakeholders involved in engineering education, including engineering students, instructors, engineering programs and institutions, employers, and accrediting bodies.

For instance, understanding the purposes of EEE can be a catalyst for actively involving students in the learning process. This aligns with research in the field of learning sciences, which emphasizes that students are most receptive to knowledge when they clearly understand their learning goals and expectations (Dotson, 2016). Moreover, the capacity to design more impactful, pertinent, and engaging learning experiences for students hinges upon instructors' understanding of the underlying purposes of EEE (Liow et al., 1993). Within the realm of engineering programs and institutions, a comprehensive understanding of the purposes behind EEE establishes a normative framework that shapes the outcomes and purposes of engineering education.

EEE has become central to most undergraduate engineering programs across the globe. When considering teaching engineering ethics to undergraduate engineering students, the question arises regarding its *modus operandi,* which includes whether an engineering ethics course is to be offered as an elective or a core module/course for an engineering program, what the learning objectives are, who shall teach engineering ethics, how engineering ethics shall be taught, and what assessment and evaluation methods should be used to evaluate students' learning. Researchers across the globe have been discussing these questions in their works. Hence, it is crucial to address these questions in this chapter as they are foundational for discussing the purposes of EEE.

A major contribution of this chapter is to construct a conceptual framework to systematically describe and compare various approaches to the purposes of EEE. It is worth noting that such a framework is inherently embedded with a tension between a normative approach and a pragmatic approach to the possible purposes of EEE. A normative approach is primarily interested in what the purposes of EEE should be. Starting from the risks and harms that engineering as a profession can give rise to and from asking how engineers can help make the world a better place, this

DOI: 10.4324/9781003464259-3

approach posits first the ends of EEE without concern for the actual means of achieving them. If the ends are clear, then the means will follow. The other approach is pragmatic and starts from the question 'What can be achieved through educational practice?' This approach considers what has been done already and the limitations inherent to any educational endeavor: time, energy, and resources (material and cognitive).

The normative approach concerns what these purposes *should be*, given the needs of the engineering profession and of society at large. The pragmatic approach concerns what these purposes *are*, in educational practice and in policy making. Such descriptive formulations usually follow complex negotiations between multiple stakeholders and will be different based on each country's own priorities. The normative approach is one in an 'ideal world' scenario, while the pragmatic one is the result of 'actual world' situated outcomes of negotiations between the stakeholders. In order to differentiate the normative from the pragmatic approaches – and to show what these have in common – we will use a conceptual framework that illustrates the wide range of these purposes, strictly philosophically speaking.

Following this introduction, this chapter will address three key themes. First, before we delve into the purposes of EEE, we discuss some fundamental questions about the nature of the purposes of EEE. These questions have been extensively discussed and debated in the EEE literature and are interrelated. The question of whether engineers' moral actions should align with their personal morality or adhere to professional ethics is pivotal for establishing the legitimacy of the field of EEE. This consideration significantly impacts the purposes of EEE, or what we, as engineering educators, expect our students to learn about engineering ethics. Such a question is also connected to another critical question in EEE: whether engineering ethics is teachable. If engineering ethics is merely an application of personal morality in engineering, then since students arrive in undergraduate classrooms, either we have little to teach them about personal morality or we cannot teach students about engineering ethics (as personal morality is closely linked to an individual's early developmental stages, it is inherently shaped by their foundational beliefs established during that period) (Abaté, 2011; Harris et al., 1996; Veach, 2006). Alternatively, if engineering ethics is construed as professional ethics, encompassing "special morally permissible standards of conduct that every member of a profession wants every other member to follow" (Harris et al., 1996, p. 93), then it implies that not only is engineering ethics a subject that can be taught, but also that understanding it requires more than an individual's personal experiences. Finally, whether teaching engineering ethics is about developing ethical knowledge and skills or developing moral habits is crucial for discussing the purposes of EEE. Further deliberation on this question extends to exploring the diverse purposes of EEE, which are captured by a conceptual framework in the following section.

Second, to capture the diversity of purposes of EEE, we constructed a conceptual framework comprising six approaches to understanding the purposes of EEE: knowledge, personal traits, actions, values in artifacts, relations, and ecosystems. Arguably, the foundational assumption of individualistic rationalism underpins the first four approaches, which perceive engineers as entirely rational and autonomous individual decision-makers (Zhu & Clancy, 2023). Moral engineers are thus those who are capable of developing moral knowledge and behavioral tendencies in engineering practice or creating technologies to exert positive moral influences on society. Conversely, the remaining two approaches adopt a holistic perspective, emphasizing the interconnected nature of the world and the impact of engineering practices on this interconnectedness.

Finally, we discuss normative analysis via two normative yet practical questions: 'Who should teach engineering ethics based on this wide variety of purposes?' and 'Who gets to decide on these purposes in practice, namely, who are the stakeholders involved in these decisions?'Both of these questions bear social and political relevance, as the qualifications, background, and disciplinary training of individuals qualified to teach engineering ethics play a pivotal role in shaping their understanding of engineering ethics, encompassing the purposes of EEE, and influencing how ethics learning outcomes are formulated (Barry & Herkert, 2015). Furthermore, more recent research has shown that decision-making concerning EEE is a socially constructed process, entailing power negotiations among diverse stakeholders (Martin et al., 2021). In recognition of the socially constructed nature of EEE and engineering education research, we begin by describing our positionality as a team of authors.

Positionality

The positionality of each co-author of this paper is shaped by our own set of personal and professional experiences. Qin Zhu is an associate professor of Engineering Education at Virginia Tech. Trained as both a materials engineer and philosopher in China, and later transitioning into the role of an engineering education researcher in the United States, he has adopted a cultural and critical perspective. From this perspective, he explores values, norms, and cultural assumptions embedded in the professional formation of engineering identity and the development and deployment of technologies, such as artificial intelligence (AI) and robotics. Lavinia Marin is a Romanian philosopher working in the Netherlands at TU Delft. In addition to her philosophical background, which is predominantly informed by the Western canon, Lavinia was trained as an electrical engineer and has worked for several years in a large state-owned company in Bucharest, Romania, which informed her vision on EEE goals in practice. Aline Medeiros Ramos is a Brazilian philosopher who received most of her postsecondary education in North America (New York and Montreal). She is now based in Canada at Université du Québec à Trois-Rivières. She specializes in medieval philosophy and ethics and has a background in classics. She teaches courses on the history of philosophy and professional ethics, primarily to engineering and medical students. Satya Sundar Sethy is a professor of Philosophy in the Department of Humanities and Social Sciences of the Indian Institute of Technology Madras. He specializes in applied ethics (engineering ethics, academic ethics), engineering education, consciousness studies, logic, and Indian philosophy. He was conferred with the 'Young Philosopher' award by the Indian Council of Philosophical Research, Ministry of Education, New Delhi, in 2017. He was also awarded a Scholar-in-Residence Fulbright Fellowship to carry out teaching and research tasks in Utah (USA) in 2022–2023.

Some fundamental questions

EEE is usually seen as essential to prepare future engineers for the complex ethical challenges they will encounter. A proper pedagogical path into this field requires the critical examination of some fundamental ethical questions. By engaging in this kind of reflection, we can gain valuable insights into the purpose and nature of EEE. Thus, we can ensure its continued relevance in rapidly evolving technologies. This section delves into some fundamental philosophical questions that determine or shape the definition and prioritization of the purposes for EEE. More specifically, it will address (1) the distinction between engineering ethics and morality, (2) the teachability of engineering ethics, and (3) the balance between theoretical knowledge, technical operationalizable skills, and moral habits in engineering ethics. Responses to these questions all have a profound impact on how the purposes of EEE are or should be articulated.

Engineering ethics, morality, and personal ethics

The first question to be explored revolves around the relationship between engineering ethics and so-called 'personal ethics.' By analyzing their similarities and differences, we can achieve a clearer understanding of the unique considerations and responsibilities that engineering ethics requires. Thus, we can articulate the distinct ethical framework required within the engineering profession.

In this chapter, we will not make any philosophical distinction between the terms 'ethics' and 'morality' (as this is outside the scope of this chapter). We will use the terms 'ethics' and 'morality' synonymously and interchangeably. Having said that, we will distinguish engineering ethics from personal ethics. While there may be some overlap between them, they differ in scope and focus in many ways. The first way they differ is in scope. Engineering ethics is a specific branch of ethics – namely of *applied* ethics – that deals with the ethical considerations and responsibilities of engineers and their professional practices. It concerns the impact of engineering decisions and actions on society, the environment, and other stakeholders. Personal ethics, on the other hand, may encompass an individual's beliefs, values, and principles that guide their personal conduct in various aspects of life, which are not limited to their professional role as an engineer (Martin, 2002). In short, while living in a society, every individual has developed personal ethics to conduct themselves in specific ways in their day-to-day life and to judge ethical matters. In contrast, engineers are people who have received specialized postsecondary education and learned engineering ethics in their educational path. They are expected to use that acquired knowledge in their ethical decision-making and evaluate whether an action (regarding an engineering task) is moral. The two also differ in context. Engineering ethics is primarily concerned with ethical dilemmas, deliberation, and decision-making reasoning within the specific context of engineering practice. It addresses issues such as professional responsibility, safety, sustainability, fairness, and the welfare of societies impacted by engineering developments. On the other hand, personal ethics applies to a person's overall behavior and choices in various contexts, including personal relationships, family, community involvement, and more. To be sure, they can – and often do – coincide, but it is not necessary that they do.

Another fundamental difference between the two regards the existence of a formal code of conduct in engineering practice, which is (often) absent in an individual's personal dealings. Engineers often adhere to a professional code, which provides guidelines and principles specific to their field. These codes are typically established by professional engineering regulatory boards and associations, outlining the responsibilities and obligations engineers should uphold in their professional practice. Thus, engineering ethics often intersects with legal and professional standards. Engineers are bound by legal obligations and regulations that govern their professional practice, and ethical misconduct can have legal consequences (Davis, 1998). On the other hand, personal ethics may overlap with legal and professional standards but are not bound by them. Personal ethics is guided by an individual's beliefs, values, and principles – which may or may not align with a specific code and are not necessarily written or institutionally upheld.

Finally, engineering ethics strongly emphasizes the needs and interests of various stakeholders affected by decisions. Engineers must consider the potential impacts of their work on public health, safety, the environment, and other societal aspects. Personal ethics may also consider the welfare of others, but it may not be as directly focused on the broader or societal consequences of specific professional actions.

The teachability of engineering ethics

Since engineering ethics differs from personal ethics, the next concern is whether it is a subject that can be taught;, that is, we must determine the extent to which engineering ethics can be cultivated and developed through pedagogical interventions. This exploration will inform the design and delivery of ethics education programs, ensuring their effectiveness in enhancing ethical competence among engineering students.

Scholars acknowledge that engineering ethics is teachable to students in engineering and technology fields (Johnson, 2020). Sethy (2017) in his research findings reported that engineering students who completed the Engineering Ethics course stated that this course offered essential and interesting information about the engineering profession. While individuals may have personal values and moral frameworks that influence their ethical decision-making, as suggested above, the specific application of ethics in engineering requires knowledge and understanding of the ethical principles and considerations relevant to the field. We contend that engineering ethics constitutes a specialized body of knowledge that necessitates deliberate, focused, and empirical acquisition. Therefore, it cannot be assumed that engineers have some inherent ethical knowledge pertaining to their profession, even if they have some knowledge of personal ethics. This is because engineering ethics requires special knowledge due to its nature as a field of applied ethics. Hence, engineering students are required to learn about engineering ethics, thus the need for EEE. (For more about requirements *vis-a-vis* professional accreditation, see Chapters 32–36.)

EEE typically involves studying ethical theories (normative theories, for the most part; see Chapter 2), exploring case studies, and examining real-world examples of ethical dilemmas that engineers may face (Herkert et al., 2020; for more, see Chapter 20). It helps engineers develop the necessary skills to identify ethical issues and analyze their implications by identifying stakeholders and relevant decision-making principles. Thus, they can make informed decisions considering society's and stakeholders' welfare. Teaching engineering ethics also involves imparting knowledge about professional codes of ethics and regulations governing engineering practice. These codes provide specific guidelines and standards that engineers should follow to ensure responsible and ethical conduct in their work (for more, see Chapter 5).

While individuals may have different personal inclinations towards ethical reasoning, the ethics education offered in engineering programs should foster a shared understanding and awareness of the ethical responsibilities inherent in engineering practice. The primary goals of EEE are to increase student sensitivity to ethical issues (i.e., the ability to perceive and evaluate moral or ethical aspects and implications in a given situation), increase student knowledge of relevant standards, improve ethical judgment, and increase ethical commitment (Herkert, 2000; Davis, 2006). Further, it may be stated that the EEE assists engineers in developing a common ethical framework and practical knowledge to navigate complex ethical situations. In other words, a significant reason to justify that engineering ethics is, in fact, teachable is to examine whether there are tools that engineering students can be taught to use and to address ethical challenges in their future careers after engaging in engineering ethics learning experiences. For instance, instructors can use the 'drawing the line' methodology to explain the situation between accepting gifts and bribes (Harris et al., 2019). Furthermore, they can consider conflict resolution as a methodology to resolve two pressing facts – obligation towards the employer and ethical responsibility towards the public – when a question arises about whose well-being an engineer must protect (Feldhaus et al., 2015).

Acquiring knowledge and skills versus developing moral habits: Balancing technical expertise and character building

To understand how engineers can apply the ethics education imparted to them to their lived experience, we must examine the balance between two critical aspects of ethical competency: the operationalizable, practical skills engineering students acquire and the habit-like tendencies cultivated in EEE. (For more on competencies and how to assess them, see Chapter 26.)

Gaining operationalizable, practical skills in engineering involves acquiring theoretical knowledge and learning specific techniques as well as how and when to apply them. It is the kind of work that requires what the virtue philosophers often call craft, craftsmanship, or skill (*ars*, in Latin, or τέχνη in Greek, whence we get words like 'artisan' and 'technique,' respectively). Now, needless to say, one can be *technically* very good at engineering without being *morally* good. We can think of someone who possesses specialized engineering knowledge and an extensive array of technical skills that could – and would – design weapons of mass destruction or a technically flawless concentration camp. The results of this person's work would be technically good but morally reproachable. This person would be good in a very restricted sense (with respect to a specific kind of technique) but not absolutely speaking (Medeiros Ramos, 2021). Being morally good, or being a good engineer to be exact, involves understanding ethical principles, theories, and frameworks that can guide decision-making in engineering practice (see Chapter 2). It also involves cultivating ethical habits and tendencies that shape an engineer's behavior and professional conduct.

Therefore, engineering ethics goes beyond mere theoretical knowledge and skill acquisition, which would only consider humans as makers or producers of artifacts or technology. It also involves the development of habit-like tendencies that influence an engineer's behavior and which considers humans as doers or agents in a broader sense. Ethical habits are ingrained patterns of behavior that are 'second nature' and reflect an engineer's character and values. These habits shape an engineer's day-to-day decision-making processes and actions, guiding them towards ethical conduct even in situations where there may be ambiguity or conflicting interests. The ethical habit (or disposition) that must be acquired (in addition to theoretical knowledge and skill) is practical reasoning (φρόνησις or *prudentia*, as it was called by ancient Greek and European medieval philosophers, respectively), and this virtue can be acquired through education and experience (Medeiros Ramos, 2022). For example, ethical principles related to practical reasoning in engineering can include a commitment to transparency and honesty in communication, a dedication to prioritizing safety and well-being, a proactive approach to addressing potential ethical concerns, and a commitment to continuous learning and improvement. Both aspects, theoretical knowledge and practical skills, as well as habit-like tendencies to act morally well, absolutely speaking, are essential in EEE and practice, as they work together to foster ethical awareness, responsible conduct, and the promotion of public trust in the engineering profession.

A conceptual framework for the purposes of EEE

The purposes of EEE are of several kinds, as reflected in the vocabulary already used for formulating learning goals in practice. In this chapter, we try to systematically conceptualize the kinds of purposes that engineering ethics pedagogy could aim for. Thus, we are looking at existing and possible purposes that the scholarship on EEE did not focus on.

In proposing this framework for classifying the purposes of EEE, we think it is more interesting to start from the normative approach. *What should we strive for, even if we currently do not have the means to achieve this in educational practice?* This approach is justified by the fact that our educational methods keep changing and improving, so the border between achievable and unrealistic keeps changing. Furthermore, if we know what we *should* strive for, we can alter our educational methods to pursue those goals or invent new methods. Yet, to see what is currently pursued and what is still missing, we first need a conceptual mapping of possible EEE goals that is as complete as possible.

The possible purposes of EEE that we identified encompass the following categories: knowledge, actions, personal traits, relationships, artifacts, and environments. We explain each category briefly and provide a summarizing table at the end. For each category, we also point at the kind of theoretical framework (in ethics or philosophy, more generally) used to posit such a goal, namely the normative grounds for having this as a purpose of EEE. When there are established pedagogical methods to achieve those purposes, we also list those.

Theoretical knowledge

This category includes all kinds of knowledge considered relevant for engineering ethics and that a student should acquire. This could contain anything from knowledge of ethical theories and principles to knowledge of values (definitions and operationalization), ethical decision-making principles, or codes of conduct. Anything that the student can memorize and learn counts here as knowledge. Knowledge can be further categorized and divided into Bloom's taxonomy of learning levels based on the complexity of the cognitive tasks required, from understanding to more sophisticated tasks such as reflection and application (Bloom et al., 1956). The theory informing a knowledge-centered purpose for EEE is moral epistemology. This theory assumes, from Plato onward, that to perform a good action, one must first know what good is in that situation (Floridi, 2013). Hence, knowledge must come before moral actions as a scaffold for them. In practice, a knowledge-focused approach is also informed by an assumption that engineers do not do what is right in certain circumstances simply because they do not know what the right thing to do is. Still, there is room left for discussion about what kind of knowledge is relevant in the ethical domain for engineers. Questions such as the following have not yet been answered in practice: *Is knowledge of general moral principles enough? Should codes of conduct be the main thing future engineers learn? Should ethical theories be taught as a plurality, or is only one theory enough? Should the focus be on theories as decision-making tools or as paradigms to think with?* Standard teaching methods in higher education will fit nicely with knowledge transfer as a purpose of EEE, such as commented texts, seminars, exams, essays, and lectures. Hence, knowledge acquisition needs to come before moral actions as a scaffold for them.

Action

This category starts from the assumption that we want to have ethics education in engineering programs to help students achieve and pursue the proper action in their specific context or, as some scholars put it, the right behaviors (Clancy & Zhu, 2023). Typically, for this class of ends, instructors would focus on contexts of action and the right actions for each context, analyzed through paradigmatic case studies. The case study–focused pedagogy (so-called microethics) falls neatly into this category. For example, one could teach starting from any engineering disaster (the Challenger explosion, the Rana Plaza collapse, the BP oil spill, the Tesla highway car crash, the VW defeat design, etc.[1]) and ask students what should have been done. There is no list of prescribed actions one should pursue, but there is a list of actions one should avoid, for example, taking bribes, overpromising features, lying, cutting corners in design or execution, and so on. Meanwhile, positive actions could be truth-telling/speaking up, whistle-blowing, sabotaging a project doomed to hurt many, and so on.

The problem with developing a list of 'positive' actions to prefer is that these are always contextual and case related. We want to avoid prescribing concrete actions in general, such as whistleblowing, which should be only a last resort tactic. We also want to promote more constructive ways of disagreeing with one's work environment. Thus, action-oriented pedagogy is highly contextualized and relies on case studies. The list of desirable and undesirable actions remains open by default and cannot be prescribed in advance.

Action-oriented purposes for EEE assume that we want future engineers to do the right thing without caring how they arrive at this decision. They could do so by deliberation but also by being nudged into it; some biases and heuristics could be used to direct them (Clancy & Zhu, 2023), and hence, moral psychology would be a fitting framework (see Chapter 10). In moral psychology, we do not assume people to be rational agents. Instead, we describe their biases and heuristics and work with these as the starting material. While the question of how to achieve the right actions in the workplace can be pursued through various means, such as environment redesign, accountability procedures, and so on, in the ethics classroom, we do not have these means. Ethics educators will rely on case studies as exemplary and discuss them, hoping that students learn from these cases and apply the lessons to similar future cases. Case-study pedagogy is relatively easy to teach and one of the favorite methods of EEE; however, it has its limitations, which have been discussed extensively (Martin et al., 2019; see Chapter 20). Another approach favoring action-taking centers on performative techniques such as role playing or improvisation (see Chapter 24). In these performative cases, students enact the behavior from their own perspective while not yet knowing the end result. From a moral psychology perspective, performative techniques have the advantage of being affect-infused, and thus more memorable for students.

Personal traits

Many learning goals for EEE are phrased in the language of competencies or skills, for example, the competencies proposed in the handbook by Ibo van de Poel and Lambert Royakkers (2011), which proposes the following: "Moral sensibility … Moral analysis skills … Moral creativity … Moral judgment skills …; Moral decision-making skills …; and Moral argumentation skills" (van de Poel & Royakkers, 2011, p. 2). The basic assumption here is that moral decision-making in EEE relies on some character traits that students do not have from the beginning and that these character traits need to be acquired and put into practice in the classroom so that students can take these along in their professional lives. Such character traits can be discussed regarding virtues, skills, and competencies. The idea is that what the right action *is* is hard to predict for each case anyway and that knowledge by itself is powerless in making students choose the right course of action, so we need to train students to become better at making moral decisions themselves by sharpening their skills in the ethical domain. Just like one cultivates and practices some skills for critical thinking, mathematical thinking, design thinking, and so on, one can be trained in the moral domain. The primary difference in this category is whether one pursues discrete skills and competencies or aims for more holistic virtues (see, e.g., Chapter 22).

Traditionally, EEE has focused on skills and competencies since these are easier to measure and, hence, to operationalize – while virtues remain a fuzzy ideal that many scholars call for (Harris, 2008; Frigo et al., 2021). However, there is a shortcoming to the skills-focused approach: one could have the competencies but fail to deploy them in the required context. Meanwhile, virtues always have end-values embedded in them; hence, if one has the virtue of honesty, one cannot but act honestly when the time comes. The theory informing this purpose is virtue ethics – be it of the Western kind, such as Aristotelian or MacIntyre-ian, or non-Western, such as Confucian or Buddhist ethics (see Chapters 2 and 8). The fundamental assumption here is that the right character traits will lead to the best decision in a particular situation. Teaching such character traits is difficult in practice because the contact hours typically allotted in EEE settings are insufficient for shaping the character traits of students in the long term. Role-playing pedagogy has shown some promise (Martin et al., 2019; Chapter 24), as it puts the student at the center of the moral scenario and asks them to act exemplarily in front of others.

Design/values in artifacts

Another possible approach is artifact-centered. In this one, we do not care about what kind of persons the engineers are, nor what actions they undertake; instead, we look at the kinds of artifacts they create – what kind of apps, vehicles, bridges, processes, structures, or infrastructures they build – and we focus the moral scrutiny on these artifacts and the values embedded in them. As with actions, it is hard to prescribe a certain kind of design for the artifacts as universally recommended. Instead, we ask students to think about the values that should be embedded in their designs; from usability and transparency to privacy and justice, a plethora of values can be achieved (see Chapter 12). The theories informing this kind of purpose are value-sensitive design (Friedman et al., 2017; Hendry et al., 2021), design for values (van den Hoven et al., 2015), ethics by design, and responsible research and innovation (RRI). The pedagogical approaches for teaching this kind of purpose rely on students building artifacts themselves or evaluating others' artifacts. Approaches such as challenge-based learning (Martin & Bombaerts, 2022) and project-based learning (see Chapter 21) suit this goal well, provided that there is a clear ethics reflection component to the project. Some recent approaches use the approach of tinkering – understood here as playing with physical materials and modifying engineering artifacts - in order to embed certain values such as inclusivity, diversity, and empathy in existing artifacts, going beyond the typical values of usefulness and efficiency which are pervasive in engineering designs (van Grunsven et al., 2024). In addition, it must be stated that students need some basic ethical knowledge (as in the first purpose, i.e., theoretical knowledge, see Chapter 2) about values and their operationalization to design with values in mind. Pedagogies centered around ethical design are increasingly popular at technical universities and are used on a wide scale in EEE pedagogy alongside case-study approaches (van Grunsven et al., 2021).

Transitioning from individualistic approaches to holistic approaches

All four kinds of purposes mentioned thus far seem to start implicitly from a sort of methodological individualism, by assuming the autonomy of engineers to decide what they do, who they are, what to know, and what to design. In positing such purposes, it is assumed that, if we teach individual students to pursue the right action, to have the right knowledge, and to acquire the right competencies, the world of engineering will benefit as a result. Yet, engineers do not act alone in the world: they function in teams, systems, and corporations. Engineering professionals interact perhaps just as much as they act as individuals. The individualistic assumption is based on how EEE pedagogy is set up practically: we teach classes of students, but we evaluate them individually. If the evaluation is always about the students as individuals, then it becomes difficult to create learning goals for groups. Granted, there are team assignments in EEE (especially the ones that are artifact-focused), but ultimately, we want each individual to play a part in the team and we strive to evaluate their performance fairly, separate from the team's. This methodological individualism limits the formulation of possible purposes for EEE.

Moving away from the individualistic approach, we acknowledge the basic fact that engineers act embedded in networks of relations, in environments, in institutional structures, and cultural contexts. These networks fundamentally constrain what engineers can do, know, design, or decide. To account for this limitation, we can map out two more kinds of purposes: relational and environmental purposes for EEE.

Relations

Engineers work in teams, are placed in corporate hierarchies, and relate to their customers, employers, and society at large. What kind of relationships and relational networks should engineers strive to cultivate or enter? The answer can be further separated into *qualities of relations* and *objects of relations*.

Qualities of relations encompass the kinds of relations that an engineer should strive to cultivate in the workplace: equal or hierarchical, collaborative, respectful, honest, multi-networked (meaning to strive to create relations that connect various networks of stakeholders), other-oriented (as opposed to self-centered, or self-serving relations), and critical (meaning that one can create networks of relations that are truth-oriented and in which peers hold each other accountable). One easily notices that there seems to be an overlap with virtue ethics here, but instead of asking what kind of virtues one should cultivate in oneself, here one asks how to cultivate meaningful relations in the workplace that promote virtues in the interaction. Thus, one need not be honest as a person, but if one's relations and networks are set up so that honesty is expected and encouraged in the interactions, the purpose is achieved. Theoretical frameworks that could inform the qualit y of relations end-goal are care ethics, Confucian ethics (Zhu, 2023), and workplace ethics (a smaller branch of business ethics; see, e.g., Chapter 11).

Objects of relations include care, maintenance, respect, equality, justice, and non-discrimination. This purpose is still relational but asks how to achieve specific group and societal values by enacting certain relations. Thus, the values are not instantiated by objects or artifacts but by relations. The relations could again be in the workplace (with one's colleagues), but they primarily involve stakeholders and broader society. For example, an engineer aiming to promote the value of maintenance in one's relations would probably think differently than an engineer focused on fostering maintenance through the objects one designs. In the latter case, one designs an object that is easy to maintain, with spare parts that can be easily found and replaced. Yet, in the former case, one thinks about the entire life cycle of the product and the people using it – how to make its usage less damaging, how to empower the end-users to engage in maintenance themselves, how to persuade society that it is easier to maintain rather than discard. All kinds of constellations of relations could be enacted to support maintenance as an end goal. A theoretical account to inform such a purpose could again be care ethics or Confucian ethics – or involve value theory.

Environments and systems

The term 'macroethics,' as coined by Herkert (2001), already encompassed how the practice of engineering has broad societal and environmental effects on our society. *Do we need then to discuss the environmental effects of engineering?* We think so because 'environment' here does not mean an ecological or natural category. Instead, it is intended to emphasize how the individual agents are connected – to each other, their artifacts, and the world around them. Environment as a concept captures this interconnectedness and mutual influence. In discussing the impact of engineering – as a profession – on the world around it, one could try to formulate purposes that do justice to this holistic view (see Chapters 7 and 9). One could strive for maintenance and non-destruction of the world but also for awareness of how one's actions affect the world. Another goal could be limiting suffering by creating systems that do not promote suffering or systemic oppression. Formulating concrete goals for this kind of purpose is notoriously difficult but worth trying. Still, one could argue that this environmental or systemic perspective should not be a goal for engineering *education*, and perhaps we should delegate it to the engineering profession as a whole by embedding it into ethical codes of conduct. Still, one should learn to think about the systemic effects of one's actions in the world, even as a student, because it may be too late to develop this awareness after graduating.

Several theories analyze the systemic nature of engineering ethics, such as science and technology studies (STS) or Luhmann's systems theory (Fuchs et al., 2023). There are several non-Western approaches particularly fitting for this scope, such as Buddhist ethics (Garfield, 2021; Bombaerts et al., 2023), *Ubuntu* philosophy (Mabele et al., 2022), and *Buen Vivir* (Estermann, 2006; Kopenawa & Albert, 2013; Viveiros de Castro, 2014). *Ubuntu* and *Buen Vivir* are non-Western traditions (from Africa and South America, respectively) centered on community – or on a more global perspective and a broader ontological scope – according to which humans, non-human animals, and nature as a whole are seen as being at the same ontological (and moral) level (Ewuoso & Hall, 2019; Mabele et al., 2022; Kopenawa & Albert, 2013; Viveiros de Castro, 2014). These two approaches are more thoroughly discussed in Chapter 8, alongside Buddhist ethics.

All six types of engineering purposes mentioned here are found at various levels of abstraction, for an analytical purpose, but these levels are not isolated in practice, as they contribute to one another and can be incremental. For example, the pursuit of moral competencies such as moral perception can have a systemic effect if one also becomes aware of one's systemic influence, or it can give rise to richer and more responsible relations. Artifacts designed with certain systemic values in mind can alter systems and shape the world we live in in ways that were not yet anticipated by their designers. There is interaction at stake between the systemic, relational, and individual levels, with the systemic level encompassing all previous ones. Table 1.1 summarizes our conceptual framework for dividing the types of goals for EEE.

Some practical questions

Who should teach engineering ethics?

In the engineering ethics literature, it has been widely debated *who* should teach engineering ethics. Exploring such a question is crucial for understanding the purposes of EEE since the positionality, academic training, and disciplinary norms all potentially impact how engineering ethics instructors define and enact these purposes. In the United States (and other countries as well), historically, the qualifications desirable for someone to teach professional ethics, including engineering ethics, have been unclear. Scholars have expressed concerns regarding both engineers and non-engineers (humanities and social sciences scholars) teaching engineering ethics. For instance, McGinn (2018) expressed concern about having engineering instructors cover ethical issues in engineering courses, as they often lack formal training in ethics – and thus, their consideration of ethical issues is likely to be "intuitive and not grounded in ethics fundamentals" (McGinn, 2018, p. 9). In addition to the concern about engineering educators lacking sufficient training in ethics, Newberry (2004) noted that most universities do not have a reward system that motivates engineering faculty members to develop the background for ethics instruction. Therefore, the question becomes whether formal training or even degrees in applied ethics or philosophy should be considered indispensable for someone to teach engineering ethics. Barry and Herkert (2015) indicated that many engineering faculty members without formal training in either engineering or philosophy have been teaching engineering ethics and trying to achieve course objectives.

There have also been concerns with philosophers or humanities scholars teaching engineering ethics. For instance, in an empirical study conducted by one of the co-authors of this chapter, a Chinese engineering faculty member summarized two significant limitations of solely relying on humanities and social sciences professors in the teaching of engineering ethics at his institution:

Table 1.1 Our conceptual framework for dividing the types of goals for EEE.

Category of purpose for EEE	*Examples from this category*	*Theoretical frameworks fitting this purpose*	*Pedagogical approaches*
Individualistic, token-oriented			
Knowledge	Moral knowledge: ethical theories, codes of conduct, values, ethical decision-making procedures.	Moral epistemology	Bloom's taxonomy of learning goals. Readings, seminars, exams, lectures.
Action	Undesirable actions: taking bribes, over-promising features, lying, cutting corners in design or execution, etc. Desirable actions: cooperation, truth-telling, double-checking, preventive maintenance, talking to stakeholders, speaking up in a team.	Moral psychology	Case-study pedagogy (microethics), deliberation. Role-playing/ improvisations.
Personal traits	Ethical skills or competencies (such as moral intuition, moral sensitivity, moral decision making, moral imagination) and virtues (honesty, determination, courage, etc.).	Virtue ethics (Western or Eastern)	Role-playing/ improvisations.
Values in artifacts	Values to be embedded in design. Negative constraints for design: dark-patterns, deceptive or manipulative design, harmful designs.	Value-sensitive design/ethics by design	Challenge-based learning. Project-based learning. Any design-centered task.
Networked and holistic			
Relationships	Qualities of relations: equal, hierarchical, collaborative, respectful, honest, multi-networked, other-oriented, critical. Objects of relations: care, maintenance, respect, equality, justice, non-discrimination.	Care ethics; Confucian ethics; workplace ethics; value theory	X
Environments and systems	World-maintenance, non-destruction of the world, awareness of impact on the system, limitation of suffering and oppression.	Buddhist ethics; *Ubuntu* philosophy; *Buen Vivir;* Systems theory; STS	X

One issue is that the number of humanities and social sciences faculty cannot fulfill the teaching requirement … Every year, our university recruits thousands of graduate students. As this [engineering ethics] course has become a required course for graduate students, even if one section can accommodate 200 students, we still need to have dozens of sections. In this sense, we are short of humanities and social sciences faculty who are able to fulfill

> this teaching requirement. The other limitation with having humanities and social sciences faculty teach engineering ethics is that they tend to overly theorize engineering ethics. Such overemphasis on theorizing engineering ethics might be okay to students with [a] good humanities and social sciences background. However, for the majority of engineering students, their humanities and social sciences background is weak. These students will have challenges learning engineering ethics and keeping pace with the instructor. Finally, there will be some negative emotion towards engineering ethics among these students.
>
> *(Zhang & Zhu, 2021)*

Barry and Herkert (2015) suggested that, in addition to engineers and moral philosophers being considered qualified for teaching engineering ethics, instructors trained in interdisciplinary fields such as the history of science and technology, technical communications, and science and technology studies should also be considered qualified – insofar as they are enthusiastic about discussing ethical issues in and the social implications of engineering. Barry and Herkert (2015) did not specify that faculty from those interdisciplinary fields need engineering expertise. However, their training programs often expect a basic understanding of engineering, ensuring these individuals are equipped for interdisciplinary exploration of themes connecting engineering and technology. Again, the question of who is qualified to teach engineering ethics is highly debatable; however, the diverse backgrounds of engineering ethics instructors will undoubtedly influence how they conceptualize ethical issues and prioritize ethical learning outcomes. These diverse backgrounds will also influence how they understand and explain EEE's general and specific purposes. In the following section, we delve into these concerns by examining the perspectives, interests, and expectations of university faculty members and other key stakeholders involved in EEE.

Who decides on the purposes of EEE?

This section reflects on the multiple stakeholders of EEE, including the broader social, cultural, and political contexts in which they are situated. These stakeholders shape how EEE is created and the purposes of EEE are formulated. More specifically, this section discusses the primary stakeholders for EEE, including their values, motivations, and needs for promoting EEE. Doing such an analysis allows us to understand better the social forces that help shape the definition and implementation of EEE purposes. By considering the potential stakeholders of EEE beyond students and teachers, we see that academic institutions, professional organizations, industry stakeholders, and ethical experts can all evaluate their respective roles and responsibilities in shaping the purposes of EEE. Various stakeholders have always advocated for engineering ethics, and EEE should thus be considered a collaborative effort.

The purposes of EEE are often considered integral to the *professional formation of engineers*, and academic institutions are typically considered responsible for transmitting theoretical knowledge and technical skills to students (at least at the outset of students' engineering careers). Engineering ethics is thus taught as an indispensable part of engineering curricula in most universities and colleges (and global accords and accreditation standards play a significant role; see Chapters 32–36). Academic institutions have a crucial role in providing foundational knowledge and offering dedicated courses or modules on engineering ethics. As noted above, faculty members with different backgrounds may perceive the purposes of EEE differently. For instance, professors and instructors in engineering departments may consider that a significant purpose of EEE is to teach students practical skills for solving ethical problems in the workplace – comparable to using scientific and technological skills to solve engineering problems. Thus, many of them incorporate ethical discussions (Chapter 25), case studies (Chapter 20), and ethical decision-making frameworks (Chapter 2) into their teaching of technical content.

In contrast, ethicists, philosophers, and professionals with similar expertise and background in ethics may consider teaching other aspects of morality – such as moral reasoning skills, moral sensitivity, and moral tendencies, which are key concepts in humanities and social sciences – to be the purposes of EEE. Their insights and knowledge can provide a deeper understanding of ethical theories, frameworks, and reasoning processes. Nevertheless, raising difficult ethical questions and teaching students how to address them should not be relegated to the 'token ethicist' or philosophy instructor but should be embraced by most – if not all – engineering instructors. Collaboration among experts can enrich the teaching of engineering ethics and ensure a comprehensive exploration of ethical issues.

There are, however, other stakeholders who play an essential part in shaping what should be the purposes of EEE. Professional engineering organizations, such as the National Society of Professional Engineers (NSPE) in the United States or the Institution of Engineering and Technology (IET) in the United Kingdom, often consider the application of professional codes of ethics as a central purpose for EEE. They develop and promote codes of ethics specific to the engineering profession. These organizations can contribute to teaching engineering ethics by offering guidance, resources, and training programs for engineers. They can – and often do – organize workshops, conferences, and seminars focused on professional ethical issues in engineering, often made explicit in codes of ethics (see Chapter 5).

In addition, employers in the engineering industry also influence and help interpret and shape the purposes of EEE. Their understanding of the purposes of EEE often emphasizes the role of engineers in reconciling their professional responsibilities, corporate social responsibility (CSR), and ethical obligations arising from their role as employees. These industry players play a crucial role in EEE by integrating ethics into their onboarding processes and continuing professional development programs. Employers can provide case studies and scenarios relevant to their specific industry to help engineers apply ethical principles to real-world situations.

All in all, collaborative efforts among academia, professional organizations, and industry should be favored as they can enhance the teaching of engineering ethics and provide diverse approaches to the purposes of EEE. Joint initiatives can be established to develop curricula, share best practices, create case studies, and facilitate discussions on ethical considerations in engineering practice.

Concluding remarks and future directions

Defining the purposes of EEE holds paramount significance for the engineering education community. It is problematic and potentially dangerous when engineering educators design ethics-learning activities without critically examining *the purposes of these activities* and assessing *whether these purposes are justified* for educating ethically and professionally competent engineers. Failing to consider the purposes of EEE can lead to adverse outcomes. At a minimum, it can undermine the effectiveness of teaching engineering ethics, manifesting as a misalignment between the curriculum and the purposes of EEE. At its most severe, the omission of certain elements from students' learning experiences can have far-reaching consequences. These consequences can range from immediate (depending upon, e.g., whether specific learning activities prioritize moral reasoning over moral empathy), to more enduring impacts (e.g., the cultivation of moral identity, values, and lifelong ethical development).

The actual purposes of EEE are closely tied to the contextual and sociocultural milieu. To put it differently, these purposes are frequently shaped by the values and ideologies of various influential collectives and entities. Accreditation bodies, professional societies, industries, engineering faculties, and individual educators each possess their own political agendas and motives for shaping these objectives. The multi-dimensional framework we constructed for this chapter (Table 1.1) can help educators reflect on the purposes and objectives embedded in the ethics learning programs designed by themselves and others – so that their programs encompass the theoretical foundations and pedagogies that enable these objectives. In other words, such a framework can facilitate educators in attaining alignment among the purposes of EEE, theoretical foundations, and pedagogies they use.

The purposes of EEE are socially constructed and vary from country to country based on unique historical, political, and cultural contexts (for more on this, see Chapters 32–34). The analytical framework we – a group of scholars hailing from diverse cultural backgrounds – have constructed and presented (Table 1.1), embraces both the individualistic and holistic aspects of EEE, incorporating perspectives from both Western and non-Western traditions. It can function as a comprehensive framework for delineating, implementing, and evaluating the purposes of EEE, catering to the needs of engineering educators and policymakers globally.

We recommend future research to explore the purposes of EEE including (1) how engineering educators and policy-makers perceive their own purposes for EEE; (2) the extent to which these purposes influence curriculum, teaching methods, and policy formulation; (3) how the diverse purposes in the engineering education ecology align with students' values and aspirations in the engineering profession; and (4) how different EEE purposes can synergize to develop well-rounded engineers capable of working across cultures.

Acknowledgments

Qin Zhu's work was supported by two National Science Foundation (NSF) grants: 2316634 and 2124984. Lavinia Marin's work has been supported by the 4TU Centre for Engineering Education (4TU.CEE) and by the research program Ethics of Socially Disruptive Technologies (ESDIT), which is funded through the Dutch Ministry of Education, Culture, and Science and the Netherlands Organization for Scientific Research (NWO grant number 024.004.031).

Note

1 Several examples of cases discussing engineering action and behavior can be found here: https://search.edusources.nl/communitys/4tu-ethics

References

Abaté, C. (2011). Should engineering ethics be taught? *Science and Engineering Ethics, 17*, 583–596. https://doi.org/10.1007/s11948-010-9211-9

Barry, B. E., & Herkert, J. R. (2014). Engineering ethics. In A. Johri & B. Olds (Eds.), *Cambridge handbook of engineering education research* (pp. 673–692). Cambridge University Press.

Barry, B. E., & Herkert, J. R. (2015). Overcoming the challenges of teaching engineering ethics in an institutional context: A U.S. perspective. In C. Murphy, P. Gardoni, H. Bashir, C. E. Harris, & E. Masad (Eds.), *Engineering ethics for a globalized world* (pp. 167–187). Springer.

Bloom, B. S., Engelhart, M. D., Furst, E. J., Hill, W. H., & Krathwohl, D. R. (1956). *Handbook I: Cognitive domain.* David McKay.

Bombaerts, G., Anderson, J., Dennis, M., Gerola, A., Frank, L., Hannes, T., & Spahn, A. (2023). Attention as practice: Buddhist ethics responses to persuasive technologies. *Global Philosophy, 33*(2), 25. https://doi.org/10.1007/s10516-023-09680-4

Clancy, R. F., & Zhu, Q. (2023). Why should ethical behaviors be the ultimate goal of engineering ethics education? *Business and Professional Ethics, 42*(1), 33–53. https://doi.org/10.5840/bpej202346136
Davis, M. (1998). *Thinking like an engineer.* Oxford University Press.
Davis, M. (2006). Engineering ethics, individuals, and organizations, *Science and Engineering Ethics, 12*(2), 223–231. https://doi.org/10.1007/s11948-006-0022-y
Dotson, R. (2016). Goal setting to increase student academic performance. *Journal of School Administration Research and Development, 1*(1), 44–46.
Estermann, J. (2006). *Filosofía andina: Sabiduría indígena para un mundo nuevo.* ISEAT.
Ewuoso, C., & Hall, S. (2019). Core aspects of ubuntu: A systematic review. *The South African Journal of Bioethics & Law, 12*(2), 93–103. http://doi.org/10.7196/SAJBL.2019.v12i2.00679
Feldhaus, C. R., Little, J., & Sorge, B. (2015). Conflict resolution and ethical decision-making for engineering professionals in global organizations. In S. S. Sethy (Ed.), *Contemporary ethical issues in engineering* (pp. 204–227). IGI Global.
Floridi, L. (2013). *The ethics of information* (1st ed.). Oxford University Press.
Friedman, B., Hendry, D. G., & Borning, A. (2017). A survey of value sensitive design methods. *Foundations and Trends in Human-Computer Interaction, 11*(2), 63–125. https://doi.org/10.1561/1100000015
Frigo, G., Marthaler, F., Albers, A., S., O., & Hillerbrand, R. (2021). Training responsible engineers: Phronesis and the role of virtues in teaching engineering ethics. *Australasian Journal of Engineering Education, 26*(1), 25–37. https://doi.org/10.1080/22054952.2021.1889086
Fuchs, L., Bombaerts, G., & Reymen, I. (2023). Does entrepreneurship belong in the academy? Revisiting the idea of the university. *Journal of Responsible Innovation, 10*(1), Article 2208424. https://doi.org/10.1080/23299460.2023.2208424
Garfield, J. L. (2021). *Buddhist ethics: A philosophical exploration.* Oxford University Press.
Harris, C. E. (2008). The good engineer: Giving virtues its due in engineering ethics. *Science and Engineering Ethics, 14*, 153–164. https://doi.org/10.1007/s11948-008-9068-3
Harris, C. E., Jr., Davis, M., Pritchard, M. S., & Rabins, M. J. (1996). Engineering ethics: What? Why? How? and When?. *Journal of Engineering Education, 85*(2), 93–96.
Harris, C. E., Pritchard, M. S., James, R. W., Englehardt, E. E., & Rabins, M. J. (2019). *Engineering ethics: Concepts and cases* (6th ed.). Cengage.
Hendry, D. G., Friedman, B., & Ballard, S. (2021). Value sensitive design as a formative framework. *Ethics and Information Technology, 23*(1), 39–44. https://doi.org/10.1007/s10676-021-09579-x
Herkert, J. R. (2000). Engineering ethics education in the USA: Content, pedagogy and curriculum. *European Journal of Engineering Education, 25*(4), 303–313. https://doi.org/10.1080/03043790050200340
Herkert, J. R. (2001). Future directions in engineering ethics research: Microethics, macroethics, and the role of professional societies. *Science and Engineering Ethics, 7*(3), 403–414. https://doi.org/10.1007/s11948-001-0062-2
Herkert, J. R., Borenstein, J., & Miller, K. (2020). The Boeing 737 MAX: Lessons for engineering ethics. *Science and Engineering Ethics, 26*, 2957–2974. https://doi.org/10.1007/s11948-020-00252-y
Johnson, D. (2020). *Engineering ethics: Contemporary and enduring debates.* Yale University Press.
Kopenawa, D., & Albert, B. (2013). *Falling sky: Words of a Yanomami Shaman.* Belknap Press.
Liow, S. R., Betts, M., & Lit, J. K. (1993). Course design in higher education: A study of teaching methods and educational objectives. *Studies in Higher Education, 18*(1), 65–79. https://doi.org/10.1080/03075079312331382468
Mabele, M. B., Krauss, J. E., & Kiwango, W. (2022). Going back to the roots: Ubuntu and just conservation in Southern Africa. *Conservation & Society, 20*(2), 92–102. http://dx.doi.org/10.4103/cs.cs_33_21
Martin, D. A., & Bombaerts, G. (2022). *Enacting socio-technical responsibility through challenge based learning in an ethics and data analytics course*. 2022 IEEE Frontiers in Education Conference. Uppsala, Sweden.
Martin, D. A., Conlon, E., & Bowe, B. (2019). The role of role-play in student awareness of the social dimension of the engineering profession. *European Journal of Engineering Education, 44*(6), 882–905. http://dx.doi.org/10.1080/03043797.2019.1624691
Martin, D. A., Conlon, E., & Bowe, B. (2021). A multi-level review of engineering ethics education: Towards a socio-technical orientation of engineering education for ethics. *Science and Engineering Ethics, 27*, Article 60. https://doi.org/10.1007/s11948-021-00333-6
Martin, M. W. (2002). Personal meaning and ethics in engineering. *Science and Engineering Ethics, 8*(4), 545–560. https://doi.org/10.1007/s11948-002-0008-3
McGinn, R. (2018). *The ethical engineer: Contemporary concepts & cases.* Princeton University Press.

Medeiros Ramos, A. (2021). Is ars an intellectual virtue? John Buridan on craft. In I. Chouinard, Z. McConaughey, A. Medeiros Ramos, & R. Noel (Eds.), *Women's perspectives on ancient and medieval philosophy* (pp. 275–301). Cham: Springer.

Medeiros Ramos, A. (2022). *John Buridan on the intellectual virtues* (Doctoral Dissertation), Université du Québec à Montréal. Retrieved May 2023. https://archipel.uqam.ca/15415/1/D4149.pdf

Newberry, B. (2004). The dilemma of ethics in engineering education. *Science and Engineering Ethics, 10*(2), 343–351. https://doi.org/10.1007/s11948-004-0030-8

Sethy, S. S. (2017). Undergraduate engineering students' attitudes and perceptions towards "professional ethics" course: A case study of India. *European Journal of Engineering Education, 42*(6), 987–999. https://doi.org/10.1080/03043797.2016.1243656

van de Poel, I., & Royakkers, L. (2011). *Ethics, technology, and engineering: An introduction.* Wiley.

van den Hoven, J., Vermaas, P. E., & van de Poel, I. (Eds.). (2015). *Handbook of ethics, values, and technological design: Sources, theory, values and application domains.* Springer.

van Grunsven, J., Franssen, T., Gammon, A., & Marin, L. (2024). Tinkering with technology: How experiential engineering ethics pedagogy can accommodate neurodivergent students and expose ableist assumptions. In E. Hildt, K. Laas, E. M. Brey, C. Z. & Miller, (eds.), *Building inclusive ethical cultures in STEM* (pp. 289–311). Springer International Publishing.

van Grunsven, J., Marin, L., Stone, T., Roeser, S., & Doorn, N. (2021). How to teach engineering ethics? A retrospective and prospective sketch of TU Delft's approach to engineering ethics education. *Advances in Engineering Ethics, 9*(2), 1–11.

Veach, C. (2006). There's no such thing as engineering ethics. *Leadership and Management in Engineering, 6*(3), 97–101.

Viveiros de Castro, E. (2014). *Cannibal metaphysics: For a post-structural anthropology.* Univocal Publishing.

Zhang, H., & Zhu, Q. (2021). Instructor perceptions of engineering ethics education at Chinese engineering universities: A cross-cultural approach. *Technology in Society, 65*, Article 101585. https://doi.org/10.1016/j.techsoc.2021.101585

Zhu, Q. (2023). Confucian ethics of technology. In G. J. Robson & J. Y. Tsou (Eds.), *Technology ethics: A philosophical introduction and readings* (pp. 93–101). Routledge.

Zhu, Q., & Clancy, R. (2023) Constructing a role ethics approach to engineering ethics education. *Ethics and Education, 18*(2), 216–229.

2
ETHICAL THEORIES

Michael Kühler, Natalie Wint, Rafaela Hillerbrand, and Ester Gimenez-Carbo

Introduction

Ethical theories analyze and justify what is morally right and wrong. Ethical theories, therefore, are *normative* theories that provide the systematic basis of arguments and viewpoints from which moral issues can be discussed. Within the context of engineering ethics education, Haws (2001, p. 226) claims that ethical theories "provide a more logical, systematic format for the resolution of ethical dilemmas." Martin and Schinzinger (2004, p. 81) define them as "attempts to provide clarity and consistency, systematic and comprehensive understanding, and helpful guidance in moral matters." Mastering at least the basics of important ethical theories is meant to enable students to reach the core aim of engineering ethics education, namely developing 'ethical competence,' that is, the ability to understand and solve ethical issues in the student's field of work based on well-reasoned ethical arguments and judgments as well as developing a corresponding personal ethical attitude for acting responsibly (Andersson et al., 2022; Franck, 2017) – although it should be noted that different conceptions of 'ethical competence' emphasize different components (Franck, 2017).

Although ethical theories provide a necessary normative framework to address moral questions, it is a matter of debate whether to use them in teaching engineering students. This debate is unfolding differently in different geographical regions and cultural traditions. In terms of engineering ethics education, the connection between cultural and ethical values (Alas, 2006) means that the degree to which ethical theories are adopted within different geographical regions depends upon various contextual factors, including educational systems and accreditation bodies. For example, within Great Britain, where the first professional engineering societies were formed, practitioners developed the first code of professional ethics, which was later used as a model for codes within the United States, and upon which the first US ethics textbooks were based (Didier, 2015). In comparison, "such codes are much less important in Europe than in North America" (Didier, 2015, p. 89). In Europe, engineering ethics is generally considered "an interdisciplinary reflection at the crossroads of professional ethics, the human and social sciences, and the philosophy of technology" (Didier, 2015, p. 87), and the development of engineering ethics was initially based on feelings of social responsibility, drawing upon insights from Science and Technology Studies (STS) (Polmear et al., 2019). Accordingly, engineers are placed within a complex system where technologies are developed in collaboration with other actors (Didier, 2015). Indeed, during their

DOI: 10.4324/9781003464259-4

cultural comparison of macro-ethics teaching practices and perceptions in engineering, Polmear et al. (2019) found that US educators primarily taught codes of ethics, ethical design, and safety – and focused on an individualistic micro-ethical approach. Likewise, Hess and Fore (2018, p. 551) highlighted "that the most common methods for integrating ethics into engineering involved exposing students to codes/standards, utilizing case studies, and discussion activities." Although it should be noted that the consistent formulation and justification of professional codes still require a philosophical framework in terms of ethical theories, it is well worth asking if it is necessary to include this framework in teaching – or if referring to professional codes is already sufficient for students to develop ethical competence and an ethical attitude.

In this chapter, we will first provide an overview of the principal ethical theories used in engineering ethics education – namely *consequentialism*, *deontology*, and *virtue ethics* – before discussing contemporary aspects of the philosophical and ethical debates surrounding them. We will then provide a brief overview of several underrepresented ethical theories and approaches, including *contractarianism and contractualism*, *care ethics*, and *discourse ethics* (for an in-depth look at *non-Western ethics*, see Chapter 8) to paint a richer picture of how ethical theories could contribute to improving ethical competence among students. We will ultimately provide a summary of common problems or barriers that engineering students encounter when learning about ethical theories. The chapter concludes by discussing the need to use ethical theories when teaching engineering students, if one sees ethical competence as the primary learning goal. We begin by describing our shared perspective as authors.

Positionality

Two philosophers specializing in applied ethics (Michael and Rafaela) and two engineering academics with backgrounds in ethics (Natalie and Ester) authored this chapter. We share an interest in fostering responsible engineering in teaching. Based on influential ethics literature and years of experience teaching ethics to engineering students, we intend this chapter to provide an overview of ethical theories and their use within engineering ethics education for engineering teachers, students, and practitioners.

Ethical theories prominently used in engineering ethics education

Consequentialism and utilitarianism

Consequentialism is an ethical theory that places the consequences of an action at the center of ethical judgments. Every time we make a choice, we should choose the option that produces the best overall consequences, and only this option can be considered morally right or obligatory (Sinnott-Armstrong, 2022). *Utilitarianism* is considered a prime example of consequentialism. According to utilitarianism, an act is morally right only if its consequences create the most 'utility' (Driver, 2022). Note that the notion of 'utility,' including the idea of 'best' consequences, is not itself morally qualified but points to a *non-moral* understanding, for example, people's happiness or the maximum fulfillment of their particular interests or preferences.

When referred to in engineering ethics education, utilitarianism is typically taught based on the first two of its three classical authors as identified by van de Poel and Royakkers (2011, pp. 78–89): Jeremy Bentham (1748–1832) (Bentham, 1789); John Stuart Mill (1806–1873) (Mill, 1861); and Henry Sidgwick (1838–1900) (Sidgwick, 1909). Classic utilitarianism's *axiology*, that is, the notion of goodness that is supposed to be maximized, is *happiness*. Hence, the classic utilitarian slogan: 'The greatest happiness for the greatest number.' Happiness is, in turn, defined in *hedonistic* terms of *feelings of pleasure*, as opposed to pain. Bentham thought pleasure and pain can be measured quantitatively, for example,concerning their intensity or duration. Mill rejected this claim and argued for a qualitative distinction between higher and lower pleasures, asserting that higher pleasures (e.g., engaging intellect, moral reasoning, or aesthetic appreciation) are necessary and preferable to lower ones (e.g., bodily sensations or sensory gratification) concerning human happiness. In any case, if happiness could be measured, it would be possible to determine the morally 'right' choice using an approach similar to a cost–benefit or risk–benefit analysis (Pantazidou & Nair, 1999, p. 206).

Yet, things are not that simple. For instance, the idea of maximizing goodness fails to address the question of how pleasure and pain should be distributed among different people, which is why utilitarianism has been criticized for failing to include a convincing notion of (distributive) justice (van de Poel & Royakkers, 2011, p. 86f). Moreover, calculating the consequences of each action individually – which would be *act utilitarianism* – easily leads to counterintuitive results. For instance, it would imply that it is *morally obligatory* to kill an innocent person if doing so avoids a worse outcome (Smart & Williams, 1973, p. 98f). In comparison, *rule utilitarianism* is intended to avoid such counterintuitive implications, as it focuses not on individual decisions or actions but on rules and the consequences of their general adoption. Accordingly, a rule that allows killing innocent persons for the greater good cannot be considered as producing the best consequences because no one could feel safe anymore. Therefore, what we morally ought to do depends on whether an option falls under a rule that, when generally adopted, produces the best overall consequences.

Deontology

In contrast to consequentialism, *deontology* rejects the moral importance of consequences. According to the most influential deontological author, Immanuel Kant (1724–1804), consequences do not matter morally at all (Kant, 1785, 1797). The moral quality of an action is instead a question of the action's inherent quality and our good 'will,' that is, our intention, which, in turn, is determined by whether we could think of, or could want, this course of action being taken up by everyone in similar circumstances. In Kant's famous words of the first formula of the categorical imperative: "act only according to that maxim through which you can at the same time will that it become a universal law" (Kant, 1785, p. 71). Put simply, the question is: '*What if everyone did this?*' Note that this question is not meant to invoke consideration for the actual consequences if everyone acted accordingly but whether the respective principle of action (*maxim*) is reasonably acceptable to everyone.

To illustrate the categorical imperative, Kant discussed the moral permissibility of lying. Imagine if one were to ask themselves whether it is morally permissible for them to lie to avoid an uncomfortable situation. If everyone were morally allowed to lie, successful lying would be impossible because lying exploits (and, therefore, depends on) honesty. Only if people assume honesty as morally required (i.e., precisely presupposing a moral obligation *not* to lie) can one successfully lure people into believing one's lie. Hence, the maxim to be allowed to lie implies both the permissibility to lie ('I want lies to be allowed') and its *im*permissibility ('I want people to believe me, so I want lies *not* to be allowed') – a logical contradiction, which is why the maxim does not stand Kant's test of the categorical imperative, and lying is, therefore, morally wrong.

Kant's ethics is typically taught with a focus on the first two formulas of the categorical imperative (van de Poel & Royakkers, 2011, pp. 89–95), where the second formula emphasizes respect for people's autonomy: "So act that you use humanity, in your own person as well as in the person of any other, always at the same time as an end, never merely as a means" (Kant, 1785, p. 87). The crucial point here is to understand that instrumentalizing others is not morally impermissible *as such* but only if it is done without respecting the others' autonomy ("*merely* as a means"). Put simply, and relating to the first formula, the question is whether everyone can autonomously (i.e., freely and reasonably) agree to the underlying principle that defines the particular social interaction (more currently, Rawls's (1971) *contractualism* has Kantian roots; see below). For instance, instrumentalizing others in the workplace is morally unproblematic as long as everyone freely and reasonably agrees to the particular conditions of working together – which is why forced labor and unfavorable working conditions that take advantage of someone's dire situation are morally wrong. Respecting people's autonomy, thus, provides an additional important aspect when applying Kant's ethics in engineering practice.

However, like utilitarianism, Kant's ethics may lead to counterintuitive results. For instance: *Would it not be morally required to lie to save an innocent person from a potential murderer?*

Consequentialism/utilitarianism and deontology revisited

When it comes to using ethical theories in teaching engineering students, referring only to their classical versions is somewhat problematic. In the case of utilitarianism, practically all its features have been subject to criticism and substantial revision (Sinnott-Armstrong, 2022). First, classic utilitarianism's *hedonism* has been criticized because feelings of pleasure and pain are notoriously hard to measure, especially for interpersonal comparison. Newer suggestions for utilitarianism's axiology propose *desire fulfillment*, *interest realization*, or *preference satisfaction*. Each of these suggestions has different implications for maximizing *goodness*. For instance, a person might *want* ice cream but would *prefer* not to gain weight, while eating a salad would be in their best *interest*.

Second, the notion of consequences needs refinement. *How much into the future should one consider the consequences of one's actions?* Moreover, the kind of consequences that one needs to consider should also be specified. As the *actual* consequences are not yet known, we can question whether to consider *intended* consequences, consequences that a person can *foresee*, or those that may be regarded as *reasonably foreseeable*.

Third, the idea of *maximizing* good consequences has also been contested. Given the complex challenges of measuring and predicting the consequences of one's actions: *Why not opt for consequences that are taken to be good enough?* Accordingly, some newer versions of utilitarianism opt for a *satisficing* principle.

Hence, following the current ethical debate, consequentialism/utilitarianism is not *one* clearly defined ethical theory but, rather, allows for a wide variety of different versions – each differing in their specification of core claims and accompanying arguments. Yet, engineering students rarely get to know this lively debate – with its intricate arguments and specifications – and are usually left with an outdated understanding.

Such shortcoming is also visible in the case of deontology. Concerning Kant's ethics, questions have been raised if there can be really no moral dilemmas, that is, conflicting moral duties without a moral solution, as Kant claimed (Kant, 1797, p. 50; McConnell, 2022). This has led to the concept of *prima facie duties* (van de Poel & Royakkers, 2011, pp. 93–95; Ross, 1930), which are weighed against each other to determine one's *actual* moral duty. The discussion of moral dilemmas is regularly used in teaching ethics and has also, for example, spun a public debate about how autonomous cars should behave in so-called *trolley cases*, in which a decision needs to be made as to who should be killed if there is no option to save everybody (Foot, 2003, p. 23; Thomson, 1986, p. 80f). Discussions in class and public debates are then usually focused on determining the 'right solution.' However, if *trolley cases* are *real* moral dilemmas, then there is, per definition, no single correct solution. Yet, it should be emphasized that *trolley cases* are used for a very different purpose in philosophical debate. Instead of seeking the 'right solution,' the philosophical goal is to analyze which aspects bear how much weight, if any, in moral analysis. For instance: *Does it make a difference in moral judgment if trolley cases involve actively killing someone or letting a person die?* The philosophical aim here is to make progress in analyzing the conceptual difference and moral weight of actions in comparison to omissions – not in finding the 'right solution.' Hence, the way these cases are typically discussed in public debate and engineering classes is misguided.

Moreover, Kant's ethics raises questions about how exactly the categorical imperative works as a universalizability test in practice, e.g., whether different specifications of one's maxim may already lead to different results (Korsgaard, 1996; O'Neill, 1989). For instance, it makes a difference if I characterize my maxim as 'I lie if it suits my needs' or rather as 'I lie if I consider the situation as socially unbearable.'

Finally, Kant had a narrow understanding of autonomy solely regarding *moral autonomy*, i.e., our ability to give ourselves a *moral law* of action. Current ethical debate has broadened the notion – now referring to *personal autonomy* – to emphasize that *respecting people's autonomy* is not confined to respecting their ability to act from a universalizable moral principle but also encompasses respecting their ability to make autonomous choices in their personal life (Christman, 2020). Most prominently, this Kantian idea has been introduced as one of four ethical principles in biomedical ethics and further explained in terms of patients needing to give their *informed consent* based on their own individual perspectives and personal values (Beauchamp & Childress, 2019, ch. 4). All these elaborated current discussions about Kant's ethics draw a much more fine-grained picture of deontological ethics than merely explaining Kant's categorical imperative.

The outcome is that engineering students typically fail to learn how critical ethical reflection and argumentation contribute to making ethical progress – both within one strand of ethical theorizing and concerning the controversy between different ethical theories. Yet, this capability of engaging in ethical argumentation and forming well-considered ethical judgments based on state-of-the-art insights is a crucial aspect of ethical competence. Accordingly, including current versions of ethical theories and debates in teaching engineering students would enable students to better understand critical ethical thinking and, thus, improve their ethical competence.

Virtue ethics

Deontological and consequentialist ethical theories share one fundamental assumption: morally right and wrong are distinguished by judging *actions*, whether these are subject to specific general standards (deontology) or based on their consequences (consequentialism). In everyday life, we often use moral judgments differently: instead of actions, we judge the acting *person*. So-called *virtue ethics* takes this as a starting point and focuses on the acting subject, that is, the person who is morally responsible. Virtue ethics is primarily concerned with the question of cultivating moral character. Its focus is on how to be a *good person* and less on defining morally *right* (or *wrong*) *actions*. Apart from care ethics (see below), other ethical theories discussed here focus on the latter. Virtue ethics thus provides a different approach to teaching engineering ethics. Its basic idea when looking at actions is that one reliably acts morally right or good once one has acquired a moral character: a person who *is* good *acts* well.

Virtue ethics approaches were popular in antiquity (with philosophers such as Plato and Aristotle) and throughout the centuries in non-Western thought (particularly in Confucian and Buddhist traditions). While virtue ethics took a backseat to other ethical theories in Western philosophy after the Enlightenment, this changed in the mid-twentieth century with influential publications by, for example, Anscombe (1958), MacIntyre (1981), and Hursthouse (1999). Virtue ethics has proven itself as a theoretical alternative to deontology and consequentialism and plays an increasing role in applied ethics. For medical, care, or business applications, virtue ethics seems to be a fairly well-established alternative to address moral questions (Oakley & Cocking, 2001). In comparison, in engineering ethics and ethics of technology, virtue-based approaches have only gained prominence more recently (Frigo et al., 2021; Steen, 2013; Vallor, 2016) and remain less utilized than utilitarian or deontological positions (Pierrakos et al., 2019).

In virtue ethics, notions like the 'good' or 'virtuous' engineer are central (Harris, 2008). Virtues refer to the characteristics of a (moral) character and are character traits in the sense of deeply rooted dispositions to direct one's actions and thoughts in a certain way. They can be trained and cultivated. Thus, a virtuous character trait develops gradually from a full study and exercise of 'right' actions. The virtues are excellences of character in their own right. Their value is not measured solely by the fact that they serve to implement the right actions. Instead, they are intrinsically valuable to those who cultivate them. It is this latter point that distinguishes virtue ethics from other ethical theories. Admittedly, all ethical theories may include the idea of virtues in the simple sense of good character traits, just like including references to consequences or rules (Hursthouse & Pettigrove, 2022). However, when utilitarianism or deontology refer to virtues, these are merely of instrumental value in ensuring that people act morally. Only virtue ethics places virtues front and center.

Virtue ethics is primarily concerned with the *how* of the morally right or good, not primarily with the *what*. Aristotle, a classic virtue ethicist, saw the highest human goal in realizing *eudaimonia*, often translated as happiness or well-being but referring to a more encompassing idea of flourishing as a human being. Similarly, in Confucian and Buddhist traditions, a full human life is tied to possessing and cultivating particular virtues. Hence, training the virtuous engineer or spelling out any virtues must give some account of a well-lived human life. Although concrete manifestations of virtues (as well as the incorporation and exercise of virtues) may change with the context, virtue ethics is rooted in a conception of human nature or human flourishing and is, therefore, decidedly non-relativistic. It is universal at its core (Vallor, 2016, p. 50).

All versions of virtue ethics involve a concept like the 'Aristotelian mean' that emphasizes the right measure for certain things. The virtuous state is an intermediate ground between two extremes, a 'golden mean' (Aristotle, EN, pp. 1106a26–b28). For instance, the classic virtue of courage lies between cowardice and rashness. Where exactly this intermediate between the two bad extremes lies depends on many contextual factors. To become virtuous, it is necessary to repeatedly identify the 'golden mean' in different situations and act accordingly. The virtuous person can do so reliably through another virtue, which Aristotle called *phronesis*, that is, a learned practical wisdom to judge situations and determine the required action correctly. Such practical wisdom also helps deal with the uncertainties of the threats and promises of technological developments (Frigo et al., 2021; Hillerbrand & Roeser, 2016).

The fact that virtue ethics must presuppose some account of a good human life is often seen as a challenge, especially in modern societies with their multitude of different ways of life and values. This presents a particular challenge for the ethics of technology when, in principle, an ethical account needs to bind present and future generations with very different cultural backgrounds all over the globe. Contemporary virtue ethicists like Vallor (2016) answer this challenge by drawing on both Western and non-Western accounts of the good life. Vallor suggests a list of technomoral virtues: *honesty*, *self-control*, *humility*, *justice*, *courage*, *empathy*, *care*, *civility*, *flexibility*, *perspective*, *magnanimity*, and *technomoral wisdom*. Before Vallor, non-anthropocentric environmental ethicists suggested virtue ethics to deal with the negative impacts of modern technologies (Sandler & Cafaro, 2005). Many of these ecological virtue ethics build on Aldo Leopold's and Henry David Thoreau's *land ethics* and have rather far-reaching implications; they are grounded in assumptions concerning the good human life (Wensveen, 2000). An anthropocentric virtue ethics for the technology era was suggested by Höffe (1993) in his synthesis of Kantian and Aristotelian approaches. Next to classical Aristotelian virtues, Höffe advocates two ecological virtues to deal with the ecological threats that modern technologies may pose:

1. The virtue of *ecological serenity* as intermediate between human hubris, which sees nature as entirely amenable for human interests, on the one side, and an acquiescence of natural threats and hazards, on the other.
2. The virtue of *ecological prudence* as an intermediate between humility and the wish for the fulfillment of each and every need.

Due to their context-sensitivity and their focus on the moral subject (i.e., the engineer, those regulating the development and use of technology, or those using technological artifacts), virtue ethics adds substantially to developing and improving ethical competence among engineering students.

Care ethics

Nair (2005, p. 695) describes *care ethics* as highlighting "the importance of responsibility, concern, and relationship over consequences (utilitarianism) or rules (deontology)." Although responsibility is central in engineering ethics, it was not until recently that care ethics was taken up in (teaching) engineering ethics and ethics of technology (Campbell et al., 2012; Frigo et al., 2023; Pantazidou & Nair, 1999).

Originally developed as an alternative to traditional Western ethical approaches by psychologist C. Gilligan and philosopher N. Noddings, contemporary care ethicists like Tronto (1993) see care ethics more as an augmentation of classical approaches. However, all versions of care ethics share a critique of traditional (Western) ethical reasoning with its focus on generalizations and abstract moral objects and subordinate elements as (at least partially) incomplete; they build on an unwarranted assumption about the nature of moral relations, namely equality. Care ethics asserts that moral relations extend beyond interactions among equals to encompass relationships between individuals with unequal power or circumstances, often involving parties who did not voluntarily enter these relationships. For instance, children find themselves with parents they did not choose. Similarly, individuals such as workers in coal mines may be dependent on their employers and industry in ways they have not autonomously chosen (Groves et al., 2021).

Care ethics takes concrete human relationships and their asymmetries as a starting point. Consequently, the moral subject "is conceived as a *relational* self, one that is constituted in part by relationships important to a person's identity" (Kittay, 2011, p. 53). Moreover, care ethicists share a sensitivity to the context-dependent features of the situation, which renders some parts of ethical reasoning *irreducibly particular* and *non-universalizable*. Every person comes with their own context and their very own specific history and identity. Care ethics contends that moral deliberation requires not reason alone but also *empathy*, *emotional responsiveness*, and *perceptual attentiveness*.

One line of integrating care ethics into the ethics of technology is via technological design. While van Wynsberghe (2013) explores the role of care in the design of robots, more recent work considers energy ethics (Frigo et al., 2023) as well. Van Wynsberghe and Frigo et al. suggest care ethics as a normative framework for *Value Sensitive Design* (VSD) (for more on VSD, see Chapter 22). Michelfelder et al. (2017) explore the concept of *Caring Design* (Flower & Hamington, 2022). Other lines of reasoning have explored the role of care ethics in *Responsible Research and Innovation* (RRI) (Pellé, 2016). Groves (2014) links responsibility to care in intergenerational perspectives, while others approach the link between care and engineering responsibility broadly (Campbell et al., 2012) and care as a guiding principle for (teaching) engineering ethics per se (Kardon, 2005; Pantazidou & Nair, 1999).

Ethics of care has been criticized for being limited in its moral scope as it seems confined to intimate settings and close-kind relationships (or at least between human beings who interact directly). However, within environmental ethics, care ethics has been shown to extend to other sentient beings (Warren, 2000). Tronto (2013), and others have argued that care ethics can tread into areas such as the political realm, especially where practices of justice are inadequate to cover a situation's contextual and narrative complexities.

Contractarianism and contractualism

A general challenge for ethical theories is that they often include controversial assumptions. For instance, utilitarianism must defend its axiology, that is, what should count as goodness. Likewise, virtue ethics must defend its account of what may count as a good human life. In contrast, *contractarianism* is an ethical theory that explicitly tries to avoid controversial assumptions and merely takes rational self-interest and bargaining power as its starting point (Cudd & Eftekhari, 2021). The classic example is Thomas Hobbes's (1588–1679) *Leviathan* (Gauthier, 1986; Hobbes, 1651). The basic idea is that only those social norms that are in the rational self-interest of individuals may be regarded as justified and, thus, morally legitimate. For instance, people have a rational interest in not getting killed against their will. To ensure that one can live safely, it is more efficient to have a social norm in place that generally forbids murder than to take care of one's protection individually. Of course, it must be ensured that everyone adheres to the norm, which is why Hobbes has added the figure of Leviathan, a supreme ruler with absolute power to guarantee people's compliance. Thinking about what is morally obligatory or permissible, then, boils down to analyzing whether one's options fall under a norm that is in people's rational self-interest and that can be sufficiently enforced. As a result, contractarianism only comprises a few basic norms that can be backed up by people's bargaining power, like prohibitions against killing and harming others at will.

Contractualism also refers to rational self-interest and includes the idea of ensuring a *reasonable* or *fair* outcome (Ashford & Mulgan, 2018). Most prominently, John Rawls (1921–2002) developed this idea regarding what constitutes a just society (Rawls, 1971, 2001). The primary (Kantian) idea is that rational, self-interested people must decide on the basic structure of society without knowing their place in it or anything about their own person (e.g., their interests, capabilities, or personal values). They must deliberate behind a *veil of ignorance* (Rawls, 1971, pp. 12–19), which ensures a fair outcome. Accordingly, moral contractualism (Scanlon, 2000) involves thinking about whether any option of how to act may fall under a *fair* (moral) rule and, thus, cannot be *reasonably* rejected.

The contractarian and contractualist ideas of *rational self-interest*, *fairness*, and the possibility of *reasonable rejection* add further aspects for engaging in critical thinking about ethical questions. Any engineering project implies drafting a 'contract' where each of the elements to be developed are perfectly defined; contractarianism and contractualism emphasize not only the importance of the contract itself, but also the importance of why it has been drafted and the conditions under which it has been drafted.

Discourse ethics

Discourse ethics may be depicted as a variant of deontology. The term refers to ethical theories that determine morally right arguments by whether they adhere to specific rules of rational discourse. Discourse ethics originated in the German-speaking world (Apel, 1990, 1999; Habermas, 1983). Both Apel and Habermas viewed discourse ethics as a shift away from Kant's philosophy of the subject, that is, of the constituents of ourselves as individual human persons, resulting in an ethics of individual conviction, towards an ethics of responsibility toward others and the world as a whole. This transition, exemplified by Jonas (1979) and increasingly influential in the ethics of technology, emphasizes the ethical implications of human actions and decision-making. Internationally, discourse ethics is best known as *argumentation ethics* (Hoppe, 1988).

Habermas (1990) formulated two core principles of discourse ethics:

> *Discourse principle (D)*: norms are only valid if they meet (or could meet) the agreement of all affected who, as such, are participants in a practical discourse.
>
> *Principle of universalization (U)*: "*All* affected can accept the consequences and the side effects its [the norm's] *general* observance can be anticipated to have for the satisfaction of *everyone's* interests (and these consequences are preferred to those of known alternative possibilities for regulation)".
>
> *(Habermas, 1990, p. 65)*

With these principles, those who participate in the discourse can, in an ideal case, determine what is morally right or wrong. Ethically permissible communication following *(D)* and *(U)* must be symmetrical; only sound arguments are allowed in this communication, and hierarchies and authorities have no place if they prevent rational communication in the form of critical arguments. As Nickel and Spahn (2012, p. 38) wrote, "The purpose and outcome of the discourse are open in a strong sense, because any party to the communication could be convinced by the other parties to change their behavior or their moral beliefs."

Nickel and Spahn (2012, p. 38) applied discourse ethics to engineering design, particularly the design of persuasive technologies, and argued that the typical *a priori* method of incorporating moral values into the design cannot fulfill the communicative standards set by discourse ethics. To achieve symmetrical communication, there must be room for those who use, or are affected by, the technology not only to co-design its technical features but also to co-design the moral values in the technological design process. The reciprocity of perspective necessitates incorporating the perspective of the other into the norm and, thereby, impartiality. In this regard, discourse ethics does without a counterfactual construction such as Rawls's *veil of ignorance* (Rawls, 1971, pp. 12–19). Though this is often cited as an advantage of discourse ethics, it is unclear how a symmetrical dialogue with future generations can be realized, even in principle, to determine principles of sustainability and intergenerational justice. This may be one reason why, despite some applications to the information technologies (Mingers & Walsham, 2010; Yetim, 2011), discourse ethics is rarely referred to explicitly in the ethics of engineering and technology. However, it plays a vital role in both the practice and teaching of engineering ethics in the guise of stakeholder discussions, deliberative technology assessments, role plays, and so on (Lennerfors, 2019). Still, engineering students could benefit from learning discourse ethics explicitly by imagining themselves in the shoes of various stakeholders and reflecting on how each stakeholder would argue in an idealized version of symmetrical communication.

Typical problems, errors, and barriers in student learning

Despite their appeal, ethical theories and their use within engineering education are not without issues. Many engineering students and educators fail to differentiate personal values from ethics, leading to a barrier in teaching ethics. Thus, they fail to grasp how ethics 'works' and what gaining 'ethical competence' means, namely the ability to understand and reflect on ethical issues based on coherent ethical arguments and principles, not subjective personal preferences. This problem is perhaps linked to faculty members' observed resistance to teaching ethics. Haws (2001, p. 227) suggests that the claim by engineering educators that "the theoretical aspect of engineering ethics is beyond our expertise" undoubtedly has consequences for both the confidence and enthusiasm with which such theories are communicated to students.

Another, perhaps more pragmatic issues associated with extensive discussion of moral theories in engineering ethics education are the amount of study needed to fully appreciate ethical theories and the limited time allocated for their teaching. Teaching ethical theories, Lawlor (2007) claims, can thus result in two undesirable consequences: either that students fail to process the nuances and complexities of each theory or that theories are simplified to the degree that they become of little use. This can cause students to dismiss their use and that of philosophical reasoning altogether. Lawlor further claims that teaching students this way can lead them to believe that ethics consists of simply picking a theory and applying it to a specific case by following it to its end.

Ironically, the problem with using ethical theories correctly is emphasized by students' familiarity with the use of *empirical* or *descriptive* theories. Such theories are meant to accurately describe and explain 'states of the world', that is, the theories need to 'fit' the world and be revised if they do not. In assuming that *normative* ethical theories can be applied in the same way, students risk erroneously choosing an ethical theory to *fit* the concrete ethical problem, for example, choosing utilitarianism because the ethical problem appears concerned with undesirable consequences. However, the 'direction of fit' of ethical theories works precisely the other way around (Anscombe, 1963; Searle, 2001). Ethical theories depict states of the world that are *not* (yet) the case but *ought* to be. Hence, if an ethical theory's content does not 'fit' the relevant 'state of the world,' it is the world that needs to change – brought about by our moral actions. What we ought to do then relies on the respective theory's ethical criteria, like maximizing utility in the case of utilitarianism or the universalizability of one's maxim in the case of deontology. Accordingly, referring to apparently undesirable consequences does not qualify as a reason to choose any ethical theory. So, the main problem for students in working with ethical theories is understanding – in direct opposition to what they know from working with empirical or descriptive theories – *not* to put the (worldly) cart before the (ethical) horse.

A related problem occurs when students commonly integrate theoretical elements of different ethical theories without being aware of the (theoretical) inconsistencies that arise by doing so. For example, utilitarianism explicitly claims that overall consequences are *all* that matters for ethical judgments. This sometimes clashes with individual rights when these are seen from a deontological perspective. Often, students, in arguing explicitly from a utilitarian standpoint, simply want to solve this by 'adding Kantian deontology' for dealing with individual rights, that is,safeguarding individual rights against a utilitarian calculation. What such students fail to see is that Kantian deontology rejects utilitarian calculation *completely* and that, therefore, they cannot hold both ethical positions at the same time without contradicting themselves. Those students thus show a lack of ethical competence and fail to reach a consistent and well-argued judgment. Such competence would consist in realizing that if one wanted to uphold a utilitarian judgment, this would imply that any individual right could only be justified as a result of a utilitarian calculation, for example, because implementing individual rights would lead to overall better consequences. Accordingly, safeguarding individual rights would only be possible against this background and would always be limited. If, on the contrary, one wanted *strictly* to uphold individual rights from a deontological perspective, this would imply dropping one's previous utilitarian standpoint that consequences are *all* that matters ethically.

To equip students with the ability to consistently justify their decisions by integrating ethical theories, it is also recommended that students actively apply their ethical competence in their daily activities. In many universities, there are possibilities for co-curricular activities outside the strict curriculum, which can increase their knowledge in this area. For example, industrial lecturers can be invited to share real-life ethical challenges and case studies with students through group discussions or short reflection papers. There are also many students involved in service-learning activities. It would be advisable for this type of work to seek a link with engineering so that the development of ethics and, in particular, the ethical behavior component could be promoted through coursework. However, such additional and partially non-mandatory activities take up students' time, adding to the challenge of how successful engineering ethics education can be integrated into student activities.

This last problem leads to the general question of how best to incorporate ethics teaching into the curricula, including teaching ethical theories. A study carried out by Walczak et al. (2010) identified five common problems in engineering schools that hinder the incorporation of ethics in students' training. In addition to the problems mentioned above (i.e., the overcrowded curricula, the limited space for ethics education, and educators needing more training for teaching ethics), two further problems were noted: inconsistency in policies and academic dishonesty.

The challenges that engineering students face when it comes to integrating ethics into their decision-making processes during the design and development of products present their flipside when it comes to *assessing* students' ethical competence. Engineering schools need to determine whether, or how well, graduates have acquired this competence. While various helpful approaches to acquiring ethical competence for future engineers can be found in the literature, the question of how to assess students' ethical competence remains a particular challenge. Some schools offer compulsory subjects within the curriculum that deal with ethics in engineering; others opt for a transversal integration with different approaches. To evaluate the effectiveness of such approaches, it would be necessary to carry out respective studies (Barry & Ohland, 2012). In any case, even if knowledge of ethical theories proved to be conducive to developing ethical competence, such knowledge alone would not guarantee that students will display an ethical attitude when making decisions related to their professional work. In other words, *knowing* what is right does not guarantee *doing* what is right. (For more on the assessment of competencies, see Chapter 26.)

From ethical theories to ethical competence?

Given not only the problems associated with teaching and learning ethical theories but also the question of whether knowing them is really necessary for students to develop ethical competence, one may argue that ethical theories could or should be entirely dismissed in teaching engineering ethics. Bouville (2008) proposes four possibilities regarding the treatment of ethical theories in teaching ethics to engineering students: (1) reject theories entirely, (2) use them without naming them, (3) mention them without justifying them, or (4) teach and justify them. Glagola et al. (1997, p. 475) thus encourage educators to "chuck out the jargon," indicating that students may feel they need to choose between ethical theories and that instead, "when we've got a moral problem, we should examine all the morally relevant considerations" (Glagola et al., 1997, p. 475). This view is "Pluralist – useful approaches may be drawn from a variety of ethical theories" (Derry & Green,

1989, p. 531). Bouville (2008) takes a similar approach, encouraging us to break ethical theories down into constituent parts and to emphasize fundamental dichotomies and elementary concepts, for example, consequences versus intentions or society as a whole versus individuals. An advantage of this, Bouville claims, is that the pairs are comparable, making it easier to find a middle ground through reasoning. Yet, doing so would only lead back to the general problem of ethical consistency we mentioned above if students came to some 'middle ground' in their ethical judgment without realizing its lack of coherent ethical justifiability. From a philosophical point of view, the uncritical suggestion of such a 'middle ground' as an implicit pluralist ethical framework is simply untenable. Accordingly, Haws (2004) suggests that the need to solve future, unprecedented ethical dilemmas indeed necessitates the inclusion of ethical theory, and that engineers who lack a foundation of strong theoretical knowledge will be unable to adapt. Such grounding is also required for engineers to justify their decisions to the broader community.

According to Newberry (2004), the purposes of teaching ethics in engineering can be classified into three categories: *particular knowledge*, *intellectual engagement*, and *emotional commitment*. Learning the main ethical theories would correspond to *particular knowledge* and would be the most easily attainable goal and the one that is often assessed in engineering schools. The second category, *intellectual engagement*, would be related to knowing how to make ethical decisions – the difficulties this presents have already been mentioned. Finally, *emotional commitment* or the desire to behave ethically would be most challenging to measure as an outcome within a curriculum – even if such an outcome would be desirable from the perspective of engineering ethics education.

Although the goal of personal ethical character building can neither be guaranteed nor properly assessed by 'merely' teaching ethics, developing an assessable ethical competence may very well be achievable by way of critically engaging with ethical theories, especially given the wide range of ethical theories with their different core ideas. If ethical competence is supposed to include the ability to reflect critically on different ethical aspects in the engineering realm, learning different ethical theories would equip students with relevant knowledge about different ethical aspects and their respective ethical roles in making well-justifiable decisions in practice. For instance, it makes a theoretical and practical difference when ethically thinking about a design problem in engineering from the perspective of, for example, utilitarianism, deontology, or care ethics, including their different ideas on how to justify ethical judgments and decisions. For example, deciding on the 'best' design of a bridge may differ if the decision is based on the overall consequences in terms of the most efficient travel connection (*utilitarianism*);or on relational aspects of the people who want to cross it using various means of transportation, including cars, bikes, and walking (*care ethics*); or on the bargaining power of those who have specific preferences for its design (*contractarianism*). Hence, students would not only gain *particular knowledge* when learning about different ethical theories but also show intellectual involvement via *critical engagement*, thereby becoming increasingly capable of justifying and critically re-evaluating their own particular ethical judgments and decisions consistently and coherently to others.

However, even if this line of argument is plausible and ethical theories may be considered conducive to developing ethical competence, one might still question whether they are necessary. Alternative teaching models encompass other factors affecting ethical decision-making and competence (Bairaktarova & Woodcock, 2017; Walczak et al., 2010). For example, Illingworth (2004) has outlined three different ways to teach applied ethics within higher education:

1. A pragmatic approach based on regulatory codes.
2. An embedded approach that makes use of reflection and role play.

3. A theoretical approach that "places an understanding of moral theory at the heart of ethics learning and teaching" and whereby "ethics of real-life or life-like situations are then presented in terms of application of that theory" (p. 10).

Conclusion

This chapter has provided an overview of the most important ethical theories as well as some theories that engineering ethics education has rather neglected. It briefly discussed whether teaching ethical theories is necessary to help students develop ethical competence – given that although knowing about ethical theories is conducive to developing ethical competence, knowledge of specific theories might not be necessary for sound engineering practice. For all practical intents and purposes, engineering ethics education may utilize other teaching models to achieve the levels of ethical awareness required (e.g., by accreditation bodies, as discussed in Chapters 32–36) – even though, from a philosophical point of view, ethical theories provide indispensable underlying frameworks for critical analysis and justification of ethical judgments. Understanding the theories discussed in this chapter can help educators and researchers achieve consistency and develop well-framed activities and materials. Understanding these seminal theories and their internal logic will also support understanding the other chapters of this handbook.

References

Alas, R. (2006). Ethics in countries with different cultural dimensions. *Journal of Business Ethics*, *69*(3), 237–247. https://doi.org/10.1007/s10551-006-9088-3

Andersson, H., Svensson, A., Frank, C., Rantala, A., Holmberg, M., & Bremer, A. (2022). Ethics education to support ethical competence learning in healthcare: An integrative systematic review. *BMC Medical Ethics*, *23*(1), 29. https://doi.org/10.1186/s12910-022-00766-z

Anscombe, G. E. M. (1958). Modern moral philosophy. *Philosophy*, *33*(124), 1–19. https://doi.org/10.1017/S0031819100037943

Anscombe, G. E. M. (1963). *Intention* (2nd ed.). Harvard University Press.

Apel, K.-O. (1990). *Diskurs und Verantwortung: Das Problem des Übergangs zur postkonventionellen Moral* (5th ed.). Suhrkamp.

Apel, K.-O. (1999). *Transformation der Philosophie: Band II. Das Apriori der Kommunikationsgemeinschaft*. Suhrkamp.

Aristotle. (EN). *Nicomachean ethics* (W. D. Ross, Trans.). The Internet Classics Archive. http://classics.mit.edu/Aristotle/nicomachaen.html

Ashford, E., & Mulgan, T. (2018, Summer). Contractualism. In E. N. Zalta (Ed.), *The Stanford encyclopedia of philosophy*. Metaphysics Research Lab, Stanford University. https://plato.stanford.edu/archives/sum2018/entries/contractualism/

Bairaktarova, D., & Woodcock, A. (2017). Engineering student's ethical awareness and behavior: A new motivational model. *Science and Engineering Ethics*, *23*(4), 1129–1157. https://doi.org/10.1007/s11948-016-9814-x

Barry, B. E., & Ohland, M. W. (2012). ABET criterion 3.f: How much curriculum content is enough? *Science and Engineering Ethics*, *18*(2), 369–392. https://doi.org/10.1007/s11948-011-9255-5

Beauchamp, T. L., & Childress, J. F. (2019). *Principles of biomedical ethics* (8th ed.). Oxford University Press.

Bentham, J. (1789). *An introduction to the principles of morals and legislation*. Clarendon Press, 1907.

Bouville, M. (2008). On using ethical theories to teach engineering ethics. *Science and Engineering Ethics*, *14*(1), 111–120. https://doi.org/10.1007/s11948-007-9034-5

Campbell, R. C., Yasuhara, K., & Wilson, D. (2012). *Care ethics in engineering education: Undergraduate student perceptions of responsibility*. 2012 Frontiers in Education Conference Proceedings, 1–6. https://doi.org/10.1109/FIE.2012.6462370

Christman, J. (2020, Fall). Autonomy in moral and political philosophy. In E. N. Zalta (Ed.), *The Stanford encyclopedia of philosophy*. Metaphysics Research Lab, Stanford University. https://plato.stanford.edu/archives/fall2020/entries/autonomy-moral/

Cudd, A., & Eftekhari, S. (2021, Winter). Contractarianism. In E. N. Zalta (Ed.), *The Stanford encyclopedia of philosophy*. Metaphysics Research Lab, Stanford University. https://plato.stanford.edu/archives/win2021/entries/contractarianism/

Derry, R., & Green, R. M. (1989). Ethical theory in business ethics: A critical assessment. *Journal of Business Ethics*, *8*(7), 521–533.

Didier, C. (2015). Engineering ethics: European perspective. In J. B. Holbrook (Ed.), *Ethics, science, technology, and engineering: A global resource* (2nd ed., Vol. 2, pp. 87–90). Acmillan Reference USA.

Driver, J. (2022, Winter). The history of utilitarianism. In E. N. Zalta & U. Nodelman (Eds.), *The Stanford encyclopedia of philosophy*. Metaphysics Research Lab, Stanford University. https://plato.stanford.edu/archives/win2022/entries/utilitarianism-history/

Flower, M., & Hamington, M. (2022). Care ethics, bruno latour, and the anthropocene. *Philosophies*, *7*(2), 31. https://doi.org/10.3390/philosophies7020031

Foot, P. (2003). *Virtues and vices. And other essays in moral philosophy* (Reprint ed.). Oxford University Press.

Franck, O. (2017). Varieties of conceptions of ethical competence and the search for strategies for assessment in ethics education: A critical analysis. In O. Franck (Ed.), *Assessment in ethics education: A case of national tests in religious education* (pp. 13–50). Springer International Publishing. https://doi.org/10.1007/978-3-319-50770-5_2

Frigo, G., Marthaler, F., Albers, A., Ott, S., & Hillerbrand, R. (2021). Training responsible engineers. Phronesis and the role of virtues in teaching engineering ethics. *Australasian Journal of Engineering Education*. https://www.tandfonline.com/doi/abs/10.1080/22054952.2021.1889086

Frigo, G., Milchram, C., & Hillerbrand, R. (2023). Designing for care. *Science and Engineering Ethics*, *29*(3), 16. https://doi.org/10.1007/s11948-023-00434-4

Gauthier, D. P. (1986). *Morals by agreement*. Oxford University Press.

Glagola, C., Kam, M., Whitebeck, C., & Loui, M. C. (1997). Teaching ethics in engineering and computer science: A panel discussion. *Science and Engineering Ethics*, *3*(4), 463–480. https://doi.org/10.1007/s11948-997-0048-9

Groves, C. (2014). *Care, uncertainty and intergenerational ethics* (2014th ed.). Palgrave Macmillan.

Groves, C., Shirani, F., Pidgeon, N., Cherry, C., Thomas, G., Roberts, E., & Henwood, K. (2021). A missing link? Capabilities, the ethics of care and the relational context of energy justice. *Journal of Human Development and Capabilities*, *22*(2), 249–269. https://doi.org/10.1080/19452829.2021.1887105

Habermas, J. (1983). *Moralbewußtsein und kommunikatives Handeln*. Suhrkamp.

Habermas, J. (1990). *Moral consciousness and communicative action* (C. Lenhardt & S. Weber-Nicholsen, Trans.). MIT Press.

Harris, C. E. (2008). The good engineer: Giving virtue its due in engineering ethics. *Science and Engineering Ethics*, *14*(2), 153. https://doi.org/10.1007/s11948-008-9068-3

Haws, D. R. (2001). Ethics instruction in engineering education: A (Mini) meta-analysis. *Journal of Engineering Education*, *90*(2), 223–229. https://doi.org/10.1002/j.2168-9830.2001.tb00596.x

Haws, D. R. (2004). The importance of meta-ethics in engineering education. *Science and Engineering Ethics*, *10*(2), 204–210. https://doi.org/10.1007/s11948-004-0015-7

Hess, J. L., & Fore, G. (2018). A systematic literature review of US engineering ethics interventions. *Science and Engineering Ethics*, *24*(2), 551–583. https://doi.org/10.1007/s11948-017-9910-6

Hillerbrand, R., & Roeser, S. (2016). Towards a third 'practice turn': An inclusive and empirically informed perspective on risk. In M. Franssen, P. E. Vermaas, P. Kroes, & A. W. M. Meijers (Eds.), *Philosophy of technology after the empirical turn* (Vol. 23, pp. 145–166). Springer International Publishing. https://doi.org/10.1007/978-3-319-33717-3_9

Hobbes, T. (1651). *Leviathan* (C. Brooke, Ed.). Penguin Classics, 2017.

Höffe, O. (1993). *Moral als Preis der Moderne: Ein Versuch über Wissenschaft, Technik und Umwelt*. Suhrkamp Verlag.

Hoppe, H.-H. (1988). On the ultimate justification of the ethics of private property. *Liberty*, *2*(1), 20–22.

Hursthouse, R. (1999). *On virtue ethics*. Oxford University Press.

Hursthouse, R., & Pettigrove, G. (2022, Winter). Virtue ethics. In E. N. Zalta & U. Nodelman (Eds.), *The Stanford encyclopedia of philosophy*. Metaphysics Research Lab, Stanford University. https://plato.stanford.edu/archives/win2022/entries/ethics-virtue/

Illingworth, S. A. (2004). *Approaches to ethics in higher education: Teaching ethics across the curriculum.* https://www.semanticscholar.org/paper/Approaches-to-Ethics-in-Higher-Education%3A-Teaching-Illingworth/6ec1f607e12785f71796efe5cae873cda11f628d
Jonas, H. (1979). *Das Prinzip Verantwortung. Versuch einer Ethik für die technologische Zivilisation* (7th ed.). Suhrkamp, 2003.
Kant, I. (1785). *Groundwork of the metaphysics of morals: A German–English edition* (M. Gregor & J. Timmermann, Eds.; Bilingual). Cambridge University Press, 2011.
Kant, I. (1797). *The metaphysics of morals* (Rev. ed.). Cambridge University Press, 1996.
Kardon, J. B. (2005). Concept of "care" in engineering. *Journal of Performance of Constructed Facilities, 19*(3), 256–260. https://doi.org/10.1061/(ASCE)0887-3828(2005)19:3(256)
Kittay, E. F. (2011). The ethics of care, dependence, and disability*. *Ratio Juris, 24*(1), 49–58. https://doi.org/10.1111/j.1467-9337.2010.00473.x
Korsgaard, C. M. (1996). *Creating the kingdom of ends.* Cambridge University Press.
Lawlor, R. (2007). Moral theories in teaching applied ethics. *Journal of Medical Ethics, 33*(6), 370–372. https://doi.org/10.1136/jme.2006.018044
Lennerfors, T. T. (2019). *Ethics in engineering.* Studentlitteratur AB. https://urn.kb.se/resolve?urn=urn:nbn:se:uu:diva-408705
MacIntyre, A. (1981). *After virtue: A study in moral theory.* University of Notre Dame Press.
Martin, M. W., & Schinzinger, R. (2004). *Ethics in engineering* (4th ed.). McGraw Hill Higher Education.
McConnell, T. (2022, Fall). Moral dilemmas. In E. N. Zalta & U. Nodelman (Eds.), *The Stanford encyclopedia of philosophy.* Metaphysics Research Lab, Stanford University. https://plato.stanford.edu/archives/fall2022/entries/moral-dilemmas/
Michelfelder, D. P., Wellner, G., & Draper, H. (2017). Designing differently. Toward a methodology for an ethics of feminist technology design. In S. O. Hansson (Ed.), *The ethics of technology* (pp. 193–218). Rowman & Littlefield.
Mill, J. S. (1861). Utilitarianism. In J. M. Robsen (Ed.), *The collected works of John Stuart Mill, volume X – essays on ethics, religion, and society* (pp. 203–259). University of Toronto Press, 1969.
Mingers, J., & Walsham, G. (2010). Toward ethical information systems: The contribution of discourse ethics. *MIS Quarterly, 34*(4), 833–854. https://doi.org/10.2307/25750707
Nair, I. (2005). Ethics of care. In C. Mitcham (Ed.), *Encyclopedia of science, technology, and ethics* (pp. 695–700). Thomson/Gale. https://repository.library.georgetown.edu/handle/10822/985341
Newberry, B. (2004). The dilemma of ethics in engineering education. *Science and Engineering Ethics, 10*(2), 343–351. https://doi.org/10.1007/s11948-004-0030-8
Nickel, P., & Spahn, A. (2012). Trust; Discourse ethics; and persuasive technology. In *Persuasive technology: design for health and safety; The 7th international conference on persuasive technology; PERSUASIVE 2012* (pp. 37–40). Linköping Electronic Conference Proceedings. https://ep.liu.se/en/conference-article.aspx?series=ecp&issue=68&Article_No=10
Oakley, J., & Cocking, D. (2001). *Virtue ethics and professional roles.* Cambridge University Press.
O'Neill, O. (1989). *Constructions of reason. Explorations of Kant's practical philosophy.* Cambridge University Press.
Pantazidou, M., & Nair, I. (1999). Ethic of care: Guiding principles for engineering teaching & practice. *Journal of Engineering Education, 88*(2), 205–212. https://doi.org/10.1002/j.2168-9830.1999.tb00436.x
Pellé, S. (2016). Process, outcomes, virtues: The normative strategies of responsible research and innovation and the challenge of moral pluralism. *Journal of Responsible Innovation, 3*(3), 233–254. https://doi.org/10.1080/23299460.2016.1258945
Pierrakos, O., Prentice, M., Silverglate, C., Lamb, M., Demaske, A., & Smout, R. (2019). *Reimagining engineering ethics: From ethics education to character education.* 2019 IEEE Frontiers in Education Conference (FIE), 1–9. https://doi.org/10.1109/FIE43999.2019.9028690
Poel, I. van de, & Royakkers, L. (2011). *Ethics, technology, and engineering: An introduction.* Wiley-Blackwell.
Polmear, M., Bielefeldt, A. R., Knight, D., Canney, N., & Swan, C. (2019). Analysis of macroethics teaching practices and perceptions in engineering: A cultural comparison. *European Journal of Engineering Education, 44*(6), 866–881. https://doi.org/10.1080/03043797.2019.1593323
Rawls, J. (1971). *A theory of justice.* Belknap Press.
Rawls, J. (2001). *Justice as fairness. A restatement.* Belknap Press.
Ross, W. D. (1930). *The right and the good.* Clarendon Press.
Sandler, R. D., & Cafaro, P. (Eds.). (2005). *Environmental virtue ethics.* Rowman & Littlefield.

Scanlon, T. M. (2000). *What we owe to each other*. Belknap Press of Harvard University Press.
Searle, J. (2001). *Rationality in action*. MIT Press.
Sidgwick, H. (1909). *The methods of ethics* (7th ed.). Hackett Publishing, 1981.
Sinnott-Armstrong, W. (2022, Winter). Consequentialism. In E. N. Zalta & U. Nodelman (Eds.), *The Stanford encyclopedia of philosophy*. Metaphysics Research Lab, Stanford University. https://plato.stanford.edu/archives/win2022/entries/consequentialism/
Smart, J. J. C., & Williams, B. (1973). *Utilitarianism: For and against*. Cambridge University Press.
Steen, M. (2013). Virtues in participatory design: Cooperation, curiosity, creativity, empowerment and reflexivity. *Science and Engineering Ethics*, *19*(3), 945–962. https://doi.org/10.1007/s11948-012-9380-9
Thomson, J. J. (1986). *Rights, restitution, and risk. Essays in moral theory* (W. Parent, Ed.). Harvard University Press.
Tronto, J. (1993). *Moral boundaries: A political argument for an ethic of care*. Routledge.
Tronto, J. (2013). *Caring democracy: Markets, equality, and justice*. New York University Press.
Vallor, S. (2016). *Technology and the virtues: A philosophical guide to a future worth wanting*. Oxford University Press.
Van Wynsberghe, A. (2013). Designing robots for care: Care centered value-sensitive design. *Science and Engineering Ethics*, *19*(2), 407–433. https://doi.org/10.1007/s11948-011-9343-6
Walczak, K., Finelli, C., Holsapple, M., Sutkus, J., Harding, T., & Carpenter, D. (2010). *Institutional obstacles to integrating ethics into the curriculum and strategies for overcoming them*. 2010 Annual Conference & Exposition Proceedings, 15.749.1–15.749.14. https://doi.org/10.18260/1-2--16571
Warren, K. J. (2000). *Ecofeminist philosophy: A Western perspective on what it is and why it matters*. Rowman & Littlefield Publishers.
Wensveen, L. van. (2000). *Dirty virtues. The emergence of ecological virtue ethics*. Humanity Books.
Yetim, F. (2011). Bringing discourse ethics to value sensitive design: Pathways toward a deliberative future. *AIS Transactions on Human-Computer Interaction*, *3*(2), 133–155.

3

INDIVIDUAL AND COLLECTIVE DIMENSIONS OF ETHICAL DECISION-MAKING IN ENGINEERING

Kari Zacharias[1], Marion Hersh, Andrew O. Brightman, and Jonathan Beever

Introduction

A foundational element of ethics in engineering practice is the complexity of the relationships between the individual and the collective in ethical decision-making (EDM). Engineers generally work as part of a team, often alongside other teams within a larger organization. Engineering also shapes and is shaped by the wider communities in which engineers participate, both professionally and personally (Davis, 2006). Norms, values, and attitudes towards ethics in these broader contexts can influence the ethical decision-making (including ethical action) of individual engineers. Furthermore, engineers and engineering work can play significant roles in large-scale sociotechnical problems, from climate change to pandemics to social injustices, which require collective decision-making and the resulting collective action[2] that extends far beyond engineering organizations. Despite this, engineering ethics research and education have largely focused on ethical decision-making by individuals, with little attention to both the complex interactions between individuals and collectives and the necessity of engineers' contributions to collective ethical action for solving large-scale sociotechnical problems.

As evidenced in the literature on the ethics of care, every individual is at least partially dependent on others for their material and emotional needs, and these dependencies can impact our ethical decision-making. Whether seen through interconnected relationships of care (e.g., Held, 2006) or autonomous decisions implemented with support (e.g., Knight, 2017), the lines connecting individual and collective aspects of decision-making become complex and, at times, blurry. In contexts ranging from biology to bioethics, others have argued that this blurriness should compel us to rethink the relationship between the individual and the collective, trading views of interconnection for views of interdependence (Beever & Morar, 2016, 2019; Sharma, 2016; Thompson, 2016). Interdependence implies that individuals can be better understood as parts of collectives as well as collectives themselves and are shaped or even constituted by these connected relationships. This view can be extended to contemporary collectives, including non-human decision-making

DOI: 10.4324/9781003464259-5

agents, wherein multi-agent systems reflect the complex interplay between engineers, programmers, operators, and machines (e.g., Awad et al., 2019).

This chapter articulates the diverse types of interactions between individuals and the collectives of which they are a part and shows how these interactions can be mutually supportive or in tension. Complex relationships point to the need for engineers to develop sensitivity to, be able to reason about, and take action on ethical issues in both micro-ethics and macro-ethics contexts as a necessary condition for effective EDM. We argue that this ethics literacy needs to include understanding the range of external factors and collective contexts that influence individual ethical decision-making and are, themselves, shaped by individual actions. Specifically, this chapter covers:

- Theoretical and practical aspects of the relationships between individuals and collectives in engineering practice
- Key concepts in individual and collective ethical decision-making
- Discussion of how current approaches to EDM in engineering education characterize individual and collective dimensions
- A case study that illustrates individual and collective dimensions of EDM

The four authors of this chapter bring perspectives from philosophy, engineering education, biomedical engineering, and science and technology studies. They are positioned across disciplinary perspectives as well as career stages and represent multiple gender, cultural, and national perspectives. Two authors are based in the United States, one in Scotland (the United Kingdom), and one in Canada. Their perspectives and positions intersect around questions of ethics and ethical decision-making in engineering, wherein each has significant pedagogical and research experience in engineering ethics education.

Relationships among individuals and collectives in engineering practice

Engineers work with and within numerous collectives throughout their careers. Collectives include the engineering teams, departments, and organizations in which engineers are typically employed, as well as the industry organizations, sociocultural contexts, and ecosystems in which engineering work is situated. Although the ability for collective action functions at a level of psychological complexity that we still struggle to understand, much of our life and work revolves around collective decisions rather than individual ones (Rachar & Salomone-Sehr, 2023). In many ways, the engineering profession depends on the ability of individual engineers to engage in collective decision-making to achieve outcomes that range from the simple levels of teamwork (e.g., partnered surveying), to more complex levels of collaboration and coordination (e.g., a multinational engineering company sourcing parts and processes from auxiliary organizations while complying with the regulations on trade and transport of different countries and applying health and safety standards and equitable treatment of personnel throughout the global organization).

In this section, we introduce the various types of collectives in which engineers are involved and discuss how acknowledging the complexity of these overlapping relationships and influences could reconceptualize the traditional views of individual decision-making, responsibility, and ethical action in engineering.

Engineers and teams

Engineers regularly work in teams of various size and composition to accomplish their project goals. Whether in small teams of independent consultants or large teams of engineering divisions in companies, the decision-making of individual engineers is often highly influenced by their team-mates, their team leaders, supervisors, and managers, as well as the ever-present collective context of that engineering group sub-culture (Jones et al., 2017). Studies have shown how group dynamics can stifle discussions of ethical issues in professional settings (Hussain et al., 2019) but also demonstrated the potential for individual 'ethical champions' to positively influence ethical decision-making within teams (Chen et al., 2020). The complex projects in which engineers engage typically require interactions among multiple perspectives, skills, knowledge, and roles. Even when a particular person is required by their role or pressure to make an independent final decision, this decision is highly contextualized. It will affect the rest of the team – in terms of both short-term actions and the long-term effects on team culture and future decisions. In addition, the other team members have individual and collective responsibilities to support and encourage the decision-maker to make ethical decisions and involve the collective in decision-making.

Engineers and organizations

Individual engineers also influence and are influenced by the cultures of the organizations they are part of, beyond the dynamics of their teams (Kim & Hess, 2022). It is much easier to act ethically in a context where the organization expects and encourages ethical behavior – whether it is a manufacturing company, academic institution, governmental agency, or non-profit organization (Mason, 2004). Conversely, when an organizational culture stresses values such as profit, market share, rapid innovation, short-term outcomes, or global influence over ethical choices, it becomes much easier for employees to act unethically.

Strongly hierarchical organizations can give rise to unethical behavior, particularly when they have a culture of blame, intimidation, or secrecy (Kranakis, 2004; Jeske, 2020; Rogal, 2020). Although the responsibility for creating and stewarding an ethical organizational culture typically falls to individual leaders or a leadership team, the decisions of every individual either support or weaken the collective culture over time. In large organizations, multiple sub-cultures can form with internal values, norms, and practices, which may or may not align well with other sub-cultures or with the overall culture of the organization (Hofstede, 1998). Such sub-cultures can increase the challenges of stewarding an organizational culture that is ethical overall. These challenges are exacerbated when the organization spans multiple national cultures, which nearly always have varied, and the often conflicting, cultural norms that influence decision-making. In addition, in multinational work, there is the risk of cultural prejudice and discrimination leading to unethical and even fatal consequences as evident in the tragic industrial disaster in Bhopal, India (Broughton, 2005).

Engineers and the profession

The nature of the engineering profession establishes that all engineers, regardless of their official status, are connected to the profession and its associated norms, values, principles, practices, and perspectives (Kasher, 2005). However, the engineering profession also includes a rich diversity of professional societies based on discipline, focus area, country, and geo-political region – including national regulators, academic societies, and informal advocacy groups. Participation in professional societies is often voluntary, although exceptions exist (such as in countries where the practice of engineering is limited to those who hold an approved professional designation; for more on professional licensure, see Chapter 34). However, even professional organizations with voluntary membership can exert significant influence on engineering ethical decision-making through guidelines, standards, codes of ethics, and informal communication (Rosenberg, 1998). While professional engineering codes of ethics have traditionally addressed a narrow scope of ethical issues, this is changing. Since the 1970s and 1980s, professional societies have begun to acknowledge broader ethical issues, such as environmental sustainability and the social impacts of engineering work, in their codes of ethics (Vesilind, 1995; for more on historic trends, see Chapter 32). These societies are also dependent on their individual members for contributions to the respective society's goals, policies, guidelines, and, ultimately, its collective engineering cultures. Thus, individual engineers should consider their roles and responsibilities in improving the ethical culture of their professional societies and, through them, the wider engineering community and profession (Davis, 1991; Lynch & Kline, 2000).

Engineers and society

Beyond their specific teams, organizations, and professional associations, engineers are responsible to the societies in which they live and work, and to the greater global society of which we are all part. Various societal and cultural influences play a role in the decisions engineers make and how they act, as both individuals and collectives. Engineers' national contexts influence the organization of their professional associations, for example, with respect to unionization and regulation (Meiksins & Smith, 1996) and the extent to which individuals and organizations prioritize collective versus individual agency, autonomy, responsibility, and accountability (Hofstede, 2011; Husted & Allen, 2007).

Sociocultural norms and values – including detrimental ones, such as normalized racism, sexism, ableism, and so on – are also reflected in the technologies that engineers create and can exert a significant negative impact on already marginalized populations (Benjamin, 2019; Eubanks, 2017). When such values and norms are deeply embedded in everyday technologies, their origins become challenging to trace and understand, even when their impacts are widely felt, particularly by people who do not fit the mold of a default user who is assumed to be white, male, non-disabled, or English-speaking. Thus, engineers face significant social concerns when making design decisions, including justice, equity, and access. Their expertise and professional status set an ethical responsibility for engineers to be involved in decision-making regarding the types of problems they work on as individuals and also contribute to influencing the types of problems that society as a whole asks engineers to solve (Hersh, 2014; Riley, 2023).

Engineers and ecosystems

Just as engineers are part of local and global societies, they are also part of larger-scale ecologies that influence and are influenced by the technologies that engineers design and produce. Thus, an additional dimension of engineering EDM involves the relationship that engineering projects have with the natural environment, including its more-than-human elements. The contemporary climate crisis makes engineering decisions about resource utilization and extraction, geographical location, habitat disruption, and energy generation vital (for more on these topics, see Chapters 6 and 11).

Local environmental considerations for engineering decisions might be handled at the collective level of an organization. However, on a global scale, engineering decisions involve many more layers of collective decision-making, including national and international governmental associations and multinational engineering companies. Engineers of all types, as both individuals and members of collectives, are part of – and have a global responsibility to – humanity, all non-human species, and the environment as a whole. Consider, for example, that the continuing development of planned large-scale oil and gas projects (Carrington & Taylor, 2022) could be influenced by individual decisions and actions (e.g., high-level intergovernmental regulation) and/or by the collective refusal of engineers and scientists to work on these projects. As stated elsewhere (Beever & Whitehouse, 2017), an 'ecosystemic' approach to decision-making is necessitated by the globalization of human communities (the global public) and the resulting anthropogenic changes to those communities' natural environments.

An engineering example

The attitudes, practices, norms, values, and cultures of the collectives described above – as well as the positioning of individual engineers with respect to them – all impact ethical decision-making. The ethical aspects of all engineering decisions are thus uniquely situated and complex. Consider the following example:[3] A recently graduated female engineer with a background in micro-fluidics from Stanford and MIT joins a startup company early in product development of a miniaturized bioanalytical device for blood testing. She quickly observes a lack of communication between the engineering and analytical chemistry departments that is jeopardizing critical system testing. Learning that an important pilot project contract for a major pharmaceutical company is about to go live, the new engineer alerts her male supervisor of the potential risk of system failure, but her report is ignored. After repeated attempts to bring attention to the problems of reliability and usability, including a direct appeal to the female CEO, the engineer is told by her supervisor to "Go find a place where you can be a big fish in a small pond." The young engineer decides to leave the company soon after.

This story could be presented as a straightforward and narrow case of the individual's conflict with management (in a way that is frequent within contemporary engineering ethics education): the new engineer has discovered a potential danger to the public, but her supervisor (a more senior engineer) and those above him disagreed. However, upon further examination, numerous collective and broader ethical aspects emerge. Power dynamics and inequalities based on gender, seniority, status, and perceived differences in knowledge are significant aspects of this case. These power relations alone make the case about more-than-individual decision-making. The younger female engineer is part of a complex collective decision-making structure contingent on various external factors. As a junior member of the engineering department, she is dependent on her supervisor for guidance, mentorship, or a potentially positive recommendation in the future. The supervisor likely depends on the engineer for her contributions or support to advance the pilot project, the success of which may affect his progression and promotion in the company. The engineer's ethical decision-making will be influenced by her relationships with and dependence on her senior colleagues, the organizational culture of the company, and potentially the perceived culture of the highly competitive and regulated healthcare technology industry.

Beyond the power dynamics within their engineering team, these two engineers are also, like all of us, members of various collectives, including families, friends, civic and cultural groups, and professional organizations, all of which have diverse values, power structures, and interpersonal dynamics. The engineers, in this case, are not *solely* engineers but also have interconnected sets of personal and professional identities. The interplay among these identities can lead to constraints or tensions in individual decision-making. In this example, the young engineer's ethical convictions, which may themselves have complex origins and influences, conflict with the organizational norms of the company she works for.

Internal pressures include what we might call 'the problem of many hats,' which can occur when we find it challenging to navigate competing identities under pressure. External constraints include the social, political, interpersonal, or institutional power dynamics that shift the conditions under which we make decisions.

The central aspect of stories like this one is that ethical decision-making is not just an internal, individual, rational, and moral capacity to be actualized. Instead, it is contingent on a wide range of relations to the communities of which we are each a part. Furthermore, our relations are not necessarily fixed. As the young engineer in this story demonstrates, one outcome of ethical decision-making can be a change in our relationships with colleagues or organizations, including resignation.

Key concepts in individual and collective ethical decision-making

Viewed through these multiple overlapping lenses of engineering collectives, we can see the two engineers from the example above in contentiously structured relationships of power based on seniority, gender, experience, training and mentoring, and leadership roles that involve tensions among their identities as individuals and as members of various collectives. The overlap between individual and collective identities requires us to consider the distinctions between individual moral agency and individual autonomy, collective agency, and external frameworks of accountability that support that agency. This section highlights these theoretical concepts as important aspects of EDM's individual and collective dimensions. They frame how individuals think and act and how collectives shape EDM outcomes, and thus are important for developing a fundamental approach to ethics education.

Agency, autonomy, responsibility, and accountability

Agency is the capacity to act. In]the context of ethical decision-making, what matters beyond mere *agency* is *autonomous* agency. Autonomy is self-governance or the ability of individuals or groups to make decisions and have those decisions implemented (Buss, 2018). Thus, autonomy can have collective dimensions since we may require support from other individuals or from a collective for our autonomous decisions to be implemented (Knight, 2017). Autonomous agency directed to moral ends is called *moral agency*. There are important questions about the scope of moral agency, for example, related to its relevance for artificial agential systems (e.g., Cervantes et al., 2020) or organizations (Watson et al., 2008). However, in engineering ethics, the scope of consideration regarding moral agency has traditionally been restricted to human individuals.

For ethical decision-making, we need to understand to what extent and in what ways moral agents have responsibility for their actions. A moral agent is responsible for their actions precisely because of their capacity for autonomous decision-making. However, ethical responsibility is rarely wholly autonomous, as it is regularly constrained and shaped by external influences of rewards, punishments, power imbalances, and life circumstances. Responsibility can also be constrained by internal factors, such as internalized beliefs about self-efficacy – the ability to implement the actions required to achieve the desired outcomes of your decisions (see Hersh & Lewoc, 2023, pp. 22–23, 26–28, 31–32). Constrained autonomy, such as being forced to obey orders, does not necessarily remove responsibility (e.g., as determined in the Nuremberg trials at the end of World War II).

Even when relatively unconstrained, responsibility is, by itself, insufficient for understanding how ethical decision-making works and why it matters. Coupled to it is *accountability*, or the external regulation of responsibility. To be held accountable is to have one's responsibility critiqued by an external mechanism (see Bivins, 2006), including through reporting mechanisms to external individuals, organizations, or constituencies (for those who are elected), or through the requirement to comply with standards. Thus, internally driven responsibility, which results from factors such as education, training, culture, and values, couples with accountability or responsibility to external relations of power, politics, and regulations, offering a more comprehensive understanding of the complex dynamics involved in ethical decision-making in engineering.

Both individuals and collectives thus shape ethical decision-making, responsibility, and accountability. We can think of ethical accountability as an accounting from an outside perspective of a multifaceted decision-making process, including "the distribution of responsibility between the actors that participate in [a collective decision-making] process" (Frasheri et al., 2022, p. 60). The young engineer and supervisor from the earlier example model a dynamic fundamental to the human condition: we have the capacity to reason and act individually, and simultaneously our reasons and actions are shaped by and shape collective decision-making actions. This capacity, called 'collective intentionality' (Schweikard, 2020), can have both positive and negative impacts on ethical decision-making, as we discuss next.

Collective decision-making and the problem of power

Positively, collective decision-making enables and empowers individuals to come together in social contracts, to share beliefs and values, and to motivate social and political responses to ethical issues. Negatively, however, this same capacity to think and value together can support implicit bias; the suppression of moral issues (Palazzo et al., 2012); the overshadowing of ethical issues by other urgencies, also known as 'ethical fading' (Rees et al., 2019); and groupthink – and has led to what Hannah Arendt famously described in 1963 as the 'banality of evil' (Arendt, 2006). 'Power dynamics' is the term generally used to discuss the interactions (between people in an organization) that are affected by differences in power. Some of these differences may be fixed, and others may change over time. There are several different factors that affect the power people have relative to others in any organization: their position in the organization; personal characteristics such as gender, race, and disability; knowledge and skills; length of time in the organization; strength or type of personality; and the ability to apply or resist pressure.

As the example of the startup company illustrates, power dynamics can make it difficult to challenge unethical behavior effectively. The younger female engineer has less power on account of her status/role in the organization, gender, and perceived lack of knowledge. These factors enable her supervisor to ignore her concerns and make it difficult for her to raise the issues. Her lower seniority in the team and different experiences result in her not being taken seriously.

Power dynamics also hold across collectives – including between organizations and nations and between organizations or nations and individuals. The power imbalance between different nations and regions or groups of people and the associated devaluation of those with lower power has contributed to ethical violations, as evidenced by the mining of uranium on sacred Aboriginal land counter to the wishes of Aboriginal people (Marsh & Green, 2020) and the Bhopal chemical disaster (Broughton, 2005). In the latter case, one of the many contributing factors was the minimal safety standards enforced by the engineering culture in Union Carbide in Bhopal, India, compared to those in its operations in the United States. The implication for engineers and particularly engineering ethics educators is to be aware of power dynamics and how collective intentionality affects decision-making, for better or worse. However, despite the risks, these individual-collective tensions in EDM are fundamental to the human condition and, therefore, need to be integrated into engineering ethics education.

Individuals and collectives in engineering ethics education

The vast majority of approaches to engineering ethics education address the interplay between the individual and the collective to some extent. In this section, we map out a landscape of EDM approaches used in engineering ethics education regarding their expressions of individual–collective relations. Our intention is not to offer a comprehensive review of EDM in engineering but to present a spectrum of relationships that demonstrate the scope of and potential for bridging the individual and the collective in engineering ethics education. By describing EDM approaches in this way, rather than, for example, categorizing them according to the 'micro' or 'macro' scale of the ethical issues that they seek to address, we draw attention to the presence of individual–collective relations in every type of engineering decision and provide support for their integration into engineering ethics education.

Engineers as individual moral agents

Many EDM models were developed mainly as tools for research – to further understand essential elements and processes involved in ethical decisions – and a few have been developed solely for educational purposes. These models range from the simple to the complex. Most of the simple models have been built upon the early work of Rest (1984, 1986), in their attempts to understand ethical reasoning in the context of moral development and compare it to similar stages of development in cognitive development. The four-stage model by Narvaez and Rest (1995) that highlighted sensitivity, judgment, motivation, and implementation has been advanced by Tuana (2007) in education for moral literacy – with components of ethics sensitivity, ethical reasoning, and moral imagination or motivation, each with a subset of components essential for the process. Tuana's conceptualization expands on the earlier models; it focuses primarily on cognitive processes to include affective (emotional and intuitive) aspects of ethical decision-making, particularly in ethics sensitivity and moral imagination. This model focuses on ethical *understanding* rather than the *action* or *implementation* that would be the behavioral outcome of EDM.

Gentile (2017) promoted a related model of EDM for educational applications (primarily business school and management training) that emphasizes the behavioral aspect of action as the goal of awareness and analysis. Gentile's ethics educational approach is designed to ensure students are prepared and practiced in completing the process of EDM – from ethics awareness (sensitivity) and ethical analysis to ethical action through repetitive practice that trains the "moral muscle memory" (Gentile 2017, p. 474).

An example of an EDM model with much higher complexity is the integrated theoretical model of Schwartz (2015). This research-focused model attempts to include the cognitive, affective, and situational components by integrating several previous EDM models into a more holistic framework (Schwartz, 2015, p. 761). This model includes both cognitive (rationalist) and affective (non-rationalist) focused aspects and emphasizes the dynamic between individual capacity and the influences of the situational context.

Each of these EDM models is oriented toward the individual as the decision-maker or moral agent, with some inclusion of the influence of the situational context (Schwartz, 2015). The models do not fully explore the complex and dynamic interactions between individuals and their collectives that influence decision-making within an engineering ecosystem, even though many engineers describe their experiences of ethics as highly influenced by their roles and responsibilities and the systems within which they are embedded (Hess et al., 2023; Fila et al., 2024). Thus, while these models of individual EDM may help educate the individual engineer, they are insufficient. Ethics education must include training to develop students' understanding of the systemic contextual and collective influences on individuals in their EDM.

Engineers as members of collectives

Many approaches to engineering ethics have been developed to address the perceived overemphasis on individual decision-making. These perspectives include 'macro' ethical approaches, influenced by organizational sociology and Science and Technology Studies (STS), which emphasize the broader contexts of engineering work. Some educators thinking in this vein have called for a complete overhaul of engineering ethics education (Bucciarelli, 2019), while others have sought to integrate macro-ethical thinking into existing structures and formats for engineering ethics instruction (Kline, 2010). Other approaches seek to integrate the 'micro' and 'macro' scales, retaining a practical focus on individual decision-making but deeply situating the engineer as decision-maker in relation to multiple collectives. Overall, these approaches view engineers' membership in collectives and their corresponding collective agency, responsibility, and accountability as fundamental to ethical decision-making.

Engineering ethics educators have attempted to frame the connections and tensions between the individual and the collective in various ways, often by naming and differentiating between different types of engineering ethics. Mclean (1993) made an early attempt to shift engineering ethics beyond a simplified focus on public safety and individual actions by introducing the three conceptual frames of technical, professional, and social ethics. Mclean considers these 'levels' of ethics to be the respective responsibilities of engineers, lawyers and managers, and politicians primarily. A strength of this framing is its awareness of the broader impacts of engineering and the importance of stakeholders who are not engineers. However, Mclean's approach is limited by its exemption of engineers themselves from considering the sociotechnical interactions inherent in engineering practice.

An alternative early framing distinguishes between the ethics of individual engineers' actions ('ethics *in* engineering') and collective ethical issues related to the role of engineers in industry, the ethics of engineering and professional engineering societies, and the ethical responsibilities of the profession ('ethics *of* engineering') (Roddis, 1993, p. 1540). Thus, for example, failure analysis should consider broader standards of engineering practice in addition to individual decisions and technical causes.

More recently, Herkert's framing of micro- and macro-ethics has gained prominence within engineering ethics communities (Herkert, 2005).[4] Herkert introduced three 'frames of reference': individual, professional, and social. Unlike Mclean, Herkert did not exclude engineers from broader social responsibility. Herkert argued for incorporating macro-ethical perspectives alongside micro-ethics in engineering ethics education. He also suggested strategies for bridging the two scales, for example, by encouraging engineering educators to foster in their students the 'moral courage' necessary to take individual ethical actions, even when those actions are limited by collective influences such as organizational cultures (Herkert et al., 2020). Subsequent scholarship builds on macro-ethics by explicitly introducing equity and justice as components of macro-ethical thinking, reinforcing the importance of engineers' membership in society-level collectives (Rottmann & Reeve, 2020). Such framings confront the fact that engineering projects are not necessarily neutral (Banks & Lachney, 2017) by teaching theories of interpersonal and structural violence and by connecting systemic violence and oppression in which engineering companies or the engineering profession are implicated in seemingly neutral decisions.

Another approach to understanding the dynamics of individual EDM within a broader context focuses on the general ethical principles that influence EDM. Many professional collectives such as engineering societies create codes of ethics that guide and stipulate acceptable engineering behaviors based on shared guiding principles. *Reflexive Principlism* is an EDM approach that connects individual reflective processes of ethical reasoning and the collective ethics of universal moral principles from bio-ethical principlism (Beever & Brightman, 2016). In Reflexive Principlism, guiding ethical principles are evaluated in light of professional codes of ethics and diverse perspectives of engineering and non-engineering stakeholders. In this approach, the ethical principles of beneficence, non-maleficence, justice, and respect for autonomy, reflexively specified and balanced in line with a specific context of application, apply at both the individual level and at the broader societal or collective levels. Other approaches to ethical decision-making outside of engineering, including Mepham's 'Ethical Matrix' (2006, 2013), have also leaned on bio-ethical principlism as their foundation.

Decentering the engineer

A third way of approaching individuals and collectives in engineering ethics moves beyond merely acknowledging the importance of the broader context toward actively decentering the individual engineer as a sole or privileged decision-maker. These approaches overlap to some extent with macro-ethical perspectives in their acknowledgment of complex relations and shared agency between designers, users, values, technologies, organizations, environments, and other actors/elements. However, decentering approaches emphasize collective responsibility and mutual accountability in addition to complexity, calling for shared decision-making processes that integrate sustained inputs from multiple stakeholders. At present, few specific EDM frameworks within engineering draw from such approaches. Yet, we see potential applications in engineering EDM for models of shared ethical decision-making that have been implemented in other fields and for critical design approaches that have been used in engineering contexts but not necessarily expressed as ethical frameworks.

Shared ethical decision-making, or SDM, involves individuals sharing information and expertise, deliberating on values, negotiating on trade-offs, collectively evaluating options, jointly articulating judgments, and collaborating on strategies for implementing action(s). It is therefore even more integrative of the individual and the collective than the approaches discussed in the previous section. This collaborative process typically involves individuals from multiple roles with a variety of experiences and expertise, including at times non-technical stakeholders. Although not yet applied directly within engineering ethics, SDM models are being developed and implemented in patient-centered healthcare (Spatz et al., 2017; Bomhof-Roordink et al., 2019). Similarly, *narrative ethics* integrates the individual and collective by giving voice to all the stakeholders, including those who are generally excluded. It has been used particularly by nurses, but a seven-stage methodology developed by Hersh (2015) shows its potential applications in engineering.

Another proposed EDM model, *guidance ethics*, emphasizes the relationality between technology and society. Guidance ethics identifies existing values of individual stakeholders around a specific technology as a means of making those values explicit and, therefore, functional. Its advocates argue that it is a bottom-up approach to ethical decision-making –avoiding principles and guidelines for the sake of stakeholders and citizens (Verbeek & Tijink, 2020). In this way, it is similar to other dialogic approaches to ethical decision-making, including the National Science Foundation–funded 'Toolbox Project' (O'Rourke & Crowley, 2013).

A focus on multiple stakeholders' values is also a core component of a range of design approaches that position the designer/engineer as only one of many humans and non-humans involved in the creation of technologies. One such approach, *value sensitive design* (VSD), shows how values (social, cultural, ethical, political) are filtered through the design work of individuals subject to collective norms and become embedded in technologies. VSD emphasizes intentionally making connections between values and choices about technological design, such that systems can be shaped by collective 'moral imagination' (Friedman & Hendry, 2019). (See Chapter 22 for more on VSD.) *Participatory design* approaches specifically include users and other stakeholders in the design process: for example, the 'Experience Lab' is a narrative-centered design approach that allows for the co-production of values that facilitate collective decision-making (Raman et al., 2017). *Design justice* practices take this participatory framework even farther, centering community needs and outcomes through processes of co-design (Costanza-Chock, 2020). Other approaches integrate non-human elements as part of the decision-making process. In Suzanne Kite's semi-speculative work 'How to Build Anything Ethically' (Kite, 2020), the author demonstrates how the Lakota people's ethical protocols, including consideration of the ontological status of stones, can be mapped onto the design of both a sweat lodge and a physical computing device.

The design approaches discussed above do not present themselves as models or protocols for engineering ethics, though they all – implicitly or explicitly – make moral arguments about 'good' technological design. We present these decentering frameworks and design approaches to identify a gap, and to offer possible ways forward, in considering relationships between the individual and the collective in engineering ethics. The examples of the diverse landscape of ethical decision-making we have offered here do not include the various other heuristics and practices that guide and shape ethics. These might include education in ethical reasoning and normative theory, academic and research integrity training, mentoring, engagement with professional codes of ethics, case-based analyses, or thought experiments. Heuristics like these can inform and guide ethical decision-making conceptualizations and approaches, scaffolding for a more significant effect. Yet they sometimes are made to stand alone, replacing effective practices with efficient practices.

Individual/collective tensions in EDM: A Boeing 737 Max case study

In this chapter, we have examined dynamic tensions between individuals and collectives in ethical decision-making from a range of perspectives and demonstrated a spectrum of approaches to EDM that take up (or fail to take up) those tensions. Here, we highlight an example case study as a means of demonstrating how these various levels of complexity can be addressed for ethics education in engineering.

Case studies can powerfully illustrate and convey lessons about ethical decision-making in relevant, real-world situations in engineering where the complexities and diverse factors of influence are at stake in importantly practical ways. Our example here is the Boeing 737 MAX incidents of 2018 and 2019. These two significant accidents involved Boeing 737 MAX passenger jets crashing a few minutes after takeoff, during which 346 people were killed. All 737 MAX planes worldwide were grounded after the second accident. There has been considerable discussion of the causes and responsibility for the accidents, including the case study by Herkert and colleagues (2020) highlighted here.

Herkert and colleagues (2020) show that the key decisions and undue haste in the design of the Boeing 737 Max were motivated by intense competition with Airbus, which had recently introduced a new aircraft. These decisions included moving the new more powerful engines to an atypical location higher and further forward on the aircraft wings and compensating for the risk of a potential stall with a software fix: the maneuvering characteristics augmentation system (MCAS) with input from only one of the two angle of attack (AOA) sensors. To speed up the certification process Boeing concealed many of their changes, including the MCAS software fix, from the Federal Aviation Authority (FAA) and its own pilots.

Some of the design changes, their concealment, and the lack of pilot training violated industry practices and regulatory norms and should have been objected to by both individual engineers and managers, and collectively. However, the replacement of engineers with business executives in management at all levels (following Boeing's merger with McDonnell Douglas in 1997) had resulted in a culture change that prioritized profit even at the expense of safety and technical competence (see Herkert et al., 2020, p. 2963; Useem, 2019). These changes made it challenging, but even more critical, for engineers to collectively raise concerns. Developing or joining collectives, such as trade unions and professional societies, within and outside companies can provide collective support for ethically-minded engineers with less risk of individual retaliation.

A further factor that downgraded the importance of safety in this safety-critical industry was the change in the FAA's approach from 2005 onwards, resulting in manufacturer self-certification becoming the norm and Boeing self-certifying 96% of its work in 2018 (Kitroeff et al., 2019a). However good the self-certification process could be, independent evaluators might identify potential problems that would otherwise be missed. Independent verification also reduces the likelihood of incomplete or false declarations, as in the case of Boeing. Thus, the collective interdependency between industry and independent certification should be considered essential for ethical engineering practice. The need for independent evaluation should have been raised by engineers in Boeing and the FAA, both individually and collectively.

The ethical failures here have been related to the 'problem of many hands' (Herkert et al., 2020), where the involvement of many different parties and complex interactions makes it difficult to determine accountability. However, engineers can be educated with a focus on shared accountability, both individually and collectively, with the potential of transforming cultures of blame. A related issue is that of 'many hats' (see section above), where the different parties involved have different disciplinary expertise. Educating engineers by providing direct experiences of working together with people from other disciplines can encourage them to see the ethical value in collective decision-making.

The case study (Herkert et al., 2020) indicated that a few Boeing engineers had raised concerns about the risks of relying on a single AOA sensor and about MCAS's erratic behavior in a flight simulator. Therefore, management was aware of the problems but did not intervene or change its behavior. If the complaints had been raised collectively, they might have been harder to ignore and could have brought the issues into the public domain. This would have made it more difficult for Boeing management to continue to focus on 'cost and schedule' rather than 'safety and quality' (Kitroeff et al., 2019b).

Further case analysis by our author team also raises several wider ethical issues which have received little attention. There seems to have been little justification for Boeing (or Airbus) to develop new aircraft in a time of climate crisis, especially designs which did not focus on maximizing safety and minimizing environmental impacts. There are good arguments for suggesting that Boeing could have instead focused innovation efforts on improving wind or wave turbines for energy generation with major positive environmental impacts. However, such large changes in corporate vision likely would have required significant collective societal pressure aligned with the internal support of environmentally-minded engineers.

The lessons for engineers starting their careers in industry sectors could include developing support networks; joining existing supportive collectives, unions, and professional societies; finding colleagues with compatible views; identifying supportive senior colleagues; and connecting with external collectives of stakeholders with related concerns. Participating in supportive collectives creates opportunities to discuss ethical concerns with colleagues and raise them in the organization and more widely if this becomes necessary. The contrast is evidenced in the ten mini-cases analyzed in McGinn's (2022) study of the Theranos startup failure. Many scientists and engineers sought to do the right thing ethically and change the culture, but nearly all failed to make a difference. They all acted independently except those who reached out as whistle-blowers to external resources. Collective participation can lead to more effective action while reducing the risks of victimization and the need for extreme moral courage by individuals. It will also mean that early-career engineers might not need to worry as much about being mocked if they raise potential ethical violations that turn out to be unfounded. Instead, they can be alert to potential ethical problems and collectively apply ethical decision-making tools and frameworks in practice.

Conclusions, recommendations for future work, and key questions

This chapter has articulated the tensions and interactions between individuals and the collectives of which they are a part as a central and often-overlooked orientation to engineering ethics. Ethical decision-making at the individual/collective intersection shapes sensitivity to ethical issues, reasoning approaches about them, and motivation for actions in response. Ethical decision-makers face both constraints and complements to their individual roles, thanks to their participation in various levels of collectives. The largely individualized approaches to engineering ethics need to include an understanding of the range of external factors and collective contexts that necessarily shape ethical decision-making. Individuals and collectives are deeply intertwined, and ethical approaches in engineering need to take that interplay seriously. In addition, we can learn more fully from EDM approaches that situate engineers differently with respect to individual/collective relations, as we demonstrate with the wider-view analysis of the 737 Max case study. Ethics literacy for engineers needs to include an understanding of context, interpersonal interactions, broader social justice, and interrelationality without overwhelming students, such that they lose a sense of their own agency. Students also need to understand both their responsibility in shaping the ethical climate of their organizations and the value of seeking support and acting collectively to challenge unethical practices while avoiding victimization. That difficult work in engineering ethics education will empower future engineers be better informed and more literate ethical decision-makers.

Notes

1 Author names are presented in reverse alphabetical order. All authors contributed equally to the work.
2 Ethical decision-making is understood comprehensively to include ethical action. We recognize that 'collective action' can be an emergent property of groups of individuals that has been studied as an important entity unto itself with even broader issues of consideration (Mayer, 2014). Expanding upon each of the components of ethical decision-making and their variations is beyond the scope of this chapter.
3 Adapted from McGinn (2022, p. 39).
4 Herkert builds on Brummer's earlier (1985) framing of micro- and macro-ethics, which distinguishes between when 'demands of conscience' of subordinates in an organization conflict with perceived occupational requirements and with superiors' setting of policy for organizations in general.

References

Arendt, H. (2006[1963]). *Eichmann in Jerusalem: A report on the banality of evil*. Penguin Books.
Awad, E., Bonnefon, J. F., Shariff, A., & Rahwan, I. (2019). The thorny challenge of making moral machines: Ethical dilemmas with self-driving cars. *NIM Marketing Intelligence Review*, *11*(2), 42–47.
Banks, D. A., & Lachney, M. (2017). Engineered violence: Confronting the neutrality problem and violence in engineering. *International Journal of Engineering, Social Justice, and Peace*, *5*(1–2), 1–12.
Beever, J., & Brightman, A. O. (2016). Reflexive principlism as an effective approach for developing ethical reasoning in engineering. *Science and Engineering Ethics*, *22*(1), 275–291.

Beever, J., & Morar, N. (2016). Bioethics and the challenge of the ecological individual. *Environmental Philosophy, 13*(2), 215–238.
Beever, J., & Morar, N. (2019). The epistemic and ethical onus of 'One Health'. *Bioethics, 33*(1), 185–194.
Beever, J., & Whitehouse, P. J. (2017). The ecosystem of bioethics: Building bridges to public health. *Jahr: European Journal of Bioethics,* 8(16), 227–243.
Benjamin, R. (2019). *Race after technology: Abolitionist tools for the New Jim Code*. Polity Press.
Bivins, T. H. (2006). Responsibility and accountability. In K. R. Fitzpatrick & C. Bronstein (Eds.), *Ethics in public relations: Responsible advocacy* (pp. 19–38). SAGE Publications.
Brummer, J. (1985). Business ethics: Micro and macro. *Journal of Business Ethics, 4,* 81–91.
Bomhof-Roordink, H., Gärtner, F. R., Stiggelbout, A. M., & Pieterse, A. H. (2019). Key components of shared decision-making models: A systematic review. *MJ Open, 9*(12), 1–11.
Broughton, E. (2005). The Bhopal disaster and its aftermath: A review. *Environmental Health, 4*(1), 6–12.
Bucciarelli, L. (2019). Ethics and engineering education. *European Journal of Engineering Education, 33*(2), 141–149.
Buss, S. (2018). *Personal autonomy*. Stanford Encyclopedia of Philosophy. https://plato.stanford.edu/entries/personal-autonomy/.
Carrington, D., & Taylor, M. (2022). Revealed: The 'carbon bombs' set to trigger catastrophic climate breakdown. *The Guardian.* https://www.theguardian.com/environment/ng-interactive/2022/may/11/fossil-fuel-carbon-bombs-climate-breakdown-oil-gas.
Cervantes, J.-A., López, S., Rodríguez, L.-F., Cervantes, S., Cervantes, F., & Ramos, F. (2020). Artificial moral agents: A survey of the current status. *Science and Engineering Ethics, 26,* 501–532.
Chen, A., Treviño, L. K., & Humphrey, S. E., (2020). Ethical champions, emotions, framing, and team ethical decision-making. *Journal of Applied Psychology, 105*(3), 245–273.
Costanza-Chock, S. (2020). *Design justice: Community-led practices to build the worlds we need.* MIT Press.
Davis, M. (1991). Thinking like an engineer: The place of a code of ethics in the practice of a profession. *Philosophy & Public Affairs, 20*(2), 295–312.
Davis, M. (2006). Engineering ethics, individuals, and organizations. *Science and Engineering Ethics, 12*(2), 223–231.
Eubanks, V. (2017). *Automating inequality: How high-tech tools profile, police, and punish the poor.* St. Martin's Press.
Fila, N., Zoltowski, C., Kerr, A., Kim, D., Loui, M., Hess, J., & Brightman, A.O. (2024). How do engineers experience ethical practice in industry?: A phenomenographic investigation. *Journal of Engineering Education.*
Frasheri, M., Struhar, V, Papadopoulos, A. V., & Causevic, A. (2022). Ethics of autonomous collective decision-making: The CAESAR framework. *Science and Engineering Ethics, 28*(6), 61–89.
Friedman, B., & Hendry, D.G. (2019). *Value sensitive design: Shaping technology with moral imagination.* MIT Press.
Gentile, M. C. (2017). Giving voice to values: A pedagogy for behavioral ethics. *Journal of Management Education, 41*(4), 469–479.
Held, V. (2006). *The ethics of care: Personal, political, global.* Oxford University Press.
Herkert, J. R. (2005). Ways of thinking about and teaching ethical problem solving: Microethics and macroethics in engineering. *Science and Engineering Ethics, 11,* 373–385.
Herkert, J. R., Borenstein, J., & Miller, K. (2020). The Boeing 737 MAX: Lessons for engineering ethics. *Science and Engineering Ethics, 26,* 2957–2974.
Hersh, M. (2015). Engineers and the other: The role of narrative ethics. *AI & Society, 31,* 327–345.
Hersh, M. (2014). Science, technology and values: promoting ethics and social responsibility. *AI & Society, 29*(2), 167–183.
Hersh, M., & Lewoc, J. B. (2023). *ICT ethics and human behaviour in ICT development: International case studies with a focus on Poland.* Springer International Publishing.
Hess, J., Kim, D., & Fila, N. (2023). Critical incidents in ways of experiencing ethical engineering practice. *Studies in Engineering Education, 3*(2), 1–30.
Hofstede, G. (1998). Identifying organizational subcultures: An empirical approach. *Journal of Management Studies, 35*(1), 1–12.
Hofstede, G. (2011). Dimensionalizing cultures: The Hofstede model in context. *Online Readings in Psychology and Culture, 2*(1), 8–34.

Hussain, I., Shu, R., Tangirala, S., & Ekkirala, S. (2019). The voice bystander effect: How information redundancy inhibits employee voice. *Academy of Management Journal, 62*(3), 828–849.
Husted, B. W., & Allen, D. B. (2007). Corporate social strategy in multinational enterprises: Antecedents and value creation. *Journal of Business Ethics, 74*(4), 345–361.
Jeske, M. (2020). Theranos: Changing narratives of individual ethics in science and engineering. *Engaging Science, Technology, and Society, 6*, 306–313.
Jones, S. A., Michelfelder, D., & Nair, I. (2017). Engineering managers and sustainable systems: the need for and challenges of using an ethical framework for transformative leadership. *Journal of Cleaner Production, 140*, 205–212.
Kasher, A. (2005). Professional ethics and collective professional autonomy: A conceptual analysis. *Ethical Perspectives: Journal of the European Ethics Network, 11*(1), 67–98.
Kim, D., & Hess, J. L. (2022). A multi-layered illustration of exemplary business ethics practices with voices of the engineers in the health products industry. *Journal of Business Ethics, 187*(1), 169–183.
Kite, S. (2020). How to build anything ethically. In J. E. Lewis (Ed.), *Indigenous protocol and artificial intelligence position paper* (pp. 75–84). The Initiative for Indigenous Futures and the Canadian Institute for Advanced Research (CIFAR).
Kitroeff, N., Gelles, D., & Nicas, J. (2019a, July 27). The roots of Boeing's 737 MAX Crisis: A regulator relaxes its oversight. *The New York Times*. https://www.nytimes.com/2019/07/27/business/boein g-737 -max-faa.html.
Kitroeff, N., Gelles, D., & Nicas, J. (2019b, October 2). Boeing 737 MAX safety system was vetoed, engineer says. *The New York Times*. https://www.nytimes.com/2019/10/02/business/boeing-737-maxcrashes.html.
Kline, R. R. (2010, Winter). Engineering case studies: Bridging micro and macro ethics. *IEEE Technology and Society Magazine*, 16–19.
Knight, J. (2017). *Cognitive accessibility 103 – A presentation at CSUN 2017*. Spaced Out and Smiling. https://spacedoutandsmiling.com/presentations/cognitive-accessibility-103-csun-2017.
Kranakis, E. (2004). Fixing the blame: Organizational culture and the Quebec Bridge collapse. *Technology and Culture, 45*(3), 487–518.
Lynch, W. T., & Kline, R. (2000) Engineering practice and engineering ethics. *Science, Technology & Human Values, 25*, 195–227.
Marsh, J. K., & Green, J. (2020). First Nations rights and colonizing practices by the nuclear industry: an Australian battleground for environmental justice. *The Exractive Industries and Society, 7*(3), 870–881.
Mason, R. O. (2004). Lessons in organizational ethics from the Columbia disaster: Can a culture be lethal? *Organizational Dynamics, 33*(2), 128–142.
Mayer, F. W. (2014). *Narrative politics: Stories and collective action*. Oxford University Press.
McGinn, R. E. (2022). Startup ethics: Ethically responsible conduct of scientists and engineers at Theranos. *Science and Engineering Ethics, 28*(5), 1–21.
Mclean, G. F. (1993) Integrating ethics and design. *IEEE Technology and Society Magazine, 12*(3), 19–30.
Meiksins, P., & Smith, C. (1996). *Engineering labour: Technical workers in comparative perspective*. Verso.
Mepham, B. (2013). Ethical principles and the ethical matrix. In J. P. Clark & C. Ritson (Eds.), *Practical ethics for food professionals: Ethics in research, education, and the workplace* (pp. 39–56). John Wiley & Sons.
Mepham, B., Kaiser, M., Thorstensen, E., Tomkins, S., & Millar, K. (2006). *Ethical matrix manual*. LEI.
Narvaez, D., & Rest, J. (1995). The four components of acting morally. In W. Kurtines & J. Gewirtz (Eds.), *Moral behavior and moral development* (pp. 385–400). McGraw-Hill.
O'Rourke, M., & Crowley, S. J. (2013). Philosophical intervention and cross-disciplinary science: The story of the toolbox project. *Synthese, 190*, 1937–1954.
Palazzo, G., Krings, F., & Ulrich, H. (2012). Ethical blindness. *Journal of Business Ethics, 109*(3), 323–338.
Rachar, M., & Salomone-Sehr, J. (2023). Collective intentions. In M. Sellers & S. Kirste (Eds.), *Encyclopedia of the philosophy of law and social philosophy* (pp. 1–8). Springer.
Raman, S., French, T., & Tulloch, A. (2017). Design-led approach to co-production of values for collective decision-making. *The Design Journal, 20*(Suppl. 1), S4331–S4342.
Rees, M. R., Tenbrunsel, A. E., & Bazerman, M. H. (2019). Bounded ethicality and ethical fading in negotiations: Understanding unintended unethical behavior. *Academy of Management Perspectives, 33*(1), 26–42.
Rest, J. (1984). Research on moral development: Implications for training counseling psychologists. *The Counseling Psychologist, 12*(3), 19–29.
Rest, J. (1986). *Moral development: Advances in research and theory*. Praeger.

Riley, D. (2023). Engineering principles: Restoring public values in professional life. In A. Fritzche & A. Santa-Maria (Eds.), *Rethinking technology and engineering: Dialogues across disciplines and geographies* (pp. 13–23). Springer Verlag Publishing.
Roddis, W. K. (1993). Structural failures and engineering ethics. *Journal of Structural Engineering, 119*(5), 1539–1555.
Rogal, L. (2020). Secrets, lies, and lessons from the Theranos scandal. *Hastings Law Journal, 72*, 1663–1685.
Rosenberg, R. S. (1998) Beyond the code of ethics: The responsibility of professional societies. In *Proceedings of the ethics and social impact component on shaping policy in the information age* (pp. 18–25). ACM Policy.
Rottmann, C., & Reeve, D. (2020). Equity as rebar: Bridging the micro/macro divide in engineering ethics education. *Canadian Journal of Science, Mathematics, and Technology Education, 20*(1), 146–165.
Schwartz, M. S. (2015). Ethical decision-making theory: An integrated approach. *Journal of Business Ethics, 139*, 755–776.
Schweikard, D. P. (2020). *Collective intentionality*. Stanford Encyclopedia of Philosophy. https://plato.stanford.edu/entries/collective-intentionality/ .
Sharma, K. (2016). *Interdependence: Biology and beyond*. Fordham University Press.
Spatz, E. S., Kurmholz, H. M., & Moulton, B. W. (2017). Prime time for shared decision-making. *JAMA, 317*(13), 1309–1310.
Thompson N. (2016). Metaphysical interdependence. In M. Jago (Ed.), *Reality making*. Oxford Scholarship Online.
Tuana, N. (2007). Conceptualizing moral literacy. *Journal of Educational Administration, 45*(4), 364–378.
Useem, V. (2019, November 20). The long-forgotten flight that sent Boeing off course. *The Atlantic*. https://www.theatlantic.com/ideas/archive/2019/11/how-boeing-lost-its-bearings/602188/.
Verbeek, P.-P., & Tijink, D. (2020). *Guidance ethics approach. Platform voor de InformatieSamenleving*. ECP.
Vesilind, P. A. (1995). Evolution of the American Society of Civil Engineers code of ethics. *Journal of Professional Issues in Engineering Education and Practice, 21*(1), 1–10.
Watson, G. W., Freeman, R. E., & Parmar, B. (2008). Connected moral agency in organizational ethics. *Journal of Business Ethics, 81*, 323–341.

4

REASON AND EMOTION IN ENGINEERING ETHICS EDUCATION

Nihat Kotluk, Johanna Lönngren, and Roland Tormey

Introduction

For decades, philosophers, sociologists, and psychologists have discussed morality, reason, and emotion, drawing on various concepts that have often been shared – albeit using different terms – across disciplines. These different fields and perspectives have affected engineering ethics and engineering ethics education. Today, many Western cultures conceptualize emotions as separated from reason and rationality (Ritzer, 2020; Roeser, 2012). This divide is particularly strong in engineering and the natural sciences (Roeser, 2020; Sinatra et al., 2014), and engineering is typically considered a highly rational and technocratic field (Roeser, 2020; Lönngren, Bellocchi et al., 2024). Consequently, students are often warned that their emotions will hinder their ability to make rational decisions (Sunderland et al., 2014). Similarly, engineering ethics education has historically emphasized the development of moral reasoning skills through opportunities to practice reasoning based on ethical principles. Conversely, emotions are typically overlooked (Hess & Fore, 2018; Kim, 2022; Morrison, 2020; Tormey, 2020). This has, however, begun to change with a growing focus on exploring the relationship between reason and emotion in engineering ethics education.

Our core argument in this chapter is that because engineering ethics education has historically been treated in primarily rationalist and individualistic terms, a fuller account of reason and emotion, therefore, needs to expand in two dimensions: (i) by seeing how reason and emotion are connected and (ii) by exploring how individual, interactional, and macro-social levels are connected. Since this work is still in its infancy in engineering, we aim in this chapter to provide a conceptual framework and some integrating concepts that will allow researchers and teachers to better conceptualize how reason and emotion are linked in engineering ethics education at a range of levels of social analysis. Our goal is not simply to summarize what has already been written about reason and emotion in this field but also to provide some directions on how to move forward.

This chapter is structured into four sections. After this introduction, the first section explores foundational frameworks for linking reason and emotion in moral judgments. Emotion is defined and the notion of *moral schema* is introduced, which we use as an integrating framework to connect reason and emotion while bridging the gap between micro-individual, meso-interactional, and macro-social levels of analysis. We then look at each level of analysis in turn. The following sec-

DOI: 10.4324/9781003464259-6

tion, therefore, explores how reason and emotion interact on the micro-individual level by looking at how individual engineers learn to make moral judgments and how individualist approaches have traditionally reinforced rationalist ways of thinking that define emotions as irrelevant or detrimental to engineering ethics. The third section focuses on the *meso-interactional level* (what sociologists would call the micro-social level) and examines how social, cultural, and emotional factors impact engineering students' moral behavior as they interact with peers. Finally, the fourth section addresses the *macro-societal level* and examines societal and cultural factors influencing moral decision-making. At this level, social institutions, social systems, and cultures shape individuals' schemas related to thoughts, feelings, and moral actions.

The authors of this chapter are researchers in engineering education. One author completed an engineering degree before moving into engineering education research. The other authors are trained in psychological and sociological learning sciences, respectively. Geographically, one author comes from Western Asia and two from Western Europe. All three were primarily trained in Western intellectual traditions, where they have worked on questions of diversity, inequality, and emotions in engineering education. The authors' backgrounds are reflected in the ideas presented in this chapter, most notably in the combination of Western psychological, interactional, and sociological perspectives on engineering ethics education.

Foundational concepts for connecting reason and emotion across different levels of social analysis

Although there is no agreement on how to define emotions, Shuman and Scherer's (2014) 'Multi-Component Model' of the definition of emotion has gained broad acceptance. It specifies five primarily biological and psychological components: (1) a cognitive component involving the evaluation of a situation, (2) a neurophysiological component involving bodily changes such as heart rate and hormonal secretion, (3) a motivational component involving action tendencies like avoidance behavior, (4) a motor expression component including facial expressions, and (5) a subjective feeling component involving affective experiences, such as feeling nervous. Sociological viewpoints build on this and emphasize that social and cultural factors and organizations shape a significant portion of those components (Bericat, 2016). Similarly, emotional expressions are often described as influenced, validated, and understood in accordance with broader societal values and culturally constructed discourses (Harré et al., 2009). This alignment of psychological and sociological understandings indicates the potential for a more integrated and collectively shared interdisciplinary understanding of emotions. Yet, there is not yet an agreed interdisciplinary definition of emotion, and differences between perspectives need to be identified, alongside shared understandings.

A helpful framework for thinking about how conceptions of emotion are changing over time is provided by Barbalet (2001, pp. 45–54), who distinguishes three approaches to theorizing reason and emotion, which appear to be common across multiple disciplines. First is the traditional or *conventional* approach, drawing on a Cartesian and Kantian understanding, theorizing emotion as a bodily disturbance that undermines reason. This approach frames good moral judgment as driven primarily by reasoning (understood as logical thought in the absence of emotional disturbance) and as being opposed to decisions based on emotion or gut instinct. This approach was dominant in moral psychology (Kohlberg, 1969) and sociology (Weber, 1930/1992) in the twentieth century. It also often underpins 'folk' theories of emotion (i.e., those used by people in everyday life). Although this traditional/Kantian approach to understanding reason and emotion remains dominant in popular culture, it is less commonly used today by emotion researchers.

A second way of theorizing reason and emotion is what Barbalet (2001) calls a *critical* approach. It sees emotion as complementary to reason by providing rapid insights and guidance on the salience of information that reason itself cannot provide. This tradition is influenced by the work of the philosopher David Hume, among others. While Barbalet presents the critical approach as coherent, other writers tend to emphasize differences within this tradition, between those who focus on emotion as informing reasoned judgments ('emotion as *moral insight*', e.g., Roeser, 2012) and those who focus on a dual-process model which sees emotion as acting alongside and potentially bypassing reasoned judgment (*social intuitionist model*, e.g., Greene et al., 2004; Haidt, 2008). These critical approaches emphasize the co-operation between reason and emotion in coming to judgment (not their conflict, which tends to be emphasized by the traditional or conventional approach).

Barbalet also identifies a third approach, a *radical* approach, influenced by the work of the psychologist William James and the sociologist Georg Simmel. Rather than seeing reason and emotion as distinct but interrelated processes, this approach sees reason and emotion as "distinct names for aspects of a continuous process" (Barbalet 2001, p. 45). From this perspective, reason is not opposed to emotion (as for Kant) or informed by emotion (as in critical perspectives). Rather, reason is seen as a way of thinking associated with particular emotional states – including trust, security, and certainty. In this sense, reason is only possible when we give ourselves emotional permission to not engage with the consequences of our actions, which might disturb our sense of certainty and security (we will return to the moral implications of this argument in the section on macro-social frameworks below).

'Schema' as an integrating concept

This chapter aims at integrating many concepts: reason and emotion on one axis and micro-individual, meso-interactional, and macro-social levels of analysis on the other. These elements are closely connected: at the micro-individual level, we explore how emotions and reasoning operate within individuals. Moving to the meso-interactional level, we examine how these aspects manifest in personal interactions. Finally, at the macro-social level, we look at the broader societal contexts and influences on emotions and reasoning. A useful integrating concept across multiple levels of social interaction is provided by the work of Firat and McPherson (2010), who use the concept of *moral schema* to integrate across these levels of analysis within a common framework.

Schemas are often described as cognitive frameworks that guide people's perceptions, interpretations of their experiences, and actions (Boutyline & Soter, 2021; c.f. Piaget, 1932/1965). According to this schema-based model, a given experience activates mental schemas, which means that in that situation, a person will automatically and intuitively perceive certain things and think and feel in particular ways – and these elements influence their judgment (Thoma & Dong, 2014, p. 56). Schemas often operate implicitly and unconsciously, and people may not always be aware of their influence. *Moral schemas* are schemas that involve automatic and implicit cognitive processes important for moral behavior as they guide people's attention to moral issues and drive moral judgment (Narvaez & Bock, 2002). While schemas have historically been conceived as purely cognitive constructs, the concept of moral schemas highlights the importance of emotions, which operate as signals "activating schemas' salience, intensity, and content in people's everyday lives" (Firat & McPherson, 2010, p. 354).

While the language in the literature on schemas primarily derives from work at the individual/psychological level, the basic idea behind the concept – that internalized frameworks from our culture and our experiences shape our perceptions and responses – is also used to make sense of patterns at interactional and macro-social levels. Yet, writers in disciplines other than psychology may use different terms to refer to this idea. In sociology, for example, the term *habitus* is often used to describe a pattern of perception, thinking, and action shaped by our past experiences (Bourdieu, 1990). As we will show in the section on the interactional level of social analysis, *discourse* is another related concept. Our goal here has been to use a single term (schema) to give clarity to the reader, not to prioritize psychological or individualistic perspectives.

Psychological perspectives and their impacts on engineering ethics education and research

Psychological perspectives on moral reasoning and emotion are explored in more detail in Chapter 10 of this handbook. Thus, we provide only a concise overview here.

Historically, rationalist (Kantian) cognitive-developmental models (what Barbalet would call conventional models) have dominated moral psychology. For example, Kohlberg's theory (1969) suggests that individuals progress through a series of stages of moral reasoning as they develop cognitively. The theory outlines six stages, divided into three levels: pre-conventional (judgment based on personal interest), conventional (judgment based on rule-following), and post-conventional (judgment based on the assessment of what is good for people in general). Moral judgment in this model is described as primarily based on reasoning and reflection. Some emotions, such as sympathy, are assumed to contribute to reasoning occasionally but are not seen as directly influencing moral judgments (Kohlberg, 1969).

However, this theory has been criticized for oversimplifying the complex nature of moral reasoning and underestimating the role of cultural (Tappan, 1997), social (Bandura, 2001), and emotional (Haidt, 2001) factors in shaping moral judgment. Within more critical approaches to understanding how reason and emotion interact, others have argued that emotions influence moral reasoning *and* crucial elements of cognitive abilities in moral judgments. For example, Hoffman (2000) emphasized emotional empathy – rather than cognition – as the primary driver of moral decision-making. However, Hoffman also connected emotion with cognition, proposing that empathy includes both cognitive and emotional processes. Haidt (2001) went even further, proposing a social intuitionist model of moral decision-making as an emotion-driven alternative to rationalist models. Haidt (2001) argued that moral judgments are guided by emotionally based intuitions that happen quickly, effortlessly, and automatically, while moral reasoning is formed slowly and with the effort exerted only after judgments have been made (*post-hoc* rationalization). Social neuroscientists have also brought a new perspective on moral judgment: the dual process theory of morality (Greene et al., 2004). According to this theory, a complex interaction between cognitive and affective psychological mechanisms influences moral judgments.

In response to the criticisms of Kohlberg's theory, Rest and his colleagues (1999) introduced the Neo-Kohlbergian approach, which retained Kohlberg's fundamental framework while integrating schema theory. This model conceptualizes moral development as a shift in the distribution of three primary 'schemas' – the *Personal Interest Schema*, the *Maintaining Norms Schema*, and the *Postconventional Schema* – rather than a linear progression through six discrete 'stages' (Thoma & Dong, 2014). Contrary to Kohlberg's assertion of the universality of moral stages, this approach recognizes that moral schemas are context-dependent and can be influenced by factors such as emotional information (Thoma & Dong, 2014). To assess individuals' levels of moral development, Neo-Kohlbergian researchers also developed the Defining Issues Test (DIT-1 and DIT-2) to detect shifts in these moral schemas (Rest et al., 1999).

The DIT has found extensive use in exploring moral reasoning development in higher education (King & Mayhew, 2002), including engineering ethics education (Hess & Fore, 2018; Hess et al., 2019). Furthermore, the DIT and DIT-like instruments, such as the Engineering and Science Issues Test (Borenstein et al., 2010), have also often been used to assess the effectiveness of engineering ethics education (Watts et al., 2017, p. 14). Thus, Kohlberg's theory (based on a conventional or Cartesian/Kantian model of the emotion–reason relationship) has played an essential role in engineering ethics education, whether implicitly or explicitly (for a comprehensive overview, see Hess, Beever, et al., [2019]). This influence significantly shapes the teaching approaches of engineering ethics educators, often placing a strong emphasis on reasoning as the primary driver of moral judgment while downplaying the role of emotions, which are often regarded as irrelevant or even harmful (Guntzburger et al., 2019; Kellam et al., 2018; Lönngren, Adawi et al., 2021; Ottemo et al., 2021; Roeser, 2020; Sunderland, 2014). For example, widely used engineering ethics textbooks such as those by van de Poel and Royakkers (2011), Fledderman (2011), and Harris et al. (2019) hardly even mention the term 'emotion' (Tormey, 2020). Similarly, engineering ethics case studies are typically designed and presented as moral dilemmas to be resolved through logical reasoning, neglecting the important emotional aspects of moral judgment (Bairaktarova & Woodcock, 2017; Martin et al., 2021; Miñano et al., 2017).

Fortunately, there is a rapidly growing interest in emotions in engineering education and engineering ethics education (Lönngren et al., 2023, preprint). For example, Roeser (2012, 2020) has emphasized the significance of emotions in technology ethics and design work. Others have shown that emotions play a crucial role in human-centered design practices (Bairaktarova & Plumlee, 2022), development of professional engineering identities (Huff et al., 2021), and engineers' ability to value moral decision-making (Hess et al., 2021). These researchers have argued that paying attention to emotions in professional engineering ethics is crucial. Others stress the importance of integrating emotions into engineering ethics education (e.g., Davis, 2015; Snieder & Zhu, 2020; Kim, 2022). Some even argue that emotions can bridge theoretical knowledge of engineering ethics and practical decision-making (Davis, 2015; Newberry, 2004). In this direction, Sunderland et al. (2014) and Sunderland (2014) examined how project-based learning may engage engineering students' emotions in engineering ethics education, finding that engagement with emotions can contribute to more meaningful ethics learning experiences for engineering students.

Another approach for introducing emotions into engineering ethics education can be using fictional films, as proposed by Hitt and Lennerfors (2022) (for more on this, see Chapter 36). According to the authors, fictional films can help engineering students relate to the ethical challenges engineers may face in their careers, empathize with the individuals affected by engineering decisions, and develop a more profound sense of moral awareness and sensitivity. Hess, Strobel, and Brightman (2017) and Hess et al. (2019) have also tested various models for incorporating emotions into teaching engineering ethics. For example, Hess et al. (2019) investigated how an emotionally engaging process can be applied to the practical examination of ethics cases. They employed a scaffolded, interactive, and reflective analysis (SIRA) approach and observed that it resulted in increased cognitive empathetic perspective-taking. Researchers have also begun to investigate the effects of including emotional information in ethics case studies on engineering students' ethics learning (Thiel et al., 2013; Watts et al., 2017), activated moral reasoning schemas (Kotluk & Tormey, 2023), and moral judgment (Higgs et al., 2020) (this is explored in more detail in Chapter 20, on 'teaching ethics using case studies').

Thus far, we have identified that there are now multiple theoretical frameworks aiming to connect reason and emotion in the ethical decision-making of individual engineers. So far, we have only explored connections between reason and emotion at the individual level. We next turn our attention to the connection of reason and emotion at the next level of social analysis – the level of social interaction.

Reason and emotion in moral social interaction

We have described how moral philosophy and psychology research have begun attending to the multiple roles that reason and emotion can play in individual moral judgment and development. Engineering, however, is an inherently *social* profession characterized by complex social relationships, distributed knowledge and production systems, unclear power structures, and conflicting value systems (Rojter, 2007; Ross & Athanassoulis, 2010). Thus, decisions on moral issues are never made by individual engineers alone – they are always also influenced by the social contexts in which engineers work and live. Similarly, engineering students learn ethics in the social contexts of educational institutions.

In this part of the chapter, we thus focus on the meso-level of social interaction among individuals and smaller groups, drawing on research from social psychology and critical discourse studies. In this type of research, emotions are not understood as residing 'inside an individual's head,' but as co-constructed, negotiated, and used as rhetorical resources for influencing social norms or power relations (Ahmed, 2014; Lönngren, Adawi et al., 2021; Pepin, 2008). In other words, social interaction researchers are not interested in whether an individual 'has' or 'experiences' an emotion. Instead, they study how people express emotions in social contexts, how social norms influence who expresses which emotions, and how the expression of emotions influences how people interact with and position themselves relative to others.

Social psychological and discourse studies employ many terms and expressions that may be new to readers trained in philosophical or psychological traditions. The most important concept for the discussion in this section is *discourse*, which refers to societal norms, rules, and practices that guide individual and collective behavior. Discourse is closely linked to schemas: we can say that people enact their schemas through discourse when they use language and other forms of communication in interaction with others to make sense of their experiences and create meaning together (Kvasny, 2005). Thus, professional ethics schemas can shape engineering ethics discourses, including discourses constructing ethics as separate from the technical subject content and irrelevant to the engineering profession (Lönngren, 2021; Martin & Polmear, 2022; Nieusma & Cieminski, 2018; Polmear et al., 2019; Tormey et al., 2015). This is important since meaning created through such discourses, in turn, influences engineers' and students' professional schemas.

Besides discourse, *power* is an essential concept in social interaction research. There are many ways of conceptualizing power, but here, we use a definition from positioning theory, where power refers to peoples' rights and/or duties to perform specific actions in a given interactional context (Harré et al., 2009; Zembylas, 2016). Thus, like emotions, power is not something individuals 'have,' but something that is dynamically assigned to people in specific situations. So far, however, there is very little research on emotions in relation to discourses and power in engineering ethics education (Lönngren, Bellocchi et al., 2024). This is unfortunate, since such research could provide important knowledge on how engineering education could position students as morally responsible agents and thus support students in developing schemas with a strong commitment to ethics in engineering practice.

To begin developing such research, scholars can draw on a growing body of social interactionist research in other educational domains, which provides many valuable concepts and theories for studying emotion and power in social practices. Hochschild's (1979) notion of *feeling rules* is an essential and highly influential concept in such research. Feeling rules are social norms about who is expected to feel and express which emotions, how to feel and express them, and in which situation. In Western cultures, feeling rules typically associate rationality with strength, authority, and power, while emotionality is associated with weakness, submission, and loss of power (Bericat, 2016). In engineering contexts, we can expect feelings rules to limit the range of emotions one can express without considerable social costs (e.g., loss of status and power; Lönngren, Adawi et al., 2021). This, in turn, may pose significant challenges for ethical decision-making since it becomes challenging to express emotions related to, for example, compassion, empathy, and care. Thus, to strengthen moral schemas, we must also challenge prohibitive feeling rules in engineering. Another critical issue is that emotionality is often associated with women, minoritized groups, and groups with lower socio-economic status. This association exacerbates inequality and unequal representation in engineering and engineering education, since groups with lower status are likely to face an even greater loss of power when they express or engage with emotions (Ahmed, 2014; Boler, 1999; Lutz, 1996). As a result, individual students and engineers have unequal opportunities to use emotions to support ethical decision-making (e.g., exploring emotions to identify and/or communicate concerns about ethical issues).

Emotional labor (Hochschild, 1983) is another crucial concept, referring to the effort professionals perform when they express emotions that are socially expected but not aligned with how they feel or when they try to change their emotions to better align with social norms and expectations. Emotional labor is pervasive in educational settings and has been studied in, for example, higher education (Lawless, 2018), science education (Zembylas, 2004), social justice education (Rivera Maulucci, 2013), and engineering education (Adams & Turns, 2020; Kotluk et al., 2023). Due to the strong rationality discourses in engineering, engineering students and professionals are often required to perform emotional labor to control and repress their emotions and to reconcile rationality discourses with demands to show empathy and care (Buzzanell et al., 2023; Lönngren, Adawi et al., 2021). Education research has also shown that members of some minoritized groups may be expected to be more empathic and caring than others (Maddamsetti, 2021; cf. Nair & Bulleit, 2020 for an introduction to care ethics in engineering education) while also having to deal with emotions related to discrimination, prejudice, and a lack of power (DeCuir-Guby et al., 2009).

Finally, *emotional capital* provides a helpful lens for exploring engineering students' and professionals' access to tools for moral decision-making. The concept is associated with Bourdieu's work and describes how organizations and cultures tend to value the *habitus* (schemas) of powerful social groups over those of others. Emotional capital has been defined as "one's trans-situational, emotion-based knowledge, emotion management skills, and feeling capacities, which are both socially emergent and critical to the maintenance of power" (Cottingham, 2016, p. 454). In educational contexts, emotional capital is unequally distributed, meaning that particular emotional schemas are valued more than others (Zembylas, 2007). This can further exasperate unequal access to emotional resources for ethical decision-making.

Much of the early work on emotions in social interaction developed as a critical response to conventional theories that saw emotion and rationality as distinct and thus sought to explain social interaction through purely rational processes (Wetherell, 2013). More recent research, however, has begun to explore more critical and radical approaches that position reason and emotion relative to each other. For example, a study of engineering students' reflections on addressing a complex sustainability issue has illustrated how students can use discourses of reason and emotion side by side (Lönngren, Adawi et al., 2021). Engineering education researchers and practitioners have much to learn from those students!

Rationality and emotion in moral action at the institutional and societal level

Previous sections in this chapter have explored how reason and emotion interact in moral judgments made at individual and meso-interactional levels. Each has highlighted an intellectual shift from more conventional understandings of the reason–emotion relationship to more critical or radical conceptions. The text has shown that these intellectual shifts are not yet well reflected in work on engineering ethics education. In this section, we focus on the macro-social level – involving social institutions and broader cultural contexts – from the perspective of sociology. This topic requires making yet another shift regarding the language used to describe phenomena related to reason and emotion.

The terms 'rationality' and 'reason' come from the same root and effectively mean the same thing (one could play a semantic game to disentangle them, but in practical terms, people use the terms interchangeably). In sociology, the term 'rationalization' is more commonly used, so it is used in this section. The key argument in this section is that rationality and emotion are important not only in understanding the ethical action of engineers at the level of the internal mental processes of individual engineers, or in terms of the discourses of reason and emotion that shape their interactions, but also because the product of engineering rationality becomes embedded in technologies and social systems that, in themselves, have an impact on the moral action of the people who live and work within them. One key sociological concept in this regard is that modern societies are characterized by a process of *rationalization* (Weber, 1930/1992).

Weber's account has been very influential and controversial in sociological analysis, particularly in Western societies (see Chapter 9 for a range of more contemporary accounts). For Weber, Western cultures that increasingly valued rational action created bureaucratic, accounting, legal, and managerial rules based on rational evaluation according to clear rules and metrics, thus removing emotion from decision-making. These rationality-based social systems and the technologies that support them are often the work of engineers. The process of division of labor, for example, in which work tasks are studied, quantified, and then subdivided and distributed across multiple workers, is based upon principles of scientific management first articulated by mechanical engineer F. W. Taylor. The principles of scientific management remain a key organizing principle in workplaces today (see Ritzer, 2020).

One of Weber's concerns was that the increasing rationalization of societies had implications for moral or ethical action (in what follows, we will apply these to engineering situations, but Weber was not, himself, particularly concerned with engineering). He saw rational social systems as characterized by rule-based and quantified behavior in which decisions are based on rules, laws, and calculations rather than on consideration of 'right' and 'wrong.' Hence, the space for moral decision-making was progressively reduced; in the language of the neo-Kohlbergian model introduced above, one could say that social organizations such as companies or other bureaucratic institutions created the conditions in which post-conventional moral reasoning was progressively replaced by decisions that were required to be taken based on conventional moral reasoning. One way this happens is through the division of labor, which implies that engineers are often charged with decisions about 'how' rather than 'what' or 'why.'

In engineering ethics education, the effects of this division of labor can be seen in the foregrounding of discussions as to what is (and is not) the ethical responsibility of the engineer. The engineering ethics textbook of van de Poel and Royakkers (2011), for example, begins with a discussion of whether the engineer who had identified potential problems with the Challenger Space Shuttle before take-off was responsible, as a salaried employee who was hierarchically below a manager, for doing more than reporting the problem to managers. From a sociological perspective, the question here is not simply whether the individual engineer has a responsibility to do more, but rather how the social system that separates the 'how' and 'what' questions had emerged in the first place, as a result of a series of rational organizational decisions.

The Challenger case is a spectacular one but, in the Weberian account, not the only one. This issue (what is the responsibility of the engineer *vis a vis* the manager) has been framed by engineering ethics educators in different ways (see, e.g., Bucciarelli, 2008; Johnson, 1992). One way of thinking about this is the distinction between the teaching of micro-ethics, which is concerned with the responsibility of the individual engineer as an employee, and macro-ethics, which is concerned with the broader responsibility of the profession as a collective (Bielefeldt et al., 2016; Dyrud, 2014; Herkert, 2001). A possible takeaway from a macro-social analysis for engineering instruction on macro-ethics is the need to problematize and analyze (rather than accept as taken for granted) the social process through which laws, rules, bureaucracy, and the division of labor in organizations reduce or expand the space for engineers to make ethical decisions.

Weber's original account of rationalization saw rationalization as implying a progressive removal of emotion as an organizing principle in social life (see, e.g., 1930/1992, p. 73). However, as Barbalet noted (2001), this conception only holds as long as a traditional, Kantian understanding of 'emotion as disturbance' is used. While rationalization could be described as aiming for the removal of disturbance from social life, the radical view of the relationship between reason and emotion recognizes that rationally working to find the optimal means towards a given end requires a calmness that comes from not feeling threatened and from trusting that a given action will result in a given outcome (Barbalet, 2001, p. 49).

Indeed, since the 1990s, there has been an increasing recognition of the importance of 'trust' in the organization of social life (Möllering, 2001). Trust has been described as a mental process involving a sense of suspension that enables favorable expectations regarding other people's actions and intentions (Möllering, 2001). In a sociological sense, trust is important because one of the features of the modern social system is that it connects people across cultures, space, and time. If we take, for example, a piece of technology like a mobile phone, the technology itself is based on using materials extracted from the earth in Central Africa (Fitzpatrick et al., 2015); built in factories in China, in which workers' rights and working conditions have been questioned (Josephs, 2013); and disposed of in countries such as China, India, Nigeria, Ghana, and Pakistan in circumstances that can lead to ground and water pollution (Awasthi et al., 2016).

In this complex sociotechnical system, a person is 'disembedded' (Giddens, 1991) from the other people they indirectly interact with across space and social distance (in the case of those from other cultures and countries who are affected by mineral extraction or electronic waste) and across time (in the case of a future generation affected by toxic waste). At a societal level, 'trust' allows us to suspend questions about the operation of social systems. Trust, therefore, reduces the impact of social complexity on individual persons (Luhmann, 1979) as it allows us to place a bracket around a whole set of relationships and to accept that these relationships will give rise to expected outcomes. Thus, while trust is generally regarded as a positive emotion, at a social level, its effects are more morally ambiguous.

Trust in disembedding social systems has another emotional impact: those affected by our decisions across cultures, space, and time are rendered psychologically invisible and emotionally distanced. In other words, if we had to interact directly with factory workers in China, they would have a face and a name and would be hard to ignore emotionally – social systems of trade disembed us from them and so enable emotional distance and make them psychologically invisible. Thus, the engineer working on an application for a mobile phone can feel no emotional connection to the Central African miner; the Chinese laborer; the Indian, Nigerian, Ghanaian, and Pakistani e-recycling worker; or those affected by the carbon footprint of their application. In this sense, rationalization produces the emotional experience of impersonality (Weber 1930/1992, p. 259) or a lack of compassion and empathy. This reinforces the lacking sense of responsibility for the socially distributed effects of engineers' actions in a rationalized world.

Conclusions and ways forward

In this chapter, first, we have presented a historical paradigm shift across multiple disciplines toward exploring the relationships between reason and emotion in moral judgment, in the moral nature of interactions and power dynamics, and in the moral problems that result from the ways in which trade, law, accounting systems, and organizations shape the effects of engineering decisions.

Considering the current evidence from moral psychology, we should focus on the role emotions play in moral decision-making and ethics learning in engineering ethics education, moving beyond the traditional focus on reason alone. For example, emotions linked to ethical issues can play a critical role in the decision-making and learning processes. Although some research has been done on the impacts of moral emotions on engineering students' moral judgment (Higgs et al., 2020; Kotluk & Tormey, 2022), more research is still needed. There is also a need to focus more on the emotional dimensions of teaching and learning ethics to ensure that future engineers have knowledge about what is ethical and the capabilities and dispositions needed to assume moral responsibility. However, engineering ethics education or research has not yet adequately investigated the impacts of moral emotions on ethics learning.

Second, we have discussed how engineering as a profession involves complex social relationships and power structures. We have shown that moral decision-making in engineering is not an individual act but is influenced by the social and cultural environments in which engineers work and live. Thus, attending to emotions in moral decision-making can have different social costs for diverse groups of engineers and engineering students. Yet, there is currently a lack of research exploring the connection between power dynamics and emotions in engineering ethics education. We suggest that a growing body of sociological and social interactionist research in other educational fields could be a foundation for such research.

Finally, taking a sociological perspective, we have examined how the institutional, emotional, and rational aspects of society (macro-level) impact engineering ethics education. Analyzing rationalization and emotion at a macro-social level has several implications for engineering ethics and engineering ethics education. First, it asks us to question conventional engineering ethics – with its focus on micro-ethical judgments, which are made within a highly constrained setting in which the engineer can only be 'responsible' for a limited number of decisions. It suggests that we should also focus more on macro-ethical questions, which take seriously the ways in which rationalized financial, legal, and organizational factors affect what gets to be 'defined in' and 'defined out' of engineer's ethical judgments. Second, it asks us to bring emotion back into ethics questions and ethics education. It highlights that we should perhaps be clearer about the role of 'trust' in bracketing off ethical components of engineering decisions and that we may need to work on building better capacities in engineers to imagine and feel emotions like compassion, empathy, or pity for those affected by the effects of engineering actions.

To sum up, this chapter has presented the arguments for why reason has historically been prioritized and why emotions have not been fully integrated into engineering ethics education on various levels – individual, interpersonal, and societal. It is crucial to note that research on the roles of reason and emotion in moral decision-making in engineering ethics education has mainly derived from moral psychology and is based on psychological processes. However, the evidence from social psychology and the perspectives presented by sociology have thus far been given little consideration. Ultimately, we suggest that engineering ethics education must engage more with social and cultural psychology and sociology to better understand how reason and emotion interact in engineering ethics education.

References

Ahmed, S. (2014). *The cultural politics of emotion* (2nd ed.) Edinburgh University Press.

Adams, R., & Turns, J. (2020). The work of educational innovation: Exploring a personalized interdisciplinary design playbook assignment. *International Journal of Engineering Education, 36*(2), 541–555.

Awasthi, A. K., Zeng, X., & Li, J. (2016). Relationship between e-waste recycling and human health risk in India: A critical review. *Environmental Science and Pollution Research, 23*, 11509–11532. https://doi.org/10.1007/s11356-016-6085-7

Bairaktarova, D., & Plumlee, D. (2022). Creating space for empathy: Perspectives on challenges of teaching design thinking to future engineers. *International Journal of Engineering Education, 38*(2), 512–524. https://www.ijee.ie/latestissues/Vol38-2/20_ijee4180.pdf

Bairaktarova, D., & Woodcock, A. (2017). Engineering student's ethical awareness and behavior: A new motivational model. *Science and Engineering Ethics, 23*, 1129–1157. https://doi.org/10.1007/s11948-016-9814-x

Bandura, A. (2001). Social cognitive theory of mass communication. *Media Psychology, 3*(3), 265–299. https://doi.org/10.1207/S1532785XMEP0303_03

Barbalet, J. M. (2001). *Emotion, social theory, and social structure: A macrosociological approach.* Cambridge University Press.

Bericat, E. (2016). The sociology of emotions: Four decades of progress. *Current Sociology, 64*(3), 491–513. https://doi.org/10.1177/0011392115588355

Bielefeldt, A. R., Canney, N. E., Swan, C., Polmear, M., & Knight, D. (2016). Microethics and macroethics education of biomedical engineering students in the United States. *Ethics in Biology, Engineering and Medicine: An International Journal, 7*(1–2). https://doi.org/10.1615/EthicsBiologyEngMed.2017018790

Borenstein, J., Drake, M. J., Kirkman, R., & Swann, J. L. (2010). The engineering and science issues test (ESIT): A discipline-specific approach to assessing moral judgment. *Science and Engineering Ethics, 16*(2), 387–407. https://doi.org/10.1007/s11948-009-9148-z

Boler, M. (1999). *Feeling power: Emotions and education*. Routledge.

Bourdieu, P. (1990). *The logic of practice*. Stanford University Press.

Boutyline, A., & Soter, L. K. (2021). Cultural schemas: What they are, how to find them, and what to do once you've caught one. *American Sociological Review, 86*(4), 728–758. https://doi.org/10.1177/00031224211024525

Bucciarelli, L. L. (2008). Ethics and engineering education. *European Journal of Engineering Education, 33*(2), 141–149. https://doi.org/10.1080/03043790801979856

Buzzanell, P. M., Arendt, C., Dohrman, R. L., Zoltowski, C. B., & Rajan, P. (2023). Engineering emotion sustainably: affective gendered organizing of engineering identities and third space. *Sustainability, 15*(6), 5051. https://doi.org/10.3390/su15065051

Cottingham, M. D. (2016). Theorizing emotional capital. *Theory and Society, 45*, 451–470. https://doi.org/10.1007/s11186-016-9278-7

Davis, M. (2015). Engineers, emotions, and ethics. In S. S. Sethy (Ed.), *Contemporary ethical issues in engineering* (pp. 1–11). IGI Golabal. https://doi.org/10.4018/978-1-4666-8130-9.ch001

DeCuir-Gunby, J. T., Long-Mitchell, L. A., & Grant, C. (2009). The emotionality of women professors of color in engineering: A critical race theory and critical race feminism perspective. In P. Schutz & M. Zembylas (Eds.), *Advances in teacher emotion research* (pp. 323–342). Springer. https://doi.org/10.1007/978-1-4419-0564-2_16

Dyrud, M. A. (2014, June). *Predatory online technical journals: A question of ethics.* Proceedings of 121st ASEE Annual Conference and Exposition, American Society for Engineering Education, paper ID #8413. http://www.asee.org/public/conferences/32/papers/8413/download.

Firat, R., & McPherson, C. M. (2010). Toward an integrated science of morality. In S. Hitlin & S. Vaisey (Eds.), *Handbook of the sociology of morality* (pp. 361–384). Springer.

Fitzpatrick, C., Olivetti, E., Miller, T. R., Roth, R., & Kirchain, R. (2015). Conflict minerals in the compute sector: estimating extent of tin, tantalum, tungsten, and gold use in ICT products. *Environmental Science & Technology*, *49*(2), 974–981. https://pubs.acs.org/doi/pdf/10.1021/es501193k

Fleddermann, C. B. (2011). *Engineering ethics* (4th ed.). Pearson Prentice Hall.

Giddens, A. (1991). *Modernity and self-identity: Self and society in the late modern age*. Stanford University Press.

Greene, J. D., Nystrom, L. E., Engell, A. D., Darley, J. M., & Cohen, J. D. (2004). The neural bases of cognitive conflict and control in moral judgment. *Neuron*, *44*(2), 389–400. https://doi.org/10.1016/j.neuron.2004.09.027

Guntzburger, Y., Pauchant, T. C., & Tanguy, P. A. (2019). Empowering engineering students in ethical risk management: An experimental study. *Science and Engineering Ethics*, *25*, 911–937. https://doi.org/10.1007/s11948-018-0044-2

Haidt, J. (2001). The emotional dog and its rational tail: A social intuitionist approach to moral judgment. *Psychological Review*, *108*(4), 814. https://psycnet.apa.org/doi/10.1037/0033-295X.108.4.814

Haidt, J. (2008). Morality. *Perspectives on Psychological Science*, *3*(1), 65–72. https://doi.org/10.1111/j.1745-6916.2008.00063.x

Harré, R., Moghaddam, F. M., Cairnie, T. P., Rothbart, D., & Sabat, S. R. (2009). Recent advances in positioning theory. *Theory & Psychology*, *19*(1), 5–31. https://doi.org/10.1177/0959354308101417

Harris, C. E., Pritchard, M. S., James, R. W., Englehardt, E. E., & Rabins, M. J. (2019). Engineering Ethics; Concepts and Cases, 6th edition. Boston, MA: Cengage.

Herkert, J. R. (2001). Future directions in engineering ethics research: Micro ethics, macro ethics and the role of professional societies. *Science and Engineering Ethics*, *7*, 403–414. https://doi.org/10.1007/s11948-001-0062-2

Hess, J. L., Beever, J., Zoltowski, C. B., Kisselburgh, L., & Brightman, A. O. (2019). Enhancing engineering students' ethical reasoning: Situating reflexive principlism within the SIRA framework. *Journal of Engineering Education*, *108*(1), 82–102. https://doi.org/10.1002/jee.20249

Hess, J. L., & Fore, G. (2018). A systematic literature review of US engineering ethics interventions. *Science and Engineering Ethics*, *24*(2), 551–583. https://doi.org/10.1007/s11948-017-9910-6

Hess, J. L., Miller, S., Higbee, S., Fore, G. A., & Wallace, J. (2021). Empathy and ethical becoming in biomedical engineering education: A mixed methods study of an animal tissue harvesting laboratory. *Australasian Journal of Engineering Education*, *26*(1), 127–137. https://doi.org/10.1080/22054952.2020.1796045

Hess, J. L., Strobel, J., & Brightman, A. O. (2017). The development of empathic perspective-taking in an engineering ethics course. *Journal of Engineering Education*, *106*(4), 534–563. https://doi.org/10.1002/jee.20175

Higgs, C., McIntosh, T., Connelly, S., & Mumford, M. (2020). Self-focused emotions and ethical decision-making: Comparing the effects of regulated and unregulated guilt, shame, and embarrassment. *Science and Engineering Ethics*, *26*(1), 27–63. https://doi.org/10.1007/s11948-018-00082-z

Hitt, S. J., & Lennerfors, T. T. (2022). Fictional film in engineering ethics education: With Miyazaki's The Wind Rises as Exemplar. *Science and Engineering Ethics*, *28*(5), 44. https://doi.org/10.1007/s11948-022-00399-w

Hochschild, A. R. (1979). Emotion work, feeling rules, and social structure. *American Journal of Sociology*, *85*(3), 551–575.

Hochschild, A. R. (1983). *The managed heart: Commercialization of human feeling*. University of California press.

Hoffman, M. L. (2000). *Empathy and moral development: Implications for caring and justice*. Cambridge University Press. https://doi.org/10.1017/CBO9780511805851

Huff, J. L., Okai, B., Shanachilubwa, K., Sochacka, N. W., & Walther, J. (2021). Unpacking professional shame: Patterns of White male engineering students living in and out of threats to their identities. *Journal of Engineering Education*, *110*(2), 414–436. https://doi.org/10.1002/jee.20381

Johnson, D. G. (1992). Do engineers have social responsibilities? *Journal of Applied Philosophy*, *9*(1), 21–34. https://www.jstor.org/stable/24353538

Josephs, H. K. (2013). Production chains and workplace law violations: The case of apple and foxconn. *Global Business Law Review*, *3*(2), 211–228. https://engagedscholarship.csuohio.edu/gblr/vol3/iss2/5

Kellam, N., Gerow, K., Wilson, G., Walther, J., & Cruz, J. (2018). Exploring emotional trajectories of engineering students: A narrative research approach. *International Journal of Engineering Education*, *34*(6), 1726–1740. https://www.ijee.ie/latestissues/Vol34-6/02_ijee3672.pdf

Kim, D. (2022). Promoting professional socialization: A synthesis of Durkheim, Kohlberg, Hoffman, and Haidt for professional ethics education. *Business and Professional Ethics Journal*, 41(1), 93–114. https://doi.org/10.5840/bpej202216115

King, P., & Mayhew, M. (2002) Moral judgement development in higher education: insights from the Defining Issues Test. *Journal of Moral Education*, *31*(3), 247–270. https://doi.org/10.1080/0305724022000008106

Kohlberg, L. (1969). Stage and sequence: The cognitive developmental approach to socialization. In D. Goslin (Ed.), *Handbook of socialization theory and research* (pp. 347–480). Rand McNally.

Kotluk, N., & Tormey, R. (2022). *Emotional empathy and engineering students' moral reasoning*. The 50th Annual SEFI Conference SEFI-2022; Towards a New Future in Engineering Education, New Scenarios that European Alliances of Tech Universities Open Up, 458–467. https://doi.org/10.5821/conference-9788412322262.1124

Kotluk, N., & Tormey, R. (2023). Compassion and engineering students' moral reasoning: The emotional experience of engineering ethics cases. *Journal of Engineering Education*, 112(3), 719–740. https://doi.org/10.1002/jee.20538

Kotluk, N., Tormey, R., Germanier, R., & Darioly, A. (2023, September). *Emotional labor experienced in team-projects: a comparison of engineering and hospitality students*. Paper presented at the Engaging Engineering Education Proceedings of the European Society for Engineering Education 51st Annual Conference (SEFI), Dublin, Ireland. https://arrow.tudublin.ie/cgi/viewcontent.cgi?article=1022&context=sefi2023_respap

Kvasny, L. (2005). The role of the habitus in shaping discourses about the digital divide, *Journal of Computer-Mediated Communication, 10*(2), JCMC1025. https://doi.org/10.1111/j.1083-6101.2005.tb00242.x

Lawless, B. (2018). Documenting a labor of love: Emotional labor as academic labor. *Review of Communication*, *18*(2), 85–97. https://doi.org/10.1080/15358593.2018.1438644

Lönngren, J. (2021). Exploring the discursive construction of ethics in an introductory engineering course. *Journal of Engineering Education, 110*(1), 44–69. https://doi.org/10.1002/jee.20367

Lönngren, J., Adawi, T., & Berge, M. (2021). Using positioning theory to study the role of emotions in engineering problem solving: Methodological issues and recommendations for future research. *Studies in Engineering Education*, *2*(1), 53–79. https://doi.org/10.21061/see.50

Lönngren, J., Bellocchi, A., Berge, M., Bøgelund, P., Direito, I., Huff, J.L., Mohd-Yusof, K., Murzi, H., Farahwahidah Abdul Rahman, N., & Tormey, R. (2024). Emotions in engineering education: A configurative meta-synthesis systematic review. *Journal of Engineering Education*, 1–40. https://doi.org/10.1002/jee.20600

Lönngren, J., Direito, I., Tormey, R., & Huff, J.L (2023). Emotions in engineering education. In A. Johri (Ed.), *International handbook of engineering education research* (pp. 156–182). Routledge. https://doi.org/10.4324/9781003287483-10

Luhmann, N. (1979). *Trust and power*. John Wiley & Sons.

Lutz, C. A. (1996). Engendered emotion: Gender, power, and the rhetoric of emotional control in American discourse. In R. Harré & W. G. Parrott (Eds.), *The emotions: Social, cultural and biological dimensions* (pp. 151–170). SAGE Publications Ltd.

Maddamsetti, J. (2021). Exploring an elementary ESL teacher's emotions and advocacy identity. *International Multilingual Research Journal*, *15*(3), 235–252. https://doi.org/10.1080/19313152.2021.1883792

Martin, D. A., Conlon, E., & Bowe, B. (2021). Using case studies in engineering ethics education: the case for immersive scenarios through stakeholder engagement and real-life data. *Australasian Journal of Engineering Education*, *26*(1), 47–63. https://doi.org/10.1080/22054952.2021.1914297

Martin, D. A., & Polmear, M. (2022). The two cultures of engineering education: Looking back and moving forward. In S. H. Christensen, A. Buch, E. Conlon, C. Didier, C. Mitcham, & M. Murphy (Eds.), *Engineering, social sciences, and the humanities. Philosophy of engineering and technology* (Vol. 42). Springer. https://doi.org/10.1007/978-3-031-11601-8_7

Miñano, R., Uruburu, Á., Moreno-Romero, A., & Pérez-López, D. (2017). Strategies for teaching professional ethics to IT engineering degree students and evaluating the result. *Science and Engineering Ethics*, *23*, 263–286. https://doi.org/10.1007/s11948-015-9746-x

Morrison, L. A. (2020). Situating moral agency: How post-phenomenology can benefit engineering ethics. *Science and Engineering Ethics*, *26*(3), 1377–1401. https://doi.org/10.1007/s11948-019-00163-7

Möllering, G. (2001). The nature of trust: From Georg Simmel to a theory of expectation, interpretation and suspension. *Sociology*, *35*(2), 403–420. https://doi.org/10.1017/S0038038501000190

Nair, I., & Bulleit, W. M. (2020). Pragmatism and care in engineering ethics. *Science and Engineering Ethics, 26*(2020), 65–87. https://doi.org/10.1007/s11948-018-0080-y
Narvaez, D., & Bock, T. (2002). Moral schemas and tacit judgment or how the Defining Issues Test is supported by cognitive science. *Journal of Moral Education, 31*(3), 297–314. https://doi.org/10.1080/0305724022000008124
Newberry, B. (2004). The dilemma of ethics in engineering education. *Science and Engineering Ethics, 10*(2004), 343–351. https://doi.org/10.1007/s11948-004-0030-8
Nieusma, D., & Cieminski, M. (2018). *Ethics education as enculturation: Student learning of personal, social, and professional responsibility*. 2018 ASEE Annual Conference & Exposition.
Ottemo, A., Gonsalves, A. J., & Danielsson, A. T. (2021). (Dis) embodied masculinity and the meaning of (non) style in physics and computer engineering education. *Gender and Education, 33*(8), 1017–1032. https://doi.org/10.1080/09540253.2021.1884197
Pepin, N. (2008). Studies on emotions in social interactions. *Introduction. Bulletin VALS-ASLA*, 88, 1–18. http://doc.rero.ch/record/11876/files/bulletin_vals_asla_2008_088.pdf
Piaget, J. (1932/1965). *The moral judgment of the child.* Free Press.
Polmear, M., Bielefeldt, A. R., Knight, D., Canney, N., & Swan, C. (2019). Analysis of macroethics teaching practices and perceptions in engineering: A cultural comparison. *European Journal of Engineering Education, 44*(6), 866–881. https://doi.org/10.1080/03043797.2019.1593323
Rest, J., Narvaez, D., Bebeau, M., & Thoma, S. (1999). A neo-Kohlbergian approach: The DIT and schema theory. *Educational Psychology Review, 11*(4), 291–324. https://doi.org/10.1023/A:1022053215271
Ritzer, G. (2020) *The McDonaldization of society: Into the digital age* (10th ed.). Sage.
Rivera Maulucci, M. S. (2013). Emotions and positional identity in becoming a social justice science teacher: Nicole's story. *Journal of Research in Science Teaching, 50*(4), 453–478. https://doi.org/10.1002/tea.21081
Roeser, S. (2012). Emotional engineers: Toward morally responsible design. *Science and Engineering Ethics, 18*(1), 103–115. https://doi.org/10.1007/s11948-010-9236-0
Roeser, S. (2020). Risk, technology, and moral emotions: Reply to critics. *Science and Engineering Ethics, 26*(4), 1921–1934. https://doi.org/10.1007/s11948-020-00196-3
Rojter, J. (2007). Engineering profession as a social practice: Impact on professional engineering education. *International Journal of Technology, Knowledge and Society, 2*(7), 135. https://doi.org/10.18848/1832-3669/CGP/v02i07/55657
Ross, A., & Athanassoulis, N. (2010). The social nature of engineering and its implications for risk taking. *Science and Engineering Ethics, 16*(2010), 147–168. https://doi.org/10.1007/s11948-009-9125-6
Shuman, V., & Scherer, K. R. (2014). Concepts and structures of emotions. In R. Pekrun & L. Linnenbrink-Garcia (Eds). *International handbook of emotions in education* (pp. 23–45). Routledge. https://doi.org/10.4324/9780203148211
Sinatra, G. M., Broughton, S. H., & Lombardi, D. (2014). Emotions in science education. In R. Pekrun & L. Linnenbrink-Garcia (Eds.), *International handbook of emotions in education* (pp. 425–446). Routledge.
Snieder, R., & Zhu, Q. (2020). Connecting to the heart: Teaching value-based professional ethics. *Science and Engineering Ethics, 26*(4), 2235–2254. https://doi.org/10.1007/s11948-020-00216-2
Sunderland, M. E. (2014). Taking emotion seriously: Meeting students where they are. *Science and Engineering Ethics, 20*(1), 183–195. https://doi.org/10.1007/s11948-012-9427-y
Sunderland, M. E., Taebi, B., Carson, C., & Kastenberg, W. (2014). Teaching global perspectives: Engineering ethics across international and academic borders. *Journal of Responsible Innovation, 1*(2), 228–239. https://doi.org/10.1080/23299460.2014.922337
Tappan, M. B. (1997). Language, culture, and moral development: A Vygotskian perspective. *Developmental Review, 17*(1), 78–100. https://doi.org/10.1006/drev.1996.0422
Thiel, C. E., Connelly, S., Harkrider, L., Devenport, L. D., Bagdasarov, Z., Johnson, J. F., & Mumford, M. D. (2013). Case-based knowledge and ethics education: Improving learning and transfer through emotionally rich cases. *Science and Engineering Ethics, 19*(1), 265–286. https://doi.org/10.1007/s11948-011-9318-7
Thoma, S. J., & Dong, Y. (2014). The defining issues test of moral judgment development. *Behavioral Development Bulletin, 19*(3), 55. https://doi.org/10.1037/h0100590
Tormey, R. (2020). Towards an emotions-based engineering ethics education. In *Engaging engineering education proceedings of the European society for engineering education 48th annual conference 2020* (pp. 1139–1148). https://www.sefi.be/wp-content/uploads/2020/11/Proceedings-DEF-nov-2020-kleiner.pdf

Tormey, R., LeDuc, I., Isaac, S., Hardebolle, C., & Cardia, I. V. (2015, June). *The formal and hidden curricula of ethics in engineering education*. 43rd Annual SEFI Conference (No. CONF). https://www.sefi.be/wp-content/uploads/2017/09/56039-R.-TORMEY.pdf

van de Poel, I., & Royakkers, L. (2011). *Ethics, technology, and engineering: An introduction*. John Wiley & Sons.

Watts, L. L., Medeiros, K. E., Mulhearn, T. J., Steele, L. M., Connelly, S., & Mumford, M. D. (2017). Are ethics training programs improving? A meta-analytic review of past and present ethics instruction in the sciences. *Ethics & Behavior*, *27*(5), 351–384. https://doi.org/10.1080/10508422.2016.1182025

Weber, M. (1930/1992). *The protestant ethic and the spirit of capitalism*. Routledge.

Wetherell, M. (2013). Affect and discourse–What's the problem? From affect as excess to affective/discursive practice. *Subjectivity*, *6*, 349–368. https://doi.org/10.1057/sub.2013.13

Zembylas, M. (2004). Emotion metaphors and emotional labor in science teaching. *Science Education*, *88*(3), 301–324. https://doi.org/10.1002/sce.10116

Zembylas, M. (2007). Emotional capital and education: Theoretical insights from Bourdieu. *British Journal of Educational Studies*, *55*(4), 443–463. https://doi.org/10.1111/j.1467-8527.2007.00390.x

Zembylas, M. (2016). Making sense of the complex entanglement between emotion and pedagogy: Contributions of the affective turn. *Cultural Studies of Science Education*, *11*(2016), 539–550. https://doi.org/10.1007/s11422-014-9623-y

5

PROFESSIONAL ORGANIZATIONS AND CODES OF ETHICS

Jeff R. Brown, Leroy Long, III, Taylor Mitchell, and Renato Bezerra Rodrigues

Introduction

Codes are prolific in engineering. Most codes are technical in nature; they make up the standards and design manuals that provide engineers with historically validated methodologies and 'best practices' that can be used to transform a basic understanding of the engineering sciences into functional systems that are deemed safe, efficient, and effective. Other codes are more aspirational, such as the codes of ethics. Undergraduate classroom instruction surrounding these codes of ethics varies drastically depending on region and pedagogical strategy, which can limit engineering students' ability to recognize how ethics will function in their daily decisions as professionals (Colby & Sullivan, 2008a). Haws (2001) even argues that a one-sided focus on codes of ethics in engineering ethics education can be harmful as codes are allegedly platitudinous and thus do not contribute to students' learning about ethics. Thus, we echo Martin et al.'s (2020) call for more curriculum instruction on the codes of ethics that govern the engineering profession, instruction that would intertwine the existing technical curriculum with more attention to legal and aspirational aspects of professional codes.

When codes of ethics do enter the curriculum, educators tend to emphasize the legal aspects of the codes, as engineers can face severe consequences for violating these codes (Hess & Fore, 2018). By legal aspects, we mean the codes that can be deemed punitive. These codes tend to represent the bare minimum of ethical behavior engineers should adhere to. For example, Chapter 471.033 of the Florida (USA) Statutes (2023) states: "The following acts constitute grounds for which ... disciplinary actions ... may be taken: Item (g) – Engaging in fraud or deceit, negligence, incompetence, or misconduct, in the practice of engineering." Violations of this statute can lead to the loss of professional licensure, fines, and/or the requirement to pay restitution to those who were adversely impacted. The American Society of Civil Engineering (ASCE) Code of Ethics (2020) also contains elements that discourage the behaviors proscribed by the Florida Statutes: "Engineers express professional opinions truthfully and only when founded on adequate knowledge and honest conviction. Engineers have zero tolerance for bribery, fraud, and corruption in all forms, and report violations to the proper authorities." This ASCE example reinforces how the organizational legal code can align with the regional statutes; the dual emphasis on fraud and corruption also makes it very clear to students that engineers – as professionals and members of

DOI: 10.4324/9781003464259-7

the discipline – must not engage in this unethical (and illegal) behavior. Students who study these legal elements of codes can easily grasp that they should not engage in corrupt behavior.

What students might not get a chance to study thoroughly, though, are the aspects of codes of ethics that are more difficult to grasp. For instance, Martin et al.'s (2020) study of Irish engineering curricula revealed that only 1% of courses included instruction on how the work of engineers is connected to society and the natural environment. We label these aspects of codes as *aspirational* to distinguish them from the technical or punitive legal aspects. We do so because they are not directly connected to regional statutes and law. They are also more abstract due to the subjective nature of the concepts. The wording of these parts of codes tends to be more abstract and, thus, needs further interpretation and agreement to be enacted. Take ASCE's other ethical tenets as an example. They state engineers shall:

- treat all persons with respect, dignity, and fairness, and reject all forms of discrimination and harassment ("1(f). Society" quoted from ASCE, 2020)
- acknowledge the diverse historical, social, and cultural needs of the community, and incorporate these considerations in their work ("1(g). Society" quoted from ASCE, 2020)
- adhere to the principles of sustainable development ("2(a). Natural and Built Environment" quoted from ASCE, 2020)
- use resources wisely while minimizing resource depletion ("2(d). Natural and Built Environment" quoted from ASCE, 2020)

Treating all persons with *respect*, *dignity*, and *fairness* is not as clear cut to students as the legal statute that engineers 'don't accept a *bribe*.' Also, these aspirational aspects of codes represent societal values that have not been widely accepted – or even understood – but point to an ethical behavior that engineers should work towards. While safety and (monetary) honesty are more understood and accepted values, issues related to equity, diversity, and inclusion are debatable and might even be considered controversial in some contexts, which we will return to later in this chapter.

By placing the aspirational elements of engineering codes of ethics alongside the more traditional legal aspects, engineering professional organizations attempt to serve as a moral compass for the profession and guide members through the myriad considerations involved in protecting the natural environment and the health, safety, and welfare of the public. There are well-documented cases of engineers actively subverting environmental laws (e.g., the 'Dieselgate' affair at Volkswagen) or disregarding the health impacts of engineering decisions when they affect economically disadvantaged communities (e.g., the water crisis in Flint, Michigan). However, there are many more cases where competing interests, politics, and moral ambiguity may leave engineers struggling to decide between the pragmatic maximization of efficiency and profit or prioritizing their aspirational commitment to public health, safety, and welfare.

We argue that one of the challenges with the aspirational aspects of codes is that they require more contextualization than the legal ones. Educators may need to weave these aspirational elements into a variety of lessons, ground them in 'real-world' experiences, and break down their concepts to make them meaningful and applicable. Students may need more time to understand the subjective and 'novel' nature of the concepts and behaviors included in aspirational codes. This can become highly complicated when one considers that different engineering disciplines practiced in different geographical and social contexts will necessarily operate under different legal and ethical obligations.

Thus, in this chapter, we contextualize the codes of ethics (hereon referred to as 'codes') and their respective professional organizations to provide a resource for engineering ethics researchers and educators who wish to use codes in their research or teaching. To do so, we use the framework of professionalism developed by Freidson (2001). After explaining Freidson's framework, we provide a brief account of the early history of codes of ethics before getting to the primary purpose of the chapter, namely to contextualize the codes in the inextricably linked histories of engineering's professional organizations, the development of the codes, and their modern implications with a geographical focus on the United States and Canada. While attention to the codes is at least present – if varied – in engineering curriculum, connections between engineering and engineering ethics are often neglected (Mitcham & Englehart, 2019). The development and promotion of engineering codes in Western societies are rooted in the modern professional organizations that began in the 19th century, for example, the Institution of Civil Engineers (ICE) in 1828, Institution of Mechanical Engineers (IMechE) in 1847, American Society of Civil Engineers (ASCE) in 1852, Institution of Electrical Engineers (IEE) in 1871, American Society of Mechanical Engineers (ASME) in 1880, and American Institute of Electrical Engineers (AIEE) in 1884. Using codes, professional organizations and their members have standardized engineering practices. Like the codes they create, professional organizations are prolific and vary by discipline and region; therefore, we provide a four-part model linked to Freidson's framework for understanding how different professional organizations guide the practice of engineering – and thus the daily decisions of engineers. We conclude by identifying the modern implications of these codes.

This model of contextualization can help engineering ethics researchers and faculty members more fully integrate the technical, legal, and aspirational dimensions of engineering, as suggested by Conlon et al. (2018). It can also help students contextualize the codes that govern their future profession and encourage them to move beyond simply solving problems to a professional who can identify problems and the broader impacts that engineering solutions have in society (Downey, 2015).

A crucial limitation of this work relates to its geographical scope. As authors, we quickly realized that a global analysis would become overwhelming; it would not be possible for us to adequately compare and contrast different codes of ethics or professional organizations from around the world. Instead, we decided to focus on North America with the systems we (authors) are most familiar with navigating. We hope the framework will prove helpful and that engineering education researchers will further refine this approach as they apply it in different contexts in different locations and organizational settings.

Positionality

The first author of this chapter, Jeff, identifies as a white male, married with two children. His traditional engineering background is in civil structures. Jeff's current research interests include ethics in engineering education and service learning in the context of international development. The second author, Leroy, identifies as a Black man (a US descendant of Black people who were enslaved) and a person of faith. He is a department chair in mechanical engineering technology and devoted community servant. Growing up in the US Midwest, Leroy earned two degrees in mechanical engineering as well as a Ph.D. in STEM education. As a graduate student, he published research and designed engineering ethics coursework. The third author, Taylor, identifies as a cisgender white woman from a family of educators. Growing up in the northeast United States, she has earned degrees in English and American Literature, focusing on gender studies and Cold War Culture. Currently, Taylor is a tenured associate professor teaching interdisciplinary literature courses and honors seminars on Cold War Culture and Environmental Literature. Her publications reflect her work in cultural studies, pedagogy, assessment, and Rhetoric and Composition. The fourth author, Renato, was born and raised in the northeast region of Brazil and is now a settler in Winnipeg, Canada. With a background in engineering and philosophy, he is now a Ph.D. candidate in engineering education – researching ways of thinking in engineering that acknowledge, identify, and respond to the sociotechnical nature of engineering practice. His general interests are in Science and Technology Studies, Emancipatory Education, and Critical Pedagogy.

Starting the conversation: sharing a framework for professionalism

A good starting point for any conversation around codes is the history of professionalism and how codes are embedded into organizations and professional practices. This can help students understand why legal and aspirational codes of ethics are critical to their future engineering profession. Even if students might think of professional ethics solely through the lens of 'responsibilities,' they should also understand the 'privileges' that accompany professional status and how these privileges are jeopardized if the public loses trust in a profession's work. Hence, scholars like Conlon et al. (2017) propose including a broad social purpose in the definition of engineering. Critical conversations (Rottman & Reeve, 2020) could then explore an engineer's 'privileges' versus 'responsibilities.' Ideally, engineers would ultimately go a step further and use ethics to take a more activist stance as professionals (Colby & Sullivan, 2008b).

A useful framework for understanding professionalism was developed by Freidson (2001). The framework identifies five critical elements that position professionalism as a unique, socially constructed mechanism for organizing a particular type of labor within the broader economy. Below are the first four:

1. Specialized work … believed to be grounded in theoretically based discretionary knowledge;
2. Exclusive jurisdiction in a particular division of labor that is created and controlled by occupational negotiation;
3. A sheltered position in both external and internal labor markets … that is based on qualifying credentials;
4. A formal training program lying outside the labor market that produces the qualifying credentials, which is controlled by the occupation and associated with higher education.

(Freidson, 2001, p. 127)

The first element, *specialized work,* should be easily recognizable to engineering students. Engineering is a distinct body of knowledge (Colby & Sullivan, 2008a). The second element, *exclusive jurisdiction,* may require additional unpacking. The distinctions between different disciplines in engineering, and their corresponding professional organizations, is a practical example of how jurisdiction over specific engineering functions has been negotiated to allow for specialization – even sub-specialization – within different engineering disciplines (see Chapters 14–18 on ethical issues in different engineering disciplines). The *sheltered position* concept is probably the least familiar to engineering students, but the basic idea is that by completing a specific degree program, obtaining specific experience, and possibly becoming a licensed professional, the status that one achieves is at least partially protected because only others who have also achieved the same status can compete for the same type of work. Finally, the *formal training program* tied to higher education is highly relevant to engineering students and should be easily recognized (Colby & Sullivan, 2008a; see Chapters 32–36 on accreditation).

These four elements of Freidson's theoretical framework form the basis for the legalistic elements of engineering codes of ethics. Because the engineer, like other professionals, is granted exclusive jurisdiction and a sheltered position within the broader market economy, the engineer is *required* to refrain from committing fraud or deceiving others through their work. Engineers are *required* to refrain from negligence, incompetence, or misconduct when executing their work. If an engineer refuses to abide by these requirements, the State or company will refuse to provide the sheltered position and/or rescind exclusive jurisdiction.

What, then, is the basis for the aspirational elements of engineering codes of ethics? The fifth element of Freidson's typology contends that professional work embodies "an ideology that asserts greater commitment to doing good work than to economic gain and to the quality rather than the economic efficiency of work" (Freidson, 2001, p. 127). Freidson builds on this assertion with three additional points that support the inclusion of aspirational elements in codes of ethics: (1) The professional ideology of service goes beyond serving others' choices. (2) Rather, it claims devotion to a transcendent value which infuses its specialization with a larger and putatively higher goal, (3) which may reach beyond that of those they are supposed to serve. Aspirational elements of engineering codes of ethics provide organizations an opportunity to articulate their 'transcendent values' while, hopefully, inspiring their members to devote themselves to these higher goals. Helping students understand why professional organizations might (or not) include references to environmental or social concerns as part of their codes of ethics is an important aspect of professional formation.

Of course, aspirational elements of codes will often have legal components defined in various locations within a legal jurisdiction's statutes. For example, legally binding building codes will often dictate minimum energy-efficiency requirements for buildings – to minimize energy use. Engineers operating in a given jurisdiction would be legally obligated to produce designs that meet these minimum requirements. *But why the need for aspirational elements in engineering codes of ethics if these obligations may be already spelled out elsewhere?* The answer can point in two directions: (1) the aspirational elements can be viewed as an encouragement for engineers to think critically about minimum requirements and attempt to improve designs beyond legal requirements, and (2) the aspirational elements can serve as a reminder to the engineer that whenever conflicts might arise, the engineer's obligation lies first with the aspirations of the profession.

Legal and aspirational parts of early codes of ethics

Concepts related to professionalism, professional responsibility, and ethical behavior have historical roots that date back centuries (Moriarty, 2001). These are concepts that can easily be introduced during class discussions, particularly the classic example of Hammurabi's Code for the Babylonian Empire in the eighteenth century BC, which established rules and guidance for a wide array of business transactions, legal proceedings, and expectations for professional services. Builders could be executed for shoddy construction if a structural failure caused the death of the building's owner or his son. Several (some gruesome) provisions were also made for physicians:

> 218. If a physician makes a large incision with the operating knife, and kill him, or open a tumor with the operating knife, and cut out the eye, his hands shall be cut off.

Fast forward roughly 1,000 years (circa eighth–fifth century BC) to when the author of Deuteronomy established several relevant demands upon the Israelites that are easily recognizable in a modern professional context: judges should "not accept a bribe, for a bribe blinds the eyes of the wise and twists the words of the righteous" (Deut. 16:19). With the focus on bribery, the language mimics the current ethical code from ASCE that mentions a 'zero tolerance for bribery.' Regardless of the profession – judge or engineer – bribes have been clearly unethical for centuries. Also, there is a call in the Old Testament to anyone engaged in homebuilding (what Freidson would call 'specialized work'): "you are to construct a railing around your roof, so that you do not bring bloodguilt on your house if someone falls from it" (Deut. 22:8). Even if the language of Deuteronomy may seem clunky to the modern reader, professors can ask students what a contemporary version of a 'code written in blood' might be. Students can compare a variety of modern codes to identify how modern codes tend to opt for the positive assertion to simply protect the safety of the public, while still threatening legal penalties or licensure removal.

Legal and aspirational aspects of codes also can be found in the Twelve Tables of Roman Law (449 BC):

> Table 7.6. The width of a road ... shall be eight feet on a straight stretch, on a bend ... sixteen feet.
>
> Table 8.1b. If anyone sings or composes an incantation that can cause dishonor or disgrace to another ... he shall suffer a capital penalty.

Comparing Table 7.6 and 8.1b can help students think critically about different community needs. Here, 7.6 is a clear-cut and policy-based code, and 8.1 is much more complex. Both give boundaries to different professions that can introduce students to the long history of professions and the changing punishments for violating professional codes. When given this context, students can situate their own engineering profession into this larger historical framework, a long view that many engineering ethics modules don't include.

Professional organizations and their codes

Students should also learn about the different types of professional organizations and how they have contributed to legal and aspirational codes to contextualize current codes better. The complicated history and current events that shape professional organizations are critical for understanding their role in codes of ethics – including legal and aspirational elements. Paralleling the way Freidson's (2001) typology of professionalism can be used in the classroom, we have identified four general categories of professional organizations: Regulatory, Body of Knowledge, Education, and Affinity Organizations. Through the graphical representation of these relationships, seen in Figure 5.1, educators, researchers, and students can better understand how these organizations form and frame professionalism and ethics in engineering.

In Figure 5.1, we situate the fifth element of the fivefold typology for professionalism – a commitment to good work – at the center to highlight the connection to codes of ethics and the ideals they aspire to achieve. The organizational types are located outside the model but closest to the elements they are generally associated with from Freidson's model. The dashed lines in the model represent modes or mechanisms that the organizations must work *through* to achieve these goals. For example, regulatory-focused professional organizations, like NCEES and Engineers Canada (through Provincial and Territorial Regulatory regulators), work not only to ensure that engineers are appropriately certified but also to develop regulatory frameworks that legislative bodies (represented by the dashed line) can enact as laws to guarantee exclusive jurisdiction and a sheltered position for engineers. Many affinity organizations, such as the Society of Women Engineers (SWE), were formed in response to persistent underrepresentation in the profession (see dashed line), and others were created to earn the power they lack but rightfully deserve.

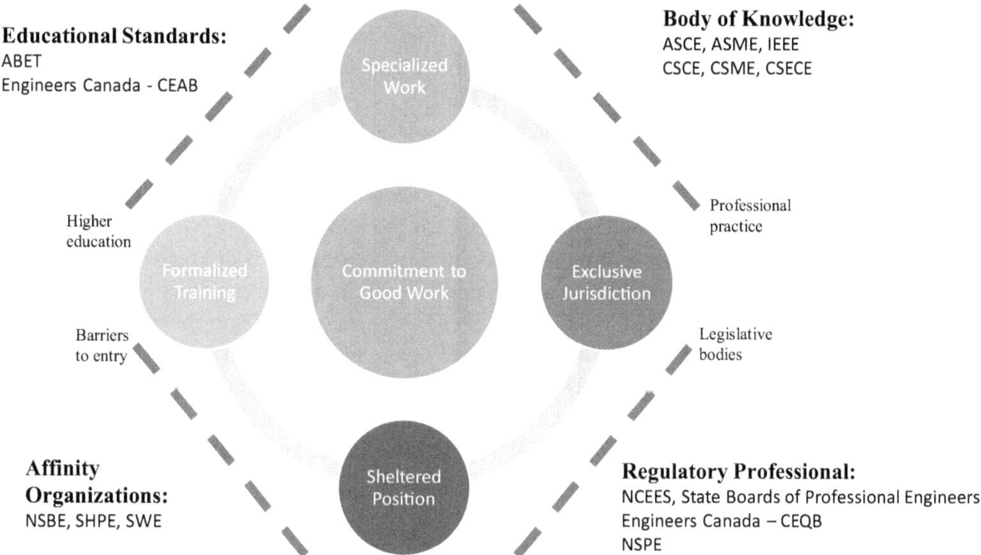

Figure 5.1 Professional organizations and their relations to Freidson's (2001) fivefold typology for professionalism.

Regulatory professional organizations

Engineers Canada and NCEES are North America's largest professional organizations dedicated to influencing how the government regulates engineering practice. Students need to understand how these organizations operate because they influence legislation and encourage ethical behavior in engineering through their specific codes.

Engineering licensure in Canada

To practice engineering in Canada, one must be licensed through a regulatory body that acts at the provincial and territorial level. At the national level, there is a 'harmonization' promoted by Engineers Canada's recommendations through guidelines, papers, and tools developed by professional engineers in collaboration with provincial and territorial regulators (Engineers Canada). However, these are not rules – and provincial and territorial engineering regulators may adopt them entirely, in parts, or not at all.

The first act to regulate the practice of engineering in Canada was passed in 1896 in Manitoba. However, it was repealed in 1913 due to a lack of administrative machinery to enforce the act. It was not until 1920, after the Great War, that six provincial regulatory bodies were formed to regulate engineering practices in their respective provinces. Canada's national engineering organization was created in 1936 to bring the provincial and territorial regulators 'into greater harmony.' Although licensure requirements vary, they generally include academic experience, work experience, professionalism and ethics, and good character. At the national level, Engineers Canada – through the Canadian Engineering Qualifications Board (CEQB) – develops guidelines and papers to promote consistent engineering practice across the country, but this board does not serve as a regulator or rule enforcer. Ethics is a core aspect of professional licensure in Canada, as candidates must pass the National Professional Practice Exam (NPPE), which covers knowledge of laws affecting the engineering profession, professional standards, engineering ethics, and other intellectual property topics (Professional Engineers Ontario, n.d.). The NPPE requirement forces all potential practicing engineers to be able to regurgitate basic legal, professional, and ethical Canadian standards, but the exam does not (nor can it) help students aspiring to be engineers to abide by legal or aspirational aspects of codes of ethics.

The engineering code of ethics in Canada has a long history. The provincial regulatory bodies developed the first engineering codes of ethics right after their creation (e.g., Engineers Geoscientists Manitoba in 1921 and Professional Engineers of Ontario in 1923). Some of these codes were included in the first version of the engineering act and, therefore, were enforceable by the regulatory body. In contrast, it took a couple of decades before other codes were included in the act through amendments (Professional Engineers Ontario, 2022). Like the ancient professionalism examples, these first codes focused mostly on the legal aspects, including bribery, underbidding, providing honest and true information, and confidentiality. In these early stages, there was no mention of aspirational elements, such as public interest or environmental consideration.

Engineers Canada (more specifically, CEQB) developed the current National Code of Ethics in consultation with provincial and territorial regulators. "Based on broad principles of integrity, truth, honesty, and trustworthiness, respect for human life and welfare, fairness, openness, competence, and accountability" (Engineers Canada, 2016, p. 2), these codes are reflected in ten tenets. Some are focused on the legal aspects (e.g., "to maintain confidentiality and avoid conflicts of interest" (Engineers Canada, 2016, p. 3), and others are more aspirational (e.g., "to hold paramount the safety, health, and welfare of the public and protect the environment" (Engineers Canada, 2016, p. 3). Engineers Canada's Code of Ethics still includes historical bribery mentions but also aspirational tenets of safety, whistle-blowing, social and environmental impacts, and equity. This shows how codes of ethics can change and evolve through time and are based on largely agreed-upon values by the professional organizations. Sharing this context with students shows where the profession came from and where it is heading.

As indicated above, in Canada, each provincial and territorial regulator is free to adopt Engineers Canada's Code of Ethics (fully, in part, or not at all) to serve regulatory and legal purposes. For instance, in Manitoba, regulators use Engineers Canada's Code of Ethics almost verbatim but do not include tenet 9 ("treat equitably and promote the equitable and dignified treatment of people in accordance with human rights legislation") or tenet 10 ("Uphold and enhance the honor and dignity of the profession") (Engineers Geoscientists Manitoba, 2018, p. #). Another example is Alberta's engineering association, whose code of ethics includes only five tenets but covers and expands on all principles from the Engineers Canada's Code of Ethics, such as an entire subsection dedicated to bias in the profession (Association of Professional Engineers and Geoscientists of Alberta, 2022). Ontario's regulators use different wording but include many legal principles outlined in the national codes. However, more aspirational principles, such as equity, diversity, sustainability, and social impacts, are not mentioned or covered in Ontario's code of ethics (Professional Engineers Ontario, 2023). Students studying in this province can and should be curious as to why this is the case – *What makes Ontario different from the other provinces? Are there political or cultural events that would keep diversity, equity, and inclusion (DEI) or social impacts out of the official principles?* Getting students to consider these larger, more complex and culturally relevant concerns about their profession can provide them with real-world context that a basic engineering ethics case study could not.

Engineering licensure in the United States (NCEES)

Individual states regulate engineering practice in the United States. Each state has a Board of Professional Engineers that is granted the authority, by state statutes, to issue Professional

Engineering (PE) licenses. All of the state Boards of Engineers are members of the National Council of Examiners for Engineering and Surveying (NCEES), which is broadly responsible for four primary functions: (1) administering the Fundamentals of Engineering Examination and the Professional Engineering Examinations; (2) developing Model Laws and Model Rules that the state Boards of Engineers can adopt; (3) promoting professional ethics; and (4) co-ordinating with other domestic and international organizations to promote professional licensure. In the United States, a Professional Engineers (PE) license is necessary and beneficial in ways that differ from the gains that result from joining an engineering organization. Some employers may only hire candidates with a professional engineer's (PE) license due to federal, state, and local laws as well as the needs of their clients. PEs are held to higher ethical and industry standards than non-licensed engineers. To become a PE, aspiring and current engineers must a) complete a bachelor's degree from an accredited engineering program, b) pass the Fundamentals of Engineering (FE) Exam, c) work for 4 years under the direction of a PE, and d) pass the PE exam.

NCEES promotes Model Laws (NCEES, 2021) and Model Rules (NCEES, 2022b), but these only become enforceable once an individual state adopts them into their state statutes. In addition to the Model Laws and Model Rules, NCEES also maintains a document containing Position Statements intended to summarize consensus positions of the state Boards of Engineers (NCEES, 2022a). These statements contain several elements that align with the elements of aspirational ethics. For example, PS 34 "encourages professional engineers and professional surveyors to incorporate in their work and lives the principles and practices of sustainability to safeguard the health, safety, and welfare of the public" (NCEES, 2022a, p. 39). Analyzing position statements like PS 34 and the conversation that spurred their appearance in the field will help students connect their professional identity to the environment and larger society impacts – something that Freidson (2001) posited in his fivefold typology. The most recent version of the Position Statements document adopted in 2022 also includes PS 32, directly addressing DEI. In this statement, NCEES encourages its member boards to "advanc[e] licensure in such a way as to be inclusive of all people for the betterment of engineering … and trea[t] its employees and volunteers of the organization in an equitable and inclusive manner with respect, dignity, and fairness that fosters participation without regard to individual differences" (NCEES, 2022a, p. 38). By sharing this PS with students, educators can ask students to wrestle with concepts of diversity and inclusion that people in some North American regions have found particularly divisive. Chapter 34 of this handbook provides additional discussion regarding processes and implications of licensure and accreditation, and case studies from the United States and Ireland.

Educational standards organizations

Students should move beyond just learning about regulatory professional organizations and understand how, through these professional organizations, specific accrediting bodies like ABET and the Canadian Engineering Accreditation Board (CEAB) have evolved. Discussing this can illustrate the connections between professional organizations, codes of ethics, and engineering education. Due to the United States' – and to a lesser extent Canada's – decentralized education system, the influence of accrediting bodies on higher education can be more profound than in other nations (Akera & Seely, 2015). Without a more powerful centralized department of education, the objectives of some degree programs in higher education end up being shaped by professional organizations and their accrediting bodies. For example, in 1929, the Society for the Promotion of Engineering Education (now the American Society for Engineering Education) published *A Comparative Study of Engineering Education in the United States and in Europe* by William E.

Wickenden (1929). Wickenden's two-volume report called for the creation of an organization to establish standards for engineering education programs and to conduct reviews of programs for compliance. Three years later, the Engineers' Council for Professional Development (ECPD) was founded to establish personal and professional development training plans, enhance recognition for the profession, and identify curricula that met specific standards (Aldridge & Cryer, 2009). By 1935, Charles F. Scott, a former president of the American Institute of Electrical Engineers (AIEE), oversaw the first evaluation visits as the ECPD chairman; the following year, ECPD completed the first evaluations of engineering degree programs. Within the decade, the professional organization had evaluated and accredited programs at 133 institutions. Almost a century later, the power of accreditation is now even more profound because to be accredited, degree programs, curriculum maps, and classroom materials need to address the codes the organizations deem necessary for engineering professions. In 1980, the ECPD became known as the Accreditation Board for Engineering and Technology (ABET), and at the turn of the century it developed EC2000. The EC2000 criteria govern ABET-accredited programs and are intended to give graduates "essential 21st century skills such as the ability to work in teams and communicate effectively" (ABET, 2021c, "Engineering Criteria, 2000" ¶ 3). Student learning outcomes (SLOs) regarding collaboration now encourage degree programs to address ways their students work together in diverse teams.

CEAB

Like ABET in the USA, the CEAB – part of Engineers Canada – accredits national undergraduate engineering programs to ensure that they meet the standards for engineering education. Engineering programs typically want their graduates to be eligible for professional licensure, and graduating from an accredited engineering program automatically meets the academic requirements for an individual to achieve licensure as a Professional Engineer in Canada (Engineers Canada, 2016). To be accredited, programs must have a minimum of 1,850 accreditation units (AU), which include basic studies, discipline-specific studies, and complementary studies, with minimum AU requirements of 420, 900, and 225, respectively. One AU equals one hour of lecture with 50 minutes of activity or two hours of laboratory activities. Basic and discipline-specific studies provide technical competence, while complementary studies ensure that professional engineers have an awareness of the broader context of engineering and include topics such as professionalism, ethics, law, communication, sustainability, project management, and engineering impact on society.

In addition to CEAB's AUs, programs must also show that their engineering graduates are competent in 12 graduate attributes; two of these are 'Professionalism' and 'Ethics and Equity' (Canadian Engineering Accreditation Board, 2022). However, these two graduate attributes represent a small portion of the AU requirements and must compete with other important topics for the 225 AU reserved for complementary studies. Based on the minimum requirements, in the best case scenario, programs would have 530 AU to teach complementary studies, which would certainly provide instructors more opportunity and students more exposure to engineering ethics and professionalism. Each degree program can determine how many AUs are dedicated to complementary studies (225–530) and how these are split among the many topics competing for space, including ethics and professionalism. Consequentially, in many situations, instructors must be selective about what they teach – and creative in how they teach engineering ethics with such limited space in the curriculum in Canada. This history (i.e., how accreditation requirements change and influence what is taught) is one aspect that should be included.

ABET

Likewise, students in the United States can be taught that ABET's student learning outcomes directly shape their higher-education curriculum and class content. Even within content-heavy curriculums, faculty members should share the SLOs and help students parse out which ones are legalistic and aspirational. For example, nestled within ABET's more measurable outcomes like SLO 3 ("an ability to communicate effectively with a range of audiences") are SLOs dedicated to ethics and professional responsibility (ABET, 2021a). In creating SLO2 and SLO4, ABET declared that students developing professional identities as engineers must expand their skill sets beyond solving complex problems.

> SLO 2. an ability to apply engineering design to produce solutions that meet specified needs with consideration of public health, safety and welfare, as well as global, cultural, social, environmental, and economic factors
> SLO 4. an ability to recognize ethical and professional responsibilities in engineering situations and make informed judgements, which must consider the impact of engineering solutions in global, economic, environmental, and societal contexts.
> *(ABET, 2021a, "Criterion 3: Student Outcomes" 1)*

Even though SLO2 doesn't directly mention ethics, the stated considerations reflect an ethical framework dependent on an intersection of needs. Asking students to produce solutions that don't just meet the bottom line reinforces ethical frameworks like the common good approach, care-based ethics, or virtue ethics. SLO4 specifies that students must recognize ethical and professional responsibilities in situations and asks that conclusions students make are informed by a variety of impacts. With these SLOs, ABET – the accrediting organization that universities must adhere to if they want their degrees to be worthwhile – frames students' education before they become professionals.

Body of knowledge organizations

Professional organizations like ASCE, ASME, and IEEE – and their global equivalents – are primarily responsible for establishing best practices and standards for their respective engineering disciplines. This ensures practicing engineers have access to relevant information and generations of accumulated knowledge and experience. These organizations also support Freidson's (2001) fivefold typology for professionalism by defining (and also limiting) which types of engineering work fall under a particular organization's purview (i.e., exclusive jurisdiction). For students, these disciplinary distinctions are critical for efficiency and organizing the body of knowledge each discipline is responsible for maintaining and developing.

These organizations also support Freidson's (2001) 'commitment to good work' through their codes. By establishing expectations for professional practice, as well as aspirational elements that point to the future these organizations are striving to create, body-of-knowledge organizations can tailor their codes to address the specific needs of their members. Even if codes from this type of organization are not legally binding, organizations can issue reprimands or other penalties that may include revoking the membership of anyone who violates the code.

American Society of Civil Engineers

Engineering students should understand several key developments in the codes for specific organizations. ASCE serves as a good example. At the time of ASCE's founding in 1852, there was some informal discussion of developing a code of ethics, but the Society adopted the position that professional ethics is a personal responsibility (Vesiland, 1995). This stance was reinforced in 1877 when a proposal was brought before the Society regarding members' conduct. The following resolution was attached to the proposal for referral to the Board of Directors: "Resolved: That it is inexpedient for this society to instruct its members as to their duties in private professional matters" (Wisely, 1977, in Vesiland, 1995, p. 5).

The development of a code of ethics was again raised in 1893 when a group of members from Cincinnati endorsed the adoption of a code and called for the appointment of a committee to draft such a code. However, no action was taken on this matter, and the idea of a code was dropped. Students should be asked why this might have happened and what events might have caused the 1902 discussion regarding the formulation of a code of ethics, which ASCE again decided against. In the words of the Society's president:

> it is, I think, safe to say that for the kindling of professional enthusiasm, and the establishment of high professional standards, the Society and its members will continue to rely, as they have done in the past, upon these vital and moral forces, and not upon the enactment of codes or upon any form of legislation.
>
> *(Wisely, 1977, in Vesiland, 1995, p. 5)*

The first code was formally adopted by ASCE in 1914. Before this established code, organizations most likely relied on the gentlemanly aspect of the profession, something like: 'there is no need for a code … we are all gentlemen here.' For example, Item 17 on Hartley's list of "100 Hints for Gentlemanly Deportment" stated that a gentleman should "Cultivate the virtues of the soul, strong principle, incorruptible integrity, usefulness, refined intellect, and fidelity in seeking for truth. A man in proportion as he has these virtues will be honored and welcomed everywhere" (Hartley, 1875, p. 192).

The most recent update to the ASCE code of ethics was adopted in October 2020 (ASCE, 2020). The intention behind the update was to make the code more accessible and to provide a hierarchy focused on stakeholders who are impacted by the practice of civil engineering: (1) Society, (2) Natural and Built Environment, (3) Profession, (4) Clients and Employers, and (5) Peers (Walpole, 2020). In the case of a conflict between stakeholders, the order in which they are listed in the hierarchy indicates priority. According to this model (ASCE, 2020, 'Natural and Built Environment'; 'Clients and Employers'), civil engineers are expected to "adhere to the principles of sustainable development" (contained in Level 2 – Natural and Built Environment) before honoring their commitment to "act as faithful agents of their clients and employers" (contained in Level 4 – 'Clients and Employers'). Of course, this does not mean that a civil engineer committed to a design alternative with a lower environmental impact is guaranteed to retain their client or employer. However, the aspirational intent is clear, and the prioritization of social and environmental impacts does provide a foothold from which present and future practitioners can continue to climb. Knowing this intent is key to engineering education on the code of ethics.

Affinity organizations

Not only do national and international organizations stipulate the legal and policy-driven goals of the engineering profession, but some also have aspirational aspects built into their mission statements and overall identities. For example, the mission of the Society of Women Engineers (SWE) is to "empower women to achieve their full potential in careers as engineers and leaders;

expand the image of the engineering and technology professions as a positive force in improving the quality of life and demonstrate the value of diversity and inclusion" ("What is SWE?", 2024, ¶ 1). In addition, the mission of the Society of Hispanic Professional Engineers (SHPE) is to change "lives by empowering the Hispanic community to realize its fullest potential and to impact the world through STEM awareness, access, support, and development" ("About SHPE", 2022) ¶ 1. Furthermore, the mission of the National Society of Black Engineers (NSBE) is to "to increase the number of culturally responsible black engineers who excel academically, succeed professionally and positively impact the community" (National Society of Black Engineers, 2023, ¶ 1).

At the time that the NSPE's Code of Ethics was being adopted, in the 1960s, only about 1% of US engineers were women (Layne, 2009) and into the 1970s, only about 1% were Black men (Westcott, 1982). SWE was founded in the 1950s, a decade before the NSPE Code of Ethics was created. Organizations like SHPE and NSBE were formed, in 1974 and 1975, over a decade after the NSPE Code of Ethics was established. Both SHPE and NSPE were founded by men, yet, a Black woman became NSBE's national chair and the first chair to serve two terms just 2 years after its formation. By comparison, it took 15 years before a Hispanic woman became the president of SHPE. Overall, SWE, SHPE, and NSBE provided safe spaces for women and people of color to enter a profession that excluded them for decades and still marginalizes them to this day. Faculty members can encourage students who identify with these affinity organizations to join them; educators can also teach all students the history of these organizations and explain why they came into existence. Students and educational ethics researchers should move beyond just recognition and explore these organizations' impact on the profession.

Modern-day implications of professional organizations and codes of ethics

We have illustrated how professional organizations strive to shape engineering practice in North America through their codes of ethics. This section will highlight a current example of the complex interactions between professional organizations, their constituencies, and the political environment that they all inhabit.

In the United States, issues of diversity, equity, and inclusion (DEI) have been controversial as institutions across society have struggled to confront the persistent injustices surrounding race, gender, and sexual orientation. The murder of George Floyd on 25 May 2020 led to nationwide protests and represented a turning point in the national conversation on race. In March 2021, deans of the colleges of engineering from all schools in the 'Big Ten Conference' (an athletic conference of ten large public research universities in the United States) signed a letter addressed to ABET in support of the inclusion of DEI as part of the General Criteria for Accrediting Engineering Programs (ABET, 2021b). Signatories included deans from Cornell University, the Massachusetts Institute of Technology, Rutgers University, and the University of Michigan. The deans cited multiple factors for their support of DEI, including ensuring that "our students will be prepared to develop technological solutions to society's most pressing problems and to combat prejudice, racism, and discrimination during their careers" (ABET, "Big 10+ Universities Deans of Engineering Letter of Support," 2021, ¶ 2). The letter also identified "the potential synergies between DEI and ethical/professional responsibilities."

In direct response to this letter, ABET proposed modifying two criteria in its overall framework for the accreditation of engineering programs:

- Criterion 5: The curriculum must include … content that ensures awareness of diversity, equity, and inclusion for professional practice consistent with the institution's mission. (ABET, 2021a, p. 6)

- Criterion 6: The program faculty must also demonstrate knowledge of applicable institutional policies on diversity, equity, and inclusion and demonstrate awareness appropriate to providing an equitable and inclusive environment for its students that respects the institution's mission (ABET, 2021a, p. 10).

First piloted during the 2023–2024 accreditation cycle, institutions are able to adopt the above criteria on a voluntary basis. After the pilot period, ABET will decide if and how to incorporate these changes related to DEI and the extent to which all engineering programs will be expected to meet these criteria (ABET, 2021b).

While these changes were underway at ABET, NCEES also began a conversation about how DEI might inform their mission regarding regulations and licensure for engineering practice. This culminated in the adoption of Position Statement 32 in August 2022, which encourages DEI (NCEES, 2022a). Recall, however, that NCEES is the umbrella organization for US Boards of Professional Engineers administered at the state level with members who are effectively political appointees. In October 2022, the Florida Board of Professional Engineers (FBPE) newsletter contained an article titled "From the Executive Director: What Exactly is DEI?" that discussed PS 32 and provided additional context about DEI that was supportive in nature. Roughly one week after this newsletter was published, FBPE issued a retraction stating, "This article does not reflect the official position and the content included therein is not supported by the FBPE" (FBPE, e-mail communication, October 13, 2022). The retraction letter further clarified that "FBPE led the opposition and voted against the DEI position and the new language," stating that "DEI is a highly politicized item in the current social and political climate, and that the NCEES organization should strive to remain politically neutral from these types of current influences" (FBPE, e-mail communication, October 13, 2022).

The political winds in Florida are clearly blowing against the aspirational goals of ABET and NCEES. The issue was further complicated in April 2023 when Florida Governor Ron DeSantis signed into law Florida Senate Bill 266 (2023), which states that public institutions of higher education in Florida may not:

> Expend any funds, regardless of source, to promote, support, or maintain any programs or campus activities that … espouse diversity, equity, and inclusion, or promote or engage in political or social activism.
>
> *(Florida Senate, 2023, p.10)*

Additional provisions state that:

> General education core courses may not suppress or distort significant historical events or include a curriculum that teaches identity politics … or is based on theories that systemic racism, sexism, oppression, and privilege are inherent in the institutions of the United States and were created to maintain social, political, and economic inequities.
>
> *(Florida Senate, 2023, p. 20)*

Surprisingly, the Florida law provided an exception for DEI-based programs that may be required "for obtaining or retaining institutional or discipline-specific accreditation with the approval of either the State Board of Education or the Board of Governors." Such an exception would appear to minimize the potential for a standoff between any public university system subject to these restrictions against DEI and ABET's ability to establish criteria for accreditation of engineering programs. Recent comments by Gov. DeSantis, made during his announcement for the US presidency in May 2023, suggest that the issue is far from settled:

> There are some tools at the federal level that we don't necessarily have at the state level. For example, some of the problems with the university and the ideological capture, that didn't happen by accident. It can trace back all the way to the accreditation cartels. Well guess what? To become an accreditor, how do you do that? You've got to get approved by the US Department of Education, so we're going to be doing alternative accreditation regimes where instead of saying, 'You will only get accredited if you do DEI,' you'll have an accreditor that will say, 'We will not accredit you if you do DEI.' We want a colorblind, merit-based accreditation scheme, and so as president controlling that agency, you can then approve other types of accreditations.
>
> *(DeSantis & Musk, 2023)*

We hope this example will spark further reflection on the significance of professional organizations and work to dispel the notion that engineering is a purely objective field. Integrating examples like this into discussions about codes of ethics in the classroom can better prepare students to engage with modern social issues that, acknowledged or not, impact engineering practice.

Concluding remarks

Haws (2001) included a stark warning to engineering educators who limit their instruction on engineering ethics to codes: "Certainly, the practice of a 'cold' distribution of the Canon of Ethics as a one-page handout deserves the level of student attention it typically receives" (p. 224). Haws was generally skeptical about codes of ethics and believed that "the Professional Engineer's Code of Ethics is platitudinous and contributes nothing to our students' understanding of either ethical systems, or the shared language in which ethical problems and solutions are couched" (p. 224). While we share some of Haws's skepticism, we conclude that engineering codes of ethics are much more meaningful and impactful when approached through a historical context and a theory of professionalism, and interpreted through the lenses of the myriad professional organizations responsible for their development. Researchers and engineering educators can apply the lens of our framework to any professional organization in different geographical regions of interest to help students assess and contextualize codes of ethics.

Research for this chapter has also revealed an interesting dynamic between conservative and progressive political forces that play out through the development of engineering codes of ethics. On the conservative side, codes of ethics seek to outline the hallmarks or necessary character traits for practicing engineers to maintain the established order and preserve exclusive jurisdiction. On the progressive side, codes of ethics provide an opportunity for professional organizations to evaluate and articulate their higher goals and ensure that the broader impacts of engineering work are considered while the work is being performed. We contend that engineering educators will be able to capitalize on this tension and help students understand that codes of ethics are living documents that, to stay relevant, must continue to evolve and adapt to the environmental, social, and economic context of their time.

References

ABET. (2021a). *Criteria for accrediting engineering programs, 2022-2023*. Retrieved January 5, 2024, from https://www.abet.org/wp-content/uploads/2022/01/2022-23-EAC-Criteria.pdf

ABET. (2021b). *Diversity, equity & inclusion*. Retrieved January 5, 2024, from https://www.abet.org/about-abet/diversity-equity-and-inclusion/

ABET. (2021c). *History*. Retrieved January 5, 2024, from https://www.abet.org/about-abet/history/

Akera, A., & Seely, B. (2015). A historical survey of the structural changes in the American system of engineering education. *International Perspectives on Engineering Education: Engineering Education and Practice in Context, 1*, 7–32.

Aldridge, M. D., & Cryer, K. (2009, August). ABET and IEEE: A history intertwined. In *2009 IEEE conference On the history of technical societies* (pp. 1–8). IEEE.

American Society of Civil Engineers. (2020). *Code of ethics*. Retrieved January 5, 2024, from https://www.asce.org/career-growth/ethics/code-of-ethics

Association of Professional Engineers and Geoscientists of Alberta. (2022). *Ethical practice*. Retrieved January 5, 2024, from https://www.apega.ca/docs/default-source/pdfs/standards-guidelines/ethical-practice.pdf

Canadian Engineering Accreditation Board. (2022). *Accreditation criteria and procedures*. Retrieved January 5, 2024, from https://engineerscanada.ca/sites/default/files/2022-11/Accreditation_Criteria_Procedures_2022.pdf

Colby, A., & Sullivan, W. M. (2008a). Ethics teaching in undergraduate engineering education. *Journal of Engineering Education, 97,* 327–338.

Colby, A., & Sullivan, W. M. (2008b). Formation of professionalism and purpose: Perspective from the preparation for the professions program. *U. St. Thomas LJ, 5*, 404.

Conlon, E., Martin, D. A., Nicolaou, I., & Bowe, B. (2018). Holistic engineering ethics? In *EESD 2018 proceedings creating the holistic engineer: International conference on engineering education for sustainable development* (pp. 52–60).

Conlon, E., Nicolaou, I. & Bowe, B. (2017). Into the deep: The role of paradigms in understanding engineering education for sustainable development. *Irish Journal of Social, Economic and Environmental Sustainability, 1(2)*.

DeSantis, R. & Musk, E. (2023). *Florida governor ron DeSantis announces 2024 presidential run on Twitter spaces with Elon Musk transcript*. Retrieved January 5, 2024, from https://www.rev.com/blog/transcripts/florida-governor-ron-desantis-announces-2024-presidential-run-on-twitter-spaces-with-elon-musk-transcript

Downey, G. L. (2015). PDS: Engineering as problem definition and solution. In S. H. Christensen, C. Didier, A. Jamison, M. Meganck, M. Mitcham, & B. Newberry (Eds.), *International perspectives on engineering education: Engineering education and practice in context* (Vol. 1, pp. 435–455). Springer.

Engineers Canada. (2016). *Code of ethics*. Retrieved January 5, 2024, from https://engineerscanada.ca/publications/public-guideline-on-the-code-of-ethics

Engineers Geoscientists Manitoba. (2018). *Code of ethics*. Retrieved January 5, 2024, from https://www.enggeomb.ca/pdf/CodeOfEthics.pdf

Florida Senate. (2023). *Senate bill 266*. Retrieved March 10, 2024, from https://www.flsenate.gov/Session/Bill/2023/266/BillText/er/PDF.z

Florida Statutes, Regulations of Professionals and Occupations, Chapter 471 (2023).

Freidson, E. (2001). *Professionalism, the third logic: On the practice of knowledge*. Chicago: University of Chicago Press.

Hartley, C. B. (1875). *The gentlemen's book of etiquette, and manual of politeness: Being a complete guide for a gentleman's conduct in all his relations towards society... From the Best French, English, and American authorities*. JS Locke & Company.

Haws, D. R. (2001). Ethics instruction in engineering education: A (mini) meta-analysis. *Journal of Engineering Education, 90*, 223–229.

Hess, J. L., & Fore, G. (2018). A systematic literature review of US engineering ethics interventions. *Science and Engineering Ethics, 24*, 551–583.

Layne, P. (2009, December). Why don't American women stay in engineering?. *PE Magazine 2009*. Retrieved January 5, 2024, from https://www.nspe.org/resources/pe-magazine/december-2009/why-dont-american-women-stay-engineering

Martin, D. A., Conlon, E., & Bowe, B. (2020, November). Exploring the curricular content of engineering ethics education in Ireland. In *2020 IFEES world engineering education forum-global engineering deans council (WEEF-GEDC)* (pp. 1–5). IEEE.

Mitcham, C., & Englehardt, E. E. (2019). Ethics across the curriculum: Prospects for broader (and deeper) teaching and learning in research and engineering ethics. *Science and Engineering Ethics*, *25*, 1735–1762.
Moriarty, G. (2001). Three kinds of ethics for three kinds of engineering. *IEEE Technology and Society Magazine*, *20*, 31–38.
National Council of Examiners for Engineering and Surveying. (2021). *Model laws*. Retrieved January 5, 2024, from https://ncees.org/wp-content/uploads/Model_Law_2021_web-2.pdf
National Council of Examiners for Engineering and Surveying. (2022a). *Manual of policy and position statements*. Retrieved January 5, 2024, from https://ncees.org/wp-content/uploads/2023/03/Policy-manual_2022_web.pdf
National Council of Examiners for Engineering and Surveying. (2022b). *Model rules*. Retrieved January 5, 2024, from https://ncees.org/wp-content/uploads/2023/04/Model_Rules_2022_web.pdf
National Society of Black Engineers. (2023). *About Us*. Retrieved January 5, 2024, from https://www.nsbe.org/about-us
Professional Engineers Ontario. (2022). *PEO turns 100*. Retrieved January 5, 2024, from https://peo.on.ca/sites/default/files/2022-06/PEOturns100-ED.pdf
Professional Engineers Ontario. (2023). *Code of ethics*. Retrieved January 5, 2024, from https://www.peo.on.ca/licence-holders/code-ethics
Professional Engineers Ontario. (n.d.). *National Professional Practice Exam (NPPE)*. Retrieved July 29, 2024, from https://www.peo.on.ca/apply/become-professional-engineer/national-professional-practice-exam.
Rottmann, C., & Reeve, D. (2020). Equity as rebar: Bridging the micro/macro divide in engineering ethics education. *Canadian Journal of Science, Mathematics and Technology Education*, *20*, 146–165.
Society of Women Engineers. (2024). About SWE. Retrieved July 29, 2024, from https://swe.org/about-swe/
The Twelve Tables (449 BC). Retrieved January 5, 2024, from https://avalon.law.yale.edu/ancient/twelve_tables.asp
Vesilind, P. A. (1995). Evolution of the American society of civil engineers code of ethics. *Journal of Professional Issues in Engineering Education and Practice*, *121*, 4–10.
Walpole, B. (2020, November 10). ASCE installs new code of ethics. *Civil Engineering Source Newsletter*. Retrieved January 5, 2024, from https://www.asce.org/publications-and-news/civil-engineering-source/article/2020/11/10/asce-installs-new-code-of-ethics
Westcott, D. N. (1982). Blacks in the 1970's: Did they scale the job ladder. *Monthly Labor Review*, *105*, *29*.
Wickenden, W. E. (1929). *A comparative study of engineering education in the United States and in Europe*. Lancaster Press.
Wisely, W. H. (1977). Professional turning points in civil engineering history. *Civil Engineering—ASCE*, *47*, 137–140.

6

A POST-NORMAL ENVIRONMENT-CENTERED APPROACH TO ENGINEERING ETHICS EDUCATION

Tom Børsen, Shannon Chance, and Gaston Meskens

Introduction

What do we mean when we use the term 'ethics' when discussing engineering? Being 'ethical' implies applying well-reasoned values and morals, and today, concepts like sustainability, equity, and diversity are increasingly associated with ethics in engineering as well (Committee on Education, 2019). Ethics extends beyond professional codes that specify what one must do as part of the engineering profession to include what one should do as a responsible and moral person (Chance et al., 2021). Developing the ability of engineers to apply ethical judgment when facing ethical dilemmas necessitates providing future engineers with education in ethics to support their moral development (see Chapter 10) and reflective practice (see Chapter 25). Engineers often associate the term 'ethics' with workplace health and safety, but today's complex environmental challenges imply embracing a broader view of health and safety to encompass the well-being of our planet and all its constituents, living and nonliving. The term 'global responsibility' is promoted by the United Nations (UN) to capture this expansive understanding of ethics. This chapter discusses how to foster global responsibility among engineers (including future engineers) and shift how they think and behave collectively and as individuals. Definitions (as fuzzy as some of them may be) are necessary for facilitating dialogue, and this chapter seeks to identify and define key terms relevant to moving forward the dialogue on what constitutes ethical engineering and how to achieve it.

In 1828, Thomas Tredgold characterized civil engineering as "the art of directing the great sources of power in nature for the use and convenience of man" (Alder, 2022, p. 2). This perspective asserts that engineers serve society by harnessing natural resources, and it undergirds many engineering sectors even today. Even in contemporary times, environmental engineering is often viewed as "improving ecological conditions," mainly to make surroundings "more suitable for humans to live" (Joshi, 2021, ¶3). However, modern leaders, like those at the United Kingdom's Institution of Civil Engineers (ICE), acknowledge "the detrimental effect that the industrial scale development which started with the Industrial Revolution [has] on our planet" (Alder, ¶3). We argue that although our early ancestors had to live with the pace of nature and struggle to circumvent its vagaries, humans today 'engineer' the natural environment to an unhealthy and unsustain-

DOI: 10.4324/9781003464259-8

able extreme. We question humans' attempts to control nature, particularly through engineering. We argue that the balance between serving humans and respecting other species, ecosystems, habitats, and so on constitutes an ethical dilemma that must be addressed.

Humans and the engineers serving them have extracted, exploited, rerouted, canalized, and otherwise 'modified' nature, ring-fencing the most dramatic features into encapsulated parks and 'natural reserves' but covering much of the rest with asphalt, concrete, brownfields, and contaminated wastelands. Now, confronted with urban heat islands, extreme weather, pollution, resource shortages, loss of biodiversity, and climate change, some propose 'extreme engineering' that is highly ambitious and employs unconventional engineering practices to address complex and severe challenges. 'Extreme engineering' practices are characterized by their aggressive groundbreaking approaches, high technological innovation, and high potential for impact and risk. Yes, the urgency to realign our relationship with nature is more pronounced than ever. Still, these extreme engineering methods are subject to growing criticism based on ethics and the risks of implementing such large-scale interventions. *How might we better respond? How can engineers achieve 'global responsibility' to people and the planet, including the non-human?*

What we need to move away from is the extractive cradle-to-grave capitalist model for producing and monetizing engineered products at the expense of the planet, including the environment, other people, and the non-human. The 'Anthropocene' is frequently invoked when questioning the prevailing extractive and human-first mindset. The National Geographic Society (2023) defines the Anthropocene as a distinct epoch "during which human activities have impacted the environment enough to constitute a distinct geological change" (¶3), exerting an overwhelming influence on the Earth's climate, geology, and ecosystems.

In this conceptual chapter, we use reflexivity – the quality of a dialogical approach to tackling complex societal problems – to assess existing normative practices and propose a framework for moving beyond them. We do this in response to the current ecological crisis; we call on engineers to help define and forward a paradigm shift (Kuhn, 1962) to transition from extractive practices and mindsets to more humble, healthy, and sustainable ones. Overall, we use three lenses: (1) reflexivity, (2) post-normal science (PNS) underpinning post-normal engineering (PNE), and (3) environmentalism.

We draw inspiration from the global conversation questioning the current status quo and calling for new and different responses, particularly the call for 'post-normal science' (Funtowicz & Ravetz, 1993). The conversation calls for a shift from the 'normal' way of doing things to a more refined 'post-normal' way of thinking and being – a concept we find helpful for repositioning engineering. Historically, societies and their groups of practitioners and thinkers (like architects and philosophers) have periodically transitioned away from paradigms once the mindset has become the status quo or normative enough to be named (e.g., modernism and structuralism, discussed by architects and philosophers alike). Critics from various groups reacted to and pushed against the boundaries of their time's existing normative ways, the status quo, to set forth via new paradigms (e.g., post-modernism and post-structuralism, respectively). They accomplished this using reflective thinking, dialogue, and rigorous debate. We experience the results of paradigm shifts when we observe paintings and sculptures, dwell in architecture, or read literature, poetry, and philosophy that integrate and seek to manifest the new mindset. Post-modernism and post-structuralism have been expressed in all these realms, and this chapter calls for engineers to embrace the emerging new post-normal paradigm and express it in their work.

We explore the idea that engineering requires a more evolved post-normal perspective regarding its role and potential. The solution isn't merely about improved models, technologies, or algorithms; it involves a collective view of engineering as an endeavor to address urgent political issues rooted in ethical and holistic thinking, transdisciplinarity, global accountability, and public participation. We contend that engineering can fully realize this aspiration only after its educational foundation is reshaped using these values. If the goal is to engineer responsible solutions to societal challenges, then policy must foster this enhanced form of education within the engineering community.

Ecological crises are exemplary for enacting post-normal approaches (such as PNS and PNE). These approaches are also relevant for addressing other crises, such as pandemics and inequality. Thus, the relationship between PNS/PNE and the environment is that the environment, with its different crises (e.g., the climate crisis or the biodiversity crisis), is used to explore aspects of PNS/PNE and the ethical frameworks that inform these practices.

This chapter aims to help bridge two realms (ethics and environmentalism), drawing from environment-centered ethical frameworks, to foster a new way of thinking about engineering. We discuss typical engineering values and practices and question what 'responsible engineering' means today. We propose a response called 'post-normal engineering' and reflect upon a range of existing normative theories, identifying some pros and cons of each approach and then proposing how engineers, engineering teachers, and future engineers might respond in a more effective post-normal way.

Positionality

The unique perspectives of each author on our team have steered the direction and scope of this chapter. We all have a foundation in design and engineering, specifically focusing on the built environment. Tom teaches technology ethics, technology assessment, public/user engagement in science, technology, engineering, and mathematics (STEM), and techno-anthropology within a department of sustainability and planning. His research emphasizes the integration of ethical judgment and participatory methods in STEM practices and education. Shannon teaches students architecture, engineering, and educational planning, emphasizing ecological principles. She advocates using site-specific, culture-specific methodologies. Gaston, a researcher in moral philosophy and science and technology studies (STS), is a philosophical activist and founder of the New Humanism Project. Our mutual interest in PNS and the desire to adopt more transformative approaches to tackle global challenges brought us together as co-authors and inspired this chapter's creation.

A post-normal approach for engineering

As we aim to propose the concept of reflexivity to rethink the 'normal' practice of engineering and its education system away from destructive and extractive practices, we believe that PNS provides inspiration and clues for how to do this. In 1993, Funtowicz and Ravetz introduced the concept of PNS as an evolved form of expertise, mainly designed for advising policy-makers during times when "facts are uncertain, values in dispute, stakes high and decisions urgent" (Funtowicz & Ravetz, 1993, p. 10). This vision now aligns with the ongoing transformation in how technoscience and engineering are perceived. Central to these evolving approaches is the realization that experts grapple with numerous uncertainties and value-laden viewpoints when shaping policies on intricate sociotechnical and ecological matters. These experts are under constant pressure from the political, public, and economic sectors to provide solutions that span multiple areas – for example, climate change, the COVID-19 response, large dam constructions, and genetically modified organism (GMO) policy-making. Consequently, engineering perspectives are transitioning. There's a growing understanding that the Anthropocene's multifaceted challenges can't be addressed with the same mindset that initially led to them; mechanizing solutions for every emerging issue isn't viable.

When introducing the idea of post-normality in 1993, Funtowicz and Ravetz (1993) identified four problem-solving strategies: core science, applied science, professional consultancy, and post-normal science. Each of these four types of scholarly activity has its equivalent in engineering. Core science in various domains is the foundation of engineering. Applied science and professional consultancy are well-described engineering practices. Traditional engineering, typified by its reliance on applied science and problem-solving, addresses routine challenges using established methods. However, these standard solutions and tools fall short regarding more intricate and unpredictable issues.

In the original writings on PNS, engineering was identified as applied science and professional consultancy. Engineering implies 'applied science' in the sense that it applies (natural) scientific theories and laws under controlled circumstances in developing new technological artifacts that can be used to make life easier for its target groups. Biotechnology and software engineering are examples of this type of engineering. Engineering can also involve 'problem-solving' where engineers address societal problems. Examples are engineering infrastructure projects (e.g., introducing central heating in major cities or constructing railway systems to connect a country or countries). This form of engineering also requires control over the context in which infrastructure is set.

Jerry Ravetz (2006) and Tom Børsen (2015) have linked PNS to technological risks and explored how the PNS framework can be applied to understanding, assessing, and managing the risks associated with technology. They argue that a broader, more inclusive approach is needed for complex and high-stakes technological issues than the approach provided within traditional scientific methodologies. Fanny Verrax (2017) also referenced PNE in a paper in *Futures*, calling experts to rethink the 'normal' engineering identity. We follow this route in part as we are concerned with how engineering can address urgent policy issues related to the environment where facts are uncertain, stakes are high, and values are in dispute. We do not perceive 'engineering' as only an applied science and client-serving consultancy (i.e., engineering must serve a good greater than the funder's request). PNE is engineering that effectively responds to post-normal times (Sardar, 2010). PNE is not (yet) defined; thus, it is one of the quests of this chapter to describe this.

At this point, we want to emphasize that other traditional and PNE practices hold value and will remain relevant. Yet, we argue that more is needed to address the current climate and environmental crises than relying solely on applied science or conventional problem-solving methodologies. Although we see immense value in the engineering professions, we also ask how engineers, as individuals and as members of professional collectives/organizations, can better tackle significant environmental challenges.

PNS develops and presents science-based advice to policy- and other decision-makers when trying to address crises through policy measures. Post-normal science-based advice portrays uncertainties at different levels – empirical, methodological, theoretical, institutional, legal, ethical, and so on (Benessia & De Marchi, 2017) – and manages conflicting stakes and ethical dilemmas through establishing extended peer communities (Meisch et al., 2022), honest brokery (Pielke Jr, 2007), and quantitative storytelling (Saltelli & Giampietro, 2017). PNE differs from PNS as PNE practitioners do not (only) provide advice; they address post-normal crises by developing sociotechnical solutions and strategies.

The ethical landscape of post-normal engineering in the Anthropocene

Delving into the 'ethical landscape' of PNE, we are inspired by a critical perspective on our current coexistence in the Anthropocene. Paul Crutzen, an atmospheric chemist, was the first to coin the term 'the Anthropocene' to describe the epoch where human actions are the dominant force impacting Earth's geology, climate, and ecosystems. Yet, the foundational beliefs and values of PNE differ from the conceptual notions associated with the Anthropocene. Crutzen (2006) proposed technical solutions, specifically geoengineering methods such as releasing sulfur compounds into the atmosphere to mitigate the sun's heat. In our view, such a proposal doesn't resonate with the post-normal emphasis on humility. It seems to overlook the potential unexpected consequences of such interventions. For instance, geoengineering might be a plausible reply if society runs out of options. However, implementing such a grand plan will require pervasive reflection, enormous assessment of unexpected consequences, and extensive discussion regarding which ethical frameworks are appropriate to consult before action can be taken. We advocate referencing PNE when discussing pressing contemporary issues like climate change. These challenges require a collaborative approach among diverse stakeholders, emphasizing humility and accountability. Addressing the intricacies of problem-solving in the Anthropocene harmonizes with the call from Jonas (1984) to prioritize the sustainability of future conditions. Jonas argued that potential negative outcomes should be given more weight than positive projections in ethical considerations.

The literature on PNS and STS provides concepts and tools to manage uncertainty, for example, the Numeral, Unit, Spread, Assessment, and Pedigree (NUSAP) approach to uncertainty (Funtowicz & Ravetz, 1990; van der Sluijs et al., 2005) and stakeholder controversies (e.g., Social Construction of Technology (SCOT)). Regarding the 'ethical values in dispute' part of the PNS one-liner (coined by Funtowicz and Ravetz and frequently repeated), 'when facts are uncertain, stakes high, values in dispute, and decisions urgent,' the literature provides little to go on, although a forthcoming special issue of *Futures* promises an investigation of relationships between 'Post-normal Science and Ethics' (Børsen & Meskens, under review).

Although we cannot and should not completely abandon the anthropocene perspective, we believe that responsible engineering originates from a deep understanding of our global challenges and an acute awareness of their ethical ramifications. This shapes how we deliberate and execute solutions to benefit present and future generations, human and non-human. The following sections identify a broad palette of ethical theories that engineers can choose from and combine when engaging with urgent political issues. We argue that there is not one ethical framework that engineers can apply in isolation to post-normal problem-solving. The engineer must reflect and discuss with self and others what ethical frameworks fit for individual (yet often complex and overlapping) issues.

Helpful ethical frameworks are covered in detail elsewhere in this handbook (e.g., virtue ethics, deontology, utilitarianism, and the common good; see Chapter 2) and continue to be relevant in PNE. Other ethical frameworks, such as the Golden Rule (do unto others as you would have them do unto you) and the Fairness approach defined by Rawls (1971), are relevant in PNE. The Fairness approach posits that a just society is one where principles are selected impartially and without bias, following two primary tenets: *basic liberties for all* and *the difference principle*, which permits inequitable responses only if they benefit the least advantaged members.

Foundational ethical perspectives like deep ecology, sustainability, and land ethics (described later in this chapter) recognize environmental systems' intricate and interconnected nature, emphasizing the importance of considering the broader ecological community in our actions and decisions. No engineering solution can fully encompass every facet of this intricate context. Invariably, certain elements will remain external to the system addressed by any solution proposed. The environment's components are intertwined, forming a holistic web where interventions in one segment inevitably impact others. Including environmental ethical frameworks in the palette of ethical frameworks for PNE is extremely important, so we describe them here. We note that Chapters 11 and 15 also provide helpful guidance for readers interested in environmental topics.

Some professions favor specific frameworks over others, but their preferred approaches may only address some of the profession's dilemmas. Engineers, policy-makers, and practitioners in many fields must learn a wider array of ethical approaches and learn when to integrate them based on the given contexts; given today's post-normal complexity, they cannot rely solely on the rules of thumb favored in their professions.

For us, PNE is characterized by a higher degree of complexity than other forms or strategies of problem-solving. PNE is embedded in a more contradictory field of interests and stakeholders than other forms of problem-solving. We are in a post-normal context where policy-makers cannot look towards normal engineering to provide adequate response.

Reflexivity

Reflexivity is a vital ethical virtue for engineers in this complex age. Traditional ways of thinking and working are evolving, and with this evolution comes a need for engineers to be deeply introspective and outwardly attuned, based on an awareness of the context in which they operate and of the values and beliefs that drive them to do what they do. We propose to understand reflexivity as this kind of awareness. If we imagine the complexity of an environmental problem as making an ethical appeal to us to deal with it 'fairly,' then we can understand reflexivity – in response to that appeal – as an *ethical attitude*, being critically aware of our own position, interests, hopes, hypotheses, beliefs, and concerns (Meskens, 2017). Chapter 25 discusses possible approaches to teaching and practicing reflexivity. It suggests that dialogue with yourself and others is the basis of reflexivity and that the dialogue should ask critical questions like: *What is the problem we face? In what way(s) is it complex? Should we do, or have done, something else? What might we be overlooking? How could we improve?* Chapters 35 and 36 ask these types of critical questions regarding the role of ethics in engineering accreditation. Chapter 31 critically probes assessment practices, confronting assumptions and biases about behavior and culture.

Important to understand is that reflexivity as an ethical attitude emerges in dialogue with others, a dialogue that – by its very form and method – is emancipatory and (respectfully) confrontational simultaneously (Meskens, 2017). It connects engineers with different views and meaningful ethical frameworks and nourishes their competence. From this perspective, we can also understand why and how reflexivity grounds other values, like precaution, transparency, accountability, protection and empowerment of the weak, and even sustainability. This kind of dialogue would stimulate sensitivity to these ethical values among all concerned and consequently enable meaningful interpretation. In this sense, it would also become an 'authoritative place' where these values could be applied as principles to inspire and steer (engineering) policy (Meskens, 2018). Dialogue with others should always involve parallel individual contemplation (this idea has religious underpinnings from, e.g., Saint Thomas of Aquinas). On the other hand, individual contemplation could and should inform and be informed by interaction with others.

We argue for rooting extended discussion of ethical and sustainable (a.k.a. post-normal) engineering in the community. This requires engaging with stakeholders. Public participation is necessary to ensure benefit to the more vulnerable. This begs the question of how to set up an extended peer community to reflect and act – to form a new and improved paradigm, advocate for change, and integrate its tenets in thought and deed. The question will be contemplated in a subsequent section, where we discuss tools for fostering public dialogue.

Ethical approaches – an overview

We now outline the ethical terrain of environment-focused engineering to explore the prospects of post-normal thinking. Engaging with classical anthropocentric (human-first) theories and newer environment-centric approaches can illuminate the engineering community's diverse and sometimes clashing beliefs. Here, we highlight ethical concerns that, in our view, could guide ethical engineering practice and education – especially if we intend to conquer contemporary environmental challenges and crises. We believe it is essential to understand all the tools we currently have for enacting global responsibility and addressing today's challenges.

Therefore, we open the toolbox of prevailing ethical approaches, considering what they offer, identifying some of their shortcomings, and suggesting how they might be integrated into the new 'post-normal' paradigm for engineering. This curated set of existing environment-focused lenses can support (future and current) engineers and engineering educators in cultivating and advocating for environmentally and socially considerate practices, policies, and mindsets. Note that the concepts we highlight are all open to interpretation. They should be topics of dialogue themselves (within political, academic, policy-making, and other professional circles, including engineering) to unveil different interpretations and interests and discern the positions of various actors. First, we apply reflexivity in identifying and briefly defining *core ethical virtues* and *procedural ethical values* that will remain valuable. Here, we use terminology proposed in the publication *The Ethical Foundations of the System of Radiological Protection* from the International Commission on Radiological Protection (2018). Then, we shift the discussion toward *responsible complexity management strategies* from business and economics. We identify *other environment-centered ethical frameworks* that provide a foundation for moving engineering ethics (and) education forward. Finally, we look at some environmentally important *policy issues*, reflect upon frameworks for *fostering public dialogue*, and consider how we can put them to work with PNE.

Core ethical virtues and procedural ethical values

Moving forward will require using many widely recognized *core ethical virtues*. These include the principles of *beneficence* and *non-maleficence*, which require doing good and avoiding harm. *Prudence* (wise and judicious decision-making) and *respect for dignity* (which values intrinsic worth) will also be important concepts to bring forward. Likewise, *openness* and *tolerance* are essential concepts, so the community of reflexive thinkers/engineers will welcome varied knowledge and opinions. *Procedural ethical values* like *accountability* (owning responsibilities and outcomes), *transparency* (ensuring clarity and openness), and *inclusiveness* (valuing and supporting diverse participants, with a particular focus on those potentially affected by engineering practices) must also be retained.

Responsible complexity management strategies

Responsible complexity management strategies that can inform PNE include *global responsibility* (upholding duties beyond borders, considering local and global impacts of our decisions and our profession), *intergenerational ethics* (considering possible consequences for future generations), and *holism* (embracing the interconnectedness of all things). *Inter-* and *transdisciplinarity* can help us break silos to achieve more holistic solutions. *Action research* provides iterative approaches to learning from experience and refining practice over time based on real-world learning and application of research. Ideals of *cosmopolitanism* can help us cultivate self-critical world-citizen perspectives (Meskens, 2022).

Many frameworks have been offered to help make activities in our current economic model more environmentally and socially sustainable. These include *corporate social responsibility* (CSR), a model for self-regulating practices within businesses and organizations (including the business side of higher education) to ensure organizations are socially accountable to themselves, their stakeholders, and the public. By practicing CSR, an organization can become more conscious of its impact on society's economic, social, and environmental realms. Via CSR, companies aim to contribute positively, often by adopting sustainable practices, engaging in philanthropy and ethical labor practices, and reducing their ecological footprint. Yet, for CSR to be effective, it needs more stringent standards, greater transparency, increased integration into core business strategies, and long-term commitment to genuine change (Christensen et al., 2021). It is largely ineffective because it is voluntary, because businesses often prioritize short-term gains and shareholder returns above long-term sustainability goals, and because CSR is frequently treated as a peripheral activity rather than a core business strategy. Global supply chains are incredibly complex and challenging to regulate, and the lack of standardization in defining CSR makes it difficult to measure and compare effectiveness. Quantifying social and environmental impact is ill-defined, and it is hard to evaluate efficacy without clear metrics. Moreover, some CSR efforts only address specific areas of concern, neglecting other important aspects of social and environmental responsibility (Scherer, 2018). Unfortunately, many companies engage in 'greenwashing' where they exaggerate or falsely claim to benefit the environment, and such abuse leads to skepticism and distrust – undermining the credibility of CSR.

The *doughnut model of economics* (Raworth, 2012) presents a framework for sustainable development, aiming to support essential human needs within Earth's ecological limits. It visualizes an ideal zone (shaped like a doughnut) that avoids both deprivation and ecological overshoot.

The *circular economy* is a significant and influential approach driven by sustainable development and resource efficiency principles. It proposes an alternative to the traditional linear economy (that follows the 'take, make, dispose' or 'cradle to grave' model) and encompasses sustainability principles, resource efficiency, and waste reduction. Proponents advocate for a closed-loop system where resources are reused, repaired, refurbished, and fully recycled. The circular economy is being implemented and practiced in various industries and by policy-makers worldwide. It provides practical strategies, demonstrating how economic activities can be restructured to minimize waste and negative environmental impact while maximizing resource efficiency. The concept has been shaped by, for example, Ken Webster and the Ellen MacArthur Foundation, Walter Stahel and the Product-Life Institute, 'Cradle-to-Cradle' concepts from William McDonough and Michael Braungart, and the European Union's Circular Economy Action Plan (part of the European Green Deal).

Another vital contribution in this realm is the *blue economy*, proposed by Gunter Pauli to complement the circular economy with solutions inspired by nature and emphasizing the sustainable use of local resources.

An investigation of economic models can only be completed by looking at concepts and tools designed with engineers and designers specifically in mind. Prominent among these are the cradle-to-cradle (C2C) design principles mentioned above (McDonough & Braungart, 2002), which seek to balance economic, environmental, and social concerns. Treating waste as a resource for another cycle is central to C2C. In contrast to recycling, which can diminish quality and introduce additional pollutants, upcycling seeks to enhance an item's value. Initial designs must be crafted to support ongoing use in various new forms. Design begins with careful material selection, avoiding 'X list' materials detrimental to humans and the environment, seeking substitutes for 'gray list' items (those presently indispensable but problematic), and always giving preference to safe, sustainable materials (from McDonough and Braungart's 'P list' of positive and healthy substances). Regenerative design minimizes harm and actively feeds and enriches the local environment. C2C designs aim to revitalize ecosystems, enhance biodiversity, and champion local communities, kindling synergies between development and ecology. Although we recognize C2C as a business-oriented approach grounded in the capitalist economy, we believe it provides some valuable concepts for students, designers, and policy-makers.

Janine Benyus introduced *biomimicry*, which is related to the C2C approach. *Biomimicry* focuses on innovation inspired by natural processes and biological systems, encouraging designers and engineers to create products and solutions that emulate nature's patterns. The approach supports sustainability, encourages a symbiotic relationship between human development and the environment, and fosters a deeper appreciation of the natural world.

Reflecting on why the techniques identified above haven't worked and why companies and institutions of higher education don't already achieve sustainability using them, we cite their voluntary nature and the ongoing hold of capitalist ideals acutely evident in the business of engineering and the built environment. There's more to the story, though. There's also a pervasive detachment from nature and a sense of technicality or instrumentality that philosophers of STS call the technical frame. Changing extractive ways of thinking firmly rooted in the engineering profession is very hard. Big organizations, including academic ones, are known for high resistance to change.

Empson et al. (2019) recommend a less opt-in approach, arguing that, in post-normal times, no design activity should be considered 'creative' that is not deeply sustainable. Nevertheless, we still see praise doled out by prize-awarding organizations for projects that lack sustainability or effectively constitute greenwashing.

Ecological economics denotes an interdisciplinary research area that advocates for an equal exchange between humans and nature. This means that when humans take from nature, they must give something back. That is the foundation for economic exchange. In a 1994 paper published in *Ecological Economics*, Funtowicz and Ravetz asked about "the worth of a songbird" (Funtowicz & Ravetz, 1994, p. 197). They argued that ecological economics requires a PNS to address the dilemma of "setting a monetary value on an irreplaceable songbird [which] forces us to be clear about what is being valued, how it is done, and indeed, what value is" (p. 198). There are no certain answers to this question. Stakes are high, and ethical values are in dispute.

Other environment-centered ethical frameworks

Looking at *other environment-centered ethical frameworks*, we draw an arc from *sustainability* (aiming for longevity and balance) through *deep ecology* (recognizing non-human entities) to *land ethics* (valuing the sanctity of the land), *material ethics* (emphasizing regenerative practices), and *values embedded* within the things we create.

Sustainability

Sustainability is a core concept. In post-normal times, living in harmony with nature is a fitting response to the conventional notion of controlling nature. The United Nations General Assembly (2022) has now stated that "to achieve a just balance among the economic, social and environmental needs of present and future generations, it is necessary to promote harmony with nature" (p. 2), but this seemingly prioritizes human needs, a flaw we see. Drawing from the ancient Greek philosopher Marcus Aurelius' (2002) insights that "all things come to their fulfillment as the one universal Nature directs" (Marcus Aurelius, 2002, Book VI, statement 9), we see a nuanced relationship: nature continues to guide the definition of harmony, but it now assumes the humble and vulnerable stance traditionally ascribed to humans.

The United Nations (UN) emphasis on living harmoniously with nature has been in its documents since at least the 1980s. The 2009 UN General Assembly Resolution 64/196 references the 1980 Resolution 35/7, the *Draft World Charter for Nature*, highlighting the dependency of life on nature's continuous processes and the dangers of excessive exploitation (United Nations General Assembly, 1980). The UN doesn't suggest reverting 'back to nature.' Instead, it advocates a balance among human economic, social, and environmental needs, aligning with the definition of sustainable development provided in the Brundtland Report (World Commission on Environment and Development, 1987). Here again, human needs have been prioritized.

Practical tools to support sustainability include *carbon calculators* (Wackernaegle & Rees, 1996) and the UN *Sustainable Development Goals* (SDGs). From a critical perspective, Seniuk Cicek et al. (2023) contend that many methods, especially those appealing to engineering mindsets like the SDGs, favor the Global North. They originate in values defined by the Global North. Moreover, when organizations in wealthy countries work toward high-level goals without drilling down into specific targets – referencing just the overarching SDG titles like *quality education* (SDG 4), *gender equality* (SDG 5), *decent work and economic growth* (SDG 8), and *industry, innovation, and infrastructure* (SDG 9) – their efforts can help to raise the standard of living locally without doing anything to help the Global South. Gains in rich countries can exacerbate inequalities between rich and poor countries. In response, Ochoa-Duarte and Peña-Reyes (2020) champion the concept of *Buen Vivir*, which, as described by Seniuk Cicek et al., is "anchored in Latin American principles and emphasizes biocentrism, postcapitalism, decolonialism, and depatriarchalization" (p. 55–56), presenting it as an alternative to address disparities they see within the SDG approach. *Buen Vivir* is also discussed in several chapters of this handbook (see Chapters 1, 8, 9, and 15).

Regarding learning and teaching sustainability in subjects including engineering, the UN Educational, Scientific and Cultural Organization (UNESCO) identified eight competencies that all students need to develop. The sustainability-related competencies are systems thinking, anticipatory, normative, strategic, collaboration, critical thinking, self-awareness, and integrated problem-solving (Didham, 2018), and they resonate with our idea of fostering reflexivity through dialogue.

Deep ecology

The term *deep ecology* was coined by the Norwegian philosopher Arne Næss (1973). It is a philosophical and ethical approach to environmentalism that emphasizes the inherent worth of all beings, regardless of their utility to human needs. Næss argued that the prevailing approach to environmental problems was too shallow, focusing on pollution and resource depletion concerning their impacts on humans. In contrast, he argued for a 'deeper' approach that recognizes the fundamental interconnectedness of all life. Ethical frameworks for post-normal times need to reflect the interconnectedness of nature more holistically.

Around the same time, Ian McHarg (1999) criticized the assumed superiority of modern humans, noting that (hu)man's presumed supremacy "lies in the inheritance of tools, information and powers from his predecessors" (p. 287). McHarg proclaimed the value of 'primitive' societies, promoting ideas of pantheists and animists that "the entire world contains godlike attributes: the relations of man to this world are sacramental. … the actions of humans in nature can affect their own fate; these actions are consequential, immediate, and relevant to life. There is, in this relationship no non-nature category" (p. 287). Hunter-gathers recognized and honored seasons and maintained balance with nature; they honored and revered the prey that sustained their lives. People living this way "could promise their children the inheritance of a physical environment at least as good as had been inherited – a claim few of us can make today" (p. 288).

Consistent with this approach, *deep ecology* asserts that humans aren't superior to other life forms. Instead of conserving the environment solely for human advantage, *deep ecology* promotes biocentric equality – valuing every living entity, from microorganisms to large mammals, for its inherent right to exist and thrive. Resonating with Marcus Aurelius' views, this perspective emphasizes that humans are just one part of the broader web of life.

Land ethics

Humans have exploited Earth's land and its constituent components, plants, and animals. Capitalist systems and economic foci have exacerbated this exploitation, but Western societies' one-way approaches to land and earth also have religious roots. Religious texts seemingly grant humans the absolute right to dominate over plants, animals, and land. Probing our languished, or absent, set of *land ethics*, brings us back to McHarg's scathing critique titled *On Values*. McHarg (1999) argued that the pronouncement in Genesis of man as "exclusively divine, given dominion over all life and non-life, enjoined to subdue the earth" (p. 288) set the tone for calamity. Islam, Judaism, and Christianity all inferred from Genesis values regarding how humans should relate to nature. In the past, Islam saw humans as stewards, entrusted to "make paradise on earth, make the desert bloom" (p. 296). In contrast, Judaism and Christianity leaned towards conquest. When the "medieval Christian Church introduced otherworldliness" (p. 296), it deepened the human perception of Earth as dangerous and impure (think of the paintings by Hieronymus Bosch of worldly, carnal sins, for example). The Western world, particularly in its more modern form as the Global North, has often viewed nature as a "crude, vile, lapsed paradise" (p. 296) and sought to conquer it. Although the West has made big achievements in social equality, McHarg acknowledges, as for the land, "nothing has changed" (p. 296). An ultimate expression of this exploitation was the urban landscape of the United States, which McHarg described as "the ransacking of the world's last great cornucopia [and] the largest, most inhumane, and ugliest cities ever made by man." This he saw as a clear example of "profound ignorance, disdain, and carelessness" (p. 298). This indictment targets architects, engineers, and financiers and helps explain why the activities of well-intentioned and often 'god-fearing' people have resulted in such low levels of sustainability.

Material ethics

As we contemplate the ethics of land use and our relationship with materials like rocks, minerals, and plant-derived resources, it is clear that we have viewed them as resources for extraction and consumption without giving much thought to long-term repercussions. We must instill a new code of *material ethics* that acknowledges the limited nature of our planet's bounty. Implementing this code across architecture, engineering, and construction sectors could drive a paradigm shift. Instead of viewing materials as endless supplies, we'd understand their limited nature and the broader implications of our cradle-to-grave consumption patterns. Adopting a lifecycle perspective would prioritize regeneration and prohibit depletion.

Engineers make key decisions regarding the selection and use of materials. They have a moral obligation to consider the implications of their choices regarding extraction. A shared code of *material ethics* would require engineers to consider not just the functional properties of materials but also their environmental impacts and the social implications of their mining, processing, and disposal. Tony Fry (1999, 2009) highlighted the role of designers and technical design in today's unsustainable world and recommended a rethought design practice that finds inspiration in intercultural, interdisciplinary, and transdisciplinary creation. Fry stated that technical design is vital for human and ecological development and observed that the dominant technological frame threatens our future. It is 'defuturing.' Thus, the central move in addressing the Anthropocene is to develop a new relation to technology.

Values embedded

Pertaining to land and material ethics alike, designers must reflect on the *values embedded* in the structures and products they create and the messages their designed outputs convey. A case study by Chance and Cole (2015) illustrates how buildings can implicitly or explicitly communicate values to their occupants. When designed with purpose, buildings can instruct new generations and guide users to recognize or assimilate lessons about environmental care, collaborative work, preservation of natural habitats, and efficient utilization of natural resources such as wind, sun, vegetation, and rainwater. This design philosophy can also be applied to engineered products in addition to buildings and structures.

Policy issues

Policy-making is an integral part of changing behavior, and engineers and engineering academics should be involved in this process. This section highlights some policy-related issues that are highly relevant for PNE: *energy justice*, *energy democracy*, and *self-imposed engineering limits*.

Energy justice describes the fair and ethical distribution of energy and alleviates the currently unequal degradation (environmentally, ecologically, and socially) caused by energy extraction. Stephens (2021) highlights the historical and racial imbalances in energy use and the mounting adverse effects of energy extraction and combustion on marginalized communities. Although getting more people access to electric power is desirable to improve living conditions, Stephens asserts that simply scaling up existing systems will inadvertently maintain disparities. Furthermore, technical approaches to counteract global warming may have unintended consequences (Stephens et al., 2021). Stephens presents the term *energy democracy* to promote social equity during the shift to electrification based on renewable energy. Considering these broad impacts when shaping and debating policy is important, and engineers should be part of this dialogue (Stephens et al., 2021).

Looking closely at the engineers' role, Lawlor and Morley (2017) postulate the necessity for engineering professional bodies to set and adhere to *self-imposed engineering limits*, concerning, for instance, carbon emissions, especially in situations where the government fails to enact adequate regulations that can keep profit-prioritizing clients in check. Lawlor and Morley assert the urgency for immediate measures to assist engineers and design teams in counteracting environmentally (and socially) detrimental design briefs. Given the plethora of interests that engineers must navigate, these professionals need more tools and policies and more reflective practices to ensure higher levels of sustainability across development, artifacts, and production processes. Within the framework of PNE, these points are pivotal. Carbon isn't just an isolated metric. Its impact should be understood from a broader perspective, and other factors that influence climate in their own way (factors like methane and nitrous oxide, as well as deforestation) should be included. Such insights are vital as engineers seek to weigh benefits against trade-offs more effectively in complex, unpredictable scenarios.

Fostering public dialogue

It is important to note existing techniques for soliciting stakeholder feedback that engineering practitioners and teachers can use to help lead change. Essential strategies include *public participation* (engaging the public in project initiations and developments), *future thinking* (projecting and planning by envisioning a range of possible scenarios), and *participatory technology assessment* (collaborative evaluations of new technologies; for more on this, see Chapter 18). These existing tools must be part of education to help a broad and diverse array of stakeholders deeply understand and reflect upon issues. Internal and external dialogue must occur for these tools to be effective. The process can articulate a new vision and/or paradigm for a healthier, more sustainable future.

The values of humility and precaution

Given the enormous complexity of the issues identified above, will it be feasible for engineering to deliver the right solutions? We contend that yes, doing so will demand recalibrating engineering to resonate with post-normal times. This will shift engineering closer to decision-making, reaching beyond conventional problem-solving. PNE is tailored to offer knowledge-driven solutions to intricate and tumultuous political challenges by the very fact of its participatory approach, involving the *extend peer community*. This isn't to say that engineering solutions are the sole answer; rather, they must work in tandem with other solutions. Therefore, in post-normal times, PNE practitioners, policy-makers, and advocates should operate humbly without harboring a singular mindset, overemphasizing their solutions, or portraying their approaches as the only path forward. As suggested before, reflexivity is essential to *fostering humility*. *Practicing reflexivity* can help engineers and the engineering community (comprised of practitioners, teachers, and students) continually assess the broader context and their role. This also holds for the choice of ethical frameworks guiding PNE in concrete circumstances. All constituents must anticipate and mitigate potential negative consequences in an unpredictable environment. Although predicting these in every instance isn't feasible, implementing strategies for ongoing monitoring, early warning detection, and timely interventions is essential within PNE.

The *precautionary principle*, sometimes called the *principle of caution*, offers a guideline for handling uncertainties and potential risks. It argues that if an action (e.g., a policy, product, or behavior) could potentially harm individuals or the environment, especially when scientific consensus is absent, we should refrain from implementing it. Tracing its origins to the *safety culture* concept, which emerged in a 1987 report focused on preventing nuclear catastrophes, the term *safety culture* now describes an institutional ethos that prioritizes safety, embedding it in every aspect of operations. Yet, in post-normal times, more in-depth scrutiny is warranted. Delving into an organization's safety culture can benefit from Schein's (1992) layered framework, which examines basic assumptions, stated values, and tangible artifacts. This framework proposes that beyond examining the explicit or professed values of engineers and their affiliated institutions, evaluating the tangible products they produce regarding environmental safety is imperative. Furthermore, it's crucial to challenge foundational beliefs about our relationship with the world, such as the perceived divides between humans and nature or mind and body.

As mentioned above, this approach to envisioning involves engaging in *future thinking* or visualizing potential future scenarios. The process employs divergent thinking to embrace uncertainty and identify many possible solutions. Traditionally, engineers have been trained with an analytical mindset, focusing on deconstructing problems and addressing them straightforwardly and efficiently.

Precaution can involve avoiding potential problems. The *problem avoidance* approach aims to solve a given problem by looking beyond or 'upstream' of the immediate, to alter the larger system, and to prevent the problem from occurring in the first place. Engineers are typically comfortable with the idea of *problem avoidance* because many engineers are attuned to convergent thinking. Convergent thinking aims to find the 'correct' solutions and minimize uncertainties. It is more aligned with traditional engineering mindsets than the divergent thinking needed to brainstorm/project/envision a vast array of possible consequences and outcomes. Incorporating divergent thinking alongside problem avoidance can enrich engineering and engineering education, fostering a more comprehensive approach to addressing complex challenges and promoting ethical decision-making.

Remediating adverse environmental impacts

Engineers often deal with remediating adverse environmental impacts that have already happened. Post-normal engineers are among those concerned with remediation, yet existing practices for addressing current anthropogenic problems sometimes lack full recognition of the causes of the anthropocentric problems. Some curricula in engineering build an understanding of anthropocentric environmental impacts through assessments and measurements, as well as dealing with the consequences of such impacts. Still, they often have false dichotomies embedded in their underlying structure that reinforce superficial notions of separation between the material and the social – and between humans and non-humans (Hawkins et al., 2017). False dichotomies can lead engineers to view environmental problems as a reality independent of cultural and societal practices. Ethical frameworks in environmental education must consider perspectives that move away from separating nature and people and stop placing humans at the fore in most problem-solving. Responsibility, environment, and climate are transversal concerns that all the different types of engineers need to think about.

Undoubtedly, engineering practice and the products engineering produces have enormous effects in multiple realms. Environmental impact procedures exist and are part of official policy in many localities. Yet, there must be more assurance that ethical frameworks (named above and detailed in Chapter 2) inform the legally required assessments. We offer this chapter to provoke more (current and future) engineers to push further and question the bounds of engineering thinking to incorporate deep reflections of an ethical nature.

The value of reflexivity in education as a fundamental ethical attitude for PNE

Recent discussions have highlighted ethical values like *precaution*, *transparency*, *openness* to diverse knowledge and viewpoints, and *accountability* as essential guides for engineering practice. Reflecting on these values in the context of PNE, we wish to underscore *reflexivity* as a pivotal ethical approach for this era.

At its core, any ethics education seeks to cultivate a heightened sensitivity toward ethical dimensions of thinking, behaving, and decision-making. With this in mind, we envision PNE as a discipline that acknowledges the inherent complexities of its practice and actively reflects upon the values of itself and others. This reflection must encompass a broad spectrum, including individual and collective rationales, interests, aspirations, beliefs, and concerns tied to specific challenges.

Reflexivity can be the bedrock for precaution, transparency, openness, and accountability. By fostering this reflexive mindset during engineers' formative years, teachers can help (future) engineers better engage with and appreciate ethical perspectives such as *deep ecology*, *sustainability*, and *land ethics*. Reflexivity represents a foundational practice to support ethical competence in thinking and being. Reflexivity doesn't emerge in a vacuum, though; it is nurtured through dialogues that probe the ethical dimensions of engineering. Cultivating the ability to reflect upon one's and others' values starts with exposure to and deliberation upon a diverse array of ethical frameworks. These might range from traditional anthropocentric theories to the more environment-centric approaches detailed in this chapter.

Engaging in dialogue and informed discussions is indispensable in engineering ethics education. Such interactions bolster ethical competence and unveil diverse (and sometimes conflicting) beliefs and viewpoints prevalent among engineers. Engaged dialogue fosters a richer, more nuanced understanding of engineering ethics than traditional teacher – pupil lecturing can.

Conclusion

This chapter has introduced PNE as an approach for developing sociotechnical strategies and solutions to urgent complex problems. Such engineering practices must be *humble* – because their intended effects and broader implications are uncertain – and *reflexive* – as different perspectives and possible ways to address crises must be discussed and considered. When working on urgent complex problems, ethical dilemmas will occur, and engineers must be able to identify dilemmas and reflect on how to transcend them. The chapter has presented a selection of frameworks and concepts that might be relevant for PNE practitioners. The frameworks and concepts introduced here highlight different ethical concerns, some of which are neglected or at least treated briefly in engineering ethics education research. Although we have aimed to be comprehensive, the 'ideas and frameworks' presented above are merely a starting point. We invite readers to join this dialogue, building upon and refining these foundational concepts in the ever-evolving domain of engineering ethics.

Overall, we have advocated PNE as a reflexivity lens and sketched an emerging vision of what PNE might look like. Fleshing out and realizing this vision is, of course, a work in progress. It will require collective and reflexive effort from a community of diverse thinkers, engineering educators, and practitioners. Essentially, via this chapter, we have launched a call for participants to join the discussion on PNE and reflexivity and use these concepts to facilitate a marked change of direction – a new paradigm for thinking and being – that draws from yet reacts to today's 'normal' engineering practices. We call you to join our community, working toward a more rigorous and reflexive way of addressing global crises through engineering and design.

Acknowledgments

The authors thank the two scholars who reviewed our chapter, Mihály Héder and Per Fors, and our theme's editor, Thomas Taro Lennerfors, for their valuable input. We also take this opportunity to thank the fabulous community of editors and authors who contributed to this handbook and continually enrich our engineering education experience through the collaborative construction of new knowledge.

References

Alder, A. (2022). *Civil engineering is constantly evolving, and so must our institution*. ICE Community blog. https://www.ice.org.uk/news-insight/news-and-blogs/ice-blogs/ice-community-blog/civil-engineering-is-constantly-evolving-and-so-must-our-institution

Aurelius, M. (2002). *Meditations*. Reprint edition, from the original translation by Maxwell Staniforth for Penguin Books Ltd. 1964. The Folio Society.

Benessia, A., & De Marchi, B. (2017). When the earth shakes… and science with it. The management and communication of uncertainty in the L'Aquila earthquake. *Futures*, *91*, 35–45.

Børsen, T. (2015). Post-Normal techno-anthropology. *Techne: Research in Philosophy & Technology, 19(*2), 233–265.

Chance, S. M., & Cole, J. T. (2015). Enhancing building performance and environmental learning: A case study of virginia beach city public schools. In T. C. Chan (Ed.), *Marketing the green school: Form, function, and the future* (pp. 54–73). IGI Global.

Chance, S., Lawlor, R., Direito, I., & Mitchell, J. (2021). Above and beyond: Ethics and responsibility in civil engineering. *Australasian Journal of Engineering Education.* Special Issue: Ethics in Engineering Education and Practice. https://doi.org/10.1080/22054952.2021.1942767

Christensen, H. B., Hail, L., & Leuz, C. (2021). Mandatory CSR and sustainability reporting: Economic analysis and literature review. *Review of Accounting Studies*, *26*(3), 1176–1248.

Committee on Education. (2019). *Civil engineering body of knowledge for the 21st century: Preparing the civil engineer for the future* (3rd ed.). American Society of Civil Engineers. https://doi.org/10.1061/9780784415221.fm

Crutzen, P. J. (2006). Albedo enhancement by stratospheric sulfur injections: A contribution to resolve a policy dilemma?. *Climatic Change, 77*(3–4), 211.

Didham, R. J. (2018). *Education for sustainable development and the SDGs: Learning to act, learning to achieve*. Policy brief of the UNESCO Global Action Programme on Education for Sustainable Development.

Empson, T., Chance, S., & Patel, S. (2019). *A critical analysis of 'creativity' in sustainable production and design*. Proceedings of the 21st International Conference on Engineering and Product Design Education.

Fry, T. (1999). *A new design philosophy: An introduction to defuturing*. UNSW Press.

Fry, T. (2009). *Design futuring*. UNSW Press.

Funtowicz, S. O., & Ravetz, J. R. (1990). *Uncertainty and quality in science for policy*. Kluwer Academic Publishers.

Funtowicz, S. O., & Ravetz, J. R. (1994). The worth of a songbird: Ecological economics as a post-normal science. *Ecological Economics*, *10*(3), 197–207.

Funtowicz, S. O., & Ravetz, J. R. (1993). Science for the post-normal age. *Futures, 25*(7), 739–755.

Hawkins, B., Pye, A., & Correia, F. (2017). Boundary objects, power, and learning: The matter of developing sustainable practice in organizations. *Management Learning, 48*(3), 292–310. https://doi.org/10.1177/1350507616677199

International Commission on Radiological Protection. (2018). Ethical foundations of the system of radiological protection. ICRP Publication 138. *Annals of the ICRP*, *47*(1), 1–65.

Jonas, H. (1984). *The imperative of responsibility: In search of an ethics for the technological age*. University of Chicago Press.

Joshi, S. (2021). *Why is environmental engineering important for human health?* https://www.g2.com/articles/environmental-engineering.

Kuhn, T. S. (1962). *The structure of scientific revolutions*. University of Chicago Press.

Lawlor, R., & Morley, H. (2017). Climate change and professional responsibility: A declaration of Helsinki for engineers. *Science and Engineering Ethics, 23*, 1431–1452.

Marcus Aurelius (2002). *Meditations*. Reprint edition, from the original translation by Maxwell Staniforth for Penguin Books Ltd. 1964. UK: The Folio Society.

McDonough, W., & Braungart, M. (2002). *Cradle to cradle: Remaking the way we make things*. North Point Press.

McHarg, I. (1999). On values. In J. M. Stein & K. F. Spreckelmeyer (Eds.), *Classic readings in architecture* (pp. 284-298). WCB/McGraw Hill.

Meisch, S. P., Bremer, S., Young, M. T., & Funtowicz, S. O. (2022). Extended peer communities: Appraising the contributions of tacit knowledges in climate change decision-making. *Futures*, 135, 102868.

Meskens, G. (2017). Better living (in a complex world) - An ethics of care for our modern co-existence. In F. Zölzer & G. Meskens (Eds.), *Ethics of environmental health* (pp. 115–136). Routledge.

Meskens, G. (2018). The politics of hypothesis – An inquiry into the ethics of scientific assessment. In F. Zölzer & G. Meskens (Eds.), *Environmental health risks* (pp. 81–106). Routledge.

Meskens, G. (2022). Cosmopolitanism and environmental health. In F. Zölzer & G. Meskens (Eds.), *Research ethics for environmental health* (pp. 32–59). Routledge.

Næss, A. (1973). The shallow and the deep, long-range ecology movement. *Inquiry, 16*, 95–100.
National Geographic Society. (2023). *Anthropocene*. https://education.nationalgeographic.org/resource/anthropocene/
Ochoa-Duarte, A., & Peña-Reyes, J. I. (2020). Work in progress: Engineering education for Buen Vivir in the context of the 4th industrial revolution. In *2020 IEEE world conference on engineering education (EDUNINE)* (pp. 1–4). IEEE.
Pielke Jr, R. A. (2007). *The honest broker: Making sense of science in policy and politics*. Cambridge University Press.
Ravetz, J. (2006). *The no-nonsense guide to science*. New Internationalist.
Rawls, J. (1971). *A theory of justice*. Belknap Press.
Raworth, K. (2012). *A safe and just space for humanity: Can we live within the doughnut?*. Oxfam.
Saltelli, A., & Giampietro, M. (2017). What is wrong with evidence based policy, and how can it be improved? *Futures, 91*, 62–71.
Sardar, Z. (2010). Welcome to postnormal times. *Futures, 42*(5), 435–444.
Schein, E. H. (1992). How can organizations learn faster?: The problem of entering the Green Room.
Scherer, A. G. (2018). Theory assessment and agenda setting in political CSR: A critical theory perspective. *International Journal of Management Reviews, 20*(2), 387–410.
Seniuk Cicek, J., Masta, S., Goldfinch, T., & Kloot, B. (2023). Decolonization in engineering education. In A. Jordi (Ed.). *International handbook of engineering education research*. Routledge.
Stephens, J. C. (2021). Electrification: Opportunities for social justice and social innovation. *MRS Bulletin*, 46(12), 1205–1209.
Stephens, J. C., Kashwan, P., McLaren, D., & Surprise, K. (2021). The risks of solar geoengineering research. *Science, 372*(6547), 1161–1161.
United Nations General Assembly. (2022). *Resolution adopted by the general assembly: 77/169. Harmony with nature*. https://daccess-ods.un.org/access.nsf/Get?OpenAgent&DS=a/res/77/169&Lang=E
United Nations General Assembly. (1980). *Resolution 35/7 draft world charter for nature*. United Nations. https://undocs.org/en/A/RES/35/7.
Van der Sluijs, J. P., Craye, M., Funtowicz, S., Kloprogge, P., Ravetz, J., & Risbey, J. (2005). Combining quantitative and qualitative measures of uncertainty in model-based environmental assessment: The NUSAP system. *Risk Analysis, 25*(2), 481–492.
Verrax, F. (2017). Engineering ethics and post-normal science: A French perspective. *Futures, 91*, 76–79.
Wackernaegle, M., & Rees, W. (1996). *Our ecological footprint: Reducing human impact on the Earth*. New Society Publishers.
World Commission on Environment and Development. (1987). *Our common future* (1st ed.). Oxford University Press.

7

ENGINEERING ETHICS EDUCATION AND ARTIFICIAL INTELLIGENCE

Cécile Hardebolle, Mihály Héder, and Vivek Ramachandran

Introduction

This chapter lies at the intersection of engineering, ethics, education, and artificial intelligence (AI). It discusses how to educate engineers about ethical issues specific to AI engineering and AI *in* engineering, and how AI may be used as a tool in the engineering ethics classroom. As with the other chapters of this handbook, we begin by describing our context, or positionality, as authors.

Positionality

Three academics have written this chapter. The first author, Cécile, is an engineer, computer scientist, and learning scientist working as a pedagogical advisor at École Polytechnique Fédérale de Lausanne (EPFL) in Switzerland. Cécile came late to ethics in the context of her work with teachers. As one of the few women during her engineering and computer science journey, she has been particularly inspired by women in AI ethics. Cécile advocates for practice-oriented, in-context approaches rooted in active and experiential learning.

Mihály is a Hungarian philosopher and computer scientist interested in engineering design, epistemology, and ethics, especially in the context of AI and other software. His career as a software engineer gave Mihály social mobility, a much-needed window to Europe and beyond, and the means to study and teach philosophy at the Department for Philosophy and History of Science at Budapest University of Technology and Economics, which has been his main occupation over the past decade.

Vivek is a non-binary roboticist, learning scientist, and lecturer educated in Asia, North America, and Europe. After completing his Ph.D. in robotics, he shifted focus toward engineering education and ethics based on his desire to emphasize the importance of societal responsibility in engineering. His research explores new ways of teaching ethics to engineers using generative AI as a pedagogical tool; his teaching focuses on developing new curricula for infusing sustainability in all aspects of engineering education.

DOI: 10.4324/9781003464259-9

What do we mean by AI?

The term 'artificial intelligence' has long been a subject of terminological debate. Perhaps the most potent force of canonization was the Russell-Norvig (1995) textbook, which offers a two-by-two matrix of definitions that we summarize thus: AI as relating to internal workings versus observable behavior; AI as performance compared to humans versus an ideal measure.

The lack of total convergence in the definitions is not only a result of the Babelian state of the human race. AI, with its boom-and-bust cycles, can be, at times, an appealing brand, capable of attracting investors and, at the same time, the subject of an increased level of scrutiny, both moral (AI-HLEG, 2019) and legal (Madiega, 2021). Although still under development, the definition we uphold in this chapter is provided by the legal efforts behind the European Union AI Act: "a machine-based system designed to operate with varying levels of autonomy and that may exhibit adaptiveness after deployment and that, for explicit or implicit objectives, infers, from the input it receives, how to generate outputs such as predictions, content, recommendations or decisions that can influence physical or virtual environments" (Council of the European Union, 2023, p. 29).

Under the hood of AI

The technology that led to the most recent developments in AI is machine learning (ML), through which software can 'learn' from data – particularly non-symbolic ML, such as artificial neural networks. Large language models (LLMs), the technology behind ChatGPT, are a recent evolution of these techniques. From an engineering standpoint, we note that non-symbolic ML generally differs from other types of software or even from older versions of AI:

1. The design process for ML software starts and *centers on data* instead of a set of fixed, human-defined rules.
2. In most cases, the obtained ML model is a *black box*, and it is hard (if not impossible) to explain how a model produces a given output.
3. The *failure modes* of ML algorithms are significantly different from those of other types of software, making it challenging to ensure the safety and security of ML-based systems.

Although not all AI technologies have the characteristics mentioned above, the ones listed here do generate specific ethical issues that engineers should be able to consider.

Engineers and AI

Concerning AI, we may simplistically consider three categories of roles for engineers: *end-users* (e.g., in AI-assisted engineering); *designers/assemblers* (e.g., designing complex AI systems, embedding AI agents into larger systems such as autonomous vehicles or robots); and *developers* (i.e., implementing AI agents). While some of these roles can be considered the domain of computer science rather than engineering, this distinction is fading as AI spreads across disciplines (e.g., mechanical engineers may contribute to developing AI agents for mechanical applications). This tendency is reflected in the introduction of AI-related courses throughout engineering curricula. Orchard and Radke (2023) report that "the use of AI is pervasive across disciplines such that whether the program majors appear to be AI related is not indicative of their students' engagement with the technology" (p. 15838). As engineering students are increasingly introduced to AI, they should simultaneously be introduced to AI ethics.

Ethical issues specific to AI arise in all of the above-mentioned roles, albeit to different extents. This is why this chapter focuses on *engineers as potential ethical actors in the AI value chain* and discusses the ethical knowledge and competencies engineers need to develop concerning AI.

State of the literature

Ethical questions with AI attract an exponential amount of interest. In December 2023, a Scopus search for 'artificial intelligence' AND 'ethics' returned 5,254 documents and showed that the annual number of publications on AI ethics has been multiplied by ten in just 5 years, from 96 papers in 2017 to 1,000 in 2022. In comparison, scholarship that looked at AI ethics education was much more limited and went from 13 annual publications to 139 over the same period. Strikingly, engineering has not been associated much with this field so far: the annual number of publications found using the query 'artificial intelligence' AND 'ethics' AND 'education' AND 'engineering' was only two for 2017 and 30 for 2022.

In this chapter, we review what exists and where development efforts are needed by considering three main questions: *Where are the ethical challenges for engineers involved with AI? What should engineers know about AI ethics? How can AI engineering ethics be taught, including the use of AI as a tool?*

AI-specific challenges for engineers

Researchers have proposed the notion of 'ethical debt' (Petrozzino, 2021) to refer to the cost generated by negative impacts resulting from ethically flawed systems, in particular AI. This cost is not only borne by system developers, designers, and end-users but also by a range of indirect stakeholders (individuals, communities, societies, and the environment), and it is generally irreversible. Multiple AI-related scandals illustrate how odious that cost may be, such as the thousands of children separated from their families in the Dutch fraud detection scandal (Sattlegger et al., 2022). As potential actors in the decision chain that leads to ethical debt, engineers may face different types of challenges depending on their role.

Engineers as AI users

One frequent claim about AI algorithms is that they can be more 'objective' or 'truthful' than humans. Even a major governmental organization like the Food and Drug Administration (FDA), which plays a major role in pharmaceutical product safety in the United States, suggested in a recent document that AI could "eliminate the subjectivity in the analysis of sophisticated counterfeits" (HHS OCAIO, 2023, p. 3). This widespread belief is contradicted by a large body of research that shows that sources of non-neutrality, subjectivity, and untruthfulness are inherent to the AI production process. For instance, Suresh and Guttag (2021) identified no less than seven different sources of bias throughout the ML life cycle. Worryingly, Griffin et al. (2023) have shown that AI developers also tend to conceptualize AI as value-neutral, with the ethical responsibility lying with the user (an issue we further detail in the following section). This is particularly problematic when AI is used in the engineering design process – ethical flaws in the design tools may induce ethical flaws in the designed products without engineers realizing it. Imagine utilizing an AI-based markerless human pose estimation tool to assess the likelihood of user injury based on the mechanical features of an electric scooter. Contingent upon the dataset it has been trained on, such a tool can be biased (LaChance et al., 2023), and its performance may be lower for specific user groups. Using such a tool in the engineering process could, therefore, result in serious safety risks for the scooter users.

It is essential that engineers assess ethical risks in AI tools they use – and exercise critical thinking about providers – amid the complex political, ideological, and financial dynamics in the AI field.

Engineers as designers/assemblers or developers of AI

Engineers' responsibility is, of course, more direct in AI designer/assembler or developer roles, where the challenges are also more numerous.

Combined ethical and technical knowledge

In their study of AI developers' agency, Griffin et al. (2023) reported that interviewees described a range of routine technical choices without realizing their ethical dimensions. They suggested that AI developers have "ethical agency 'veiled' as technical agency" (Griffin et al., 2023, p. 6). While this implies that some technical choices in AI entail ethical dimensions, the opposite is also true: some ethical choices in AI entail technical dimensions. For instance, a dedicated field of study researches the fairness of AI algorithms, which has resulted in the development of a range of fairness metrics to assess model fairness as well as technical solutions to try to improve it (Pessach & Shmueli, 2022). As we elaborate later, nearly all dimensions pertaining to the ethics of AI involve some combination of ethical and technical knowledge. Without this combination, engineers will find it challenging to assess and mitigate ethical issues.

Dilemmas

The AI domain is also full of ethical dilemmas disguised as technical dilemmas. Decisions made by AI-powered autonomous vehicles in life-or-death situations, popularized by the Moral Machines project at the Massachusetts Institute of Technology (MIT, n.d.), are a well-known example. Other less visible but perhaps more impactful ethical dilemmas arise in the design decisions engineers make when building AI systems – usually called 'trade-offs' in the AI literature. One example is the fairness–accuracy trade-off: currently, methods that improve the fairness of a model usually decrease its overall accuracy (Pessach & Shmueli, 2022). Many other examples can be found in Sanderson et al. (2023).

Some of these dilemmas not only need recognition and resolution but also re-evaluation. Regarding new technologies, we are usually presented with trade-off situations (Héder, 2021), which often appear to be either using the technology and risking harm or not using it and risking missing out on economic progress. The literature on technological determinism warns us that these first takes are almost always wrong and driven by a misguided 'technological imperative.' Engineers, business owners, and beneficiaries of technological advancements often hastily accept risky features as inherent to technology, implying that society must tolerate these risks. These are false trade-offs, which can be ultimately prevented at marginal, sometimes completely trivial cost – or even no cost at all – with better policies (Héder, 2021, p. 127).

Modularization

Engineers increasingly work with modules that they assemble instead of developing models from scratch (Widder & Nafus, 2022). Generic models such as 'foundation models' can be reused and fine-tuned for specific applications. Modularity introduces what Widder and Nafus call 'dislocated accountability.' Their interview-based research found that "acknowledgement of harms was consistent but nevertheless another person's job to address, almost always at another location in the broader system of production, outside one's immediate team" (Widder & Nafus, 2022, p. 1). While modularization is not specific to AI, it creates additional challenges in the case of AI because of AI's black-box nature.

Topics in AI ethics

We now present a selection of ongoing conversations in AI ethics that can provide inspiration for AI ethics curricula for engineers. In doing so, we highlight existing controversies and debates, and identify knowledge and skills for engineers to develop. We are not aiming for exhaustiveness in the themes we cover (see Hagendorff, 2022 and Kazim & Koshiyama, 2021 for more complete overviews).

Fairness and bias

Avoiding bias in any kind of system, including AI, is a central concern. It is widely recognized that a biased automated sociotechnical system can cause extreme levels of harm: the well-researched case of the algorithm for analyzing Dutch child benefits (i.e., signaling risks for biased reasons), together with inadequate bureaucratic processes, resulted in tens of thousands of wrongfully canceled child benefit cases (Sattlegger et al., 2022).

Training data is one major source of bias in AI. In applications where generating data for AI training requires some form of human involvement, the process is exposed to cognitive biases. Three are especially prevalent – selection bias, conformity bias, and exposure bias – but there are several more (Chen et al., 2023). Bias can also arise from the model itself, even with unbiased data. For instance, a model may over-generalize from some data points and under-generalize based on others as a result of applying various heuristics that do not have much to do with the semantics of the data. Other sources of bias arise from choices in the model development process (Suresh & Guttag, 2021).

A biased system is unfair, and can take several forms. It may exhibit the Matthew effect, discriminate based on protected attributes (e.g., ethnicity, religion), or exhibit error rates that differ significantly among groups. Current methods to address unfairness issues at the algorithmic level include intervening on the training data, the model, or its output (Pessach & Shmueli, 2022). However, identifying and addressing bias is not always a straightforward statistical exercise. Although some methods can shed light on causal relationships in unfairness issues (Dubber et al., 2020), the definition of bias in certain edge cases requires elaborate philosophical or political discussions (e.g., see Coeckelbergh, 2022, chap. 3, p. 86).

Fairness is among the most widely addressed topics in AI ethics syllabi (Garrett et al., 2020), most frequently introduced through a review of existing fairness metrics with mathematical definitions. While contradictory to each other (Pessach & Shmueli, 2022), these metrics still allow students to perform calculations on example datasets and models and are often used to introduce the philosophical notion of fairness. Fairness evaluation and algorithm auditing are essential skills for engineers to develop – alongside bias mitigation design.

Safety and the alignment problem

Safety is quite a central consideration in AI ethics because of the scale at which these systems can be deployed; even small error proportions can have massive consequences. A significant challenge is to cope with AI's black-box nature and specific failure modes. Latent errors, hard-to-predict modes of failure, and model drift are examples of the numerous difficulties with safety in AI systems such as self-driving cars (Cummings, 2023). Traditional software safety methods such as testing and code audits are very hard – if not impossible – to use, and public news is rife with examples of AI systems with worrying safety issues, including fatalities (Raji et al., 2022). AI safety risks can be considered at different time scales (Sætra & Danaher, 2023), which is the subject of a raging debate between advocates of the long-term risks – in particular, 'existential risks' (also called 'x-risks') that threaten human existence – and those arguing that more attention should be paid to demonstrated short-term risks that are already affecting populations and the environment.

Autonomy in AI systems, which implies a capacity to make (im)moral decisions, raises a specific safety risk called the 'alignment problem': ensuring that the values manifested in an AI's decisions and acts are aligned with human society. The problem of AI alignment is twofold, according to Gabriel (2020): (1) whose values should an AI be aligned with and (2) how to do the alignment. If the question of selecting the values (to align with) in a pluralistic world is evidently and intrinsically perilous, its implementation is far from trivial as it involves operationalizing the selected values (which will again give rise to debate at another level).

Beyond the performance measures currently central in AI curricula, engineers should be given a practical understanding of AI safety, drawing attention to the potential negative impacts on humans and the environment, both at the micro (individuals) and macro (societies) levels. Evaluating these impacts requires risk assessment methods – an approach also used in the EU AI Act. Finally, an introduction to values and their role in the design process and skills with methods such as value-sensitive design or VSD (Friedman & Hendry, 2019) seem particularly relevant (for more on teaching using such approaches, see Chapter 22).

Transparency and explainability

While traditional AI-leveraged methods were essentially self-explanatory (using logical rules, decision trees, and semantic technologies), these turned out to have less success and more modest capabilities than ML, which, in turn, has a tendency to produce black boxes. An active system that we don't understand – one that makes decisions for us instead of us – naturally raises concerns. The problem is epistemic and the idea is that opacity (Héder, 2023a) takes away our control and our sense of intellectual oversight (Héder, 2023b). On the other hand, transparency can be a way to build trust in the system. The notion of transparency or explainability is, therefore, the most common feature of regulation (Hagendorff, 2020).

However, the fact that explainability and transparency build trust should not be accepted without challenge. Some findings indicate that this effect may present itself only occasionally and may even decrease trust (Scharowski, 2020; Schmidt et al., 2020). Naturally, the transparency of a system to an individual greatly depends on the *a priori* knowledge of that person about how AI works, as well as the person's level of exposure to the system. Therefore, the draft standard in this question (P7001, Winfield et al., 2021) distinguishes between expert, user, and bystander roles.

Human agency

The increasing presence of AI systems in our lives raises questions regarding our agency (Prunkl, 2022): *Is our agency augmented or antagonized by increasing AI autonomy?*

AI's impact on human agency is a double-edged sword. While AI tools have contributed to improving people's quality of living, often by employing their data to provide tailored recommendations (Logg et al., 2019), there are concerns about how the data is obtained, stored, and utilized – and controversies regarding manipulation and surveillance (Floridi et al., 2021; Ienca, 2023). AI tools like chatbots may seem impressive at mimicking human interactions that seemingly display feelings of empathy (Stark & Hoey, 2021), and that characteristic can add to the automation bias problem – where humans overly trust AI recommendations – that undermines critical thinking and accountability (Ienca, 2023; Suresh et al., 2020). Moreover, AI can also perpetuate falsehoods and contribute to the illusory truth effect (i.e., the propensity for humans to believe misinformation as truth by dint of repetition).

This highlights the ethical responsibility in engineering to engage (as users and creators) with AI systems in a way that respects user boundaries, maintains transparency, and upholds ethical interaction standards. A balanced approach is essential in classroom discussions, examining the potential benefits *and* the ethical issues AI systems pose – as this will determine how human empowerment and agency are protected and strengthened. For a more detailed critique of how AI autonomy affects human agency, we refer readers to Mhlambi and Tiribelli (2023).

Sustainability

Currently vastly under-addressed in typical AI curricula, sustainability questions materialize a central dilemma: AI offers some potential for addressing some of the complex climate change issues (Larosa et al., 2023) while at the same time requiring colossal amounts of resources, including energy, data, hardware, and human labor (Bender et al., 2021). The complex cost–benefit questions related to AI should not be left out of current efforts to introduce sustainability into engineering programs.

While the environmental impacts of AI in general remain massively undocumented, recent studies on LLMs tend to show that both the carbon and the water footprints of these systems are significantly larger than for other IT systems (Li et al., 2023; Luccioni et al., 2023; Patterson et al., 2021). In addition to parameters related to cloud infrastructure, the size of the datasets and models, but more importantly their architecture, seem to increase the impact at the time of both training and use. The GPT models (e.g. ChatGPT) seem to have a particularly high environmental impact, which is concerning given the attention they generate in (engineering) education.

Unfortunately, AI also presents other sustainability issues (Crawford, 2021). Researchers have investigated the questionable labor practices behind AI (Hagendorff, 2022), exemplified by the Kenyan workers who made ChatGPT less toxic, reducing the amount of violent, racist, and sexist outputs for end-users by reviewing and labelling harmful content manually. On the hardware side, although many sustainability issues are not AI-specific but cloud-computing related, the exponential increase in dataset and model size leads to a race for optimized hardware. In addition to the catastrophic environmental impact of hardware production (Crawford, 2021), these impact reduction efforts are likely to be counteracted by increasing demand (rebound effect, see Grubb, 1990).

Although more research is needed, engineers should be introduced to these issues as early as possible and develop skills for evaluating AI systems' carbon and water footprints. The systemic nature of these issues also calls for macro approaches encompassing the whole AI life cycle and including questions of resources and labor dynamics at a large scale. In particular, engineers need to develop systems thinking skills and practice with methods such as life-cycle assessment.

Regulation of AI

The successes of AI in the 2020s provoked a wave of soft laws and regulations (Héder, 2020). One major challenge is assigning responsibility for unfortunate or unwelcome events. Since responsibility is closely associated with decision-making, AI automated decision-making has created a responsibility gap (Matthias, 2004): a system making a decision is not a legally accountable agent, unlike the human being it replaced. Therefore, responsibility needs to be assigned elsewhere, but this redistribution is far from trivial. Another issue is the vast potential of AI for technology lock-in because, as with any software, once developed at significant expense, the margin cost of reproduction is minimal. The fact that software can be reused cheaply and infinitely removes the incentive for creating a completely new one at high capital expenditure. The lack of serious new computer operating system projects illustrates this point quite well. In this case, the decisions made in the early stage, lacking information and foresight, may have long-lasting consequences. Finally, generative AI challenges existing copyright and intellectual property frameworks.

In addition to theoretical background – on how norms are created or studying certification materials and reports of actual systems – mock evaluation sessions and simulated certification processes (e.g., where one team of students act as the product owners while others as the certifying body) can provide engineers with a pragmatic understanding of regulatory issues. Yet, the rapid evolution of AI regulations will make it challenging to keep educational material up to date.

Pedagogical methods

We now turn to the pedagogical methods that could be used to teach engineers about AI ethics. Although still few, there are some reviews of AI ethics syllabi, mainly in the United States (Garrett et al., 2020; Raji et al., 2021; Saltz et al., 2019; Tuovinen & Rohunen, 2021). They tend to show that the range of pedagogical methods used in AI ethics is quite diverse and has much in common with engineering ethics education (EEE) methods – the overall topic of this handbook. The following subsections provide an overview of existing approaches and identify avenues for future work related to EEE and AI. We will discuss the specific case of how AI could be used as a tool for EEE.

General engineering ethics methods

Readings followed by class discussions are among the most frequently used methods in AI ethics classes (Garrett et al., 2020; Raji et al., 2021; see also Chapter 25 on reflective and dialogical approaches to teaching EEE). Reading lists generally include research papers and news articles that help relate course content to current events. Although academic readings may provide insights into the multidisciplinary nature of AI ethics (Raji et al., 2021), the vocabulary used may create difficulties for students, and engineering students generally have little experience with these methods, especially at the undergraduate level (Tuovinen & Rohunen, 2021).

Pedagogical methods in AI ethics also include *case studies* (e.g., see Alam, 2023; and Chapter 20) for students to practice assessing ethical dilemmas and ethical decision-making. Unfortunately, there's no shortage of opportunities to build cases on AI-related real-world events. Cases are frequently used with other techniques such as *role plays* (Hingle & Johri, 2023; and Chapter 24) and *debates* (Alam, 2023; and Chapter 25). Some AI ethics courses make use of *science fiction* (Burton et al., 2018; and Chapters 13 and 24) to equip "students with skills to cope also with the unforeseen ethical issues in their future work" (Tuovinen & Rohunen, 2021, p. 21). *Games* are another pedagogical tool used in AI ethics, whether in digital or physical form (Alam, 2023; Hardebolle et al., 2022).

Experiential practice-based approaches

Practice-based approaches are a central component of AI courses for engineers. Exercises and projects (see Chapter 21 on problem-based learning in EEE) provide opportunities to experience the AI development process, which, besides aiding future AI developers, also enhance engineers' understanding of the underlying mechanisms of AI. Below, we discuss how such activities can provide opportunities to teach and learn AI ethics in context.

Exercises with data

The data on which AI systems rely provides valuable opportunities for ethics education. Public datasets such as the COMPAS (Angwin et al., 2016) are frequently used for bias analyses. The 'AI and Equality Toolbox' (AI and Equality, n.d.) provides a Jupyter Notebook (an interactive document including modifiable code) for exploring biases within the German Credit Dataset (Hofmann, 1994). Other fairness-related datasets can be found in a review by Pessach and Shmueli (2022).

Another pedagogical intervention worthy of attention had students use 'datasheets' (i.e., structured documents that provide contextual information about a dataset) when working on an ML problem (Boyd, 2021). The study found that participants using datasheets identified ethical issues earlier and more often than those without. Similar documents called 'model cards' exist for ML models (Mitchell et al., 2019). Introducing engineers to such tools could potentially help address the dislocated accountability issue related to modularity. A key challenge for educators is that these tools are in their infancy and will likely evolve.

Exercises with models

Having students train an AI model themselves can create interesting conditions for ethical reflection, as suggested by Ko and colleagues: "have students train basic machine learning models, and then reflect on the application and limitation of those models to particular contexts, such as admissions and financial aid decisions" (Ko et al., 2023, chap. 15).

AI models for classification and prediction can be the object of fairness analysis exercises by having students compute and interpret fairness metrics (Pessach & Shmueli, 2022). The 'Human Contexts and Ethics' program of Berkeley (Berkeley CDSS, n.d.) proposes Jupyter Notebooks that include programming tasks using fairness assessment libraries, which can also produce visualizations (Quedado et al., 2022). Thanks to the notebook format, the exercises integrate ethical reflection questions related to the limits of fairness metrics and the contextual nature of fairness. Such approaches could be applied to other ethical issues reviewed previously.

However, as far as we know, evaluation of such methods in terms of impact on student learning is lacking. When reporting on a survey of engineering students that included an AI fairness case study, Orchard and Radke (2023) commented: "students are often able to identify and suggest actions for mitigating the [fairness] issue from a technical standpoint but rarely connect it with broader ethical and societal implications" (p. 15834). Educators and researchers should take this preliminary result as a warning about the limits of addressing ethical concepts such as fairness solely through mathematical and technical lenses. More research is needed to identify how to combine experiential approaches with broader philosophical approaches.

Engineering projects

A central experiential component of engineering education, engineering projects provide evident opportunities to integrate AI ethics. In their 'simplest' form, interventions in projects can build on ethical reflection tools such as questions (Saltz et al., 2019). Assessing students' ethical reflection can be a difficulty for engineering educators, but can be overcome with appropriate pedagogical support (e.g. grading rubrics). Involving ethicists and social scientists in projects is another way to integrate ethics into AI engineering (Tigard et al., 2023), provided that students receive appropriate training and support for interdisciplinary teamwork to ensure a positive experience. Finally, projects involving a human research component can help students develop research ethics skills and methods from the human sciences fields (Williams et al., 2020).

Projects can also provide opportunities for students to practice specific engineering ethics methods applied to AI, a type of intervention we were not able to find in existing publications. We suggest in particular value-sensitive design (Friedman & Hendry, 2019; see Chapter 22); participatory design (Gerdes, 2022; see Chapter 23); ethical risk assessment (Hardebolle et al., 2023); technology assessment (Børsen, 2021); and life-cycle assessment (Ligozat et al., 2022). The main challenge, in this case, is to involve trained specialists of these methods in the design and supervision of the projects and/or to train the teaching teams. This challenge is also an asset – shared with the approaches that we discuss in the next section.

Curriculum-wide interventions

Harvard University implements a curriculum-wide program called "Embedded EthiCS" (Grosz et al., 2019) wherein philosopher-designed ethics modules involving case studies with analytic methodologies and small group discussions are embedded into computer science courses. While some evaluations have been conducted in terms of students' interest and self-efficacy toward ethical issues (Horton et al., 2022), more research is essential to assess the impact of such interventions, particularly for engineers.

Northeastern University chose to embed 'Values Analysis in Design' modules into AI-related courses (Kopec et al., 2023). While the pedagogical methods used are mostly similar to those mentioned above, its specific focus on value analysis builds on prior engineering ethics work with value-sensitive design (Friedman & Hendry, 2019). An evaluation showed significant changes in students' attitudes with respect to values and ethically responsible design (Kopec et al., 2023). While further evaluation is needed, such approaches could also be applied to programs teaching AI to engineers. See Chapter 22 on VSD and Chapter 12 on engineering design for further discussions.

Overall, the contextualization of ethical concerns in practical settings provided by experiential practice-based and embedded approaches seems promising. However, they have their detractors (e.g. Raji et al., 2021), and the evidence is still extremely limited.

AI as a tool for teaching ethics

Applications of AI as a teaching and learning tool are almost as old as the field itself, but the LLM boom has now heightened interest and fears. It would be beyond the scope of this chapter to elaborate on the use of AI for general education, or even for engineering education, and so we focus only on applications to ethics education, a domain which seems under-explored. However, we want to make clear that we by no means consider AI as a silver bullet for this task, not least because it comes with concerning ethical issues that we explore at the end of this section.

AI in case-based learning

Previous work as explored the use of AI for on-the-spot assistance but also for preparatory training in moral decision-making (O'Neill et al., 2022). For instance, AI could be used to provide interactive, personalized, step-by-step guidance in case study analysis. In utilitarian calculus applications, additional information (e.g., background, stakeholder preferences, and probabilities) could be interactively provided to learners. Alternatively, learners could be presented with similar cases to compare since, unlike case law, in ethics precedents are not binding. O'Neill et al. (2022) have, however, flagged critical ethical risks associated with this use – such as unintended influence – and others to which we return later.

Students as critics of AI output

Students' experience with publicly available generative AI could be leveraged for ethics interventions. For instance, students could be asked to create text, images, or videos and analyze the output in terms of the kind of values or biases they present (e.g., political bias, see Narayanan, 2023) or to identify instances of 'plausible non-sense' in AI chatbot outputs (Hardebolle & Ramachandran, 2023) and reflect on how much such systems should be trusted. While we found academic work doing this type of analysis (e.g., Srinivasan & Uchino, 2021), we were not able to find studies on educational interventions. One challenge is that methodologies for performing such evaluations rigorously can be quite complex. It is worth highlighting that some studies found that students may be reluctant to use generative AI tools even when encouraged to do so (Prasad et al., 2023).

For classification or prediction models, students could be guided to use one such model to make a decision and then reflect on how they made the decision, particularly in terms of their own cognitive biases. Such an activity could provide an introduction to the challenges of AI-assisted decision-making, particularly the issue of automation bias (Suresh et al., 2020). The effectiveness and challenges of such interventions remain to explore.

Students as 'subjects' of AI processing

Prior research has explored activities where students have worked with data on themselves to increase learning engagement. Shapiro et al. (2020) showed how such activities can support critical reflection and help students develop an ethics of care. Although they are preliminary, these results seem promising as "students were confronted with the idea that they are the 'other' within systems that use and may exploit personal data and as a result, began to consider what care they desire or demand from these systems" (Shapiro et al., 2020, p. 9).

A similar 'making it personal' approach has been explored in AI-type tasks reported by Register and Ko (2020). Students implemented a model that predicts a student's grade based on self-reported measures of interest in courses and input their own data. Although the intervention was limited to very simple models, students seemed to pay more attention to the teaching material and appeared better able to explain underlying ML mechanisms. We see potential interest in these methods for students to empathize with end-users, realize what AI-assisted decision-making means in practice, and get a better understanding of transparency issues. Still, these hypotheses would need to be tested.

Even AI tools that may be generally considered 'harmful' or 'unethical' could be used as pedagogical instruments to explain the consequences of their irresponsible use on unwitting stakeholders. For instance, Ramachandran et al. (2023) examined the effect of using deepfakes as a pedagogical tool to foster students' empathy towards victims of this technology (as in the case of non-consensual deepfakes, pornographic ones in particular). This topic appeals to the students' sense of responsibility as potential creators of AI tools. Engaging in such discussions prompts students to confront similar ethical quandaries they may encounter as professionals in the future, enhancing their moral sensitivity, motivation, and reasoning.

Another way of making it personal is to have an AI assess students' productions (such as essays) and guide students to reflect on the process and its results afterward. While evaluations of the potential of AI for this type of use exist, we did not find interventions that make use of it for teaching ethics. Such an intervention could provide opportunities for discussing the ethics of automated evaluation, trustworthiness, transparency, and empowerment questions, as well as the role of emotions in ethics. However, instructors should exercise caution and assess both the ethics and legality of this type of setup in their own context.

Overall, the balance between benefits and risks of all the interventions mentioned in this section should be carefully evaluated, a point we address in the next section.

Ethics of using AI for EEE

In this section, we examine the ethical risks associated with the use of AI in EEE by successively adopting the point of view of the five 'ethical lenses' of the 'Digital Ethics Canvas' (Hardebolle et al., 2023), a methodological tool designed for teaching ethical risk assessment to engineers.

Sustainability

Encouraging AI usage in universities raises systemic environmental risks as the rising energy consumption and resultant carbon footprint from server operations per user is immediately multiplied by large numbers of students. As we have seen, the environmental impact of generative AI is much higher than most other types of software or digital tools (Luccioni et al., 2023). Instructors ought to evaluate the necessity of using AI systems for specific educational tasks, and consider alternatives that have a lower impact. More generally, environmental impact and labor practices should be treated as essential criteria when selecting an AI system.

Privacy

Student privacy and data security is of prime importance in educational contexts. The collection of student data for AI use in ethics education is a real risk since it may include sensitive information about student values and morality (O'Neill et al., 2022). The potential re-use of student data for AI training is also of concern since training data can be retrieved from models (Carlini et al., 2023). While European institutions are particularly attentive to institutional use with the General Data Protection Regulation (GDPR), it is imperative to sensitize students to data consent and its consequences, especially with US-hosted tools such as ChatGPT.

Fairness

Students should not incidentally be subjected to unfair treatment or outcomes while using AI for ethics education. Two aspects to consider for fair treatment are access and accessibility. Although free accounts can facilitate access, they often lead to problematic differences in privacy treatment. Accessibility considerations (e.g., interface, language) are often not considered in software interfaces, and AI is no exception. Regarding outcomes, although demonstrating the biases in AI-generated output (Abid et al., 2021) can be helpful as an educational exercise, instructors should not underestimate the emotional response or even trauma that exposure to biased information can generate and must take appropriate measures.

Non-maleficence

Significant attention has been drawn to the potential adverse effects of generative AI on human learning, even though some argue that this issue dates back to the invention of writing (see Plato, Phaedrus 14, pp. 274–275). The impact on human skills generally requires more research. An open question regards whether AI harms the learning assessment process: on one hand, it interferes with students' writing; on the other hand, it can be used by instructors to ease the tedious process of analyzing textual productions (which comes with other risks, as discussed earlier). In addition, we should not lose sight of the harms that arise at a more macro/global level, among which we can cite content stolen from authors and artists – and information pollution on a large scale.

Empowerment

The 'plausible nonsense' (also called 'hallucinations,' see Huang et al., 2023 for a review) unpredictably generated by LLMs might offer intriguing exercises for practicing critical thinking. However, aggravated by the lack of information provided to users on the unreliability of the output, it remains problematic in numerous scenarios (e.g., searching for information). When available, AI tools that provide ways for users to evaluate output quality are generally preferable, particularly in educational settings. In addition to dis-empowerment risks relating to the black-box nature of AI and the associated "inescapability of outside influence" (O'Neill et al., 2022, p. 9), some interface designs can also increase the human tendency to anthropomorphize these systems, which can lead to serious consequences (manipulation, in particular, emotional manipulation and dependency), particularly for vulnerable groups.

With our review of the ethical risks above (which is not exhaustive), we hope that we have illustrated how critical reflective practice can be applied to the case(s) of using AI tools in ethics education. Beyond AI, our use of digital tools in education should be driven by our values – an exercise that is challenged by the pressure of productivity and the strong push from tool vendors.

Conclusions

This chapter has grappled with a unique set of challenges and opportunities. Although the topics of AI ethics and AI in education are rich and constantly evolving, pedagogical methods are still nascent, particularly within the context of engineering education. We navigate the inherent complexities of this field by adopting an interdisciplinary view that balances our varying opinions. However, we are simultaneously unwavering in our commitment to addressing the broad spectrum of ethical issues that arise when AI is used in education. One of the limitations of this chapter, and a challenge for EEE practitioners and researchers, is the temporality of our conclusions – AI and AI ethics evolve at lightning speed as new technologies, policies, and ethical dilemmas emerge.

Acknowledgments

Special thanks to Jules Ronné for the biomechanical engineering examples.

References

Abid, A., Farooqi, M., & Zou, J. (2021). Large language models associate Muslims with violence. *Nature Machine Intelligence*, *3*(6), 461–463. doi: 10.1038/s42256-021-00359-2

AI and Equality. (n.d.). *A human rights toolbox.* Retrieved 2024-01-08, from https:// aiequalitytoolbox.co m/

AI-HLEG. (2019). *Ethics guidelines for trustworthy ai* (Tech. Rep.). European Commission, B-1049 Brussels: European Commission.

Alam, A. (2023). Developing a curriculum for ethical and responsible AI: A university course on safety, fairness, privacy, and ethics to prepare next generation of AI professionals. In G. Rajakumar, K.-L. Du, & A. Rocha (Eds.), *Intelligent communication technologies and virtual mobile networks* (pp. 879–894). Springer Nature. doi: 10.1007/978-981-99-1767-9_64

Angwin, J., Larson, J., Mattu, S., & Kirchner, L. (2016, May). Machine bias. *ProPublica.* Retrieved 2023-12-20, from https://www.propublica.org/article/ machine-bias-risk-assessments-in-criminal-sentencing

Bender, E. M., Gebru, T., McMillan-Major, A., & Shmitchell, S. (2021). On the dangers of stochastic parrots: Can language models be too big? In *Proceedings of the 2021 ACM conference on fairness, accountability, and transparency* (pp. 610–623). ACM. doi: 10.1145/3442188.3445922

Berkeley CDSS. (n.d.). *Human contexts and ethics.* Retrieved August 1, 2024, from https://data.berkeley.edu /human-contexts-and-ethics

Børsen, T. (2021). *Using technology assessment in technical study programs as a means to foster ethical reflections on the societal effects of technologies and engineering solutions.* Proceedings of the SEFI 49th Annual Conference. Berlin, Germany, 685–694.

Boyd, K. L. (2021). Datasheets for datasets help ML engineers notice and understand ethical issues in training data. *Proceedings of the ACM on Human-Computer Interaction*, *5*(CSCW2), 438:1–438:27. doi:10.1145/3479582

Burton, E., Goldsmith, J., & Mattei, N. (2018). How to teach computer ethics through science fiction. *Communications of the ACM* , *61*(8), 54–64. doi:10.1145/3154485

Carlini, N., Hayes, J., Nasr, M., Jagielski, M., Sehwag, V., Tramèr, F., Balle, B., Ippolito, D., & Wallace, E. (2023). *Extracting training data from diffusion models.* arXiv. doi: 10.48550/ arXiv.2301.13188

Chen, J., Dong, H., Wang, X., Feng, F., Wang, M., & He, X. (2023). Bias and debias in recommender system: A survey and future directions. *ACM Transactions on Information Systems*, *41*(3), 1–39.

Coeckelbergh, M. (2022). *The political philosophy of ai: an introduction.* John Wiley & Sons.

Council of the European Union. (2023, November). *Artificial* intelligence act—preparation *for the trilogue. interinstitutional file 2021/0106(COD), Document 15688/23.* Council of the European Union.

Crawford, K. (2021). *Atlas of AI: Power, politics, and the planetary costs of artificial intelligence.* Yale University Press.

Cummings, M. L. (2023, July). *What self-driving cars tell us about ai risks - IEEE spectrum.* Retrieved October 24, 2023, from https://spectrum.ieee.org/ self-driving-cars-2662494269

Dubber, M. D., Pasquale, F., & Das, S. (2020). *The Oxford handbook of ethics of AI.* Oxford Handbooks.

Floridi, L., Cowls, J., Beltrametti, M., Chatila, R., Chazerand, P., Dignum, V., Luetge, C., Madelin, R., Pagallo, U., Rossi, F., Schafer, B., Valcke, P., & Vayena, E. (2021). An ethical framework for a good AI society: Opportunities, risks, principles, and recommendations. In L. Floridi, (Eds.), *Ethics, Governance, and Policies in Artificial Intelligence*, vol 144, Springer, Cham.

Friedman, B., & Hendry, D. (2019). *Value sensitive design: Shaping technology with moral imagination.* The MIT Press.

Gabriel, I. (2020). Artificial intelligence, values, and alignment. *Minds and Machines*, *30*(3), 411–437.

Garrett, N., Beard, N., & Fiesler, C. (2020, February). More than "if time allows": The role of ethics in AI education. In *Proceedings of the AAAI/ACM conference on ai, ethics, and society* (pp. 272–278). ACM. doi: 10.1145/ 3375627.3375868

Gerdes, A. (2022). A participatory data-centric approach to AI ethics by design. *Applied Artificial Intelligence*, *36*(1), 2009222. doi: 10.1080/08839514.2021.2009222

Griffin, T. A., Green, B. P., & Welie, J. V. M. (2023, January). The ethical agency of AI developers. *AI and Ethics*. doi: 10.1007/s43681-022-00256-3

Grosz, B. J., Grant, D. G., Vredenburgh, K., Behrends, J., Hu, L., Simmons, A., & Waldo, J. (2019). Embedded EthiCS: Integrating ethics across CS education. *Communications of the ACM* , *62*(8), 54–61. doi: 10.1145/3330794

Grubb, M. J. (1990). Communication Energy efficiency and economic fallacies. *Energy Policy*, *18*(8), 783–785. doi: 10.1016/0301-4215(90)90031-X

Hagendorff, T. (2020). The ethics of AI ethics: An evaluation of guidelines. *Minds and machines*, *30*(1), 99–120.

Hagendorff, T. (2022). Blind spots in AI ethics. *AI and Ethics*, *2*(4), 851–867. doi: 10.1007/s43681-021-00122-8

Hardebolle, C., Kovacs, H., Simkova, E., Pinazza, A., Di Vincenzo, M. C., Jermann, P., Tormey, R., & Dehler Zufferey, J. (2022). *A game-based approach to develop engineering students' awareness about artificial intelligence ethical challenges*. Proceedings of the 50th SEFI Annual Conference. Barcelona, Spain, 2276–2281. doi: 10.5821/conference-9788412322262.1283

Hardebolle, C., Macko, V., Ramachandran, V., Holzer, A., & Jermann, P. (2023). *The digital ethics canvas: A guide for ethical risk assessment and mitigation in the digital domain*. Proceedings of the 51st SEFI Annual Conference 2023. Dublin, Ireland.

Hardebolle, C., & Ramachandran, V. (2023, October). *Plausible nonsense and carbon footprint: Micro and macro Ethics of generative AI in the classroom*. Retrieved October 31, 2023, from https://www.sefi.be/2023/10/14/plausible-nonsense-and-carbon-footprint-micro-and-macro-ethics-of-generative-ai-in-the-classroom/

Héder, M. (2020). A criticism of AI ethics guidelines. *Információs Társadalom*, *20*(4), 57–73. doi: 10.22503/inftars.XX.2020.4.5

Héder, M. (2021). AI and the resurrection of technological determinism. *Információs Társadalom*, *21*(2), 119–130. doi: 10.22503/inftars.XXI.2021.2.8

Héder, M. (2023a). The epistemic opacity of autonomous systems and the ethical consequences. *AI & SOCIETY*, *38*, 1819–1827. doi: 10.1007/s00146-020-01024-9

Héder, M. (2023b). Explainable AI: A brief history of the concept. *ERCIM NEWS*, *134*, 9–10.

HHS OCAIO. (2023). *Artificial intelligence use cases – FY2022*. Retrieved December 20, 2023, from https://www.hhs.gov/sites/default/files/hhs-artificial-intelligence -select-use-cases.pdf

Hingle, A., & Johri, A. (2023). *Recognizing principles of AI Ethics through a role-play case study on agriculture*. ASEE Annual Conference and Exposition, Conference Proceedings.

Hofmann, H. (1994). *German credit data (statlog))*. UCI Machine Learning Repository. doi: 10.24432/C5NC77

Horton, D., McIlraith, S. A., Wang, N., Majedi, M., McClure, E., & Wald, B. (2022). Embedding Ethics in computer science courses: Does it work? In *Proceedings of the 53rd ACM* technical symposium *on* computer science education (Vol. 1, pp. 481–487). ACM. doi: 10.1145/3478431.3499407

Huang, L., Yu, W., Ma, W., Zhong, W., Feng, Z., Wang, H., Chen, Q., Peng, W., Feng, X., Qin, B., & Liu, T. (2023). *A survey on hallucination in large language models: Principles, taxonomy, challenges, and open questions* (arXiv:2311.05232). arXiv. https://doi.org/10.48550/arXiv.2311.05232

Ienca, M. (2023). On artificial intelligence and manipulation. *Topoi*. doi: 10.1007/ s11245-023-09940-3

Kazim, E., & Koshiyama, A. S. (2021). A high-level overview of AI ethics. *Patterns*, *2*(9), 100314. doi: 10.1016/j.patter.2021.100314

Ko, A. J., Beitlers, A., Wortzman, B., Davidson, M., Oleson, A., Kirdani-Ryan, M., Druga, S., & Everson, J. (2023). *Critically conscious computing: Methods for secondary education*. Retrieved October 23, 2023, from https://criticallyconsciouscomputing.org

Kopec, M., Magnani, M., Ricks, V., Torosyan, R., Basl, J., Miklaucic, N., Muzny, F., Sandler, R., Wilson, C., Wisniewski-Jensen, A., Lundgren, C., Baylon, R., Mills, K., & Wells, M. (2023). The effectiveness of embedded values analysis modules in computer science education: An empirical study. *Big Data & Society*, *10*(1), 20539517231176230. doi: 10.1177/20539517231176230

LaChance, J., Thong, W., Nagpal, S., & Xiang, A. (2023). *A case study in fairness evaluation: Current limitations and challenges for human pose estimation*. AAAI 2023 Workshop on Representation Learning for Responsible Human-Centric AI.

Larosa, F., Hoyas, S., García-Martínez, J., Conejero, J. A., Fuso Nerini, F., & Vinuesa, R. (2023). Halting generative AI advancements may slow down progress in climate research. *Nature Climate Change*, *13*(6), 497–499. doi: 10.1038/s41558-023-01686-5

Li, P., Yang, J., Islam, M. A., & Ren, S. (2023). *Making AI Less "thirsty": Uncovering and addressing the secret water footprint of AI models*. arXiv. doi: 10.48550/ arXiv.2304.03271

Ligozat, A.-L., Lefevre, J., Bugeau, A., & Combaz, J. (2022). Unraveling the hidden environmental impacts of AI solutions for environment life cycle assessment of AI solutions. *Sustainability*, *14*(9), 5172. doi: 10.3390/su14095172

Logg, J. M., Minson, J. A., & Moore, D. A. (2019). Algorithm appreciation: People prefer algorithmic to human judgment. *Organizational Behavior and Human Decision Processes, 151*, 90–103.
Luccioni, A. S., Jernite, Y., & Strubell, E. (2023). *Power hungry processing: Watts driving the cost of AI deployment?* arXiv. doi: 10.48550/arXiv.2311.16863
Madiega, T. (2021). Artificial intelligence act. *European Parliament: European Parliamentary Research Service*.
Matthias, A. (2004). The responsibility gap: Ascribing responsibility for the actions of learning automata. *Ethics and Information Technology, 6*, 175–183.
Mhlambi, S., & Tiribelli, S. (2023). Decolonizing AI ethics: Relational autonomy as a means to counter AI harms. *Topoi, 42*(3), 867–880. doi: 10.1007/s11245-022-09874-2
MIT. (n.d.). *Moral machine.* Retrieved January 8, 2024, from http://moralmachine.mit.edu
Mitchell, M., Wu, S., Zaldivar, A., Barnes, P., Vasserman, L., Hutchinson, B., Spitzer, E., Raji, I. D., & Gebru, T. (2019). Model cards for model reporting. In *Proceedings of the conference on fairness, accountability, and transparency* (pp. 220–229). doi: 10.1145/3287560.3287596
Narayanan, A. (2023, March). *Does ChatGPT have a liberal bias?* Retrieved November 2, 2023, from https://www.aisnakeoil.com/p/does-chatgpt-have-a-liberal-bias
Orchard, A., & Radke, D. (2023). An analysis of engineering students' responses to an AI Ethics scenario. In B. Williams, Y. Chen, & J. Neville, (Eds.), *Proceedings of the 37th AAAI conference on artificial intelligence, AAAI 2023* (Vol. 37, pp. 15834–15842).
O'Neill, E., Klincewicz, M., & Kemmer, M. (2022). Ethical issues with artificial ethics assistants. In C. Véliz (Ed.), *The Oxford handbook of digital ethics*. Oxford University Press. doi: 10.1093/oxfordhb/9780198857815.013.17
Patterson, D., Gonzalez, J., Le, Q., Liang, C., Munguia, L.-M., Rothchild, D., So, D., Texier, M., & Dean, J. (2021). *Carbon emissions and large neural network training.* arXiv. doi: 10.48550/arXiv.2104.10350
Pessach, D., & Shmueli, E. (2022). A review on fairness in machine learning. *ACM Computing Surveys, 55*(3), 51:1–51:44. doi: 10.1145/3494672
Petrozzino, C. (2021). Who pays for ethical debt in AI? *AI and Ethics, 1*(3), 205–208. doi: 10.1007/s43681-020-00030-3
Prasad, S., Greenman, B., Nelson, T., & Krishnamurthi, S. (2023). Generating programs trivially: Student use of large language models. *Proceedings of the ACM Conference on Global Computing Education, 1* . doi: https://doi.org/10.1145/ 3576882.3617921
Prunkl, C. (2022). Human autonomy in the age of artificial intelligence. *Nature Machine Intelligence, 4*(2), 99–101.
Quedado, J., Zolyomi, A., & Mashhadi, A. (2022). A case study of integrating fairness visualization tools in machine learning education. In *Extended abstracts of the 2022 CHI conference on human factors in computing systems.* ACM. doi: 10.1145/3491101.3503568
Raji, I. D., Kumar, I. E., Horowitz, A., & Selbst, A. (2022). The fallacy of AI functionality. In *2022 ACM conference on fairness, accountability, and transparency* (pp. 959–972). ACM. doi: 10.1145/3531146.3533158
Raji, I. D., Scheuerman, M. K., & Amironesei, R. (2021). You can't sit with us: Exclusionary pedagogy in AI ethics education. In *Proceedings of the 2021 ACM conference on fairness, accountability, and transparency* (pp. 515–525). New York, NY, USA: ACM. doi: 10.1145/3442188.3445914
Ramachandran, V., Hardebolle, C., Kotluk, N., Ebrahimi, T., Riedl, R., & Jermann, P. (2023). *A multimodal measurement of the impact of deepfakes on the ethical reasoning and affective reactions of students.* Proceedings of the 51st SEFI Annual Conference 2023. Dublin, Ireland.
Register, Y., & Ko, A. J. (2020). Learning machine learning with personal data helps stakeholders ground advocacy arguments in model mechanics. In *Proceedings of the 2020 ACM conference on international computing education research* (pp. 67–78). ACM. doi: 10.1145/3372782.3406252
Russell, S. J., & Norvig, P. (1995). *Artificial intelligence a modern approach.* Prentice hall.
Sætra, H. S., & Danaher, J. (2023). Resolving the battle of short- vs. long-term AI risks. *AI and Ethics*. doi: 10.1007/s43681-023-00336-y
Saltz, J., Skirpan, M., Fiesler, C., Gorelick, M., Yeh, T., Heckman, R., Dewar, N, & Beard, N. (2019). Integrating ethics within machine learning courses. *ACM Transactions on Computing Education, 19*(4), 32:1–32:26. doi: 10.1145/3341164
Sanderson, C., Douglas, D., & Lu, Q. (2023). Implementing responsible AI: Tensions and trade-offs between ethics aspects. In *2023 International joint conference on neural networks (IJCNN)* (pp. 1–7). doi: 10.1109/IJCNN54540.2023.10191274

Sattlegger, A., van den Hoven, J., & Bharosa, N. (2022). Designing for Responsibility. In *DG.O 2022: The 23rd annual international conference on digital government research* (pp. 214–225). ACM. doi: 10.1145/3543434.3543581

Scharowski, N. (2020). *Transparency and trust in AI* (Unpublished doctoral dissertation). Institute of Psychology.

Schmidt, P., Biessmann, F., & Teubner, T. (2020). Transparency and trust in artificial intelligence systems. *Journal of Decision Systems, 29*(4), 260–278.

Shapiro, B. R., Meng, A., O'Donnell, C., Lou, C., Zhao, E., Dankwa, B., & Hostetler, A. (2020). Re-shape: A method to teach data ethics for data science education. In *Proceedings of the 2020 CHI conference on human factors in computing systems* (pp. 1–13). ACM. doi: 10.1145/3313831.3376251

Srinivasan, R., & Uchino, K. (2021). Biases in generative art: A causal look from the lens of art history. In *Proceedings of the 2021 ACM conference on fairness, accountability, and transparency* (pp. 41–51).

Stark, L., & Hoey, J. (2021). The ethics of emotion in artificial intelligence systems. In *Proceedings of the 2021 ACM conference on fairness, accountability, and transparency* (pp. 782–793). Association for Computing Machinery. https://dl.acm.org/doi/proceedings/10.1145/3442188

Suresh, H., & Guttag, J. (2021). A framework for understanding sources of harm throughout the machine learning life cycle. In *Equity and access in algorithms, mechanisms, and optimization* (pp. 1–9). ACM. doi: 10.1145/3465416.3483305

Suresh, H., Lao, N., & Liccardi, I. (2020). Misplaced trust: Measuring the interference of machine learning in human decision-making. In *12th ACM conference on web science* (pp. 315–324). ACM. doi: 10.1145/3394231.3397922

Tigard, D. W., Braun, M., Breuer, S., Ritt, K., Fiske, A., McLennan, S., & Buyx, A. (2023). Toward best practices in embedded ethics: Suggestions for interdisciplinary technology development. *Robotics and Autonomous Systems, 167*, 104467. doi: 10.1016/j.robot.2023.104467

Tuovinen, L., & Rohunen, A. (2021). Teaching AI ethics to engineering students: Reflections on syllabus design and teaching methods. In *Proceedings of AAAI'23/IAAI'23/EAAI'23* (pp. 15834–15842). doi: 10.1609/aaai.v37i13.26880

Widder, D. G., & Nafus, D. (2022). *Dislocated accountabilities in the AI supply chain: Modularity and developers' notions of responsibility.* arXiv. doi: 10.48550/arXiv.2209.09780

Williams, T., Zhu, Q., & Grollman, D. (2020, April). An experimental ethics approach to robot ethics education. *Proceedings of the AAAI Conference on Artificial Intelligence, 34*(09), 13428–13435. doi: 10.1609/aaai.v34i09.7067

Winfield, A. F., Booth, S., Dennis, L. A., Egawa, T., Hastie, H., Jacobs, N., Muttram, R. I., Olszewska, J. I., Rajabiyazdi, F., Theodorou, A., Underwood, M. A., Wortham, R. H., & Watson, E. (2021). IEEE P7001: A proposed standard on transparency. *Frontiers in Robotics and AI, 8*, 665729.

SECTION 2

Interdisciplinary contributions to engineering ethics education

Roland Tormey

This section of the handbook aims to provide readers with a conceptual toolbox that can enhance the reading of ideas and concepts that emerge in other sections and later chapters. This reflects that many of the ideas and directions of travel in engineering ethics education come from disciplines as diverse as philosophy, sociology, psychology, organizational and management studies, engineering design, and law, as well as from the practices and epistemologies of different engineering disciplines, such as mechanical, chemical, electrical, software or civil engineering, among others (this issue is explored in more depth in Section 3 of this handbook). It also reflects the idea that some of the most productive conversations in engineering ethics education are those that can transcend the assumptions and methods of a single discipline.

Section 1 has dealt with some foundational issues and debates in engineering ethics education, but here in Section 2, the term 'foundations' is used in a slightly different way. The curriculum for the education of professionals (such as engineers or teachers) is often based on several different 'foundational' disciplines for the profession in question. For example, teachers start by learning some psychology, sociology, history, and philosophy, alongside learning the subject content they will teach, before seeing how to apply these in teaching that subject. Similarly, engineers learn physics and mathematics (and perhaps now computer science too), alongside learning the specific technical knowledge and practices of their engineering domain, before seeing how these all apply in their specific engineering discipline. In recent decades there have been moves toward more integrated curriculum models such as problem-based learning or the 'Conceive, Design, Implement, Operate' (CDIO) syllabus, designed to help students learn these theoretical foundations alongside learning to apply them in practice. However, even in such models, there remains an implicit sense that the foundational disciplines must be learned to become a professional (PBL and CDIO curriculum models are explored in Chapter 21).

In this section, we try to map out the 'foundational' disciplines for the interdisciplinary professional practice that is engineering ethics education. As we noted in the handbook's introduction, this implies understanding different subjects, learners and learning, and teaching methods. Reflecting the subjects underpinning how engineering ethics is taught in different places, these foundations certainly include philosophy, sociology, critical theory, organizational studies and law. Since our focus involves understanding teaching and learning as well as understanding ethics content, moral and social psychology also has a significant contribution to make (alongside sociology

DOI: 10.4324/9781003464259-10

and philosophy, which also have important contributions to make to understanding the educational process). And, since we are exploring how ethics is taught in engineering programs, engineering design is, itself, a foundational discipline that needs to be understood to effectively teach engineering ethics.

To call these disciplines 'foundations' is perhaps misleading because engineering ethics education is not a stable structure that rests on a solid and unchanging base (like a building's foundation). These 'foundational' disciplines are constantly changing, and these shifts in the foundations change the face of engineering ethics education itself. When van de Poel and Royakkers (2011) wrote the introduction to their engineering ethics textbook, for example, they noted that their text included many ideas that would be less familiar from older texts: a focus on engineering design, sustainability, and the social nature of engineers' work. When recently writing a new edition, they again found new dimensions to include, because engineering ethics education has not ceased to develop. Any account of the disciplinary foundations of engineering ethics must also address where it is going and where it is and has been. In the chapters that follow, several ideas emerge as common themes.

Positionality

In common with the chapters across this handbook, making explicit our positionality as editors is useful. Doing so can help readers assess how each editor's positionality impacts the decisions that helped shape their respective sections of the handbook. The editor of this section, Roland Tormey, is a sociologist and a learning scientist who taught for a decade and a half in teacher education before switching focus to engineering education. It is important to him, therefore, that this section does not just explore ideas but also asks how they impact on how we teach and what we teach. Roland's research work has often been at the intersection between sociology and psychology, and this has certainly influenced the focus on interdisciplinary perspectives in this section. He has worked closely with colleagues in several African countries (notably Rwanda and Uganda), and much of his research work has focused on diversity, equity, and inclusion issues in education; this is certainly reflected in the focus on colonization, power, and exclusion which runs throughout the chapters in this section. Finally, much of his research over the past decade and a half has been on emotion in teaching and learning, including in the learning of engineering ethics. This, too, has certainly colored the choices which gave rise to this section.

Chapter topics

In Chapter 8, 'Engineering Ethics Education through a Critical View: Some Philosophical Foundations,' Cristiano Cordeiro Cruz, Aline Medeiros Ramos, and Jie Gao begin by questioning the way in which engineering ethics is traditionally framed by exploring the relationship between secular ethics and religious thinking. In doing so, they raise questions about the way in which the mythology that Western ethics creates for itself (as secular, rational, and developed) serves a colonial agenda. They introduce seven distinct ethical systems, with a particular focus on three non-Western paradigms: South American *Buen Vivir*, African *Ubuntu*, and Asian Confucianism. In doing so, they raise questions about how hegemonic Western ethical theories (consequentialism, deontology, virtue ethics) frame questions of connectedness, power, and liberation in particular ways, and how these things can be seen differently.

Robert Braun, John Kleba, and Richard Randell pick up on these themes in Chapter 9, 'Sociological, Postcolonial, and Critical Theory Foundations of Engineering Ethics Education.' Building on the work of Mitcham, they note that engineers shape the world but do so in ways that are built on what feels like 'common sense' but are rather an unquestioned and taken-for-granted set of practices, ethics, ontologies, epistemologies, and political philosophies. They note that "Ontologies – at least in the Global North, with its universalist and hegemonic ambitions to explain how the world ostensibly actually is – are normative, they aim to determine what is real and what is not." Recognizing this leads to questioning the ways in which material objects and systems become hegemonic (Western) signifiers in technoscientific modernity. Critical theory, postcolonial theory, and Science and Technology Studies (STS) are all presented as being lenses that can help us to reflect on ways in which power asymmetries are engineered into everyday life and as providing insights through which alternative worlds can be imagined.

Inês Direito, Curwyn Mapaling, and Julianna Gesun in Chapter 10, 'Psychological Foundations of Engineering Ethics Education,' shift the gaze from the external social world to the internal mental world of the ethical engineer as decision-maker. Just as Chapter 8 identifies the hegemonic power of Western normative theories, this chapter explores the parallel hegemonic power of the Kohlbergian tradition in psychological research focused on understanding and measuring moral reasoning. Alongside this, the authors identify alternative important research traditions such as those focused on empathy and care, on moral intuitions and, more recently, on positive psychology. They call for an increased focus on students' thriving through an integration of positive psychology approaches to thinking about engineering ethics education. This chapter connects to work on emotion and reason in Section 1 (Chapter 4) as well as raising issues about how and what we measure when assessing the impact of engineering ethics education, topics that re-emerge in Section 6.

As with previous chapters in this section, Chapter 11, 'Organizational Studies and Engineering Ethics Education,' by Silvia Bruzzone and Silvia Gherardi, shifts the focus from describing ethics as being something an individual does to describing ethics as something that emerges in the situated practice of being an engineer. Rather than ethics involving one person being responsible for making the right decision, the authors frame ethical questions as involving multiple actors in different roles, each of whom potentially understands what is happening in different ways and who together engage in discursive practices to negotiate understandings of a situation. Since ethics is always situated, they argue, learning from situated cases and stories "would allow engineering students to immerse themselves in professional practices and situations in which the complexity of 'everyday dilemmas' requires them to think about situations, thus training their capacity of responsiveness." They illustrate this with three cases that tease out what the concepts of situated doings, collective knowing, sociomateriality, and response-*ability* can mean when applied to engineering ethics education. The focus on situatedness and perspective taking resonates, for example in later chapters in Section 3 on ethics pedagogies, notably the chapters on case studies (Chapter 20) and service learning (Chapter 23).

In Chapter 12, 'Ethics and Engineering Design Foundations,' Diana Bairaktarova, Natalie Wint, and Mauryn C. Nweke also engage in themes that have resonated throughout this section, exploring the inherently social nature of the engineering design process. They explore the ways in which the engineering design process is traditionally described, identifying how ethical considerations enter into each stage of this process. They consider how a range of approaches, such as empathic design, value-sensitive design, and human-centered design, work to make visible the (often hidden) social nature of the process (an idea picked up again in Chapter 20). Linking to themes explored in Chapter 9, they also identify how STS can enrich the discourse on ethics in engineering design. The themes and topics raised in this chapter resonate throughout both Section 4 (exploring ethics pedagogies) and Section 3 (exploring how ethics can be integrated in different engineering disciplines).

Chapter 13, 'Law in Engineering Ethics Education: An Exploration,' by Andreas Kotsios, Thomas Taro Lennerfors, and Mikael Laaksoharju, in common with other chapters in this section, begins by questioning some of the taken-for-granted assumptions that are part of everyday discourse on ethics – in this case, the relationship between what is ethical and what is legal. They note that while the ethical and the legal are often framed as being in opposition (as evidenced by debates as to whether engineers should be learning more about ethics or law), an exploration of the relationships and tensions between the two frameworks provides an opportunity for thinking about right, wrong, responsibility, the role of prior understanding of consequences in decision-making, and the ways in which arguments are formulated and supported in relation to a decision. This harks back to ideas that appear in Chapter 1 of this handbook. These questions are explored through a number of different case studies of integrating law and ethics in engineering ethics.

Chapter topics, trends, and implications

In these chapters, several ideas emerge as common themes. The question "Whose Reality Counts?" was posed by Robert Chambers (1997) in the title of a book that assessed how postcolonial power relations shaped what was held to be 'real' and how this, in turn, meant Western technological expertise typically ignored the indigenous knowledge of communities in the Global South. This often resulted in 'solutions' that didn't work, a waste of resources, and negative impacts for the communities affected. The question of how social power affects what is deemed 'knowledge' is one that reoccurs throughout these chapters. In Chapter 11, 'Organizational Studies and Engineering Ethics Education,' Silvia Bruzzone and Silvia Gherardi ask how contemporary organizational studies provide conceptual tools to comprehend, for example, how 'fire risk' is understood, by whom it is understood, and what consequences result from how fire safety technologies are developed and deployed.

Noting that engineers are tacit sociologists in that their work is predicated on (often implicit) theories regarding the nature of social life, Robert Braun, John Kleba, and Richard Randell explore how technologies are built upon and embed taken-for-granted notions about social relationships as well as relationships with the natural world in Chapter 9, 'Sociological, Postcolonial, and Critical Theory Foundations of Engineering Ethics Education.' The authors discuss a range of concepts for making sense of these relationships, including sociotechnical systems, socio-materiality, lyseology, and ontopolitical power, to name a few.

In Chapter 8, 'Engineering Ethics Education through a Critical View: Some Philosophical Foundations,' Cristiano Cordeiro Cruz, Aline Medeiros Ramos, and Jie Gao turn the question on the discipline of philosophy itself, asking how the assumptions of individualist technoscience are supported by the exclusion or marginalization within engineering ethics of normative ethical theories other than the 'big three' normative frameworks of White, European men: consequentialism, deontology, and virtue. They explore how including a broader set of normative perspectives allows us to rethink the basis of engineering ethics. Inês Direito, Curwyn Mapaling, and Julianna Gesun (Chapter 10) similarly question how particular voices – in their case, a 'feminine voice' – have been heard or silenced in moral psychology, while Diana Bairaktarova, Natalie Wint, and Mauryn C. Nweke (Chapter 12) bring the question to the heart of engineering practice and ask how users – or even 'humans' – are 'centered' or 'de-centered' in the design process. Andreas Kotsios, Thomas Taro Lennerfors, and Mikael Laaksoharju (Chapter 13) explore the diverse ways in which different positions can be juxtaposed or reconciled in both law and in ethics, noting that the rules for coming to a decision and taking responsibility are different in both, and that each reproduces – in different ways – established power relations.

A second theme that cuts across all chapters in this section is the decentering of the individual engineer in ethics education. While it has been common for engineering ethics education to prioritize micro-ethical questions (Lönngren, 2021; Polmear et al., 2019; Swan et al., 2019), these chapters locate engineers and engineering in a network of relationships with the social and natural world which requires thinking about engineering ethics *in context* (which is to say, in the lived reality of specific issues in particular places and times), rather than as the sovereign authors of individual ethical decisions. This is evident in Chapter 8, which directs our attention to relational normative theories like the ethics of care, *Buen Vivir*, *Ubuntu*, and Confucianism (picking up themes first raised in Section 1). It is also evident in the prominence Direito, Mapaling, and Gesun give to ethics of care in Chapter 10, 'Psychological Foundations of Engineering Ethics Education,' including moral psychology foundations.

Chapters 9 (by Braun, Kelba, and Randall), 11 (by Bruzzone and Gherardi), and 13 (by Kotsios, Lennerfors, and Laaksoharju) all focus on social practices in which individuals do not make decisions alone but in which power is shaped, negotiated, and obscured in the relations between a range of actors, including engineers, farmers, scientists, managers, accountants, politicians, regulatory and policy departments, research centers, legislators, lawyers, advertisers, and others. This power shapes and is shaped by the framing of what decisions are possible and what becomes taken-for-granted and unquestioned. In this context, Chapter 11 (by Bruzzone and Gherardi) cites the organizational sociologist Stewart Clegg, who defined ethics as "the social organizing of morality, the process by which accepted and contested models are fixed and refixed, by which morality becomes ingrained in the various customary ways of doing things" (Clegg et al. 2007, p. 111).

A third shift evident across these chapters is from purely cognitive and rationalist accounts of ethics to accounts that view ethics in a more 'whole person' way. This builds on ideas exploring 'intuitive' and 'reflective' ethical decision making which were raised in Section 1 and foreshadows questions of intuition (thinking fast) and reflection (thinking slow) which are returned to in Chapter 25. As noted in the Clegg definition (cited in Chapter 11), ethics is as much about 'customary ways of doing' as it is about ways of thinking. Hence Bruzzone and Gherardi propose that engineering ethics education must engage more proactively with "sensorial and embodied types of knowledge (not just cognitive knowledge)." Similarly, the focus on moral intuitionism in Chapter 10 (Direito, Mapaling, and Gesun) directs our attention from the cognitive to the pre-cognitive, while the ethics of care approach – described in both the philosophical (Chapter 8) and psychological (Chapter 10) foundations chapters – draws our attention to the role of emotional relationships and empathy in pro-social behavior.

This theme is further elaborated in the description of empathic design in engineering practice in Chapter 12, 'Ethics and Engineering Design Foundations,' by Bairaktarova, Wint, and Nweke. An underlying theme across several of these chapters is that "emotions cannot be avoided in engineering ethics … [so] teachers and researchers need to be much more deliberate in addressing emotion. Ignoring them will not make them go away" (Kotluk & Tormey 2023, p. 736).

Conclusions from the editor of this section

The chapters in this section have a great deal of commonality: they all start by questioning the common-sense or hegemonic ways in which engineering and/or engineering ethics are understood. In doing so, they raise questions about power, and about whose voices are heard and whose are excluded in debates around engineering ethics education. Concepts like empathy, postcolonial theory, and social practice re-emerge in multiple chapters and act as connecting threads for the section.

The reader will note, however, that despite their connections, writing styles vary radically from chapter to chapter, reflecting the differences in writing styles and types of argumentation typical of different disciplines. Each authoring team has explicitly addressed their positionality to aid readers in making sense of this difference in styles (just as we editors have done). Reading any of the chapters in this section alone will undoubtedly be fruitful. But reading multiple chapters will give the reader more than the sum of the parts.

References

Chambers, R. (1997). *Whose Reality Counts? Putting the First Last*. Rugby, Warwickshire: Practical Action Publishing.

Clegg, S., Kornberger, M., & Rhodes, C. (2007). Business Ethics as Practice. *British Journal of Management, 18*(2), 107–122. https://doi.org/10.1111/j.1467-8551.2006.00493.x

Kotluk, N., & Tormey, R. (2023). Compassion and engineering students' moral reasoning: The emotional experience of engineering ethics cases. *Journal of Engineering Education,112(3)*,719–740. https://doi.org/10.1002/jee.20538740

Lönngren, J. (2021). Exploring the discursive construction of ethics in an introductory engineering course. *Journal of Engineering Education, 110*, 44–69. https://doi.org/10.1002/jee.20367

Polmear, M., Bielefeldt, A., Knight, D., Canney, N., & Swan, C. (2019). Analysis of macroethics teaching practices and perceptions in engineering: A cultural comparison. *European Journal of Engineering Education*, 44(6), 866–881. https://doi.org/10.1080/03043797.2019.1593323

Swan, C., Kulich, K., & Wallace, R. (2019). A review of ethics cases: Gaps in the engineering curriculum. *Paper presented at the ASEE Annual Conference, Tampa, FL*. Retrieved from https://peer.asee.org/31988

van de Poel, I. & Royakkers, L. (2011). *Ethics, Technology, and Engineering; An Introduction*. Chichester: Wiley-Blackwell.

8

ENGINEERING ETHICS EDUCATION THROUGH A CRITICAL VIEW

Some philosophical foundations

Cristiano Cordeiro Cruz, Aline Medeiros Ramos, and Jie Gao

Introduction

This chapter presents some fundamental philosophical and religious ideas that serve as the background for thinking about ethics and morality. It also sketches some elements that stress how engineering and technology are shaped by and shape how we live individually and collectively, as well as how we make sense of ourselves, life, and reality as a whole. This should help our readers understand why studying ethics – and philosophy of engineering or technology – is important, especially if engineering is to be used to empower and liberate marginalized persons or communities and construct other possible social arrangements and meanings for life.

The chapter is divided into four sections and relies not only on texts that are part of the 'canon' of philosophy and the social sciences but also on less commonly read sources that we think are worth integrating into the engineering ethics mainstream if we want to have a forward-looking approach more in tune with critical perspectives.

Where are this chapter's authors writing and thinking from? Cristiano Cruz is a Brazilian researcher with a background in engineering and philosophy who currently investigates emancipating engineering interventions aiming at helping decolonize engineering practice and education as well as the philosophical reflection on technical design and technology. He is a member of the Brazilian network of popular/grassroots engineering – teaching, and doing research and extension, at two Brazilian engineering schools. Aline Medeiros Ramos, a Brazilian philosopher based in Canada, specializes in medieval philosophy and ethics She has a background in classics, and she teaches courses on the history of philosophy and professional ethics, especially to engineering and medical students. Jie Gao is a doctoral candidate based in Switzerland. With a background in philosophy of mind and social sciences, she is now conducting interdisciplinary research in learning science, specifically within the context of sustainability education, on sense-making and emotional development. In addition to her research, Jie actively contributes to teaching and research in the Humanities education of engineers at her university.

DOI: 10.4324/9781003464259-11

Why is studying/discussing ethics important for engineers?

Why should engineers study or discuss ethics in general and the philosophy of technology/engineering in particular? One straightforward answer is: 'Higher education should help engineers become better citizens and better human beings, in addition to giving them the training they need to engage in their professional activities.' Compelling as this answer is, it is not the only one, even if we recognize that students are usually still developing their morality at the time they enter university (Clancy & Zhu, 2022).

Technology, society, and worldview: mutual shaping and supporting

A less common yet notable answer to the question of why engineering students should study ethics and philosophy of technology is that both engineering and designing or creating technology are crucial in

1) Supporting or creating any ethical-political order that can be less or more hierarchical, participatory, conservative, diverse, respectful, and so on (Feenberg, 2010, 2017), or
2) Emulating any cosmology or worldview that, for instance, can be individualistic and take everyone and everything as resources or consider everything as interrelated parts of an integrated whole of which one must take care (Hui, 2016, 2017; Hui & Lovink, 2017).

In other words, the reality in which we live is simultaneously social and technical (i.e., sociotechnical), meaning that not only is technology shaped by society and the interests, values, and/or strategies of powerful groups, but technology also, and conversely, shapes society. For example, bridges built low on purpose to prevent buses from using the highway below them (e.g., by Robert Moses in New York City, an idea translated elsewhere) emulate a racially segregated reality intended in the first place by the racial prejudices of their designer (Winner, 1986). Even though non-accessible cities are usually not a consequence of any intended strategy, they still replicate a reality wherein people with a physical disability find much trouble navigating (Winner, 1986). In this case, like in many others, long-lasting socially sustained values and prejudices or preconceptions keep unwittingly driving designers' choices. The reality these technologies help mirror and perpetuate is an oppressing one, supported by such values and preconceptions; an example is the construction of racist algorithms and digital technologies (Poster, 2019).

Further, even the basis upon which we humans make sense of society and everything else (and, consciously or not, take as the fundamental guiding for our living and acting in the world) – that is, our cosmology or worldview – shapes and is shaped by technology. As Arturo Escobar says, "Give me a maloca [i.e., indigenous longhouse], and I will raise a relational world (including the integral and interdependent relations between humans and non-humans); conversely, give me a suburban home, and I will raise a world of de-communalized individuals, separated from the natural world" (Escobar, 2018, p. 111). This understanding holds in many other cases, for instance, if 'maloca' is replaced with 'agroecology' and 'suburban home' with 'modern, mechanized agriculture.'

That is why Hui (2016, 2017) coined the term *cosmotechnics* to highlight the mutual and supporting relation between cosmology or worldview and technology. Hui argues that any cosmology needs specific technologies to be emulated (like the South American Indigenous relational and caring worldview needs a *maloca*) and that any cosmotechnic (like the mainstream one of which American suburban houses are one materialization) builds or emulates the cosmology it draws on (like the one underpinning the American suburban houses, which takes the individual as a being both de-communalized and separated from the natural world).

Hui's cosmotechnic allows us to see and/or recognize some important things. Hui helps us see that the dominant Western, modern, capitalist cosmotechnic, which emulates a world of natural and human resources to be exploited for profit maximization, is one among *many* other possible cosmotechnics. Such dominant cosmotechnic is shaped by and shapes the dominant cosmology that is individualistic, racist, sexist, and specist (this last term meaning that it places the highest value on humans that are thus taken as entitled to dispose of other species and nature as a whole as humans see fit). Another realization from Hui is that if we want to build different worlds – ones that mimic relationality, solidarity, care, and so on – we must construct other cosmotechnics (starting by appropriating or changing the already available one).

In sum, there is no such thing as neutral technology that can be used to advance any ethical-political order or any cosmology. Engineering is never politically neutral; it either works within, reinforces, and re-creates a political/social and/or cosmological status quo, and is thus conservative, or it challenges it, and is thus progressive, empowering, emancipating,[1] or decolonial)[2]. In other words, being 'politically or cosmologically neutral' in engineering is not a choice, for it is impossible. Doing engineering will always and inevitably be either sustaining or changing reality.

Progressive, empowering, emancipating, or decolonial engineering

To engage in progressive, empowering, emancipating, or decolonial engineering, to help socially and cosmotechnically co-construct any other possible world, requires practicing engineering differently from the mainstream or dominant form (Cruz 2021a, 2021b; see also Chapter 6 of this handbook). Such a practice, and the knowledge systems that support it, can only be achieved by somehow considering or incorporating into engineering the ethical-political and/or cosmological values and fundamentals we want to see respected or served in the world we want to help build (Cruz, 2021a).

It thus seems correct to say that venturing into non-Western or non-dominant cosmologies and developing engineering practices and technical solutions that support them can allow us to work alongside marginalized groups and communities – which nurture or cherish such cosmologies – in the construction of the sociotechnical reality they want. It also means widening our capacity for 'doing' engineering and for developing technology, even if we stick to our worldview, whatever that worldview might be (Cruz, 2021a). Yet we also stress the need to relearn and examine the worldviews and world histories we encounter, as a first step toward open dialogue allowing for a plurality of worldviews (for more on this, see Chapter 6).

There are countless ways to widen (or decolonize) Western, capitalist, dominant engineering. One option is using Scandinavian participatory design (Simonsen & Robertson, 2013) in its emancipatory strand (Robertson & Simonsen 2013). Another, which arose in Latin America, is *popular engineering* (PE) – meant as *grassroots* engineering – and named after Paulo Freire's 'popular education.' It is an educative process aimed at helping to emancipate people (Freire, 1970) that is taken as a guiding principle for *popular engineers*. PE draws on *action research* (Coghlan, 2021) and social technology's *sociotechnical adequacy* (Dagnino et al., 2004) to both help the supported group and community dream the world(s) they might find worth building and sociotechnically or cosmotechnically build this (these) world(s) alongside them. As part of that process, a dialogue of knowledge is established between the supported group or community and the technical team. Both 'sides' teach and learn, thereby enriching each other's capacity to know, be, and act. This dialogue widens (or decolonizes) engineering (Cruz, 2021b).

PE seeks to help empower/emancipate the supported group or community as much as possible. Empowerment through sociotechnical intervention or design has at least eight dimensions, ranging from *sociotechnical inclusion* (e.g., giving people access to a service that improves their basic conditions for living well) to *political emancipation* (i.e., community capacity and support for self-determination aimed at its members' flourishing and not harming anyone else) (Kleba & Cruz, 2021). The more these dimensions are addressed, and the more caring and critical or questioning this process is, the more empowering and emancipating its outcomes (Kleba & Cruz, 2021). PE aims at the highest possible emancipation.

Accomplishing that level of emancipation is far from easy; it demands much more than just well-established methods. To practice PE, engineers need training complementary to the traditional, technocratic education they usually receive at the university. In undergraduate courses, such training can be obtained through socially and environmentally committed extension activities. That is the main form PE takes today in Brazil. Indeed, many of Brazil's most successful PE teams are found in extension centers and are formed by teachers, techno-administrative employees, and graduate and undergraduate students (Cruz, 2021b).

Progressive engineering, ethics, and the remainder of the chapter

The remainder of this chapter reflects deeply on ethics to illustrate how diverse the human ethical and cosmological landscape is and can be. These sections can help denaturalize the mainstream or dominant ways of conceiving rightness and fulfillment and making sense of reality, allowing us to critically question what might seem unquestionable. This is to help the reader build skills and the ability to imagine other possible ways of being and flourishing. Such de/re-construction is a fundamental first step toward any progressive, empowering, emancipating, or decolonial engineering (or engineer).

Ethics as a tug of war between philosophical and religion traditions

Ethics and religion: a brief historical perspective

Throughout human history, religion and ethics have closely intertwined. Many ethical systems have been based on metaphysical conceptions of nature and human beings, aligned with specific religious beliefs. Although ethics education is often presented as secular (i.e., separate from religious influence and beliefs), it can be influenced by religious beliefs and prior moral experiences (such as powerful experiences that students have via informal learning contexts like study abroad, service learning, social groups, etc.).

The relationship between religion and moral philosophy has a long history. In ancient Greek philosophy, piety was considered a moral issue, as seen in Plato's *Euthyphro* (2017). During the European Middle Ages, the teaching of ethics and religious doctrine was interconnected due to the intimate association of theology and philosophy.[3] Some of what is now taught in secular ethics classes was taught as part of the theology curriculum at the first universities (Marenbon, 1990).

Religious creeds have had significant impact on ethical systems. Christians, Jews, and Muslims, for example, all rely on some form of the Divine Command Theory of meta-ethics (Hare, 2015). But the boundaries between religious beliefs and ethical reasoning are not always clear. The *Euthyphro dilemma*, which makes one wonder whether a certain way of acting is right because the gods command it or rather if the gods command it because it is right (Plato, 2017, *Euthyphro*,10a), has been taken up in secular, philosophical discussions of ethics. In the European Middle Ages, natural law theories added a theistic aspect to Aristotle's theory of the four causes:[4]

by referring to a belief in a god involved in the creation and workings of the universe and who thus influences human life and experience, these theories have claimed that natural law is not merely descriptive but also prescriptive, because God is the ultimate source and final cause of creation. The Decalogue, often referred to as 'the Ten Commandments' that form a significant part of the religious and moral foundation of Abrahamic religions (Exodus 20:1–17; Deuteronomy 5:6–21), finds a parallel in deontological ethical systems such as Kant's (Sandberg, 2013). The so-called 'Golden Rule' (i.e., 'treat others as you would like to be treated') found in many religious traditions across the globe, for instance, finds parallels in many philosophical moral systems (Blackburn, 2001, p. 101) and even in evolutionary psychology (Hare, 2015, 2019; Greene, 2013). Numerous philosophical theories have emerged from contemplation of religion and doctrine, and conversely, religious and doctrinal thought has also been influenced by philosophy (Hare, 2019).

However, not all ethical theorists ground their systems in religious claims. Some make a point of distancing themselves from religious beliefs, declaring that they rely on reason alone. This is often the case with consequentialist theories. Jeremy Bentham criticized religion and its institutions (Bentham, 1818/2011, 1822, 1823/2013). John Stuart Mill proposed a moral theory that was not grounded in religious beliefs, advocating a purely scientific or philosophical approach, a "religion of humanity" (Mill, 1874/1974, pp. 69–124).

Ethics: a current perspective

Ethics, also called moral philosophy, is nowadays taken to be the philosophical discipline concerned with distinguishing between right and wrong or good and bad, regardless of religious beliefs. It encompasses the study of moral systems, beliefs, and practices. It requires higher-level thinking to engage in reflection, critical thinking, argument building, justification, and application of moral beliefs, ideas, and systems (Kaurin, 2018). Ethics is not reducible to personal opinions or preferences, and is often understood as a normative discipline that deals with the obligations individuals have towards themselves and others, including future generations, non-human animals, living beings, supernatural entities, and ancestors' souls. Ethical discussions often include providing reasons for our choices and considering "what it means to be a conscientious moral agent" (Rachels & Rachels, 2018, p. 13).

While scholars in fields such as biology, economics, and cognitive science have tried to describe and explain morality, the capacity for a moral sense in humans is believed to have arisen through an interplay of biology and culture. Although foundational ethical beliefs, such as the proscription of the murder of innocent persons, have remained constant and consistent throughout human history, how ethical standards are interpreted and applied can differ over time and across geographical contexts, and may be influenced by religious views and cultural variables, such as history, institutional regulations, and social ecologies. Every ethical system is based on and reflects a particular cosmology or worldview, which includes beliefs about human nature, the ontological and moral status of other beings, and the essence of reality. As noted earlier in this chapter, some people may view reality as solely material and mechanistic, while others view it as an interconnected web of living entities or a sacred whole. These assumptions can lead to different attitudes toward the natural world. Some may see themselves as exceptional beings, superior to other species and entitled to dominate and exploit nature for their own benefit. Other people may view themselves as part of a tightly interrelated and interdependent reality, responsible for its well-fbeing, and possessing a nature that is not fundamentally different from that of other living beings.

The relationship between ethics and religion has been briefly discussed above, and the remainder of this chapter will focus on 'properly' philosophical ethics, independent of religion, unless otherwise stated.

Engineering ethics: then and now

As a philosophical inquiry, ethics is nowadays divided into three main fields: (1) *meta-ethics*, which deals with the nature and meaning of ethical terms such as 'the good,' 'rights,' and 'obligations'; (2) *normative ethics*, which prescribes norms upon which ethical action ought to be based; and (3) *applied ethics*, which involves the application of moral philosophy, often of a normative nature, to practical issues. The latter is where engineering ethics often finds itself.

Historically, ethics, or moral philosophy, was less compartmentalized than it is today. In Classical Greek philosophy and its Latin medieval development, ethics was concerned with all questions regarding morality and the virtues. Take the case of craft (τέχνη or *tékhnē* – whence we get terms such as 'technique' and 'technology,' so important in engineering), known to ancient and medieval people in the Aristotelian tradition as the intellectual virtue of production (1934, *Nicomachean Ethics* VI.5 1140a *et passim*). Unlike the other intellectual virtues (namely, prudence, understanding, knowledge, and wisdom), which are concerned with the ability to reason and make correct decisions, *tékhnē* involves a practical dimension in the sense that it is the ability to produce something according to a pre-established set of rules and which is in accordance with a pre-established goal (1934, *Nicomachean Ethics* VI.5). Aristotle's distinction between craft and the other intellectual virtues is relevant to engineering ethics in that it highlights the unique ethical challenges posed by the production of technology. While the other four intellectual virtues are concerned either with the individual's theoretical intellect or with guiding moral choice, *tékhnē* involves the creation of artifacts that can have a significant impact on society and the environment. Just as there is virtue in producing conclusions from premises, there is virtue in producing something out of something else (like a statue out of a piece of marble, or a building from stones or bricks). An important caveat, however, is that Aristotle and the tradition that followed did acknowledge that it was possible for someone to be good with regards to this kind of production – that is, to have the *virtue of craft* – without necessarily being wise or being good *absolutely* (1934, *Nicomachean Ethics* VI.5 1140b; Medeiros Ramos, 2021). This raises important questions about the responsibilities of engineers and their obligations to consider the broader ethical consequences of their work, not simply the aptness or 'fit' of what they produce or design. Aristotle's argument that one can be skilled in craft without being wise or good *absolutely* underscores the importance of developing a comprehensive understanding of the ethical dimensions of technology beyond mere technical expertise and the need for investing in engineering ethics education.

Moreover, nowadays, we tend to look for sources beyond the 'Western canon' to inform our practices. Philosophical foundations from various parts of the world offer a diverse tapestry of ethical frameworks. In Asia and parts of the Middle East, sociotechnical systems and practices have evolved in contexts where philosophical and religious traditions such as Confucianism, Buddhism, Taoism, Hinduism, and Islam have been dominant. While the mutual shaping between these traditions and technological developments may not always be evident, understanding these traditions can offer insights into the rich cultural, ethical, and societal milieu in which sociotechnical systems and practices operate. The same is true for South American and African traditions, as discussed below.

Ethical systems and their presuppositions

Western and non-Western

As we have seen in the previous section and as some philosophers have noted more thoroughly, ethics and religious traditions share many foundational beliefs (Hare, 2019). In this section, we will explore some commonalities shared by some religious doctrines – both Western and 'non-Western' – and philosophical moral theories.

Western and the so-called 'non-Western' ethical traditions are rich and diverse, reflecting the historical, cultural, and philosophical influences that have shaped them. While there are some differences between these traditions, it is important to recognize that some ethical values and principles (such as compassion, justice, and respect for human dignity) can be seen as somewhat universal and can be found across most – if not all – cultures and societies.

Western ethical traditions have been influenced by various philosophical and religious perspectives, such as ancient Greek and Roman philosophy, Christian theology, and Enlightenment rationalism. These traditions have emphasized the importance of individual autonomy, reason, and human rights, among other values. In the contemporary Western context, secularism and liberalism have also played key roles in shaping ethical values and principles, to the point where we can no longer "look to Aristotle for any elucidation of the modern way of talking about 'moral' goodness, obligation, etc." (Anscombe, 1958, p. 2).

'Non-Western' ethical traditions, on the other hand, have been influenced by various philosophical and religious perspectives, including Confucianism, Buddhism, Taoism, Hinduism, and Islam. These traditions often emphasize communal and collective values, such as harmony, social order, and respect for authority. Like Western moral philosophy, 'non-Western' ethical traditions are also often closely tied to religious practices and beliefs. Confucianism, Taoism, and Buddhism are often regarded as the fundamental pillars that underpinned the social fabric of ancient Chinese society. Elements from these traditions can be seen as intertwined; for centuries they have co-existed and interacted. Individuals may exhibit reverence and adherence towards all three traditions simultaneously. As philosophies and religions, they had an impact not just on matters of spirituality, but also on domains such as governance, science, arts, and social structure. In recent academic discourse, the cultural and social traditions of East Asian societies, communities, and individuals are sometimes represented under the concept of the 'Global East,' a concept that moves beyond mere geographical boundaries and Euro-centric or North-Atlantic-centric understandings to encapsulate the essence of East Asian thought, its diaspora, and its interactions with the Global Community (Yang, 2018).

There is a common misconception that Western ethical traditions are philosophical and secular, while 'non-Western' traditions are primarily religious and thus second-rate. This misleading and oversimplified view ignores the rich and complex ethical traditions of 'non-Western' cultures. This view ignores that non-European cultures have also developed complex philosophical and ethical systems and that both traditions have been shaped by their respective religious and philosophical contexts. For instance, Western ethical thought, as we have seen, has been deeply influenced by the works of philosophers such as Aristotle, Kant, Bentham, and Mill, often nourished by or read through religious lenses. In contrast, so-called 'non-Western' ethical traditions, such as *Buen Vivir* and *Ubuntu*, as we shall see below, have their own philosophical and religious sources.

Another prevailing misconception, and the reason why we have used 'scare quotes' to talk about 'non-Western' ethics, is that non-European ethics is homogenous. Non-European cultures are often viewed as monolithic and lacking diversity, leading to oversimplified generalizations, while there actually exists a wide range of ethical traditions across different non-European cultures, each with its own unique characteristics and philosophical influences. Grouping them under the generic 'non-Western' label grossly reduces their richness and diversity to a generic form of dissension.

Antithetically, some wrongly see European ethics as universally applicable, while non-European ethics is seen as culturally specific or limited. However, both are shaped by cultural, historical, and philosophical factors and cannot be generalized without understanding their context. On another misinterpretation, non-European ethics is viewed as 'primitive' or outdated, suggesting that non-European cultures are less developed. However, non-European cultures have complex, relevant ethical systems and supporting worldviews on par with European ones.

Seven ethical traditions

Neither European nor non-European ethical traditions are static; rather they evolve and change over time in response to new social, political, and philosophical contexts. We will now examine some Western and 'non-Western' ethical traditions side by side, considering some of their religious or cosmological assumptions. Virtue ethics, deontology, and consequentialism are presented in some more detail in Chapter 2 in this handbook. Since *Buen Vivir*, *Ubuntu*, and Confucianism are typically less familiar to Western readers, we will allocate space here to these three 'non-Western' traditions.

Virtue ethics can be traced back to Ancient Greece, specifically to Aristotle's inquiry in the *Nicomachean Ethics* about what virtues, or traits of character and intellect, make a person good. Aristotle draws on the classical Greek worldview according to which the fulfillment of whatever exists – human beings included – has to do with its fitting into its natural place or realizing its natural *telos* (i.e., the ultimate end or purpose towards which something is directed or aimed), with the human *telos* being flourishing or *eudaimonia* (εὐδαιμονία).

Virtue ethics is not a normative ethical theory per se, for it does not establish a norm that ought to be followed in our deliberation regarding actions. Instead of focusing on actions, virtue ethics focuses on the development and improvement of character through considering our motivations and reasons for acting, as well as our intended goals (Hursthouse & Pettigrove, 2023). Virtue ethics continued to be the standard moral theory in Europe through the Middle Ages, when it was made to accommodate the context of Divine Command Theory and the so-called '*law* conception of ethics' (Anscombe, 1958). It was also in the Middle Ages that the theological virtues of charity, faith, and hope[5] were added to Aristotle's scheme of moral and intellectual virtues. Virtue ethics became less popular after the Renaissance (Grönum, 2015) and, from the seventeenth century on, ethics became more concerned with properly normative theories, such as deontology and consequentialism, which are action focused. Since the mid-twentieth century, however, a revival of virtue ethics has been underway, propelled mainly by Catholic philosophers like Elizabeth Anscombe (especially her 1958 article, "Modern moral philosophy," which was germinal to this revival) and Alasdair MacIntyre.[6] These twentieth-century contributions have sparked ongoing debate and brought virtue ethics back on the map as a theory worth considering in contemporary reflections, regardless of religious beliefs. Proponents of virtue ethics like Philippa Foot and Rosalind Hursthouse, for instance, came to virtue ethics from outside the framework of Catholicism, and their ideas are now ubiquitously taught in ethics classes (see, e.g., Foot, 1978; Hursthouse, 1999, 2007).

Deontology can also seem to share some common ground with religious beliefs (Hare, 2019). Since it is a duty-based ethical theory, it relies on the idea that, at least on some base level, we all owe each other something. The most famous proponent of deontology, Immanuel Kant, is a thinker from the Enlightenment who considers reason to be the most elevated and distinguishing characteristic of human beings. He takes human autonomy, which derives from reason, to be a requirement for the fulfillment of human nature (Kant, 2019). He also takes human reason and autonomy to be the foundations of the universality of the moral law. Kant's version of deontology, famously expressed through the categorical imperative (Kant, 2019), is grounded on reason (or the universalization of Western, modern reason). He argued that it is a person's reasoning and motives for acting that make an action morally right or wrong – and never the consequences of a given action. There are, however, other versions of deontology that are compatible with *Divine Command Theories* (Alexander & Moore, 2021, section 7). In those cases, it is not human rationality but God's authority that establishes the covenant that binds us to ethical obligations. Deontological ethical theories often serve as a foundation for constructing professional *codes of deontology* or *codes of conduct* in various fields, including engineering. These codes outline the ethical duties and responsibilities that individuals within a specific profession should uphold. They provide a framework that articulates the inherent obligations and rules governing professional conduct, acting as a compass for professionals to navigate ethical challenges and ensure adherence to moral principles.

Consequentialism is a kind of teleological moral theory according to which the criterion for determining the moral value of an action lies solely on the consequences of that action (Sinnott-Armstrong, 2023). Its famous maxim that 'the end justifies the means' dates back to Antiquity, but it is Chinese Mohism which is usually credited with being the earliest recorded form of consequentialist reasoning found in a religious tradition (Fraser, 2022), with its emphasis on impartiality and on the production of beneficial consequences of actions. Modern Western consequentialism traces its origins to Bentham and Mill,[7] classical utilitarians who sought to establish the basis of morality by calculating the positive and negative consequences of actions in order to identify the course of conduct that embodies the principle of utility, that is, the act that maximizes pleasure and minimizes pain for the greatest number of individuals. Modern Western consequentialism takes pleasure and pain as the fundamental drivers for human action and values each individual life equally. A classic example of utilitarian reasoning is seen in cost–benefit analyses (Audi, 2005). Such analyses are often used to tackle issues like the trolley problem, a famous thought experiment proposed by Foot (1978), in which an individual is faced with a moral dilemma: they must choose between switching a lever to divert a trolley and save five people at the cost of killing one person on an alternate track, or doing nothing and allowing the trolley to continue and kill the five people while sparing the one on the alternate track. Utilitarians would typically argue in favor of pulling the lever. This decision is based on the belief that sacrificing one life to save five results in a net gain in overall happiness, as the greater number of lives saved contributes to a more favorable outcome from a utilitarian perspective.

Ethics of care is a relatively new normative ethical theory but one that is fast-growing in popularity. It first sprung from feminism but soon developed into a more general and comprehensive account of both individual and political morality (Engster, 2007; Gilligan, 1982; Held, 1993; Noddings, 1984; Slote, 2007; Tronto, 1993, 2010). It is grounded on a relational understanding of life, either human or non-human, and on our responsibility toward people and nature around us. Although *ethics of care* is usually grouped together with *virtue ethics* because both are non-principial ethical systems, care is a practice more fundamental than cultivating a virtue. It has been argued that without care, there will be no justice, for human development and flourishing hinge fundamentally on the care that those needing it receive. In contrast to the dominant Kantian and utilitarian ethics, which require universality and impartiality in the application of moral principles – and take them as achievable – ethics of care is sensitive to contextual nuances of concrete situations, the web of relationships a person finds themselves in, and the interrelatedness of the interests of carers and cared-for. Emotions such as empathy, compassion, sensitivity are appreciated; they are relational capabilities that enable morally concerned persons. However, "we need an *ethics* of care, not just care itself," argues Virginia Held: "The various aspects and expressions of care and caring relations need to be subjected to moral scrutiny and *evaluated*, not just observed and described" (Held, 2006, p. 11).

One example of non-Western ethics is Andean *sumak kawsay*, a Quechuan expression translated into Spanish as *Buen Vivir* (Good Living). With some variations, this ethics continues to be practiced and advanced by many Indigenous peoples in South America. *Buen Vivir*'s supporting cosmology presents and enacts reality as a deeply interconnected whole governed by four main principles: relationality, correspondence, complementarity, and reciprocity (Estermann 2006, pp. 125–147). Humans are not exceptional beings detached from nature or superior to other animals. In fact, according to its perspectivism, other animals see themselves as humans and other animals as non-humans (Viveiros de Castro, 2014, chap. 2). Humans, though, have a specific role in the South American Indigenous world. We must act as cosmic shamans, mediating not only the conflicts created by our misconduct, excesses, or disturbance of natural balance but also other beings' misconduct, excesses, or imbalances (Estermann, 2006, pp. 214–215; Kopenawa & Albert, 2013, chap. 2).

We share with other beings – for example, animals, plants, mountains, rivers – the similar spirit and capacity for agency (Viveiros de Castro, 2014, chap. 2). That is why these other beings can also act wrongly. Even though the essential paradigmatic relationship among all beings (humans included) is that of hunter and prey, that does not lead us into a Hobbesian war of all against all because there is a natural tendency toward balance or cosmic order and because breaking the natural laws or balances leads to punishment (e.g., drought, flooding, lack of prey, disease) (Kopenawa & Albert, 2013).

Buen Vivir is thus closer to Stoicism than to Aristotle's virtue ethics or to Kant's deontology. "The moral order as a system of reciprocal relationships corresponds to the cosmic order as a system of complementary and corresponding relationships. Therefore, [*Buen Vivir*] is not so much a reflection on the normativity of human behavior but on its 'being' within the holistic whole of the cosmos" (Estermann, 2006, p. 246). Then, *Buen Vivir* is "both teleological and deontological ethics: the purpose of acting ethically (*telos*) is the conservation of the [cosmic] order, which at the same time is the fulfillment of a normativity felt as a duty" (Estermann, 2006, p. 252).

Buen Vivir's uniqueness, which makes it worth presenting in this handbook, primarily concerns its supporting cosmology and the distinctive way South American Indigenous peoples have lived for centuries and keep living when allowed to do so, compared to the hegemonic Western, capitalist, urban ways of life. Based on *Buen Vivir*'s principle to "Act in such a way that you contribute to the conservation and perpetuation of the cosmic order of vital relationships, avoiding disorders thereof" (Estermann, 2006, pp. 51–52), South American Indigenous peoples have shown how our ways of living can be more than only harmless to 'nature,' they can help it grow stronger, more diverse, resilient, and complex (Cunha & Almeida, 2004).

Nothing could be more appealing to us today than to (re)learn ways of living that promote nature instead of destroying it. Agroecology and malocas, as mentioned earlier, are but two examples of technology that help us get closer to fulfilling this ideal. They are two versions of a South American Indigenous cosmotechnic, drawing on and enacting or supporting the relational and interdependent cosmology they possess and allowing them to structure their collective and individual lives accordingly. An engineering practice capable of producing or improving these and other versions of South American Indigenous cosmotechnics seems highly desirable not only for the sake of working with these peoples in sociotechnical projects of their interest but also to increase non-Indigenous peoples' capacity to conceive and construct, for instance, 'nature-improving' sociotechnical solutions.

Crossing the Atlantic, we arrive in Africa, the homeland of *Ubuntu*, a unique ethics found in communities of virtually every Sub-Saharan country and rooted in notions of communitarianism, reconciliation, relationality, and interdependence (Mabele et al., 2022, p. 2). *Ubuntu*'s cosmology, like that of *Buen Vivir*, presents and enacts reality as a deeply interconnected and interdependent whole. Therefore, it makes no sense to consider an individual as an autonomous being or to take human beings as exceptional (or superior) and detached (or different in their nature/essence) from every other living being, as many Western ethics do (Mabele et al., 2022, p. 8). Unlike *Buen Vivir*, though, the central stage is not occupied by a naturally ordered cosmos and by the continuous duty of harmonizing unbalances (or injustices) caused by this cosmos' constituents, that is, human and non-human beings. Instead, *Ubuntu* focuses on the community, which starts with the community of other human beings that a person belongs to but also encompasses the person's ancestors and descendants (current and future), every other living being, and the gods (Mabele et al., 2022, p. 5; Ewuoso & Hall, 2019, p. 99). Humanness is not something a person possesses in themselves but something they accomplish through – and as – caring and life-fostering relationships with all the members of their (widened) community (Le Grange, 2019, p. 325; Ewuoso & Hall, 2019, p. 98).

The centrality of community does not mean the negation of oneself. Instead, taking the good of the (widened) community or well-being as one's primary duty means acknowledging that one cannot be well if one's community is suffering and, conversely, a community is not well if it causes suffering to (some of) its members. In other words, if one causes harm to the community, they cause harm to themselves; if one seeks good for the community, they benefit themselves (Dju & Muraro, 2022, p. 248; Ewuoso & Hall, 2019, p. 97). Therefore, *Ubuntu* does not stand for resignation concerning possible social injustices but rather for always having in mind the affirmation of others' lives and humanness as part of our search for affirming our own lives and humanness (or the conjugation of others' needs with the search for fulfilling one's own needs) (Ewuoso & Hall, 2019, p. 96; Mabele et al., 2022, p. 6). In sum, "The struggle for individual freedom, social justice and environmental sustainability is one struggle" (Le Grange, 2019, p. 325). For *Ubuntu*, "the morally right action is one that connects, rather than separates" (Ewuoso & Hall, 2019, p. 99). Thus, for some scholars, *Ubuntu* is a form of ethics of duty, while for others, it is a virtue ethics (Dreyer, 2015, p. 199; Le Grange, 2019, p. 324; Metz, 2007, p. 383).

As with *Buen Vivir*, *Ubuntu* brings forward aspects frequently forgotten in average dominant Western ways of life, so obsessed with individual autonomy and happiness. In the case of *Ubuntu*, core aspects have to do with the necessary commitment to a community's well-being, the sacredness of life-fostering relationships (to all living beings and to our descendants, ancestors, and gods), and the centrality of life and life fulfillment in our existences.

When it comes to technological development, *Ubuntu* was and still is taken as a paradigm for producing technologies attuned to, or supporting, other possible sociotechnical realities (or cosmotechnical orders). That is the case – or at least was at the beginning – of the free software Ubuntu. Many free software communities also (claim to somewhat) draw on the *Ubuntu* philosophy and ethics, understanding that "while a single company is responsible for all enhancements in a program, free software is not only free but there is a community ready and willing to improve it and distribute these improvements" (Augusto-Vieira, 2016, p. 44). More recently, *Ubuntu* has also been taken for the enrichment of artificial intelligence (AI) governance, emphasizing processes of co-operation and social harmony through the inclusion of communities most affected by AI's potential harms (Mhlambi, 2020).

Transitioning from the communal ethos of *Ubuntu*, we now journey to East Asia, exploring the tradition of *Confucianism* (also explored briefly in Chapters 32 and 33 in this volume, on accreditation of engineering ethics). This ancient yet enduring philosophy offers another relational perspective on ethics, emphasizing the role of individuals within society and the cultivation of virtues that nurture both personal and societal harmony.

Confucianism is one of East Asia's foundational ethical and philosophical systems. This tradition emphasizes a relational ontology wherein individuals exist within a dense web of duties and responsibilities. These duties are defined by various societal roles, whether familial, as seen in parent–child dynamics, or societal, as observed in friend–friend and ruler–subject interactions. Central to Confucian thought are virtues like *Ren* (仁, often translated as benevolence or humanity), *Yi* (义, righteousness or justice), *Li* (礼, ritual propriety), Zhi (智, wisdom), and Xin (信, trustworthiness). *Ren* is regarded as one of the highest values incorporating kindness and human-heartedness. According to Confucius, it is achieved through "loving others" (1998, Analects 12.22) and "overcoming oneself and returning to ritual propriety" (1998, Analects 12.1), meaning that *ren* operates within a web of virtues, rooted in traditional familial and social networks.

Contemporary Confucian society places a strong emphasis on *he* (harmony). Confucian harmony is both a metaphysical and a moral concept. It is not about uniformity but rather about the co-existence of different diverse elements, about working through creative tensions and establishing favorable relationships among them. In comparative philosophy, Confucian harmony has been discussed as a cosmic, personal, and social virtue, an ideal in relation to nature (Bell & Metz, 2011; Li & Düring, 2022). According to one of the most comprehensive scholarly introductions to Confucian harmony. authored by Li Chenyang, *he* consists of a dynamic process of harmonization instead of conformity to a pre-set order (Li, 2013). The human world, composed of individuals, families, communities, and societies, is not naturally harmonious. In this sense, the development of moral character and the attainment of harmonization may coincide within the Confucian person-making philosophy; a person of *ren* is capable of harmonizing within oneself, with others, and with the world.

When Confucianism is juxtaposed with *Ubuntu*, we see that both ethical systems clearly underscore relational ethics. However, while *Ubuntu*'s spirit 'I am because we are' underscores interconnectedness and mutual respect for one another, Confucianism delves deeper into the structure and dynamics of social roles and the duties arising from them. In a comparative analysis with Western philosophical traditions, it becomes evident that both Aristotle's virtue ethics and Confucianism champion the cultivation of virtuous character and self-examination (or *zixing*). However, the virtues which Aristotelian ethics seeks have a teleological basis: contributing to individual flourishing or *eudaimonia*. Confucian philosophy lacks teleology in the sense of a preconceived cosmic design. While Aristotle endeavors to offer an account of human relationships in the context of justice and friendship, his emphasis remains on the individual (Sim, 2007). In contrast, in Confucian philosophy, the development of virtues is contingent upon social interactions, ritual observance, and the emulation of exemplars (Lai & Lai, 2023). A lifelong process of "learning to be human," as neo-Confucian scholar Tu Wei-Ming puts it, occurs through the "creative tension" between our social context and our potential for self-transcendence (Wei-Ming, 1985, p. 15).

In the Chinese context, many academics have used historical traditions to explain a Chinese philosophy and ethics of engineering and technology. Li Bo-cong (2002) first brought the Dao–Qi relation to the forefront. Here, the Dao symbolizes the heavenly pattern and the natural laws, while Qi embodies the material, the tangible, and the instantiation of Dao. This dialectic hints at a tripartite system where science is for understanding, technology for creating, and engineering for application. Similarly, Pak-Hang Wong (2012) proposed reconstructing the Confucian notions of

Dao, harmony, and personhood, so that an alternative ethics of technology based on the Confucian tradition may then offer an antidote to the atomistic view of humans. Such an interpretation causes engineering ethics to transcend the bounds of professional ethics, as it necessitates the identification of both harmony and discord in technology–society relations, and a closer examination of the nature of affected social roles and the responsibilities attached to them.

Closing remarks

As stated at the beginning, this chapter intended to provide some philosophical foundations that could help us – engineering teachers, researchers, and engineers – not to take the ethical-political and cosmological bases of the dominant engineering practices for granted but to consider them critically. As seen, those bases are, to a non-negligible extent, contingent, particular, local, and non-universalizable. What is more, not only is engineering practice (and the technology produced via engineering practice) shaped by the dominant ethical-political and cosmological values and understandings, but it also supports or emulates a reality that reinforces both aspects to the benefit of the powerful who profit somehow with them and to the detriment of a vast majority of disempowered people(s). That is why one cannot be politically or cosmologically neutral when engaged in engineering, for the activity and its outputs either support or confront the status quo. The only real choice is between being (or trying to be) conservative or progressive.

Whatever choice one makes, that choice will be free, informed, and/or justified if it is not based on illusions or misconceptions but on serious, supported critical reflection. With this chapter, we aimed to offer an opportunity for our readers to become acquainted with or go deeper into some well-founded, up-to-date thoughts on engineering, technology, and ethics. Hopefully, such reflections can help you be better positioned to choose how you will practice, teach, or do research on engineering in a more informed way – or in better accordance with your worldview, political perspective, ideals, and so on.

Throughout the chapter, even though we acknowledged systemic forms of power (like capitalist structures and religious actors and institutions) that force the world (and engineering with it) to be one way or another, much emphasis was given to individual and local disruption (like progressive engineering and popular engineering) as though they could be achieved without any constraint, as the result of a mere acknowledgment of how reality is. That is deceiving. There are no individual superheroes capable of overcoming oppression, of freeing or emancipating any given marginalized group or community. But there can be collective initiatives, even small ones, that manage to face these systemic forces and, if only locally and for some time, succeed. Even when they do not last much longer, their success is (or can be) a powerful reminder that other worlds, with these other forms of engineering they demand, still are – as they have always been – possible.

Notes

1 Empowerment can be defined as the "multi-dimensional social process that helps people gain control over their own lives" by fostering power in people and groups/communities to operate the changes they may want in their own lives, territories, and society (Page & Czuba, 1999). Empowerment is liberative or emancipatory whenever it allows individuals and groups to improve their lives – i.e., "being more fully human" (Freire, 1970, chaps. 1–2) – and/or fight for their rights or for building other possible social realities and/or ways of living, without dwarfing other people's and groups' rights or legitimate search for self-determination.

2 'Decolonial' and 'emancipating' can reasonably be taken as synonyms. For more on Decolonial Theory, see Chapter 9 of this handbook.

3 At that time, philosophers from different traditions considered philosophy to be subordinate to theology. Al-Ghazâlî, writing in the philosophical tradition of the Islamic world, upheld such a belief, as did Bonaventure, a Franciscan friar and professor at the University of Paris. But not all philosophers in the Middle Ages held the same view. An obvious counterexample is Ibn-Rushd's retort to Al-Ghazâlî (de Libera, 2019).

4 In short, Aristotle's theory explains that everything has four fundamental aspects (called 'causes,' αἴτῐᾰ): *material cause* (what a thing is made of), *formal cause* (what it is, its structure or form), *efficient cause* (what caused it to be or where its change comes from), and *final cause* (what its good, purpose or goal is) (Falcon, 2023).

5 In reference to Paul of Tarsus' letters, such as 1 Thess, 1:3, 1 Thess. 5:8, and 1 Cor. 13.

6 See, e.g., MacIntyre's (1981/2013) renowned book *After Virtue*.

7 Although Bentham and Mill used somewhat different criteria in their calculations, both of their approaches remain influential in contemporary ethical theory.

References

Alexander, L., & Moore, M. (2021). Deontological ethics. In: E. N. Zalta (Ed.), *The Stanford encyclopedia of philosophy*. https://plato.stanford.edu/entries/ethics-deontological/

Anscombe, G. E. M. (1958). Modern moral philosophy. *Philosophy*, *33*, 1–19

Aristotle. (1934). *Nicomachean ethics* (H. Rackham, Trans.). Harvard University Press.

Audi, R. (2005). The ethical significance of cost-benefit analysis in business and the professions. *Business & Professional Ethics Journal*, *24*(3), 3–21. http://www.jstor.org/stable/27801385

Augusto-Vieira, G. (2016). Ubuntu, the philosophy of complete happiness. *Atlantico*, *2/6*, 40–44. https://ibraf.org/wp-content/uploads/2017/09/atlantico06.pdf

Bentham, J. (1818/2011). *Church–of–Englandism and its catechism examined* (J. E. Crimmins & C. Fuller, Eds.). Oxford University Press.

Bentham, J. (1822). *An analysis of the influence of natural religion on the temporal happiness of mankind*. R. Carlile.

Bentham, J. (1823/2013). *Not Paul, but Jesus*. Bentham Project. https://www.ucl.ac.uk/bentham-project

Bell, D. A., & Metz, T. (2011). Confucianism and ubuntu: Reflections on a dialogue between Chinese and African traditions: Confucianism and ubuntu. *Journal of Chinese Philosophy*, *38*, 78–95. https://doi.org/10.1111/j.1540-6253.2012.01690.x

Blackburn, S. (2001). *Ethics: A very short introduction*. Oxford University Press.

Bo-cong, L. (2002). *An introduction to philosophy of engineering*. Daxiang Publishing Press.

Clancy III, R., & Zhu, Q. (2022). Global engineering ethics: What? Why? How? and When? *Journal of International Engineering Education*, *4*(1). https://digitalcommons.uri.edu/jiee/vol4/iss1/4

Coghlan, D. (2021). Action research. In: V. P. Glăveanu (Ed.), *The Palgrave encyclopedia of the possible* (pp. 9–16). Palgrave Macmillan. https://doi.org/10.1007/978-3-319-98390-5_180-1

Confucius. (1998). *The original analects: Sayings of confucius and his successors* (E. Bruce Brooks & A. Taeko Brooks, Trans/Eds.) (Translations from the Asian Classics). Columbia University Press.

Cruz, C. (2021a). Decolonizing philosophy of technology: Learning from bottom-up and top-down approaches to decolonial technical design. *Philosophy & Technology*, *34*, 1847–1881. https://doi.org/10.1007/s13347-021-00489-w

Cruz, C. (2021b). Brazilian grassroots engineering: A decolonial approach to engineering education. *European Journal of Engineering Education*, *46*(5), 690–706. https://doi.org/10.1080/03043797.2021.1878346

Cunha, M., & Almeida, M. (2004). Traditional populations and environmental conservation. In: A. Veríssimo, A. Moreira, D. Sawyer, I. Santos, & L. P. Pinto (Eds.), *Biodiversity in the Brazilian Amazon* (pp. 182–191). Editora Estação Liberdade.

Dagnino, R., Brandão, F., & Novaes, H. (2004). Sobre o marco analítico-conceitual da tecnologia social. In: A. E. Lassance Jr., C. J. Mello, E. J. S. Barbosa, F. A. Jardim, F. C. Brandão, H. T. Novaes, ... S. M. P. Kruppa (Eds.), *Tecnologia social: Uma estratégia para o desenvolvimento* (pp. 15–64). Fundação Banco do Brasil.

Dju, A., & Muraro, D. (2022). Ubuntu como modo de vida: contribuição da filosofia africana para pensar a democracia. *Trans/Form/Ação*, *45*(Edição Especial), 239–264.

Dreyer, J. (2015). Ubuntu: A practical theological perspective. *International Journal of Practical Theology*, *19*(1), 189–209. https://doi.org/10.1515/ijpt-2015-0022

Engster, D. (2007). *The heart of justice: Care ethics and political theory*. Oxford University Press.
Escobar, A. (2018). *Designs for the pluriverse: Radical interdependence, autonomy, and the making of worlds*. Duke University Press.
Estermann, J. (2006). *Filosofía andina: Sabiduría indígena para un mundo nuevo*. ISEAT.
Ewuoso, C. & Hall, S. (2019). Core aspects of ubuntu: A systematic review. *South African Journal of Bioethics Law* 12(2), 93–103. https://doi.org/10.7196/SAJBL.2019.v12i2.679
Falcon, A. (2023). Aristotle on causality. In E. N. Zalta & U. Nodelman (Eds.), *The Stanford encyclopedia of philosophy*. https://plato.stanford.edu/archives/spr2023/entries/aristotle-causality/
Foot, P. (1978). *Virtues and vices and other essays in moral philosophy*. Basil Blackwell.
Feenberg, A. (2010). *Between reason and experience: Essays in technology and modernity*. MIT Press.
Feenberg, A. (2017). *Technosystem: The social life of reason*. Harvard University Press.
Fraser, C. (2022). Mohism. In: E. N. Zalta (Ed.), *The Stanford encyclopedia of philosophy*. https://plato.stanford.edu/archives/spr2022/entries/mohism/
Freire, P. (1970). *Pedagogy of the oppressed*. Continuum.
Gilligan, C. (1982). *In a different voice: Psychological theory and women's development*. Harvard.
Greene, J. (2013). *Moral tribes*. Penguin.
Grönum, N. (2015). A return to virtue ethics: Virtue ethics, cognitive science and character education. *Verbum et Ecclesia*, *36*(1), 1–6. https://dx.doi.org/10.4102/VE.V36I1.1413
Hare, J. (2015). *God's command*. Oxford University Press.
Hare, J. (2019). Religion and morality. In: E. N. Zalta (Ed.), *The Stanford encyclopedia of philosophy*. https://plato.stanford.edu/archives/fall2019/entries/religion-morality/
Held, V. (1993). *Feminist morality: Transforming culture, society, and politics*. University of Chicago Press.
Held, V. (2006). *The ethics of care: Personal, political, and global*. Oxford Academic, https://doi.org/10.1093/0195180992.003.0002
Hui, Y. (2016). *The question concerning technology in China: An essay in cosmotechnics*. Urbanomic Media.
Hui, Y. (2017). On Cosmotechnics: For a renewed relation between technology and nature in the anthropocene. *Techné: Research in Philosophy and Technology*, *21*(2–3), 319–341. https://doi.org/10.5840/techne201711876
Hui, Y., & Lovink, G. (2017). For a philosophy of technology in China: Geert Lovink interviews Yuk Hui. *Parrhesia*, *27*, 48–62.
Hursthouse, R. (1999). *On virtue ethics*. Oxford University Press.
Hursthouse, R. (2007). Environmental virtue ethics. In R. L. Walker & P. J. Ivanhoe (Eds.), *Working virtue* (pp. 155–172). Oxford University Press.
Hursthouse, R., & Pettigrove, G. (2023). Virtue ethics. In E. N. Zalta & U. Nodelman (Eds.), *The Stanford encyclopedia of philosophy*. https://plato.stanford.edu/archives/fall2023/entries/ethics-virtue/.
Kant, I. (2019). *Groundwork for the metaphysics of morals* (C. Bennett, J. Saunders, & R. Stern, Eds. & Trans.). Oxford: Oxford University Press.
Kaurin, P. S. (2018). *Ethics: Starting at the beginning*. Wavell Room. https://wavellroom.com/2018/08/23/ethics-starting-beginning/
Kleba, J., & Cruz, C. (2021). From empowerment to emancipation: A framework for empowering sociotechnical interventions. *International Journal of Engineering, Social Justice and Peace*, *8*(2), 28–49. https://doi.org/10.24908/ijesjp.v8i2.14380
Kopenawa, D., & Albert, B. (2013). *Falling sky: Words of a Yanomami Shaman*. Belknap Press.
Lai, Y.-Y., & Lai, K. (2023). Learning from exemplars in confucius' *Analects*: The centrality of reflective observation. *Educational Philosophy and Theory*, *55*(7), 797–808. https://doi.org/10.1080/00131857.2022.2132936
Le Grange, L. (2019). Ubuntu. In A. Kothari, A. Salleh, A. Escobar, F. Demaria, & A. Acosta (Eds.), *Pluriverse: A post-development dictionary* (pp. 323–326). Tulika Books.
Li, C. (2013). *The Confucian Philosophy of Harmony*. Routledge.
Li, C., & Düring, D. (2022). Harmony as a virtue. In C. Li, & D. Düring (Eds.), *The virtue of harmony* (pp. 21–42). Oxford University Press.
de Libera, A. (2019). *La philosophie médiévale* (3re éd.). Presses Universitaires de France.
Mabele, M., Krauss, J., & Kiwango, W. (2022). Going back to the roots: Ubuntu and just conservation in Southern Africa. *Conservation and Society, AOP* special issue, pp. 1–11. https://doi.org/10.4103/cs.cs_33_21
MacIntyre, A. (1981/2013). *After virtue*. Notre Dame Press.

Marenbon, J. (1990). The theoretical and practical autonomy of philosophy as a discipline in the Middle Ages. In: M. Asztalos, J. E. Murdoch, & I. Niniluoto (Eds.), *Knowledge and the sciences in medieval philosophy* (pp. 262–274). Yliopistopaino.

Medeiros Ramos, A. (2021). Is *ars* an intellectual virtue? John Buridan on craft. In: I. Chouinard, A. McConaughy, A. Medeiros Ramos, & R. Noël (Eds.), *Women's perspectives on ancient and medieval philosophy* (pp. 275–301). Springer.

Metz, T. (2007). Ubuntu as a moral theory: Reply to four critics. *South African Journal of Philosophy*, *26*(4), 369–387.

Mhlambi, S. (2020). *From rationality to relationality: Ubuntu as an ethical & human rights framework for artificial intelligence governance*. Carr Center Discussion Paper. https://carrcenter.hks.harvard.edu/publications/rationality-relationality-ubuntu-ethical-and-human-rights-framework-artificial

Mill, J. S. (1874/1974). *Three essays on religion: Nature, the utility of religion, and theism.* Longmans, Green, Reader and Dyer.

Noddings, N. (1984/1995). Caring [1984]. In V. Held (Ed.), *Justice and care: Essential readings in feminist ethics* (pp. 7–30). Routledge.

Page, N., & Czuba, C. (1999). Empowerment: What is it? *Journal of Extension*, *37*(5), 1–5.

Plato. (2017). *Euthyphro Apology Crito Phaedo* (Christopher Emlyn-Jones & William Preddy, Ed. & Trans.). Harvard University Press.

Poster, W. (2019). Racialized surveillance in the digital service economy. In R. Benjamin (Ed.). *Captivating technology: Race, carceral technoscience, and liberatory imagination in everyday life* (pp. 133–169). Duke University Press.

Rachels, J., & Rachels, S. (2018). *The elements of moral philosophy*. McGraw Hill.

Robertson, T., & Simonsen, J. (2013). Participatory design - An introduction. In J. Simonsen & T. Robertson (Eds.), *Routledge international handbook on participatory design* (pp. 1–17). Routledge.

Sandberg, J. (2013). Deontology. In A. L. C. Runehov & L. Oviedo (Eds.), *Encyclopedia of sciences and religions*. Springer. https://doi.org/10.1007/978-1-4020-8265-8_703

Sim, M. (2007). Remastering morals with Aristotle and Confucius. Cambridge University Press.

Sinnott-Armstrong, W. (2023). Consequentialism. In: E. N. Zalta & U. Nodelman (Eds.), *The Stanford encyclopedia of philosophy*. https://plato.stanford.edu/archives/win2023/entries/consequentialism/

Slote, M. (2007). *The ethics of are and Empathy*. New York: Routledge.

Tronto, J. (1993). *Moral Boundaries, a Political Argument for an Ethic of Care*. New York: Routledge.

Tronto, J. (2010). Creating caring institutions: Politics, plurality, and purpose. *Ethics and Social Welfare, 4*(2), 158–171. doi:10.1080/17496535.2010.484259

Viveiros de Castro, E. (2014). *Cannibal Metaphysics: For a Post-Structural Anthropology*. Minneapolis: Univocal Publishing.

Wei-Ming, T. (1985). *Confucian Thought: Selfhood as Creative Transformation*. New York: SUNY Press.

Winner, L. (1986). Do artifacts have politics? In L. Winner (Ed.), *The Whale and the Reactor: A Search for Limits in an Age of High Technology* (pp. 19–39). Chicago: Chicago University Press.

Wong, P.-H. (2012). Dao, harmony and personhood: Towards a confucian ethics of technology. *Philosophy & Technology*, *25*(1), 67–86. https://doi.org/10.1007/s13347-011-0021-z

Yang, F. (2018). Religion in the Global East: Challenges and opportunities for the social scientific study of religion. *Religions*, *9*(10), 305. https://doi.org/10.3390/rel9100305

9
SOCIOLOGICAL, POSTCOLONIAL, AND CRITICAL THEORY FOUNDATIONS OF ENGINEERING ETHICS EDUCATION

Robert Braun, John Kleba, and Richard Randell

Introduction

Traditionally, engineering ethics has been viewed as belonging exclusively to the domain of philosophy. In this chapter, it is argued that engineering ethics has much that can be learned from sociological approaches. This is especially important as all engineers are also tacit sociologists; they form an opinion about the social world in which they dwell, socialize, and work – and into which they imagine their engineered artifacts will be deployed. A greater understanding of formal sociology enables engineers to contextualize their practices, understand problems, and generate engineering ideas in a more interdisciplinary and multi-dimensional way. Sociology also helps us understand how and why technology ethics (and the role of engineers) change over space (culturally) and time (historically), along with the structural changes of social systems and the history of ideas. Sociology gives us tools to deconstruct simplistic views when working with students, such as technological determinism and the belief technical design is value neutral. This chapter presents what we regard as the three most crucial sociological approaches and their potential contributions to engineering ethics education: critical theory, postcolonial theory, and Science, Technology, and Society (STS) studies.

Critical theory is associated with numerous intellectual traditions seeking human emancipation. In respect to its implications for engineering, if the potential liberating powers of technology are to be realized, this will only occur through human-designed social change based on a larger dialogue about goals and values (Mitcham & Briggle, 2009). Postcolonial studies provide a decentered, diasporic rewriting of earlier nation-centered imperial grand narratives of technoscientific modernity. STS offers a critique of technological determinism and solutionism and their correlate deficit logic, and of artifacts that are assumed to be void of socio-political agential powers. All these intellectual and research traditions provide resources for reflecting on and following ethical pathways in engineering and engineering education.

This chapter suggests ways to conceptualize the 'self-knowledge' of engineers, focusing on the social, political, epistemological, and ontological aspects of common sense and the frequently unarticulated, taken-for-granted social practices and ethics of engineering (Mitcham, 2014). At the same time, we ask whether other worlds, ways of life, social imaginaries, and material practices

DOI: 10.4324/9781003464259-12

are possible, and how such potential futures could be realized with the help of a reflective engineering education and practice.

Before providing an overview of the three sociological approaches, we open the chapter with positionality statements and remarks on the link among ethics, engineering, and society. The two last sections expose paths to integrate sociological approaches in the theory and practice of engineering ethics education.

Positionality

The first author, Robert Braun, was born, raised, and educated in Hungary during "socialist times" (the Soviet/Russian occupation of Eastern Europe, 1948–1989). He comes from an assimilated Jewish academic family. Coming of age in the late 1980s, he was involved in activism – fighting for the human rights of Roma people in Hungary – and also in the emerging political movements around the opening up of local politics and the collapse of the Soviet Union. Robert was involved in the budding left-liberal parties that emerged; he participated in the democratic transition. He left for the United States in 1992 to do Ph.D. coursework and a dissertation (Rutgers University) on a fellowship offered by the Soros Foundation; his academic carrier started at Eötvös Lorand University, Department of Jewish Studies thereafter. He later joined the philosophy department of Corvinus University, the leading economics and social science university in Budapest. Parallel to his academic work, Robert remained active in politics and held a number of public offices and various (mostly founding) positions in technology and politics-oriented business enterprises. In 2015, he moved with his family to Vienna, Austria, where he joined the faculty of the Institute for Advanced Studies. His research moved into the direction of Science and Technology Studies with a focus on the ontological politics of technology transitions and quantum social science. His specific interest is in the Anthropocene, not as a geological epoch but as a political meta-apparatus of world-making. He has researched and published extensively on one of the core apparatuses of the Anthropocene, automobility and its politics; his current research moves more in the direction of applying quantum theory to understand accident events in automobility.

The second author, John B. Kleba, was born in Brazil during the dictatorship (1964–1984). In 1984, he began studying Social Sciences at the Federal University of Santa Catarina, South Brazil, protesting for democracy among one million citizens nationwide and engaging in the ecological movement. John's postgraduate work was directed to the critique of 'development.' At that time, he spent one month living within and studying the Landless Workers' Movement. Shifting between disputing sociological theories and clashing streams of activism, he learned that an open mind is essential when striving for a proactive attitude toward social change. In 1992, he moved to Germany earning a Ph.D. in Science and Technology Studies in Bielefeld and working in Bremen as a research assistant in law and society. He investigated issues such as the access and benefit-sharing regimes (genetic engineering, conflicting worldviews, regulatory frameworks) and pollution double standards in the trans-national chemical industry. In 2005, he moved with his family to Brazil to work at the Aeronautics Technological Institute (ITA). His research included working with Indigenous Peoples, Quilombolas (slave-descendant communities) and other social movements, especially related to the privatization of the commons and the clash of traditional medicinal versus Western knowledge systems. The invisibility of people made vulnerable and marginalized by colonialist structures, their ways of knowing and existing, and the multitude of critiques raised in the Global South have been constantly present in his reflections. At ITA, he established the Citizenship and Social Technologies Lab (LabCTS), which has engaged hundreds of engineering students in sociotechnical interdisciplinary projects in partnership with civil society organizations and public schools.

A descendant of nineteenth-century European colonial settlers, the third author, Richard Randell, was born in Melbourne, Australia. Richard completed his school education in Adelaide, South Australia. His high school years coincided with the last years of Australia's involvement in the Vietnam War. Out of approximately 600 students, 6 opposed the war. Participation in anti-war demonstrations was his first contact with an alternative politics. During school vacations, he visited family in a small town 800 kilometers west of Adelaide. Several times a week, indigenous peoples visited the town by bus from two nearby mission stations administered by the Lutheran Church. Only as an adult did he discover that many of those visitors were refugees from British atomic tests that were conducted to the north, between 1956 and 1963, in Maralinga, where their people had lived for 65,000 years. His current research interest focuses on the Anthropocene, and more specifically on the various sociotechnical apparatuses with which the colonial powers have transformed much of the planet into a space of exception, where everything is permitted, and nothing is considered a crime. Maralinga is such a space of exception, created by one such sociotechnical apparatus that was brought into being by the work of scientists and engineers. After completing his BA degree at Flinders University of South Australia, he attended the University of Wisconsin-Madison, where he completed a Master of Science (MS) degree and a Ph.D. in sociology. Following a long break from academia, he returned to teaching and only later to research in the field of mobility studies and critical automobility studies.

On ethics, engineering, and society

At least since the Enlightenment, the debate on moral principles is not only about what choices are morally right or wrong or which virtues we should encourage – it is also about demanding, justifying, negotiating, and designing new forms of social co-operation today and toward better futures (Cohen, 2009; Mannheim, 1985; Wright, 2010). *How should this techno-ethical debate inform the social design in law and public policies in all current technological controversies (the regulation of artificial intelligence, autonomous mobility, and climate change policies, among others)?* Justifying, negotiating, and designing new forms of social co-operation and forging or experiencing relations are not optional. They are built into the very fabric of the world we inhabit, a world that largely has been constituted through efforts that may be placed under the umbrella term 'engineering.'

Like engineering, sociology is an ethical enterprise. Even producing accurate research findings about social reality leads to the question of which data are to be collected, analyzed, and re-arranged – which also expresses ethical-political choices. For example, policies that address inequality require data about structural racism and gender inequality. Studying social life in all its varied manifestations is the goal of sociology. Therefore, sociological studies encompass ethical goals and activities (Lybrand & Randell, 2022). They are contributing factors to social development even if they are themselves the symptoms and effects of social circumstances. Because they are working toward improvements – betterment not only of technology but also of social life as a whole – engineers are also compelled to establish, albeit usually implicitly, social theories about the social world they intend their artifacts to be written into.

Sociology and engineering are intimately interconnected, even if the connections are largely unacknowledged. Engineers are not only tacit sociologists, they are also tacit ethicists. They have specific and strong views on the social (which they perceive as lacking), and also about rights and wrongs of the social order (which they perceive as in need of betterment). Much of this chapter may be read as a description of the tacit social theory (and lack of awareness thereof) of engineering. Whether through commission or omission, ethics is *not* optional, contingent, only occasionally relevant. Indeed, ethics is always relevant. Not only is ethics built into material artifacts and their relations, but also we exist within a world where what should be done or not done is frequently taken to be common sense. Yet what we take to be common sense is a common sense that has been *constructed* and *disseminated* by agents with their own interests. The world is neither 'out there' nor is it free of agency. The world is not a subject-independent container in which subjects and objects 'interact,' into which artifacts are engineered and deployed.

The world (or worlds) is/are constructed by the material-discursive practices of agents that create entities and their relations and categorize them into kinds: subjects and objects, living or dead, agential or void of agency. Constructs such as these are mobilized and enacted by common sense, a specific way of seeing. To the degree such agents successfully convince us of their view of 'common sense,' that too is 'engineering' – in this case, the engineering of the social world. *To develop and build or not to develop and build?* If the former is chosen, *what* and *how* it should be done always raises ethical issues, as will the latter choice. What is common sense in one given time and space is also a way of imposing particular worldviews, social hierarchies, and forms of lives over others. Amongst the agencies that construct and disseminate common sense is engineering itself.

Paraphrasing an essay on poetry by Percy Bysshe Shelley (1840, p. 57), the engineering ethicist Carl Mitcham (2014, p. 19) described engineers as the "unacknowledged legislators of the world" who, "by designing and constructing new structures, processes, and products, [influence] how we live as much as any laws." Engineers are not only unacknowledged legislators who *regulate* the world we inhabit and are a part of, they are also *co-creators* of it. Seen from the vantage point of the social, reflecting on engineering ethics is engaging with the co-creation of the world by science and its applications (Jasanoff, 2004).

Engineering is perceived by many, engineers included, to be the field *par excellence* to develop solutions to the major challenges of our time and to design and construct *desirable* (i.e., 'ethically good') possible futures. We may call this 'lyseology' – mobilizing science and knowledge production to convince policy-makers and the general public that the present possesses some form of lack that should be addressed with a new technology brought to life and offered as a solution (Braun, 2024). It is a neologism from the Greek word *lysi* (solution) and *logos* (knowledge). Lyseology is the use and misuse of science to suggest that it is in the future, populated by new but not yet existing engineered artifacts, that a better world is believed to lie. It is a modified version of agnotology (Proctor, 2008) – the use and misuse of science to produce ignorance in support of corporate interests. (Chapter 6 discusses similar topics and may interest readers of this chapter.) Support for science, technology, engineering, and mathematics (STEM) education and research at the expense of other fields is symptomatic of this belief.

Martin Heidegger (1977, p. 4) once observed that "we are delivered over to [technology] in the worst possible way when we regard it as something neutral; for this conception of it, to which today we particularly like to do homage, makes us utterly blind to the essence of technology." It is an observation that engineers would do well to reflect on, not just in the abstract but with respect to the moral choices and consequences that follow from each and every artifact that their work has contributed to realizing (Braun & Randell, 2022). Engineering is a social (material-discursive) practice that is embedded in what is commonly understood to be the social; it is part of the social, and simultaneously creates, reproduces, and sustains the social. Engineering is also embedded in what is generally termed as the natural, not only through, for example, biochemicals or radioactive materials that impact, influence, and alter life and its various forms, but also through engineering artifacts that impact life on Earth and the Earth's ecosystem. Engineering, in short, is embedded in and constitutive of the socionatural – the hybridization of reciprocal intermingling of the natural and the social (Arias-Maldonado, 2015). Engineering ethics asks us to be cognizant of and reflect on such embeddedness and embedding.

Critical theory

'Critical theory' has both a narrow and a broader meaning in the social sciences, humanities, and philosophy. In the narrow sense, the term designates the tradition associated with the Frankfurt School. According to the School founders, a critical theory is distinguished from 'traditional theories' by pursuing human emancipation and liberation in all circumstances of domination and oppression (Horkheimer & Adorno, 1973). In its broader sense, critical theory (CT) encompasses a variety of approaches, often in association with social movements with a similar agenda, that seek to identify the dimensions of injustices, power asymmetries, and exploitation, such as gender studies, critical race theory, class analysis, postcolonial studies, and posthumanism (Bohman et al., 2023). CT combines philosophy with empirical social scientific research and is aimed at *critique,* explanation, understanding, and also changing the current state of affairs. It is *practical* – seeking emancipation – in the ethical sense of the term.

This multidisciplinary field focuses on how knowledge is formed and how power underlies these formations. As a mode of social analysis, it is concerned with language and discourse (written texts, visual images, and other discursive forms) and the relationship between power and discourse. Its central interest is the political, and its primary assumption is that the political pervades the world we inhabit, not just within discourses that claim to be non-political (Esposito, 2021) but the very materiality of the world. Artifacts themselves, as much of the STS scholarship has demonstrated, are political and enact politics (Winner, 1986). Bridges, roads, airports, computers, data, and so forth contain both the ethics and politics of their architects and operators (Jasanoff & Kim, 2015).

One way to define and delimit critical theory is to ask what it is not – to ask what a non-critical theory might be. For the subject at hand, one answer might be 'engineering.' The point is not that engineers should cease being engineers and train in another field. Rather, it is the degree to which what they develop, research, and construct – what they 'engineer' – is pursued with (or without) reflection on how what *has been* or *will be* engineered fits into or (re)constructs an irremediably political world – what the possible political, social, and co-constructive consequences might be. It is not (just) a question of telling oneself to take ethical issues into consideration. What is required are intellectual tools to do so, including familiarity with the theoretical and disciplinary fields that are the subject of this chapter, as well as an awareness of how such tools may be acquired by reflection and education. This is why we argue that sociological, postcolonial, and critical theory foundations should be part of all engineering education, under the heading of ethics or elsewhere in the curriculum. Discourse, a strategic apparatus of "the said as much as the unsaid" (Michel Foucault, 1980, pp. 195–195), creates and reproduces the social as well as the material, together with its mechanisms of power. One of those mechanisms of power is 'technology.' Perhaps the most obvious technology into which are built hierarchies of power, control, and ownership is the assembly line, whether it be the Fordist assembly line of an automobile or smartphone factory, or the kitchen of a fast-food chain (Ritzer, 2021) (see also Chapter 4 on reason and emotion). Technologies tend to reproduce already existing social hierarchies. Reflecting on power and ideological bias – for example, in relation to class, race, and gender – provides a way to identify how the politics and the ethics of engineering are intertwined with these hierarchies. Science, technology, and innovation co-produce and reinforce already existing structures of social injustice, violence, and social exclusion (Braun & Randell, 2022).

Science, Technology, and Society (STS)

STS focuses on technology–society relationships and is critical of approaches that assume that technological development follows its own logic, independently of the social world in which it is embedded. Much of the STS scholarship is critical of technological determinist accounts of technological development. The main tenet of STS is that technology and engineering are shaped by a variety of social factors and forces, and vice versa. Technologies are reflexively embedded in and embed social practices, norms, processes, conventions, discourses, and institutions. These make up what is commonly seen in sociology as building blocks of the social (social change/stability, structure/agency, cultural diversity/hegemony, etc.). This is what in STS is called technology being *co-produced* (Jasanoff, 2004) by numerous human (people) and non-human agents (norms, institutions, artifacts) and 'becoming-with' in a multi-species world (Haraway, 2008).

Innovation is understood to be embedded in a network of social institutions, forming what in STS is called a 'sociotechnical system.' It is a 'system' (e.g., a patterned network of relationships constituting a coherent, dynamic whole that exists between individuals, groups, institutions, and artifacts) composed of practices, organizations, and logics, which is the intertwined social context, composed of engineering practices and technologies. That context includes the economy, business strategies, government policies, everyday habitual practices, complex perceptual lifeworlds, and local and national cultures. If technology is rooted in the social, we must go beyond social construction accounts of technology. We need to zoom out from sociotechnical systems and focus our attention on 'the world,' of which 'the social' (e.g., the network of human subjects) is one aspect.

STS is concerned with, and, for the most part, critical of (a) the hegemonic assumption that there is one single universal world and (b) the ways in which it has been discursively constructed. This construction has involved the conversion of matter into what is commonly referred to as 'nature'; the conversion of nature into what economists traditionally call 'resources'; the transformation of the materiality of entire domains of the inorganic and the non-human into matter that can be possessed, transformed, and extracted; and the linking of matter and worlds to markets to generate growth (Escobar, 2020).

Projects to transform the sociomateriality of the world create and reproduce a modernist capture: a conversion process with a desire to solve problems. This is what we have referred to as lyseology – the use and misuse of science to suggest that, in the present, the world is populated with problems while the future could be bettered by substituting problems with (engineered) solutions. This capture is manifested in the climate emergency, which will affect humans and non-humans in myriad ways (Escobar, 2019). From an STS perspective, the world we currently inhabit that is so co-constructed can provisionally be called the world of modernist technoscience. So conceptualized, the following questions arise: *How did this hegemonic world come into existence? What are its component elements? How does it sustain and reproduce itself? What are its contradictions and contested features? Are there other (cultural, ethical-political) worlds disputing alternative developments, and with which sociotechnical consequences? And, for the subject at hand, where do engineers and engineering fit into this?* To these questions we turn in the next section.

From critical theory and STS to a critical ontology of engineering

'Ontology' traditionally has been understood to be a field of metaphysics, institutionally and intellectually located primarily, but not exclusively, in the discipline of philosophy. This is one way of thinking about ontology, as a discourse grounded in metaphysics that aims to establish the properties and boundaries of an ostensibly independently existing reality. An alternative way of thinking about ontology is as a set of practices through and by which worlds are created, not by philosophers but by members of society – engineers, for example – in and through their routine, mundane activities. Ontologies – at least in the Global North, with its universalist and hegemonic ambitions to explain how the world ostensibly actually is – are normative; they aim to determine what is real and what is not, what counts as a thing, a signified; what counts as a representation, a signifier. Mundane practices that dwell in and enact relations in the world always tacitly construct and reproduce assumptions regarding, and reflections on, actual and possible worlds and the kinds of entities that exist within that world. For the subject at hand, this is the world of technoscientific modernity, a world that engineering, its knowledge of and assumptions about science, as well as material practices of technology, is and has been instrumental in constructing, reproducing, and sustaining. If engineers, as Mitcham intimated, are the unacknowledged legislators and makers of not just any world but the world of modernist technoscience, *what are the implications and socio-ethical consequences of this?*

Accounting for world-making requires attending to what can be called 'ontology work' (Braun & Randell, 2023). Ontology work constitutes the mundane, everyday, professional, and lay efforts that are directed to the construction and reproduction of a world – an ontology. It is the work routinely performed by human agents engaged in the continual effort at imagining, creating, and sustaining objects, artifacts, infrastructures, networks, connections, and relations that populate our everyday world. Beyond creating the artifacts that humans have no choice but to engage with, humans themselves are subjectified (Michel Foucault, 2006). We also are constructed, as specific kinds of selves with certain beliefs, ethics, and desires, selves whose desires revolve around consumption, for example.

"How and by whom," C. B. Jensen (2021, p. 101) has asked, are "such worlds ... performed, maintained, challenged, transformed, or destroyed"? *How can this routine, everyday ontology work be complemented with a correlate ethics? What would the source of such ethics be?* Technoscientific modernity is a world that has been constructed and routinely sustained by shared, stabilized, and publicly performed visions about desirable futures. These visions are enacted by a myriad of human and non-human agents that form a sociotechnical 'system' or 'network' (Jasanoff, 2015). Ontopolitical power is shaped by and within these practices (A. Mol, 1999). It is exercised by those responsible for the reproduction and administration of technoscientific modernity: engineers, manufacturers, (repressive) state apparatuses (Althusser, 2014) such as regulatory and policy departments and research centers, advertising agencies, and so forth.

Technoscientific modernity, the world we inhabit, is an example of what the Situationist writer Raoul Vaneigem (1983 (1967)) called a factory of collective illusion. The allusion to the world being a factory relates to it being constructed by desires (of consumption, of lyseology, of happiness). Illusions on this account are not ideas, thoughts, or dreams that are in peoples' heads; they are as real as anything in the world. It is a factory that has created a world full of stuff, not only physical entities but also other agential powers: deterritorialized networks and relations (culture, money, media, etc.); manifold hierarchies (of material inequalities, of knowledges and beliefs, of access); desires and wants (sexual, consumerist, colonial); subjects (disciplined, controlled individuals and groups); and so forth. What is relevant here is that all these entities acquire agency that impacts the world and all of us. Imaginaries are hegemonic. They are comprised not only of visions, images, and discourses, but also the ostensibly material and physical, technological artifacts such as machines, of modernity. As Paul Virilio put it in a different context, "to invent the train is to invent the rail accident of derailment ... to invent the family automobile is to produce the pile-up on the highway" (Virilio, 2007, p. 10). "Derailment," "pile-up" – what Virilio calls the "integral accident" – are as real as the technological artifacts that populate our world. It is a world captured and converted by the manifold agencies engaged in ontology work. The entity created by such work is what Timothy Morton has called a 'hyperobject' (Braun & Randell, 2021; Morton,

2016): objects massively distributed in time and space relative to humans, in which humans are trapped inside. It is the late-modern, global, capitalist world we inhabit.

It is one of the factories within what the Frankfurt School critical theorists Max Horkheimer and Theodor Adorno (1973) called 'the culture industry.' It is an industry, to extend Horkheimer and Adorno's metaphor to the subject at hand, that produces ontology. That ontology, that world (assumed to be universal, causal, deterministic), is the everyday (Western, Eurocentric) lifeworld. It is comprised of dynamic and complex relations between people and stuff that are inscribed in artifacts and their material-semiotic networks (J. Law, 1986), which create, through a continuously unraveling process, the One-World-World in which we all dwell (John Law & Lien, 2018). The activity that goes by the name 'engineering' is a form of what Heidegger called human and violent *thinging* (Heidegger , 2002, p. 7), which rests on 'rational' thinking. Western thinging not only creates objects as individual entities, but also imaginaries (shared, stabilized, and publicly performed visions about desirable futures), hyperobjects (invisible objects massively distributed in time and space), and scapes (interconnected, globalized, and hegemonic transformations into resources).

Technoscientific modernity is a deficit ontology, wherein the present is perceived as imperfect and deficient (Dewandre, 2018) but rectifiable through the unending task of techno-political lyseology. Lyseology not only suggests solutions in the future, but inscribes lack, the missing object of techno-scientific desire, into the present. The central assumption and conviction of modernist engineering is that the world possesses a lack; something is absent and needs to be added or fixed. However, what is lacking, usually subsumed in the concept of innovation, is defined uncritically, veiling, and erasing crucial dimensions of the social reality. And the idea of an amorphous 'we' does not account for sharp social differences (class, nation, culture, and rights, among others). For example, facial recognition using artificial intelligence can hide issues of structural racism (Raji et al., 2020).

It is in this corrected future that a better world, full of new and improved technologies, is believed to lie. New goods and services are assumed to improve general well-being. Challenges are typically reduced to and understood in terms of technical properties (e.g., improving efficiency). By bringing into being new artifacts, entities, connections, and networks, engineers tacitly do 'ontology work.' However, this ontology work typically lacks the self-awareness that 'thinking' creates not only of artifacts and their relations *in the world,* but also 'things,' 'entities,' 'beings,' 'agencies,' and '(intra)connections' that the world is made of (Barad, 2007). Modernist techno-science is a worldview wherein it is assumed that the future in the present can be controlled by humans, provided they possess adequate knowledge of mathematics, physics, biology, chemistry, and other (natural) sciences, and have adequate engineering skills and the means to bring that future into being. Such a worldview is modernist in that it creates and upholds binaries of nature/culture and natural/social, as well as visions of human exceptionalism, Cartesian object/subject dualism, and Newtonian physical determinism. A critical ontology (of engineering), based on work in the critical social sciences in recent decades (Latour, 2000; John Law & Lien, 2018; Annemarie Mol, 2014; Woolgar & Lezaun, 2013), calls into question such basic modernist assumptions, especially in light of current controversies related to the ontological capture discussed above by the use of, for example, nuclear energy (Jasanoff & Kim, 2009), geoengineering (Shapiro, 2021), fracking (Howell et al., 2019), and autonomous mobility (Braun & Randell, 2020), to name some contemporary techno-ethical debates surrounding emerging technologies.

Postcolonial studies – views from the Global South

Colonialism is the historical process of European (and later, also American) violent dispossession and political conquest of the rest of the world. Contemporary postcolonial studies examine how patterns in power/knowledge and power/violence reproduce dominance over peoples, raising issues of identities, narratives, and inequalities. Fanon denounces racialized subjectivities and the foundational violence of colonialism (Fanon, 2021). Edward Said (2019) started the post-structuralist critique of Western epistemology by undermining the ideological belief of value-free knowledge, revealing that 'knowing the subaltern' (in the way this knowing has been historically established) is part of subjugating it. Recognizing the coloniality of a specific assemblage of power and knowledge as well as processes of power/violence, one manifestation of which is modernist technoscience and its ontology, *ipso facto* is to denounce how it continues to destroy community-based livelihoods, cultural diversity, and lifeworlds based on human–non-human co-existence.

Western neocolonialism acts in at least three ways against other cultural worlds: disparaging, erasing, and making it invisible. First, it downgrades 'inferior' and 'primitive' non-modern cultures, requiring of them the acceptance of a specifically Western idea of progress. That is why it is so crucial to build a critique of Western development (Escobar, 2015; Kleba & Reina-Rozo, 2021). The UN Sustainable Development Goals represent an advancement, but not enough from the point of view of postcolonial critiques (Hidalgo-Capitán et al., 2019). Second, Western neocolonialism obliterates minorities, their languages, and their living spaces, accelerating the extinction of cultures and biodiversity. Finally, by making invisible and speechless the representatives of non-Western cultures (Santos, 2011), this hegemony hinders their political articulation towards 'other possible worlds' (Castro-Gómez & Grosfoguel, 2007; Mignolo, 2018).

Today, postcolonial studies encompass numerous approaches, including milestone contributions of the Global South,[1] such as the Pluriverse (Kothari et al., 2019), the *Buen Vivir*, and the epistemologies of the South (Santos, 2011). The pluriverse and *Buen Vivir* are polysemic concepts; depending on which interpretation we accept, they either converge or diverge. Pluriversality establishes a communication between theories, social movements, and social actors of the social and political periphery engaged in a critical intercultural dialogue (Dussel, 2012, p. 26). Rooted in the idea that we live in a world with plural cosmologies and worldviews, the pluriverse encompasses a collage of anti-systemic traditions of the global South (*Buen Vivir* in Latin America, *Ubuntu* in South Africa, *Tazkijah* in the Islamic culture, *Swaraj* in India, and *Kongsi* in China, among others), along with intellectual movements of the Global North such as degrowth and ecofeminism (Kothari et al., 2019) (see also Chapter 8 on the philosophical foundations of engineering ethics education). Pluriversality is a critique of the project of a 'Western' (Euro-American, colonial, Cartesian, Newtonian) way of looking (Kuhn, 1962) that conceives of the world as being 'out there' – external and independent as well as anterior of human or non-human actions and perceptions, complete with knowable and definite, universal forms and relations of stuff that (are assumed to) populate it. The One-World World (OWW) of Western onto-epistemology, *in which* and not *of which* entities and their politics are performative (John Law, 2015; John Law & Lien, 2018), is occluding and suppresses potential alternative ontologies and subaltern indigenous subjects.

Buen Vivir (BV) (in Quechua *Sumak Kawsay* and in Aymara *Suma Qamana*), by contrast, has originated in the political struggles of indigenous peoples of the Andes region, spreading to Latin America. This intellectual and political movement is divided into three main political strands: (1) community socialism, combining local traditions with twenty-first-century socialism in governmental programs (in Ecuador and Bolivia) (Garcia Linera, 2015); (2) cultural-ancestral indigenism (Blanco & Aguiar, 2020), which opposes the Western appropriations of BV (Hidalgo-Capitán & Cubillo-Guevara, 2014); and (3) the 'pluriverse,' as explained above, which is linked to post-development (Beling et al., 2021; Escobar, 2015) and non-Eurocentric perspectives of knowing (epistemologies) (Mignolo, 2018; Reiter, 2018). Challenging misconceptions and stereotypes around some understandings of BV (Walsh-Dilley, 2017, p. 515) and the pluriverse has provided a powerful tool for counter-hegemonic struggles.

Linking this debate to the field of ethics, the pluriverse and BV represent other ways to comprehend 'the right' and 'the good,' as well as the world(s) that they enact. Both must be considered in both research and sociotechnical projects. Both concepts strongly value reciprocity, communalism, conviviality, and redistribution, as well as priority to the commons instead of private property (Chuji et al., 2019; Kothari et al., 2019). There is a strong defense of the common good and 'building community,' in opposition to capitalism and individualism. An additional line of the pluriverse critique stands for the feministic and new leftist ethics of care (Cohen, 2009; Puig de la Bellacasa, 2017) and the rights of nature (Escobar, 2011), opposing neoliberal development, extractivism (Gudynas, 2009) and technocracy (Feenberg, 1999, p. 4).

There is strong empirical evidence that economic reciprocity involving joint work (*mutirões*) and non-monetary exchanges of Andean indigenous populations, such as those described by Acosta (2016), have an essential symbolic and ethical character for the reproduction of social ties and the consolidation of community identity. At the same time, the empirical realities of non-Western cultures are far more diverse and problematic than some postcolonial discourses tend to represent. *So, how can engineers integrate postcolonial critique in their ways of thinking and practicing?* In the following sections, we are going to explore this intricate question.

Towards critical engineering

How does the moral economy that engineers are a part of look and what is the role and potential of engineers in this process? A postcolonial and STS analysis of engineering offers us a chance to decenter conventional accounts of optimistic hegemonic and global technoscience. It may reveal and complicate durable dichotomies produced under and by colonial regimes. Dichotomies can help clarify general trends as analytical tools. However, they often disseminate oversimplified accounts blurring empirical realities that are much more hybrid, complex, and 'messy.' These binaries usually operate in terms of global/local, first-world/third-world, Western/Indigenous, modern/traditional, developed/underdeveloped, and so forth. Postcolonial and STS approaches help understand how ideas about difference – racial (white/other), temporal (modern/traditional), class (elite/subaltern), knowledge (science/knowledge systems) – are enacted, stabilized, and/or disturbed in the performance of technoscientific modernity and Western hegemonic ways of seeing and doing engineering.

Many of these binaries originate in a foundational, Cartesian, and Newtonian thinking that lies at the core of engineering – imagining the all-knowing engineering subject educated in and trained by elders who present a world seen from an observation deck constructed by a neo-positivist European scientific ethos. This *thinking* and *thinging* is political – it refers to the politics involved in the practices that shape the world that has come to possess a deficit, and to assigning subjects and objects that populate the world. Engineering in its current form is a dominantly colonialist project: it sees maniform *terrae nullius* – the surface of the Earth, sea, the mass under the surface, space, other planets, the body, the virtual, and so forth as belonging to no-one and open to (re) population and appropriation by engineering artifacts and networks. A critical, STS-inspired and postcolonial approach to engineering and engineering ethics questions the hegemonic ambitions of a European deficit ontology and its accompanying technoscientific epistemology. It opens up possibilities to engage with alternative indigenous and/or scientific ontologies (like quantum theory inspired agential realism (Barad, 2007) or Everettian Many Worlds Interpretation (Everett, 1957, 2012) and reflect on what such alternative ontologies – worlds – might be.

Critical engineering in theory and practice (education)

Taking up the challenge of the Brazilian educator Paulo Freire,[2] *what would be the point of fostering critical thinking in engineering education if there is no connection between theory and practice? Is another way of practicing and teaching engineering, one that looks at technology as ontologically and ethically biased, even possible? If so, which conditions, means, and tools would be required for such a transformation? Most importantly, for the purposes of this handbook, what would an alternative engineering teaching and practice look like?* This is a broad issue for which we do not intend to provide recipes; rather we offer a few pointers, keeping in mind that any initiative must be situated, adapted, and experimented in its local cultural, social, and institutional context.

The question of what could be done differently in training engineers is vital. Concerning critical thinking, engineering students must move "beyond deterministic models of technology and decontextualized models of engineering, where engineering decisions are understood to be 'purely' technical and without inherent social ... implications." (Nieusma, 2011, p. 22.609.7). However, the deconstruction of such naïve views of engineering students often collides with cognitive bias and is perceived as "troublesome" (Kabo, 2010, pp. 4–5). So, in this endeavor, we may need the help of learning tools such as threshold theory active methodologies (Kabo et al., 2009) and action research (Argyris & Schön, 1989).

Considering the three core dimensions of education, the 'know-what' (theories, critical thinking, reflection) should be able to connect in meaningful ways with the dimensions of 'know-how' (action, abilities, and competencies in practice) and 'attitudes' (ethical behavior and values) (Varela, 1999). In this sense, a milestone approach in engineering education is represented in the intellectual and political movements of 'engaged programs' such as Humanitarian Engineering (Smith et al., 2019) (see also Chapter 23 on Humanitarian Engineering), Engineering for Social Justice (Baillie et al., 2021; Nieusma & Riley, 2010) and Engaged and Grassroots Engineering (Cruz, 2021; Cruz et al., 2021).

Particularly in the university engineering formation, such 'engaged programs' provide us with a possible path to integrate critical thinking and postcolonial critique in theory and practice. Engaged extension programs can be implemented, including curricular and extra-curricular activities (Kleba & Cruz, 2020; Smith et al., 2019), encouraging students to work in sociotechnical hands-on projects with social movements, organizations of civil society, and communities (Timmermans et al., 2020). Such projects should respond to real needs and give priority to the vulnerable and the needy (Schneider et al., 2009). A complete project cycle can be worked on, inspired by concepts such as 'design thinking' (immersion, ideation, prototyping, testing, and implementation) (Brown & Wyatt, 2010). In working with communities, the first step is to understand local problems and possible solutions from 'the inside,' from local singularities and local knowledge. The whole project should be co-constructed with the stakeholders, following participatory research in practice (Simonsen & Robertson, 2013; Braun et al., 2022). An interepistemic approach must be assured, in which science enters into dialogue with other knowledge systems, such as indigenous and small farmer knowledge (Fúnez-Flores, 2022). The multi-dimensional aspects of sociotechnical interventions should be considered, especially thinking about the related processes of empowerment/disempowerment (Kleba & Cruz, 2021). Competencies and abilities closely related to (ethical) values such as caring (ethics of care), empathy, and listening must be trained in practice. Otherwise, they risk being ideals with no connection with social change and agency.

We should also question systemic changes 'from above' in technology and society regarding what can be done differently in government and business. Advances in legislation and policies may allow engineers to engage in critical ethical-political agendas. Ways to move environmental and social corporative governance (ESG) forward may be explored. For instance, Colorado School of

Mines offers a minor in Corporate Social Responsibility (as part of the Humanitarian Engineering Faculty), striving "to work for communities' wellbeing inside of corporate settings" in a critical way (Braun, 2019; Lucena & Kleine, 2021, p. 100).

Engineers working in the Third Sector at the crossroads with public policies show they can act as game-changers in fostering social innovation (Avelino, 2019, p. 197). Amongst a multitude of possible examples, 'Techo' works with civil engineering projects of infrastructure and participative community housing in slums in Latin America (Melo et al., 2021), and 'AlterMundi' promotes internet community networks in Argentina, at the same time politically mobilizing to steer information technologies policies towards the public interest (Prato et al., 2021).

Our starting argument in this chapter was that engineering ethics should engage with sociological approaches. Engineers, we suggest, are also architects of the social: they form a professional opinion about the social world and shape, even create, this world by engineering its semiotic-material furniture. The toolkit that sociology and its cognates – critical and postcolonial theory as well as STS studies – offer enables engineering students to think critically about artifacts, sociotechnical systems, and the sociotechnical imaginaries we inhabit. More importantly, by better understanding social mechanics, they also apprehend that technoscience (the complex practice of creating scientific knowledge, technical systems, and artifacts) creates our social reality as much as the representations of these realities. With the help of sociological approaches, engineers can develop competencies to work with the complexity and multiplicity of the social, to critically enact alternative worlds and their appliances, to be aware of values and desires – including dimensions of social justice and democracy – embedded in technology, and to design more reasonable responses to the urgencies of the present world.

Notes

1 The Global South is understood here not geographically but drawing on the line separating the world citizens who enjoy high living standards and those destitute and marginalized wherever they live.

2 Freire criticizes the separation between theory and practice in Western thought and what he called 'banking education.' For him, action and reflection blend into 'praxis,' which is an essential part of the liberating dynamic from oppressive tendencies (Freire, 1970).

References

Acosta, A. (2016). O Buen Vivir: uma oportunidade de imaginar outro mundo. In C. M. Sousa (Ed.), *Um Convite à Utopia* (Vol. 1, pp. 203–233). EDUEPB.

Althusser, L. (2014). *On the reproduction of capitalism: Ideology and ideological state apparatuses*. Verso.

Argyris, C., & Schön, D. A. (1989). Participatory action research and action science compared: A commentary. *American Behavioral Scientist, 32*(5), 612–623.

Arias-Maldonado, M. (2015). *Environment and society; Socionatural relations in the anthropocene*. Springer. https://link.springer.com/chapter/10.1007/978-3-319-15952-2_4

Avelino, F. (2019). Transformative social innovation and (Dis)empowerment. *Technological Forecasting and Social Change, 145*, 195–206. doi:10.1016/j.techfore.2017.05.002.

Baillie, C., Byrne, C., Haralampides, K., Riley, D., & Arif, S. (2021). Engineering, social justice and peace: the journey towards a movement. In C. Alvear, C. Cruz, & J. Kleba (Eds.), *Engenharia e outras práticas técnicas engajadas – vol.1: Redes e movimentos* (pp. 107–134). EDUEPB. https://doi.org/10.5281/zenodo.4908523

Barad, K. (2007). *Meeting the universe halfway: Quantum physics and the entanglement of matter and meaning*. Duke University Press.

Beling, A. E., Cubillo-Guevara, A. P., Vanhulst, J., & Hidalgo-Capitán, A. L. (2021). Buen vivir (good living): A "glocal" genealogy of a Latin American Utopia for the world. *Latin American Perspectives, 48*(3), 17–34. doi:10.1177/0094582x211009242

Blanco, J. P., & Aguiar, E. P. (2020). El Buen Vivir Como Discurso Contrahegemónico. Postdesarrollo, Indigenismo Y Naturaleza Desde La Visión Andina. *Mana, 26*. doi:10.1590/1678-49442020v26n1a205.

Bohman, J., Flynn, J., & Celikates, R. (2023). Critical Theory, Stanford encyclopedia of philosophy (Fall 2023 ed.), E. N. Zalta & U. Nodelman (Eds.). https://plato.stanford.edu/archives/fall2023/entries/critical-theory/
Braun, R. (2019). *Corporate stakeholder democracy*. CEU University Press.
Braun, R. (2024). Radical reflexivity, experimental ontology and RRI. *Journal of Responsible Innovation, 11*(1). https://doi.org/10.1080/23299460.2024.2331651
Braun, R., Loeber, A., Vinther Christensen, M., Cohen, J., Frankus, E., Griessler, E., . . . Starkbaum, J. (2022). Social labs as temporary intermediary learning organizations to help implement complex normative policies. The case of responsible research and innovation in European science governance. *The Learning Organization*. doi:https://doi.org/10.1108/TLO-09-2021-0118
Braun, R., & Randell, R. (2020). Futuramas of the present: The "driver problem" in the autonomous vehicle sociotechnical imaginary. *Humanities and Social Sciences Communications*. https://doi.org/10.1057/s41599-020-00655-z
Braun, R., & Randell, R. (2022). The vermin of the street: The politics of violence and the nomos of automobility. *Mobilities, 17*,1, 53–68.
Braun, R., & Randell, R. (2023). The political ontology of automobility. *Mobility Humanities*, 2(1), 7–22. https://doi.org/10.23090/MH.2022.07.1.2.007
Brown, T., & Wyatt, J. (2010). Design thinking for social innovation. *Development Outreach, 12*(1), 29–43. doi:10.1596/1020-797x_12_1_29.
Castro-Gómez, S., & Grosfoguel, R. (Eds.). (2007). *El giro decolonial: reflexiones para una diversidad epistémica más allá del capitalismo global*. Siglo del Hombre Editores.
Chuji, M., Rengifo, G., & Gudynas, E. (2019). Buen Vivir. In A. Kothari, A. Salleh, A. Escobar, F. Demaria, & A. Acosta (Eds.), *Pluriverse: A post-development dictionary*. Tulika Books.
Cohen, G. A. (2009). *Why not socialism?* Princeton University Press.
Cruz, C. (2021). Brazilian grassroots engineering: A decolonial approach to engineering education. *European Journal of Engineering Education, 46*(5), 690–706. doi:10.1080/03043797.2021.1878346
Cruz, C., Kleba, J., & Alvear, C. (Eds.) (2021). *Engenharias e outras práticas técnicas engajadas* (Vol. 2). EDUEPB. https://eduepb.uepb.edu.br/download/engenharia-e-outras-praticas-tecnicas-engajadas-vol-2/?wpdmdl=1836&masterkey=618ed68a15375
Dewandre, N. (2018). Political agents as relational selves: Rethinking EU politics and policy-making with Hannah Arendt. *Philosophy Today, 62*, 493–519.
Dussel, E. D. (2012). Transmodernity and interculturality: An interpretation from the perspective of philosophy of liberation. *Transmodernity: Journal of Peripheral Cultural Production of the Luso-Hispanic World, 1*(3). doi:10.5070/t413012881.
Escobar, A. (2011). Sustainability: Design for the pluriverse. *Development, 54*(2), 137–140. doi:10.1057/dev.2011.28.
Escobar, A. (2015). Degrowth, post-development, and transitions: A preliminary conversation. *Sustainability Science, 10*(3), 451–462.
Escobar, A. (2019). Thinking-feeling with the earth: Territorial struggles and the ontological dimension of the epistemologies of the South. In B. de Sousa Santos & M. Meneses (Eds.), *Knowledges born in the struggle: Constructing the epistemologies of the global south*. Routledge.
Escobar, A. (2020). *Pluriversal politics: The real and the possible*. Duke University Press.
Esposito, R. (2021). *Instituting thought: Three paradigms of political ontology*. Polity Press.
Everett, H. I. (1957). "Relative state" formulation of quantum mechanics. *Review of Modern Physics, 29*, 454.
Everett, H. I. (2012). *The everett interpretation of quantum mechanics: Collected works 1955-1980 with commentary* (J. A. Barrett & P. Byrne, Eds.). Princeton University Press.
Fanon, F. (2021). *The wretched of the earth* (60th anniversary ed.). Grove Press.
Feenberg, A. (1999). *Questioning technology*. Routledge.
Foucault, M. (1980). *Power/Knowledge: Selected interviews and other writings* (C. Gordon Ed.). Pantheon Books.
Foucault, M. (2006). *The order of things: An archaeology of the human sciences*. Vintage Books.
Freire, P. (1970). *Pedagogy of the opressed*. The Seabury Press.
Fúnez-Flores, J. I. (2022). Inter-epistemic dialogues with decolonial thought from Latin America and the Caribbean. *Educational Studies, 58*, 5–6. doi:10.1080/00131946.2022.2132397.
García Linera, Á. (2015). *Socialismo comunitario: un horizonte de época*. Vicepresidencia del Estado, Presidencia de la Asamblea Legislativa Plurinacional.

Gudynas, E. (2009). Diez tesis urgentes sobre el nuevo extractivismo. In C. L. A. E. S. Caap (Ed.), *Extractivismo, Política y Sociedad* (pp. 187–225). CentroAndino de Acción Popular. Centro Latino Americano de Ecología Social.
Haraway, D. J. (2008). *When species meet*. University of Minnesota Press.
Heidegger, M. (1977). *The question concerning technology, and other essays*. Harper & Row.
Heidegger, M. (2002). *Off the beaten track* (J. Young & K. Haynes Eds.). Cambridge University Press.
Hidalgo-Capitán, A. L., & Cubillo-Guevara, A. P. (2014). Seis Debates Abiertos Sobre El Sumak Kawsay. *Íconos - Revista de Ciencias Sociales, 48*, 25. doi:10.17141/iconos.48.2014.1204.
Hidalgo-Capitán, A. L., García-Álvarez, S., Cubillo-Guevara, A. P., & Medina-Carranco, N. (2019). Los Objetivos del Buen Vivir. Una propuesta alter-nativa a los Objetivos de Desarrollo Sostenible. *Iberoamerican Journal of Development Studies*, 8(1), 6–57. https://doi.org/10.26754/ojs_ried/ijds.354.
Horkheimer, M., & Adorno, T. W. (1973). *Dialectic of enlightenment*. Allen Lane.
Howell, E. L., Wirz, C. D., Brossard, D., Scheufele, D. A., & Xenos, M. A. (2019). Seeing through risk-colored glasses: Risk and benefit perceptions, knowledge, and the politics of fracking in the United States. *Energy Research & Social Science, 55*, 168–178. doi:https://doi.org/10.1016/j.erss.2019.05.020
Jasanoff, S. (Ed.) (2004). *States of knowledge: The co-production of science and the social order*. Routledge.
Jasanoff, S. (2015). Future imperfect: Science, technology, and the imaginations of modernity. In S. Jasanoff & S. H. Kim (Eds.), *Dreamscapes of modernity: Sociotechnical imaginaries and the fabrication of power* (pp. 1–33). University of Chicago Press.
Jasanoff, S., & Kim, S.-H. (2009). Containing the atom: Sociotechnical imaginaries and nuclear power in the United States and South Korea. *Minerva, 47*, 119–146. doi:10.1007/s11024-009-9124-4
Jasanoff, S., & Kim, S.-H. (2015). *Dreamscapes of modernity: Sociotechnical imaginaries and the fabrication of power*. University of Chicago Press.
Jensen, C. B. (2021). Practical ontologies redux. *Berliner Blätter, 84*, 93–104.
Kabo, J. (2010). *Seeing through the lens of social justice: A threshold for engineering* (Phd Thesis.). Caroline Baillie. Queen's University, Kingston Canada.
Kabo, J., Day, R., & Baillie, C. (2009). Engineering and social justice: How to help students cross the threshold. *Practice and Evidence of Scholarship of Teaching and Learning in Higher Education, 4*, 126–146.
Kleba, J. B., & Cruz, C. C. (2020). *Building engaged engineering in curriculum - a review of Brazilian and Australian cases*. ASEE Virtual Conference 22–26 June 2020. Retrieved from https://peer.asee.org/building-engaged-engineering-in-curriculum-a-review-of-brazilian-and-australian-cases.pdf
Kleba, J. B., & Cruz, C. C. (2021). Empowerment, emancipation and engaged engineering. *International Journal of Engineering, Social Justice, and Peace, 8*(2), 28–49. doi:10.24908/ijesjp.v8i2.14380.
Kleba, J. B., & Reina-Rozo, J. D. (2021). Fostering peace engineering and rethinking development: A Latin American view. *Technological Forecasting and Social Change, 167*, 120711. doi:10.1016/j.techfore.2021.120711
Kothari, A., Salleh, A., Escobar, A., Demaria, F., & Acosta, A. (2019). *Pluriverse: A post-development dictionary*. Tulika Books.
Kuhn, T. (1962). *The structure of scientific revolutions*. University of Chicago Press.
Latour, B. (2000). *Pandora's hope: Essays on the reality of science studies*. Harvard University Press.
Law, J. (1986). On the methods of long distance control: Vessels, navigation, and the portuguese route to India In J. Law (Ed.), *Power, action and belief: A new sociology of knowledge?* (32nd ed., pp. 234–263). Routledge, Henley.
Law, J. (2015). What's wrong with a one-world world? *Distinktion: Journal of Social Theory, 16*(1), 126–139. doi:10.1080/1600910X.2015.1020066
Law, J., & Lien, M. E. (2018). Denaturalizing nature. In M. d. l. Cadena & M. Blaser (Eds.), *A world of many worlds* (pp. 131–171). Duke University Press.
Lucena, J., & Kleine, M. S. (2021). Colorado school of mines humanitarian engineering program: Negotiating the technical/social divide to create 'engineering as it should be'. In C. Cruz, J. Kleba & C. Alvear (Eds.), *Engenharias e outras práticas técnicas engajadas* (Vol. 2). EDUEPB. https://eduepb.uepb.edu.br/download/engenharia-e-outras-praticas-tecnicas-engajadas-vol-2/?wpdmdl=1836&masterkey=618ed68a15375
Lybrand, S., & Randell, R. (2022). Possible in sociology. In V. P. Glăveanu (Ed.), *The Palgrave Encyclopedia of the possible* (pp. 1178–1184). Springer International Publishing.
Mannheim, K. (1985). *Ideology and utopia: An introduction to the sociology of knowledge*. Harcourt Brace Jovanovich.
Melo, Y. V. C., Espitia, I., & Costa, J. (2021). Desenvolvimento do capital social comunitário em assentamentos vulneráveis: A experiência da organização Teto (Techo) na Colômbia e no Brasil. In C. Alvear, C.

Cruz, & J. Kleba (Eds.), *Engenharia e outras práticas técnicas engajadas – vol.1: Redes e movimentos* (pp. 219–250). EDUEPB.

Mignolo, W. (2018). Epistemic disobedience and the decolonial option: A manifesto. *Transmodernity: Journal of Peripheral Cultural Production of the Luso-Hispanic World, 1*(2). doi:10.5070/t412011807.

Mitcham, C. (2014). The true grand challenge for engineering: Self-knowledge. *Issues in Science and Technology, 31*(1), 19–22.

Mitcham, C., & Briggle, A. (2009). The interaction of ethics and technology in historical perspective. In A. Meijers (Ed.), *Philosophy of technology and engineering sciences* (pp. 1147–1191). North-Holland.

Mol, A. (1999). Ontological politics: A word and some questions. *The Sociological Review, 47*, 74–89. doi:10.1111/j.1467-954X.1999.tb03483.x

Mol, A. (2014, January 13). Other words: Stories from the social studies of science, technology, and medicine. *Theorizing the Contemporary, Fieldsights*. Retrieved from https://culanth.org/fieldsights/other-words-stories-from-the-social-studies-of-science-technology-and-medicine

Morton, T. (2016). HYPEROBJECTS. *CSPA Quarterly, 15*, 7–9. Retrieved from http://www.jstor.org/stable/90000650

Nieusma, D. (2011). Engineering, Social Justice, and Peace: Strategies for Pedagogical, Curricular, and Institutional Reform. 2011 ASEE Annual Conference & Exposition Proceedings, 22.609.1-22.609.12. https://doi.org/10.18260/1-2--17890

Nieusma, D., & Riley, D. (2010). Designs on development: Engineering, globalization, and social justice. *Engineering Studies, 2*(1), 29–59. doi:10.1080/19378621003604748.

Prato, A., Weckesser, C., & Segura, M. (2021). Las redes comunitarias de Internet en Argentina. AlterMundi y una red extendida durante la pandemia. In C. Alvear, C. Cruz & J. Kleba (Eds.), *Engenharia e outras práticas técnicas engajadas – vol.1: Redes e movimentos* (pp. 285–320). EDUEPB. https://doi.org/10.5281/zenodo.4908523

Proctor, R. N. S. L. (2008). *Agnotology: The making and unmaking of ignorance*. Stanford University Press.

Puig de la Bellacasa, M. (2017). *Matters of care*. University of Minnesota Press.

Raji, I. D., Gebru, T., Mitchell, M., Buolamwini, J., Lee, J., & Denton, E. (2020). *Saving face: Investigating the ethical concerns of facial recognition auditing*. AIES '20: Proceedings of the AAAI/ACM Conference on AI, Ethics, and Society, 145–151. https://doi.org/10.1145/3375627.3375820

Reiter, B. (2018). *Constructing the pluriverse*. Duke University Press.

Ritzer, G. (2021). *The McDonaldization of society: into the digital age* (10th ed.). SAGE.

Said, E. W. (2019). *Orientalism*. Penguin Books.

Santos, S. (2011). Boaventura de. *Utopía y Praxis Latinoamericana, 16*(54), 17–39.

Schneider, J., Lucena, J., & Leydens, J. A. (2009). Engineering to help. *IEEE Technology and Society Magazine, 28*(4), 42–48. doi:10.1109/MTS.2009.935008

Shapiro, J., & McNeish, J.-A. (2021). *Our extractive age: Expressions of violence and resistance*. Routledge.

Shelley, P. B. (1840). A defence of poetry. In M. Shelley (Ed.), *Essays, letters from Abroad, translations and fragments* (pp. 1–57). Edward Moxon.

Simonsen, J., & Robertson, T. (2013). *Routledge international handbook of participatory design*. Routledge.

Smith, J., Tran, A. L. H., & Compston, P. (2019). Review of humanitarian action and development engineering education programmes. *European Journal of Engineering Education, 45*(2), 249–272. doi:10.1080/03043797.2019.1623179

Timmermans, J., Blok, V., Braun, R., Wesselink, R., & Nielsen, R. Ø. (2020). Social labs as an inclusive methodology to implement and study social change: The case of responsible research and innovation. *Journal of Responsible Innovation, 7*(3), 410–426. doi:10.1080/23299460.2020.1787751

Vaneigem, R. (1983 (1967)). *The revolution of everyday life*. Left Bank Books and Rebel Press.

Varela, F. J. (1999). *Ethical know-how: Action, wisdom, and cognition*. Stanford University Press.

Virilio, P. (2007). *The Original Accident* (J. Rose, Trans.). Polity Press.

Walsh-Dilley, M. (2017). Theorizing reciprocity: Andean cooperation and the reproduction of community in highland bolivia. *The Journal of Latin American and Caribbean Anthropology, 22*(3), 514–535. doi:10.1111/jlca.12265

Winner, L. (1986). Do artifacts have politics? In L. Winner (Ed.), *The Whale and the reactor: A search for limits in an age of high technology* (pp. 19–39). Chicago University Press.

Woolgar, S., & Lezaun, J. (2013). The wrong bin bag: A turn to ontology in science and technology studies? *Social Studies of Science, 43*(3), 321–340. doi:10.1177/0306312713488820

Wright, E. O. (2010). *Envisioning real utopias*. Verso.

10

PSYCHOLOGICAL FOUNDATIONS OF ENGINEERING ETHICS EDUCATION

Inês Direito, Curwyn Mapaling, and Julianna Gesun

Introduction

Traditionally, psychology has studied morality in three aspects: cognition, affect, and behavior (Narváez & Rest, 1995). The psychological foundations of ethics are rooted in theories and frameworks of moral development, cognitive and affective processes, and socialization. Moral psychology started as a branch of developmental psychology – which studies the influences of *nature* and *nurture* on humans' physical, cognitive, and behavioral changes throughout their lifespan. Moral psychology extended this and focused on the development of moral reasoning. It has been driven by developmental questions such as 'How does a person develop a moral sense?' and 'At what age does a person understand moral choices and consequences?'

Developmental psychology has had a profound impact on ethics education by offering insights into the development of moral reasoning and by creating frameworks that guide the development and teaching of ethical reasoning skills. In the 1980s, the field started to change its focus from reasoning – as the critical way to acquire and perform moral knowledge – to affect and emotion(s), sociocultural contexts, and automatic processes (Haidt, 2013a). In addition to the dominant 'ethic of autonomy' that characterized secular Western justice-based principles in judging morality, cross-cultural studies by Schweder et al. (1997) proposed two other ethics – an 'ethic of community,' based on communal values and hierarchical social order; and an 'ethic of divinity,' based on concepts such as purity and holiness.

Teaching and research regarding ethics in engineering education have been shaped and informed predominantly by cognitive models of moral development and based on the assumption that ethical reasoning translates automatically to ethical behaviors (Bairaktarova & Woodcock, 2017). However, long-term ethical behaviors should be the goal of engineering ethics education (Clancy & Zhu, 2023). Engineering actions have a tremendous impact on society and the environment, but discerning what these impacts are and deciding on what it means to be ethical in certain situations can be highly subjective and influenced by cultural norms and values (Haidt & Joseph, 2004).

Teaching and researching ethics requires knowledge of theories focusing on human moral development and an understanding of how psychological frameworks have informed teaching and research in engineering ethics education. This chapter draws on literature in psychology to describe the conceptual grounding of the study of moral development, offers an overview of dif-

DOI: 10.4324/9781003464259-13

ferent theoretical perspectives in psychology and how these informed methods and instruments are used to assess moral reasoning in engineering education, and discusses current and emerging issues in engineering ethics education and research.

Positionality statement

We present a combined positionality statement for this work to highlight the impact of our group identity, which transcended our differentiated individual identities. While writing this chapter, we found that our collaboration as a group of scholars took on an identity and positionality that was more than the sum of our individual backgrounds and expertise (Baumeister et al., 2016). While we all conduct research in engineering education and have backgrounds in the social sciences, we each approach the research from unique perspectives given our different disciplinary fields (Clinical Psychology, Educational Psychology, Engineering Education). Thus, writing this chapter required acknowledging our shared educational research interests based on positive psychological perspectives, human-centered approaches to teaching and learning, and non-cognitive variables (such as growth mindsets, thriving, and resilience) and value. Together, we bring more collective awareness of different communities of scholarship regarding these shared interests. Similarly, we merged cultural perspectives of working in academia, sharing personal and professional perspectives from three continents (Africa, Europe, and North America) and four countries (Portugal, South Africa, the United Kingdom, and the United States). Moreover, as a group of authors, we acknowledge that the ways and frameworks we were taught are a product of the so-called 'WEIRD' societies – Western, educated, industrialized, rich, and democratic (Henrich et al., 2010). We also acknowledge that many of the psychology theories and frameworks that founded and shaped the study of moral development embraced universalist conceptions of the human mind and behavior (for more on universalist conceptions, see Chapter 9). Furthermore, we recognize the significance of intellectual humility in our approach (for more on intellectual humility, see Chapter 6). The overview of the foundational theories in moral psychology presented in this chapter, while extensive, is not exhaustive. We acknowledge the value of non-WEIRD perspectives, such as the practice of *Ubuntu* in humanistic psychology (Hanks, 2008, also Chapters 8 and 9), in enriching engineering ethics education and social responsibility (Munir, 2020). We aim to embrace these perspectives in theory and in practice in current and future work through collaborative work with scholars across different continents and cultures.

The study of moral development in psychology

In this section, we describe some of the major theoretical models and concepts of moral psychology, explaining how they shaped and informed engineering ethics education, particularly regarding the development of engineering-specific instruments to assess individual moral reasoning. Different theories and approaches are presented chronologically to provide a historical and contextual overview of this field.

Moral development and moral reasoning

Kohlberg's theory of moral development

Developmental psychology is a branch of psychology that studies the influences of nature and nurture on humans' physical, cognitive, and behavioral changes throughout their lifespan. The initial focus of developmental psychology was on infant and children's cognitive development and implications for education. A well-known example is Jean Piaget's theory of cognitive development, which identified four developmental stages (sensorimotor, pre-operations, concrete operations, and formal operations). Although Piaget explored how morality principles developed in children, Lawrence Kohlberg's cognitive-developmental approach to moral education, developed in the 1960s and 1970s, dominated both the study of morality within developmental psychology for decades (Rest et al., 2000) and the shaping of educational programs to teach professional ethics (Pritchard, 1999). Kohlberg's stages of moral reasoning have been commonly used to assess the ethics education of undergraduate engineering students in the United States (Bairaktarova & Woodcock, 2017).

Kohlberg's theory of moral development states that moral reasoning progresses through six hierarchical stages of value-orientation that are grouped into three broader levels: the pre-conventional level, conventional level, and post-conventional level (Kohlberg, 1981). According to Kohlberg, most adults never reach the post-conventional level.

- At the pre-conventional (pre-moral) level, individuals are responsive to cultural rules of good and bad, but there is no true sense of right or wrong. Moral judgments are based on physical consequences (Stage 1: Punishment and obedience orientation) and rewards (Stage 2: Instrumental relativist orientation) of obeying/disobeying authority figures.
- At the conventional level, individuals begin to internalize societal norms, rules, and conventions. Moral judgments are based on conformity to the expectations of others' families or groups (Stage 3: The interpersonal concordance or 'good boy-nice girl' orientation) and duty to maintain social order (Stage 4: The 'law and order' orientation).
- At the post-conventional (autonomous or principled) level, individuals abide by their own moral values and principles. Moral judgments consider the relativism of personal values (Stage 5: The social-contract, legalistic orientation) and the universality of abstracts and ethical principles (Stage 6: The universal-ethical-principle orientation).

This theory of moral development was widely tested in moral education programs in schools through the presentation of moral dilemmas – hypothetical situations in which moral principles conflict (see Kohlberg's widely-known Heinz dilemma, 1969) – to generate debate (Blatt & Kohlberg, 1975). Much of the discussion was left to the participants themselves, with researchers intervening to clarify, summarize, or present a view of their own that was a stage above that of most of the class. This speaks to one of Kohlberg's key ideas as to how individual moral reasoning progresses through these stages, encountering viewpoints that challenge their thinking and stimulate them to formulate better arguments (Kohlberg, 1981). This method of inducing cognitive conflict epitomizes Piaget's equilibration model, an ongoing process allowing for the balance between *assimilation* (the incorporation of new information into an already existing cognitive structure) and *accommodation* (the transformation of pre-existing cognitive structures to fit new knowledge) in an individual's transition from one major developmental stage onwards to the next. Cognitive structures are mental frameworks, or schemas, that individuals use to process and organize information.

Kohlberg claimed that these stages of moral reasoning proceed in a culturally universal invariant sequence (Kohlberg, 1973, p. 630), where each stage provides a progressively broader framework for dealing with moral issues, being more cognitively adequate than the previous stage. He maintained that his model is universal due to the stages referring to the underlying modes of reasoning, not specific cultural beliefs. However, it is now well-documented that individuals from different cultures move through Kohlberg stages at disparate rates and reach different endpoints (Worthy et al., 2020).

Despite these claims and limitations, Kohlberg's work in developmental psychology was fundamental in demonstrating that cognition (reasoning) plays an important role in moral judgment. Although a primitive innate sense of morality is found in humans (Bloom, 2013), this morality is limited – it can change and evolve during growth and improve through education. Students' cognitive and psychosocial development continues in university or college years (Chickering & Reisser, 1993; Perry, 1970), including their ethical reasoning.

In engineering ethics education, Magun-Jackson (2004) proposed the adoption of the model developed by Hersh et al. (1979), which adopted a Kohlbergian approach, to stimulate cognitive conflict and perspective-taking in engineering students with the use of ethical dilemmas, dialogues, and role-playing. This model highlights the need to understand that different students might be in different developmental stages (Chickering & Reisser, 1993), might bring different cultural perspectives to ethical issues, and might have different verbal abilities to express and discuss their ethical reasoning (Clancy et al., 2022). This is particularly relevant in engineering education settings with international students.

Neo-Kohlbergian model of moral judgment

Following on the tradition of Kohlberg, but addressing concerns raised by other psychologists and philosophers, Rest et al. (2000) proposed a neo-Kohlbergian model of moral judgment and developed a psychometric test called the Defining Issues Test (DIT) (Rest et al., 1999a). The 'Kohlbergian part' (Rest et al., 2000, p. 383) of this model comprises the following features: the focus of morality research is on cognition, personal experiences and understanding, and the development of higher morality. As in Kohlberg's approach, adulthood is characterized by the shift from conventional to post-conventional moral thinking. While adopting a developmental framework, the neo-Kohlbergian approach structures developmental changes in moral schemas, rather than moral stages. Schemas are cognitive structures that represent knowledge of certain concepts; they can guide perception and help in the interpretation of new information and in problem-solving (Rest et al., 2000). Moreover, moral schemas are defined as being context dependent (Thoma & Dong, 2014) rather than abstract universal principles, which contradicts Kohlberg's claims of the universalism of moral stages. These schemas are influenced by multiple contexts, including the immediate environment and cultural, religious, and historical contexts.

According to Narváez and Rest (1995), four psychological processes are involved in acting morally – the Four Component Model (FCM) considers moral sensitivity, moral judgment, moral motivation, and implementation. Narváez and Rest argue that each process (or component) integrates both cognitive and affective elements. Moral sensitivity "involves the ability to interpret the reactions and feelings of others" (Bebeau, 2002, p. 283). This requires empathy and perspective-taking. In professional settings, it involves not only considering others' perspectives but also knowing the codes and regulations of the profession – which can be described as 'ethical sensitivity' (Bebeau, 2002). Moral judgment is the component most researched in psychology. It "involves deciding which of the possible actions is most moral" (Narváez & Rest, 1995, p. 386). This requires weighing the pros and cons of possible and anticipated routes of action. Moral motivation involves "prioritizing moral values over other personal values" (Bebeau et al., 1999, p. 22). Identity development toward a shared code of professional expectations and values, through normative socialization, is essential to moral motivation and ethical orientation in professional settings (Bebeau, 2002). Moral implementation "presupposes that one has set goals, has self-discipline and controls impulse, and has strength and skill to act in accord with one's goals" (Bebeau et al., 1999, p. 22). This component relates to the importance of character and agency to responsible ethical practice.

Instruments for assessing moral reasoning

The DIT was developed as an alternative test to Kohlberg's moral judgment interviews. Like these interviews, the DIT uses stories to focus the participant on a moral dilemma and elicit their construction of moral reasoning. However, whereas in the moral judgment interview participants are asked to respond to a moral dilemma, the DIT is a recognition test where participants are asked to rate and rank a set of statements representing features of moral dilemmas (Rest et al., 1999b; Thoma & Dong, 2014). Most importantly, the development of the DIT in 1974 introduced a theoretical departure from Kohlberg's framework. Instead of assessing stages of moral reasoning linearly, the neo-Kohlbergian approach organizes moral reasoning in three moral schemata: pre-conventional (or Personal Interest Schema), conventional (or Maintaining Norms Schema), and post-conventional (Rest et al., 2000). The DIT is, thus, a tool for activating moral schemas and assessing those schemas. The DIT-2 (Rest et al., 1999b) is a refined version of the DIT and comprises a set of five moral dilemmas. Each dilemma is described in one paragraph and is followed by 12 statements representing three levels of moral reasoning schemas. Participants are asked to rate the moral importance of each statement on a 5-point Likert-type scale (from great importance to no importance) and then to rank the four statements that best describe their understanding of how the protagonist ought to solve the respective dilemma.

The DIT has been used to assess the effectiveness of educational interventions (Drake et al., 2005). As a response to the call for contextual and profession-specific tools to measure moral judgment (Bebeau, 2002), Borenstein et al. (2010) adapted the DIT-2 and developed the Engineering and Science Issues Test (ESIT). The ESIT has been used in case studies in civil engineering (Murzi et al., 2019) and compassion-induced engineering ethics cases (Kotluk & Tormey, 2023). Another instrument developed to assess individual ethical decision-making in project-based design teams is the Engineering Ethical Reasoning Instrument (EERI), also based on Kohlberg's moral development theory (Zhu et al., 2014).

Ethics of care, empathy, and pro-social behavior

Whereas most research and practice in engineering ethics education has been influenced by (neo) Kohlbergian theories and instruments of moral judgment, the concept of care and associated constructs, such as empathy and social responsibility, has been increasingly explored and integrated into engineering education (Strobel et al., 2011). In particular, Hess and colleagues' work is focused on empathy's role in engineering ethics education (Hess et al., 2017; Hess et al., 2021).

Ethics of care

Gilligan (2003) developed the concept of ethics of care to complement Kohlberg's theory of moral development. In Govrin's words (2014, p. 8), "Ethics of care was the first theory to challenge the Kohlbergian-Kantian view that moral judgment is determined by rational psychological processes. In moral psychology, it was the first theory to present a model of moral judgment based on emotions."

Gilligan's work brought to the fore the 'feminine voice,' and the role of empathy and compassion, when making morality-based decisions. This voice was previously overlooked in initial theories of moral development that were male-oriented and focused on logic and individualism (Gilligan, 2003). Gilligan's research suggested that the act of caring was an intrinsic part of moral development. This ethics of care "emphasizes interpersonal relationships, connectedness, and self-awareness rather than abstract or decontextualized moral reasoning" (Hess et al., 2016, pp. 236–

237). It is "based on the daily activity of caring rather than on abstract principles" (Pantazidou & Nair, 1999, p. 207).

Other scholars, such as Noddings (1984, 2013) and Tronto (1994), further developed theories of care. In Nodding's care theory, the basic feature of ethics is the relationship between people, and the aim of moral education is preparing students to engage in caring relationships, to "care-for those they encounter directly and to care-about the suffering of people at a distance" (Noddings, 2013, p. 394). Moral education can be promoted through modeling, dialogue, practice, and confirmation (Noddings, 2002). Unlike Gilligan and Noddings, Tronto conceptualized care as a gender-neutral practice. She described integrity of care as a combination of four moral elements: (1) attentiveness (caring about), (2) responsibility (care taking), (3) competence (care giving), and (4) responsiveness (care receiving).

In engineering education, the work by Pantazidou and Nair (1999) described engineering and care as both a response to a need and an orientation towards action. These authors explored how Tronto's framework was suitable for guiding teaching care and practicing care in engineering ethics education and matched Tronto's four moral elements to different phases of engineering design (Dieter, 1991): (1) the identification of a societal need (attentiveness of the care giver); (2) design conceptualization to respond to the need (responsibility of the care giver); (3) feasibility analysis and production (competence of the care giver); and (4) acceptance of design product (responsiveness of the care receiver).

For an overview of other feminist theories and engineering ethics, see Riley (2013).

Empathy and pro-social behavior

In Hoffman's theory of pro-social moral development, empathy is core to ethical reasoning, and empathic perspective-taking to ethical decision-making (Hoffman, 2000). According to Hoffman's research, empathy can be divided into two types: cognitive and affective empathy. Cognitive empathy involves understanding someone else's perspective, while affective empathy involves feeling and sharing someone else's emotions.

An empathic response requires "the involvement of psychological processes that make a person have feelings that are more congruent with another's situation than with his own situation" (Hoffman, 2000, p. 30). Central to this response is empathic distress, defined as a prosocial motive – "one feels distressed when observing someone in actual distress" (Hoffman, 2000, p. 63). Empathic distress is associated with helping. It precedes helping, and individuals feel better after helping. However, empathic distress does not always lead to helping behaviors. A classic phenomenon in psychology is the 'bystander effect,' where individuals are less likely to help a victim, and feel less responsible, in the presence of other people/bystanders (Darley & Latané, 1968).

Hoffman's work highlights the importance of empathy in promoting pro-social behavior, which can be cultivated through education, modeling, and other forms of socialization. He posits that socialization, which enables individuals to experience a range of emotions, enhances their capacity for empathy. Adults exposed to role models demonstrating pro-social behavior, during childhood and thereafter, are more likely to develop and exhibit similar pro-social and helping behaviors. Induction to pro-social behavior, the voluntary attention to someone else's perspectives and distress, is also a key element in enhancing empathic potential (Hoffman, 2000).

An important element of empathy is perspective-taking. Batson (2009) distinguishes two types of perspective-taking: imagine-other perspective and imagine-self perspective. The former is about "imagining how another is thinking and feeling." It "is not so much what one knows about the feelings and thoughts of the other but one's sensitivity to the way the other is affected by his or her situation" (Batson, 2009, p. 7). This latter part is about "imagining how one would think and feel in the other's place." This is aligned with the concepts of 'role-taking' and 'role-playing' as teaching methods in engineering ethics education, which are expanded in Chapter 20 on teaching ethics using case studies.

Intuition and ethics

In psychology, most research on moral development has been dominated by cognitive and rationalist frameworks, which consider that moral judgment is an outcome of moral reasoning. Contrary to this trend, in his Social Intuitionism Model, Haidt argues that intuition comes first and strategic reasoning comes second (Haidt, 2013a). Strategic reasoning (process) and judgment (outcome) are interrelated mechanisms. Moral judgments are often made quickly and automatically, based on gut feelings and emotions rather than logical reasoning – "moral judgment is caused by quick moral intuitions and is followed (when needed) by slow, *ex post facto* moral reasoning" (Haidt, 2001, p. 817).

As a result of their research exploring morality and culture, Haidt and Joseph developed the concept of intuitive ethics – "an innate preparedness to feel flashes of approval or disapproval toward certain patterns of events involving other human beings" (Haidt & Joseph, 2004, p. 56) – with implications for moral education and moral diversity. These authors propose four moral patterns, or moral foundations, shared across different cultures – suffering, hierarchy, reciprocity, and purity – with associated emotions and virtues.

Haidt's work on intuitive ethics has significant implications for our understanding of moral psychology and political ideology. Haidt argues that people's political beliefs are often based on their moral intuitions rather than logical reasoning (Haidt, 2013b). Haidt's research on intuitive ethics has also shed light on the role of moral emotions, such as disgust and anger, in shaping moral judgments. For example, research by Rozin et al. (2008) has shown that people's moral judgments can be influenced by disgust, with morally relevant stimuli that elicit disgust being more likely to be judged morally wrong. On the other hand, elevation, the emotional response to "witnessing acts of virtue or moral beauty" (Algoe & Haidt, 2009, p. 106), a reaction to 'moral excellence,' motivates pro-social behavior and volunteering (Cox, 2010), as well as ethical leadership in organizations (Vianello et al., 2010). For a detailed account of how engineering ethics education relates to emotion, see Chapter 4.

In our everyday lives, intuition plays a role in shaping our decision-making processes in numerous crucial ways (Haidt, 2013b). In engineering ethics education, understanding the role of intuition and moral intuitions in ethical decision-making is essential to preparing future engineers to deal with complex ethical issues. In the next section, we will discuss the role of intuition and cognitive biases in engineering ethics education. Suggested guidelines on how to integrate intuition into engineering ethics education are provided are provided in the last sections of the chapter.

Generally speaking, intuition is the ability to understand or grasp something immediately without conscious reasoning. Ethical decisions are often guided by intuition, which manifests as a 'gut feeling' or a sense of right or wrong. Even though intuition may be influenced by personal experiences, cultural background, and one's own values, it can potentially provide valuable insights when making ethical decisions.

Cognitive bias in engineering ethics education

A complex ethical dilemma may involve multiple stakeholders and long-term consequences when making ethical decisions in engineering. Engineers can make appropriate decisions based on intuitive judgments if they can identify potential ethical issues quickly. However, sole reliance on intuition may result in biases and cognitive errors. Therefore, when making ethical decisions, a balance must be struck between intuition and analysis.

It is important to recognize the limitations and potential pitfalls of intuition when making ethical decisions. Various cognitive biases may affect intuitive judgments (Berthet, 2021; Caviola et al., 2014). All humans have cognitive biases. A study by Steele et al. (2016) in US institutions showed that home and international graduate students are prone to different cognitive biases. In the following paragraphs, we discuss confirmation bias, anchoring bias, and groupthink in greater detail.

Confirmation bias refers to the tendency for individuals to seek, interpret, or remember information in a way that confirms their pre-existing beliefs, ideas, or hypotheses (Nickerson, 1998). Often, this bias can lead to errors in judgment or flawed conclusions, influencing decision-making and reasoning processes. As humans, we naturally tend to look for patterns in the world around us and make sense of them. There are several ways in which confirmation bias can manifest itself, namely selective exposure, selective perception, and selective recall.

The purpose of Table 10.1 is to examine how confirmation bias can manifest itself by applying an engineering-related vignette that we developed for this chapter – '*Nothing better than steel': Engineers may experience confirmation bias when developing a new product or system. Imagine that an engineer is developing a new type of suspension system for a car, and they strongly believe that a particular material (e.g., steel) is the best choice for suspension components.*

Even in the face of strong evidence, confirmation bias can lead to overconfidence, poor decision-making, and reluctance to change one's beliefs. Being aware of this bias, seeking out diverse sources of information, engaging in critical thinking, and being open to updating beliefs in light of new information are important.

In engineering education, the *anchoring bias*, as detailed by Berg and Moss (2022), poses a significant barrier to effective perspective-taking. When engineering educators anchor their judgments on initial information, especially in the case of uncertainty or insufficient data, they may unintentionally limit their openness to new methodologies or diverse viewpoints. This bias towards initial impressions or familiar concepts can lead to a reliance on outdated teaching methods, restricting the educators' ability to adapt to new insights and perspectives in the rapidly evolving field of engineering.

Moreover, the impact of anchoring bias can be exacerbated by priming effects, where prior exposure to certain stimuli, such as conventional engineering concepts, shapes future decisions and judgments. This phenomenon further emphasizes the importance of engineering educators actively seeking varied information sources and continuously updating their judgments with new evidence. By acknowledging and actively working to overcome this bias, engineering educators can more effectively foster an inclusive and dynamic educational environment, promoting diverse perspectives and better preparing students for the complexities of modern engineering challenges.

Groupthink, a psychological phenomenon extensively studied in fields like social psychology and organizational behavior, occurs when a group suppresses dissenting opinions in favor of consensus and harmony, often at the cost of critical thinking and objective decision-making. This concept, explored in works by scholars such as Janis (1972) and Turner and Pratkanis (1998), highlights how the desire for group conformity can overshadow the evaluation of information, risks, and potential outcomes. In such settings, ethical and moral reasoning can be compromised, as the group prioritizes agreement over the thorough examination of moral implications and ethical standards.

Addressing groupthink requires fostering a culture where open communication, critical thinking, and diverse viewpoints are valued. Psychological research, including studies on moral

Table 10.1 Application of types of confirmation bias explanations to our engineering vignette

Type of confirmation bias	*Explanation*	*Engineering Scenario*	*Ethical Implications*
Selective exposure	It is possible that people prefer to consume information that adheres to their beliefs and avoid information that does not. An echo chamber effect may result, where individuals surround themselves with people and sources who are similar to them, reinforcing their perspectives and beliefs.	It is possible for the engineer to only search for and read research papers, case studies, or articles that demonstrate the benefits of using the specified steel for suspension systems, avoiding or dismissing evidence that suggests alternatives (e.g., aluminum, carbon fiber, or titanium) are more effective.	This bias can lead to ethical issues in engineering due to the lack of thorough examination of all relevant data and alternatives. Engineers have a responsibility to consider diverse perspectives and information to ensure the most effective and efficient solutions are chosen, not just those that align with pre-existing beliefs.
Selective perception	People may disregard or downplay contradictory evidence when presented with ambiguous or conflicting information.	Engineers may overemphasize or ignore the disadvantages of steel, such as its higher weight or increased cost, if they encounter a study that compares the performance of various materials. It is also possible for them to emphasize the negative aspects of alternative materials while ignoring their positive aspects.	By overemphasizing or ignoring certain aspects of data, engineers risk compromising the integrity and safety of their designs. Ethical engineering requires a balanced and fair assessment of all available information, especially in cases where public safety and welfare are at stake.
Selective recall	When people have information that confirms their beliefs, they may remember it more vividly and accurately than when they have information that challenges their beliefs. It may be difficult for them to consider alternative viewpoints if their opinions are further entrenched as a result of this.	In the case of a specific steel, the engineer may remember positive testimonials or success stories more vividly than any negative feedback or failure cases. Consequently, the material's suitability for the application may be overestimated.	Favoring information that confirms pre-existing beliefs can lead to overconfidence in certain materials or designs, potentially overlooking risks or flaws. Ethical engineering practice necessitates a critical and objective evaluation of all experiences and data, both positive and negative, to ensure the most reliable and safe engineering solutions.

development by Kohlberg (1981) and on ethical decision-making by Rest (1986), suggests that encouraging a questioning attitude and constructive criticism can significantly enhance ethical reasoning in group settings. Leaders in the engineering industry play a crucial role in mitigating groupthink by seeking dissenting opinions and challenging group assumptions. Techniques like including external viewpoints, assigning a 'devil's advocate,' and dividing the group into smaller, independent subgroups are effective strategies to reduce the risk of groupthink, as they introduce a variety of perspectives and critical evaluations.

The concept of groupthink can be exemplified through a scenario where a team of engineers, led by a directive leader, works on a safety system for a plant. The leader's strong influence could inadvertently steer the team towards a unanimous approach, potentially overlooking critical safety and ethical considerations. This scenario underscores the importance of ethical and moral reasoning in engineering decision-making, particularly in collaborative environments where the risk of groupthink is prevalent. These and other cognitive biases should be addressed in engineering ethics education, and strategies should be taught to mitigate their impact on decision-making.

Engineering ethics education through Greene's dual process theory (2015) and Kahneman's *Thinking, Fast and Slow* (2012) highlights the interplay between intuitive and analytical thinking. There are many ethical dilemmas that engineers face daily which require not only technical expertise but also sound moral judgment. The dual-process theory can be incorporated into engineering ethics curriculums to demonstrate that intuition can be susceptible to cognitive biases and shortcuts that may not necessarily lead to ethical decisions. Engineers can be better prepared to navigate difficult ethical situations by promoting self-awareness of these biases and developing critical thinking skills. As a result, engineers can strike a balance between fast, automatic, and emotionally driven thought processes that may drive initial reactions and the slow, deliberate thinking required for making ethical, thoughtful decisions (Greene, 2015). In addition to enhancing engineering ethics education, such an approach also contributes to developing more socially and ethically responsible engineers.

Personal and cultural perspectives should also be highlighted in engineering ethics education. People may make intuitive judgments differently based on their cultural backgrounds, personal values, and life experiences. Research on intuitive ethics underscores the importance of acknowledging these differences in judgment (Haidt & Joseph, 2004). It is the responsibility of educators to foster an environment that encourages open discussions and respects the perspectives of all students.

It is important to use intuition when making ethical decisions in engineering, but this is often overlooked. Incorporating intuition into engineering ethics education, as Haidt and Joseph suggest (2004), can facilitate the development of responsible and ethical engineers by providing students with a more comprehensive and nuanced understanding of ethical dilemmas. Some suggestions on incorporating intuition into engineering ethics education are presented in the last sections of this chapter.

Positive psychology

Positive psychology has emerged as a significant area of interest in recent years, focusing on the promotion of well-being and human flourishing by understanding, nurturing, and harnessing the strengths and virtues of individuals. Traditionally, it involves the study of the processes and conditions that enable individuals, but also groups and institutions, to function at their optimum level (Gable & Haidt, 2005). Research on the psychological aspects of what makes life worth living has been overshadowed by research on disorders and damage carried out before the 2000s. Positive psychology began as a result of the recognition of this imbalance in the early twenty-first century, a movement that was motivated by a desire to conduct research and engage in areas that had previously been neglected and 'one-sided.' In this spectrum of positive psychology, the psychobiological basis of morality is one of many areas explored. Positive moral emotions can uplift and transform people (Keyes & Haidt, 2003).

It is important to note that the term *positive psychology* does not necessarily imply that other psychological studies outside of its scope are negative. In fact, most of the academic work in positive psychology is neutral in nature, aiming to address the full spectrum of the human experience. Importantly, positive psychology seeks to broaden the focus of research to include topics that have traditionally been examined primarily through their negative aspects. For example, much-published research on morality has been concerned with negative moral emotions, such as anger and disgust when others do wrong, as well as shame and guilt when one does wrong. Conversely, positive moral emotions like gratitude and admiration are much less well-studied empirically, according to Gable and Haidt (2005). Therefore, the emergence of positive psychology was largely motivated by a recognition that the science of psychology could benefit from a more balanced approach that includes researching and understanding 'what goes right' with individuals, families, groups, and institutions.

In the context of engineering education, positive psychology can help cultivate an ethically grounded and emotionally intelligent generation of engineers. The final sections of the chapter will further delve into the importance of integrating positive psychology into engineering education.

How moral psychology has been used in engineering education and research

Engineering ethics education – current and emerging issues

Researchers have documented several methods to study ethics in engineering education in the classroom. For example, instructors have created dedicated engineering ethics courses where engineering students read and rank statements corresponding to moral foundations, ethical codes/standards, or case studies (Clancy et al., 2022; Hess & Fore, 2018). In other cases, students are also invited to respond to ethical reflection prompts given in courses that may or may not be focused on ethical reasoning (Hess et al., 2021; Hashemian & Loui, 2010).

Additional studies have integrated ethics into existing technical engineering courses (Davis, 2006; Hess et al., 2019). Despite these different methods of incorporating ethics and moral psychology into engineering classrooms, there are no established 'best' practices for teaching and studying ethics in the classroom (Hess & Fore, 2018). Thus, instructors are encouraged to incorporate whichever format of ethics instruction makes sense for their class. Overall, many opportunities remain for researchers to investigate and evaluate practices for teaching ethics in engineering classes, as we outline below.

Studies on empathy and care have been gaining attention in engineering education and research since 2011 (Strobel et al., 2011). Recent works have focused on the relationship between empathy and empathic perspective-taking in engineering ethics (Hess et al., 2017, 2021), empathic communication (Sochacka et al., 2020), and empathy as a learnable skill, a practice orientation, and a professional way of being (Walther et al., 2017).

Outside the classroom, service learning has emerged as a way for engineering students to develop an understanding of ethical responsibility and morality by working in local communities. Prior research documents several methods to include service learning in engineering curricula. For example, engineering courses (at the undergraduate and graduate levels) have integrated elements of service work with local community projects, Habitat for Humanity, Campus Connect, and medical clinics (Tsang, 2000). Additionally, service learning in engineering could stem from student-initiated efforts across the engineering curriculum (Pritchard, 2000). Reflection is a crucial aspect of service learning that distinguishes it from volunteering, as student reflections demonstrate and assess an understanding of ethical responsibility. This process of reflection enables students to consider their value ideals more deeply (Pritchard, 2000). (Chapter 6 also discusses reflection and reflexivity. See Chapter 23 for more on ethics in service-learning and humanitarian engineering education, and Chapter 25 on reflective and dialogical approaches in engineering ethics education.)

As studies of engineering ethics education continue to grow, so do its critiques. One major critique focuses on the disconnect between ethics education and behavior change. For example, prior interventions focus on building ethical awareness yet "pay little attention to how well ethical awareness predicts ethical behavior" (Bairaktarova & Woodcock, 2017, p. 1129). Large-scale survey research on undergraduate engineering students revealed that students experienced many opportunities for high-quality ethics education, but ethical knowledge and behaviors varied (Finelli et al., 2012). Regarding engineering professional practice, there seems to be agreement that developing ethical behaviors and understanding the connection between ethical awareness and behavior change is critical (Bairaktarova & Woodcock, 2017). One explanation for the disconnect between ethical knowledge and behaviors is that moral educational approaches in engineering have been 'rule-based' with a focus on 'negative' consequences, which can lead to negative outcomes such as moral schizophrenia (Han, 2015; Harris, 2008; Stovall, 2011).

Recent trends in positive psychology emphasize that "moral education should serve for students' flourishing and authentic happiness" (Han, 2015, pp. 441–442). The moral education of engineering students should also prioritize their physical and mental well-being. In the next sections, we call for more integration of positive psychology and intuition in engineering ethics education.

Integrating positive psychology into engineering ethics education

A positive psychology approach can be integrated into an engineering ethics curriculum by engineering educators focusing on the following strategies:

- Ensure that self-reflection is encouraged: Students should be provided opportunities to reflect on their emotions and emotional responses to ethical dilemmas. As a result, they can develop self-awareness and understand how emotions influence their decision-making processes.
- Teach techniques for regulating emotions: Teach students effective methods for managing their emotions, such as mindfulness cognitive restructuring and stress-reduction techniques (Huerta et al., 2021).
- Develop empathy and the ability to take a perspective: Foster empathy and social awareness among students by incorporating exercises that encourage them to consider other people's emotions and perspectives (Sochacka et al., 2020).
- Promote positive mindsets: Educate students about the importance of growth mindsets, emphasizing learning from mistakes and viewing challenges as opportunities for improvement (Campbell et al., 2021).
- Develop collaborative problem-solving skills by facilitating group discussions: Promote a supportive learning environment by providing students with opportunities to work together on ethical dilemmas and to develop relationship management skills.

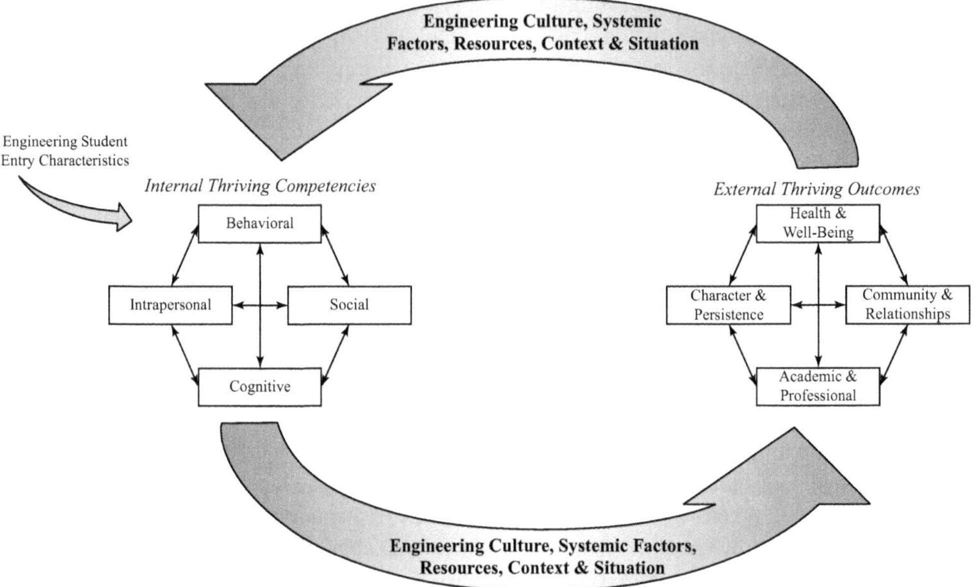

Figure 10.1 Model of engineering thriving (Gesun et al., 2021).

An important goal of engineering education is to cultivate students' ability to find solutions that serve diverse societies. Based on a model of engineering thriving (Gesun et al., 2021), all competencies are highly interrelated and ought to be studied alongside other competencies (see Figure 10.1). It is a moral imperative for education, particularly in engineering, to facilitate students' thriving, equipping them with the skills to confront ethical problems and exercise their ethical responsibilities (Mapaling, 2023). This approach involves making students aware of ethical problems, understanding their impact as moral agents, and developing solutions to these problems. Engineering schools have responded by incorporating ethics education through freestanding courses and/or integrating ethics across the curriculum, employing various problem-solving activities such as role-playing, computer simulations, and group projects. Refer to the section of this handbook titled "Teaching Methods in Engineering Ethics Education" for a comprehensive mapping of established and emerging methods utilized in teaching engineering ethics.

Incorporating intuition into engineering ethics education

Educators should consider the following strategies when incorporating intuition into engineering ethics education:

- Promote self-reflection: It is important for students to reflect upon their intuitions and to consider the factors that may influence these judgments. Developing self-awareness and learning to recognize potential biases can be facilitated through this process.
- Enhance students' ethical sensitivity: Educators can sharpen students' ethical sensitivity through case studies, role-playing activities, and other interactive activities to help them recognize ethical dilemmas more instinctively.
- Create an atmosphere of moral imagination: Rules and codes are important, but not sufficient, for ethical decisions. To enhance students' intuitive understanding of ethical dilem-

mas, educators can cultivate students' moral imaginations by encouraging them to imagine different perspectives, stepping outside cultural constraints, and potential consequences, and make use of stories and metaphors (Umbrello, 2020). Moral imagination "demands that individuals be able to see this cultural reality as a perspective and to transcend it to the point where an individual can understand alternative stories arising from different cultures and contexts" (Mehalik & Gorman, 2006, p. 296).
- Maintain a balance between intuition and analysis: A balanced approach to engineering ethics should be emphasized during engineering ethics education. Students can make well-informed decisions if educators provide the tools and frameworks to critically evaluate their intuitive judgments.

To date, the focus on the psychological foundations of engineering ethics education has been heavily constrained by cognitive theories and rational approaches. This limited perspective can be detrimental to engineering ethics education, as "expertise in moral reasoning does not seem to improve moral behaviors, and it might even make it worse" (Haidt, 2013b, p.104). Haidt's observation signals a critical challenge in engineering ethics education: the gap between theoretical moral reasoning and practical ethical behavior. The recognition of this gap necessitates a shift in focus towards applying moral psychology in ways that promote actual ethical behavior among engineering students and professionals.

Promising directions for future work include fostering long-term ethical behaviors in engineering students. As Clancy and Zhu (2023) highlight, it is the behavior of professional engineers and the products they design that impact society. Thus, behavior change is more pivotal than merely understanding what is 'right' and what is 'wrong.' Additionally, exploring cross-cultural variations in ethical reasoning is vital. Preliminary results from Clancy et al. (2022) indicate cultural differences in moral intuitions among engineering students, suggesting a rich area for further study.

The emergent field of emotions in engineering education also offers new perspectives. Researchers like Kim (2022) and Kotluk and Tormey (2023) are exploring the role of emotions and emotional empathy in moral behavior, an area ripe for further investigation.

Moral psychology, as a foundational discipline in ethics education, encompasses diverse research traditions. These traditions, from developing and evaluating moral reasoning to exploring pro-social behavior, care, and empathy, are critical pillars for engineering ethics education. In recent times, frameworks such as moral intuitionism and positive psychology have notably enriched the discourse within moral psychology, offering new perspectives and methodologies.

This chapter has delved into the psychological foundations of engineering ethics education, with a particular focus on cognitive approaches to moral reasoning. Although these approaches remain predominant, the chapter has also introduced and examined diverging perspectives, including care, empathy, pro-social behavior, and moral intuitionism. These diverse approaches underline the complexity of ethical reasoning and the importance of a multifaceted approach to ethics education.

Considering these discussions, the chapter advocates for the integration of positive psychology into the curriculum of engineering ethics education. Positive psychology, with its emphasis on fostering well-being and success, presents a complementary approach that can significantly enhance traditional methods of teaching ethics. By nurturing moral and ethical reasoning, attitudes, and behaviors, positive psychology offers a holistic framework that prepares engineering students to face ethical challenges and promotes their overall well-being.

As we look towards the future of engineering ethics education, embracing this integration will be pivotal. Converging cognitive approaches with care principles, empathy, pro-social behavior, moral intuitionism, and positive psychology can create a more robust and comprehensive educational experience. Such an approach can not only equip students with the necessary tools to navigate ethical dilemmas but also cultivate a generation of engineers who are both ethically minded and oriented towards contributing positively to society.

References

Algoe, S., & Haidt, J. (2009). Witnessing excellence in action: The other-praising emotions of elevation, admiration, and gratitude. *Journal of Positive Psychology, 4*(2), 105–127.

Bairaktarova, D., & Woodcock, A. (2017). Engineering student's ethical awareness and behavior: A new motivational model. *Science and Engineering Ethics, 23*, 1129–1157.

Batson, C. D. (2009). These things called empathy: Eight related but distinct phenomena. In J. Decety & W. Ickes (Eds.), *The social neuroscience of empathy* (pp. 3–15). MIT Press.

Baumeister, R., Ainsworth, S., & Vohs, K. (2016). Are groups more or less than the sum of their members? The moderating role of individual identification. *Behavioral and Brain Sciences, 39*, E137.

Bebeau, M. (2002). The defining issues test and the four component model: Contributions to professional education. *Journal of Moral Education, 31*(3), 271–295.

Bebeau, M., Rest, J., & Narváez, D. (1999). Beyond the promise: A perspective on research in moral education. *Educational Researcher, 28*(4), 18–26.

Berg, S. A., & Moss, J. H. (2022). Anchoring and judgment bias: Disregarding under uncertainty. *Psychological Reports, 125*(5), 2688–2708.

Berthet, V. (2021). The impact of cognitive biases on professionals' decision-making: A review of four occupational areas. *Frontiers in Psychology, 12,* Article 802439.

Blatt, M. M., & Kohlberg, L. (1975). The effects of classroom moral discussion upon children's level of moral judgement. *Journal of Moral Education, 4*(2), 129–161.

Bloom, P. (2013). *Just babies: The origins of good and evil.* The Bodley Head

Borenstein, J., Drake, M. J., Kirkman, R., & Swann, J. L. (2010). The engineering and science issues test (ESIT): A discipline-specific approach to assessing moral judgment. *Science and Engineering Ethics, 16*(2), 387–407.

Campbell, A., L., Direito, I., & Mokhithi, M. (2021). Developing growth mindsets in engineering students: A systematic literature review of interventions. *European Journal of Engineering Education, 46*(4), 503–527.

Caviola, L., Mannino, A., Savulescu, J., & Faulmüller, N. (2014). Cognitive biases can affect moral intuitions about cognitive enhancement. *Frontiers in Systems Neuroscience, 8,* Article 195.

Chickering, A. W., & Reisser, L. (1993). *Education and identity.* Jossey-Bass Publishers.

Clancy, R., & Zhu, Q. (2023). Why should ethical behaviors be the ultimate goal of engineering ethics education? *Business & Professional Ethics Journal, 42*(1), 33–53.

Clancy, R., Zhu, Q., Gammon, A., & Thorpe, R. (2022). *Exploring the relations between ethical reasoning and moral intuitions among first-year engineering students across cultures.* Proceedings of the 2022 ASEE Annual Conference & Exposition, Minneapolis, MN. https://peer.asee.org/41487

Cox, K. S. (2010). Elevation predicts domain-specific volunteerism 3 months later. *The Journal of Positive Psychology, 5*(5), 333–341.

Darley, J. M., & Latané, B. (1968). Bystander intervention in emergencies: Diffusion of responsibility. *Journal of Personality and Social Psychology, 8*(4), 377–383.

Davis, M. (2006). Integrating ethics into technical courses: Micro-insertion. *Science and Engineering Ethics, 12*(4), 717–730.

Dieter, G.E. (1991). *Engineering design: A materials and processing approach.* McGraw-Hill.

Drake, M. J., Griffin, P. M., Kirkman, R., & Swann, J. L. (2005). Engineering ethical curricula: Assessment and comparison of two approaches. *Journal of Engineering Education, 94*(2), 223–231.

Finelli, C. J., Holsapple, M. A., Ra, E., Bielby, R. M., Burt, B. A., Carpenter, D. D., Harding. T. S., & Sutkus, J. A. (2012). An assessment of engineering students' curricular and co-curricular experiences and their ethical development. *Journal of Engineering Education, 101*(3), 469–494.

Gable, S., & Haidt, J. (2005). What (and why) is positive psychology? *Review of General Psychology, 9*(2), 103–110.

Gesun, J. S., Pitman-Gammon, R., Berger, E. J., Froiland, J. M., & Godwin, A. (2021). Developing a consensus model of engineering thriving using a Delphi process. *International Journal of Engineering Education*, *37*(4), 939–959.

Gilligan, C. (2003). *In a different voice: psychological theory and women's development.* Harvard University Press.

Govrin, A. (2014). From ethics of care to psychology of care: reconnecting ethics of care to contemporary moral psychology. *Frontiers in Psychology, 5*, Article 1135. https://doi.org/10.3389/fpsyg.2014.01135

Green, J. D. (2015). Beyond point-and-shoot morality: Why cognitive (neuro)science matters for ethics. *The Law & Ethics of Human Rights, 9*(2), 141–172.

Haidt, J. (2001). The emotional dog and its Rational tail: A social intuitionist approach to moral judgment. *Psychological Review, 104*(4), 814–834.

Haidt, J. (2013a). Moral psychology for the twenty-first century. *Journal of Moral Education, 42*(3), 281–297.

Haidt, J. (2013b). *The righteous mind: why good people are divided by politics and religion*. Penguin Books.

Haidt, J., & Joseph, C. (2004). Intuitive ethics: How innately prepared intuitions generate culturally variable virtues. *Dædalus, 133*(4), 55–66.

Han, H. (2015). Virtue ethics, positive psychology, and a new model of science and engineering ethics education. *Science and Engineering Ethics, 21*(2), 441–460.

Hanks, T. L. (2008). The ubuntu paradigm: Psychology's next force? *Journal of Humanistic Psychology, 48*(1), 116–135.

Harris, C. E. (2008). The good engineer: Giving virtue its due in engineering ethics. *Science and Engineering Ethics, 14*(2), 153–164.

Hashemian, G., & Loui, M. C. (2010). Can instruction in engineering ethics change students' feelings about professional responsibility? *Science and Engineering Ethics, 16*(1), 201–215.

Henrich, J., Heine, S., & Norenzayan, A. (2010). The weirdest people in the world? *Behavioral and Brain Sciences, 33*(2–3), 61–83.

Hersh, R. H., Paolitto, D. P., & Reimer, J. (1979). *Promoting moral growth: From Piaget to Kohlberg.* Longman.

Hess, J. L., Beever, J., Zoltowski, C. B., Kisselburgh, L., & Brightman, A. O. (2019). Enhancing engineering students' ethical reasoning: Situating reflexive principlism within the SIRA framework. *Journal of Engineering Education, 108*(1), 82–102.

Hess, J. L., & Fore, G. (2018). A systematic literature review of US engineering ethics interventions. *Science and Engineering Ethics*, *24*(2), 551–583.

Hess, J. L., Miller, M., Higbee, S., Fore, G. A., & Wallace, J. (2021). Empathy and ethical becoming in biomedical engineering education: A mixed methods study of an animal tissue harvesting laboratory. *Australasian Journal of Engineering Education, 26*(1), 127–137.

Hess, J. L., Strobel, J., & Brightman, A. O. (2017). The development of empathic perspective-taking in an engineering ethics course. *Journal of Engineering Education, 106*(4), 534–563.

Hess, J. L., Strobel, J., & Pan, R. (Celia) (2016). Voices from the workplace: practitioners' perspectives on the role of empathy and care within engineering. *Engineering Studies*, *8*(3), 212–242.

Hoffman, M. L. (2000). *Empathy and moral development: Implications for caring and justice.* Cambridge University Press.

Huerta, M. V., Carberry, A. R., Pipe, T., & McKenna, A. F. (2021). Inner engineering: Evaluating the utility of mindfulness training to cultivate intrapersonal and interpersonal competencies among first-year engineering students. *Journal of Engineering Education, 110*(3), 636–670.

Janis, I. L. (1972). *Victims of groupthink: A psychological study of foreign-policy decisions and fiascoes.* Houghton Mifflin.

Kahneman, D. (2012). *Thinking fast and slow*. Penguin Books.

Keyes, C. L. M., & Haidt, J. (Eds.). (2003). *Flourishing: Positive psychology and the life well-lived.* American Psychological Association.

Kim, D. (2022). Promoting professional socialization: A synthesis of Durkheim, Kohlberg, Hoffman, and Haidt for professional ethics education. *Business and Professional Ethics Journal, 41*(1), 93–114.

Kohlberg, L. (1969). Stage and sequence: the cognitive-development approach to socialization. In D. Goslin (Ed.), *Handbook of socialization theory and research* (pp. 347–480). Rand McNally.

Kohlberg, L. (1973). The claim to moral adequacy of a highest stage of moral judgment. *The Journal of Philosophy, 70*(18), 630–646.

Kohlberg, L. (1981). *The philosophy of moral development: Moral stages and the idea of justice*. Harper & Row.
Kotluk, N., & Tormey, R. (2023). Compassion and engineering students' moral reasoning: The emotional experience of engineering ethics cases. *Journal of Engineering Education, 112*(3), 719–740.
Mapaling, C. (2023). *Academic resilience of engineering students: A case study* (Unpublished doctoral thesis). Nelson Mandela University.
Magun-Jackson, S. (2004). A psychological model that integrates ethics in engineering education. *Science and Engineering Ethics, 10*(2), 219–224.
Mehalik, M. & Gorman, M. (2006). A framework for strategic network design assessment, decision making, and moral imagination. *Science, Technology, & Human Values, 31*(3), 289–308.
Munir, F. (2020). Ubuntu in the engineering workplace: Paying it forward through mentoring. *Commonwealth Youth and Development, 18*(1). https://hdl.handle.net/10520/ejc-cydev-v18-n1-a7.
Murzi, H., Mazzurco, A., Pikaar, I., & Gibbes, B. (2019). Measuring development of environmental awareness and moral reasoning: A case-study of a civil engineering course. *Europen Journal of Engineering Education, 44*(6), 954–968.
Narváez, D., & Rest, J. (1995). The four components of acting morally. In W. Kurtines & J. Gewirtz (Eds), *Moral behavior and moral development: An introduction* (pp. 385–400). McGraw-Hill.
Nickerson, R. S. (1998). Confirmation bias: A ubiquitous phenomenon in many guises. *Review of General Psychology, 2*(2), 175–220.
Noddings, N. (1984). *Caring: A feminine approach to ethics and moral education*. Cambridge University Press.
Noddings, N. (2002). *Educating moral people*. Teachers College Press.
Noddings, N. (2013). *Caring: A relational approach to ethics & moral education*. University of California Press.
Pantazidou, M., & Nair, I. (1999). Ethic of care: Guiding principles for engineering teaching & practice. *Journal of Engineering Education, 88*(2), 205–212.
Perry, W. G., Jr. (1970). *Forms of intellectual and ethical development in the college years: A scheme*. Jossey-Bass.
Pritchard, M. S. (1999). Kohlbergian contributions to educational programs for the moral development of professionals. *Educational Psychology Review, 11*(4), 395–409.
Pritchard, M. S. (2000). Service-learning and engineering ethics. *Science and Engineering Ethics, 6*(3), 413–422.
Rest, J. R. (1986). *Moral development: Advances in research and theory*. Praeger.
Rest, J. R., Narváez, D., Bebeau, M. J., & Thoma, S. J. (1999a). A neo-Kohlbergian approach: The DIT and schema theory. *Educational Psychology Review, 11*(4), 291–324.
Rest, J. R., Narváez, D., Thoma, S. J., & Bebeau, M. J. (1999b). DIT2: Devising and testing a revised instrument of moral judgment. *Journal of Educational Psychology, 91*(4), 644–659.
Rest, J. R., Narváez, D., Thoma, S. J., & Bebeau, M. J. (2000). A neo-Kohlbergian approach to morality research. *Journal of Moral Education, 29*(4), 381–395.
Riley, D. (2013). Hidden in plain view: Feminist doing engineering ethics, engineers doing feminist ethics. *Science and Engineering Ethics, 19*, 189–206.
Rozin, P., Haidt, J., & McCauley, C. R. (2008). Disgust. In M. Lewis, J. M. Haviland-Jones & L. F. Barrett (Eds.), *Handbook of emotions* (pp. 757–776). Guilford Press.
Shweder, R. A., Much, N. C., Mahapatra, M., & Park, L. (1997). The "big three" of morality (autonomy, community, divinity) and the "big three" explanations of suffering. In A. Brandt & P. Rozin (Eds.), *Morality and health* (pp. 119–169). Routledge.
Sochacka, N. W., Walther, J., Miller, S. E. & Youngblood, K. M. (2020). *Facilitating empathic communication modules in undergraduate engineering education: A handbook.* Version 3, March 2020. https://eeti.uga.edu/wp-content/uploads/2019/02/Empathy-Modules-Workbook_2020_v3.pdf
Steele, L.M., Johnson, J.F., Watts, L.L., MacDougall, A. E., Mumford, M. D., Connelly, S., & Williams, T. H. L. (2016). A comparison of the effects of ethics training on international and US students. *Science and Engineering Ethics, 22*, 1217–1244.
Stovall, P. (2011). Professional virtue and professional self-awareness: A case study in engineering ethics. *Science and Engineering Ethics, 17*, 109–132.
Strobel, J., Morris, C. W., Klingler, L., Pan, R., Dyehouse, M., & Weber, N. (2011). *Engineering as a caring and empathetic discipline: Conceptualizations and comparisons.* Proceedings of the Research in Engineering Education Symposium 2011, Spain, 603–616.

Thoma, S. J., & Dong, Y. (2014). The defining issues test of moral judgement development. *Behavioral Development Bulletin, 19*(3), 55–61.
Tronto, J. (1994). *Moral boundaries: A political argument for an ethic of care.* Routledge.
Tsang, E. (Ed.). (2000). *Design that matters: Service-learning in engineering.* American Association of Higher Education.
Turner, M. E., & Pratkanis, A. R. (1998). Twenty-five years of groupthink theory and research: Lessons from the evaluation of a theory. *Organizational Behavior and Human Decision Processes, 73*(2–3), 105–115.
Umbrello, S. (2020). Imaginative value sensitive design: Using moral imagination theory to inform responsible technology design. *Science and Engineering Ethics, 26*, 575–595.
Vianello, M., Galliani, E. M., & Haidt, J. (2010) Elevation at work: The effects of leaders' moral excellence. *The Journal of Positive Psychology, 5*(5), 390–411.
Walther, J., Miller, S. E., & Sochacka, N. W. (2017). A model of empathy in engineering as a core skill, practice orientation, and professional way of being. *Journal of Engineering Education, 106*(1), 123–148.
Worthy, L. D., Lavigne, T., & Romero, F. (2020). *Culture and psychology. How people shape and are shaped by culture*. MMOER. https://open.maricopa.edu/culturepsychology/
Zhu, Q., Zoltowski, C. B., Feister, M. K., Buzzanell, P. M., Oakes, W. C., & Mead, A. D. (2014). *The development of an instrument for assessing individual ethical decision-making in project-based design teams: Integrating quantitative and qualitative methods*. Proceedings of the 2014 ASEE Annual Conference & Exposition, Indianapolis, IN. https://peer.asee.org/23130

11

ORGANIZATION STUDIES AND ENGINEERING ETHICS EDUCATION

Response-*able* engineering and education, situating ethics-*in*-practice

Silvia Bruzzone and Silvia Gherardi

Introduction

This chapter explores the ways in which the recent perspectives on ethics that have been a subject of discussion in management and organization studies might contribute to a specific reformulation of education on engineering ethics. Drawing our inspiration from posthumanist practice theory (Gherardi, 2022; Gherardi & Laasch, 2021), we suggest that the focus of attention should be shifted away from what managers and engineers actually do toward managing and engineering practices that are conceived as local entanglements of heterogeneous elements (humans, technologies, other non-humans, discourses) and the loci of ethical doings. The proximity of these two areas is nothing new. Engineering is performed within complex organizational settings and practices (Vinck, 2003) and is connected to management and decision-making processes. In addition, as engineers are primarily employed by public or private organizations of all sizes, business and management ethics are inevitably intertwined with engineering ethics. In recent decades, urgent issues that have been brought to the fore by global challenges such as sustainable transition, climate change, and the inequalities associated with the neoliberal economy, combined with irresponsible management practices on the part of business actors in the neoliberal era (Mintzberg & Laasch, 2020), have led to an upturn in the interest being paid to engineering ethics and responsible management. As a result, the debate around ethics and corporate responsibility is increasingly becoming a constitutive element of engineering and management education. For this reason, it is important to look at how the conversation on ethics between the two areas – management/organization studies and engineering – has unfolded and at the potential synergies between them.

The efforts to introduce a moral and ethical dimension into management on the one hand and engineering on the other have followed a similar path, with the development of a mass of external tools, such as codes and rules of conduct and ethics, principles, reporting practices, and corporate social responsibility (CSR), in order to support – and hopefully guide – 'correct' behavior. This approach to ethics – which has been called ethics-as-technology (Boening-Liptsin, 2022) – is tied to the image of a technologist (a manager, an engineer, or a decision-maker) who acts based on an

DOI: 10.4324/9781003464259-14

individualistic understanding of morality in society. More profoundly, however, as we will see, it entails a specific understanding of the loci of morality – of the ethics of ethics – associated with an independent, self-determined moral subject and a 'trait' of an individual's personality.

Engineering ethics has traditionally been developed by the introduction of ethical theories into engineering education and by the application of philosophical concepts such as utilitarianism and virtue ethics to situations engineers had to deal with in the hope that they would become sensitized to them and act ethically (Johnson & Wetmore, 2008). In management and organization studies, it was believed that the introduction of principles of responsible management education and tools such as CSR or codes of conduct would produce similar results. Despite the good intentions and the increasing amount of space devoted to ethics in business and engineering education, observers have been highly critical of the ability of these approaches and tools to inspire more ethical behaviors and to reform the immoral practices of contemporary capitalism (Rasche & Gilbert, 2015).

It must be acknowledged, however, that in recent decades, the literature on organizational and engineering ethics has enormously expanded its efforts to bridge the gaps between classical ethics theory and the formalism of ethics-as-technology, on the one hand, and the complexity of the ethical challenges currently being faced by the engineering and management sectors, on the other. The advances in technological innovation in general, and in big data in all areas of society in particular – such as biomedicine, cybersecurity, food production, smart cities, and sustainable transitions – raise new ethical issues and compel new action. It is no chance that the debate on ethics in artificial intelligence (AI) is an especially lively one (Boening-Liptsin, 2022; Johnson & Wetmore, 2008), and that the need to move away from traditional forms of ethics (and from ethics-as-technology) is particularly urgent. The dark side of digitalization may be seen as a contemporary phenomenon, as part of the posthuman societal condition: algorithmic control now revolves less around human managers and more around employee interaction with a non-human algorithm, which leads to a "disintermediation of managers" (Kellogg et al., 2020, p. 387).

The objective of the attempts to supersede the traditional approaches is to draw closer to the context and level of the lived experience and focus on the situated practices from which ethics emerges. Reference to local dynamics and 'practices' has become central to these developments in both management and engineering. The mobilization of ethics of care (Gilligan, 1982; Tronto, 1993) (for more, see Chapter 10) in these areas marks a move away from formal rules to a focus on the values that emerge *in* practices. The impetus for the shift from abstract and decontextualized principles to an observation of how ethical and moral principles are generated in situated practices has a direct effect on how future engineers will be educated. Students will be able to learn about ethics when they are confronted with moral dilemmas that arise in working practices relating to their own and other professions. Professionals will be able to develop a practical reflexive sensibility (Hibbert & Cunliffe, 2013) for how morality is enacted in situated professional circumstances, and it will then be possible to bring their experience back into education and into induction practices for newcomers to the particular profession.

The article is organized as follows. First, we review some of the main contributions to the current debate on management studies – in particular ethics-as-practice and ethics-*in*-practice – that claim that ethics cannot be reduced to norms and principles, or to something that one *has*, or to a characteristic of the self. Instead of being an attribute of a moral, independent individual, ethics is what emerges from our everyday connections to others and their needs. With this in mind, we take inspiration from the concept of response-*able* managing (Gherardi & Laasch, 2021), which decenters the locus of ethics from managers to managing practices, to introduce the concept of 'response-*able* engineering' in the second section of the chapter. In the third section, we use three vignettes that illustrate examples of situated engineering practices to consider ordinary engineering practices as an arrangement of humans, non-humans, and the world from which responsible engineering emerges. Finally, we discuss the implications of response-*able* engineering for education as a form of training that requires creativity, imagination, and an ability to learn from ordinary situations.

From normative ethics to ethics-as-practice and ethics-*in*-practice

We will first summarize how the conversation on ethics has been shaped in management and organization studies and how it has been debated of late. The idea that corporations – and organizations more broadly – need to think about ethics and responsibility has emerged as a reaction to the irresponsible and unethical conduct of business and corporate interests in liberal economies. The need to reform the sector had already been raised in the 1970s, and it became urgent in the 1990s, when major financial scandals, environmental disasters, and the global exploitation of workforces raised a general alarm. Business schools were also accused of playing an important role in educating people to act irresponsibly rather than for the common good (Painter-Morland, 2015).

Organizational ethics has predominantly been approached in two ways (Hancock, 2008). On one hand, in the form of deontological ethics based on Kantian thought, which introduces a normative or legislative approach in which compliance with formal codes of conduct and the adoption of various tools – CSR, life-cycle analysis (LCA), reporting practices and others – have been favored. On the other, based on Aristotelian thought, ethics has also been seen to be connected less with the 'codifying of moral imperatives' and more with virtues, that is, with the personal traits of the subject, which might be an individual or an organization. This has triggered the development of solutions for raising awareness and educating individuals on moral behavior.

This position began to be criticized in the 1990s, especially in critical management studies (Clegg et al., 2007b; Painter-Morland, 2011; Parker, 2003), for which ethics is acknowledged to be more complex than norms, something that cannot be delegated to formal models that are supposed to guide complex decisions and real-life situations. A margin of interpretation and negotiation is always needed. On the one hand, despite the attention and the literature on ethics and responsibility, normative and codification approaches are considered not to have met the challenge or brought about a reform of deviant business practices. The risk of over-relying on codes and codification (Bevan & Corvellec, 2007) and the connection to the specific personal traits of people and organizations would promote the status quo, or business as usual, if not even glorify business (Rhodes & Pullen, 2017). On the other hand, the practice turn in organization studies (Gherardi, 2000; Nicolini, 2012; Orlikowski, 2000; Sandberg & Tsoukas, 2015; Schatzki et al., 2001; Shove et al., 2012) has focused new attention on the contextual and everyday life of ethics and on the choices made by individuals in actual situations and the definition of *ethics-as-practice* (Clegg et al., 2007a). Ethics has been defined as "the social organizing of morality, the process by which accepted and contested models are fixed and refixed, by which morality becomes ingrained in the various customary ways of doing things" (Clegg et al., 2007a, p. 111). Ethics-as-practice looks at how ethics is enacted, with a specific focus on discursive practices and discursive sense-making: "discourses that make sense of behavior and often retrospectively categorize practices as more or less ethical, where discourse is considered as a source that legitimizes behavior and construct frameworks (including vocabulary) to justify practices" (Clegg et al., 2007a, p. 113).

The authors formulated a research agenda for ethics-as-practice that includes five points (Clegg et al., 2007a, pp. 118–119): (a) it analyses the precise points at which a way of behaving becomes seen as problematic, but it is less concerned with finding a solution and more interested in how behavior becomes an ethical problem; (b) it focuses on the complex heterogeneous web that makes organizations work and how operative ethical discourses are mobilized; (c) it comes up at the level of actual practices in use, and not 'grand narratives'; (d) rather than focusing on the ultimate values embodied in discourses and discussing whether they are good or bad, it asks how these values come into being, and why they should be 'better' than others; and (e) by placing the emphasis on the context and the embeddedness of ethics, it refrains from making generalized judgments, focusing on the local meaning and sense-making practices that constitute ethics.

The adoption of a research agenda for ethics-as-practice is a significant move toward grounding a methodological framework for a study of ethicality in managerial practices and countering imposed ethical universalism. However, we have identified two weak points in the research agenda of ethics-as-practice: the centrality that is still attributed to humans as the exclusive locus of morality and decision-making; and the separation of ethics from politics. With regard to the former, we will show how a posthumanist practice theory conflicts with theories of practice centered around 'humans and their practices,' and in the case of the latter, we acknowledge the need to link ethics and politics, as some authors have done under the label of critical business ethics (McMurray et al., 2011, Rhodes & Pullen, 2017; Wray-Bliss, 2009).

While ethics-as-practice invites researchers to investigate how ethics is enacted, a posthumanist practice theory conceptualizes ethics as emerging from the entanglement of humans, non-humans, and materialities. In other words, ethics emerges *in*-practices rather than being enacted by human beings alone.

This shift is marked by a passage from responsibility – formal roles or personal traits – to a capacity for responsiveness: that is, a commitment to changing environments (Painter-Morland, 2011) and an affective response to the worldly connections of which we are a part. Feminist posthumanism (Barad, 2007; Braidotti, 2013) has influenced management studies and provided an even more radical departure from the ethical subject – that is, centered around the human – and has expanded ethical subjectivity as a commitment to human, nonhuman, and worldly relationships. We are therefore focusing on ethics-*in*-practice in order to stress the importance of situated practices within which the capacity to provide responses is performed and produces effects.

Barad (2007) is often cited on the subject of ethics because of her description of posthumanist theory as ethico-onto-epistemological, which implies that ethics is inseparable from being and understanding. Barad argues that researchers do not uncover pre-existing facts about independently existing things as they existed frozen in time, like little statues positioned in the world. Rather, we learn about phenomena – about specific material configurations of the world's becoming, of which we too are a part. Which practices we enact matter – in both senses of the word. Making knowledge is not simply about making facts but about making worlds, engaging materially as part of the world in giving it a specific material form. The term response-*ability* (Barad, 2007; Haraway, 2008) captures a different understanding of ethics-*in*-practice. Here, response-ability is not about 'being responsible', and it cannot be established in advance; rather, it implies being open to the call of others (Despret, 2016). In this sense, ethics cannot be dissociated from politics, that is, from what is included and excluded from ethical doings.

The fact that we are a part of the world is not merely a critique of anthropocentrism; it also means adopting a research agenda in which posthumanist theory in many disciplines is a generative tool "to help us re-think the basic unit of reference for the human in the bio-genetic age known as 'anthropocene,' the historical moment when the Human has become a geological force capable of affecting all life on this planet" (Braidotti, 2013, p. 5). The idea of a posthuman societal condition reflects a situation in which humans and non-humans are increasingly folded into one another: more-than-human assemblages of digital cultures, emerging biotechnologies, algorithmic automation, and various cyborg formations are deeply enmeshed.

The definition of our contemporary epoch as 'Anthropocene' has shown great promise in sustainability studies. However, it is an extremely heated debate, even among scholars who share a deep concern for the planet, because the term 'Anthropocene' is understood differently depending on the disciplinary priorities (Calás et al., 2018; Gibson-Graham, 2011). We need to bear in mind, however, that this planet is not ours to ruin or save, and so when we mobilize the concept of Anthropocene it is more appropriate to ask, as Ulmer (2017, p. 6) does, "What the Anthropocene might *do* in research". This is the starting point for a conversation in which some have suggested that the Anthropocene produces awareness and sustainable living practices, or an understanding that we live in an interconnected world, while others have pointed to the term as the core of the problem, as it reminds us of everything that is excluded – women, the Global South, people with disabilities, non-humans, and all types of externalities (human and non-human) involved in a linear economy (Sperling, 2019) – and points to the future a post-Anthropocene might imagine in the form of alternative economic systems and environmental policies. Ulmer's (2017) answer to this is that it might do several things: "situate research within a particular time period; support inquiries that include aspects of in/non/human life; and highlight the purpose and significance thereof. Anthropocenic thinking invites scholars to refine their political commitments both in and to research" (Ulmer 2017, p. 6). From her response, we see how in doing research on (and teaching) sustainability, responsibility, and ethics cannot be separated, and this insight lies at the core of the capacity of responsiveness, that is, a commitment to an affective and effective response to the worldly connections of which we are a part.

Response-*able* engineering

Our aim is to develop a theoretical-methodological framework for an empirical study of response-able engineering as a phenomenon emerging in and through practices. We draw a parallel with the emergent field of responsible management, in which the three areas of ethics, responsibility, and sustainability are all connected (Laasch et al., 2020), and propose an exploration of responsible engineering as a means of integrating sustainability, responsibility, and ethics into engineering practices. The objective here is to develop a framework that is suited not only to ethics, responsibility, and sustainability, but also – and especially – to the interactions among them.

When we approach responsible management and responsible engineering as practice (Gherardi & Laasch, 2021), we make an epistemological shift from ontological issues (what an object 'is') to onto-epistemological issues, or how an object (response-able engineering in this case) is made, and how it comes to be accomplished within situated practices. As we focus on this, we draw on four principles (Gherardi & Laasch, 2021; Price et al., 2020) for an ethics-*in*-practice analysis – situatedness, sociomateriality, collective knowledgeable doings, and texture of practices – as we will explain in greater detail below.

First, the principle of *situatedness* brings to the fore how a practice can be conceived as an *agencement*[1]) of humans, non-humans, tools, technologies, rules, and discourses, and how practices happen in time and space and in association with other practices. The focus here is on how a particular practice might be performed differently, and on how that practice may be enacted with different social effects. Accordingly, when we consider practices of response-able engineering, we ask: *How is response-able engineering accomplished* in situ*? What activities are performed within the practice we describe, and with what consequences in terms of sustainability, responsibility, and ethics?*

We apply a posthumanist practice approach (Gherardi, 2019) and focus on the here and now of a way of ordering humans, non-humans, tools technologies, rules, and discourses that produce (or do not produce) responsible effects in the dynamics of sustainability, responsibility, and ethics.

Second, practice as a *collective knowledgeable doing* is the operating definition for approaching knowledge as a situated activity, a form of knowing that takes place while it is being practiced. It is not enough to look at practice as a set of sayings and doings, because they are knowledgeable activities. Thus, 'knowing-*in*-practice' connects doing with knowing as a way of ordering heterogeneous elements into a cohesive whole by making aesthetic, ethical, and political judgments that are socially recognized and collectively sustained as professional deontology and as knowledgeable doing *in situ*. When we consider situated practices of response-able engineering, we ask: *What counts as legitimate knowledge within the institutionalized body of knowledge called engineering? Whose knowledge counts as knowledge? How is knowing-in-practice actually made?* Beyond questions such as these, there is the concern with power and with forms of knowing in an ethically accountable manner.

The third principle is *sociomateriality*, which means that the social and the material (corporeal, technical, and digital) form an ecology of knowing, and that the introduction of a new artifact, technology, or tool produces a realignment of practices that is both material and cultural (that is, sociomaterial). The concept of sociomateriality enables us to pose questions such as: *How is knowing materialized in tools that are kept, innovated, or discarded? How is this 'knowing' embedded in specific artifacts and devices in use in order to create accountability? What are the effects of the adoption of certain devices of monitoring activities on organizing for response-ability?* It might be said that artifacts have policy since sustainability, responsibility, and ethics are embedded in the way humans, non-humans, and discourses form an *agencement* that achieves agency from the entangled elements.

The fourth principle – *a texture of practices* – assumes that practices of response-able engineering will co-exist with others in a texture of practices, with an interconnection that links each one to the other, interconnected, practices. They can, therefore, cut across boundaries, whether they be organizational, institutional, community, or professional. Organizational practices can also extend beyond organizational boundaries, just as social practices extend across into organizations. We have already mentioned how what is called the Anthropocene might be seen as the effect of interconnected practices. The term 'texture' denotes 'connectedness in action,' that is, how each practice is interdependent and interwoven with others in an endless series of relationships that continually move into each other. This enables us to ask questions like: *How is the connection of one practice to others achieved, maintained, or disrupted? How do we trace and map the relationships that are formed between a 'here and now' practice and its connection and manifestation within other practices that may be far away in time and space?*

To conclude, we might say that response-able engineering can be seen as an *agencement* of humans and non-humans more than just as something humans alone are doing and/or the discourses they are engaging in, and as such, *agencement* embraces both the social and the material. This principle of 'agentivity' means that we do not study engineers as individuals as collectives, or just engineering work and activities, or management discourse in isolation, or education or learning as separate activities. Rather, we study the social and material effects of their interconnections.

In the sections that follow, we will use three vignettes to illustrate how the proposed framework might be put to work and how it might be used in the context of education for reflecting on ethics, responsibility, and sustainability within professional engineering knowledge.

Response-able engineering practices (through vignettes)

We will now show how the theoretical framework on response-able engineering can be used to interpret three stories that lie at the crossroads between engineering innovation and management.

Two are taken from research conducted by one of the authors (on wildfire and water management connected to climate change). The third – on waste management – comes from the work of two other researchers. In line with posthumanist thought and post-qualitative research (Lather & St. Pierre, 2013), we understand that the role of the researcher – of us as authors – is not as an outsider in the research process but as an agentic actor in the research question, process, and results. This being the case, we make the claim that our position – as women researchers who are committed to feminist posthumanist practice thought, who both have a background in organizational studies, and one of whom (the first author) co-ordinates courses in the social sciences as part of a civil engineering program, as well as being a practitioner of art-based methods for sustainable transitions – matters in the way we frame the contribution (epistemological background, choice of cases, and analysis), and is assumed to open up new possibilities (results) in understanding, researching, and educating on engineering/managing responsibility. The choice of these cases is motivated by a desire to show how ethical issues emerge from (engineering) doings and do not pre-exist them, and how it is these doings that should be acknowledged.

As we mentioned above, engineering is not treated here as being restricted to what engineers do, but rather as an ensemble of practices involving a multiplicity of actors (including engineers and other professionals), rules, procedures, materialities, and discourses. Engineering never happens in a void; it takes place in relation to specific situations that we refer to here as 'stories'.

Vignette 1. What is wildfire and how to respond to it? Environmental caring between policing and collective doings

The increase in the number of wildfires is a widely acknowledged consequence of climate change, and in particular of the droughts that are being recorded in different parts of the world, even in Northern countries. The need to anticipate this type of risk has become a priority. In this case, we focus on the practice of firefighting to highlight how the sociomaterial enactments of a forecasting practice let emerge different modes of existence of fires and different ethos in terms of how to respond to them.

It is based on an ethnographic study by one of the authors (Bruzzone, 2019) on the introduction of forecasting technology into wildfire fighting in a Mediterranean region and the shift from a reactive to a preventive approach to fire based on forecasting the meteorological and vegetation conditions that may turn a small outbreak into a devastating wildfire. The practice involves the foresters (who are responsible for firefighting), the forecasters (who produce the forecasts), the forecasting data and maps, the volunteers (who patrol the areas that might be impacted), and forest in all its forms.

It emerged from a study of the way foresters use forecasting maps that they are just one of the elements that contribute to the foresters' decision as to whether or not to activate the patrolling service. As key informants suggested, the decision relies on "a constellation of factors" (Bruzzone, 2019, p. 58) that are mostly based on the foresters' professional knowledge and the policing activity of investigating arson (human-caused fires).

The conceptualization aligning wildfires with arson competes with a second conceptualization – that held by the forecasters, who believe that fires are mainly accidental and, therefore, that maps can contribute to raising awareness and knowledge of fire and should be distributed as widely as possible across the territory, with the intention of improving the forecasts. This would mean securing systematic data on their use by the foresters to take preventive action and on the actual number of outbreaks of fire in relation to forecasts. Instead, they only receive general aggregate information on a monthly basis, which allows them to build on this information and improve the forecast to a limited extent.

In this case, the specific sociomaterial entanglements linked to the use of technology suggest that the forecasting technology is not neutral, but rather produces – or reproduces – a certain understanding of fire as mostly a criminal activity, and therefore an understanding that wildfire prevention mainly involves avoiding criminal acts (while providing the public with general information about the risk). In this context, forecasts are considered to potentially coincide with criminal plans by providing information about the times and zones when and where a fire might have the most devastating effect. Their diffusion and use should, therefore, be restricted as much as possible (basically, this means restricting them to the authorities in charge of policing).

While the various sociomaterial entanglements that give shape to the forecasting practice mean that wildfire emerges as a multiple (there are at least two types: wildfire as an accident versus wildfire as a criminal act), each with a different ethos of action and response, the specific configuration of power between human and non-human actors means that the specific ontology of fire as a criminal act prevails, and that the response is mainly framed in terms of policing. In other words, in the transition from reaction to prevention, control of the definition of risk and the power to relate the risk remain unchanged.

This conflict of understandings has important implications, as it excludes other potential uses of the forecasting maps and preventive actions addressed to the public, particularly those that encourage responsible action among the population. The capacity of response-ability in the case of wildfire is therefore restricted to just a few actors who have the power and ability to respond, and a more collective response-ability toward fire is held back.

In this case, we can see what an ethico-onto-epistemology approach implies. There is nothing – such as fire – that pre-exists the situated practices of firefighting. It is the emergence of knowledge of fire and its ontology from the specific sociomaterial alignments that define fire as a crime. As we have seen, the introduction of the forecasting technology does not alter this practical understanding because it is embedded in the previous approaches to fire. This reaffirms what and whose knowledge of fire is legitimate.

However, the forecasting technology opens new potential sociomaterial entanglements – which then extend to other humans and non-humans, to citizens and media channels, and so on. This enacts a new collective knowing and being of fire as an accident. In Barad's (2007) words, fire emerges from these two "intra-acting" (p. 33) cuts, which bring about different becomings of fire.

Ethics is not separate from these doings and ways of knowing; it also emerges from the intra-activity of all the entangled elements as specific responses to an ongoing situation. In the case of fire, ethics refers to deciding both what matters – what fire is and what counts as firefighting – and what is excluded from being relevant as a potential action. In this case, what matters is the specific alignment of fire as a crime. It is therefore possible to say that the question of response-ability engages a variety of contradictory choices and mattering in environmental caring between policing activity and collective doings.

Vignette 2. Competing sociomaterial moralities emerging in a circular economy

In the second vignette, we discuss how value alternatives of responsible recycling do not pre-exist but emerge from the different sociomaterial practices at play in the name of the circular economy. We focus on an example reported by Laser and Stonewall (2020), in which the dominant practice and morality of disassembling iPhones to recover materials silence alternative kinds of practice, expertise, and work among those who are engaged in giving iPhones a second life.

Laser and Stonewall (2020) analyze how a high-tech company (Apple) 'takes responsibility' for the social and environmental degradation associated with e-waste. Apple has recently introduced two recycling robots that are used to recover precious components – such as gold, platinum, silver, and copper that can be recycled – from iPhones, and as a means of reducing an exponential rise in e-waste. The initiative is part of their 'GiveBack' campaign, which aims to recycle in a 'safe and clean' way.

Laser and Stonewall (2020) use their research on informal recyclers in the United Kingdom, Germany, and India to analyze the types of value generated in the Apple example compared to what they observed in the informal sector. They argue that in the case of Apple's robots, the specific alignment of humans and non-humans is problematic because it is intended to relieve consumers of guilt and to encourage them to buy new iPhones with a clear conscience. In addition, the type of algorithmic knowledge embedded in the robots hides and silences the informal knowledge and skills of those who perform and support repair and reuse practices. While disassembling/shredding and 'GiveBack' practices are becoming dominant across the world, this specific way of constructing value and responsibility has come under attack from environmental activists – using the slogan 'Don't give it to him' [the recycling robot] – as it conflicts with a culture of repair and long-term use of products and supports a logic of planned obsolescence.

In this case, the competing alternative actions and values in relation to what responsible recycling is do not pre-exist: rather, they occur within a sociomaterial entanglement of activities and practices. "Creating value is a process of joining together: classifying, grouping, combining, making, re-forming. Yet it is also a process where persons, things, parts of bodies, or landscapes are disentangled, abandoned, dismissed, or corrupted" (Greeson et al., 2020, p. 5).

Whereas the centrality of morality is normally addressed to humans, we see in this discourse that other beings – in this case robots – are also supposed to be the locus of morality. A technocratic solution is the means used by the corporation to take responsibility. However, this reaffirms a concept that says that ethics and responsibility are connected to a single company that is willing to propose responsible/sustainable products, and to an individual consumer who, by choosing responsible or sustainable products, may feel free to continue consuming, or at least less guilty about doing so. In this sense, as Rhodes and Pullen (2017) have noted, this understanding of responsibility – as codified, or in this case by the use of 'responsible technologies' – serves the status quo and business-as-usual.

If we apply posthumanist ethics instead, and move away from individual ethics – of an individual or a corporation – to ethics as they emerge *in* sociomaterial practices, we might ask what kind of response to the problem of e-waste emerges from the technocratic entanglement based on the robotic recycling infrastructure, and what it silences or excludes. In this sense, "the new high-tech infrastructure in fact does not simply replace the previous recycling and repair efforts (but it might make their work more difficult and expensive)" (Laser & Stowell 2020, p. 185). Repairing practices are especially an aspect of the informal sector and know-how (or bricolage) of both the Global South and the Global North and involve environmental groups and diverse materialities engaged in repair and 'second-life' practices. In this sense, ethics and response-ability are connected to a texture of practices that are scattered across time and space and which focus on the prevention (of e-waste) and combating product obsolescence. This case raises a clear issue in terms of politics and democratic processes, as "the robots are representative of Apple's control, relying upon algorithms that 'black box' decisions about which recycling practices are considered 'optimal'" (Laser & Stowall, 2020, p. 186). In this way, it also 'black boxes' the response, which is *de facto* delegated to the technocratic solution (the robot), a practice one might even define as a form of *de*-respons-*ability*. As the authors suggest, the complexity of e-waste and the patchy texture of the practices it mobilizes call for broader democratic debates on alternative ways to respond to e-waste.

Vignette 3. From hydroengineering to situated water engineering

The third vignette describes a case of water engineering as a response to climate change and the rise in sea levels. Water is an essential element of urban design and dynamics, and it is key factor of climate change; increasingly, storm events and the rise in sea levels are viewed as major threats to cities that experience repeated flooding and require adaptive responses.

The traditional approaches to water engineering, which focus on hard defensive infrastructures, are no longer considered to be sufficient in the long run, and more sustainable solutions are being sought. New 'soft engineering' approaches are introducing a new ethos in the name of 'living with water' or 'giving room to water,' in contrast to the traditional logics of 'containment,' 'repulsion,' 'defence,' 'separation' from water, and other standardized solutions.

The vignette we present, which is taken from a study conducted by one of the authors (Bruzzone, 2012), shows the broader implications of paradigmatic shift such as this. It means rethinking engineering beyond what *engineers do* and pure technical solutions within a texture of sociomaterial practices and through the development of a situated water ethos.

In the 1990s, after many years of employing a well-established tradition of defensive infrastructures, the Flemish regional water authority decided to counteract the effects of the rise in sea levels by developing a huge project of managed retreat that consisted in a controlled flooding of upstream low-lying farming areas to prevent downstream urban areas from being flooded.

The regional water authority was accustomed to developing defensive infrastructures and, where necessary, as in this case, to carrying out expropriations to compensate farmers and enable projects to move forward. On this occasion, however, the project encountered huge opposition from a variety of sources: from the powerful local mayor, who opposed interference by the state; from environmentalists, because the project would destroy a protected area for birds; from farmers, because the compensation they were to receive would not make up for the loss of the generous Common Agricultural Policy (CAP) subsidies; and from local inhabitants, who were alarmed by a project that seemed to go in the opposite direction – creating an area that would be flooded! – to centuries of defensive infrastructures designed to protect Flanders against flooding. There were also fears of an invasion of mosquitoes in the area. The project placed local support for the government in jeopardy, and in the end, it was dropped.

The project team – which included a young lead engineer – realized that the traditional top-down approach of state engineering would not work as a way of legitimizing the intervention.

A new opportunity for the project emerged when it was revised and took shape within a new texture of practices. The story, which has been reported following an Actor-Network Theory (ANT) narrative (Bruzzone, 2012), highlights the process of 'interessement' (Callon, 1986) and the sociomaterial reconfigurations that led to the new project. The loss of the bird sanctuary was compensated by the development of another protected zone for birds in a different area of the city. Some of the farmers were allowed to stay on and dedicate themselves to new farming activities while doing some maintenance work. Through a strong communications campaign and the creation of an on-site information point, the local inhabitants were reassured about the safety of the project, and new eco-tourist and leisure activities were included in the project and developed as part of it, with the collaboration of environmental groups.

This case highlights the change in the ethos of the project from a top-down water engineering approach that excluded and silenced all human and non-human actors to a new project integrated into a multiplicity of *in situ* sociomaterial practices. In modern times, water has mostly been treated as an abstraction, a calculation, and a de-territorialized *materia* (Linton, 2010; Neimanis, 2017). Pierre Lascoumes (1994) has proposed the term 'eco-power' (*éco-pouvoir*) to refer to the normative and knowledge mechanisms of control and regulation – what Foucault calls bio-power (1976, 1977–1978) – which are extended to all living beings (and not just to humans). State engineering has been one of the most powerful incarnations of eco-power, entailing a power to 'tell the risk' (of flooding, in this case) by means of a variety of mechanisms (maps, technical tools, and expertise) which define the legitimate knowledge about the risk and the legitimate response to it.

The current water crisis is, in the first place, a "social crisis" of our imaginaries of water as something that is disconnected "from social and ecological relations" (Linton, 2010, p. 14). In our example, we move from an imagination of water as being "out there" (Neimanis, 2017, p. 21) to an understanding that water and the project are not separate from the location and all the human and non-human actors and that they are intra-acting within situated practices.

In this reframing of the imaginary associated with water, engineering can be understood as a texture of sociomaterial practices scattered across time and place that takes account of a multiplicity of human and nonhuman actors engaged in the context: it is about farming, maintenance, bird inhabitants to be taken care of, the development of new natural areas outside the perimeter of the project, and inhabitants and local groups involved in co-creating a situated response to the risk of flooding.

In other words, it means making a shift from universal, standardized engineering solutions and power to what we call 'situated engineering', which is enacted through multiple sociomaterial *agencements*. In this sense, response-able engineering means moving away from providing a response *to* (a situation or a risk) to engaging in responding *with* a plurality of human and non-human actors, knowledge, practices, and discourses. In other words, a relational ethics of engineering entails overcoming traditional separations (such as nature versus society, or legitimate knowledge and power versus recipients of knowledge and power) that emerge from local collective knowings and doings. Finally, if, as we have claimed, ethics emerges from practices, and if, to paraphrase Latour (1984), engineering is politics by other means, then we acknowledge the lack of separation between engineering ethics and politics.

Concluding remarks: implications for response-able engineering education

We began this chapter by proposing a theoretical framework that mobilizes response-able engineering as emerging from *situated doings*, *collective knowing*, *sociomateriality*, *and a texture of practices*. We then used this framework to relate three stories and show how response-able engineering emerges (or does not emerge) from these situations. We will now conclude by reflecting on what the approach we have called 'ethics-*in*-practice' means in terms of engineering education.

Despite the differences in content and the issues at stake, we can observe some common patterns in these stories that refer to the elements and questions raised in the theoretical framework. In each of them, we see how engineering unfolds as a specific *agencement* of humans, technology, rules, and discourses that do not take place abstractly but rather in a specific time and space. Just as engineering practice is situated, so is the engineering ethics that emerges from this *agencement* and response-able engineering, which is accomplished *in situ*. In each case, we see that the definition of the problem, the knowledge of it, and the ethical questions associated with it are not pre-existing but are generated by the *agencement* of human and non-human actors, technology, and discourses. Ethics is an emerging question about the different ways to handle a problem and the possible responses to it, which might exclude other possibilities and capacities to respond. It may involve new definitions of a problem such as wildfires and the responses that emerge from the introduction of forecasting technology, and which may question established power relations; competing ways of taking responsibility for e-waste between technocratic solutions that promote the status quo (and business-as-usual) and more collective preventive actions; or the need for engineering to move away from technical problem-solving to situated engineering-*with* and response-ability.

In all these cases, knowledge is produced both in situated practices and from what counts as legitimate knowledge. Knowledge of fire and how to prevent it is developed out of the forecasting practice and is also based on established prior forestry knowledge. In the second case, knowledge about e-waste and the possible solutions and ethical implications are developed within parallel and competing networks in which one is more legitimate than the other. In the third case, a method for dealing with flooding is also produced *in situ* in the form of a shift in what can be considered legitimate knowledge from a standardized technical solution to deal with risk toward a more composite and situated solution developed *with* the territory.

In all these cases, sociomateriality is a central element of the *agencement* of the practice in question and plays a key role in the manner in which response-ability is organized: in the first case, the forecasting tool is at the heart of possible alternatives for responding to wildfires (*should it only be used for policing activities or should it be shared with the wider public to raise awareness and promote collective care for the territory?*). In the second case, the robot is an artifact that black-boxes e-waste issues and all the decisions about what e-waste is and how to respond to it. The composite materiality of the last case – a 1,500-acre flood-control area made up of a mix of dikes, pumps, sluices, arable land, wild flora and fauna, animals and plants, cycle paths, and so on – plays a core role in the shifting response: *should the situated heterogeneity be taken into account when preparing the response (a response-*with*) or should it be disregarded in the name of the* raison d'Etat *of which standardized engineering knowledge is one of the most powerful expressions?*

Finally, all the stories emerge as connections of different practices, between previous 're-active' practices and new preventive practices that engage a plurality of actors in the first case, and consumer practices, disassembling and value chains for re-valued materials, and a plurality of invisible practices of recycling scattered across the informal circuits of the Global South and North in the second. We have shown that the texture of practice is the same as the one response-ability is built upon and through which it should be acknowledged.

Response-able engineering education should, therefore, focus on a pedagogical perspective in which learning is not a matter of reproducing previously provided knowledge (Barad, 2007; Lenz Taguchi, 2011; Zembylas, 2018) but rather develops new knowledge via a greater affective involvement in and commitment to studies on the part of students. Learning from situated cases – or stories, as illustrated in this chapter – would allow engineering students to immerse themselves in professional practices and situations where the complexity of 'everyday dilemmas' requires them to think about situations, thus training their capacity for responsiveness. Therefore, collaboration between practitioners and academia should be enhanced, as fieldwork practices may provide the living material to train and develop forms of responsiveness among students. This would enable filling the gap between professional practices and engineering education and enable cross-fertilization.

By extending certain pedagogical approaches from the field of organization and management to ethics in engineering education (Bruzzone, 2022), we have outlined several principles that may be a source of inspiration for grounding ethics in professional practices. First, students should be encouraged to bring their everyday ethical dilemmas and experiences into their curriculum and learning processes. For teachers, the same principle means viewing pedagogy as a sociomaterial assemblage and the locus where specific possibilities and impossibilities arise. For example, they might consider the extent to which their educational methods are inclusive and speak to student diversity.

Second, an approach to engineering education grounded in ethics as situated and emergent in both professional and interprofessional practices would shift the focus away from 'what engineers do' – an individual approach – to how response-able engineering can emerge from a texture of everyday situated practices. As a further principle, we also suggest breaking down the barriers around the discipline and multiplying the capacity for vision by borrowing methodological approaches from other areas; aesthetic learning processes (Burman, 2014; Styrke, 2015), for example, address sensorial and embodied types of knowledge (not just cognitive knowledge) that have also been proven to be very valuable in engineering education (Bruzzone & Stridsberg, 2023). Lastly, as we have claimed and illustrated in our three vignettes, there is no separation between ethics and politics, and engineering education should focus more carefully on the political and moral implications of engineering knowing and doing, as well as on its collective and distributed character in society. This is not an option for the future; it is a way of responding to today's posthuman condition and to the call of others.

Note

1 *Agencement* has been used as a philosophical term by Deleuze and Guattari (1987) with the sense of 'in connection with,' which gives a first good approximation of the term. The problem, however, is that its translation into English as 'assemblage' has changed the original meaning. The French term, in fact, has a processual connotation – the idea of establishing or forming an assemblage. It focuses on process and on the dynamic character of the inter-acting between the heterogeneous elements of the phenomenon. While a certain use of the term 'assemblage' risks rigidifying the concept into the thingness of final or stable states, the French term *agencement* works as an evocation of emergence and heterogeneity.

References

Barad, K. (2007). *Meeting the universe halfway: Quantum physics and the entanglement of matter and meaning.* Duke University Press.

Bevan, D., & Corvellec, H. (2007). The impossibility of corporate ethics: For a Levinasian approach to managerial ethics. *Business Ethics: A European Review*, *16*(3), 208–219. https://doi.org/10.1111/j.1467-8608.2007.00493.x

Boenig-Liptsin, M. (2022). Aiming at the good life in the datafied world: A co-productionist framework of ethics. *Big Data & Society*, *9*(2), 205395172211397. https://doi.org/10.1177/20539517221139782

Braidotti, R. (2013). *The posthuman.* Polity Press.

Bruzzone, S. (2012). Climate change and re-organizing of land use. Flood-control area as network effect. *International Journal of Urban and Regional Research, 36*, 2001–2013. doi:10.1111/j.1468-2427.2012.01162.x

Bruzzone, S. (2019). Wildfire forecasting: Between criminal act and unintentional events. *Culture and Organization, 25*(1), 52–64. doi:10.1080/14759551.2016.1213730

Bruzzone, S. (2022). A posthumanist research agenda on sustainable and responsible management education after the pandemic. *Journal of Global Responsibility, 13*(1), 56–71. doi:10.1108/JGR-05-2021-0045

Bruzzone, S., & Stridsberg, H. (2023). Dancing urban waters. A posthuman feminist perspective on arts-based practice for sustainable education. In M. Cozza & S. Gherardi (Eds.), *The* posthumanist epistemology *of practice theory. Re-imagining method in organization studies and beyond* (pp. 123–148). Palgrave.

Burman, A. (Ed.). (2014). *Konst och lärande: essäer om estetiska lärprocesser*. Södertörns Högskola.

Calás, M. B., Ergene S., & Smircich, L. (2018). Becoming possible in the Anthropocene? Becoming social entrepreneurship as more-than-capitalist practice. In P. Dey & C. Steya (Eds.), *Social entrepreneurship: An affirmative critique* (pp. 264–293). Edward Elgar. doi:10.4337/9781783474127.00027

Callon, M. (1986). Éléments pour une sociologie de la traduction. La domestication des coquilles Saint-Jacques et des marins-pêcheurs dans la baie de Saint-Brieuc. *L'année sociologique*, *36*, 169–208.

Clegg, S., Kornberger, M., & Rhodes, C. (2007a). Business ethics as practice. *British Journal of Management*, *18*(2), 107–122. https://doi.org/10.1111/j.1467-8551.2006.00493.x

Clegg, S., Kornberger, M., & Rhodes, C. (2007b). Organizational ethics, decision making, undecidability. *The Sociological Review*, *55*(2), 393–409. https://doi.org/10.1111/j.1467-954x.2007.00711.x

Deleuze, G., & Guattari, F. L. (1987). *A thousand plateaus: Capitalism and schizophrenia*. University of Minnesota Press.
Despret, V. (2016). *What would animals say if we asked the right questions?* Minnesota University Press.
Foucault, M. (1976). *Histoire de la sexualité I. La volonté de savoir*. Gallimard.
Foucault, M. (1977–78 [2004]). *Sécurité, territoire, population. Cours au Collège de France*. Gallimard/Seuil.
Gherardi, S. (2000). Practice-based theorizing on learning and knowing in organizations. *Organization*, *7*(2), 211–223. https://doi.org/10.1177/135050840072001
Gherardi, S. (2019). *How to conduct a practice-based study: Problems and methods* (2nd ed.). Edward Elgar.
Gherardi, S. (2022). A posthumanist epistemology of practice. In C. Neesham, M. Reihlen, & D. Schoeneborn (Eds.), *Handbook of philosophy of management*. Springer. https://doi.org/10.1007/978-3-319-48352-8_53-1
Gherardi, S., & Laasch, O. (2021). Responsible management-as-practice: Mobilizing a posthumanist approach. *Journal of Business Ethics*. https://doi.org/10.1007/s10551-021-04945-7
Gibson-Graham, J. K. (2011). A feminist project of belonging for the Anthropocene. *Gender, Place & Culture*, *18*(1), 1 21. https://doi.org/10.1080/0966369x.2011.535295
Gilligan, C. (1982). *In a different voice: Psychological theory and women's development*. Harvard University Press.
Greeson, E., Laser, S., & Pyyhtinen, O. (2020). Dis/Assembling value. *Valuation Studies*, *7*(2), 151–166. https://doi.org/10.3384/vs.2001-5992.2020.7.2.151-166
Haraway, D. (2008). *When species meet*. Minnesota University Press.
Hancock, P. (2008). Embodied generosity and an ethics of organization. *Organization Studies*, *29*(10), 1357–1373. https://doi.org/10.1177/0170840608093545
Hibbert, P., & Cunliffe, A. (2013). Responsible management: Engaging moral reflexive practice through threshold concepts. *Journal of Business Ethics*, *127*(1), 177–188. https://doi.org/10.1007/s10551-013-1993-7
Johnson, D. G., & Wetmore, J. M. (2008). STS and ethics: Implications for engineering ethics. In J. Hackett & O. Amsterdamska (Eds.), *The* handbook *of* science *and* technology studies (3rd ed., pp. 567–582). MIT Press.
Kellogg, K. C., Valentine, M. A., & Christin, A. (2020). Algorithms at work: The new contested terrain of control. *Academy of Management Annals*, *14*(1), 366–410. https://doi.org/10.5465/annals.2018.0174
Laasch, O., Suddaby, R., Freeman, R. E., & Jamali, D. (2020). (Eds.), *The research handbook of responsible management*. Edward Elgar. https://doi.org/10.4337/9781788971966.00006
Lascoumes, P. (1994). *L'éco-pouvoir. Environnements et politiques*. La Découverte.
Laser, S., & Stowell, A. F. (2020). Thinking like Apple's recycling robots: Towards the activation of responsibility in a postenvironmentalist world. *Ephemera: Theory and Politics in Organization, 20*(2), 163–194. https://eprints.lancs.ac.uk/id/eprint/138476
Lather, P., & St. Pierre, E. A. (2013). Post-qualitative research. *International Journal of Qualitative Studies in Education*, *26*(6), 629–633. https://doi.org/10.1080/09518398.2013.788752
Latour, B. (1984 [1988]). *The Pasteurization of France Followed by Irreductions*. Harvard University Press.
Lenz Taguchi, H. (2011). Investigating learning, participation and becoming in early childhood practices with a relational materialist approach. *Global Studies of Childhood*, *1*(1), 36–50. https://doi.org/10.2304/gsch.2011.1.1.36
Linton, J. (2010). *What is water? The history of a modern abstraction*. UBC Press.
Mintzberg, H., & Laasch, O. (2020). Mintzberg on (ir)responsible management. *Research Handbook of Responsible Management*, 73–83. https://doi.org/10.4337/9781788971966.00010
McMurray, R., Pullen, A., & Rhodes, C. (2011). Ethical subjectivity and politics in organizations: A case of health care tendering. *Organization*, *18*(4), 541–561. https://doi.org/10.1177/1350508410388336
Neimanis, A. (2017). *Bodies of water*. Bloomsbury Publishing.
Nicolini, D. (2012). *Practice theory, work, and organization: An Introduction*. Oxford University Press.
Orlikowski, W. J. (2000). Using technology and constituting structures: A practice lens for studying technology in organizations. *Organization Science*, *11*(4), 404–428. https://doi.org/10.1287/orsc.11.4.404.14600
Painter-Morland, M. (2011). Rethinking responsible agency in corporations: Perspectives from Deleuze and Guattari. *Journal of Business Ethics*, *101*(S1), 83–95. https://doi.org/10.1007/s10551-011-1175-4
Painter-Morland, M. (2015). Philosophical assumptions undermining responsible management education. *Journal of Management Development*, *34*(1), 61–75. https://doi.org/10.1108/jmd-06-2014-0060

Parker, M. (2003). Introduction: Ethics, politics and organizing. *Organization, 10*(2), 187–203. doi:10.1177/1350508403010002001

Price M. O., Gherardi, S., & Manidis, M. (2020). Enacting responsible management: A practice-based perspective. In O. Laasch, R. Suddaby, R. Freeman, & D. Jamali (Eds.), *The research handbook of responsible management* (pp. 392–409). Edward Elgar. doi:10.4337/9781788971966.00035

Rasche, A., & Gilbert, D. U. (2015). Decoupling responsible management education. *Journal of Management Inquiry, 24*(3), 239–252. https://doi.org/10.1177/1056492614567315

Rhodes, C., & Pullen, A. (2017). Critical business ethics: From corporate self-interest to the glorification of the sovereign pater. *International Journal of Management Reviews, 20*(2), 483–499. https://doi.org/10.1111/ijmr.12142

Sandberg, J., & Tsoukas, H. (2015). Practice theory: What it is, its philosophical base, and what it offers organization studies. In R. Mir, H. Willmott & M. Greenwood (Eds.), *The Routledge companion to philosophy in organization studies* (pp. 184–198). Routledge. doi:10.1093/oso/9780198794547.003.0013

Schatzki, T. R., Knorr-Cetina, K., & Von Savigny, E. (2001). *The practice turn in contemporary theory*. Routledge.

Shove, E., Pantzar, M., & Watson, M. (2012). *The dynamics of social practice: everyday life and how it changes*. Sage.

Sperling, A. (2019). Anthropocene. In R. T. Goodman (Ed.), *Bloomsbury handbook of 21st-century feminist theory* (pp. 311–324). Bloomsbury. doi:10.5040/9781350032415.ch-022

Styrke, B. M. (2015). *Kunskapande i dans: om estetiskt lärande och kommunikation*. Liber.

Tronto, J. C. (1993). *Moral boundaries: A political argument for an ethic of care*. Routledge.

Ulmer, J. B. (2017). Posthumanism as research methodology: Inquiry in the Anthropocene. *International Journal of Qualitative Studies in Education, 30*(9), 832–848. doi:10.1080/09518398.2017.1336806

Vinck, D. (2003). *Everyday engineering. An ethnography of design and innovation*. MIT Press.

Wray-Bliss, E. (2009). Ethics: Critique, ambivalence, and infinite responsibilities (unmet). In M. Alvesson, T. Bridgman, & H. Willmott (Eds.), *The Oxford handbook of critical management studies* (pp. 267–285). Oxford Academic. https://doi.org/10.1093/oxfordhb/9780199595686.013.0013

Zembylas, M. (2018). The entanglement of decolonial and posthuman perspectives: Tensions and implications for curriculum and pedagogy in higher education. *Parallax, 24*(3), 254–267. https://doi.org/10.1080/13534645.2018.1496577

12
ETHICS AND ENGINEERING DESIGN FOUNDATIONS

Diana Bairaktarova, Natalie Wint, and Mauryn C. Nweke

Introduction

Engineering design is often considered central to engineering practice (van Gorp & van de Poel, 2001). According to Archer (1992) engineering design is "directed towards meeting a particular need, producing a practicable result and embodying a set of technological, economic, marketing, aesthetic, ecological, cultural and *ethical* values determined by its functional and social context" (p. 8). Devon and van de Poel (2004) claim that design is "quintessentially an ethical process" (p. 461), contending that "ethics is not an appendage to design but an integral part of it" (p. 461). The decisions made during engineering design are thus critical in determining an engineering artifact, process, or technology's impact on society. When examining modern design more broadly from a philosophical perspective, Parsons (2016) identifies three aspects of the process involving 'design ethics': (1) when designers face ethical issues applying norms and rules during design, (2) when choices are made regarding what is designed, and (3) when designs modify or change existing conceptions of ethics.

The social form of inquiry involved in design and the ill-structured nature of the problems has led to design being defined as a reflective practice (Schön, 1987), highlighting the importance of considering societal impacts and ethics in engineering artifact, process, and technology development. Engineering training, however, does not appear to adequately prepare students to assume professional and ethical responsibility for the societal impacts of technology; students and recent graduates often have difficulty connecting social consciousness with user needs. Considering user needs reflects a commitment to designing solutions prioritizing user well-being and satisfaction, ensuring that benefits are distributed equitably. This ethical stance fosters a sense of social responsibility. However, "engineering education has been described as characterized by a 'culture of dis-engagement' in which ethical and societal concerns are constructed as different from, and less important than, purely technical concerns" (Lönngren, 2020, p. 44). There is evidence that the nature of programs can diminish students' inclination toward ethical discourse. For instance, empirical analyses have revealed a decline in students' interest in public welfare as they progress through their education (Cech, 2014). Moreover, in a study by Tormey et al. (2015), the moral reasoning of Swiss engineering students appeared to diminish during a period of ethics instruction – something the research team attributed to a hidden curriculum that encourages students to adopt an epistemology-based application of established principles and laws.

DOI: 10.4324/9781003464259-15

The approach taken by students when identifying customer needs is often limited to conducting surveys and organizing focus groups early in the process (Bairaktarova et al., 2016). Consequently, students tend to treat needs as a checklist of requirements to use as inputs in their design processes. To address this situation, there is a need for interventions that enable students to cultivate a more socially conscious level of understanding. This shift towards considering user needs aligns with essential engineering ethical principles such as honesty, integrity, and fairness in the design process.

While we recognize that the impact of engineering work has led to an emphasis on broader issues such as sustainability, social responsibility, and ethics, our chapter focuses explicitly on ethics in engineering design, acknowledging that these ethical considerations are a vital and distinct component of the broader societal concerns within the field of engineering. We outline the definition of engineering design before introducing ways to classify engineering design. The stages of the engineering design process are explained so that we may understand how ethics relates to each stage. We then discuss the ethical implications of the social nature of engineering design. The subsequent sections focus on teaching ethics within engineering design and various educational models we may use. Finally, we share insights from Science and Technology Studies (STS) and end the chapter with concluding remarks, recommendations, and future directions.

Before that, we outline the ways our positionality impacts our work. Our perspectives and insights, shared in this chapter, are enriched by our backgrounds and experiences. Having all come from engineering and science backgrounds, we acknowledge the need for engineering researchers to challenge the myth of objectivity in research. We, therefore, took part in an exercise to surface our interpretive lens, and a summary is shared to inform the interpretation of the chapter presented.

Diana, with nearly 15 years of experience as a design engineer and a decade dedicated to engineering education and research, embodies a strong commitment to addressing the ethical dimensions of engineering and nurturing a holistic perspective in the field. Throughout her industry career, she often encountered ethical dilemmas that the engineers involved seemed ill-equipped to navigate effectively. Diana's experience revealed that engineering solutions frequently fail to prioritize user needs; her realization of this led Diana to transition to academia, pursuing an advanced degree in engineering education. Her mission is unwavering – to educate the next generation of engineers with a strong ethical foundation and an innate ability to empathize with the end user. She seeks to address challenges impacting people's everyday lives, using her expertise to nurture and elevate the engineering profession.

Natalie has been an engineering academic for 4 years. During her academic career, she has started to question *who* and *what* engineering is for and who benefits from and suffers from the cost of engineering decisions. She takes a broadly philosophical approach and is inspired by STS ideas. She enjoys incorporating the social sciences into teaching, which partly results from a conflict between personal and professional identity. She now leans towards qualitative research approaches. Her motivation to participate in this work was to collaborate with those with different experiences and develop her knowledge further. She considers herself an expert neither in engineering nor ethics but is trying to navigate the interface between them.

Mauryn has been an engineering academic for 7 years. She has worked on refining the content and pedagogical approaches involved in teaching engineering design and professional skills, particularly to first-year undergraduate engineering students. She has considered philosophical and practical approaches to integrating themes around ethics, social responsibility, equity and inclusion, and responsible innovation into engineering design. Having completed a Ph.D. in engineering sciences, she sought to develop competence in education and social science research techniques to deepen her understanding of research-informed approaches to her work. This led to her undertaking a postgraduate engineering education degree, and she now uses the research skills she gained to support educational developments in the engineering curriculum. Although her expertise is broader than the focus presented in this chapter, it has served as a great learning opportunity and a valuable interfacing of knowledge between herself and her fellow authors.

Our chapter benefits from these perspectives, fostering a holistic understanding of ethics in engineering design that encompasses practical, philosophical, and ecological dimensions, thus providing a comprehensive view of the subject.

Engineering design: types of design and the design process

Design, according to Petroski (1998), is what most distinguishes engineering from science: "Design is a process through which one creates and transforms ideas and concepts into a product that satisfies certain requirements and constraints" (p. 5). Brey (2022) provides a fuller understanding of design, emphasizing that it is an all-encompassing term and a core activity of society, pointing to fields such as craft and applied arts, fine arts, architecture, and applied social sciences. In differentiating engineering design, he relies on the ABET definition: "the process of devising a system, component, or process to meet desired needs. It is a decision-making process (often iterative), in which basic science and mathematics and engineering sciences are applied to convert resources optimally to meet a shared objective" (ABET, 2018, p. 5). Engineering design is thus considered an activity carried out only with specialized training (e.g., ideation through free-hand sketching, CAD modeling, prototyping), knowledge, and methods for applying this knowledge.

Types of engineering design

The engineering design process can manifest in several ways, resulting in various, although sometimes overlapping, types of design. According to van Gorp and van de Poel (2008), how engineers address ethical issues depends on the type of design process used, and it is thus useful to consider these typologies. Vincenti (1990, 1992) categorizes engineering design processes using two dimensions: hierarchy and type. Concerning the former, the degree of external constraint is larger for design processes lower in the design hierarchy as the higher levels pose constraints (e.g., dimensional constraints, or constraints concerning functionality) on lower levels (Vincenti, 1990). An example of this is provided by van de Poel and van Gorp (2006), who describe piping and equipment design for (petro)chemical plants as being at the lower levels of the hierarchy. The design process and the product and chemicals involved are at higher levels and are also within the control of the petro(chemical) company. Engineering firms contracted to design piping and equipment need to adhere to economic and practical (e.g., space constraints) requirements as well as safety codes, regulations, and standards, all of which place external constraints on the design. Fulfilling multiple requirements imposed by such constraints thus results in ethical questions such as: *What is safe enough?*

A design can then be considered either normal or radical. In normal design, both operational principle (Polayni, 1962), how the design works, and the normal configuration or "the general shape and arrangement that are commonly agreed to best embody the operational principle" (Vincenti, 1990, p. 209) remain the same as in previous designs. In contrast, in radical design, the operational principle and/or normal configuration are unknown. According to van Gorp and van de Poel (2008), most decisions made during normal design are based on regulative frameworks (e.g., minimum safety requirements). Frameworks can also be used during radical design; however, the absence of normal configurations and operational principles may mean they are less applicable. In these latter cases, decisions are primarily made based on design team norms. The DutchEVO has been described (van de Poel and van Gorp, 2006) as a radical design. Its lightweight design meant a standard configuration could not be used and this led to questions and discussions regarding how safety could be operationalized. Through such examples, the importance of such distinctions becomes clear in that they shift responsibility from the engineering community and society involved in the formulation of regulations to the individual design engineer, thus having implications for trust in engineering and its products.

Brey (2022) describes a similar dichotomy between routine design and innovative or creative design. He defines routine design as "design that proceeds within a well-defined state space of potential designs, where all variables, their applicable ranges, and the knowledge to compute their values are directly instantiable from existing design prototypes" (Brey, 2022, p. 32). This is contrasted with innovative or creative design, which ventures beyond these established parameters.

We can also categorize engineering design in several ways, such as original, adaptive, redesign, selection, product, and industrial (Dieter & Schmidt, 2021). Original design, which incorporates the use or application of novel technologies, requires the result (be that a tangible or intangible artifact) to be unique. In contrast, adaptive design modifies an existing solution to fulfill different requirements or applies it in a novel way. Re-design aims to significantly improve an existing design and tends to result in enhanced service, function, and/or capability. Many tangible designs use manufacturer-supplied components with specified properties, performance attributes, quality, and cost, with components selected based on required properties. While the terms 'product design' and 'industrial design' tend to be used interchangeably depending on the engineering discipline, their target markets differ. Both designs aim to design a consumer product to be sold – but whereas this is a primary goal of 'product design,' 'industrial design' focuses more on the interface between the consumer and the product. For example, in industrial design, the end user may not be the typical customer in the public market as its primary goal is to create enhanced designs for manufacturers. When these products are commercialized, the target market is a niche one, which, in turn, may have broader access to sell to the public.

We will now focus on the stages of the engineering design process and investigate how ethical principles apply at each stage.

The engineering design process

The engineering design process has several distinct stages with unique characteristics and challenges. It begins with identifying the problem and progresses through the development of solutions, prototyping, and testing. Van de Poel (2000) distinguished the following five points of ethical relevance during the engineering design process:

1. The formulation of goals, design criteria, and requirements and their operationalization.
2. The choice of alternatives to be investigated during a design process and the selection among those alternatives later in the process.
3. The assessment of trade-offs between design criteria and decisions about the acceptability of particular trade-offs.
4. The assessment of risks and secondary effects and decisions about the acceptability of these.
5. The assessment of scripts and political and social visions that are (implicitly) inherent in a design and decisions about the desirability of these scripts.

(p. 3)

The first stage involves choices regarding *what* problems we decide to solve and *who* benefits – decisions that Chan (2018) claims must "presume some fundamental ideas of what is a good or worthwhile life," with ideals "usually (being) tempered within the parameters of some acceptable rules or obligatory norms" (p. 186). Multiple stakeholders are involved in formulating design criteria, both by detailing specific requirements and through general legislation (van Gorp & van de Poel, 2001). While some see the role of engineers as morally neutral, their being responsible for finding the best possible technological solution within constraints, van Gorp and van de Poel (2001) argue that how something is designed influences who will use it and for what purpose. Again, making use of the DutchEVO example, van Gorp and van de Poel describe how design will determine the person's physical ability to drive. They explain that the relationship between the user and a product impacts emotional sustainability. Therefore, if people enjoy their car, they might use it more often, leading to unsustainable behavior.

Later stages of the process include concept design during which creativity can bridge opposing moral values (van de Poel & Royakkers, 2011). During this stage, we may use what Johnson (1993) refers to as moral imagination, "an ability to imaginatively discern various possibilities for acting in a given situation and to envision the potential help and harm that are likely to result from a given action" (Johnson, 1993, p. 13). For example, van de Poel and Royakkers (2011) describe a plan to close the Eastern Scheldt in the Netherlands after a flood disaster. Environmentalists and fishermen opposed the closure, and thus, ecological care and safety values were posed against one another. A storm surge barrier allowing water through, but that was to be closed when a flood threatened, was posed as a creative compromise balancing safety and environmental concerns.

The simulation stage involves ensuring concept designs meet design requirements and consideration for the desirable or acceptable level of reliability in predictions (van de Poel & Royakkers, 2011). Although prediction reliability is a methodological issue, the reliability of predictions is considered a moral concern depending upon what is at stake. For example, in the case of nuclear power plants, for which failure is catastrophic, reliability would be considered more important than in simulations associated with everyday devices.

The decision stage involves analyzing simulation results alongside original requirements to compare concept designs and determine trade-offs and compromises that need to be made (van de Poel & Royakkers, 2011). This is particularly significant when considering the multiplicity of stakeholders involved, which will be discussed further when considering trade-offs and introducing Constructive Technology Assessment (CTA).

During detailed design, the selected design is elaborated, and such ethical questions as the choice of materials and their associated risks and health/environmental impacts are addressed (van de Poel & Royakkers, 2011). The subsequent prototype development and testing involves moral judgments about the extent to which tests represent circumstances in which designs are eventually used.

Finally, the manufacture and construction stages involve considering ethical issues such as labor conditions, emissions, and use of hazardous materials (van de Poel & Royakkers, 2011). Engineers must consider the moral issues raised by risks and hazards of designs and make decisions concerning the acceptability of these risks. This process typically involves attempts to characterize the risks involved, for example by conducting a risk assessment that considers factors such as failure modes, exposure, consequences, and probability, followed by the need to answer an ethical question regarding the acceptability of risk. Van de Poel and Royakkers (2011) outline four potential ethical considerations: informed consent, whereby risks are seen as more acceptable if those at-risk consent to involvement in the relevant activity (e.g., an experiment); assessing whether advantages outweigh disadvantages; the availability of alternatives for the best available technology; and, finally, the distribution of risks and benefits. These methods assume risks can be, at least to some extent, predicted. However, there is an increasing need to focus on cases with uncertain hazards associated with new technology. In such cases, the precautionary principle, which originated from the Rio Declaration (United Nations, 1993), is proposed. Sandin (1999) defines its four dimensions as threat, uncertainty, action, and prescription. Ethical considerations are also relevant when understanding the degree to which designs solve the original problem and address user needs. Findings often lead to adjustments in the design process, such as in response to engineering disasters (e.g., Ermer, 2008), or inform future designs, especially when engineering innovations are misused or not applied as intended, as documented in other cases (e.g., Leydens & Lucena, 2018; Lucena et al., 2010; Riley, 2008).

The ill-structured nature of design problems means that not all design criteria can be met simultaneously, and there is a need for compromises throughout the design process. Decisions about which trade-offs are acceptable are normative in nature (van Gorp & van de Poel, 2001), with ethical decisions being made when moral values, such as safety and sustainability, are at stake. These trade-offs can be determined in a variety of ways, three of which are listed below.

Cost–benefit analysis: alternatives are compared based on their advantages and disadvantages expressed in monetary terms. Contingent validation, an approach used to express non-economic values (e.g.,, safety) in monetary terms, can be seen as problematic because of the incommensurable nature of values (van de Poel & Royakkers, 2011). One issue is how a choice is made once analysis is carried out. For example, it may be that the option with the highest net value is chosen or that all options having an overall advantage are eligible for selection based on other ethical criteria.

Thresholds: these are commonly used in technical codes and standards defining the minimum level of a design criterion that should be met.

Multi-criteria analysis: involves scoring and comparing options based on specific criteria. It assumes various design criteria can be measured using the same scale and that an ethical decision can be made based on relative weightings.

Although these methods introduce a somewhat systematic approach, what happens in reality can be different. Such trade-offs involve numerous stakeholders with varying opinions about what constitutes an acceptable one (van de Poel & Royakkers, 2011), and design is typically considered to be a social process, as discussed next.

Design as a social process: exploring individual and co-design dynamics

Engineers navigate a space consisting of "conflicting goals … non-engineering success standards, non-engineering constraints, unanticipated problems, distributed knowledge, and collaborative activity systems" (Jonassen et al., 2006, p. 139). Van Gorp (2005) argues that design should be considered a social process, saying that "choices are made in, and by groups of people. During the design process, communication, negotiation, argumentation, (mis)trust between engineers and power differences between engineers influence the design" (p. 29). These social arrangements for making decisions in a design process are referred to as social ethics, and project management structures may ensure the correct processes are in place for ensuring that moral values such as safety remain paramount.

As the social nature of design suggests, it is difficult to discern the responsibilities of an individual engineer. Chilvers and Bell (2018) assert that increasing focus on normative goals "focuses attention on the agency of engineers within sociotechnical networks where these same networks can both enhance and constrain engineers' capacities to contribute to positive change" (p. 205). However, Devon and van de Poel (2004) suggest ways to improve the design process, highlighting three example issues that may be considered from a societal ethics perspective. First, the division of design tasks and allocation of roles and responsibilities have implications for the distribution of responsibility. Second, how decision-making takes place and what opportunities exist to revise decisions can influence the likelihood of ethical choices. Finally, the degree to which various stakeholders, including those affected by the product, are included (or excluded) from the design process can affect the decisions made.

In many ways, the ethical decision-making process parallels the engineering design process. This comparison was supported by Bero and Kuhlman (2010); they drew parallels between the general engineering design process defined by Dominick et al. (2001) and the ethical decision-making process defined by Martin and Schinzinger (1996); see Table 12.1.

One of the key stakeholders in the design process is the user, whose needs, preferences, and experiences must be carefully considered to create products and systems that effectively meet their requirements and expectations.

Understanding users through design

Engineering designers need to consider not only how their design presents a solution to a consumer problem but also what aspects of the design may cause concerns for customers. For this reason, the approaches and tools used by designers have changed dramatically over the last three decades: engineering designers now focus on user needs in each phase of the process to create highly usable, accessible, and valuable products. These different design approaches have evolved in response to the changing landscape of technology, societal values, and the recognition of the pivotal role of users in the design process. They have unique historical roots and disciplinary applications, allowing designers to select the most appropriate approach depending on the specific context and objectives of design projects.

User-centered design

User-centered design (UCD), sometimes referred to as human–computer interaction (HCI), traces its roots back to the mid-twentieth century when the field of computer science recognized the need to accommodate human capabilities and limitations. Pioneers like Don Norman and Jakob Nielsen

Table 12.1 Comparison of the stages of engineering design and ethical decision-making (adapted from Bero and Kuhlman, 2010)

Step	*Engineering design process*	*Ethical decision-making process*
1	Problem identification	Identification of moral factors relevant to the case
2	Definition of the constraints	Identification of conflicting moral factors and definition of dilemmas
3	Ideation	Ranking of moral theories
4	Initial design of potential design solutions	Generation of various options for action and potential consequences of each action
5	Design selection and detailed design	Making the decision
6	Implementation of final design	Enacting the decision

significantly influenced the development of UCD. In this approach, engineering designers focus on the user's inherent way of doing things. According to the literature, UCD focuses on four main activity phases (Harte et al., 2017):

1. Specify the user and the context of use.
2. Specify the user requirements.
3. Produce design solutions.
4. Evaluate designs against requirements.

UCD emphasizes iterative design, incorporating user feedback throughout. It is often applied in software development and digital interfaces, where user experience is critical – as applied in engineering, UCD tends to focus on developing products and related experiences that are functional, valuable, useful, and usable.

Human-centered design

Human-centered design (HCD) shares historical origins with UCD but has a broader application, encompassing a variety of design domains. It acknowledges that human interactions extend beyond the digital realm. HCD has evolved to include architectural design, industrial design, and other fields. Sometimes referred to as participatory design, this approach focuses on incorporating users' thinking, behavior, and emotions into design to better understand their needs. It is often associated with engineering disciplines related to social impact, including biomedical engineering (e.g., prosthesis design) and biochemical engineering (e.g., manufacture of vaccines). HCD incorporates the steps used in UCD in addition to considering ways to connect with the customer:

- Understanding the end-user
- More clearly defining the problem
- Brainstorming potential solutions
- Creating prototypes
- Testing and refining with a particular focus on minimizing risk and maximizing safety

HCD is characterized by its holistic approach, considering both the user and the broader human context, including cultural and societal factors. It finds application in diverse industries, from product design to urban planning.

Empathic design

Empathic design, also known as empathetic design, was introduced by industrial designer Roger Martin in the 1980s. It emerged as a response to the limitations of traditional, functionalist design approaches.

Design practitioners (Koskinen et al., 2003; Mattelmäki & Battarbee, 2002; Suri, 2003) argue that empathy is a human quality that designers need to develop and enhance to meet customer needs by creating products that are useful and practical yet meaningful. Battarbee et al. (2002) suggest that for designers to empathize with the users, they should extend their perspectives by putting themselves in the users' shoes. Koskinen et al. (2003) and Fulton Suri (2003) argue that being an empathic designer involves engaging in specific activities to imagine being in the users' position.

Empathic design is often employed where emotional connections with the user are paramount, such as healthcare or product design. Considering that empathy – as described by the psychological literature – includes both affective and cognitive components, Kouprie and Visser (2009, p. 445) propose an empathic design framework comprised of four phases that can help designers develop and apply techniques and tools in design:

1. *Discovery* – Entering the user's world; achieving willingness.
2. *Immersion* – Wandering around in the user's world; taking the user's point of reference.
3. *Connection* – Resonating with the user; achieving emotional resonance and finding meaning.
4. *Detachment* – Leaving the user's world; designing with a user perspective.

Value-sensitive design

Value-sensitive design (VSD) emerged in the early twenty-first century as a response to the growing importance of ethics and societal values in technology and product development. This approach considers the relevant ethical values in a systematic manner (Friedman et al., 2006). It involves both consideration of evidence regarding the experiences and values of those affected by designs, making trade-offs among these values, and technical investigations analyzing designs to determine the extent to which they meet morally relevant values.

VSD is inherently ethical, aiming to ensure that design choices align with fundamental human values. It has found prominence in domains such as information technology and artificial intelligence, where ethical considerations are critical to ensuring that technology aligns with societal values. For example, during VSD, empirical investigations can be conducted to determine the role various values play in influencing behavior, and technical investigations to determine the degree to which technology supports or discourages specific values.

While all these theoretical frameworks and types of design provide valuable insights into how design is conducted, it is equally important to explore how these approaches, including ethical considerations, are effectively taught and integrated into design education to empower the next generation of designers. For more on VSD teaching methods, see Chapter 22.

Ethics of design in engineering education

In the context of engineering education, the act of design takes on a profound ethical dimension. Designing is inherently transformative, with each change to the environment raising questions of responsibility, values, and consequences. As designers engage in this process, they alter not only their surroundings but also their own ethical awareness. This ethical interplay is underpinned by the concept of situatedness, closely linked to the work of John Dewey, and constructive memory (Newman & Holzman, 1997). These ideas form the ethical foundation upon which engineering students ground their knowledge within the dynamic situations they construct during their design interactions.

Situatedness and constructive memory thus provide the conceptual bases for grounding the knowledge of designers in the situation being constructed by their interactions with the environment (Greeno et al., 1996). Situated theories have purposefully been used to understand learning as context-specific social processes by characterizing cognition as being socially shared (Clancey, 2012; Newman & Holzman, 1997). These ideas are rooted in John Dewey's early objections to stimulus-response theory (Newman & Holzman, 1997). For more on situatedness, see Chapter 11.

Within the landscape of engineering education, it is crucial to acknowledge a significant dissonance. Fry (2009) astutely notes that design ethics is "massively underdeveloped and even in its crudest forms remains marginal in design education" (p. 34). The predominant focus within engineering ethics education has traditionally centered on individual ethical considerations and applying normative ethical theories. This approach, however, falls short of addressing the intricate web of social ethics inherent in the design process, which relies on social relations and communal decision-making structures (Devon & van de Poel, 2004). In response to this disparity, advocates for a more holistic approach to design ethics, such as Kirkman et al. (2017), propose a paradigm shift. They advocate for design ethics courses that immerse students in complex problem situations, equipping them with tools and guidance to navigate and make sense of scenarios. Kirkman et al. see these tools as the "very idea of an ethical value, along with schemas and appropriate vocabulary for framing and reframing problem situations, developing options, and sorting and connecting ethical values implicated in those options" (p. 3). Such approaches bridge the gap between engineering education and the encompassing reality of ethical considerations in practice, preparing students to tackle the intricate, multi-dimensional challenges they will face.

Educational models

Several efforts have focused on incorporating existing design frameworks in the classroom. Below we provide some educational models, including curricula and program reform related to ethics of design in engineering education.

Design for sustainability – Product Realization for Global Opportunities is a hybrid course offered to students in the United States and Brazil to initiate collaboration on projects focusing on designing products that could improve housing, living conditions, and personal security. Undergraduate engineering and business students applied sustainability framework in designing new technologies (Mehalik et al., 2008).

Eco-design – Delft University of Technology (TU Delft) offers six bachelor's and master's-level courses focused on eco-design (Boks & Diehl, 2006). The leadership at TU Delft modified existing courses to include corporate social responsibility (CSR) principles, further emphasizing the importance of including a holistic sustainability perspective.

Empathic design – Empathic techniques have been integrated into capstone design (Guanes et al., 2021) and product design distance-learning courses for geographically distributed master's-level engineering students (Bairaktarova et al., 2016) to provide immersive design experience in ill-structured problems and design decision making. Other scholars have used a more formal method – the Empathic Experience Design (EED) Method (Genco et al., 2011; Johnson et al., 2014) – during the conceptual design phase of the design process with senior-level students.

Design for development – Nieusma and Riley (2010) showcased several universities in Sri Lanka and Nicaragua partnering with US universities that apply specific development interventions considering engineering as a professional activity with social justice goals.

Human-centered design – Engineering Projects in Community Service (EPICS), founded at Purdue University in the late 1990s, is a service-learning design program where multidisciplinary teams of undergraduate students partner with community organizations at local and global levels to address human, community, and environmental needs.

Both HCD and service-learning programs have been considered as "rich sites for exploring engineering students' ethical decision-making" (Corple et al., 2020, p. 264). In their work on understanding ethical decision-making in engineering design, Corple and colleagues applied the four principles of Beever and Brightman's (2016) framework of reflexive principlism: beneficence, providing benefits to society; non-maleficence, avoiding causing harm; autonomy, respecting the agency of individuals in decision-making; and justice, distributing risks, benefits, and costs equitably among all individuals. In so doing, they examined students' descriptions of (a) engineering design decision-making to determine (b) where the principles emerged and (c) how they shaped students' sense-making regarding beneficence, or what is good, throughout the design process. Corple et al. found that students demonstrated more ethical sensitivity when working closely with project partners and users throughout the engineering design process, particularly during the last phase when they delivered their product to partners and users and were required to question if their decisions were ethical. In comparison, students distanced from project partners did not incorporate user/partner needs and concerns into their engineering design decisions intentionally or consistently. However, Corple et al. (2020) highlighted that a strong emphasis on users might lead to students over-associating ethical concerns with user concerns at the cost of considerations for secondary stakeholders or environmental impacts – something they describe as outsourcing ethical decision-making to users and project partners and not considering the breadth of potential ethical implications of decisions. They conclude by suggesting that educators use 'reflexive principlism' to identify and display students' intuitive ethical decision-making during engineering design, enabling them to help students apply ethical frameworks in prescriptive ways. Corple et al. (2020) offer the following suggestions to educators:

1. Have students write down how they navigated ethically challenging engineering design decisions and what their decision-making processes were and why.
2. Use Beever and Brightman's (2016) text to teach students the four principles of reflexive principlism and describe how they may materialize in engineering contexts.
3. Ask students to examine their written decision-making processes to identify if and where the moral principles appear in their thinking.

According to Bowers (1998), although the ethics of technology is often included in specialized courses in Science and Technology Studies (STS), this topic is generally not found in general engineering courses. Dyrud (2017) suggests introducing cases that focus on artifacts developed by engineers – for example, illustrating the non-neutrality of technology using the example of the IBM tabulator that read punch cards that stored detailed information about Jews in Nazi Germany. Concerning the inclusion of CSR in the design curriculum, researchers suggest that CSR directly influences the ethical behavior of engineers (Hutchins & Sutherland, 2008).

Insights from Science and Technology Studies

As demonstrated throughout this chapter, particularly in the section on design as a social process, engineering design involves the decisions of individuals within groups. Such ideas are linked strongly to those from STS and the philosophy of technology. According to Manzini and Cullers (1992), design ethics within engineering and technology focuses on choices about the side effects and risks posed by modern technology as opposed to everyday design decisions and practice. Similarly, Verbeek (2011b) reminds us that the products of engineering and technological design mediate actions and experiences of users, irrespective of the degree to which designers engage in ethical reflection.

Technological mediation (Ihde, 1990; Latour, 1992; Verbeek & Crease, 2005) concerns how technological artifacts (co)shape human perception and behavior. It can influence moral decisions (Ihde, 1991) and "can have systematic tendencies to promote values of tendencies to promote or benefit values, such as privacy and sustainability, as well as harm or detract from them" (Brey,

2022, p. 408). Firstly, mediation of perception involves influencing people's sensory experience of reality, for example, looking through a virtual reality (VR) headset. Such technology changes what is considered 'real' and, therefore, what contributes to ethical decision-making. For example, ultrasound has allowed us to make decisions about unborn children (Verbeek, 2011a). Secondly, mediation of action is based on the idea that designs are inscribed with scripts (Akrich, 1992; Latour, 1992, 1994) that shape human action. For example, it can be considered that the microwave encourages us to eat quickly and alone. Verbeek (2011b), thus, refers to designing as "materializing morality" (p. 90), proposing two ways designers can incorporate the mediating role of technology into the design process. First, they may assess and reflect upon the degree to which actions resulting from technological mediation are morally justified. Second, they may choose to explicitly design desirable forms of technological mediation (e.g., a speed bump encourages drivers to slow down) – something which Achterhuis (1995, 1998) refers to as the moralization of technology, and which has been criticized for risking human freedom and democracy. At this point, designers may consider which values and norms are to be embodied and the ways in which these may be materialized. Verbeek (2011b) compares this process to the conceptual investigations conducted in VSD, analyzing the values supported by a design (Friedmann et al., 2006). Similarly, Vallor (2016) takes a 'technomoral' virtue ethics approach, claiming that technology can promote virtues (e.g., honesty and empathy) and vices (e.g., dishonesty, carelessness). Drawing upon the work of Fletcher (2012), Brey (2022) highlights the need to differentiate between a design which is prudentially good, meaning good or bad for something, for example "goodness for persons and goodness for society" (p. 404), from that which is morally good. For example, taking someone's money may be prudentially good for a person or an organization, but it would be morally wrong to accept it if not given with free will.

However, the mediating role of design depends not only on decisions made by designers but also on users and the unforeseen ways in which design mediates actions. Verbeek (2011b) warns of what Tenner (1996) refers to as a rebound effect, whereby technology is used in a way different from that intended, is not used at all, or when there is a difference between designer and user expectations. Thus, engineers must actively imagine how their designs might be used and expand the design process to consider a broader range of actors, values, and interests. Verbeek (2011a, 2011b) refers to Jelsma's (2006) approach to redesign, which involves analyzing user logic (how users interpret and adapt the design), script logic (how the design influences behavior), and the impact of these factors on determining design outcomes.

Verbeek (2011b) suggests a design process that allows for responsible, intentional intervention. He proposes anticipating technological mediation, which he suggests can be done using imagination – following the Constructive Technology Assessment (CTA), a systematic method that generates variations of designs based on feedback from stakeholders and thus focuses on how technology emerges from a specific context – and using scenarios and simulations focusing on using a design in specific situations in various ways. While processes like CTA are considered a means for the democratization of design, thus removing the fear of unknown technology, they are limited in that little attention is paid to non-human actors.

Anticipating technological mediation will result in further complexities, trade-offs, and questions, for example, the degree to which potential mediation is justified. Including a desirable mediating characteristic can negatively affect other design features. To illustrate, Verbeek (2011a) describes automatic speed restrictions in cars that come at the cost of freedom and experience. Assessment of mediation involves stakeholder analysis that considers moral arguments associated with designs, including reflecting on, for example, the morals of intended and implicit mediations, the form of mediation (e.g., forcing action versus nudging), and the outcome of mediation in society (Verbeek 2011b). In discussing mediation assessment, Verbeek (2011b) considers the issues of responsibility, freedom, and democracy – initially questioning the extent to which humans can be held responsible for actions mediated by technology. He distinguishes between causal and moral responsibility, proposing that technology (co)shapes moral responsibility by mediating human action and contributing to causal responsibility. Thus, responsibility is shared between humans and technology, with the designer taking some responsibility for design decisions. Such factors become increasingly important when considering technology such as autonomous vehicles. Similarly, in his discussion of freedom, he draws upon the work of Foucault (1988), who argues that freedom results from the situated nature of human life, meaning it cannot exist in an absolute sense. Many forms of mediation do not force specific actions; instead, they inform them (e.g., through persuasion), meaning technology is then considered a coauthor of what we do. Foucault thus argues that technological mediations are limited and questions whether we should allow technology that allows no room for freedom of individuals (e.g., by oppressing or limiting human behavior). The value placed on individual freedom links directly to the way in which technology can be perceived as a threat to democracy, with Verbeek (2011b) suggesting that democratic processes such as CTA be used to encourage democratization of the design process. Technology can also be used to promote democratic values, for example, the 'good life.'

Concluding remarks

In this chapter, we discussed engineering design foundations through the lens of ethics, providing a synopsis of emerging issues in ethics in engineering design education. We presented the development of engineering design thinking in problem-solving, particularly related to engineering ethics, to demonstrate that engineering design is as much about the process as it is about providing a service or product intended for human use. Designers nowadays focus on the user and their needs in each phase of the design process to create highly usable, accessible, and valuable products. Regardless, there is a fundamental tension rooted in the structure of society (Conlon & Zandvoort, 2011), which still results in a lack of interaction with the externalist approach that engineering ethics takes to technology, as well as a focus on products as opposed to processes (van de Poel & Verbeek, 2006).

In the last several decades, we witnessed that the more advanced technology becomes, the more humanity is exposed to unanticipated side effects and risks of harnessing technology (Wolin, 2001). Many scholars echo calls from Conlon and Zandvoort (2011) to examine the different resources available to engineers to help them have a voice in public policy, including consideration for the values and beliefs of professional engineering bodies and individual engineers. Given the impact of engineering design in society, engineering educators have an instrumental role in emphasizing the importance of ethics in the engineering profession, including design, and in continuing to integrate ethics as a core competency in the engineering curriculum.

Engineering design education should incorporate a robust ethical component throughout the curriculum. It should encompass ethical principles and considerations relevant to design, encouraging students to recognize and address the ethical dimensions of their work. It can incorporate real-world problem scenarios by emphasizing complex, real-life problems in design education. These scenarios should expose students to multifaceted challenges, integrating technical, social, and ethical dimensions, mirroring the intricacies of professional engineering practice.

Further, fostering interdisciplinary learning and collaboration and encouraging students to work with experts from diverse fields – including ethics, sociology, and policy – can provide a more holistic, comprehensive, and socially responsible approach. Educators could shift the focus of engineering ethics education from individual ethics and normative theories towards social ethics methods (see, e.g., Chapter 3), encouraging students to explore social relations and decision-making processes that underpin design, aligning more closely with the iterative nature of design. They can encourage students to continually question assumptions, assess consequences, and engage in reflective dialogues about the ethical implications of design.

Engineering educators could consider empowering engineering students to advocate for public policy (see Chapter 6) that aligns with ethical and societal values; engineering design education must continually adapt to evolving technological and societal landscapes, and flexibility and adaptability should be core principles to keep curricula relevant. Professional engineering bodies should play an active role in promoting and supporting ethics education. They should provide resources, guidelines, and forums for discussing ethical challenges in engineering design.

By implementing recommendations and recognizing the implications, engineering design education can better prepare students to navigate complex ethical dimensions of the field, fostering socially responsible, thoughtful engineers well-equipped to address the challenges of our ever-evolving technological world.

References

ABET. (2018). *Criteria for accrediting engineering programmes, 2018-2019*. https://www.abet.org/accreditation/accreditation-criteria/criteria-for-accrediting-engineering-programs-2018-2019/

Achterhuis, H. J. (1995). De moralisering van de apparaten [The moralization of apparatuses]. *Socialisme en Democratie*, *52*(1), 3–12.

Achterhuis, H. J. (1998). *De erfenis van de utopie* [The legacy of utopia]. Ambo.

Akrich, M. (1992). The de-scription of technological objects. In W. E. Bijker & J. Law (Eds.), *Shaping technology/building society* (pp. 205–224). MIT Press.

Archer, B. (1992). The nature of research in design and design education. In B. Archer, K. Baynes, & P. Roberts (Eds.), *The nature of research in design and technology education: Design curriculum matters* (pp. 7–14). Department of Design and Technology, Loughborough University.

Bairaktarova, D., Bernstein, W., Reid, T., & Ramani, K. (2016). Beyond surface knowledge: An exploration of how empathic design techniques enhances engineers understanding of users' needs. *International Journal of Engineering Education*, *32*(1A), 1–12.

Battarbee, K., Baerten, N., & Hinfelaar, M. (2002). Pools and satellites: Intimacy in the city. In *Proceedings of the conference on designing interactive systems DIS'00* (pp. 237–245). ACM Press.

Beever, J., & Brightman, A. O. (2016). Reflexive principlism as an effective approach for developing ethical reasoning in engineering. *Science and Engineering Ethics*, *22*, 275–291. https://doi.org/10.1007/s11948-015-9633-5

Bero, B., & Kuhlman, A. (2010). Teaching ethics to engineers: Ethical decision-making parallels the engineering design process. *Science and Engineering Ethics*, *17*(3), 597–605. doi:10.1007/s11948-010-9213-7

Boks, C., & Diehl, J. C. (2006). Integration of sustainability in regular courses: experiences in industrial design engineering. *Journal of Cleaner Production*, *14*, 932–939: doi:10.1016/j.jclepro.2005.11.038

Bowers, C. A. (1998). The paradox of technology: What's gained and lost? *NEA Higher Education Journal*, *14*, 49–57.

Brey, P. (2022). Understanding Engineering design and its social, political, and moral dimensions. In S. Vallor (Ed.), *The Oxford handbook of philosophy of technology* (pp. 395–416). Oxford University Press.

Cech, E. (2014). Culture of disengagement in engineering education? *Science, Technology and Human Values, 39*(1), 42–72.

Chan, J. K. H. (2018). Design ethics: Reflecting on the ethical dimensions of technology, sustainability, and responsibility in the Anthropocene. *Design Studies*, *54*, 184–200.

Chilvers, A., & Bell, S. (2018). Professional lock-in; Structural engineers, architects and the disconnect between discourse and practice. In B. Williams, J. Figueiredo, & J. Trevelyan (Eds.), *Engineering practice in a global context* (pp. 205–221). CRC Press.

Clancey, W. J. (2012). Scientific antecedents of situated cognition. In P. Robbins & M. Aydele (Eds.), *The Cambridge handbook of situated cognition* (pp. 11–34). Cambridge University Press.
Conlon, E., & Zandvoort, H. (2011) Broadening ethics teaching in engineering: Beyond the individualistic approach. *Science and Engineering Ethics, 17*, 217–232. https://doi.org/10.1007/s11948-010-9205-7
Corple, D. J., Zoltowski, C. B., Feister, M. K., & Buzzanell, P. M. (2020). Understanding ethical decision-making in design. *Journal of Engineering Education, 109*(2), 262–280. https://doi.org/10.1002/jee.20312.
Devon, R., & van de Poel, I. (2004). Design ethics: The social ethics paradigm. *International Journal of Engineering Education, 20*(3), 461–469.
Dieter, G., & Schmidt, L. (2021). *Engineering design* (6th ed.). McGraw Hill.
Dominick, P. G., Demel, J. T., Lawbaugh, W. M., Freuler, R. J., Kinzel, G. L., & Fromm, E. (2001). *Tools and tactics of design*. Wiley.
Dyrud, M. A. (2017). *Ethics and artifacts*. Paper presented at 2017 ASEE Annual Conference & Exposition, Columbus, OH. doi:10.18260/1-2--28298
Ermer, G. (2008, June 22–25). *Understanding technological failure: Ethics, evil, and finitude*. Paper presented at 2008 ASEE Annual Conference & Exposition. Pittsburgh, PA. doi.org/10.18260/1-2—4161
Fletcher, G. (2012). Resisting buck-passing accounts of prudential value. *Philosophical Studies, 157*, 77–91. https://doi.org/10.1007/s11098-010-9619-8
Foucault, M. (1988). *Power knowledge*. Random House.
Friedman, B., Kahn, P. H. Jr., & Borning, A. (2006). Value sensitive design and information systems. In P. Zhang & D. F. Galletta (Eds.), *Human-computer interaction in management information systems: Foundations* (pp. 348–372). M.E. Sharpe.
Fry, T. (2009) *Design futuring: Sustainability, ethics, and new practice*. Berg.
Genco, N., Johnson, D., Ho¨ ltta¨-Otto, K., & Seepersad, C. C. (2011, January). A study of the effectiveness of empathic experience design as a creativity technique. In *International design engineering technical conferences and computers and information in engineering conference* (Vol. 54860, pp. 131–139).
Greeno, J. G., Collins, A. M., & Resnick, L. B. (1996). Cognition and learning. In D. C. Berliner & R. C. Calfee (Eds.), *Handbook of educational psychology* (pp. 15–46). Prentice Hall.
Guanes, G., Wang, L., Delaine, D., & Dringenberg, E. (2021). Empathic approaches in engineering capstone design: Student beliefs and reported behavior. *European Journal of Engineering Education*, 47(3), 429–445. https://doi.org/10.1080/03043797.2021.1927989
Harte, R., Glynn, L., Rodríguez-Molinero, A., Baker, P. M., Scharf, T., Quinlan, L. R., & ÓLaighin, G. (2017). A human-centered design methodology to enhance the usability, human factors, and user experience of connected health systems: A three-phase methodology. *JMIR Human Factors, 4*(1), e8.
Hutchins, M. J., & Sutherland, J. W. (2008). An exploration of measures of social sustainability and their application to supply chain decisions. *Journal of Cleaner Production, 16*, 1688–1698. https://doi.org/10.1016/j.jclepro.2008.06.001
Ihde, D. (1990). *Technology and the lifeworld*. Indiana University Press.
Ihde, D. (1991). *Instrumental realsim: The interface between philosophy of science and philosophy of technology*. Indiana University Press.
Jelsma, J. (2006). Designing 'moralized' products. In P. P. Verbeek & A. Slob (Eds.), *User* behavior *and* technology *development* (Vol. 20). Springer. https://doi.org/10.1007/978-1-4020-5196-8_22
Johnson, D. G., Genco, N., Saunders, M. N., Williams, P., Seepersad, C. C., & Holtta-Otto, K. (2014). An experimental investigation of the effectiveness of empathic experience design for innovative concept generation, *Journal of Mechanical Design, 136*(5), 051009-1–051009-12.
Jonassen, D. G., Strobel, J., & Lee, C. B. (2006). Everyday problem solving in engineering: Lessons for engineering educators. *Journal of Engineering Education, 95*(2), 139–151. https://doi.org/10.1002/j.2168-9830.2006.tb00885.x
Johnson, M. (1993). *Moral imagination: Implications of cognitive science for ethics*. University of Chicago Press.
Kirkman, R., Lee, B., & Fu, K.K. (2017). Teaching ethics as design. *Advances in Engineering Education, 6*(2), n2.
Koskinen, I., Battarbee K., & Mattelmäki, T. (Eds.). (2003). *Empathic design, user experience in product design*. IT Press.
Kouprie, M., & Visser, F. S. (2009). A framework for empathy in design: Stepping into and out of the user's life. *Journal of Engineering Design, 20*(5), 437–448.
Latour, B. (1992). Where are the missing masses? The sociology of a few mundane artifacts. In W. E. Bijker & J. Law (Eds.), *Shaping technology/building society* (pp. 225–258). MIT Press.

Latour, B. (1994). On technical mediation – Philosophy, sociology, genealogy. *Common Knowledge*, *3*, 29–64.

Leydens, J. A., & Lucena, J. C. (2018). *Engineering justice: Transforming engineering education and practice*. IEEE Press.

Lönngren, J. (2020). Exploring the discursive construction of ethics in an introductory engineering course. *Journal of Engineering Education*, *110*, 44–69.

Lucena, J., Schneider, J., & Leydens, J. A. (2010). *Engineering and sustainable community development*. Morgan and Claypool.

Manzini, E., & Cullars, J. (1992). Prometheus of the everyday: The ecology of the artificial and the designer's responsibility. *Design Issues*, *9*(1), 5–20. https://doi.org/10.2307/1511595

Martin, M. W., & Schinzinger, R. (1996). *Ethics in engineering* (3rd ed.). McGraw-Hill.

Mattelmäki, T., & Battarbee, K. (2002). Empathy probes. In T. Binder, J. Gregory, & I. Wagner (Eds.), *Proceedings of the participatory design conference* (pp. 266–271). CPSR.

Mehalik, M. M., Lovell, M., & Shuman, L. (2008). Product realization for global opportunities: Learning collaborative design in an international setting. *International Journal of Engineering Education*, *24*(2), 357–366.

Newman, F., & Holzman, L. (1997). *The end of knowing: A new developmental way of learning*. Routledge.

Nieusma, D., & Riley, D. (2010). Designs on development: Engineering, globalization, and social justice. *Engineering Studies*, *2*(1), 29–59.

Parsons, G. (2016). *The philosophy of design*. Polity.

Petroski, H. (1998). Polishing the gem: A first-year design project. *Journal of Engineering Education*, *87*(4), 445.

Polanyi, M. (1962). *Personal knowledge*. University of Chicago Press.

Riley, D. (2008). *Engineering and social justice*. Morgan and Claypool.

Sandin, P. (1999). Dimensions of the precautionary principle. *Human and Ecological Risk Assessment*, *5*(5), 889–907.

Schön, D. A. (1987). *Educating the reflective practitioner: Toward a new design for teaching and learning in the professions*. Jossey-Bass.

Suri, F. J. (2003). Empathic design: Informed and inspired by other people's experience. In I. Koskinen, K. Battarbee, & T. Mattelmäki (Eds.), *Empathic design, user experience in product design* (pp. 51–57). IT Press.

Tenner, E. (1996). *Why things bite back: Technology and the revenge of unintended consequences*. Alfred A. Knopf.

Tormey, R., LeDuc, I., Isaac, S., Hardebolle, C., & Cardia, I. V. (2015). *The formal and hidden curricula of ethics in engineering education*. Proceedings of the 43rd Annual SEFI Conference, Orléans, France.

United Nations Conference on Environment and Development Rio de Janeiro, Brazil. (1993). *Agenda 21: Programme of action for sustainable development ; Rio Declaration on environment and development ; Statement of forest principles: The final text of agreements negotiated by governments at the United Nations Conference on Environment and Development (UNCED), 3-14 June 1992*. United Nations Department of Public Information.

Vallor, S. (2016). *Technology and the virtues: A philosophical guide to a future worth wanting*. Oxford Academic. https://doi.org/10.1093/acprof:oso/9780190498511.001.0001.

van Gorp, A. (2005). *Ethical issues in engineering design; Safety and sustainability* (Simon Stevin Series in the Philosophy of Technology Vol. 2). TU Eindhoven & TU Delft.

van Gorp, A., & van de Poel, I. (2001). Ethical considerations in engineering design processes. *Technology and Society Magazine, IEEE, 20*, 15–22. doi:10.1109/44.952761

van Gorp, A., & van de Poel, I. (2008). Deciding on ethical issues in engineering design. In P. E. Vermaas, P. Kroes, A. Light, & S. A. Moore (Eds.), *Philosophy and design: From engineering to architecture* (pp. 77–90). Springer.

van de Poel, I. R. (2000, September 6–8). Ethics and engineering design. University as A Bridge from Technology to Society. Proceedings of the IEEE International Symposium on Technology and Society. La Sapienza University, Rome, pp. 187–192.

van de Poel, I. R., & Royakkers, L. (2011). *Ethics, technology, and engineering an introduction*. Wiley-Blackwell.

van de Poel, I. R., & van Gorp, A. C. (2006). The need for ethical reflection in engineering design. *Science, Technology and Human Values, 31*(3), 333–360.

van de Poel, I. R., & Verbeek, P. P. (2006). Ethics and engineering design. *Science, Technology & Human Values, 31*, 223–236.
Verbeek, P. P. (2011a). Designing morality. In I. van de Poel & L. Royakkers (Eds.), *Ethics, technology, and engineering an introduction* (pp. 198–216). Wiley-Blackwell.
Verbeek, P. P. (2011b). *Moralizing technology: Understanding and designing the morality of things*. University of Chicago Press.
Verbeek, P. P., & Crease, R. P. (2005). *What things do: Philosophical reflections on technology, agency, and design*. Penn State University Press.
Vincenti, W. G. (1990). *What engineers know and how they know it: Analytical studies from aeronautical history*. The John Hopkins University Press.
Vincenti, W. G. (1992). Engineering knowledge, type of design, and level of hierarchy: Further thoughts about what engineers know. In P. Kroes & M. Bakker (Eds.), *Technological development and science in the industrial age* (pp. 17–34). Kluwer.
Wolin, R. (2001). *Heidegger's children; Hannah Arendt, Karl Lowith, Hans Jonas, and Herbert Marcuse*. Princeton University Press.

13

LAW IN ENGINEERING ETHICS EDUCATION

An exploration

Andreas Kotsios, Thomas Taro Lennerfors, and Mikael Laaksoharju

Introduction

The television series *Perry Mason* (aired 2020–2021) is a fictional legal drama based on a book series by Erle Stanley Gardner. In Chapter 5 of the first season, there is an exchange between two of the main characters – Perry Mason, a scruffy private investigator turned defense attorney, and his associate Della Street – regarding arranging a suicide to look like a natural death for insurance reasons. Della remarks that it appears very easy for Perry to break rules, to which Perry responds: "Well … the way I see it, there's what's legal and there's what's right."

Later in the episode, Della reuses the phrase to justify to her girlfriend that she is hiding sensitive case files in her home. The imagery in the scene suggests that she is also alluding to the, at the time, illegal nature of their relationship.

The narrow interpretation of this phrase seems to suggest that there is (sometimes) a conflict between law and ethics. However, at the same time, this phrase seems to indicate that there is a strong relationship between law and ethics, not only because a given scenario can be examined from both perspectives but also because this relationship seems to contain an inevitable element of interaction: acting ethically affects our perception of law, while acting in accordance with law affects our perception of ethics.

Although in legal studies, the relationship between law and ethics is extensively discussed,[1] this relationship has not been equally examined within the domain of engineering ethics education (EEE) – even though we can find several research papers within this domain that mention ethics and law, added to the range of textbooks about engineering ethics that bring up law. Our view is that the importance of law in relation to engineering ethics and, consequently, its importance as a subject worth including in EEE remains vague in this literature and deserves further discussion.

The primary purposes of this chapter are, thus, to present how the relationship between law and ethics has been depicted in the literature of engineering ethics education, assess this relationship, and encourage the inclusion of law – understood not as mere legal rules but as a system of norms – in engineering ethics education. Our central point is that ethics and law are often presented in antagonistic ways, but they can also be mutually reinforcing in engineering ethics education. To achieve this, law cannot be regarded as a mere *matter of fact*, as something immutable that pre-

DOI: 10.4324/9781003464259-16

sides irrespective of ethical discourse (and, as something that Perry Mason believes to be in conflict with ethics), but as a *matter of concern*, namely as something in need of critical examination in relation to ethics (Latour, 2004). Simply put, the point in engineering ethics education when it comes to its relation to law should not be to present law as just something to be complied with, as something to be critiqued as being unethical, or as something deficient concerning ethics, but to genuinely include law as an additional layer in ethical discussions.

Similarly, and considering that, to our knowledge, only a few papers explicitly describe a possible law and ethics integration within the engineering ethics classroom, we will highlight how law can be integrated into teaching topics related to engineering ethics based on our own experiences. We hope this initial sharing can serve as an impetus to create a discussion between EEE researchers and teachers about teaching practices.

Before we start this outlined endeavor, we should address a possible counter-argument, which we often hear from some students, namely the idea of functional specialization – that lawyers should do law and ethicists ethics. *When developing a system, why could engineers not just follow the guidelines and recommendations by lawyers and ethicists – or develop a system that lawyers and ethicists scrutinize* post hoc *to ensure that it complies with relevant laws and ethical values?*

Now, in a handbook on EEE, it is probably relatively uncontroversial to claim that ethics should be taught in engineering education. To the arguments why ethics should be studied, we want instead to add arguments for why engineers should also study – at least some aspects of – law together with ethics. This is to understand that their professional behavior and choices are not only passively affected by legislation but also positively affect the interpretation and formation of such legislation. By doing so, engineers could affect ethical values as well as laws in society, fulfilling what has been seen as their macro-ethical role. In other words, they would not only be seeing ethics from the individual perspective but also perceiving the ethical obligations that the engineering profession carries or could potentially carry.

In line with the purposes of the chapter, the next section will focus on the relationship between ethics and law and how it has been constructed within a range of EEE literature, both research papers and educational material. While presenting this literature – the review of which reveals the recognition of multiple fruitful connections between ethics and law while, at the same time, making apparent the lack of an in-depth discussion related to these connections – we will also provide our own reflections on it, and highlight possible gaps in this literature. More importantly, although it is marginally present in previous literature, we introduce the idea that law needs to be seen as fundamentally an interpretative endeavor. In that sense, the main goal of teaching law in the context of engineering ethics education is not to promote the perception of law as a matter of mere compliance, as an existing body of enforceable norms that engineering students are expected to more or less memorize and learn to follow in their future profession. Instead, the point of teaching law as part of engineering ethics education is to demystify what 'applying the law' means and critically reflect on the processes that form law and its meaning, that is, why the law is created, interpreted, and applied the way it is.

After that, and in line with our second aim, we will argue that it is not clear from established literature how law and ethics are combined in courses and classrooms. Therefore, we will present some options based on our experience as well as the driving forces and barriers for each of the options: one case on autonomous vehicles, one on ethics and law in medical technology development, one about sustainability in global industrial companies, and one about data, ethics, and law. Consistent with other chapters in this handbook, we begin by introducing ourselves as authors so that readers can assess how our experiences shape the perspectives we share and the topics we identify as central.

Positionality

Andreas Kotsios is an associate professor in Commercial Law at Uppsala University. His research focus has been on law and new technologies and, more specifically, on different aspects of fairness within this area. He, therefore, has always been interested in the relationship between law and ethics, especially in the domains of consumer law, data protection law, and artificial intelligence (AI) law. Regarding teaching engineering ethics, he was invited to develop the course 'Data, Ethics, and Law' in 2019 with (the other two authors) Thomas and Mikael. In the beginning, he envisioned focusing on compliance questions for data scientists and engineers so that the ethics teacher would then use this material to discuss ethical questions. However, it became apparent to him that even non-lawyers can better understand compliance issues when seen through the lenses of ethics.

Thomas Taro Lennerfors, a professor of industrial engineering and management at Uppsala University, is interested in ethics and philosophy. For many years, he has educated engineering and management students in ethics (and written papers and books about it). Previously, he did not think about the relationship between law and ethics, except in superficial ways (perhaps even contributing to the gap by defending ethics against law), perhaps because he thought there was quite a big gap between the areas. Since 2018, he has been interacting with legal scholars in his ethics teaching and realized that many commonalities and shared premises exist. He wanted to reflect on his experiences and understand more about the relationships between ethics and law, and this chapter has been a vehicle for furthering his thinking.

Mikael Laaksoharju holds a Master of Science in Engineering and a Ph.D. in Human–Computer Interaction and is currently working as associate professor in Human–Computer Interaction. He has been teaching interaction design, full-stack programming, and ethics in relation to computer science topics for more than a decade, and his teaching focuses on developing engineering students' skills to handle engineering problems in ways that respect human values. When he met Andreas in 2017, as a fellow participant in a panel debate, he realized that ethics and law have more in common than he had previously imagined, especially when considered as processes of identifying what is right.

We bring our expertise and our diverse perspectives together in team teaching and leverage our differences to help spark critical reflection and dialogue. The author team is quite interdisciplinary, spanning technology, social sciences, and humanities. However, we all work at the same university, and our team is thus not diverse in terms of institutional belonging, which can hinder us from seeing all the ways of integrating law in engineering ethics education in diverse institutional settings. The literature we reviewed is written in English and represents viewpoints from Europe and North America, which can also blind us to perspectives from other continents. We hope that the paper will spark debate and engage scholars from all parts of the globe. In this first section, we present our thoughts and our exploration of relevant literature. Then, we provide four examples – based mainly on our experiences – illustrating ways to include law within engineering ethics education.

Positioning law within EEE literature

In this section, we draw on unsystematic but intensive searches combining the concepts of law, legal, engineering ethics, and EEE, using the keywords 'engineering ethics education,' 'legal,' and 'law.' On Scopus in January 2024, 'engineering ethics education' AND 'law' generated nine hits, and 'engineering ethics education' AND 'legal' generated eight hits. We excluded papers that only mentioned either of the keywords in passing. In addition, we queried Google Scholar to find literature that may not have been indexed with the specific keywords. To our surprise, the discussion about law in engineering ethics education is quite scarce.

There are plenty of connections between law and ethics within EEE that are hidden in plain sight – in other words, where there is an implicit connection between ethics and law – which are rarely discussed in-depth. A superficial similarity is that the case-based method that is often used to illustrate ethical principles and problematize practices ultimately derives from the legal tradition (Giraudou et al., 2018; Rottmann & Reeve, 2020) (see Chapter 20 on case studies). It is also repeatedly stated in engineering ethics textbooks that engineering is not merely a job or a set of tasks but a well-defined *profession* that requires extensive training and implies specific responsibilities, in a way similar to the legal profession, although the arguments and comparisons between these two professions are without exception absent in textbooks (e.g., Lennerfors, 2019). Some engineering ethics educators despairingly recount how engineers who are neither well-versed in nor sympathetic toward engineering ethics education often conflate it with how to make engineers comply with the law. "Of course it is important that engineers follow the law" is an utterance that Taebi (2021, p. xi) reports hearing often. Another interesting connection between ethics and law involves how it is legally demanded within the Swedish context (where we, the authors, are active) that students develop skills to handle ethics. Through legislation, some common engineering practices have also become the normative way of performing engineering, for instance, documentation practices to ensure transparency. The ethical and legal implications of novel technological solutions are sometimes difficult to understand. Nevertheless, there is an expectation, and even a direct mandate, on engineers to deliver products that are 'legal by design' and 'ethical by design.'[2]

Weil's (1984) research argues that at its very birth, engineering ethics was primarily shaped by engineering and philosophy scholars, but other domains also contributed, including law, social, and management sciences. Apart from this mention, however, it is unclear how legal scholars participated in the shaping of the field.

Moving onwards, the only direct debate about the role of law in engineering ethics education that we have found is Davis' (2006) discussion with Zandvoort et al. (2000). In a quite passionate piece, Davis seems to have had a heated discussion with Zandvoort et al. (2000) about EEE in the United States versus EEE in Europe. Part of the argument from the European perspective seems to have been that the US focus is quite micro-ethical (focusing on the individual decision maker), while the European perspective is also harboring "the actual and possible role of law, organizations and procedures for collective decision-making" (p. 298). Zandvoort et al. (2000) argue that engineers need to understand the broader context for their own decision-making and be interested in helping to reshape that context if necessary. Because of this aim of EEE in the view of Zandvoort et al. (2000), the teaching needs to consider the role of law, as mentioned in the preceding quote.

Davis responds to the article, seemingly in affect, but also perplexity, as he cannot fully grasp how one could claim that the US tradition of engineering ethics does not pay attention to law. For Davis (2006), this is more a question of the amount of time dedicated to law. He writes:

> I agree that we should devote some time in engineering ethics to explaining the structure of large organizations and the legal constraints under which they must operate. … Students should also know something about product liability, the Federal Occupational Safety and Health Administration, patent law, and other ways in which government constrains what engineers can legally do. … But, again, the question is not whether we who teach engineering ethics should do something about such things. The question is how much we should do – or, rather, whether we are not already doing enough.
>
> *(Davis, 2006, pp. 227–228)*

The discussion did not continue (at least not in writing) after this exchange of opinions, which could have led to a state of openness regarding how law should be integrated, who is doing it and who is not, and how much should be integrated. Given that, to our knowledge, there is no current debate about these issues (which we judged from reading the papers that have cited the articles mentioned in this debate), we herein review the literature that we have found on the relationship between law and ethics within EEE. Often the argumentation in the texts is very brief, which has required us to make some assumptions when interpreting the literature.

Ethics and law as different domains

A common way to portray the relationship between law and ethics is through a Venn diagram, that is, with the domains having partly overlapping and partly exclusive areas. Although this understanding is mainly based on a textbook of business ethics (Treviño & Nelson, 2021), the non-complete overlap between ethics and law is also brought up in the textbook by van de Poel and Royakkers (2011) when they present a logical fallacy they call "confusion of ethics and law" (p. 128). Therefore, and in addition to conversations we had with some teachers in engineering ethics, we believe that this is a reoccurring way in which the relationship between ethics and law is conceptualized – also within engineering ethics. The idea is that there is a great overlap between what is legal and what is ethical; however, some practices can be illegal but ethical, while some others unethical but legal. Treviño and Nelson (2021) discuss how having an affair with someone who reports to you might be legal but considered unethical. Many practices in the 2008 financial crisis were also legal but considered unethical.

Often, responsibility is used to discuss differences between law and ethics. Van de Poel and Royakkers (2011) describe legal responsibility as liability. They map a few differences between legal and moral responsibility. One is that legal responsibility is established through law in different jurisdictions, while moral responsibility is theoretically grounded. Second, liability is established in an official and regulated process in court, while moral responsibility is established more informally. Third, liability often involves the obligation to repair damages, which is not always the case for moral responsibility. Fourth, liability is backward-looking, while responsibility can be both backward- and forward-looking. Another important difference that we may add here is that ethics, or moral responsibility, follows individuals regardless of where the person is. It is thus not limited to certain jurisdictions like law.

Law as the minimum requirement – ethics as going beyond the law

The basic understanding of Treviño and Nelson (2021) is that law reflects society's minimum requirements on business behavior, while ethics goes far beyond it. This is also the case when we study Carroll's (1991) pyramid of corporate social responsibility (CSR), which states that the company in the base has legal and economic obligations, on top of which there are ethical and philanthropic functions. This clearly positions law as the minimum requirement that needs to be transcended through ethics. Coeckelbergh (2006) similarly argues that legal regulation has some advantages for engineers as well as for wider society. As he puts it, "it provides a floor to performance: everybody has to reach a certain level" (p. 239). This also leads to the position that society knows what to expect and it makes behavior much more predictable, as we mentioned above. It also leads to certainty for engineers that they are actually meeting the regulations, and do not think about any other fuzzy dimension related to broader considerations of risk and public perceptions of their practice.

Law lags behind – ethics less so (differences of temporality)

Treviño and Nelson (2021) also discuss the fact that laws change (citing how racial discrimination was legal for a long time in the United States), which is also echoed in an engineering ethics textbook (Lennerfors, 2019). In Taebi (2021), as well as in Royakkers and van de Poel (2011), it is argued that laws are lagging behind technological development, which increases the need for ethical standards to ensure good technology within legally void spaces.[3] Interestingly, van de Poel and Royakkers (2011) indicate that corporations in which technological development takes place might be more equipped than governments to foretell ethical implications of new technology, and thus might have an ethical obligation to do so.

One should of course remember that ethical discussions also lag behind technological development, something that is often referred to as the 'cultural lag' (Ogburn, 1922), but in our interpretation of the existing literature, ethics seems to lag less behind than law.

Law as an insufficient motivating force

In a discussion about CSR, van de Poel and Royakkers (2011) write that laws might not have the same motivational power to ensure good behavior as ethics. We interpret this to also mean that a sole focus on the legal perspective could mean that everything that is legal is also ethical, which might even degenerate into a posture that 'everything that we can get away with from a legal perspective is OK.' Coeckelbergh (2006) similarly urges a need to move "away from regulation, control, and liability as legal and imposed responsibility and toward responsibility as a feeling within the individual as an autonomous professional" (p. 245). Lennerfors (2019) also raises the issue of enforcement, namely that many laws are not followed, which connects back to the motivational problems that laws might have. The fact that many laws related to engineering are not followed – at least in some areas, such as cybersecurity – is also backed up by empirical data (Selzer et al., 2021). For example, for small and medium size enterprises (SMEs), the risk of having to pay a fine based on some cybersecurity regulation is extremely low, while the cost for implementing a cybersecurity strategy is rather high. This has been regarded as one possible reason why many SMEs are not eager to invest in their cybersecurity. If law is considered without reference to ethics, it would seem rational to examine law through a cost–benefit analysis, which in many cases would lead to the conclusion that it makes more sense to pay the potential fines than to follow the law.

How law can be used for or against ethical values (applications of law)

Zandvoort (2005) argues that not all laws are conducive to responsible engineering practice. Rather, he argues that there are a range of aspects of legal systems that directly obstruct engineers' possibility to take responsibility. Mostly he discusses the incompatibility between employment laws and the responsibility that engineers have toward their employers, and how that can lead to clashes with ethical obligations as stated in engineering codes of conduct. This leads to practices such as whistle-blowing becoming salient.[4]

Coeckelbergh (2006), however, highlights some positive ways that laws can be used to take responsibility:

> Rather than being 'in the way,' regulation can help engineers in a variety of ways. For example, legal requirements may help engineers to resist managerial pressure and in this way support ethical behavior. If the engineer's claims for safety have to survive in a context

> dominated by competition for money and power, regulation with an ethical content may be the engineer's ethical life jacket.
>
> *(p. 250)*

At the same time, Treviño and Nelson (2021) argue that corporations can exploit differences in legal standards and enforcement in different countries (e.g., environmental pollution laws). This practice, similar to the phenomenon of ethics dumping – relocation of (research) practices to countries with laxer legal and ethical standards and subsequently importing back the results (Floridi, 2019) – is attributed to a lack of ethics, and reading the law to the letter. This can also concern engineering work, where differences in different jurisdictions may lead to enterprises relocating certain parts of technological development to places where laws might not be as strict.[5]

Law as socio-political and not only ethical

Although it is expected or assumed in some accounts that laws are based on ethical values, Lennerfors (2019) argues that laws are socially and politically constructed and that ethics might have a more complicated relationship to law. In other words, laws can result from political processes that often need to consider specific ethical values and disregard others. Furthermore, laws can be a result of power differences in society. However, while laws can be a result of unequal power positions, it is also argued that we can contribute to changing laws that do not cohere with our ethics, which is also what Zandvoort et al. (2000) argue when they state that engineering as a profession needs to contribute to legislative processes (for more on this topic, see Chapter 6).

Law and compliance

Since laws are often seen as the minimum requirement and much of the above literature discusses positive laws rather than the legal process, the discussion often turns to compliance, which resounds with the utterance from engineers mentioned by Taebi (2021) above. Within this stream of thought, ethics is thus about something more than compliance. However, also within the domain of EEE, ethics is sometimes seen as compliance, not in the research literature but in teaching practice as witnessed by Holsapple et al. (2012):

> I'm still struggling to find a better way, rather than threatening them, to get them to appreciate the importance, but I just can't come up with a better solution other than just sort of describing the worst case scenario and sort of motivating them to be honest.
>
> *(p. 178)*

Sometimes, ethics education and students' interpretation of it are very similar to law, focusing on codes of conduct that are expected to be complied with (cf. Sunderland, 2019).

Furthermore, Lennerfors (2019) argues that it is not so easy to follow laws – they must be interpreted in courts and everyday practice – and thus, there is a "need for a reflective, critical attitude rather than one of pure submission" (p. 22). However, there is no more argumentation about the need to see law as an interpretative endeavor.

The pedagogical nature of law

Lastly, Lennerfors (2019) presents the argument that law can teach us what is ethical. In other words, especially if there is a lack of critical thinking, people might learn part of what is ethical through learning what is illegal. This is yet another relationship that law and ethics could have, highlighting the need for critical reflexivity about the law.

Our assessment of the above

The above shows how scholars in engineering ethics think about law in relation to ethics education. Even though the relationship between law and ethics is not denied – even when it is adversarial – it is mainly left untouched, keeping a distinct border between these two domains. While we do not deny the difference between law and ethics as somewhat separate domains of knowledge, this does not mean that they must be kept separate in EEE. Instead, our central point is that ethics and law can mutually reinforce engineering ethics education.

For example, what is not presented in the literature on EEE is that both ethics and law are about developing arguments for and against specific claims. The sources that are legitimate to use are, admittedly, different. To develop an ethical argument, combining empirical material with any subset of ethical theories and even using logical reasoning based on examples to reach a conclusion is perfectly fine. The criterion for a strong argument is that it is difficult to refute, given agreement about context and premises. In law, the goal is the same: to find arguments that are difficult to refute. Yet the process is somewhat more formalized – even if not carved in stone.

Within law, the persuasiveness of an argument is often related to what is referred to as the hierarchy of legal sources.[6] This hierarchy may look different in different jurisdictions, but for the sake of this chapter, we may argue that the main source of law is the legislative text – simply put: *what do the written laws claim?* In many cases, the preparatory work that led to a piece of legislation is considered as an authoritative interpretation of the provisions of this legislation while precedents – the previous decisions of the highest court when defining the meaning of statutory provisions – find themselves also high up on this hierarchy. We may also find here what is called 'legal doctrine,' namely the corpus of legal research aiming to provide "a systematic exposition of the principles, rules and concepts governing a particular legal field or institution and analyzes the relationship between these principles, rules and concepts with a view to solving unclarities and gaps in the existing law" (Smits, 2017, p. 210). Of course any other argument – be it from within or outside the legal realm, for example, ethical or political considerations – can be used, since even if the hierarchy of legal rules can be defined, the doctrine of legal sources cannot, but the more we move down this hierarchy or outside of it, the probability for an argument to be accepted by lawyers is decreased – even if it is a good argument – in the case where another opposing argument comes from the higher levels of this hierarchy.

At first thought, it may appear strange that not all possible arguments have equal weight in court. You may even have read about cases that have created public outrage because the courts' decisions have violated common-sense judgment. However, in order to retain at least some kind of legal certainty,[7] in the sense that people should have the possibility to assess what the law expects from them – a value that is deeply founded in modern societies – a system of norms where 'anything goes' could be rather problematic.

Yet another point that can reinforce EEE regards the consequences of violating law versus ethics. Legal consequences are ideally known in advance, or, as it is commonly stated, they are prescribed by law. The exact consequences are, of course, decided in each case by a court or authority, but overall, the range of possible consequences is determined by the legislator. When it comes to ethics, on the other hand, no such delimitation is to be found. Instead, the public reacts to violations against ethical mores, and the punishment may range from mild disapproval to life-changing and life-lasting consequences, like exclusion from a social sphere.

Positioning law within EEE practice

Like the absence of debate around ethics and law, there is a marked absence of literature that describes how law and ethics are integrated into engineering teaching. Respondents stated legal elements as a prevalent means of teaching this in an empirical study regarding teaching practices related to fulfilling ethics-related outcomes conducted by Martin et al. (2020). Those authors did not report, however,[8] whether such legal perspectives were combined with ethics or taught standalone.

Furthermore, in a literature review of engineering ethics interventions in the US context conducted by Hess and Fore (2018), there is no mention of law. This could indicate that it is not prevalent in practice, that law is taught in courses that are not related to engineering ethics, or that law is taught together with engineering ethics but that no scientific investigations have been reported in the US-based literature regarding the effect of such integration on engineering ethics learning outcomes.

Therefore, we will provide examples of *how* and *why* to integrate law and ethics and will reflect upon their pros and cons. We present these examples (primarily drawn from our own experience) in order, based on their level of integration between the domains of law and ethics from less to more integrated.

Ethics and law as disconnected domains – the autonomous car trolley problem

This first example shows how ethics and law can play a role in engineering ethics education, but through temporarily suspending the other domain. This is based on an interpretation of a short glimpse of how the trolley problem is used in EEE, based on a chapter by Nyholm (2022). Nyholm explains that teachers frequently present the trolley problem during ethics classes. The students then sometimes ask whether "it wouldn't be the case that one would go to jail if one pushed a large person off a bridge to his death in order to save five people on the tracks, or even if one redirected a train onto a side track where one person is hit and killed by the trolley" (Nyholm, 2022, p. 222). Usually, philosophy teachers then argue that we should suspend those considerations and instead think about the trolley problem from a theoretical perspective, focusing on the best choice to make irrespective of the current legal regime.

It is easy to imagine that such an exercise could co-exist with legal discussions about real-world crashes with self-driving cars. These discussions would focus on liability and accountability for algorithmic decisions. Yet, despite a common framing of the decision problem of self-driving cars as a version of the trolley problem, few, if any, sources reviewed integrated ethical concerns into the legal discussion. Even after legal responsibility has been established, there are also concerns that the companies have not been appropriately held accountable, signifying the difference between law and ethics.

The point of this example – and we must say that we are not sure whether there is actually any course combining the autonomous car trolley problem with real-world crashes – is that ethics and law could be used to discuss the same scenario by suspending the connection between them temporarily and then re-integrating them.

Discussing and going beyond law from an ethical perspective – MedTech, ethics and law

In a course about medical technologies (MedTech), ethics, and law at Uppsala University, Lennerfors taught the ethics-related segments. He had limited interaction with the other teachers in the course, who taught legal perspectives, but he viewed their recorded lectures before his own lectures. Two lectures of the course were dedicated to ethics, and Lennerfors presented some of the above-mentioned relationships between ethics and law, bringing them closer to the students' technical domain. The Poly Implant Prothèse (PIP) scandal can be seen as a situation where the legal perspectives were not enough, highlighting how laws can be reinterpreted based on our view of (failing) patent safety. Lennerfors mentioned ethical ramifications of law, for example, the Medical Device Regulation (MDR), which has been criticized for imposing increased administration that pushed smaller actors out of the market, contributing to a concentration of power. He explored the laws and regulations studied in the course in addition to MDR, including the International Organization for Standardization (ISO) standards for MedTech products and what such sources say about ethics. Concerning regulations, Lennerfors situated ethics as also including the more positive side of doing good, while law was positioned as avoiding the bad. Within this domain, health technology assessment (HTA) often goes beyond what is legally mandated. However, HTA has also been criticized for taking a perspective that prioritizes economics too much (Hofmann, 2020) to the exclusion of ethical issues. Therefore, ethics need to be studied, and the rest of the lectures were based on the framework of ethics in engineering presented in Lennerfors (2019).

In reflection, the teachers of this course noted the lack of integration among themselves, which was limited to a short initial meeting. The ethical interpretation of the laws and regulations studied in the course was made by the ethics teacher alone; he was not in class with the other teachers. This approach, where law is included in an ethics discussion conducted by an ethicist, is a positive starting point. It is probably one of the most resource-efficient ways to introduce law in engineering ethics education. However, the ethicists' lack of legal expertise limits the discussion quality. As we pointed out in the beginning, law as a domain of knowledge is not only about what rules exist, but it also contains questions of how these rules came to be, what their purpose is, how they are seen within the system of law, what processes can change their understanding, what interpretation has been given, what alternatives exist, and so on. Even though an ethicist could speculate about some of these questions, the lack of a legal expert can lead to somewhat arbitrary discussions without connection to the realities of the legal processes in question. Now, these lectures aimed not to integrate ethics and law but to present ethics as another perspective apart from the other teachers' legal perspective. However, the ethics teacher still felt the need to engage with what the legal experts had said to show what concerns ethics brings up that might not appear in discussions about law.

Ethics and law as parallel perspectives – sustainability in global industrial companies

Another example of how law and ethics can be used together will be given from the industrial engineering and management domain, namely the course 'Managing Sustainability in Global Industrial Companies.' Although this might seem different from engineering ethics education, most of the students were engineers pursuing a Master's degree in management and innovation studies. Another puzzling idea might be that the course was about sustainability, not ethics. However, within the business domain, much of what was earlier called CSR or business ethics is increasingly going under the concept of corporate sustainability, and scholars have called for an integration of the discussions in these two related domains (Lennerfors & Murata, 2023). This course has already been described in previous literature (Fors et al., 2023; Giraudou et al., 2018; Lennerfors et al., 2020), and what follows are Lennerfors' reflections on the 2018 round of the course, after which another teacher took over as the primary responsible teacher.

A collaboration with a legal scholar in Japan, who co-taught the course, was also initiated within this course. With this legal scholar's presence, the course aimed to include a legal approach to the corporate sustainability perspective foundational to the course. In contrast to the MedTech course, the law teacher was part of the teaching and a recurring part of the course, providing legal perspectives. The students were presented with cases where they needed to discuss the ethical and legal perspectives of corporate responsibility. The legal scholar was there to discuss the legal aspects, which meant that the teachers were working side by side, although they were not always engaging in discussions in front of the students about the differences between ethics and law, nor directly encouraging students to reflect on such differences critically. There were discussions amongst teachers before or after classes about the differences between ethics and law, reflecting on both subject areas. The initial reaction from the corporate responsibility teachers was that the discussion about corporate responsibility within legal studies seemed to be markedly similar, and the teachers indicated that the space between ethics and law was much smaller than the abyss they initially perceived.

This second approach, where law and ethics are examined in parallel by ethicists and lawyers together, seems to provide a better understanding of the relationship between law and ethics. It starts making apparent that law and ethics are not as distant from each other as many prominent narratives seem to point out. Additionally, the participation of a legal scholar helped the other teachers and the students understand what the law asks from corporations concerning CSR – not as a mere matter of compliance but more as a starting point for the ethical discussion that can be built upon this knowledge.

As we will show in the next part, however, this approach is still somewhat limited in that it does not integrate law into ethics but merely regards them as parallel connected domains. This course did not include a deep discussion of how law is formed. There was also a clear division of labor between the teachers, which could have contributed to sedimenting this separation between ethics and law.

Integration of law and ethics in EEE – the case of data, ethics, and law

As our final example, we present the course we (authors) teach together, called 'Data, Ethics and Law,' which aims to present law and ethics as mutually dependent matters of interest that can be approached deliberatively. Instead of providing students with mere information related to codes of ethics, ethical misconduct, rules, and breaches, the teachers ask the students to reflect on the *function* of ethics and law in society – as the two main societal mechanisms for attributing responsibility – and to develop an understanding of why ethical norms arise and why laws are created, formulated, and interpreted the way they are.

The course includes a dozen lectures on ethics and law topics relevant to engineers and computer scientists working with data in any form within their future professions. The lectures are complemented by seminars in which students discuss selected readings on these topics. To encourage deliberation rather than blind acceptance, the topics are introduced through common phenomena in data processing and computing rather than as moral and legal obligations. The teachers' central policy has been not to dismiss students' ethical positions and instead encourage students to inquire why they have come to consider something as ethical or unethical and why they believe that something is deemed to be legal or illegal. Through this approach, the students learn to see that both ethical sentiments and laws are social in the sense of steering society in a direction that is considered good for its members.

It is not the point of this section to go through all seminars of the course, which vary from issues related to intellectual property and access to information to the trustworthiness of AI. Still, we will provide some insights into how we approach teaching about data protection and the General Data Protection Regulation (GDPR) – very important matters for most computer and data engineers and scientists.

GDPR is one of the most important laws worldwide regulating the processing of personal data. Non-compliance can lead – and has led – to extremely high fines. Therefore, as a data engineer, knowledge of this law is definitely of added value. *Yet, how can we claim that a few lectures on the GDPR can lead to knowledge of data protection law if we only focus on the provisions of this legislation? Moreover, how is it even possible for engineers (who have probably never before worked with legal texts or used legal methods) to be able to understand what is expected of them from a text that is so large and complicated that even lawyers have difficulties navigating it (i.e., the 99 articles, 173 preambles, numerous guidelines by European and Member States authorities, decisions by the Court of Justice of the European Union (EU), national courts, and data protection authorities, to name a few)? Is it adequate to provide slides, tables, and so on with simplified interpretations of this law, to say that these students have at least some knowledge on the matter?* Suppose the point is to examine whether students can answer simplistic questions regarding specific data protection legal rules. In that case, one may argue that, yes, this is adequate. However, this is an elementary approach, especially in engineering ethics, when the goal is to prompt students to reflect upon their future actions. In such a context, the main reason for introducing law should not be about following a blueprint for being compliant but rather about being able to add something to the discussion about how engineers should behave.

The idea in this course is, therefore, to provide students with some basic information on what is contained in this piece of law, but, most importantly, encourage them to understand what this law is aiming to achieve, what processes affect its understanding – and its creation – and how its goals may change over time. The main point of our course is not just to provide a static picture of what the law *is* – something that even legal scholars may have difficulties with – but to have students reflect on how their behavior constitutes an act of interpretation of a piece of legislation. The idea is to have students realize that they are not merely passive subjects of law but part of a society that constantly affects how the law should and should not be understood.

This integrated approach seems to fit in better with a new trend in EU law concerning regulating complex systems (e.g., areas related to regulating AI, achieving sustainability, and the like), where ethics is understood as an integral part of the legislative process, meaning that ethics both affect the justification of why there is a need to regulate specific phenomena and inform the desired behavior of the actors that are – or are to be – regulated. The latest example of such considerations is probably the proposed AI Act in the EU, which is based on a text named 'Ethics Guidelines for Trustworthy AI,' and in parallel, asks the actors involved in the development and distribution of AI systems to do so by following 'ethical standards.'[9]

The literature reviewed in the previous part of this chapter has shown that law and ethics interconnect. However, this interconnection can no longer be understood only in a common-sensical way, namely that law and ethics influence each other – one may ask what is not influenced by ethics – but, instead, that law and ethics seem, now more than ever, to be connected institutionally. Law aims at the creation of ethical technologies, and at the same time, the discourse in ethics directly affects the official legislative and regulatory processes and outcomes. Thus, ethics is regarded as a tool for interpretation and regulation.

Therefore, by integrating law into ethics, it is possible a) to make apparent to students that when law becomes part of an ethical analysis, what 'society' thinks becomes an integrated part of the reasoning; b) to strengthen the students' understanding of general ethical issues; c) to show that legislation and law are generally affected by underlying ethical considerations (e.g., the importance of privacy); and d) to promote the ideas of law as 'normative ethics' and law as a social dynamic that is corresponding to the needs of society as well as the idea that existing laws are not mere 'frozen politics' but a result of historical social needs that are still relevant to varying degrees. The main point is precisely to make apparent that law is not mere compliance, something separate from ethics, but that instead, it is an integral part of socio-ethical considerations.

Of course, the approach has some challenges, especially related to misconceptions about legal methodology or legal argumentation. Students may erroneously believe, for example, that a piece of law such as the GDPR is applied to all cases related to data processing, even if no personal data are included, or that ethical argumentation (without being supported by legal argumentation) can be used in courts and so on. Such issues may nevertheless appear when law is not taught in an extensive and in-depth manner. More importantly, these are problems that can be addressed through discussions in the classroom. Ultimately, when the question is how to help students develop into thoughtful professionals, a discussion of ethics that integrates law can only enrich ethics discourse and enable students to have legal discussions, even with limited legal knowledge.

This thoughtfulness is also depicted in an outcome that we often witness in this course: even when students disagree with a specific piece of legislation and how it has been interpreted and applied until now, they nevertheless seem to adopt a more empathetic view towards law in general when they start regarding law as a process of discussion in which ethical motives and trade-offs between values may have led to a particular understanding of law becoming the prevailing one. When law is seen as a discussion process, it is easy to imagine that alternative understandings can become dominant in the future, depending on the public discourse, and that the students *can* actively participate in shaping these understandings. Our goal is precisely to equip the students for this epiphany.

Concluding remarks

In this chapter, we have aimed to explore how the relationship between ethics and law is constructed within engineering ethics education and some possible ways to connect ethics and law in engineering ethics education. Our central point has been that ethics and law can be mutually reinforcing in engineering ethics education, particularly when law is seen as a fundamentally interpretative endeavor. This implies that the goal of teaching law in the context of engineering ethics education is not to promote the perception of law as a matter of mere compliance but to demystify what 'applying the law' means and critically reflect on the processes that form law and its meaning (i.e., *why* law is created, interpreted, and applied the way it is), which are fundamentally ethical issues.

To illustrate how to integrate law within engineering ethics education, we have provided four examples. Although our argument has favored the most integrated approach (the one presented last), we also see benefits in connecting ethics and law in other ways. Ultimately, we acknowledge that the combination of ethics and law depends on the bricolage of available competencies and resources at a given institution. We hope this chapter will stimulate and re-ignite a discussion about the role of law within EEE.

Acknowledgments

The authors are grateful to the anonymous reviewers as well as the editors, Roland Tormey and Shannon Chance, for their valuable comments and suggestions. Special thanks are extended to the

Knut and Alice Wallenberg Foundation for funding Andreas' and Mikael's contributions within the DDLS – WASP-HS project 'Trustworthy AI-based Medical Treatment.'

Notes

1 For an overview see Mark C. Murphy, Philosophy of Law: The Fundamentals (2006) Wiley-Blackwell with references that range from legal positivists that negate morality within the system of law to natural law theorists that aim to define the universal moral standards in law.
2 For a discussion on legal by design see Hildebrandt (2020) and on ethics by design European Commission (2021).
3 For some critique of this, see Aspray and Doty (2023).
4 At the same time we should not forget that at least some jurisdictions have lately introduced legislation that aims at protecting whistleblowers, see Directive (EU) 2019/1937 of the European Parliament and of the Council of 23 October 2019 on the protection of persons who report breaches of Union law.
5 On forum shopping in relation to data protection see for example Brandão (2023).
6 For a discussion on the so-called hierarchy of law and the doctrine of legal sources see Samuelsson (2012).
7 See however the Critical Legal Studies movement and the idea of legal indeterminacy (Fenwick & Wrbka, 2016).
8 In the paper, there are no insights into this, which can be because respondents did not provide such details or because the authors chose to focus on other issues despite respondents providing such details.
9 More specifically, since the end of 2000's, with the introduction of the Charter of Fundamental Rights and especially after the discussions that initiated the GDPR, the rhetoric of ethics has become increasingly more apparent. Ethical values are to affect the production of legislation as well as its interpretation and application. For example the Commission set up the High-Level Expert Group that published the so-called "Ethics Guidelines for Trustworthy AI" which in turn affected the proposed AI Act, one of the main legislation for the regulation of AI in the EU. Similarly, in the proposal of the AI Act we find that the goal of this piece of law is to lead to – among other things – ethical AI. Such a *telos* has to be taken into consideration when we try to interpret and apply the AI Act in the future – or for what is worth any piece of law regulating aspects of AI systems.

References

Aspray, W., & Doty, P. (2023). Does technology really outpace policy, and does it matter? A primer for technical experts and others. *Journal of the Association for Information Science and Technology, 74*, 885–904.

Brandão, D. M. (2023), The one-stop-shop and the European data protection board's role in combatting data supervision forum shopping. *International Data Privacy Law, 13*, 313–330.

Carroll, A. B. (1991). The pyramid of corporate social responsibility: Toward the moral management of organizational stakeholders. *Business Horizons, 34*, 39–48.

Coeckelbergh, M. (2006). Regulation or responsibility? Autonomy, moral imagination, and engineering. *Science, Technology, & Human Values, 31*, 237–260.

Davis, M. (2006). Engineering ethics, individuals, and organizations. *Science and Engineering Ethics, 12*, 223–231.

European Commission. (2021). *Ethics by design and ethics of use approaches for artificial intelligence*, version 1.0. European Commission.

Fenwick, M., & Wrbka, S. (2016). The shifting meaning of legal certainty. In M. Fenwick & S. Wrbka (Eds.), *Legal certainty in a contemporary context* (pp. 1–6). Springer.

Floridi, L. (2019). Translating principles into practices of digital ethics: Five risks of being unethical. *Philosophy & Technology, 32*, 185–193.

Fors, P., Lennerfors, T. T., & Woodward, J. (2023). Case hacks in action: Examples from a case study on green chemistry in education for sustainable development. *Digital Chemical Engineering, 9*, 100129.

Giraudou, I. J., Lennerfors, T. T., & Woodward, J. R. (2018). Shouldn't we expect more from case-based learning? The transformative potential of multidisciplinary frameworks in sustainability education. *Transformative Dialogues: Teaching and Learning Journal, 11*, 1–13.

Hess, J. L., & Fore, G. (2018). A systematic literature review of US engineering ethics interventions. *Science and Engineering Ethics, 24*, 551–583.

Hildebrandt, M. (2020). *Law for computer scientists and other folk.* Oxford University Press.

Hofmann, B. M. (2020). Why ethics should be part of health technology assessment. In A. L. Caplan & B. Parent (Eds.), *The ethical challenges of emerging medical technologies* (pp. 33–39). Routledge.

Holsapple, M. A., Carpenter, D. D., Sutkus, J. A., Finelli, C. J., & Harding, T. S. (2012). Framing faculty and student discrepancies in engineering ethics education delivery. *Journal of Engineering Education, 101*, 169–186.

Latour, B. (2004). Why has critique run out of steam? From matters of fact to matters of concern. *Critical Inquiry, 30*, 225–248.

Lennerfors, T. T. (2019). *Ethics in engineering.* Studentlitteratur.

Lennerfors, T. T., Fors, P., & Woodward, J. R. (2020). Case hacks: Four hacks for promoting critical thinking in case-based management education for sustainable development. *Högre Utbildning, 10*, 1–15.

Lennerfors, T. T., & Murata, K. (2023). Ethics and sustainability in digital cultures: A prolegomena. In T. T. Lennerfors, K. Murata (Eds.), *Ethics and sustainability in digital cultures* (pp. 1–16). Routledge.

Martin, D. A., Conlon, E., & Bowe, B. (2020). Exploring the curricular content of engineering ethics education in Ireland. In *2020 IFEES world engineering education forum-global engineering deans council (WEEF-GEDC)* (pp. 1–5). IEEE.

Murphy, M. C. (2006). *Philosophy of law: The fundamentals.* John Wiley and Sons Ltd.

Nyholm, S. (2022). Ethical accident algorithms for autonomous vehicles and the trolley problem. In H. Lillehammer (Ed.), *The trolley problem* (pp. 211–230). Cambridge University Press.

Ogburn, W. F. (1922). *Social change with respect to culture and original nature.* BW Huebsch, Incorporated.

Rottmann, C., & Reeve, D. (2020). Equity as rebar: Bridging the micro/macro divide in engineering ethics education. *Canadian Journal of Science, Mathematics and Technology Education, 20*, 146–165.

Samuelsson J. (2012). *Technology and the doctrine of legal sources* (Working Paper 2012:4). Uppsala University

Selzer, A., Woods, D., & Böhme, R. (2021). Practitioners' corner. An economic analysis of appropriateness under article 32 GDPR. *European Data Protection Law Review, 7*, 456–470.

Smits, J. (2017). What is legal doctrine?: On the aims and methods of legal-dogmatic research. In R. Van Gestel, H. Micklitz, & E. Rubin (Eds.), *Rethinking legal scholarship: A transatlantic dialogue* (pp. 207–228). Cambridge University Press.

Sunderland, M. E. (2019). Using student engagement to relocate ethics to the core of the engineering curriculum. *Science and Engineering Ethics, 25*, 1771–1788.

Taebi, B. (2021). *Ethics and engineering: An introduction.* Cambridge University Press.

Treviño, L. K., & Nelson, A. (2021). *Managing business ethics: Straight talk about how to do it right.* John Wiley & Sons.

van de Poel, I., & Royakkers, L. (2011). *Ethics, technology, and engineering: An introduction.* John Wiley & Sons.

Weil, V. (1984). The rise of engineering ethics. *Technology in Society, 6*, 341–345.

Zandvoort, H. (2005). Good engineers need good laws. *European Journal of Engineering Education, 30*, 21–36.

Zandvoort, H., Van de Poel, I., & Brumsen, M. (2000). Ethics in the engineering curricula: Topics, trends and challenges for the future. *European Journal of Engineering Education, 25*, 291–302.

SECTION 3

Ethical issues in different engineering disciplines

Tom Børsen

Teaching ethics to engineering students often occurs in classes that enroll students from many engineering disciplines. The content in such interdisciplinary engineering classes is general and often not perceived as relevant for students specialized in particular engineering disciplines. There are, as we will see in the chapters in this section, issues in engineering ethics that are transversal to most, if not all, engineering disciplines and relevant for most, if not all, engineering students. This section of the handbook investigates how transversal ethical issues can be linked to specific engineering disciplines and, in this way, be perceived as relevant by specialized engineering students. We, the editors of this handbook, selected five engineering disciplines – civil engineering, aerospace and mechanical engineering, electrical and electronic engineering, chemical engineering, and software engineering – and invited prominent scholars of each discipline to author chapters where they identify, analyze, and discuss the ethical issues that they perceive relevant for students of their specific engineering area to become acquainted with during their education. Thus, this section contains five chapters coining ethical content of engineering ethics teaching in civil, mechanical, and aerospace, chemical, software, and electrical and electronic engineering.

Positionality

As a way to understand why this section appears as it does, it might be useful to make explicit the positionality of the section editor. Tom Børsen was originally educated as a chemist with a minor in philosophy, who after graduation engaged in research in science and engineering education with a focus on how to promote ethical and socio-ecological responsibility among technical experts. This led to several curriculum development projects bridging social science and humanities (SSH) and science, technology, engineering, and mathematics (STEM) disciplines. Most predominantly, Tom led the establishment of the BSc and MSc degrees in Techno-Anthropology at Aalborg University. He has also taught ethics at different engineering programs. Thus, after a process of self-reflection, Tom has identified two ways in which his positionality can have influenced the content of this theme: He has experienced that the proximity between engineering ethics content and the engineering discipline in which it is situated is imperative for engineering students' acceptance and sense-making of EEE. The section shows that Tom is not the only one who has had this experience. The second way the section editor's positionality may have had implications

DOI: 10.4324/9781003464259-17

for the section is reflected in his engagement in promoting engineers and other technical experts to take responsibility for the broader implications of their work. This concern aligns well with how the chapters in the section are conceptualized.

Chapter topics

Here we provide a synopsis of each chapter in this section, highlighting the unique contributions of each.

Chapter 14, 'Ethical Considerations in Civil Engineering' by Irene Josa, Ester Gimenez-Carbo, and Christina Nick, overviews ethical issues inherent in civil engineering, a field vital for providing essential infrastructure such as water access, shelter, transportation, and communication. While distinct from military engineering, this branch of engineering often intersects with it, particularly in areas like transport infrastructures. The chapter underscores the global trend towards a more holistic and socially conscious approach to civil engineering, embraced by professional associations worldwide. Furthermore, it emphasizes the need for civil engineers to excel in technical skills and *also* demonstrate leadership and ethical decision-making abilities, navigating complex dilemmas by considering social, cultural, and philosophical dimensions. The authors stress the importance of a comprehensive perspective that encompasses the entire life cycle of infrastructure projects, from design and planning to decommissioning, in determining the ethical responsibilities of individual professionals.

The authors present a framework and set of tools (e.g., applicable case studies) to identify content for ethics education suggested to be delivered to civil engineering students (Table 14.1). They present ethical issues in civil engineering – professionalism, social responsibility, sustainability, health and safety, EDI (equality, diversity, and inclusion), and decolonization – and relate them to each stage of the life cycle of civil engineering projects. The chapter showcases a range of existing case studies and discussion topics for teachers wanting to integrate ethics teaching into their civil engineering courses.

The authors of Chapter 15, 'Ethical Issues in Mechanical and Aerospace Engineering,' Aaron W. Johnson, Corin L. Bowen, Cristiano Cordeiro Cruz, and Renato Bezerra Rodrigues, examine ethical issues within mechanical and aerospace engineering, encompassing the application of scientific and mathematical principles to develop mechanical systems, including aircraft and spacecraft. The chapter adopts a 'social justice' lens, viewing mechanical and aerospace engineering as potential avenues for empowering marginalized communities.

A vital aspect of this chapter is its deliberate utilization of a critical, emancipatory framework inspired by activist scholars (Freire, 1970; hooks, 1994). Departing from the traditional portrayal of mechanical and aerospace engineering as politically neutral domains, the authors challenge the prevalent notion by advocating for a shift towards macro-ethical considerations alongside micro-ethics. They argue for cultivating students' critical consciousness, emphasizing the importance of addressing systemic oppression perpetuated by engineering practices. While confronting mainstream perceptions of engineering ethics, this approach encourages readers to engage with discomfort, explore relevant literature, and get involved in discussions to broaden their understanding of these perspectives.

In Chapter 16, 'Ethical Issues in Electronic and Electrical Engineering,' Susan M. Lord and John E. Mitchell delve into the ethical dimensions within electronic and electrical engineering (E&EE), emphasizing the crucial alignment of ethical education with technical content. They address the scarcity of case studies specific to E&EE contexts, attributing it to the nature of E&EE components being often integrated into larger engineered systems and less visibly public-facing compared to other engineering disciplines. The investigation by Lord and Mitchell highlights a recurring trend whereby electrical technology remains peripheral in traditional engineering ethics case studies, which lack focus on E&EE contexts. These case studies often drift outside the discipline or address irrelevant professional issues. The authors advocate for explicitly integrating ethics relevant to E&EE into technical studies and emphasize the importance of engineers teaching these topics. They explore various approaches in ethics education for E&EE students, ranging from discussions on professional responsibility to standalone ethics courses and integration across modules in a course.

Additionally, Lord and Mitchell examine the challenges of teaching ethics in E&EE, identifying opportunities to address ethical issues within the culture of electronic and electrical engineering. This includes scrutinizing offensive technical jargon and controversial imagery like the Lena photograph used in image processing. By identifying and confronting these persistent ethical challenges, the authors advocate for proactive measures to assist students in navigating the intricate ethical landscape of electronic and electrical engineering.

Chemical engineering encompasses the vast domain of large-scale chemical synthesis, production, transportation, storage, and diverse industrial and consumer product applications. In Chapter 17, 'Ethics in Chemical Engineering,' Jan Mehlich, Tom Børsen, and Dayoung Kim draw upon historical and contemporary ethical case studies from research journals to outline ethical issues. They emphasize using varied case studies in ethics education to foster different dimensions of responsibility among chemical engineering students.

Mehlich, Børsen, and Kim probe ethical issues pertinent to chemical engineering education, exploring concerns about the unpredictable nature of chemicals and the potential misuse of chemical engineering processes. The chapter advocates for reflections on the broader implications of chemical engineering projects as a fundamental aspect of ethical practice in chemical engineering. By examining moral, legal, and institutional responsibilities, as well as individual and collective responsibilities, the authors provide valuable insights for educators designing ethics education curricula in chemical engineering. They propose linking case studies with various forms of responsibility to enhance the effectiveness of teaching materials and encourage critical discourse among students.

In Chapter 18, 'Ethical Issues in Software Engineering,' Stephanie J. Lunn, Isis Hazewindus, Prajish Prasad, and Vivek Ramachandran aim to cultivate ethical awareness among students, teachers, and software engineers. They stress the importance of recognizing responsibilities in the creation, dissemination, and maintenance of software – emphasizing the need for educational environments that foster ethical learning alongside technical skills – and for continually reassessing ethics and ensuring accountability and transparency as technologies like artificial intelligence and machine learning dynamically evolve. Embracing 'Ethics by Design,' they propose tools, discussions, and assignments to promote ethical reflection and aid in problem-solving.

In exploring ethical issues in software engineering, the authors advocate for cultivating ethical mindsets throughout the software development and data life cycles. Linking these two cycles into a brand new 'ETHOS' model for facilitating ethical reflection, the authors provide a tool that can help scaffold dialogue and help people analyze ethical issues in software engineering practice as well as in engineering ethics classes.

Trends and implications

Transversal ethical issues are exemplified within the concrete discipline, as the observant reader of all five chapters in this section will notice, and there are overlaps among the different ethical issues perceived as relevant by the authors of the chapters. This is not to say that all ethical issues are present in all chapters, because they are not; however, most issues are identified as central in three or four chapters. Few, if any, ethical issues are limited to only one engineering discipline.

Professional codes of ethics are one common ethical topic described in several chapters. Yet, authors across this section argue that codes cannot be the only content of engineering ethics education – additional topics must be included. Chapter 16 on electrical and electronic engineering underscores that students need to know the ethical codes of their engineering discipline *and* also be able to relate them to their own projects and technical work. Many engineering companies have business ethical schemes that students can apply in parallel to the professional ethical codes of conduct, for example, during an internship.

The broader implications of engineering work within different non-technical or socio-ecological spheres (e.g., the environment, human health, users' privacy, employees' well-being, vulnerable groups, and society as a whole) constitute another transversal ethical issue that appears in several chapters and is considered relevant in many engineering disciplines. Chapter 14 on ethical issues in civil engineering names this ethical topic 'social responsibility.' Socio-ecological responsibility is an extended predicate one can put on this transversal ethical issue. Chapter 17 on chemical engineering bridges this ethical issue with the previous one by suggesting that engineering ethical codes must include macro-ethical concerns regarding, for example, environmental and human health implications. Chapter 18 on ethical issues in software engineering encourages engineering solutions to actively strive for improvements in these non-technical spheres, bridging the following ethical issues.

Sometimes, harmful effects on workers, technology users, or local communities are or should have been known and could have been prevented. Sometimes, the implications for humans are unforeseen and emerge due to complex systems that are difficult to control. Chapter 17 on chemical engineering asserts this distinction as central for taking responsibility for the implications. Engineers and company managers must be held responsible for their neglect if harmful implications were or could have been known (e.g., were reasonably foreseeable). When engineering design, solutions, and innovation are new and emerging, and implications uncertain and difficult to predict, engineers and other actors can and should exercise precaution. Chapter 18 on software engineering contains scenarios of both intentional misuse and unforeseen implications. Chapter 15 on mechanical and aerospace engineering argues that the harmful implications often follow from hegemonic power structures. One way to gain knowledge about the human implications of technology is to involve those affected by engineering solutions and design. Engineering that engages employees and users is denoted 'human-centric engineering.'

Sustainable design is a transversal ethical issue related to the issues identified above, focused on the dual influences of engineering activities on the environment: engineering can contribute to sustainable development, but it can also pollute the environment. Chapter 14 on civil engineering argues that engineering projects must actively strive to minimize the use of resources and emissions of greenhouse gasses. Chapter 16 on electrical and electronic engineering suggests incorporating environmental concerns in technical design. Chapter 15 on mechanical and aerospace engineering warns that developing 'green' products (in a narrow sense) is not sustainable if the products enable or encourage unsustainable practices. The difference between this ethical issue and the two previous ones is that socio-ecological responsibility focuses on identifying the implications, human-centric engineering focuses on co-creation, and sustainable design creates socio-ecological beneficial consequences. Several chapters point toward cases where engineers and engineering solutions have had very positive implications for human communities; they assert that engineers in technical design should actively seek beneficial implications.

Some chapters mention military applications as an ethical issue to be discussed and reflected upon in engineering ethics education. Chapter 15 on mechanical and aerospace engineering draws attention to the military-industrial complex and notes that military equipment, which has been used in armed conflicts around the world, kills innocent civilians in high numbers. Chapter 17 on chemical engineering includes cases from World War I and the Vietnam War.

Social justice is the final transversal ethical issue highlighted in this section's introduction. It covers an array of topics: Chapter 14 on civil engineering discusses both 'equality, diversity and inclusion' and 'decolonization' as topics to be included in engineering ethics education. Chapter 16 on electrical and electronic engineering provides examples from the academic and professional engineering culture (such as terminology and jargon) that can adversely reduce the sense of belonging among women and students of color in engineering. Engineers' use of an image of Lena Forsén in image processing, an image originating from the sexist context of *Playboy* magazine, is one of several examples the authors critique and pose for in-class and professional dialogue.

None of the chapters in this section suggest that students must be able to provide clear-cut answers to ethical challenges by being acquainted with some of the ethical issues presented in this section's chapters. It is essential to dialogue, reflect, and act on these issues when they appear in students' disciplinary and cross-disciplinary contexts. Chapter 18 on software engineering includes a sophisticated reservoir of dialogical methods and reflection tools that can be applied in other engineering disciplines. We find connectivity among the different ethical issues presented by the authors of this section: discussions, reflections, and analyses should be part of engineering ethics education as they can probe the various issues and identify the relationships among the different ethical topics.

The section highlights some general aspects of EEE, some of which are discussed in this section of the handbook, and others in previous sections. The general ethical issues must be translated into discipline-specific cases, questions, exercises, and design projects. Thus, we find relationships between this section and the previous two ones in the way that the first section highlights the importance of evoking normative frameworks in disciplinary-specific EEE content, and the second section emphasizes that thorough descriptions of the emotional, social, organizational, cultural, and legal context must be included as parts of disciplinary-specific EEE content. Such an affinity to the concrete sociotechnical context is a central way to root the more general normative frameworks and transversal ethical issues in specific engineering disciplines. After this section has reflected on the content of disciplinary EEE, one next step can be to identify appropriate teaching and assessment methods, which are the topics addressed in the following sections.

Conclusions from the section editor

As the editor responsible for leading this section, Tom Børsen expects that the selection of chapters will also be relevant to read for engineering ethics teachers, students, practitioners, and researchers from other disciplines than those directly addressed in the chapters. They are not intended to only inspire those directly involved in the discipline discussed. A common denominator in all the chapters is a suggested balance between – on the one hand – disciplinary exercises, examples, case studies, and scenarios deeply rooted within the specific discipline and – on the other hand – more general or transversal ethical issues such as professional and socio-ecological responsibilities, human-centric engineering, sustainable design, military applications, and social justice. Our rationale is that the transversal ethical issues identified in these chapters can be translated into engineering disciplines other than those from which they originate. Moreover, the chapters outline the importance of translating transversal ethical issues into specific engineering disciplines, rather than presenting ethical issues to students in a generic or general engineering context. The disciplinary proximity of engineering ethics education comes through disciplinary exercises, examples, cases, scenarios, and designs that illustrate more general engineering ethical issues but in ways that clearly link to the student's chosen discipline of focus.

Tom hopes readers will enjoy assessing how the above briefly presented transversal ethical issues – including professional codes of ethics, socio-ecological responsibility, human-centered engineering, sustainable design, military applications, and social justice – unfold in specific disciplinary contexts. That said, Tom and the larger editorial team recognize that the section does not provide a final and complete overview of ethical issues in different engineering disciplines; therefore, they welcome readers' comments on the following chapters and encourage additional research in the realm of EEE for specific engineering disciplines.

References

Freire, P. (1970). *Pedagogy of the oppressed.* Continuum.
hooks, b. (1994). *Teaching to transgress: Education as the practice of freedom.* Routledge.

14

ETHICAL CONSIDERATIONS IN CIVIL ENGINEERING

Irene Josa, Ester Gimenez-Carbo, and Christina Nick

Introduction

Civil engineering is the oldest branch of engineering, with civil engineering practices dating back to ancient times, with the establishment of the earliest urban settlements. The roads, canals, viaducts, and other public works developed by the Roman Empire are still visible today. Other notable examples of ancient civil engineering were constructed in Egypt, Greece, Mesopotamia, China, the Inca Empire, and Persia – thus illustrating the global impact and diversity of ancient civil engineering achievements. This predates the formal establishment of professional societies, which emerged later to organize and advance various engineering disciplines. Civil engineering was founded to serve and benefit society. Since its inception, the civil engineering profession has supported the development of societies and improved people's quality of life by providing access to basic needs such as water, shelter, transport, and communications. Over time, civil engineers began to formalize the scope, definition, and responsibilities under their remit, and today, professional bodies specific to civil and structural engineering practices proliferate worldwide. They include, for example, the US-based American Society for Civil Engineering (ASCE), the UK-based Institution of Civil Engineers (ICE), the Japan Society of Civil Engineers (JSCE), the International Association of Structural Engineers (IABSE), and the UK-based Institution of Structural Engineers (IStructE). Many subfields of civil engineering host professional organizations as well, such as the International Society for Soil Mechanics and Geotechnical Engineering (ISSMGE), the American Concrete Institute (ACI), and the Society of Fire Protection Engineers (SFPE).

In recent years, increasing attention has been paid to ethical issues within this profession – see, for instance, the 'vision of the 2025 engineer' by one of current society's most active professional engineering bodies, the ASCE (ASCE Steering Committee, 2006). Civil engineers' knowledge, skills, and work have great potential for enhancing humanity's quality of life and solving some of today's most pressing global challenges (Koehn, 1991). Nonetheless, undesired outcomes can arise if citizens' health, safety, and welfare are not placed at the center of civil engineers' work. Several controversial issues related to infrastructure have filled the news over the past decades, including corruption cases and infrastructure collapses.

DOI: 10.4324/9781003464259-18

To avoid adverse societal and environmental effects resulting from civil engineers' work, future professionals must be trained to understand, identify, and solve ethical issues related to their work (Banik & Gouranga Banik, 2011). Engineering ethics education is vital in equipping engineers with the necessary knowledge and skills to navigate complex ethical dilemmas that may arise throughout their careers. In addition to the technical education essential for civil engineers, an effective education program should include a comprehensive exploration of ethical questions, case studies highlighting real-world ethical challenges civil engineers face, and opportunities for ethical decision-making and reflection.

Considering the above, this chapter aims to provide an overview of the ethical issues, dilemmas, and challenges that are fundamental in the education of civil engineering students. The chapter also addresses how the discipline's students and practitioners might address these issues, dilemmas, and challenges.

Positionality

How we approach our discipline – the questions we ask, the methodologies and case studies we select – are shaped by our positionality, so it is vital to reflect on this briefly.

Two of this chapter's authors are civil engineers, while the third is an applied ethicist with several years of experience teaching engineering ethics. The chapter is the result of an interdisciplinary exchange in which the authors have tried to bring together ethical theories and concepts with concrete applications relevant to the day-to-day practices of civil engineers.

All three authors identify as white and come from and currently work in Western European countries. As such, we have no lived experience of the racial oppression and global inequalities faced by many people. All three authors identify as cis heterosexual women. Although this makes us particularly attuned to the gendered dimensions of civil engineering, it also means that we have no first-hand experience of the marginalization faced by queer people. Finally, one of the authors identifies as disabled. This has certainly heightened our awareness of the impact that our choices have on disabled people. Still, the first-hand knowledge of a single person will necessarily provide only a limited guide to the lived experiences of a large and heterogeneous group.

The danger coming from our positionality is that, through our writing, we could re-center the voices of those in positions of social advantage. Attempting to mitigate this, we have consciously tried to problematize areas where civil engineering currently works to reinforce oppression and marginalization. We have placed emphasis on equality, diversity, and inclusion (EDI) throughout the chapter.

Overview of the chapter

The chapter starts by contextualizing the civil engineering discipline. The historical framework of the profession is described, as are the subdisciplines that make up civil engineering.

Two crucial aspects must be factored in when discussing ethics in civil engineering. One involves the different areas for ethical considerations, which we have grouped under the themes of (1) professionalism; (2) social responsibility; (3) sustainability; (4) health and safety; (5) equality, diversity, and inclusion (EDI); and (6) decolonization. Secondly, civil engineering projects have different life-cycle stages: (1) design and planning, (2) development and construction, (3) operation and maintenance, and (4) decommissioning. This chapter introduces the most important aspects of the different life-cycle stages that require consideration for each thematic area. Nevertheless, readers of this chapter might want to explore the contents according to either thematic groups or according to the life-cycle stages. To help the reader go through the contents in the order they find most convenient, the different contents of this chapter have been summarized in Table 14.1. This table also includes additional content that readers can explore independently.

- Ethical issues explored
  - Case studies presented in this chapter
    - Other potential case studies

This chapter offers conceptual insights into ethical concerns and provides practical examples through case studies and discussion exercises for enriching students' understanding of ethical issues (Harris et al., 2014; Whitbeck, 2011). The selected case studies portray real-world situations, covering a diverse spectrum of civil engineering contexts and subdisciplines, aligning with the chapter's themes.[1]

The case studies serve as catalysts for student discussion. They encourage students to explore various scenarios, anchoring their arguments in established professional ethics codes while striving for solutions that align with these ethical guidelines. Teachers unfamiliar with this teaching approach can examine the discussions' structure, alignment with ethical codes, personal experiences in similar situations, and the resulting conclusions. (For more detailed advice on teaching via case studies, please see Chapter 20, and via discussion and reflection, Chapter 25.)

We suggest discussion questions that can be used in engineering classrooms to prompt discussion of core issues – but we encourage readers to consider these suggestions *alongside advice on applying case study teaching methods* as presented in Chapter 20 and *recommended methods for teaching engineering ethics using reflective and dialogical approaches* as presented in Chapter 25. The case study examples we draw from below are listed in Table 14.1, where key concepts, location within the construction process, the ethical issues explored, and other cases to be explored are also highlighted.

Lastly, civil engineering encompasses six primary subdisciplines: construction and buildings, energy infrastructure, environmental technology, water, transportation, and urbanism. This chapter addresses ethical concerns in each area, emphasizing specific issues where relevant. Each subdiscipline is represented with at least one case study, providing a comprehensive view of ethical challenges in civil engineering.

Contextualization

In 2006, the ASCE convened a summit on the 'Future of Civil Engineering' to establish a global vision for the profession in the twenty-first century. The resulting document, 'The Vision for Civil Engineering in 2025' (ASCE, 2006), outlines the role of civil engineers as stewards of the natural environment and resources, innovators and integrators of ideas and technology across sectors, and leaders in shaping public policy.

This vision has been embraced by civil engineering associations worldwide, recognizing the profession's responsibility in transforming the built and natural environments to achieve the '2030 Agenda' regarding the United Nations Sustainable Development Goals (SDGs). ASCE adopted a policy statement in support of the United Nations Sustainable Development Goals (ASCE, 2023), and ICE (2020) published the 'UN75 Sustainable Engineering in Action' to highlight civil engineering's role in this agenda.

The call for a more holistic and socially conscious approach to civil engineering is a global phenomenon that professional associations in many countries have embraced. Various professional engineering organizations worldwide have developed codes of conduct that establish ethical standards for civil engineers. These codes, such as the Code of Conduct of the European Federation

Table 14.1 Overview of possible contents for civil engineering ethics education (non-exhaustive list)

	Key Concepts	*Design and Planning*	*Development and Construction*	*Operation and Maintenance*	*Decommissioning*
Professionalism	• Codes of conduct • Registration, accreditations, charters	• Bribery and corruption • Whistle-blowing		• Conflicts of professional duties	
		- Maharashtra Irrigation Scam		○ NSPE Board of Ethical Review Case 93-13	
Social Responsibility	• Corporate social responsibility (CSR)	• Unforeseen consequences of innovation • Distribution of responsibility	• Working in foreign settings • Global supply chains	• Impact on users, local communities, and the environment • Maintenance of infrastructure to avoid disasters • Project failure responsibility	
		- Walkie Talkie Skyscraper London	○ Workers' Rights on Saadiyat Island Abu Dhabi	- Morandi bridge - L'Ambiance Plaza	- Energy site redevelopment in Italy
Sustainability	• Circular economy • Obsolescence	• Design with sustainability criteria • Sustainable use of materials • Ethically sourced materials • Minimisation of waste • Energy efficiency			
		- Dutch-style cycling lanes in Manchester - Sanergy, Kenya	- Green concrete and BioMason	○ Powerhouse Kjørbo - Shanghai Tower	- Dam Removal Europe

	Key Concepts	*Design and Planning*	*Development and Construction*	*Operation and Maintenance*	*Decommissioning*
Health and Safety	• Risk–Benefit Analysis	• Safety issues consideration in early stages • Calculating risk - I-35W Bridge Collapse - Grenfell Tower	• Workers' safety and protection - Panama Canal and contract/foreign workers	• Users' safety • Preventative and corrective maintenance - Secret repairs to Citicorp Building	• Hazardous materials ○ La Scala Opera House Asbestos deaths
Equality, diversity and inclusion (EDI)	• Inclusive design • An inclusive profession	• User-centered design ○ Trans-inclusive sanitation in Nepal		• Considering socio-historic contexts • Energy poverty • Universal access to water - Leeds City Council TIBB Project	• Impact on communities - Demolition as urban policy in the American Rust Belt
Decolonization	• Colonialism and neo-colonialism	• Working in foreign settings • Global power imbalance - Water access for nomadic pastoral communities	• Resource extraction • Modern slavery and global exploitation	• Considering socio-historic contexts ○ The Central Corridor Tanzania	• Waste and environmental racism - Construction waste sent to landfills abroad

of National Engineering Associations (FEANI), the ethical code of the Federation of African Engineering Organizations (FAEO), the guidelines for ethical conduct of JSCE (Japan's society), and the Code of Ethics of Engineers Australia, share common themes of promoting responsibility to society and the environment. These examples illustrate that professional associations around the world are recognizing the need for civil engineers to be responsible stewards of the built and natural environment and to act with integrity and social awareness in their professional practice.

Therefore, the training of civil engineering graduates must include not only technical knowledge but also the development of ethics skills essential for their professional practice. In addition to being excellent technicians, civil engineers need the ability to lead, influence, integrate, and weigh up the various social issues that underpin optimal approaches regarding planning, design, and construction. Engineers can more effectively navigate complex ethical dilemmas by cultivating ethics skills and assessing and balancing social, cultural, and philosophical aspects while making decisions. If students do not learn to appreciate elements of science such as its history, its relationship to culture, religion, differing worldviews, commerce – and its philosophical, epistemological, ontological, and methodological assumptions – then the vast opportunity for science and engineering to enrich culture and human lives cannot be fully realized (Matthews, 1994).

It is therefore essential to ask ourselves what role universities play in training twenty-first-century civil engineers and what civil engineering model to use. Universities are responsible for cultivating a comprehensive educational experience that extends beyond technical knowledge. They should aim to equip future civil engineers with a broad skill set, including critical thinking, ethical reasoning, interdisciplinary collaboration, and a holistic understanding of their work's societal and environmental impacts. By incorporating these elements into their curriculum, universities can foster a new generation of civil engineers who are technically proficient, ethically conscious, socially responsible, and capable of addressing the complex challenges of our time. In doing so, universities can play a pivotal role in shaping the civil engineering profession and ensuring its alignment with the needs and aspirations of society.

Key ethical considerations for civil engineers

Professionalism

Civil engineers are part of a profession; their first and foremost responsibility is to the public good. The profession is regulated through several relevant professional bodies. They have a key role in supporting and guiding engineers to help them abide by professional and ethical standards. This kind of association serves several purposes, such as providing career support for members (Grigg, 1998), helping steer the direction of education (Hildreth & Gehrig, 2010), and offering a platform for exchanging ideas. These associations exist for most engineering disciplines, as in civil engineering.

Crucially, professional bodies lay out the professional responsibilities of their members in relevant codes of ethics (for more on professional organizations and codes of ethics, see Chapter 5). Professional ethics is essential for appropriately governing behavior among professionals of different disciplines. It differs from personal morality, which refers to those standards of conduct that apply to everyone in society rather than to members of a particular group. Instead, professional ethics are those standards of conduct that all members of a profession should follow (Harris et al., 2014). As such, civil engineering ethics applies primarily to civil engineers. Although the separation between professional and personal responsibility is commonly accepted, other authors believe that professionals must acknowledge personal responsibility for the consequences of professional conduct (Christians & Nordenstreng, 2004). This perspective highlights the importance of considering the ethical implications of professional actions on a personal level.

The construction of civil works usually involves large projects in which substantial amounts of public money are invested, and social agents, public administrations, and private companies are involved. For this reason, cases where the engineers in charge lack professionalism tend to receive much media attention. Such lack of professionalism is widespread during the work's planning, design, development, and construction phases. It ranges from bribes to influence the location of a particular infrastructure to the construction of unnecessary engineering works that allows construction companies to enrich themselves at the expense of public money (see the Maharashtra irrigation scam for a real-life example of this) to the choice of suppliers who are in some way linked to people who are to be favored financially.

Codes of conduct in civil engineering identify and describe various professional responsibilities, such as declaring conflicts of interest, maintaining client confidentiality, refusing bribery, and reporting corruption (ICE, 2022). These codes primarily prioritize the safety, health, and welfare of the public, along with values like honesty and objectivity (Zhang & Wang, 2018).

The responsibilities that are laid out by professional associations, however, cannot be adopted unthinkingly. Codes of conduct can only provide general advice as a starting point for engineers. For example, they do not guide what should be done when two or more professional responsibilities clash. Instead, they require engineers to interpret them (Davis, 1991). Common conflicts of duties that may arise are, for example, having to weigh up the duty of confidentiality to one's client with the impact on the public if any safety issues are found during maintenance or with the potential adverse environmental impacts of decisions taken during the decommissioning stage of civil engineering projects.

Case study: National Society of Professional Engineers Board of Ethical Review Case 97-13 (National Society of Professional Engineers Board of Ethical Review, 2008)

A civil engineer is subcontracted to inspect the current state of a bridge as part of a more comprehensive maintenance project. The scope of the engineer's work is restricted to identifying any existing pavement damage and reporting this to the client.

While conducting the work, the engineer notices a defect in the bridge's wall. Not long ago, a car driver had lost control of their vehicle, crashed through a part of the wall not far away from where the engineer is now spotting the defect, fell into the river, and died. The engineer believes that this defect could have contributed to the fatal crash.

Although it is outside of the scope of the work the engineer had been hired for, the engineer puts this detail into their notes and verbally informs the client of their suspicion, who then forwards this information to the public agency overseeing the maintenance project. The public agency contacts the engineer and asks them not to mention this supposition in their final report as it did not fall within the scope of their work. The engineer agrees to keep their notes but not to include the information in the final report. They do not raise their suspicion with anyone else.

Discussion activity: This case presents a clash between client confidentiality and public safety. *Do you think that the engineer was correct in retaining the information in their notes but not to include them in the final report? Do you think they were correct in not raising this information with anyone else?*

Social responsibility

Civil engineers bear a primary duty to serve the public good. One way this social responsibility plays out during the design and planning process is that they must scrutinize the potential consequences of their innovations for society and the environment. *But what happens when unintended negative consequences occur despite their best efforts because of their innovation?* It is helpful here to distinguish between unforeseen and unforeseeable harms. An unforeseen harm is a negative consequence that the engineer neither anticipated nor intended. Yet, in principle, an unforeseen harm could have been envisioned if, for example, more detailed work and consideration had gone into the design and planning process. In contrast, unforeseeable harms are those in which the engineer would not reasonably be expected to have known that a particular harm could occur as a result of their design, for example, because the necessary scientific knowledge was not available at the time (Sucklin et al., 2021). We tend to hold civil engineers responsible for unintended but foreseeable consequences. In contrast, we tend to resist doing the same in cases where the consequences were unforeseeable during the planning process.

This is further complicated when we consider that civil engineering projects tend to require input from professionals from various disciplines, such as architecture or project management. This brings benefits as teams are more interdisciplinary, but also raises questions regarding the distribution of tasks and accountabilities in areas where responsibilities overlap. If a project fails, civil engineers may be held responsible for any design or construction errors that contributed to the failure. This responsibility may be legal or ethical, and civil engineers may face consequences such as lawsuits, loss of professional licenses, or damage to their professional reputation. However, in cases where failures result from a range of factors, including inadequate project management, lack of resources, changes in project requirements, or unforeseen environmental or economic conditions, responsibility may be shared among multiple stakeholders, and a thorough investigation may be required to determine the causes of the failure and the parties responsible.

During the development and construction phase, a particularly important aspect to consider is the relationship with the local inhabitants of the areas where the construction is carried out. This includes the contracts made with and treatment of local suppliers and the workers who carry out the construction, especially in countries that do not have strong protection for workers' rights. It is common for large engineering projects to be carried out by multinational companies that are based in the Global North. It is not permissible for these companies to exploit a lack of legal regulations and protections when contracting and negotiating the working conditions of local workers or when purchasing anything from local producers. A vital framework in this context is that of corporate social responsibility (CSR), according to which companies should integrate environmental and social considerations into their business operations and interactions with their stakeholders. According to stakeholder theory, businesses have duties not only to their shareholders but also to anyone who may be affected by their business, such as employees, users, local communities, and the environment. As such, CSR activities are seen as ethically justifiable, and indeed required, even when they do not directly maximize profits (Goodpaster, 1991).

Operation and maintenance also represent crucial stages in the life cycle of infrastructure in terms of social responsibility. During the operation phase, civil engineers are responsible for ensuring the efficiency and safety of the functioning of the infrastructure. For this, regular inspections, maintenance activities, and addressing any operational issues that may arise are fundamental.

In the last stages of the life cycle of infrastructure, civil engineers are responsible for ensuring that these structures are decommissioned or dismantled in a socially and environmentally responsible manner when they reach their serviceability limit. This includes minimizing the risk to human health and the environment, ensuring that materials and components of the infrastructure are disposed of or recycled in an environmentally responsible manner, and ensuring that the decommissioning or dismantling process does not cause unnecessary disruption to the local community.

Case study: Saadiyat Island Abu Dhabi and worker rights

Saadiyat Island, Abu Dhabi, hosts prominent cultural institutions like the Louvre Abu Dhabi and the Guggenheim Abu Dhabi. However, concerns arose regarding working conditions and workers' rights during its development (Human Rights Watch, 2009). Reports revealed exploitative conditions for migrant workers: extended work hours, low wages, and poor living conditions. Many workers paid fees to recruiters, leaving them vulnerable and in debt. Their passports were confiscated and their movement was restricted, effectively trapping them.

This situation prompted protests and calls for action, eventually leading cultural institution developers to partner with the International Labour Organization for improved worker conditions and compliance with labor standards. This case underscores the social responsibility of companies, urging them to prevent exploitation and human rights abuses. Civil engineers must promote ethical practices and safeguard worker rights during projects.

Discussion activity: *What are the social responsibilities of companies and organizations involved in large-scale development projects? How can civil engineers ensure that their projects are ethical and socially responsible? What measures can be taken to prevent exploitation and protect worker rights in large-scale development projects?*

Sustainability

Civil engineering is crucial for society's development, but its activities have significant environmental impacts, like resource depletion and greenhouse gas emissions (see also Chapters 6 and 11). As a result, civil engineers bear great responsibility for enhancing infrastructure while safeguarding the environment (Ramírez & Seco, 2012). Most codes of conduct and ethics established by professional associations place high importance on protecting the environment (Byrne, 2012). Zhang and Wang (2018) note that these codes acknowledge environmental protection as the top responsibility of engineers. Ethics and sustainability are closely linked, as sustainability challenges involve various stakeholders with conflicting needs (Curren & Metzger, 2017). Balancing societal demands and ethical considerations in finding solutions can be challenging. Hence, an understanding of ethics can be critical when deciding what techniques or strategies need to be employed (Biedenweg et al., 2013).

Designing and planning projects with sustainability in mind is of the utmost importance for civil engineers. Priorities include minimizing the use of energy, materials, and resources, and embracing circular design principles to reduce waste and regenerate resources. A thorough analysis of the supply chain is also essential to gauge the project's true environmental impact (Engineering Council, 2021). Relevant considerations for civil engineers can include, for example, the suitability of a particular location (e.g., *Would the project interfere with a wildlife corridor?*), the materials to be used (e.g., *Will the building supplies be sourced locally?*), and energy efficiency (e.g., *Does the design use natural light effectively?*). More fundamentally, however, sustainability ought to be a crucial guiding factor when we decide which projects to work on in the first place. Sustainability should guide project selection, favoring initiatives like expanding safe bike lanes over new motorways, even if the latter were designed sustainably.

The construction stage of a project offers opportunities to increase the quality and sustainability of the project; excessive ground disturbance can be avoided, the use of recycled materials and the re-use of resources can be increased, and waste can be appropriately managed while meeting the needs of all project stakeholders from designers to suppliers, workers, and users (Vanegas, 2003). It is important to remember that manufacturing and construction processes typically use non-renewable resources. To address this, civil engineers must raise the use of recycled or regenerative materials and view demolition waste as a resource for recycling and reintegration into production (here again, Chapter 6 holds relevance).

Sustainability extends beyond these initial stages, remaining vital in the maintenance and operation of engineering projects. The operation and use of buildings is a key source of global greenhouse gas emissions (UN Environment Programme, 2022). Thus, prioritizing energy efficiency in new designs and retrofitting older buildings with energy-saving measures is critical. Notably, there is a distinction between nearly zero-energy buildings (NZEB) with high energy efficiency and a significant share of renewable energy, zero-emissions buildings (ZEB) achieving high energy efficiency entirely from local or on-site renewables, and net-positive buildings that generate more renewable energy than they consume throughout their life cycle (Cole & Fedoruk, 2015).

To enhance building sustainability, typical upgrades include improved insulation, shifting from gas central heating to ground-source heat pumps, and replacing incandescent bulbs with LEDs. When designing energy-efficient new buildings, civil engineers should prioritize natural light through well-placed windows and effective use of solar energy. At the same time, they need to consider the ability to effectively use solar energy to heat the building during colder months while avoiding overheating during warmer ones. Ways of using natural ventilation or the creation of green roofs and facades to aid in this process can be incorporated. Designers (including teams of civil engineers and architects) should also consider to what extent the building can be designed to create the energy it uses, for example, via solar panels.

Sustainability's role isn't confined to the project's initiation, operations, and maintenance but also includes decommissioning. Civil engineers often need to manage substantial volumes of construction and demolition waste; these rank among the most significant waste sources.

In the context of sustainability and ethics, circular economy practices aim to minimize waste and optimize resource utilization. The circular economy is based on two cycles: the technological cycle and the biological cycle (McDonough & Braungart, 2009). In the technological cycle, materials are repaired, re-used, re-purposed, and recycled; in the biological cycle, any nutrients are returned to nature. With the introduction of circular economy practices, decommissioning is no longer restricted to demolition but to any other process aimed at recirculating the material of the decommissioned infrastructure (Ellen McArthur Foundation, n.d.). Yet it is worth noting that critics of the circular economy argue that, no matter how circular our economy is, unless we stop aiming for constant economic growth, we will always exceed planetary boundaries (Hickel, 2020).

Infrastructure obsolescence is a critical factor to consider in sustainability efforts. Obsolescence occurs when an infrastructure component or system is no longer effective or efficient in meeting current demands. Civil engineers can take several approaches to confront obsolescence, including planning for obsolescence by using materials that can be easily dismantled, re-used, recycled, or returned to the natural environment and planning to delay obsolescence by carefully selecting materials – encouraging a shift from a use-and-throw-away mindset to one that prioritizes re-use and repair (Lawlor, 2015). Civil engineers can limit the effects of infrastructure obsolescence by incorporating long-term planning and life-cycle cost analysis into their work.

Case study: Powerhouse Kjørbo

Between 2013 and 2014, two 1980s-era office buildings on Oslo's outskirts underwent complete renovation, resulting in Powerhouse Kjørbo (UNFCCC, n.d.). This project became one of the world's first renovations to generate more renewable energy than it consumes throughout its lifetime (encompassing construction, materials, maintenance, operation, and eventual demolition).

Preserving the old structure while enhancing insulation, cladding, and windows maximized energy efficiency. Solar energy generation was enabled via photovoltaic systems, and old building materials were repurposed, transforming external glass panels into interior office walls. Ground-source heat pumps are used to supply energy. Furthermore, the renovation encourages sustainable behaviors by providing bike storage and showers to promote cycling to work, and parking spaces equipped for electric vehicles.

Class activity: In groups, select a campus building and plan upgrades to achieve a desired energy efficiency level.

Health and safety

Health and safety are paramount in civil engineering projects throughout their life cycle. During the planning phase, engineers must carefully assess risks and benefits. This includes considering the nature of the potential harm, the affected parties' awareness and consent, the project's overall value, the availability of alternative and less risky approaches, and opportunities to take reasonable precautions to mitigate the potential harm. Engineers often face trade-offs between cost-effectiveness and safety, but they must prioritize ethical concerns alongside economic considerations (Toole, 2007).

During the construction phase, risks exist for both workers and local communities. Although zero risk is unattainable, most accidents are preventable. Occupational risk prevention should be integrated into projects. Strict adherence to health and safety measures and the provision of personal protective equipment are essential. Neglecting safety regulations or failing to provide necessary measures is unethical and may be illegal. Neglecting workers' well-being, including working hours and rest periods, can lead to exhaustion-related accidents. For instance, during Qatar's 2020 World Cup construction, over 38,000 workers suffered accidents, resulting in 37 deaths (Re & Tunon, 2021).

Ensuring occupational health and safety during operation and use must be a top priority (Lukhele et al., 2023). Given the longevity of these projects, continuous maintenance plays a vital role in sustaining user safety. Maintenance encompasses preventive and corrective aspects.

Preventive maintenance aims to anticipate and address issues before they arise, typically on a scheduled basis. It involves creating a maintenance plan that factors in the product's expected lifespan, material degradation rates, downtime costs, resource availability, and user safety. The goal is to maintain reliability and safety without wasting time or resources. Corrective maintenance addresses specific problems that arise during regular maintenance or in response to incidents (Stenström et al., 2016). Insufficient or infrequent preventive maintenance often leads to such issues. However, routine maintenance may not always revisit initial design choices, limiting maintenance personnel's ability to detect health and safety issues that are rooted in design flaws.

At the end of an infrastructure's life cycle, whether dismantling for re-use or demolition, potential associated risks must be considered. Decommissioning can involve handling hazardous materials like asbestos, lead, or radioactive substances, posing health risks to workers and the public. Civil engineers must ensure safe decommissioning in compliance with health and safety regulations. Environmental impacts, such as proper waste disposal and site remediation, also require attention.

Worker safety during decommissioning is critical, involving proper training, equipment, and safety protocols for handling hazardous materials. Long-term community effects should not be overlooked, hazards should be effectively communicated to the public, and measures should be put in place to mitigate potential health risks.

Case study: La Scala Opera House asbestos deaths

In 2001, six retired workers from La Scala Opera House in Milan, Italy, died due to asbestos exposure (ACTS FACTS, 2016). Their exposure occurred while handling insulation materials containing asbestos fibers in the theatre's workshops, highlighting the grave risks of working in neglected older structures.

Asbestos, once valued for its insulation and fire-resistance properties, later proved detrimental to human health – causing lung cancer, asbestosis, and mesothelioma. Despite Italy's 1992 ban on asbestos, many older buildings still harbor asbestos-containing materials (ACMs). La Scala Opera House failed to safeguard its workers, failing to provide protective gear, professional removal, and education regarding health risks.

Discussion activity: *What are the potential health and safety risks associated with working in older buildings? What measures can civil engineers take to protect workers and occupants from hazardous materials like asbestos? How can communication and training be improved to raise awareness of health and safety risks in the workplace?* Class discussion of this topic can benefit from consideration of individual and collective dimensions of ethical decision-making in engineering as presented in Chapter 3.

Equality, diversity, and inclusion (EDI)

Although infrastructure is considered essential to satisfy the most basic human needs, civil engineering has inherently incorporated biases in its processes and methodologies, which have contributed to injustices. Civil engineers, therefore, must acknowledge how their projects impact people with different characteristics such as gender, age, socio-economic status, health status (Lowe et al., 2015), or location (Maswime, 2021). Not considering these characteristics can further perpetuate existing vulnerabilities (Field et al., 2022; Hao et al., 2023).

A civil engineer's responsibility to "promote equality, diversity and inclusion" (Royal Academy of Engineering and Engineering Council, 2017, p. 1) clearly applies to how they design and plan projects. Structures and products should be accessible to as many people as possible and ought not to discriminate unjustifiably based on characteristics such as the ones mentioned above. The primary principle that civil engineers should follow to achieve this, is to put users' needs at the center of their design (for more on this please see Chapter 12 on the foundations of engineering design). The planning process should start with considering who the users will be, identifying their different characteristics and needs, and embracing this diversity. Sometimes, this means that design solutions must offer users different options to engage with a product or space. The more flexibility and the ability to adapt to people's needs is built into the design, the more inclusive the final product will be. It is important that inclusion for civil engineers is not just about physical access to spaces but also includes emotional and intellectual factors such as, for example, whether a space is signposted and enjoyable to use. Furthermore, design should not only be inclusive but also avoid hostile features such as, for example, sectioned benches intended to deter rough sleeping.

The need to make inclusive design choices comes up in various civil engineering projects. Examples that readily spring to mind often relate to designs that increase accessibility for disabled users, such as, for example, integrating ramps into building plans to make them wheelchair accessible or incorporating tactile paving in city planning to help blind and visually impaired pedestrians navigate streets safely. Yet inclusive engineering can also include issues such as, for example, the provision of public toilets in urban planning to address existing gender inequalities (Greed, 2019).

Another factor that impacts the inclusiveness of our design choices regards the makeup of the profession itself. In particular, the number of men practicing civil engineering is still significantly higher than the number of women; for example, in 2022 in the United States, only 17.1% of civil engineers were women (U.S. Bureau of Labor Statistics, 2022). This trend will not be corrected in the coming years since the number of women who access relevant university courses is still lower than that of men; for example, in 2017–2018 in the United States, only 25.9% of civil engineering undergraduate degrees were awarded to women (Roy, 2018).

To better understand just how widespread the idea is that civil engineering is a profession primarily practiced by men, let us briefly consider the well-known case of the Citicorp Center building in Manhattan. It was about to collapse due to a miscalculation during the project phase, which made it weak against forces caused by winds. Luckily, a Princeton student who was analyzing the structure of the building for their studies noticed the miscalculation and alerted the engineers' office in New York. The building was secretly repaired, and the truth only became publicly known, by chance, 20 years later. While this case is often used in ethics classes, it is interesting to note that for a long time, the student's name was not known and that everyone simply assumed he was a man. In fact, the Princeton student who discovered the calculation error and presented it in her undergraduate thesis was a woman: Diane Hartley.

When engineering projects are created, they are inevitably affected by the broader socio-political context at the time. This means that the end product may reflect existing inequalities and biases within the profession. This can happen both implicitly and explicitly. Regarding the former, we tend to design projects with people like us in mind. Even if we do not set out to favor one group at the expense of another, our biases mean that we can inadvertently create projects that do so. In a profession like engineering that historically has been dominated by men, for instance, many projects have not been created with women in mind. We can see this, for example, in urban transport systems, which usually do not account for the gendered dimensions of travel. For example, women have been found to make more non-commuting trips relating to domestic responsibilities and travel more during off-peak hours (Ng & Acker, 2018). Sometimes, however, existing inequalities and biases are used explicitly to justify creating discriminatory projects. There is, for example, a well-documented history of urban planning and infrastructure projects that were influenced by racist attitudes to explicitly serve white communities at the expense of minoritized ethnic groups (Reft et al., 2022). When thinking about the operation and maintenance of existing projects, it is therefore essential for civil engineers to bear the relevant socio-political context in mind and aim to counter unjustified discrimination through appropriate changes and upgrades.

While the end-of-life stage of infrastructure is sometimes seen as an opportunity to develop new infrastructure which might enhance society's quality of life, it can come at the cost of forcing dwellers to move or depriving users of a service that they were previously utilizing (e.g., a community center, a school). Demolition can particularly affect disadvantaged communities. Civil engineers can promote inclusive practice through transparency and stakeholder engagement in decommissioning. A successful example of EDI being effectively included at the project's end-of-life is the Kibera public space project (in Nairobi, Kenya), which involved rehabilitating a former dumping site to turn it into a public space.

Case study: trans-inclusive sanitation

Since 2011, Nepal has recognized a third gender category for those who don't identify as strictly male or female. In 2012, the first gender-inclusive public toilet opened, allowing use by any gender. A different approach emerged later, featuring gender-segregated toilets with spaces for women, men, and third-gender individuals. Feedback from the third-gender community was positive, as they gained access to appropriate sanitation facilities, reducing the need for open defecation. Having a designated third-gender toilet also raised awareness and acceptance among the public. However, this approach raised concerns among trans individuals. People often assume all trans individuals identify as being third-gender, potentially leading to abuse and violence when trans men and women use gender-specific toilets. Balancing the needs of third-gender and trans individuals remains a challenge in ensuring equitable access to sanitation (Boyce et al., 2018).

Class activity: Consider inclusive design principles in a diversity impact assessment for both gender-inclusive and gender-segregated public toilets in Nepal. Discuss the best way to accommodate the diverse needs of various groups.

Decolonization

The global history of colonialism has shaped many existing engineering projects. During European colonialism, European powers occupied foreign territories to violently extract labor and natural resources for the sole benefit of the colonizers. They portrayed indigenous populations as backward and in need of civilization through European-based science and technology. This has left long-lasting legacies and created power imbalances between the Global North and the Global South, which have been reflected in development policies and practices. The aim of decolonization is to undo these harmful legacies by fundamentally questioning and undoing the unfair privileges held by the Global North and the systems it has created to uphold them (Maldonado-Torres, 2016). In this way, the aim of decolonization goes beyond equality, diversity, and inclusion initiatives; whereas the latter wants the inclusion of marginalized groups in the current system, the former wants to question and undo that very system. For more on this topic, please see Chapter 9 titled 'Sociological, Postcolonial, and Critical Theory Foundations of Engineering Ethics Education.'

Civil engineering is deeply intertwined with the project of colonialism – after all, it was civil engineers who built the ports, roads, and railways that enabled the slave trade and theft of natural resources from colonized countries (Muller, 2018). When engineers are designing and planning their projects, they need to be aware of how they are connected to this wider global context and how they may, even if only inadvertently, further entrench global inequalities.

When planning to work on projects in foreign settings, especially in countries in the Global South, civil engineers must be mindful of their positionality in the broader socio-political context of (neo-)colonialism. For example: *Does the project have local involvement or is it administered through a top-down approach driven by the interests of a multi-national organization? Are local collaborators meaningfully involved in the project's design, or are they relegated to physical or unskilled work? Are local collaborators sufficiently remunerated and given acknowledgment for the work that they are contributing?* More broadly, engineers need to ask themselves to which extent the projects they are designing are buying into unhelpful 'white savior' and 'deficit' narratives in which skilled and educated professionals from the Global North are conceptualized as coming to fix the problems experienced by the Global South (Eichhorn, 2020; Noxolo, 2017).

The influence of colonialism on engineering projects also extends to the construction phase. During colonial times, civil engineers were involved in building infrastructure that facilitated the exploitation of resources and the control of colonized territories. Today, even when working within a domestic context, the reliance on global supply chains means that civil engineering projects will take on global dimensions in which engineers must be cognizant of this historical context of resource extraction and the potential perpetuation of unfair power dynamics. For example: *Does our design rely on materials created abroad by workers in unsafe conditions who are not paid a fair wage? Or are we planning on using materials from abroad that are causing environmental degradation and endangering local ecosystems and livelihoods?*

Initiating a process of decolonization also relates to the operation and maintenance of already existing structures. To see this, consider how colonizers created a false narrative that presented the building of infrastructure projects such as railways, roads, and ports as a benefit to local populations – a way of integrating them into the European world of enlightenment and progress (Carneiro et al., 2000). While many occupiers left once the former colonies gained independence, the infrastructure projects that they had built remained – and to this day crucially shape aspects of transport and the economy in formerly colonized countries. When working on maintaining and upgrading such structures, it is therefore essential for engineers to consider whether their continued use is sustaining colonial policies and ideas, albeit in a slightly different guise, today.

Many life-cycle stages in civil engineering, especially the decommissioning phase of projects, involve producing large amounts of waste. Here, civil engineers should be particularly aware of how the disposal of industrial waste can contribute to environmental racism by placing disproportionate environmental burdens on indigenous communities and communities of color through the positioning of landfills and toxic waste disposal facilities on their land or in their neighborhoods (Bullard, 1993), or by sending waste produced in the Global North to countries in the Global South (Okafor-Yarwood & Adewumi, 2020). What happens to the waste that civil engineers create can contribute to reinforcing racial disparities and global inequalities.

Case study: The Central Corridor, Tanzania

In the early twentieth century, German colonial rulers constructed a central railway line spanning Tanzania to facilitate the exploitation and export of the nation's natural resources. Today, the Tanzanian government is investing in the Central Corridor, which aligns with this historic railway. The Corridor comprises a transportation network featuring motorways, railways, ports, and pipelines, coupled with an economic focus on industries such as oil, gas, fishing, agriculture, and tourism. The government contends that industrialization along the Corridor will generate jobs and alleviate local poverty.

However, the new Corridor follows the footprint of colonial-era infrastructure, initially built to support German interests in an export-oriented economy. Similarly, the Central Corridor aims to boost Tanzanian exports, integrating the nation into the global market dominated by former colonial powers. While it enhances the movement of goods and capital, it does not necessarily increase the mobility of the rural population as urbanization is actively discouraged in order to maintain a sufficient workforce along the corridor. Thus, the colonial railway line serves as a conduit for carrying colonial policies, albeit with different manifestations, into the present day (Enns & Bersaglio, 2020).

Class activity: Choose a structure in your area with colonial history, describe its past and current use, and analyze whether its present function perpetuates colonial agendas.

Conclusions

This chapter has provided an overview of some ethical issues related to different areas and life-cycle stages that can be included in civil engineering education. However, it is essential to note that the topic of ethics in civil engineering is vast and complex, and it is impossible to cover everything in a single chapter. Nonetheless, we hope that this chapter has served as a starting point for readers to reflect on the ethical responsibilities that come with the practice of civil engineering and that engineering education should be prepared to address.

We have highlighted the importance of considering the full life cycle of infrastructure projects, from design and planning to development and construction, to operation and maintenance, to decommissioning. We have also discussed some key aspects that should be considered to facilitate determining the ethical responsibilities of individual professionals. It is well worth also considering and actively engaging in discussion regarding the individual and collective dimensions of ethical decision-making in engineering (Chapter 3), because civil engineers need to learn to work together to shape their collective approach to ethics in our constantly changing world, in addition to monitoring their own individual behavior and the behavior of the firms and projects where they operate (see also Chance et al., 2021).

However, it is important to acknowledge that the considerations outlined in this chapter only scratch the surface of a complex and multifaceted topic. Ultimately, the responsible practice of civil engineering requires a deep understanding of the ethical challenges and considerations that arise throughout the project life cycle. We hope that this chapter will inspire readers to continue exploring these issues in greater depth and to approach their work with a commitment to ethical and sustainable engineering practices.

Note

1 More case studies from a variety of engineering disciplines can be found at the following website https://epc.ac.uk/resources/toolkit/ethics-toolkit/ethics-toolkit-case-studies/

References

ACTS FACTS. (2016). *La Scala Workers hold memorial for asbestos victims, The Monthly Newsletter from Arts, Crafts and Theater Safety (ACTS)* (Vol. 30, No. 1). ACTS FACTS.

ASCE (2006). *The vision for civil engineering in 2025.* https://doi.org/10.1061/9780784478868

ASCE (2023). The role of the civil engineer in sustainable development. https://www.asce.org/advocacy/policy-statements/ps418---the-role-of-the-civil-engineer-in-sustainable-development

Banik, G., & Gouranga Banik, P. E. (2011). *AC 2011-2906: Ethics: Why it is important and how we can teach it for engineering and construction students?* American Society for Engineering Education.

Biedenweg, K., Monroe, M. C., & Oxarart, A. (2013). The importance of teaching ethics of sustainability. *International Journal of Sustainability in Higher Education, 14*(1), 6–14.

Boyce, P., Brown, S., Cavill, S., Chaukekar, S., Chisenga, B., Dash, M., . . . Thapa, K. (2018). Transgender-inclusive sanitation: Insights from South Asia. *Waterlines, 37*(2), 102–117.

Bullard, R. D. (1993). The threat of environmental racism. *Natural Resources & Environment, 7*(3), 23–26, 55–56

Byrne, E. P. (2012). Teaching engineering ethics with sustainability as context. *International Journal of Sustainability in Higher Education, 13*(3), 232–248.

Carneiro, A., Diogo, M. P., Simões, A., & Troca, M. (2000). Portuguese engineering and the colonial project in the nineteenth century. *Icon, 6*, 160–175.

Chance, S., Lawlor, R., Direito, I., & Mitchell, J. (2021). Above and beyond: Ethics and responsibility in civil engineering. *Australasian Journal of Engineering Education, 26*(1), 93–116.

Christians, C., & Nordenstreng, K. (2004). Social responsibility worldwide. *Journal of Mass Media Ethics, 19*(1), 3–28.

Cole, J. R., & Fedoruk, L. (2015). Shifting from net-zero to net-positive energy buildings. *Building Research & Information, 43*(1), 111–120.

Curren, R., & Metzger, E. (2017). Living well now and in the future: Why sustainability matters. MIT Press.
Davis, M. (1991). Thinking like an engineer: The place of a code of ethics in the practice of a profession. *Philosophy & Public Affairs*, *20*(2), 150–167.
Eichhorn, S. J. (2020). How the west was won: A deconstruction of politicised colonial Engineering. *Political Quarterly*, *91*(1), 204–209.
Ellen McArthur Foundation. (n.d.). *What is a circular economy?* https://ellenmacarthurfoundation.org/topics/circular-economy-introduction/overview
Engineering Council. (2021). *Guidance on sustainability for the Engineering profession.* https://www.engc.org.uk/media/3555/sustainability-a5-leaflet-2021-web_pages.pdf
Enns, C., & Bersaglio, B. (2020). On the coloniality of "new" mega-infrastructure projects in East Africa. *Antipode*, *52*(1), 101–123.
Field, C., Sutley, E., Naderpajouh, N., van de Lindt, J. W., Butry, D., Keenan, J. M., Smith-Colin, J., & Koliou, M. (2022). Incorporating socioeconomic metrics in civil Engineering projects: The resilience perspective. *Natural Hazards Review*, *23*(1), pp. 1–13.
Goodpaster, K. E. (1991). Business ethics and stakeholder analysis. *Business Ethics Quarterly*, *1*(1), 53–73.
Greed, C. (2019). Join the queue: Including women's toilet needs in public space. *The Sociological Review*, *67*(4), 908–926.
Grigg, N. S. (1998). Universities and professional associations: Partnerships for civil engineering careers. *Journal of Management in Engineering*, *14*(2), pp. 3–68.
Hao, H., Bi, K., Chen, W., Pham, T. M., & Li, J. (2023). Towards next generation design of sustainable, durable, multi-hazard resistant, resilient, and smart civil engineering structures. *Engineering Structures*, 277, 115477.
Harris, C. E., Pritchard, M. S., James, R. W., Rabins, M. J., & Englehardt, E. E. (2014). Engineering ethics: Concepts and cases. Wadsworth Cengage Learning. ISBN: 9781133935209
Hickel, J. (2020). What does degrowth mean? A few points of clarification. *Globalization*, *18*(7), 1105–1111.
Hildreth, J. & Gehrig, B. (2010). AC 2010-1584. *A body of knowledge for the construction engineering and management discipline.*
Human Rights Watch. (2009). *"The Island of happiness" - exploitation of migrant workers on saadiyat Island, Abu Dhabi*. Human Rights Watch.
ICE (2020). UN75: Sustainable Engineering in Action.
Institution of Civil Engineers. (2022). *Code of professional conduct*. Institution of Civil Engineers.
Koehn, E. (1991). An ethics and professionalism seminar in the civil engineering curriculum. *Journal of Professional Issues in Engineering Education and Practice*, *117*(2), 96–101.
Lawlor, R. (2015). Delaying obsolescence. *Science and Engineering Ethics*, *21*(2), 401–427.
Lowe, A. W., Partington, D. R., & Richardson, S. G. (2015). Achieving inclusive design: Consultation with disabled people. *Proceedings of the Institution of Civil Engineers: Municipal Engineer*, *168*(1), 45–53.
Lukhele, T. M., Botha, B., & Mbanga, S. (2023). Exploring the nexus between professional ethics and occupational health and safety in construction projects: A case study approach. *International Journal of Construction Management, 23*(12), 2048–2057.
Maldonado-Torres, N. (2016). *Outline of ten theses on coloniality and decoloniality*. Frantz Fanon Foundation.
Maswime, G. V. (2021). The enduring relationship between civil engineering and spatial injustice. In H. H. Magidimisha-Chipungu & L. Chipungu (Eds.), *Urban inclusivity in Southern Africa* (pp. 21–37). Springer International Publishing.
Matthews, M.R. (1994). Science teaching: The role of history and philosophy of science. New York: Routledge.
McDonough, W., & Braungart, M. (2009). *Cradle to cradle*. Random House.

Muller, M. (2018). Decolonising engineering in South Africa – Experience to date and some emerging challenges. *South African Journal of Science*, *114*(5/6), 1–6.
National Society of Professional Engineers Board of Ethical Review (2008). Case No. 07-11
Ng, W.-S., & Acker, A. (2018). *Understanding urban travel behaviour by gender for efficient and equitable transport policies*. International Transport Forum.
Noxolo, P. (2017). Decolonial theory in a time of the re-colonisation of UK research. *Transactions of the Institute of British Geographers*, *42*, 342–344.

Okafor-Yarwood, I., & Adewumi, I. J. (2020). Toxic waste dumping in the Global South as a form of environmental racism: Evidence from the Gulf of Guinea. *African Studies*, *79*(3), 285–304.
Ramírez, F., & Seco, A. (2012). Civil engineering at the crossroads in the twenty-first century. *Science and Engineering Ethics*, *18*, 681–687.
Re, F., & Tunon, M. (2021). *One is too many – The collection and analysis of data on occupational injuries in Quatar*. International Labour Organisation.
Reft, R., Phillips de Lucas, A. K., & Retzlaff, R. C. (2022). *Justice and the interstates: The racist truth about urban highways*. Island Press.
Roy, J. (2018). *Engineering by the numbers*. American Society for Engineering Education.
Royal Academy of Engineering, & Engineering Council. (2017). *Statement of ethical principles*. Royal Academy of Engineering, & Engineering Council.
Stenström, C., Norrbin, P., Parida, A., & Kumar, U. (2016). Preventive and corrective maintenance – cost comparison and cost –benefit analysis. *Structure and Infrastructure Engineering*, *12*(5), 603–617.
Sucklin, J., Hoolohan, C., Soutar, I., & Druckman, A. (2021). Unintended consequences: Unknowable and unavoidable, or knowable and unforgivable? *Frontiers in Climate*, *3*, pp. 1–12.
Toole, M. T. (2007). Design engineers' responses to safety situations. *Journal of Professional Issues in Engineering Education and Practice*, *133*(2), 121–131.
UN Environment Programme. (2022). *2022 Global status report for buildings and construction*. UN Environment Programme.
United Nations Framework Convention on Climate Change (UNFCCC). (n.d.). *Powerhouse Kjørbo - Norway*. https://unfccc.int/climate-action/momentum-for-change/activity-database/powerhouse-kj%25C3%25B8rbo
U.S. Bureau of Labor Statistics. (2022). *Labor force statistics from the current population survey*. https://www.bls.gov/cps/cpsaat11.htm
Vanegas, J. A. (2003). Road map and principles for built environment sustainability. *Environmental Science and Technology*, *37*, 5363–5372.
Whitbeck, C. (2011). Ethics in Engineering Practice and Research. 2nd ed. Cambridge University Press.
Zhang, M., & Wang, S. (2018). The content and structure of code of ethics for engineers in China. In M. E. Auer & K.-S. Kim (Eds.), *Engineering education for a smart society* (pp. 46–56). Springer International Publishing.

15

ETHICAL ISSUES IN MECHANICAL AND AEROSPACE ENGINEERING

Aaron W. Johnson, Corin L. Bowen, Cristiano Cordeiro Cruz, and Renato Bezerra Rodrigues

Introduction

This chapter discusses ethical issues in the related fields of mechanical and aerospace engineering from a critical perspective. These disciplines use mathematical and scientific principles to design, analyze, and construct mechanical systems. Mechanical engineering, which focuses on a wide variety of mechanical systems, is one of the oldest, most common, and broadest degree programs to be offered by colleges of engineering (American Society for Engineering Education, 2022; Dixit et al., 2017; Grayson, 1980). Mechanical engineers design commercial power-producing and power-using machines such as medical devices, kitchen appliances, industrial robots, and automobiles (US Bureau of Labor Statistics, 2023a). Aerospace engineering covers many of the same concepts but applies them to designing aircraft and spacecraft systems for commercial and military purposes (US Bureau of Labor Statistics, 2023b). As aerospace engineering can be considered a specific focus area within mechanical engineering, we address both fields in this chapter.

Undergraduate mechanical and aerospace engineering curricula cover a wide range of technical concepts such as statics and dynamics, strength of materials, thermodynamics, combustion, fluid mechanics, controls, machine design, and manufacturing. In addition, undergraduate mechanical and aerospace programs also commonly emphasize design processes and professional skills, such as communication, leadership, and ethics. Across the world, the structure of degree programs varies, as does the approach to the integration of professional skills into the curriculum. However, the incorporation of 'ethics,' by some definition, is common globally, particularly in the 23 countries that are signatories of the Washington Accord (International Engineering Alliance, 2021). The various conceptualizations of ethics education generally agree on the necessity of considering interactions between engineering technology and the people who both affect it and are affected by it. However, there are vastly differing approaches to *what* exactly 'ethics' encompasses and *how* ethics is taught – from presentations of case studies of engineering disasters to emphasis on the importance of communication with stakeholders to analyses of the perpetuation of colonization through technology. Accreditation documents, like the Washington Accord, do not address these questions. We argue that ethics education must engage mechanical and aerospace engineering students in deeper conversation to critically analyze the sociotechnical impacts of their fields and

DOI: 10.4324/9781003464259-19

develop the critical consciousness (Freire, 1970) necessary to build toward social justice. (For more on reflective and dialogical approaches in engineering ethics education, see Chapter 25.)

In this chapter, we deliberately employ a critical, emancipatory framework as framed by activist scholars such as Paulo Freire (Freire, 1970) and bell hooks (hooks, 1994). In doing so, we aim to push against the Western[1] conceptualization of mechanical and aerospace engineering as apolitical, neutral fields (Cech, 2013). This typical framing of engineering ethics focuses on *micro-ethics*, the individual responsibilities of engineers within the established guidelines of their profession, rather than *macro-ethics*, the collective social responsibility of the profession (Herkert, 2005). (For more on individual and collective dimensions of ethical decision-making in engineering, see Chapter 3.) Instead, we argue that developing students' critical consciousness (as per Freire, 1970) should be a primary objective of engineering ethics education. To this end, we address how mechanical and aerospace engineering have historically facilitated oppression and inequity through their embrace of apolitical framings, their avoidance of macro-ethical topics, and their lack of development of critical consciousness in education. Our analyses focus very explicitly on systems of power as we attempt to address root causes of systemic oppression (thus complementing Chapters 8, 9, 35, and 36, which also adopt critical perspectives and provide helpful context beyond the fields of mechanical and aerospace engineering). This approach may be challenging to some readers because it confronts mainstream realities in engineering practice and critically questions the development of engineering identity (Cech, 2013). We encourage the reader to lean into this discomfort, explore some of the literature cited herein, and discuss it with colleagues, peers, and the authors of this chapter to understand this 'unfamiliar' perspective.

To structure this chapter, we present and dissect four topics central to mechanical and aerospace engineering: the military-industrial complex, automation, sustainability efforts, and humanitarian engineering. While these topics involve engineers from many disciplines, they directly relate to positions of employment commonly held by aerospace and mechanical engineers. These four topics involve various socio-political issues that render them useful for students' reflection and discussion within a critical ethics education. They serve as concrete examples that can be taught in mechanical and aerospace engineering courses to help students understand the macro-ethical implications of their work. Furthermore, this type of critical systemic analysis can be applied to other topics in mechanical and aerospace engineering. Critical education on these and other issues is imperative to build toward emancipatory sociotechnical orders[2] – not to undermine existing approaches and perspectives on ethics education, but to add a perspective that is commonly ignored and overlooked in engineering. To close the chapter, we include a discussion of a critical question for the future of the fields: *What does a conceptualization of mechanical and aerospace engineering education committed to social justice look like?* To do this, we highlight examples of mechanical or aerospace engineering practice and education committed to social justice and the empowerment of marginalized communities.

Positionality, frameworks, and approach

We first outline the positionality of the author team, which consists of our social backgrounds and an acknowledgment of their impact on our perspectives on and approach to engineering ethics education. Corin Bowen is a white cisgender woman from a rural community in the United States with a background in structural engineering and mechanics. She works as a tenure-track professor of engineering education research. Cristiano Cordeiro Cruz is a white Brazilian cisgender male with a background in philosophy and engineering who works as a visiting researcher at two Brazilian universities. He is a member of the Brazilian Network of Popular Engineering. Aaron Johnson is a white, cisgender male from the United States with a background in aerospace engineering who conducts engineering education research as a tenure-track professor in an aerospace department. Renato Bezerra Rodrigues is a white Brazilian cisgender male with a background in engineering and philosophy who is currently pursuing a Ph.D. in engineering education. We all recognize the ways in which we each come from a privileged background, and we are all committed to exploring emancipatory approaches to our work as engineering researchers and teachers. Due to our personal experience with engineering education in the United States, Canada, and Brazil, we primarily analyze engineering and engineering education systems within the context of these settings and our own social positions within them.

In this chapter, we take an emancipatory approach to engineering ethics education. We recognize that the field of engineering and the educational processes within it enable, support, and perpetuate colonial,[3] cis-heteropatriarchal, racist, capitalistic, and militaristic systems (Cech, 2013). We assert that engineering education should prepare students to apply their engineering skills and knowledge not only in the service of mainstream capitalist beneficiaries within the domain of engineering industry, but also in the construction of new lived realities at both individual and community scales (Freire, 1970). In this way, engineering agency can be redirected toward social justice and decolonizing efforts toward equitable societal change (Alvear et al., 2021; J. A. Leydens & Lucena, 2018).

Technology and engineering are not politically or ontologically neutral (Cruz et al., 2024). They support, emulate, and/or create social orders and ways of living, such as the individualistic, consumerist, and Christian ones that are dominant in the Global North. Thus, any change in how we (want to) live and make sense of our lives – both individually and collectively – requires widening or decolonizing engineering practice, supporting knowledge, and research. Alternative framings exist that describe alternative ways of living and being, such as the Amerindian *Buen Vivir* (Acosta & Abarca, 2018; Estermann, 2006) and African *Ubuntu* (Ewuoso & Hall, 2019; Le Grange, 2019), which are centered upon values such as collective wellness, humanity, and care. Alternative framings for engineering and engineering education are capable of leading to the development of different technologies that bring about the construction of such different social arrangements and ways of living. Within engineering fields, if ethics education is to facilitate such emancipatory sociotechnical arrangements, then mechanical and aerospace engineering students in our programs must:

1. Realize that the hegemonic sociotechnical order we live in is constructed and moldable.
2. Be aware of the possibility of alternative sociotechnical arrangements.
3. Be able (to learn how) to construct the technological infrastructure of other possible worlds.

These are not trivial endeavors; they require a complete reframing of both engineering more generally and engineering labor and practice more specifically. As such, considerable dialogue will be needed to initiate these efforts. This chapter aims to contribute to and encourage such dialogue, especially within educational spaces. For this reason, we lean heavily into the domain of macro-ethics and attempt to challenge existing Western norms of acceptable macro-ethical behavior. We also advocate for sociotechnical education in which macro-ethics is integrated into technical engineering coursework rather than compartmentalized as 'separate' from the role of an engineer (e.g., Benham et al., 2021). This helps combat the prevailing engineering culture, which purports engineering to be purely technical, objective, and apolitical (Cech, 2013; Nieusma, 2015).

Sociotechnical issues in mechanical and aerospace engineering

The military-industrial complex

The military-industrial complex refers to the capitalistic enterprise of weapons research, development, and manufacturing that supplies Western imperial governments and military forces, a lucrative endeavor that frequently occurs at the expense of people and communities in the Global South (Christiansen, 2020). The fields of mechanical and aerospace engineering in the Global North are closely tied to the military-industrial complex in terms of financial support, research directions, and career prospects. The US aerospace industry is often referenced monolithically as the 'aerospace and defense' industry (*2023* Aerospace and Defense Industry Outlook, n.d.), conflating commercial and military aerospace technology. The design and optimization of the manufacture of weaponry are also undertaken mainly by mechanical engineers (*Aerospace Engineering*, n.d.; *How to Become a Firearm Engineer*, 2018). Given the career prospects of aerospace and mechanical engineers in Western countries, the relationship between engineering practice and colonial militaries is a common macro-ethical dilemma for students in these fields (Strehl et al., 2023). However, this relationship is not frequently acknowledged nor discussed within educational spaces. Instead, the connection is generally accepted (particularly within aerospace engineering) without critical questioning or discussion by engineers or engineering educators. Thus, we argue the need for aerospace and mechanical engineering programs in Western countries to directly confront the macro-ethical issue of the military-industrial complex in their programs and classrooms through critical questioning, intentional reframing, and dialogue.

Throughout modern history, imperial forces have utilized engineering and technological skills and development to advance the efficiency of weapons in producing mass death and destruction. The success of this agenda has been illustrated by the escalating efficiency of weaponry in terrorizing marginalized populations, typically perpetuated by powerful entities in the Global North at the expense of people and communities in the Global South. The growth and expansion of global capitalism, dominated and controlled by a wealthy elite in Western countries, is inextricably intertwined with ongoing processes of colonialism. Put simply, war is a profitable business. Arms sales from the five biggest 'defense' companies in the world (all of which are located in the United States) totaled over 191 billion US dollars in 2021 (SIPRI, 2021). An economic report in 2012 demonstrated that, within the for-profit corporate model, the defense industry is actually *more* profitable than other areas of industry (Wang & San Miguel, 2012), generating income that is directed to a wealthy elite consisting of corporate owners and shareholders while simultaneously producing weapons that are used to decimate populations in colonized regions. These private 'defense' corporations employ mechanical and aerospace engineers to use their technical training to produce weapons systems. An estimated 30%–60% of engineers employed in the United States work on projects sponsored by military funding sources (Meiksins & Smith, 1996; Papadopoulos & Hable, 2008).

In Western countries, mechanical and aerospace engineering educational spaces and programs fail to challenge this reality. Both *what* we teach and *how* we teach it are tied to military goals, practices, and structures (Lucena, 2011). Military entities and the for-profit corporations that supply them are lucrative sources of income for universities (McLaren & Farahmandpur, 2001). As such, the systemic structures of militarism and capitalism effectively feed on one another to drive forward engineering efforts to enact violence against oppressed peoples, including in spaces explicitly designated for educational purposes. Introducing macro-ethical content into mechanical – and, particularly, aerospace – engineering curricula should directly confront this reality through open acknowledgment and critical discussion.

One recent example that serves as a valuable case study of the macro-ethics of aerospace engineering is Raytheon's production of missiles and other weaponry that have been used by the Saudi Arabian government to commit genocide[4] in Yemen (Bachman, 2019; Wilken & Kane, 2020; Drysdale, 2022). Since 2014, the United States has been supportive of the Saudi-led campaign in Yemen, stemming from conflict between the political interests of the Saudi and Iranian governments. Yemen, the poorest nation in the Arab region, is experiencing ongoing genocide as a result of the conflict, with 233,000 people believed to have died since the start of the conflict (*UN humanitarian office*, 2020). From a macro-ethical perspective, the lived experience of the Yemeni people necessitates critical reflection on the role of the US military-industrial complex in the mass murder of Yemenis. Raytheon has sold billions of US dollars' worth of bombs and missiles to the Saudi Arabian government since the onset of the conflict (LaForgia & Bogdanich, 2020). The company has vehemently denied any wrongdoing, echoing arguments to evade responsibility or blame supported by the prevailing 'apolitical' framing of Western engineering (Cech, 2013). However, from a critical, macro-ethical perspective, engineers at Raytheon would certainly bear part of the responsibility for work that directly results in the murders of tens – if not hundreds – of thousands of innocent people.

Resistance to the military-industrial complex already exists within the body of Western engineering workers, for example, through organized groups such as Science for the People (Rullán, 2021). This movement was formed in the 1960s by students and academics in direct response to research activity that supported the Vietnam War (Wisnioski, 2003). If, throughout their education as mechanical and aerospace engineers, more students become exposed to the critical issue of militarism in engineering and have the opportunity to see resistance work as a possible *part* of being an engineer, this would open up new corridors for organized masses of engineers to radically transform the worlds they exist within, rather than solely using their engineering skills and knowledge to supply the violence of a continued militaristic and capitalistic world order.

Automation

Since the Industrial Revolution of the late eighteenth century, mechanical engineers have designed and built mechanical devices – and later computers – to fully or partially complete tasks without human assistance. Whatever the form, mechanical and aerospace engineers have and will continue to play a large role in developing automated machines like industrial robots or flight management systems. Thus, our educational programs are tasked with presenting to students which considerations and whose voices are worthy of influence within the design process.

A common argument supporting automation is that it has a net benefit for society by increasing safety, efficiency, and the amount of time humans have for leisure. However, history has shown that automation changes the nature of work rather than removing work (Autor et al., 2020; Bainbridge, 1983) and leads to deskilling (the loss of manual skill) in the long term (Feenberg, 2009; Wiener & Curry, 1980), while the benefits of automation are not equitably distributed across society, contributing to and widening the impacts of existing power relations.

As a case study of the societal-level impacts of automation, we look at the introduction of the mechanical tomato harvester in California, the United States, in the 1960s (Winner, 1980). Between 1961 and 1967, the number of harvesters in use grew from 25 to 1,000 (Schmitz & Seckler, 1970), and today nearly all tomatoes grown in the United States for processed food are harvested mechanically. While the tomato harvester was a technical success in the mechanical engineering and agricultural science fields, it significantly impacted agricultural labor in California. The machine displaced nearly a half-million person-hours of work in 1965 alone

(Schmitz & Seckler, 1970), and an estimated 32,000 jobs were lost by the late 1970s (de la Peña, 2013). Most of these jobs were previously filled by Mexican-American workers, who reasonably viewed the harvester as yet another manifestation of racial capitalism. The harvester also led to a consolidation of the tomato-growing industry, in which the number of tomato growers decreased by approximately 85% after the machine's introduction, with smaller farms closing because they did not have the land nor resources to make a profit given the steep price of a mechanical harvester (Carlisle-Cummins, 2015; de la Peña, 2013). This consolidated wealth into the hands of a few powerful owners and led to disproportionate effects on small, rural farming communities.

The tomato harvester is a prominent case study in Winner's (1980) foundational article on the politics of artifacts. Winner does not claim that the mechanical tomato harvester was a plot to benefit large private corporations and hurt small farmers and agricultural workers. Yet, he argues that the mechanical harvester had inherent politics that favored specific social interests (private corporations) while harming others (farm laborers and small farmers). Winner also points to mechanization projects that *were* "almost conspiratorial" in their intentions (Winner, 1980, p. 125). In the mid-1880s, the McCormick Harvesting Machine Company in Chicago spent much money on new pneumatic molding machines. These machines did not simply increase manufacturing efficiency; they also concentrated power with the factory owners by weeding out the skilled workers who were engaged in labor organizing and replacing them with unskilled workers to operate the molding machines. As Winner writes, "After three years of use the machines were, in fact, abandoned, but by that time they had served their purpose – the destruction of the union" (Winner, 1980, p. 125).

Labor automation will continue to displace jobs and political power with inequitable results, concentrating more wealth within a very small subset of people who form the capitalist elite. For example, autonomous vehicles will put more than three million commercial drivers out of work, many of whom have less education (Autor et al., 2020). One common counter-argument to these large-scale job losses is that automation also creates new jobs. However, as the Massachusetts Institute of Technology (MIT) *Future of Work* report points out, the benefits of automation are only achieved equitably in our existing capitalist framework if governments make large-scale investments to retrain displaced workers for new jobs and support communities built around disappearing industries (Autor et al., 2020). This support has rarely materialized as automation has increased.

Alternatively, another solution to the inequity of automation is an emancipatory approach in which communities work together with engineers to determine the automation that *they themselves* want and need. In this way, the focus is not only on the technical aspects of automation; instead, it is on the sociotechnical impacts and emancipatory power of automation to elevate the lives of everyone. *What might it look like to provide structural power to marginalized communities so they can access the benefits of automation without incurring the harms? And what role do engineers play in this process?*

Creating automation technology is a prominent aspect of many mechanical and aerospace engineers' careers. Therefore, students in these fields need to understand the full impact that these technologies have on people and systems. This requires reflection on one's own positionality and power. Once students have engaged in this self-reflection, they can work to understand the perspectives of others – particularly of those with less power who are significantly impacted by automation technology, such as the farm workers in the case of the tomato harvester. Students can learn how automation affects minoritized people and society through critical discussions and investigation of historical case studies and current events (the teaching methods chapters on teaching ethics using case studies, Chapter 20, and reflective and dialogical approaches in engineering ethics education, Chapter 25, provide detailed explanations regarding how to use these approaches).

Sustainability efforts

The choices we have made about technological development, use, and disposal in our recent history have impacted planetary systems to a point where the environmental conditions that enable human survival have become a primary societal concern. It is particularly critical now that future mechanical and aerospace engineers be made aware of the potential impact of their work on society and the environment. Sustainability efforts focused on reducing emissions from automobiles and airplanes are a prominent topic in these disciplines, but this is a narrow focus. All phases of a technology life cycle can impact the environment; sustainable engineering design must consider them all while acknowledging that each phase has its own set of environmental impacts and requires specific sustainable approaches and decisions.

In mainstream engineering terms, sustainable engineering involves improved efficiency, optimization of resources, and disposal. However, it is equally important to encourage students to examine the purpose of the use of the technology and think critically about why the technology exists in the first place and whom it serves functionally, financially, and politically. A specific technology's production, use, and disposal can be considered 'sustainable' according to engineering professional standards by using recyclable materials, applying zero-emission manufacturing processes, and/or being decomposable. However, if that technology's very existence and purpose foster an unsustainable sociotechnical order, we argue that this type of engineering is unsustainable and unethical. For example, even if machines such as mining excavators, harvesting equipment, and oil drill rigs are developed following all current sustainable processes and standards, we argue that these technologies are unsustainable because their purpose is to enable unsustainable practices and processes. This furthers the degradation of our planetary conditions, which has been shown to disproportionately impact oppressed populations such as those in the Global South (Ngcamu, 2023; *Social Dimensions of Climate Change*, n.d.; Suri, 2023).

Take the commonly used case study of the 2015 Volkswagen emissions scandal (Atiyeh, 2019), in which engineers developed and installed software in diesel engines to cheat emissions tests run by regulatory agencies. Using this software was clearly unsustainable, unethical, and illegal, but this case presents an overly simplistic picture of unethical practices in mechanical and aerospace engineering. It gives the impression that unethical (and unsustainable) behavior is easily identifiable and that simply following codes, laws, and regulations makes engineering practice sustainable and ethical. In the case of vehicle emissions, reducing emissions can certainly decrease negative impacts on the environment and people's health. However, a sole focus on transitioning to low-emission and electric vehicles serves to mask our current unsustainable sociotechnical order. Low-emission and electric automobiles and aircraft tackle only the 'use' stage of technology. The make-to-stock manufacturing of vehicles still requires enormous extraction of natural resources – including the lithium needed for batteries for electric vehicles extracted from Indigenous communities in Africa and South America – and ever-increasing landfills for car disposal.

Additionally, these vehicles still foster a car-centric culture that furthers individualistic and consumerist behavior. A sustainable solution would be to provide a public infrastructure that is both environmentally sustainable and structurally accessible (Marx, 2022). Increasing efficiency and reducing emissions can create a 'less unsustainable' society, but it is far from contributing to a truly sustainable and just sociotechnical order.

Engineering sustainability should not only consider the land; it must also consider people's relationship to land. A prominent example is the effect of 'development' on the livelihood – and existence – of *quilombolas* in Brazil, who are descendants of runaway enslaved African and Indigenous peoples. The *quilombola* communities on Boipeba Island, which is under federal environmental protection, are currently threatened by a mega real-estate tourist development that encompasses approximately 20% of the island and includes a resort, residential properties, a golf course, a harbor, and a small airport (*MPF cobra revogação de autorização para megaempreendimento na Ilha de Boipeba, em Cairu (BA)*, 2023; Uzêda & Sabrina, 2023). Mechanical engineers are implicated in this project through their work on the machinery's design, construction, and use; the project's water distribution systems; its buildings' heating, ventilation, and air-conditioning systems; and the manufacturing of materials and parts necessary to make this endeavor a reality. Similarly, *quilombola* people living near the Alcântara Satellite Launch Center on Brazil's northern Atlantic Coast are threatened by a proposed expansion from a joint Brazilian-United States technology agreement in 2019 that would allow US companies to launch rockets from the facility (Fox, 2021; McCoy & Traiano, 2021). The location is attractive to spaceflight companies because of its proximity to the equator, which reduces the fuel a rocket needs to boost a satellite to orbit. However, this expansion would forcibly displace hundreds of people from their *quilombola* communities for the purpose of increased profits for the capitalist elite.

Employing an emancipatory lens, it becomes irrelevant how efficient the buildings are, how 'green' the concrete is, or how non-polluting the vehicles are, as unsustainable projects are defined by their negative impact on people and the environment.

It is essential that students critically analyze cases that do not have a clear-cut answer to whether they are considered ethical or sustainable. If we want to educate engineers capable of making truly ethical and sustainable decisions, students must start asking questions that go beyond the immediate use of technology and consider what types of sociotechnical orders they enable through their practice. *Are the engineers making a project come to life holding paramount the protection of environment – as professional societies (Engineers Canada, 2016; NSPE, 2019) have agreed upon as an ethical responsibility of engineers? Can it be in any way ethical to consider 'profit' concurrent with 'people' and 'planet' as the 'triple bottom line' of sustainability?* If we understand sustainability as the balance between human societies' 'needs' (mostly wants) and the planetary conditions that enable human survival (and advocate for justice and emancipation), then engineering practice must be revisited. Therefore, we suggest that ethical and sustainability issues presented to mechanical and aerospace engineering students go beyond life-cycle analysis and focus on *how* and *if* the technology promotes sustainable sociotechnical orders.

Humanitarian engineering

As Mitcham and Muñoz (2010) explain, humanitarian engineering is "the artful drawing on science to direct the resources of nature with active compassion to meet the basic needs of all – especially the powerless, poor, or otherwise marginalized" (p. 35). It is rooted in the broader humanitarian movement, focusing on relieving human suffering in situations of natural disasters (e.g., earthquakes, tornadoes/hurricanes/tsunamis, and extended droughts) or other calamities (e.g., wars, epidemics, and human-caused disasters).

Humanitarian engineering – sometimes referred to within educational spaces as 'service learning' – offers an alternative framework for applying mechanical and aerospace technologies for explicitly 'humanitarian' purposes. (How to teach using service-learning and humanitarian engineering education is the dedicated focus of Chapter 23.) Many mechanical and aerospace engineering students participate in humanitarian engineering projects as a part of their undergraduate curricular or co-curricular experience. At Michigan State University, for example, 'humanitarian projects' within their mechanical engineering department "aim to make daily tasks such as processing food, retrieving water, caring and storing crops, and simplifying other manual tasks to make them more efficient" (Michigan State University Department of Mechanical Engineering, n.d., p. 3). As another example, the Michigan Sustainability Applications for Aerospace Vehicle Engineering program has worked with Air Serv International to develop an uncrewed aircraft explicitly for use in humanitarian missions (*Air Serv and University of Michigan Partner to Improve Aerial Support in Humanitarian Aid*, n.d.).

The most visible perspective within Western engineering education on humanitarian engineering efforts is one that supports these initiatives as attempts to redirect resources to support those in need. However, from an emancipatory lens, the poor and oppressed deserve both the agency to determine what sociotechnological solutions will best serve their communities and the power and resources to enact these solutions. Humanitarian engineering, on the other hand, typically relies on models of charity and philanthropy, in which powerful entities decide on behalf of marginalized groups 'what these people really need' and then decide how much they would like to provide from the position of a savior. This has race, class, and gender connotations, thus fitting into (rather than challenging) the existing oppressive model of engineering practice (Keshavarz, 2020).

The worldwide and heavily resourced organization of Engineers Without Borders (EWB) is one case study through which we can consider the implications of humanitarian engineering. EWB collaborates on "long-lasting and sustainable infrastructure solutions" through more than 370 "community-driven projects" in 40 countries (EWB-USA, n.d., ¶ 3). The most visible perspective on the organization and its efforts is one of sharing resources to make progress toward meeting basic human needs using engineering technology. From a more critical perspective, EWB provides elite students in highly resourced Western institutions with the valuable opportunity to do hands-on work that doesn't often provide meaningful support to the communities in which they work and which people in those communities do not have agency over. Complaints from target communities suggest that projects often inadvertently create additional work for the people and deplete additional resources rather than providing viable support (Thompson et al., 2022). Meanwhile, Western institutions benefit significantly from the positive publicity generated by EWB, especially within the neoliberal capitalist context in which publicity is monetized. The organization itself began to publicly grapple with its problematic positioning through a short report released in 2019 on the dangers of 'voluntourism' (EWB-SCU, n.d.), and its participants and leaders have also published their own critically reflective analyses (Partida, 2019; Thompson et al., 2022).

The complexity of the macro-ethical implications of humanitarian engineering is significant. Existing critiques describe the personal conflict between a desire to enact positive change in the world and awareness of problematic positioning as a savior. A more profound macro-ethical critique that engages directly with marginalized populations will be required to build structural solutions that serve emancipatory purposes. Emancipatory solutions must provide agency to marginalized populations that allows them to set the goals, determine the methods, and decide how to assess the success of engineering projects themselves. This will significantly affect how mechanical and aerospace engineering students participate in service-learning and community engagement projects, as described in the following section.

Mechanical and aerospace engineering education for social justice and community emancipation

We conclude this chapter by articulating how instructors can bring these conversations into mechanical and aerospace engineering classrooms. First, acknowledging the realities of oppression and power in engineering is a non-trivial step towards ethical and just behavior. Faculty members should make space for students – and themselves – to learn about and consider these issues. Case studies, which prominently feature in micro-ethics education, can also be used for this macro-ethics education in *any* mechanical or aerospace engineering course – especially those usually considered purely 'technical.' Instructors can present their students with articles and information about the examples discussed in this chapter – the sale of United States arms to Saudi Arabia, the mechanical tomato harvester, tourist and spaceport development in Brazil, and voluntourism – and others. Students can use these case studies to learn about minoritized perspectives, reflect on the role of engineering in oppression and emancipation, and engage in constructive dialogue with their peers about how to consider social justice and emancipation in their future engineering careers. For example, Aaron Johnson (an author of this chapter) has designed and implemented one-day dialogue-based macro-ethics lessons for aerospace engineering courses (Benham et al., 2021, 2022; Ennis et al., 2023). These interventions start with a brief lesson introducing students to concepts like positionality, rights-holders, and ethical lenses. Students then read a short issue brief or article about a macro-ethical issue in aerospace engineering, such as the military-industrial complex, orbital debris, or space settlement and resource utilization. Finally, students engage in small- and large-group discussions in which they consider the power of and impact on various stakeholders, also sharing their own perspectives. Importantly, students are encouraged to reflect on their own positionality in relation to the sociotechnical systems they are analyzing.

Beyond just acknowledging and engaging students in dialogues about macro-ethical topics in mechanical and aerospace engineering, it is also important to teach students about ways in which mechanical and aerospace engineering can *empower* people, groups, or communities marginalized by current sociotechnical systems. Empowerment can be defined as the "multi-dimensional social process that helps people gain control over their own lives" (Page & Czuba, 1999, p. 11) by fostering power in people and groups/communities to operate the changes they may want in their own lives, their territories, and in their society. Empowerment is emancipatory when it allows individuals and groups to improve their lives – that is, 'being more fully human' (Freire, 1970) – fight for their rights, or build other possible social realities and/or ways of living, without dwarfing other peoples' rights and search for self-determination.

Ways of using engineering for empowerment range from sociotechnical inclusion (i.e., giving people/groups access to some essential or desired service, such as clean water) to political emancipation (i.e., allowing/encouraging the community to advance its political potentialities and build alliances with other actors that enlarge its potential to enact desirable and agentic changes) (Kleba & Cruz, 2021). Other ways of cultivating a community's empowerment through engineering interventions and designs include acknowledging, cultivating, or sharing cultural differences, qualitative relationships, technical competencies, investigative competencies, social and economic emancipation, and environmental awareness (Kleba & Cruz, 2021). Notably, an emancipatory outcome is usually as dependent on the intervention process as on the sociotechnical product created.

Within the United States, programs and organizations dedicated to using mechanical and aerospace engineering for social justice can serve as examples for students. The research conducted by the Space Enabled Research Group at MIT seeks to advance justice on Earth through applications of space technology. Examples of such research include using satellites to monitor invasive species in Benin (Onuoha, 2019) and developing decision support tools to help local leaders in Indonesia manage coastal flooding (Lombardo et al., 2022). From an advocacy perspective, The JustSpace Alliance (2022) works for more inclusion and ethical consideration in space exploration; in 2022, they campaigned for renaming the James Webb Space Telescope, as James Webb was an administrator in the US Department of State in part responsible for the implementation of government policy to remove employees who were not heterosexual (Prescod-Weinstein et al., 2021; Witze, 2022).

Lastly, there exist frameworks for helping students to intentionally consider social justice and community emancipation in their engineering practice. In Brazil, a movement known as 'popular engineering' (PE) – meant as grassroots engineering – aims to contribute to emancipating marginalized groups and communities through interventions performed by teams of students and teachers linked to a university extension center (Cruz, 2021a). The approach of these extension centers draws on Paulo Freire's emancipatory approach to popular education (Freire, 1970) and on action-research designing/intervention methodology (Greenwood & Levin, 1998) in order to construct – or (re)appropriate – existent technologies alongside (i.e. *with*, rather than *for*) the local community. The devised sociotechnical solutions embody the group's knowledge and values, make sense to the group's members, and can be operated, fixed, and improved upon by the community. Further, these projects take the ideal of solidarity economy and self-management as the intended broader sociotechnical horizon (Cruz, 2021a, 2022).

The construction of a flour mill in a rural community in the northeast of Rio de Janeiro state can serve as a valuable case study of PE for community emancipation. This project was pursued alongside a community of 63 *quilombola* families. The community has lived since 2014 in a settlement for landless rural workers. The community was already accustomed to autonomously organizing its collective work and internal functioning, and it was the community that initially sought support from the Federal University of Rio de Janeiro (UFRJ), Macaé for the mill's construction (Laricchia et al., 2021). In the design and construction of the flour mill, engineering support was provided by undergraduate students and teachers from mechanical, civil, and production engineering programs at UFRJ (Laricchia et al., 2021). Given the specific conditions of the settlement – such as lack of access to electricity – the community and the engineering team decided to use a bicycle to power the cassava shredder. To determine technical aspects of the shredder, the operator was assumed to be a middle-aged woman, since this assumption was realistic for the community's intended use of the mill. To increase safety and promote social interaction, the shredder was designed to allow the two people (the person who feeds it with cassava and the one pedaling it) to work face-to-face (Laricchia et al., 2021). The intentional design of this system enforces conviviality and allows for the inclusion of aged people and people with some disabilities into the productive process. The community determined all these characteristics of the built sociotechnical solution in an agentic manner and, once they were embodied in the designed artifact, these values emulated a sociotechnical reality that supports and strengthens their community (Laricchia et al., 2021).

Engineering practices employing a PE framework seem to align with the framework defined in Western academic spaces of Engineering for Social Justice (J. Leydens et al., 2014). This orientation for engineering practice is committed to fostering social justice and community empowerment through a co-constructed, bottom-up approach to technical design and technology to support realizing marginalized peoples' and communities' imagined alternative social, political, and economic orders. In so doing, it takes engineering itself as something to be widened, demanding expanded definitions of engineering knowledge, leading to alternative ways of conceiving reality and the field of engineering itself. This reframing of mechanical and aerospace engineering fields, then, would impact every step of the educational process, from content delivery and assessment to student projects and career pathways. PE, or Engineering for Social Justice, is an enactment of an alternative conceptualization of engineering.

Concluding remarks

In this chapter, we have employed a critical perspective of engineering ethics to push back against the hegemonic conceptualization of engineering as apolitical and neutral. We showed how mechanical and aerospace engineering have facilitated oppression and inequity and argued that developing students' critical consciousness should be a primary objective of engineering ethics education. The four topics central to mechanical and aerospace engineering presented here – the military-industrial complex, automation, sustainability efforts, and humanitarian engineering – provide examples of how critical perspectives can be applied to analyze the oppressive effects of the fields of mechanical and aerospace engineering and therefore are essential for the education of students in these fields. Educators can engage students in dialogues about these case studies using the summaries and references presented in this chapter. However, these are certainly not the only macro-ethical issues essential to present to mechanical and aerospace engineering students, and we encourage educators to apply a similar critical analysis to other applications that their students will likely contribute to in their engineering careers. Lastly, students need to understand how to use their future positions as engineers to work toward social justice. To help do this, we encourage educators to build on our discussion of mechanical and aerospace engineering education for social justice and community emancipation. Nurturing students' understanding of oppression caused and perpetuated by mainstream mechanical and aerospace engineering – and ways in which these same fields can be reconfigured to help oppressed peoples emancipate themselves – can develop the critical consciousness necessary to empower students to develop and use their skills as decolonial engineers. Such activity can foster a widened and decolonized engineering community capable of supporting the (co-)creation of more just and equal sociotechnical orders.

Notes

1 This chapter uses 'Western' and 'the Global North' interchangeably to signify those countries that engaged in, and benefitted from, colonization at the expense of the Global South (Braff & Nelson, 2022).

2 'Sociotechnical order' is a concept that highlights how 'technology' and 'society' are intertwined, not existing apart from one another (Feenberg, 2010; Karwat et al., 2015). That means that there is no technological reality totally separated from a social order (or, conversely, a social order separated from a technological one) in a way that any change in one of them does require or produce a change in the other. An *emancipatory* sociotechnical order is one in which overcoming oppressions of any kind (economic, racial, religious, sexual, etc.) is actively sought to allow people, individually and collectively, to flourish, or, as Paulo Freire says it, to *be more fully human* (Freire, 1970).

3 Coloniality and the liberation from it, *decolonization*, are meant here as they are defined by the Decolonial Theory. The hegemonic oppression faced worldwide has as its main bases capitalism (and the mainstream politico-military tools to enforce it) – *coloniality of power* – Western technoscience – *coloniality of knowledge* – and Western worldviews, ideals, and values (which articulate Christianity, racism, masculinity, and so on) – *coloniality of being*. Thus, coloniality as a whole is an offspring and continuation of the European colonial domination of the sixteenth to the twentieth centuries, but usually without military occupation. Each coloniality demands and supports the other two, so overcoming any of them requires overcoming all the three. Since engineering is a central player in the coloniality of knowledge, building another possible world (with alternative social arrangements and possibilities of living and making sense of life) demands enlarging the construct of engineering itself. For more on Decolonial Theory, see Chapter 9. For more on decolonizing engineering, see Cruz (2021a, 2021b, 2022).

4 In October 2022, Genocide Watch, a Washington D.C.-based non-governmental organization, referred to the conflict in Yemen as "the world's most severe humanitarian crisis" and declared the conflict an instance of genocide (Drysdale, 2022, p. 1). The fact that many of the deaths are caused by food scarcity, disease, extreme poverty, and the internal displacement crisis (Bachman, 2019; Drysdale, 2022; Wilken & Kane, 2020) – consequences of the violence rather than the violence itself – allows many governments and organizations to avoid characterization of the conflict as a genocide (Bachman, 2019). However, Bachman's

socio-political analysis explains that the Saudi-led coalition (supported by and including Western nations) "is conducting an ongoing campaign of genocide by a 'synchronised attack' on all aspects of life in Yemen" by targeting the medical and economic infrastructure of Yemen's civilian population (Bachman, 2019, p. 1).

There are significant political implications – as well as even more dire possible consequences – of refraining from characterizing the mass death in Yemen as genocide. The same is true in the case of the genocide occurring in Gaza at the time of writing (*Nearly 25,000 Palestinians killed*, 2023). See Bachman (2019) for discussion of the conceptualization of genocide and the application of the term in the case of the Saudi-led Coalition's genocide of the people of Yemen.

References

2023 Aerospace and Defense Industry Outlook. (n.d.). Deloitte United States. Retrieved October 19, 2023, from https://www2.deloitte.com/us/en/pages/manufacturing/articles/aerospace-and-defense-industry-outlook.html

Acosta, A., & Abarca, M. M. (2018). Buen Vivir: An alternative perspective from the peoples of the Global South to the crisit of capitalist modernity. In V. Satgar (Ed.), *The climate crisis: South African and global democratic eco-socialist alternatives* (pp. 131–148). Wits University Press.

Aerospace Engineering. (n.d.). Careers in aerospace. https://www.careersinaerospace.com/roles/engineering/aerospace-engineering/

Air Serv and University of Michigan Partner to Improve Aerial Support in Humanitarian Aid. (n.d.). Air Serv International, Inc. Retrieved October 19, 2023, from https://www.airserv.org/air-serv-and-university-of-michigan-partner-to-improve-aerial-support-in-humanitarian-aid/

Alvear, C., Cruz, C., & Kleba, J. (Eds.). (2021). *Engenharias e outras práticas técnicas engajadas: Vol. 1: Redes e Movimentos*. eduepb.

American Society for Engineering Education. (2022). *Profiles of engineering and engineering technology, 2021*.

Atiyeh. (2019, December 4). Everything you need to know about the VW diesel-emissions scandal. *Car and Driver*. https://www.caranddriver.com/news/a15339250/everything-you-need-to-know-about-the-vw-diesel-emissions-scandal/

Autor, D., Mindell, D. A., & Reynolds, E. B. (2020). *The work of the future: Building better jobs in an age of intelligent machines*. Massachusetts Institute of Technology. https://doi.org/10.7551/mitpress/14109.001.0001

Bachman, J. S. (2019). A 'synchronised attack' on life: The Saudi-led coalition's 'hidden and holistic' genocide in Yemen and the shared responsibility of the US and UK. *Third World Quarterly*, *40*(2), 298–316. https://doi.org/10.1080/01436597.2018.1539910

Bainbridge, L. (1983). Ironies of automation. *Automatica*, *19*(6), 775–779.

Benham, A., Callas, M., Fotherby, R., Jones, M., Chadha, J., Dobbin, M., & Johnson, A. W. (2021). *Developing and implementing an aerospace macroethics lesson in a required sophomore course*. 2021 IEEE Frontiers in Education Conference (FIE), 1–9. https://doi.org/10.1109/FIE49875.2021.9637172

Benham, A., Fotherby, R., Johnson, A. W., & Bowen, C. L. (2022). *Student perspectives of aerospace engineering macroethics issues and education*. 2022 IEEE Frontiers in Education Conference (FIE), 1–5. https://doi.org/10.1109/FIE56618.2022.9962654

Braff, L., & Nelson, K. (2022). The Global North: Introducing the region. In K. Nelson & N. T. Fernandez (Eds.), *Gendered lives: Global issues* (p. Ch. 15). State University of New York Press.

Braun, R., Kleba, J., & Randell, R. (2024). Sociological, postcolonial and critical theory foundations. In S. Chance, T. Børsen, D. A. Martin, R. Tormey, T. T. Lennerfors, & G. Bombaerts, (Eds.), *The Routledge international handbook of engineering ethics education handbook*.

Carlisle-Cummins, I. (2015, July 15). From ketchup to California cuisine: How the mechanical tomato harvester prompted today's food movement. *Civil Eats*. https://civileats.com/2015/07/15/from-ketchup-to-california-cuisine-how-the-mechanical-tomato-harvester-prompted-todays-food-movement/

Cech, E. A. (2013). The (Mis)Framing of social justice: Why ideologies of depoliticization and meritocracy hinder engineers' ability to think about social injustices. In J. Lucena (Ed.), *Engineering education for social justice* (Vol. 10, pp. 67–84). Springer Netherlands. https://doi.org/10.1007/978-94-007-6350-0_4

Christiansen, I. (2020). Linkages between economic and military imperialism. *World Review of Political Economy*, *11*(3). https://doi.org/10.13169/worlrevipoliecon.11.3.0337

Cruz, C. C. (2021a). Brazilian grassroots engineering: A decolonial approach to engineering education. *European Journal of Engineering Education*, *46*(5), 690–706. https://doi.org/10.1080/03043797.2021.1878346

Cruz, C. C. (2021b). Decolonizing philosophy of technology: Learning from bottom-up and top-down approaches to decolonial technical design. *Philosophy & Technology*, *34*(4), 1847–1881. https://doi.org/10.1007/s13347-021-00489-w

Cruz, C. C. (2022). Decolonial approaches to technical design: Building other possible worlds and widening philosophy of technology. *Techné: Research in Philosophy and Technology*, *26*(1), 115–146. https://doi.org/10.5840/techne202253156

Cruz, C. C., Gao, J., & Ramos, A. M. (2024). Philosophical and religious foundations of Engineering ethics in a global perspective. In S. Chance, T. Børsen, D. A. Martin, R. Tomey, T. T. Lennerfors, & G. Bombaerts (Eds.), *Routledge International Handbook on Engineering Ethics Education*.

de la Peña, C. (2013). Good to think with: Another look at the mechanized tomato. *Food, Culture & Society*, *16*(4), 603–631. https://doi.org/10.2752/175174413X13673466711769

Dixit, U. S., Hazarika, M., & Davim, J. P. (2017). *A brief history of mechanical engineering*. Springer International Publishing. https://doi.org/10.1007/978-3-319-42916-8

Drysdale, L. (2022, November 2). Yemen: Genocide emergency. *Genocide Watch*. https://www.genocidewatch.com/single-post/yemen-genocide-emergency

Engineers Canada. (2016). *Public guideline on the code of ethics*. Engineers Canada. https://engineerscanada.ca/publications/public-guideline-on-the-code-of-ethics#-the-code-of-ethics

Ennis, M., Strehl, E., Johnson, A. W., Bowen, C. L., & Jia-Richards, O. (2023). *Work in progress: Implementing an orbital debris macroethics lesson in a junior-level spacecraft dynamics course*. 2023 ASEE Annual Conference & Exposition Proceedings. 2023 ASEE Annual Conference & Exposition, Baltimore, MD.

Estermann, J. (2006). *Filosofía andina: Sabiduría indígena para un mundo nuevo*. ISEAT.

EWB-USA. (n.d.). *International community program*. Engineers Without Borders USA. https://www.ewb-usa.org/programs/international-community-program/

EWB-SCU. (n.d.). *Voluntourism*. Engineers Without Borders USA, Santa Clara University Chapter. https://ewbscu.com/voluntourism/

Ewuoso, C., & Hall, S. (2019). Core aspects of ubuntu: A systematic review. *South African Journal of Bioethics and Law*, *12*(2), 93. https://doi.org/10.7196/SAJBL.2019.v12i2.679

Feenberg, A. (2009). Critical theory of technology: An overview. In G. J. Leckie & J. E. Buschman (Eds.), *Information technology in librarianship: New critical approaches* (pp. 31–46). Libraries Unlimited.

Feenberg, A. (2010). *Between reason and experience: Essays in technology and modernity*. The MIT Press. https://doi.org/10.7551/mitpress/8221.001.0001

Fox, M. (2021, February 16). Hundreds of Black families in Brazil could be evicted to make way for space base expansion. *The World*. https://theworld.org/stories/2021-02-16/hundreds-black-families-brazil-could-be-evicted-make-way-space-base-expansion

Freire, P. (1970). *Pedagogy of the oppressed*. Continuum.

Freire, P. (1978). Extension or communication. In *Education for critical consciousness* (pp. 81–145). Sheed and Ward.

Grayson, L. (1980). A brief history of engineering education in the United States. *IEEE Transactions on Aerospace and Electronic Systems, AES*, *16*(3), 373–392. https://doi.org/10.1109/TAES.1980.308907

Greenwood, D., & Levin, M. (1998). *Introduction to action research: Social research for social change*. Sage Publications.

Herkert, J. R. (2005). Ways of thinking about and teaching ethical problem solving: Microethics and macroethics in engineering. *Science and Engineering Ethics*, *11*(3), 373–385. https://doi.org/10.1007/s11948-005-0006-3

hooks, b. (1994). *Teaching to transgress: Education as the practice of freedom*. Routledge.

How to Become a Firearm Engineer. (2018, June 27). Houston Chronicle. https://work.chron.com/become-firearm-engineer-28076.html

International Engineering Alliance. (2021). *Graduate attributes & professional competencies*.

Karwat, D. M. A., Eagle, W. E., Wooldridge, M. S., & Princen, T. E. (2015). Activist engineering: Changing engineering practice by deploying praxis. *Science and Engineering Ethics*, *21*(1), 227–239. https://doi.org/10.1007/s11948-014-9525-0

Keshavarz, M. (2020). Violent compassions: Humanitarian design and the politics of borders. *Design Issues*, *36*(4), 20–32. https://doi.org/10.1162/desi_a_00611

Kleba, J. B., & Cruz, C. C. (2021). From empowerment to emancipation—a framework for empowering sociotechnical interventions. *International Journal of Engineering, Social Justice, and Peace*, *8*(2), 28–49.

LaForgia, M., & Bogdanich, W. (2020, May 16). Why Bombs Made in America have been killing Civilians in Yemen. *The New York Times*. https://www.nytimes.com/2020/05/16/us/arms-deals-raytheon-yemen.html
Laricchia, C., Oliveira, M., & Costa, R. (2021). Tecnologia social e educação popular o desenvolvimento de uma casa de farinha em um assentamento de reforma agrária. In D. Sansolo, F. Addor, & F. Eid (Eds.), *Tecnologia social e reforma agrária popular* (Vol. 1, pp. 199–242). Cultura Acadêmica Editora.
Le Grange, L. (2019). Ubuntu. In A. Kothari, A. Salleh, A. Escobar, F. Demaria, & A. Acosta (Eds.), *Pluriverse: A post-development dictionary* (pp. 323–326). Tulika Books.
Leydens, J. A., & Lucena, J. C. (2018). *Engineering justice: Transforming engineering education and practice*. IEEE Press.
Leydens, J. A., Lucena, J. C., & Nieusma, D. (2014, June). *What is design for social justice?* 2014 ASEE Annual Conference & Exposition. http://peer.asee.org/23301
Lombardo, S., Israel, S., & Wood, D. (2022). *The environment-vulnerability-decision-technology framework for decision support in Indonesia*. 2022 IEEE Aerospace Conference (AERO), 1–15. https://doi.org/10.1109/AERO53065.2022.9843544
Lucena, J. (2011). What is engineering for? A search for engineering beyond militarism and free-markets. In G. L. Downey & K. Beddoes (Eds.), *What is global engineering education for?* (pp. 361–383). Springer International Publishing. https://doi.org/10.1007/978-3-031-02125-1_6
Marx, P. (2022). *Road to nowhere: What silicon valley gets wrong about the future of transportation*. Verso.
McCoy, T., & Traiano, H. (2021, March 26). A story of slavery—And space. *Washington Post*. https://www.washingtonpost.com/world/interactive/2021/brazil-alcantara-launch-center-quilombo/
McLaren, P., & Farahmandpur, R. (2001). Teaching against globalization and the new imperialism: Toward a revolutionary pedagogy. *Journal of Teacher Education*, *52*(2), 136–150. https://doi.org/10.1177/0022487101052002005
Meiksins, P., & Smith, C. (Eds.). (1996). *Engineering labour: Technical workers in comparative perspective*. Verso.
Michigan State University Department of Mechanical Engineering. (n.d.). *Humanitarian projects*. Engineering for Global Improvement of Human Life. https://web.archive.org/web/20220328093429/https://me.msu.edu/humanitarian-projects
Mitcham, C., & Muñoz, D. (2010). Humanitarian engineering. In B. Caroline (Ed.), *Humanitarian engineering* (pp. 27–35). Springer International.
MPF cobra revogação de autorização para megaempreendimento na Ilha de Boipeba, em Cairu (BA). (2023, March 14). Ministério Público Federal. https://www.mpf.mp.br/ba/sala-de-imprensa/noticias-ba/mpf-cobra-revogacao-de-autorizacao-para-megaempreendimento-na-ilha-de-boipeba-em-cairu-ba
Nearly 25,000 Palestinians killed during 70-day Israeli genocide in Gaza. (2023, December 15). Euro-Med Human Rights Monitor. https://euromedmonitor.org/en/article/6035/Nearly-25,000-Palestinians-killed-during-70-day-Israeli-genocide-in-Gaza
Ngcamu, B. S. (2023). Climate change effects on vulnerable populations in the Global South: A systematic review. *Natural Hazards*, *118*(2), 977–991. https://doi.org/10.1007/s11069-023-06070-2
Nieusma, D. (2015). Analyzing context by design: Engineering education reform via social-technical integration. In S. H. Christensen, C. Didier, A. Jamison, M. Meganck, C. Mitcham, & B. Newberry (Eds.), *International perspectives on engineering education* (Vol. 20, pp. 415–434). Springer.
NSPE. (2019). *Code of ethics for engineers*. National Society of Professional Engineers. https://www.nspe.org/sites/default/files/resources/pdfs/Ethics/CodeofEthics/NSPECodeofEthicsforEngineers.pdf
Onuoha, O. (2019, November 18). Ufuoma ovienmhada of MIT media lab space enabled group is using satellite data to curb the effects of invasive species In Africa. *Space in Africa*. https://africanews.space/ufuoma-ovienmhada-of-mit-media-lab-space-enabled-group-is-using-satellite-data-to-curb-the-effects-of-invasive-species-in-africa/
Page, N., & Czuba, C. E. (1999). Empowerment: What is it? *Journal of Extension*, *37*(5), 1–5.
Papadopoulos, C., & Hable, A. (2008). *Including questions of military and defense technology in engineering ethics education*. 2008 Annual Conference & Exposition Proceedings, 13.725.1–13.725.15. https://doi.org/10.18260/1-2--4345
Partida, S. (2019, October 31). Can engineers without borders build the bridge between STEM and social awareness? *Columbia Daily Spectator*. https://www.columbiaspectator.com/the-eye/2019/10/31/can-engineers-without-borders-build-the-bridge-between-stem-and-social-awareness/
Prescod-Weinstein, C., Tuttle, S., Walkowicz, L., & Nord, B. (2021, March 1). The James webb space telescope needs to be renamed. *Scientific American*. https://www.scientificamerican.com/article/nasa-needs-to-rename-the-james-webb-space-telescope/

Rullán, C. (2021, December 27). Se réapproprier la science. *Science for the People Magazine, 90*. https://magazine.scienceforthepeople.org/online/se-reapproprier-la-science/

Schmitz, A., & Seckler, D. (1970). Mechanized agriculture and social welfare: The case of the tomato harvester. *American Journal of Agricultural Economics, 52*(4), 569–577. https://doi.org/10.2307/1237264

SIPRI. (2021). *SIPRI arms industry database | SIPRI*. Stockholm International Peace Research Institute. https://www.sipri.org/databases/armsindustry

Social Dimensions of Climate Change. (n.d.). [Text/HTML]. World Bank. Retrieved October 23, 2023, from https://www.worldbank.org/en/topic/social-dimensions-of-climate-change

Strehl, E. A., Ennis, M., Johnson, A. W., & Bowen, C. L. (2023). *Work in progress: Undergraduate student perceptions of macroethical issues in aerospace engineering*. 2023 ASEE Annual Conference & Exposition Proceedings. 2023 ASEE Annual Conference & Exposition, Baltimore, MD.

Suri, S. (2023, June 28). *It's time for climate justice- A Global South perspective on the fight against the climate crisis*. Observer Research Foundation. https://www.orfonline.org/research/a-global-south-perspective-on-the-fight-against-the-climate-crisis/

The JustSpace Alliance (Director). (2022, July 5). *Behind the name: James webb space telescope*. https://www.youtube.com/watch?v=jqrZ0Pl-KjQ

Thompson, L., Chan, A., Cannon, J., & Lehr, J. (2022, June). *Deconstructing the white savior model through engineers without borders student chapters: An unlikely intervention*. 2022 ASEE Annual Conference & Exposition.

UN humanitarian office puts Yemen war dead at 233,000, mostly from 'indirect causes' | UN News. (2020, December 1). UN News. https://news.un.org/en/story/2020/12/1078972

US Bureau of Labor Statistics. (2023a, February 6). *What mechanical engineers do*. https://www.bls.gov/ooh/architecture-and-engineering/mechanical-engineers.htm#tab-2

US Bureau of Labor Statistics. (2023b, February 21). *What aerospace engineers do*. https://www.bls.gov/ooh/architecture-and-engineering/aerospace-engineers.htm#tab-2

Uzêda, A., & Sabrina, F. (2023, March 23). Boipeba: Dono da globo quer construir condomínio de luxo em área federal comprada de empresário acusado de tomar terras. *The Intercept Brasil*. https://www.intercept.com.br/2023/03/23/boipeba-dono-da-globo-quer-construir-condominio-de-luxo/

Wang, C., & San Miguel, J. (2012). The excessive profits of defense contractors: Evidence and determinants. *Journal of Public Procurement, 12*(3), 386–406. https://doi.org/10.1108/JOPP-12-03-2012-B004

Wiener, E. L., & Curry, R. E. (1980). Flight-deck automation: Promises and problems. *Ergonomics, 23*(10), 995–1011. https://doi.org/10.1080/00140138008924809

Wilken, D., & Kane, S. (2020, December 7). Timestream: The yemen crisis. *Genocide Watch*. https://www.genocidewatch.com/single-post/timestream-the-yemen-crisis

Winner, L. (1980). Do artifacts have politics? *Daedalus, 109*(1), 17.

Wisnioski, M. (2003). Inside "the system": Engineers, scientists, and the boundaries of social protest in the long 1960s. *History and Technology, 19*(4), 313–333. https://doi.org/10.1080/0734151032000181077

Witze, A. (2022, November 18). NASA really, really won't rename Webb telescope despite community push-back. *Nature*. https://www.nature.com/articles/d41586-022-03787-1

16

ETHICAL ISSUES IN ELECTRONIC AND ELECTRICAL ENGINEERING

Susan M. Lord and John E. Mitchell

Introduction

As is evident throughout this handbook, ethics education in engineering takes many forms. In this chapter, we focus on the discipline of electronic and electrical engineering (E&EE) with the aim to provide examples of various discipline-specific approaches for the classroom. We hope E&EE educators may be inspired to try one of the approaches or develop their own. This, and the other discipline-specific chapters (Chapters 14, 15, 17, and 18 of this handbook) showcase the importance of including options for students that engage their disciplinary interests rather than only generic 'engineering ethics' where some of the examples might not resonate with students. For instance, the examples of failures of bridges or hotels may not seem as relevant to electrical engineering students as examples drawn from consumer electronics. It is important for E&EE students to conceptualize ethics as not merely an academic exercise but something relevant to their lives and professional practices. In addition, engineering ethics is an inherently interdisciplinary field, spanning multiple engineering as well as social sciences disciplines, and including ethics in engineering curricula helps students see the value of knowledge in other disciplines.

In this chapter, after describing our positionality as authors, we then compare ethics education in electronic and electrical engineering to other engineering disciplines to highlight why we believe that ethics has not had as much prominence in E&EE as it has had in some other disciplines. Following this, we discuss the importance of ethics within E&EE education. We consider the literature on ethics specific to electronic and electrical engineering education, including discussions of professional responsibility connected to codes of ethics and E&EE courses. We provide examples, from the literature and our own work, of integrating ethical considerations into specific technical E&EE courses. We demonstrate that although there are ethics-related case studies available, relatively few relate directly to electronic and electrical engineering; those relevant for E&EE mostly cover general engineering considerations or codes of conduct rather than E&EE scenarios. We propose that instructors draw from the teaching models and ethical issues and concerns presented below; various educators and researchers have identified these concerns and used these models to integrate ethical issues within the delivery of technical content. The presented set of E&EE-related concepts and models is not exhaustive; rather, we aimed to identify broad topics and identify some specific problematic examples from E&EE culture that instructors can use to

DOI: 10.4324/9781003464259-20

facilitate in-class discussions. We believe these can be integrated within the context of technical content delivery.

Positionality

We, the two authors of this chapter, are electrical engineering educators with decades of experience in academia. We have held leadership positions in the IEEE Education Society. Susan Lord is a white cisgender woman with undergraduate and graduate degrees in electrical engineering who is a full professor at a US university focusing on teaching. Her research is in engineering education. Her experiences of marginalization as a woman in E&EE have contributed to her desire to change the culture of E&EE to be more welcoming and inclusive. John Mitchell is a white cisgender man with undergraduate and graduate degrees in electronic and electrical engineering and a full professor at a research-intensive UK university. His career started with research focused on communications systems but has developed to focus mainly on engineering education, particularly curriculum design. Both have been involved in developing integrated programs where technical and transferable skills, such as ethics, are combined within the curriculum. We acknowledge that our positions of privilege have informed our approach to our work, including the writing of this chapter. Thus, our examples are drawn from the published Western literature that we know best.

Ethics in E&EE compared to other disciplines

Fleddermann (2000) stated in the opening of his paper on ethics case studies that "Rarely is electrical technology at the focus of the classic case studies used in engineering ethics courses and textbooks" (p. 284). Our research for this chapter has demonstrated that this continues to be the case two decades later. Although ethics is undoubtedly taught in electrical engineering courses, the case studies tend to be situated outside the discipline or to address professional issues where the setting is electrical and electronic engineering but the E&EE context is not central to the ethical issue at hand (e.g., whistle-blowing in semiconductor manufacturing that could be in any manufacturing process). Although research (e.g., Barry & Whitener, 2011) has suggested that electronic and electrical engineers are reasonably well prepared to handle ethical issues, this may be because ethics classes have typically considered generic, professional ethics rather than issues directly linked to the discipline (Bielefeldt et al., 2018).

We argue that it is important for students to grapple with ethical considerations (a) explicitly relevant to their discipline, (b) integrated into their technical studies, and (c) taught by engineers. To accomplish this, it is crucial to understand why electrical engineering case studies are rare. We hypothesize that while other branches of engineering produce infrastructure that is directly public-facing – civil engineering (buildings), chemical engineers (chemical plants), mechanical engineers (cars and planes) – electronic engineers especially (but also electrical engineers) build components that are within all these engineered systems or products. This degree of separation means that direct ethical considerations related to electrical and electronics engineering are less obvious than those within other engineering disciplines. Of course, this is just a perception – and one that in many areas has never been entirely true and becomes less so with computer systems and electronic control being at the heart of the modern world. This view is encompassed in the growing refrain that modern engineering and modern engineering graduates must be equipped to grapple with the ethical and social aspects of engineering and the technical aspects. To do this effectively, they must perceive ethical considerations as central to the problem-solving process of their discipline, recognizing that a 'good' solution must be ethical, sustainable, and inclusive just as much as technically feasible, manufacturable, and financially viable.

Although many of these ethical considerations will be social or professional in nature, some will be technical and can be quantified with calculations and technical arguments. The next section describes enacted and proposed approaches for bringing ethics into the electrical engineering classroom.

Importance of ethics in E&EE education

For many programs, the teaching of ethics within electronic and electrical engineering is based on codes of conduct or codes of ethics (see Chapter 5). This is unsurprising, as many of these codes are linked to professional bodies – and often to the accrediting bodies that will evaluate the content of courses (see Chapters 32–36 for more on accreditation). As such, these codes provide an interesting starting point for discussing ethics within the electrical engineering curriculum.

In our realm, the best known of these is the IEEE Code of Ethics (IEEE, 2020), which provides a normative ethical framework of both consequentialist (e.g., those relating to health and safety) and deontological (bribery and corruption) positions. These are spelled out in even more detail in examples from the National Society of Professional Engineers (NSPE) Code of Ethics for Engineers (NSPE, n.d.). Chapter 2 identifies and defines a wide range of applicable theoretical frameworks.

Many professional engineering institutions have produced similar statements of ethics and professional values worldwide. However, they typically apply to the engineering profession in general rather than to electronic and electrical engineering specifically. The Association of Computing Machinery (ACM) (2018) emphasizes several issues of particular importance to those in computing systems – a topic that has considerable overlap with E&EE (e.g., issues related to privacy, issues when modifying or retiring systems, and systems that become integrated into the infrastructure of society) – and identifies relevant ethical and professional codes. In the UK, the Institute of Engineering and Technology (IET) (2019), formerly the Institute of Electrical Engineers) also produces a rules-of-conduct document and in it (in keeping with rules of conduct), ethics and professional codes are combined. Codes in the E&EE realm frequently include considerations for upholding the image and reputation of the engineering profession; for example, the IET (2019, p. 3) specifies that "Members shall neither advertise nor write articles (in any medium) for publication in any manner that is derogatory to the Institution or to the dignity of their profession."

Although professional bodies are often considered the bastions of professional standards, they face their own ethical challenges. For example, a lively debate ensued when the IEEE announced that they would ban Huawei scientists as reviewers in their publications in response to legislation within the United States (Mervis, 2019). This highlights that while we usually consider that codes encompass moral binarism, stating clear and delineated positions on professionalism and professional ethics, applying and upholding them is far more complex. As we will explore later in this chapter, some of the most interesting applied moral discussions invoke an element of moral relativism in forming personal ethics. (For more on moral development theories, see Chapter 10. For more on relativism, see Chapter 28 on epistemological development.)

Although the professional bodies oversee frameworks for the engineering profession, it falls to accrediting bodies to distill these codes into required learning of professionally accredited programs. These typically follow the specifications set out by the Washington Accord (International Engineering Alliance, 1989) and are all relatively similar in scope and coverage, as seen in Chapter 32.

Approaches to ethics in E&EE education

Historically, electrical and electronic engineers have been involved in ethics education. Many engineering ethics textbooks have been written by electrical engineers, including those by Martin and Schinzinger (1996; Martin is in philosophy and Schinzinger in E&E engineering), Unger (1994), Fledderman (2008), and Baura (2006). Joseph Herkert, another E&E engineer, collaborated on a review of engineering ethics that included a section on education (Barry & Herkert, 2015). Herkert also edited a volume on *Social, Ethical, and Policy Implications of Engineering* (2000) that drew from work conducted in the IEEE Society for the Social Implications of Technology. The Society publishes the *IEEE Technology and Society* magazine, which often includes work related to ethics (IEEE Technology and Society, n.d.).

Colby and Sullivan (2008) published a review of approaches to engineering ethics education that focused on standalone courses in ethics (taught by engineering or other faculty), discussion of professional responsibility (often tied to codes of ethics), and modules (typically delivered within two or three class periods). Colby and Sullivan based their work on in-person visits to undergraduate programs in mechanical and electrical engineering at universities in the United States. We use Colby and Sullivan's categorization of discussion of professional responsibility tied to codes of ethics, standalone courses, and modules to frame our review of the literature in this section.

Discussion of professional responsibility tied to codes of ethics

In their teaching, faculty members in E&EE have incorporated discussions of professional responsibility tied to codes of ethics (often the IEEE Code of Ethics). They often incorporate these codes when discussing case studies that fit into one class period. This type of discussion, or content presentation, could be facilitated at any level of study; therefore, we provide examples below from Master's courses and (first-, third-, and fourth-year) undergraduate courses.

The literature shows that discussions of professional responsibility may be specifically focused on the IEEE Code of Ethics. In a project-based course titled 'Electric Power Engineering' for Master's students at Chalmers University in Sweden, Ehnberg et al. (2022) helped students tie the IEEE Code of Ethics to their specific projects. Students used the code as a tool to identify ethical dilemmas or risks for their project and then explored ways to avoid these dilemmas in their final reports. To illustrate, in a project that was designing an electrical brake for a wind turbine, students considered the ethical canon of "To hold paramount the safety, health, and welfare of the public, to strive to comply with ethical design and sustainable development practices, to protect the privacy of others, and to disclose promptly factors that might endanger the public or the environment" (Ehnberg et al., 2002, p. 2) from the IEEE Code of Ethics. Students identified a risk/dilemma of the brake system failing in strong winds and described preventive action of making sure to be well aware of the safety issues related to installation; students highlighted these in their report. After this experience, the students reported that they believed the IEEE Code of Ethics was relevant to their projects and future careers. Ehnberg et al. (2022) stress the importance of focusing on discussion with real-world examples. An interesting aspect of this research is that the interventions were done in-person, online, and pre-recorded; all three approaches had similar outcomes in terms of student engagement and responses to questions about the relevance of the codes to their projects, suggesting that instructors have flexibility in how they choose to implement such work.

Some discussions of professional responsibility focus on case studies and use the codes of ethics to suggest ways to analyze or move forward. Clancy et al. (2005) developed a module for one 3-hour laboratory period for a first-year module on circuits at Worcester Polytechnic Institute in the United States. Students examined case studies on engineering ethics drawn from internet resources (National Institute for Engineering Ethics, n.d.). Case studies were chosen to be relevant and easily understood by first-year students but not to be specifically relevant to E&EE topics. Students learned about and used the IEEE Code of Ethics to suggest action plans. Data collected from focus groups indicated enhanced student awareness of ethical issues. The authors provided useful advice for others interested in incorporating ethics into their classes. They believe that the adoption of the program by other faculty members was made easier due to the availability of an unused laboratory period in the first week of the term. Materials were also provided, and one of the authors assisted the instructor during the first offering of the program.

Ekong (2015) developed a module for teaching ethics that could be incorporated into any E&EE undergraduate course using case studies from electrical and computer engineering practice and exposure to codes of ethics from the IEEE, IEEE Computer Society, Association for Computing Machinery (ACM), and National Society for Professional Engineers (NSPE). Ekong implemented this intervention at Mercer University in the United States in two 50-minute class periods in a third-year microcontroller class with introductions, case studies, assignments, and quizzes. Ekong stressed the importance of having an electrical engineering faculty member lead the instruction because "a student is more likely to relate to the topic, e.g. Engineering Ethics, if that topic is taught by a professor in the student's discipline" (p. 1). Ekong chose E&EE-related case studies because the case studies "in ethics courses are usually taken from mechanical and civil engineering disciplines. Electrical and Computer Engineering (ECE) students may have difficulties relating to these cases" (p. 1). The case studies from NSPE focus on E&EE topics: software security, quality of products (defective chips), copyright (using unlicensed proprietary software to create a new software product), and compliance with ADA guidelines. Students must connect the specific NSPE code of ethics to the case study (NSPE, n.d.). (See Chapter 18 on ethical issues in software engineering for more on this realm.)

Another example of case studies tied to the IEEE Code of Ethics has been implemented in several capstone design courses within electrical engineering. Motivated by the desire to incorporate achievement of ABET (2000) Student Outcomes related to ethical responsibility and contemporary issues, Jiménez et al. (2006) developed a 2-hour module on 'Social and Ethical Implications of Engineering Design,' which aims to help students "reflect on issues and challenges associated with (i) professional integrity, (ii) engineering, industry and social responsibility, and (iii) technological impact, societal and global awareness" (p. 1). This module was incorporated in capstone courses in various areas of electrical and computer engineering taken in the last year of undergraduate study at the University of Puerto Rico-Mayagüez. The 2-hour module included an introduction to ethics, a discussion of professional integrity in engineering, and a consideration of ethical frameworks, including the IEEE Code of Ethics. Each course featured a case study relevant to the topics of that capstone course in the discussion of integrity in engineering. For example, in the 'Communication System Design: Signal Processing' course, the ethics case study centers on digital rights management (DRM) and the tension between consumer rights and the right of companies to protect their digital content. In the 'Communication Systems Design: Circuits and Antennas' course, the ethics case study considers the health hazards of electromagnetic waves – focusing on the high-power radio transmission towers in Cesano, Italy, and their impact on the local community (Hellemans, 2005). Jiménez et al. (2006) reported that students who responded to surveys at the end of the courses were positive about the impact of these modules on their own development. All students said it was very important for 'ethics to be integrated into engineering courses.' Over 80% indicated that the module had a high impact (5/5) on their 'willingness to be guided by ethical principles in professional work' and 'to be alert to and sensitive to ethical problems.' However, only 60% strongly agreed that others would be motivated to act ethically due to this module. In their comments, students expressed a desire for more opportunities to practice using ethical principles in their electrical engineering designs and curricula.

Standalone courses on ethics taught by E&EE faculty members

Passino (1998) provided an example of using Martin and Schinzinger's (1998) textbook to support the teaching of professional and ethical aspects of E&EE to a class of 120 students at Ohio State University in the United States. This example is primarily an ethics class taught by an E&EE faculty member but not focusing specifically on E&EE topics. The course was intentionally placed in the final year of the curriculum in hopes that most students had already worked on an engineering job and could bring this experience to the discussions. The main topics were "(1) safety and risk with case studies; (2) engineering as social experimentation and its link with design with case studies; and (3) professionalism and organizational issues with case studies" (Passino, 1998, p. 274). The author recommended avoiding:

> spending too much time on ethical theories at the expense of getting the students to debate case studies (engineering students tend to identify much more closely with case studies and become convinced of the importance of the material easier than via ethical theories). It is important to make some connections between technical design issues in engineering and safety, risk, ethical, and professional issues. Certainly, some time should be spent on codes of ethics.
>
> *(Passino, 1998, p. 274)*

At the University of Illinois at Urbana-Champaign in the United States, Michael Loui created an elective course, 'Engineering Ethics,' for third- and fourth-year students studying electrical and computer engineering (Loui, 2005). The course emphasizes "ethical issues in engineering including professionalism, responsibility, confidentiality, conflict of interest, risk and safety, relationships between engineers and managers, loyalty, whistle-blowing, codes of ethics, licensing, and choosing a vocation" (Loui, 2005, p. 384). Loui specifically chose case studies related to E&EE, such as the Bay Area Rapid Transit (BART) case, where technical problems arose from the electronic sensors, electrical signaling, and software controls. The BART case is historically important because it was the first and only time that IEEE filed an amicus brief in support of whistle-blowing engineers (Unger, 1973). Loui found that students benefited from cases of actual incidents and activities that included discussions with diverse perspectives.

Modules integrating ethics in electrical engineering curricula

Colby and Sullivan's recommendations for engineering programs to better prepare students for "the ethical-professional dimensions of their work" (2008, p. 335) include (1) defining ethics and professional responsibility broadly specifically going beyond codes of ethics; (2) integrating with other learning goals; and (3) using active pedagogies, since professional responsibility includes skills and habits in addition to knowledge. Current efforts to incorporate ethics into electrical engineering curricula follow these recommendations; they integrate ethical issues related to electrical engineering into required courses where the ethics are tied to technical learning outcomes using active pedagogies and not directly tied to the IEEE Codes of Ethics. This is consistent with the pioneering work of Donna Riley, who developed a handbook on thermodynamics that provided specific examples of tying social and ethical content to technical content (Riley, 2011).

Below, we summarize such efforts in E&EE for the standard electrical engineering courses of 'Introduction to Circuits' and 'Controls Systems.' We also provide some guidelines for doing this type of work and examples of student responses.

Circuits class

The 'Introduction to Circuits' class is typically the first course that students in electrical engineering that students majoring in EE&E encounter. It is a required course for students in other engineering disciplines as well. It is often taken in the second year of the curriculum. Finding ways to incorporate ethics into this course has powerful implications for students seeing the relevance of ethics to the field of E&EE.

Conflict minerals: Lord et al. (2018) incorporated a module on conflict minerals into the 'Introduction to Circuits' course at the University of San Diego in the United States. Conflict minerals include tantalum, tin, tungsten, and gold mined in areas such as the Democratic Republic of the Congo (DRC) where the money from their production supports armed conflict. The module was designed to connect conflict minerals' ethical implications to capacitors, a typical topic in this course. Learning objectives included:

- Analyzing capacitors as electrical devices
- Defining conflict minerals and describing at least two social issues surrounding them
- Describing where conflict minerals are used
- Describing potential options for engineers concerned with societal implications of conflict minerals

Before the module, students completed calculations about tantalum (Ta) in capacitors and cell phones and identified where Ta is mined. During the in-class module, the instructor defined and introduced some history about conflict minerals. Students discussed conflict minerals and their societal and ethical implications and brainstormed ways to reduce reliance on conflict minerals as engineers. For homework after the module, students were each assigned a well-known company. They researched the company's conflict-minerals policies and presented their findings to classmates in a subsequent class. Students were asked to highlight social implications and concerns about these strategies. Additional modules explored electronics recycling and sustainable innovation (Lord et al., 2018).

*EV batteries and circular economy:*In another module for 'Introduction to Circuits,'students explore electric vehicle (EV) batteries, tying them to the technical topic of voltage dividers which is typically covered in this class and the concept of the circular economy, where products are reused or recycled for as long as possible (Judge et al., 2022; Lord & Finelli, 2023). Learning objectives include:

- Designing a voltage divider for a DC source to illustrate repurposing EV battery packs
- Estimating energy available in end-of-life EV batteries
- Describing societal risks introduced by recycling EV batteries that could be alleviated by applying circular economy principles

Instructional activities include listening to a podcast about the circular economy and answering questions, estimating the energy demand existing end-of-life EV batteries could meet, discussing how the circular economy relates to circuits' concepts and EV batteries, and discussing ways to use the circular economy to repurpose batteries. For example, batteries no longer suitable for EVs could be used in residential solar energy systems. Exercises for homework include designing a voltage divider to provide a specific output voltage from a repurposed EV battery and exploring the effect of energy degradation on EV battery repurposing. The module helps students explore the ethical issues of the circular economy as an alternative to the traditional economy, how engineering design is connected to the end of product life, and electric vehicles and sustainability.

Lord and Finelli (2023) are working on a US National Science Foundation (NSF) grant to incorporate sociotechnical modules, including ethical considerations, into the 'Introduction to Circuits' course. They will implement the conflict minerals and EV batteries modules in larger classes and develop more modules (Finelli & Lord, 2023). They are recruiting partners who teach 'Introduction to Circuits' in other universities to implement these modules elsewhere.

Controls class

The concept of social justice can be an interesting way to incorporate ethical considerations into electrical engineering courses. Researchers at the Colorado School of Mines in the United States incorporated social justice concepts in an 'Introduction to Feedback Controls' (IFCS) course/module (Johnson et al., 2015; Leydens et al., 2021). The course is for electrical and mechanical engineering students and is taken as a technical elective in students' third or fourth year of undergraduate study. The interventions related to social justice included a guest lecture by a faculty member from social sciences focused on the 'Engineering for Social Justice' (E4SJ) criteria (Leydens & Lucena, 2018) and a reading from Riley's (2008) *Engineering and Social Justice* book on mindsets in engineering. As Johnson et al. (2015) state, "Social justice defies a universal definition, but is related to the vision that people and communities have the right to equality (in various senses), to health, to dignity, and to opportunities" (p. 1). Thus, social justice considerations for engineering often explore ethical questions of impact. The authors provide an example of modern agricultural machinery where:

> Advanced control systems have made crops more affordable. These same systems have reduced the sustainability of the family farm, significantly changing the agricultural lifestyle, which has had far-reaching implications on rural communities and their economies. Thus, for engineering practice, a social justice framework encourages exploration of the following questions: In the short and long-term, from engineering designs, models, and other interventions, who benefits? Who does not benefit? Who suffers?
>
> *(Johnson et al., 2015, p. 2)*

In the controls class, instructors presented what they called 'social justice' examples that connected the controls' topic of resonance to the safety implications of an unbalanced washing machine and issues related to wind energy systems and active prosthetics. Homework problems were rewritten to include social considerations. For example, a problem about a water tank with no context was rewritten to explicitly ask students to balance protecting the pump, ensuring enough water for a village's needs, and not wasting water by overfilling. Using focus groups, the researchers investigated students' responses to incorporating this content in courses where the social justice content was made visible (Group A) compared to the traditional approach where the social justice content is not visible (Group B). Their results indicated that some students valued incorporating ethical topics into this technical class while others did not. Many students focused on social justice as related to "individual ethics or responsibilities but did not recognize the obligations of the broader group of professional engineers to society" (Leydens et al., 2021, p. 740). The findings highlight the importance of integrating ethical considerations into multiple classes since one class is insufficient to help students learn to deal with the complexity of ethical issues and develop robust ethical knowledge and ideas.

Pedagogical guidelines and student responses

Incorporating ethical issues effectively in electrical engineering classes is challenging for many reasons. The instructor's identity can come into play because those educated with a deep technical focus may not feel prepared to engage in discussions about ethics or venture into challenging topics related to race, gender, and so on. Instructors need some degree of comfort in dealing with ambiguity, which differs from how most technical work is taught (for more on this, see Chapter 28). For example, there is a 'right' answer to calculating the voltage at the middle of a voltage divider but no 'right' answer to the question: *Should I get a new cell phone every year now that I know the impact on people's lives?* Learning these ethical concepts by doing only mathematical problems is impossible. On the other hand, ethical issues can be incorporated using collaborative methods such as discussions and active learning (Colby & Sullivan, 2008; Loui, 2000). (Please see Chapter 25 on reflective and dialogical approaches in engineering ethics education.) It is also important for instructors to follow good pedagogical practices, including having clear learning objectives and assessments (see Chapter 29). Gelles and Lord (2021) proposed a framework for integrating sociotechnical content into engineering classes that might also help teachers consider integrating ethical content, which is sociotechnical. The (Gelles & Lord, 2021) framework involves the following steps:

0. Identify possible sociotechnical collaborators.
1. Identify a salient course topic that has broader social and environmental implications.
2. Identify, add, or update existing course learning objectives and/or the ABET student outcome(s) that this sociotechnical course topic aligns with.
3. Create learning objectives for specific sociotechnical modules.
4. Create modules by designing activities for homework before and/or after class session(s) and class session(s) that integrate technical content and calculations students are familiar with and social and environmental context.
5. Include low stakes assessment for the module (e.g., homework) and consider including sociotechnical questions on exams.
6. Conduct formative assessment and/or engineering education research on sociotechnical modules to get student input and improve module offerings in the future.
7. Refine modules and identify possible sociotechnical collaborators for the next course offering.

In contrast to some efforts to include ethics, which are met by resistance among engineering students, this integrated approach with modules tied to technical content has been shown to be appreciated by students. Students see ethics as 'real world' engineering and beneficial for their learning. For example, after the conflict minerals module, students participated in surveys and interviews (Lord et al., 2018, 2019). In the survey, all students who responded said the topic mattered to them as engineers. In the interviews, several students pointed out that these topics were not typical for engineering classes and that they found them engaging and would like to see more of this.

> *I thought it was a really interesting topic that has larger social consequences. It was cool to get away from the stigma of engineers only worrying about math and showing that engineering is able to have effect in other disciplines.*
>
> *(Lord et al., 2019, p. 5)*

The experience helped some students see connections between their personal and professional lives as ethical engineers.

> *Prior to this class, I did not even know what conflict minerals were or where they were used. The in-class group presentations on this topic were especially relevant because I researched on Samsung, while having a Samsung phone. I learned that non-conflict tantalum and tin are used in the circuitry of my cellphone. In the future, I can apply this knowledge as an engineer (in design) and as a consumer (in purchasing from companies that have specific procedures for dealing with conflict minerals).*
>
> *(Lord et al., 2019, p. 11)*

In the interviews, students indicated that they found modules to be well-integrated into the class. When asked about engineering as a field, every student brought up the modules, emphasizing that they saw the ethical context as important for developing their sense of engineering in the real world and its potential. It did, however, challenge some students' definition of 'engineering' as indicated below:

> *Obviously, we looked at a lot of stuff that wasn't engineering including the conflict minerals, and the Sunshine Box which I thought was really cool. And that was very clearly ... I mean it was engineering but at the same time it was very clearly like looking at it from different angles.*
>
> *(Lord et al., 2019, p. 12)*

One measure of the success of these modules is that a student specifically recommended keeping them and even expanding them because they were beneficial for learning as "*real examples of how the things worked*" (Lord et al., 2019, p. 13).

In interviews that Lord and Finelli conducted after the EV battery module, students said that they found the module interesting, impactful, and relevant.

> *I came in hating electrical engineering, like it was just not for me. So I think like actually doing the voltage divider and using that for like sustainability purposes and the circular economy was really cool to like actually be like, okay, the stuff we're learning is like being used for something ... I liked that part of it.*
> *... we are a part of the issue if we don't decide to fix it.*
>
> *(Lord & Finelli, 2024, p. 4)*

Opportunities for exploring ethical issues in electronic and electrical engineering

In considering ethics in E&EE, there are many opportunities to explore issues that have not been explored in the literature. In this section, we begin with some issues that could easily be included in a technical class to align ethical discussions with the delivery of electronic and electrical engineering content. As highlighted above, although there are some publications in these areas, they have not all focused on the pedagogical implementation of discussion of such issues in the electrical engineering classroom. In the subsequent section, we identify topics specific to the culture of electronic and electrical engineering that may provide useful talking points for instructors who are comfortable leading discussions of sensitive ethical issues. Both sections provide instructors with ideas regarding ethical considerations they could use with students. We encourage electrical engineering educators to take up the challenge of doing this important work and explore how to directly connect consideration of ethics with the technical electronic and electrical engineering content they are teaching and the culture in which they are embedded.

Broad issues for future E&EE curricula

This section provides some examples of ethical issues that could be covered in electrical or electronic engineering classes in Table 16.1. For each, we include an example of a potential reading that might be introduced in the class. These are examples for future development, not citations of work reporting curricular interventions.

Ethical issues in the culture of electronic and electrical engineering

In addition to tying to technical issues, we encourage readers to look locally and consider the culture of E&EE and how implicit and explicit aspects of the academic and professional culture can impact students in the E&EE classroom. *What messages do engineering educators send by the language that we use?* Here, we discuss some terms prevalent in E&EE that can adversely impact the sense of belonging for many people, including women and students of color: the language of master/slave, male/female connectors, resistor codes, and the 'Lena' photograph used in digital image processing (an area within electronic engineering). These are opportunities for E&EE instructors to identify their own potential blind spots and explore relevant ethical issues with their students in classes. This is probably more challenging than the sociotechnical integration we call for in the previous section. We suggest a pedagogical approach for having these discussions at the end of this section as well as some important considerations for educators who want to take on the challenge of doing this work to change culture.

Technical jargon

Taheri (2020) explains the importance of language as

> one of the most powerful tools we have as humans that incorporates personal assumptions, social norms, and cultural ideologies. It is therefore important to consider language critically and to watch for biases in usage. Language reflects the world it is used in, but it is also active in maintaining or redesigning that world. It can be a tool of discrimination or of empowerment.
>
> *(Taheri, 2020, p. 151)*

Taheri goes on to review some discriminatory and non-inclusive language in science, technology, engineering, and mathematics (STEM) fields to raise awareness and suggest alternatives. Two of these examples are particularly relevant to electrical engineering and are discussed further here.

Master/slave: Historically in E&E engineering, the phrase 'master/slave' is used for digital designs where one section of software, hardware, or firmware, the 'master,' controls another, the

Table 16.1 Ethical topics that could be integrated into E&EE classes

E&EE Topic/ Class	*Summary of ethical issue*	*Supporting material/ case material*
Electro-magnetics	There have been numerous news reports expressing concerns about the abundance of electromagnetic energy sources, primarily for communication but also power lines, and their effect on human health. Although the risk is low, there is evidence that it is not entirely non-existent. Engineering class discussions might consider what level of risk is acceptable, who should decide, and potentially even link to technical evaluations such as the power from a mobile phone mast at a distance over the power from the phone itself.	Examples – Ghorbani et al. (2018); Reading – Hardell and Sage (2008)
Integrated circuit design	Instructors could structure a case study (Chapter 20) and facilitate reflection and discussion (Chapter 25) around the responsibility to disclose to customers information on flaws in the design of an integrated circuit (IC), the Intel Pentium Processor. Students can be prompted to consider the company's responsibilities and ethical position on releasing potentially damaging details to the users of its products.	Example Fleddermann (2000); Reading Crothers (1994)
Electronic design	Consumer electronics have developed to a position where typically, repair is not possible. If a significant fault occurs, replacement and disposal are the only options. Recently, a global movement backed by legislation on the 'right to repair' has sought to reverse this. The aim is to require manufacturers to design in the possibility for repair where possible. However, doing this is not always easy and may have unintended consequences. Cunningham and Hobbs (2023) may be used for a debate on how far 'right to repair' should go and where it is appropriate.	Reading – Cunningham and Hobbs (2023); O'Reilly (2020)
Off-grid, smart-grids, and renewable energy	The topic of electricity supply is a rich area for discussions that links ethical and sustainability considerations with technical evaluation. Exercises may ask groups to consider energy supply in a remote area, where considerations about the users are as central to the engineering design solution as the technologies available. The design could be supported by energy usage monitoring and discussions of available materials to minimize environmental impact.	Example – Louie (2018)
Image technology and privacy	With the growth of imaging and security technologies come potential privacy issues. Harrington (2014) highlights the level of intimate detail that can be observed. Engineering class discussions could consider how privacy and security could be balanced, and if alternative technologies might be possible to increase security while preserving privacy.	Reading – Harrington (2014)
Weapons systems	Many electronic engineering employment opportunities can be found in the military and defense sector and its supply chains. This is an ideal opportunity for a discussion on personal ethics and how each person must use their own moral compass to determine where they feel comfortable. Some may feel perfectly fine working on autonomous weapon systems, whereas others would refuse this type of work. Chapter 15, on ethical issues in mechanical and aerospace engineering, discusses such dilemmas in detail.	Reading – Hersh (2022)
Wireless systems class	Apple Airtags have a wide range of useful and innocuous applications, yet they have also raised security and privacy concerns. *What responsibility do technology companies have for the unintended uses of their products? Is it on firms to consider the implications and produce 'fixes' that reduce potential harm?*	Reading – Roth et al. (2022)

'slave.' The use of this terminology began in 1904. Eglash (2007) explored the term's history, its relationship to racialized social connotations, and why it is so popular in engineering, and then posed recommendations for alternatives. In 2004, this term was listed by the Global Language Monitor (Reuters, 2004) as one of the most politically incorrect terms of the year. In the wake of the 2020 murder of George Floyd in the United States and the heightened awareness of systemic racism, stronger calls have emerged for the elimination of this terminology including by students (Steele, 2020), EE professionals (Ellis, 2020), and the media (Canales, 2020).

> Santiago Gomez, a graduate student in computer engineering, was so perturbed when he encountered the terminology in a textbook – *Digital Design, 6th Edition* – for his Logic Design course ... that he wrote to the textbook's publisher, Pearson, calling for the language to be changed.
>
> *(Steele, 2020, p. 2)*

Alternatives have been proposed and adopted. For example, in 2021, Microchip issued a product change notification that listed "1) Replaced terminology "Master" and "Slave" with "Host" and "Client" respectively" (Microchip, 2021, p. 1). It is interesting that the reason for change stated was "To Improve Manufacturability" (p. 1). Other proposed alternatives include 'leader–follower,' 'primary–secondary,' 'writer–reader,' 'primary–replica,' and 'coordinator–worker' (Taheri, 2020, p. 153).

Danowitz and colleagues (2021) investigated the importance of this terminology for student inclusion and belonging at an undergraduate predominantly white university in the United States. They found that "42% of students surveyed either agree or strongly agree that use of master–slave terminology is problematic, including 100% of female and 100% of African American students, and that the use of the terminology may create conditions to evoke Stereotype Threat" (Danowitz et al., 2021, p. 1). The authors also critiqued the usefulness and accuracy of the terminology for learning.

Male/Female connectors: Traditionally, connectors used throughout E&EE are referred to as 'male' and 'female,' referring to whether they have a plug or a socket. As described by Wikipedia, "the female connector is generally a receptacle that receives and holds the male connector" (Wikipedia Gender of Connectors, n.d., ¶ 1). Wikipedia has a long entry on this topic, which conflates gender and sex, does not critique or suggest problematic aspects of this terminology, and provides detailed descriptions of hermaphroditic connectors and gender changers. This terminology can be seen as reductionist, reducing gender to physical attributes as well as being uncomfortable in terms of its focus on sexual intercourse with the joining of these connectors called 'mating.' This terminology also can be seen as unnecessarily sexualizing and enforcing a heteronormative narrative. Wikipedia lists plug, pin, and prong as options for 'male' connectors, and receptacle, socket, and slot for 'female' connectors (Wikipedia Gender of Connectors, n.d.). Various creative alternatives to 'male' and 'female' have been proposed such as 'worm–apple,' 'pen–cap,' 'bottle–cork,' and 'sword–sheath' (Eveleth, 2015) or 'outie–innie' (Pearlstein, personal communication, 2019).

Resistor codes

Another problematic aspect of the culture of E&EE is the mnemonic code or phrase used for remembering the resistor color code. In 1961, Alan Dundes wrote a paper exploring the richness of mnemonic devices as examples of folklore in various fields where he flatly states the following for 'resistor code reminder' (1961, p. 41). Although the language is horrifying, it is sometimes used by educators today, and to show what students over the past half century have been exposed to, we quote Dundes' explanation:

> This device gives the resistance in ohms of a coded resistor. The code is: Black (0); Brown (1); Red (2) ; Orange (3); Yellow (4) Green (5); Blue (6); Violet (7); Gray (8); and White (9). 0 1 2 3 4 5 6 7 8 9 (ohms) B B R O Y G B V G W (colors). Thus if a resistor had a red, a green and a black band, the resistance would be 2, 5, and 0, that is, 250 ohms.

The mnemonic device:

> Bad Boys Rape Only (Our) Young Girls But Violet Gives Willingly.
> Black Boys Rape Our Young Girls But Violet Gives Willingly.
> Bad Boys Run Our Young Girls Behind Victory Garden Walls.

There is no commentary in the Dundes (1961) article on the sexist or racist nature of these devices. One example of the impact of such codes is depicted in *Violet Gives Willingly*, a documentary film about a woman "confronting troubling memories of her short-lived career as an electrical engineer" (International Documentary Association, n.d., ¶ 3) in 1974 (Sanford, 2022). As a woman studying E&EE in the 1980s, one of us (Susan) was horrified by the codes she heard as an undergraduate student and did not want to commit them to memory due to the references to misogyny and rape.

Yet Dundes' (1961) mnemonic persists. A recent search we authors conducted for 'resistor color codes' using Quora[1] turned up a range of possibilities including some of the ones above. A more recent website suggests using a less offensive but still gendered pneumonic device "Bright Boys Rave Over Young Girls But Veto Getting Wed" (WikiHow, 2023). A response to the query we ran using ChatGPT in February 2024 on "What are several mnemonics for the resistor color code," provided "Bad Boys Ravage Our Young Girls But Violet Generally Wins" as the first option and the one with "Violent Gives Willingly" fifth.

Recently, the first author of this chapter, Susan, briefly referenced the resistor mnemonics in class, and several students (white women and men of color) stopped her and asked for more explanation. Developing a discussion around this topic could be a rich experience for E&EE students and faculty. However, we recognize that not all educators will have the skill to facilitate such an explicit discussion with undergraduates, but they can work to develop such skills to help E&EE education and culture evolve.

Image processing

Sometimes, what might seem innocuous and commonplace in a technical community may have a more problematic backstory. One such example is the Lena (or Lenna) image. If you have ever looked into work on image processing or image compression standards such as JPEG, you will likely have seen this image – a 512 x 512-pixel image of the face and upper arm of a young woman in a straw hat with a blue feather turning to look over her shoulder towards the camera. The image is of Swedish model Lena Forsén (Söderberg), but what is perhaps not so commonly known is that the complete image (of which a cropped area from the top portion of the origin photograph is the widely used test image) first appeared as a nude centerfold in the November 1972 issue of 'Playboy' magazine. The image was used by the Signal and Image Processing Institute (SIPI) at the University of Southern California for comparative tests between compression algorithms but has now become ubiquitous within the image processing community. It is argued that the widespread use of an image from what many consider to be a pornographic source only serves to reinforce the underrepresentation of women in computer sciences and engineering (Culnane & Leins, 2019), including by Lena herself, as she stated in the 2019 short film *Losing Lena* (Bartley, 2019).

It is of particular interest how the male-dominated culture of the technology and computing industry has served to not only normalize the use of such an image over many years but continues to do so even to this day. As a journal editor, the second author, John, recently found himself in an altercation with an eminent professor of computer science at a top US institution over his suggestion that alternative images would be far more appropriate. This exposes a potentially interesting debate with students on several fronts, but mainly around why the provenance of such an image matters and how it impacts the drive for diversity and inclusion with engineering and tech. It is now common for journals to advocate for the use of alternative (and arguably technically better images) such as 'Cameraman,' 'Mandril,' or 'Peppers,' and some, such as 'Nature Nanotechnology,' have stated that they will automatically reject submissions using the 'Lena' image (Nature Nanotech, 2019).

In November 2023, decades after the image first appeared, the IEEE officially adopted this policy citing ethical reasons.

> IEEE's diversity statement and supporting policies such as the IEEE Code of Ethics speak to IEEE's commitment to promoting an inclusive and equitable culture that welcomes all. In alignment with this culture and with respect to the wishes of the subject of the image, Lena Forsén, IEEE will no longer accept submitted papers which include the "Lena image".
>
> *(IEEE Author Center Magazines, 2024)*

Moving forward

Hopefully, this offensive language is not taught to students in today's E&EE classrooms. We might prefer not to state language such as the resistor codes at all in hopes that they will die out. Yet, today, the tradition is still alive online, and addressing it as an ethical issue is essential to changing the culture. Discussing such issues with a critical eye as to how they have contributed to and been reflective of the culture of E&EE is difficult but important work. It is worth noting that E&EE continues to have a very low representation of women compared to other disciplines in both the United States (Lord et al., 2015, 2019) and the United Kingdom (Bellingham et al., 2023). Educators need to do more to identify and address cultural factors pushing women and people of color away from E&EE studies and careers. To confront ethical issues in E&EE related to language, belonging, and discrimination, we encourage E&EE educators to reflect on their own language choices. Educators can develop skills in discussing these sensitive topics by educating themselves; talking with colleagues outside of engineering in fields such as sociology or ethnic studies or education; discussing these issues with colleagues, friends, and graduate students; and then introducing them to undergraduate students. We recognize that some dangers exist for raising sensitive topics in class, particularly for students and instructors from vulnerable minoritized groups. Instructors must carefully navigate these discussions, ensuring a safe and inclusive learning environment while addressing the historical and contemporary issues in the field. Chapter 25 on reflective and dialogical teaching approaches may be a helpful resource.

The University of Washington Information Technology group has a website with an Inclusive Language Guide, which contains information regarding problematic terminology in information technology, including 'master–slave,' 'male–female' connectors, and others such as 'blacklist-whitelist' and 'blackhat/whitehat.' This is a good resource for exploring technical terminology and why it can be considered problematic (UW-IT, 2023).

A pedagogical strategy that might be helpful for educators who want to lead discussions of these ethical issues is the confront-address-replace (CAR) framework. According to Asfaw and colleagues,

> The CAR Strategy is meant to be a proactive and modern pedagogy which encourages discussion and thought on whether or not we should replace questionable aspects within engineering. The CAR Strategy does not force students to replace 'master–slave' or any terminology from their vernacular – it simply welcomes it.
>
> *(Asfaw et al., 2021, p. 25)*

Specifically, this CAR strategy involves *confronting* the historical significance of a term such as 'master–slave.' The strategy then moves to *addressing* the inaccuracies of this problematic terminology. Finally, the discussion turns to recommendations for alternatives to *replace* the problematic language, although these are suggestions rather than prescriptions. Researchers have studied the impact of a CAR strategy as a teaching framework in computer engineering at a predominantly white undergraduate university focusing on 'master–slave' terminology (Asfaw et al., 2021). Over two-thirds of the students surveyed in that study agreed that this was an effective framework for eliminating offensive terminology and that they would like to see it used in other engineering classes where it is applicable. The researchers suggest that this strategy could be used for other examples of problematic E&EE terminology.

Conclusions

In this chapter, we have identified the importance and historic scarcity of case studies that are directly and technically relevant to students of electronic and electrical engineering. We have highlighted example topics where instructors can go beyond the typical professional subject of ethics and codes of conduct to introduce ethical issues. We have suggested discussion points and case studies that can be integrated in the core curricula and classes that form electronic and electrical engineering. We have also demonstrated that within electronic and electrical engineering there exists a number of contemporaneous case studies and issues that would make excellent discussion points within the technical classes of an electronic and electrical engineering program. These include sociotechnical cases where ethics trade-offs are a central consideration and issues of the culture and language within electronic and electrical engineering which can be directly linked to engineering content and bring the discussion of ethics into technical classes. We have presented cases that can provide active learning activities to breakup lecture classes by introducing participatory debates on topics directly related to the theoretical concepts under discussion. We consider that this explicit connection of ethical concerns with technical content is vital to engage students and allow them to see that ethics is not an afterthought or an irrelevant and abstract concern but central to good electronic and electrical engineering practice and best practice in electronic and electrical engineering design. We hope that instructors will find these cases useful and will be able to integrate them into their classes.

Acknowledgments

We thank Tom Børsen for inviting us to write this chapter. We acknowledge the work of Erin Henslee and Pinar Denis Tozun, who worked on another draft of this chapter that they were not able to complete. We thank Michael Loui for his careful reading of a draft.

Note

1 https://www.quora.com/Whats-the-best-way-to-remember-resistor-colour-codes

References

ABET. (2000). *Criteria for accrediting engineering programs, 2001 – 2002.* https://www.abet.org/accreditation/accreditation-criteria/criteria-for-accrediting-engineering-programs-2022-2023/

ACM. (2018). *ACM code of ethics and professional conduct.* Retrieved January 12, 2024, from https://www.acm.org/code-of-ethics

Asfaw, A. F., Randolph, S., Siaumau, V., Aguilar, Y. R., Flores, E., Lehr, J. L., & Danowitz, A. (2021, July). *Statistical analysis of the CAR (Confront, Address, Replace) strategy and its efficacy when applied to master-slave terminology*. 2021 ASEE Virtual Annual Conference Content Access.

Barry, B. E., & Herkert, J. R. (2015). Engineering ethics. In A. Johri & B. M. Olds (Eds.), *Cambridge handbook of engineering education research* (pp. 673–692). Cambridge University Press.

Barry, B. E., & Whitener, J. C. (2011, June). *How do civil, electrical, and mechanical engineering students compare?: Ethically speaking.* 2011 American Society for Engineering Education Annual Conference Proceedings, pp. 22–781.

Bartley K. (Director). (2019). *Losing Lena*, [Film]. Finch.

Baura, G. D. (2006). *Engineering ethics: An industrial perspective*. Elsevier.

Bellingham, K., Mitchell, J. E., Guile, D., & Direito, I. (2023). *Exploring the variation in gender balance on undergraduate engineering courses in UK universities.* Proceedings of the SEFI Annual Conference.

Bielefeldt, A. R., Polmear, M., Knight, D., Swan C., & Canney, N. (2018). *Education of electrical engineering students about ethics and societal impacts in courses and co-curricular activities.* IEEE Frontiers in Education Conference (FIE), San Jose, CA, pp. 1–9

Canales, K. (2020, July 2). Twitter is dropping coding terms like 'master' and 'slave' after 2 engineers led an internal effort to press for change. *Business Insider*. https://www.businessinsider.com/twitter-dropping-coding-terms-master-slave-2020-7

Clancy, E. A., Quinn, P., & Miller, J. E. (2005). Assessment of a case study laboratory to increase awareness of ethical issues in engineering. *IEEE Transactions on Education, 48*(2), 313–317.

Colby, A., & Sullivan, W. M. (2008). Ethics teaching in undergraduate engineering education. *Journal of Engineering Education, 97*(3), 327–338.

Crothers, B. (1994, November 18). Pentium woes continue, *Infoworld, 16*(48), 1–18.

Culnane, C., & Leins, K. (2019). *It's time to retire Lena from computer science, engineering and technology.* University of Melbourne.

Cunningham, R., & Hobbs, D, (2023). The evolution of the right to repair. *Antitrust, 37*(3), 43–47.

Danowitz, A., Asfaw, A. F., Benson, B., Hummel, P., & McKell, K. C. (2021, January). *Assessing the effects of master slave terminology on inclusivity in engineering education.* 2021 CoNECD.

Dundes, A. (1961). Mnemonic devices. *Midwest Folklore, Autumn, 11*(3), 139–147.

Eglash, R. (2007). Broken metaphor: The Master-Slave analogy in technical literature. *Technology and Culture, 48*(2), 360–369.

Ehnberg, J., Lundmark S. T., & Lundberg S. (2022). *Introducing ethics by IEEE code of ethics in international electrical power engineering education.* 2022 31st Annual Conference of the European Association for Education in Electrical and Information Engineering (EAEEIE), Coimbra, Portugal, 2022, pp. 1–6.

Ekong, D. U. (2015, April 9–12). *An instructional module for teaching ethics within an ECE curriculum.* SoutheastCon 2015, Fort Lauderdale, FL, USA, pp. 1–4.

Ellis, L. (2020 June 18). *It's time for IEEE to retire master-slave.* https://www.eetimes.com/its-time-for-iE&EE-to-retire-master-slave/

Eveleth, R. (2015). *A modest proposal for re-naming connectors and fasteners.* The Last Word on Nothing. https://www.lastwordonnothing.com/2015/11/27/a-modest-proposal-for-re-naming-connectors-and-fasteners/

Finelli, C. J., & Lord, S. M. (2023, September). *Integrating sociotechnical issues in the introduction to circuits course*. 2023 European Society for Engineering Education (SEFI) Conference, Dublin, Ireland.

Fleddermann, C. B. (2000). Engineering ethics cases for electrical and computer engineering students, *IEEE Transactions on Education, 43*(3), 284–287.

Fleddermann, C. B. (2008). *Engineering ethics* (2nd ed.). Pearson Education.

Gelles, L. A., & Lord, S. M. (2021). Pedagogical considerations and challenges for sociotechnical integration within a Materials Science class. *International Journal of Engineering Education*, *37*(5) 1244–1260.

Ghorbani, M., Maciejewski, A. A., Siller, T. J., Chong, E. K., Omur-Ozbek, P., & Atadero, R. A. (2018, June). *Incorporating ethics education into an electrical and computer engineering undergraduate program*. American Society for Engineering Education Annual Conference Proceedings, Salt Lake City, UT.

Hardell, L., & Sage, C. (2008). Biological effects from electromagnetic field exposure and public exposure standards. *Biomedicine & Pharmacotherapy*, *62*(2), 104–109.

Harrington J. E. (2014). *Dear America, I saw you naked*. https://www.politico.com/magazine/story/2014/01/tsa-screener-confession-102912/

Hellemans, A. (2005, October). *Sins of transmission: Vatican's radio high-power antennas stand accused of causing cancer*. IE&EE Spectrum, pp. 12–15.

Herkert, J. R. (Ed.). (2000). *Social, ethical, and policy implications of engineering: Selected readings*. Wiley-IEEE.

Hersh, M. A. (2022). Professional ethics and social responsibility: military work and peacebuilding. *AI & Soc*, *37*, 1545–1561.

IEEE. (2020). *IEEE code of ethics*. https://www.iE&EE.org/about/corporate/governance/p7-8.html

IEEE Author Center Magazines. (2024, February 7). Article Submission Requirements, https://magazines.ieeeauthorcenter.ieee.org/create-your-ieee-magazine-article/article-submission-requirements/

IEEE Technology and Society Magazine. (n.d.). https://technologyandsociety.org/technology-and-society-magazine/

Institute of Engineering and Technology. (2019). *IET rules of conduct*. https://www.theiet.org/about/governance/rules-of-conduct/

International Documentation Association. (n.d.), *Violet Gave Willingly*. https://www.documentary.org/project/violet-gave-willingly

International Engineering Alliance. (1989). *Washington accord, international engineering alliance*. https://www.ieagreements.org/accords/washington/

Jiménez, L. O., O'Neill-Carrillo, E., Frey, W., Rodríguez-Solis, R., Irizarry-Rivera, A., & Hunt, S. (2006, October). *Social and ethical implications of engineering design: A learning module developed for ECE capstone design courses*. IEEE Frontiers in Education. 36th Annual Conference

Johnson, K., Leydens, J. A., Moskal, B., Silva, D., & Fantasky, J. S. (2015). *Social justice in control systems engineering*. 2015 American Society for Engineering Education Annual Conference Proceedings, Seattle, WA.

Judge, G., Lord, S. M., & Finelli, C. J. (2022). *A social responsibility module exploring electric vehicle batteries for a circuits class*. 2022 American Society for Engineering Education Annual Conference Proceedings, Minneapolis, MN. https://peer.asee.org/40831

Leydens, J. A., Johnson, K. E., & Moskal, B. M. (2021). Engineering student perceptions of social justice in a feedback control systems course. *Journal of Engineering Education*, *110*(3), 718–749.

Leydens, J. A., & Lucena, J. C. (2018). *Engineering justice: Transforming engineering education and practice*. Wiley-IEEE Press.

Lord, S. M., & Finelli, C. J. (2023, October). *Work-in-progress: Sociotechnical modules for the introduction to circuits course*. Frontiers in Education (FIE) Conference, College Station, TX. https://doi.org/10.1109/FIE58773.2023.10343488

Lord, S. M., & Finelli, C. J. (2024, June). *Integrating sociotechnical issues in electrical engineering starting with circuits: Year 1*. 2024 American Society for Engineering Education Annual Conference, Portland, OR.

Lord, S. M., Layton, R. A., & Ohland, M. W. (2015). A multi-institution study of student demographics and outcomes for electrical and computer engineering students in the USA. *IEEE Transactions on Education*, *58*(3), 141–150. https://doi.org/10.7227/IJMEE.42.1.5

Lord, S. M., Ohland, M. W., Layton, R. A., & Camacho, M. M. (2019). Beyond pipeline and pathways: Ecosystem metrics. *Journal of Engineering Education*, *108*(1), 32–56. https://doi.org/10.1002/jee.20250

Lord, S. M., Przestrzelski, B., & Reddy, E. (2018, November). *Teaching social responsibility: A conflict minerals module for an electrical circuits course*. Proceedings of the 2018 WEEF-GEDC Conference, Albuquerque, NM. https://ieeexplore.ieee.org/document/8629755

Lord, S. M., Przestrzelski, B., & Reddy, E. (2019, June). *Teaching social responsibility in a circuits course*. 2019 American Society for Engineering Education Annual Conference, Tampa, FL. https://peer.asee.org/33354

Louie, H. (2018). *Off-grid electrical systems in developing countries*. Springer.
Loui, M. C. (2000). Fieldwork and cooperative learning in professional ethics. *Teaching Philosophy, 23*(2), 139–155.
Loui, M. C. (2005). Ethics and the development of professional identities of engineering students. *Journal of Engineering Education, 94*(4) 383–390.
Martin, M. W., & Schinzinger, R. (1996). *Ethics in engineering* (3rd ed.). McGraw Hill.
Mervis, J. (2019). Update: In reversal, science publisher IE&EE drops ban on using Huawei scientists as reviewers. *Science Insider*. https://www.science.org/content/article/update-reversal-science-publisher-ieee-drops-ban-using-huawei-scientists-reviewers
Microchip. (2021). *Product change notification for SYST-22XRSR883*. https://media.digikey.com/pdf/PCNs/Microchip/SYST-22XRSR883.pdf
Nature Nanotech. (2019). A note on the Lena image. *Nature Nanotechnology, 14*, 728. https://doi.org/10.1038/s41565-019-0490-2
National Institute for Engineering Ethics. (n.d.). http://www.niee.org
National Society of Professional Engineers. (n.d.). *NSPE code of ethics for engineers*. https://www.nspe.org/resources/ethics/code-ethics
O'Reilly, K. (2020, January 28). Engineering a repairable world. *IEEE Spectrum*. https://spectrum.ieee.org/engineering-a-repairable-world
Passino, K. M. (1998). Teaching professional and ethical aspects of electrical engineering to a large class. *IEEE Transactions on Education, 41*(4), 273–281.
Reuters. (2004, Decēmber 2). "Master/slave" named most politically incorrect term. *Seattle Post Intelligencer*. https://www.seattlepi.com/national/article/master-slave-named-most-politically-incorrect-1161133.php
Riley, D. (2008). *Engineering and social justice*. Morgan & Claypool.
Riley, D. (2011). *Engineering thermodynamics and 21st century energy problems: A textbook companion for student engagement*. Morgan & Claypool.
Roth, T., Freyer, F. Hollick, M., & Classen, J. (2022, May 26). *AirTag of the clones: Shenanigans with liberated item finders*. IEEE Security and Privacy Workshops (SPW), San Francisco, CA, USA.
Sanford, C. (Director). (2022). *Violet gave willingly*. https://violetgavewillingly.com/
Steele, M. (2020, July 16). *Striking out racist terminology in engineering*. https://violetgavewillingly.com
Taheri, P. (2020). Using inclusive language in the applied-science academic environments. *Technium Social Sciences Journal, 9*, 151–162.
Unger, S. H. (1973). The BART case: Ethics and the employed engineer. *IEEE CSIT Newsletter, 1*(4), 6–8. https://doi.org/10.1109/CSIT.1973.6498744
Unger, S. H. (1994). *Controlling technology: Ethics and the responsible engineer* (2nd ed.). John Wiley & Sons.
University of Washington Information Technology (UW-IT). (2023). *IT īnclusive language guide*. https://itconnect.uw.edu/guides-by-topic/identity-diversity-inclusion/inclusive-language-guide/
Wikihow. (2023, August 24). *How to remember electrical resistor codes*. https://www.wikihow.com/Remember-Electrical-Resistor-Color-Codes#:~:text=Here's%20one%20mnemonic%20%22Bright%20Boys,%2C%207%2C%208%2C%209.&text=Best%20method%20to%20memorize%20is,Great%20Britain%20Veto%20getting%20wed
Wikipedia. (n.d.). Gender of connectors and fasteners. https://en.wikipedia.org/wiki/Gender_of_connectors_and_fasteners#:~:text=The%20terms%20plug%2C%20pin%2C%20and,used%20for%20%22female%22%20connectors

17
ETHICS IN CHEMICAL ENGINEERING

Jan Mehlich, Tom Børsen, and Dayoung Kim

Introduction

In this chapter, the ethical dimensions of chemical engineering (ChE) activities, including their effects on the environment and human lifeworld, will be addressed. This description aims to identify possible content for ethics education geared toward chemical engineering students. After pointing out the specific and characteristic features of chemical engineering in contrast to other fields of engineering and to academic chemistry, various cases and scenarios of chemical engineering efforts with ethical implications are presented. These are contextualized along the lines of different domains of responsibility. The underpinning idea is that engineering ethics education should prepare chemical engineers to exercise different forms of responsibility in their forthcoming careers and that familiarity with matching content will enable this pedagogical objective (Barry & Herkert, 2015; Bielefeldt et al., 2019; Børsen et al., 2020; Shallcross, 2010; Shallcross & Parkinson, 2006). Presenting historical and contemporary cases will allow the ChE student to understand their quintessence and apply this orientational knowledge to assessing other cases. Since the practical application of engineering ethics competence plays out in discourses in interdisciplinary team assemblages, the chapter presents a short practical guide on the role of chemical engineers in such professional settings. Hopefully, this chapter will inspire educators who design ethics courses or teach ethics in ChE to choose content that ChE students and their future employers find relevant in exercising professional and ethical responsibilities.

The author team includes a female Korean engineering educator with a chemical engineering background now living and working in Virginia (United States), a Caucasian male Danish chemist-turned-techno-anthropologist, and a Caucasian male German chemist-turned-ethicist. All three work as academic scholars/researchers at universities and are primarily trained in Western intellectual traditions in their disciplines, yet with different cultural (Asian and European) upbringings. We acknowledge that these personal backgrounds have undoubtedly shaped the choice of content, perspectives, comments, and conclusions presented in this chapter. For example, all authors are formed by Western ethical frameworks and share a belief that chemical engineers have both epistemic and social responsibilities.

DOI: 10.4324/9781003464259-21

Chemical engineering and its specific features

The demarcation of ChE against other engineering fields and against the academic scientific discipline of *chemistry* is important for understanding the specific ethical dimensions, and it explains why ChE, with its specificities, deserves its own chapter in this book. ChE graduates employ scientific knowledge of chemical substances, their characteristics and reactions, and their effect on organisms and ecosystems, for the elaboration of large-scale synthesis and production processes, transportation and storage methods and technologies, and various applications for industrial and consumer products. The chemical industry, as the main employer of chemical engineers and as the primary locus of research and development (R&D) with a strong innovative driving force, is an economically significant key industry in most developed countries around the globe. This translation of knowledge about new and synthetic substances and reactions sustains the creative character of the chemical sciences and, more often than not, enters uncharted territory with new developments and designs. Thus, the ethical relevance of *chemical* engineering springs not only from its sheer quantity of impactful output but also from its inherent newness and multi-leveled risk potential.

Chemical engineering and chemistry as an academic science

Chemistry investigates the background of the science encompassing aspects of the organic, inorganic, analytical, physical, and bio-chemistry. ChE is more multidisciplinary and practical, applying engineering science to problems relating to heat transfer, fluid dynamics, equipment design, and so on. Most chemists work in laboratories as research scientists, analytic chemists, pharmacologists, biochemists, toxicologists, or forensics, whereas typical industries for chemical engineers are energy, mining, food and drink, pharmaceuticals, wastewater, pulp, paper, and so on (Edwards & Shelley, 2018).

The ethical dimensions of these two domains' work are different. Since most of the work that chemists do has only little direct impact on people and the environment, codes of professional conduct and good scientific practice are at the center of science ethics. While there are certain codes of good engineering practice that apply to chemical engineers, there are external responsibility dimensions to engineering ethics that concern the impact of engineering products on individuals and societies as well as the ecosphere. The urgency of the societal and environmental impact is, on the one hand, caused by the mere scale of the output. While chemists work in labs with relatively small amounts of chemicals in closed experimental set-ups, most engineering work is done at the industrial scale with processes that involve a large throughput of substances and energy. On the other hand, it arises from the nature of ChE that a large group of stakeholders – people for whom something is *at stake* because of ChE work output – is entangled in ChE compared to academic chemistry. ChE stakeholders include entrepreneurs, managers, workers, regulators, clients and customers, consumers, and civil society. Therefore, the work of chemical engineers can more directly involve macro-ethical issues, which distinguishes it from academic chemistry (where work involves more micro-ethical issues like research ethics; see below for further details). Overall, chemical engineers have more touch points with interdisciplinary perspectives given their work's natural engagement of multiple stakeholders.

Chemical engineering and other engineering domains

Simply put, ChE deals with chemicals while other branches of engineering do not. This fact needs to be reflected in ethics teaching for ChE students. Chemical engineers design and manufacture materials and products using scientific principles from chemistry, biology, mathematics, and physics. They may come up with innovative processes to use and transform energy. They can also work with microorganisms, food, pharmaceuticals, and fuels. ChE encompasses:

- Biochemical, biomolecular, cellular, and microbial engineering
- Biomedical, pharmaceutical, and tissue engineering
- Biotechnological and genetic engineering
- Food, textile, and paper engineering
- Materials and molecular engineering
- Metallurgical, welding, and corrosion engineering
- Petroleum, plastics, and polymer engineering
- Process engineering

In all these fields, working with substances and materials is central to daily activities (Denn, 2011; Green & Southard, 2018; Shallcross, 2017). Chemical engineers are experts in exploiting knowledge about material properties and characteristics for applications in devices and processes. These materials, their conversion or transformation products, or their waste products impact people, society, and the environment. Given the large scale of industrial processes in which ChE expertise is employed, these impacts can potentially be disastrous. Thus, an increased responsibility for overseeing and controlling the impacts of the material output of professional work may be formulated for ChE.

Given the scope of this chapter, we will not introduce details of each subdiscipline within ChE listed above. Instead, we propose contents and teaching strategies that can be jointly taught across different chemical engineering subdisciplines, focusing on common issues and themes applicable to various subdisciplines.

Cases for teaching ethics in ChE

This section presents prominent and insightful historical and contemporary ethical case studies relevant to chemical engineers. The compiled cases are extracted from different sources: a collection of ethical case studies in a special issue of the *International Journal for Philosophy of Chemistry HYLE* (re-issued in Schummer & Børsen, 2021), engineering ethics textbooks (Bowen, 2014; Harris et al., 2018; Johnson, 2020; Whitbeck, 2011), and searches in the academic journals *Science and Engineering Ethics*, *Science, Technology, & Human Values*, and *Journal for Business Ethics*. The selection of ethical case studies was then analyzed, and those considered most ethically relevant for ChE were selected. These are presented below to illustrate the normative implications of chemical engineering activity and possible solution pathways for conflicts encountered in professional contexts. As each case is linked to different forms of responsibility chemical engineers need to consider, educators can select which cases to use based on their class topic.

Ethical issues related to the unpredictable nature of chemicals

As outlined above, a main characteristic feature of ChE is the production, processing, or application of substances and materials for harvesting their specific properties for certain purposes. Inherently, as ChE innovation brings about new substances and materials, or new conditions under which they are processed, transformed, transported, stored, or recycled, one of the main sources of accidents, problems, and conflicts is the unpredictable nature of chemicals. Chemical accidents and other adverse effects of ChE products in the environment and society can be attributed to chemical engineers' inability to foresee each and every factor of the complex network of chemicals, facilities, people, and system conditions that affect the safety and operability of larger engineered systems. Adverse effects are not necessarily caused by irresponsible conduct or someone's intentional wrong behavior – they may be regarded as the result of system failure or the unmanageable material complexity of ChE work.

One example is the unexpected side effects of chemical drugs. In 1957, a drug, Thalidomide, was launched in Germany by the medicinal company Grünenthal (Ruthenberg, 2016). It was considered a harmless drug to mitigate pregnant women's morning sickness but was ultimately shown to have serious implications for babies conceived. Thalidomide, when taken during certain stages of pregnancy, could lead to infants being born with limb deformities. Testing for unintended effects over longer terms was not conducted early on, and it was therefore impossible to foresee those outcomes. Adverse effects were not linked back to Thalidomide until the drug had been in use for several years. A lesson of the Thalidomide case is that we cannot expect fully effective but entirely harmless drugs. While modern medical practices now report long-term effects of drugs, establishing direct links between conditions and specific medications remains challenging. Thus, in designing and processing chemicals, potential adverse outcomes of the products need to be seriously considered. In other words, chemical engineers hold a so-called 'external' or 'social responsibility' to deliberately think of the potential outcomes of the products they design and produce. In this case, the implications regarded unintended harmful health effects.

Beyond direct impacts on patients' health, chemical drugs can also have unintended effects on human culture. An example involves psychotropic drugs such as anti-depressive medicines. These chemicals affect emotions, mood, motivation, and behavior. These drugs can help many people with diagnoses such as depression, obsessive-compulsive disorder (OCD), anxiety, post-traumatic stress disorder (PTSD), and other mental illnesses. According to Klavs Birkholm (2016), they also influence our culture if used intensively – by changing our perception of illness and health. Francis Fukuyama and others argue that humans strive for recognition and that psychotropics' effects on the state of mind of healthy humans might challenge this quest (Fukuyama, 2003), lead to passivity and apathy, and create false expectations of constant happiness through the misleading vision of perfect control over one's mental states. These sociocultural implications resemble ethical discussions on human enhancement. They are related to works of science fiction like Huxley's (2010) *Brave New World* and their warning against drugging the population to maintain an inhuman society.

The following case concerns the false assumption that chemical waste can be easily decomposed. From 1942 to 1952, Hooker Electrochemical Company dumped chemical waste at a place called Love Canal, which later developed into a settlement (Fjalland, 2016; Levine, 1982). A school was built close to the waste dump. Inhabitants complained about the odor. Public authorities initiated an investigation to establish whether it was unhealthy to live in Love Canal. They concluded that living at some (specified) distance from the waste dump was safe. This did not align with the inhabitants' registered occurrences of illness. Eventually, it was found that the model used by the authorities was wrongly chosen. Love Canal was declared unhabitable and was abandoned. Lessons learned here include that assuming that chemical waste can be stored safely is not straightforward and public authorities must listen to early warnings from concerned experts and local citizens. The case calls for participatory approaches to uncertainty, and related ethical issues deal with the science–society or expert–laypeople relationships and their respective epistemic authority.

Unexpected consequences of chemicals occur, at times, in proximity to the exposure, for example, to those taking drugs or their offspring or to those living next to a waste facility. However, this is not always the case. Sometimes, the implications are global.

The intensive use of dichlorodiphenyltrichloroethane (DDT) is an illustrative example of how the work of chemical engineers can turn into a global environmental issue (Børsen & Nielsen, 2017; Bouwman, 2013). DDT is part of a cluster of chemicals called persistent organic pollutants (POPs) that are hard to break down, are attracted to and soluble in lipids or fats (lipophilic), and can disperse over long distances. DDT disrupts the hormonal system and is recognized for its detrimental effects on reproductive development. It has toxic effects on internal organs as well as on the neural system. It is suspected to be carcinogenic. It builds up in the food chain because of its long breakdown halftimes. When DDT is used extensively, resistance becomes apparent, leading to the eventual loss of its desired effects. During and after World War II, DDT was used as an agricultural insecticide to control crop pests and as a tool to combat insect-borne diseases such as malaria. Its use rose exponentially until its residue was registered all around the biosphere, even at the poles. Its use in agriculture is now banned and its use to combat malaria is restricted. The ethical debate surrounding DDT revolves around balancing its effectiveness in disease control with its detrimental environmental and health effects.

Other examples where chemicals have caused global environmental pollution are those of Bisphenol-A (Martin et al., 2021; Resnik & Elliott, 2015), other POPs (Godduhn & Duffy, 2003), and PFAS (Kwiatkowski et al., 2020). If chemicals are used on a global scale, global environmental challenges can merge, where everybody on Earth is affected by chemical substances and their unforeseen and undesired implications. When chemical engineering has such powers, responsibility means stewardship for the whole Earth and is driven by a view of humanity as part of the world and the environment (for more on this, see Chapter 6).

Quintessentially, the ethical issues arising due to the unpredictable nature of chemicals as introduced in this section illustrate that chemical engineers' good professional practice does not only mean absence of misconduct but also paying as much attention as possible to what could go wrong and to the validity and legitimacy of factual claims.

Ethical issues related to misuse of chemical engineering

This section describes a different kind of ChE-related problems. Rather than the unpredictability of material effects, the cases in this section have in common that they are the result of irresponsible decision-making in research and innovation contexts that involve ChE and in which chemical engineers are part of a bigger collective of agents including managers, businesspeople, regulators, workers, civil society, and others. In these settings, interests might diverge and corrupt each other. In the following cases, ChE expertise involved activities beyond purely technical or scientific professional conduct.

Military application of chemistry is one area where ChE or ChE-enabled technologies are purportedly misused. There are several illustrative examples of military uses of ChE, including the use of poison gas in World War I, for example, as well as military applications of Napalm and Agent Orange during the Vietnam War. In this context, 'misuse' denotes the deliberate exploitation of chemical substances originally intended for specific applications, such as the textile industry or agriculture, with the intent to cause harm or produce destructive outcomes. This underscores the dual nature inherent in the use of numerous chemical and ChE innovations, as discussed by Tucker (2012). Many questions can be asked concerning the internal and external forces that drive engineers and other actors into unethical behavior:

- *Is it always wrong to develop or use chemical weapons?*
- *What ethical values are in play when assessing chemical weapons?*
- *What are the responsibilities of chemical engineers?*
- *To what extent were the implications of chemical weapons known?*
- *How do the responsibilities of chemical weapons engineers relate to the responsibilities of governmental representatives, politicians, military people, merchants, and the management of the weapons industry?*
- *What forces drive chemical engineers to misuse their knowledge, skills, and competencies?*

An illustrative example of misuse of ChE appears in the Fritz Haber case, which is thoroughly introduced in journal articles (e.g., Schummer, 2018) and books (Charles, 2011; Stoltenberg, 2004). The case literature is extensive. It regards the implementation of ChE in the development of chemical weapons such as chlorine gas and mustard gas that were applied during World War I. The case also illuminates the backdrop of weapons development, shedding light on the interconnected academic-military-industrial-governmental complex. Haber personified all spheres as he was a university professor, a captain in the army, involved in the chemical industry, and appointed to serve in the German war department. The case shows that Haber himself explicitly chose not to follow established ethical values. As described by Schummer (2018):

> First, he argued that he had never cared about the Hague Convention [an international agreement banning gas weapons]. Second, he was convinced that in times of war ethical standards are to be replaced by patriotism, such that warfare engagement becomes a moral duty for scientists … Third, he was fully aware that his weapons program initiated an arms race among the enemies.
>
> *(p.13)*

The details and discussions of the case in the literature show that the development and use of chemical gas weapons cannot be justified by any major ethical framework (see Chapter 2). Discussing the case with students gives them ample opportunity to reflect on the limits of what may be ethically justified and on current practices in honoring influential persons in their field of study.

A related case is that of Agent Orange. Arthur Galston discovered a chemical compound, 2,3,5-triiodobenzoic acid (TIBA), that inhibits the growth of leaves and, instead, increases the number of floral buds on soybean plants. In higher concentrations, though, it leads to defoliation (abscission of leaves) and the death of the plant. Galston learned later that the US army, interested in this chemical's defoliation effects, produced derivates of this compound and used the most powerful ones (known as Agents Orange, White, and Blue) in the Vietnam War, with disastrous effects on the ecosystem, food chains, and aboriginal culture and lifestyle. Galston remained extremely concerned about the implications of his work, not shying away from confrontation with the US government. He became a public voice of science that reminded fellow scientists of the inherent dual-use potential of every scientific discovery. According to Galston, "The only recourse for a scientist concerned about the social consequences of his work is to remain involved with it to the end!" (Galston, 1972, p. 223). Certainly, the same applies to engineers. It is important to note that the issue is not Galston's responsibility as causal contributor to the United States' use of Agent Orange in the Vietnam War. He was not able to predict that a fertilizer that has defoliation effects in high concentrations would be exploited deliberately as a warfare product. The responsibility attributions of him to himself and, as he hopes, all scientists and engineers to themselves, refers to being aware of what happens with research and innovation output and, if necessary, to actively seeking participation in its discourse (Jacobs & Walters, 2005).

These cases open discussions of how to set up institutional regimes to prevent the development and application of chemical weapons by referring to the Chemical Weapons Convention – an international treaty that prohibits chemical weapons (Frank et al., 2018).

Perhaps counterintuitively, most so-called *chemical incidents* fall into this category, too. According to the World Health Organization (WHO, 2023), chemical incidents refer to:

> An explosion at a factory that stores or uses chemicals, contamination of the food or water supply with a chemical, an oil spill, a leak from a storage unit during transportation, deliberate release of chemicals in conflict or terrorism, or an outbreak of disease that is associated with a chemical exposure.
>
> *(¶ 2)*

Only in a few cases do chemical incidents occur due to the unpredictable nature of chemicals. Most, indeed, can be reconstructed as the result of a chain of questionable decisions. They are viewed as misuse if commercial interests overshadow the safety concerns of workers and civil society. They are viewed as the result of (unavoidable) uncertainty and risk amidst the complexity of chemical knowledge and expertise. One cannot generally judge whether a chemical incident reflects misuse or the unpredictable nature of chemistry. Chemical accidents that happen due to neglect of ChE expertise (and, perhaps, ChE warnings) in the pursuit of commercial or profit interests reflect a form of misuse, especially when public health is put at risk.

In December 1984, an accident happened at a factory producing the insecticide ‘Sevin’ in Bhopal in central India. Between 3,000 and 10,000 people were killed and more than 100,000 were injured. The company that owned the plant claimed that vandalism committed by a dissatisfied employee was the course of the accident. This accusation was never proven, and Eckerman and Børsen (2018) present an alternative explanation. Different safety standards were installed and operated in India than in a similar facility in Virginia (United States): The manufacturing pathway – including the toxic compound methyl isocyanate (MIC) that explodes when in contact with water – was risky, but in Bhopal it was chosen over more expensive, but safer, alternatives. The MIC was stored in huge tanks and not in several smaller tanks. The safety system was under-dimensioned and manually managed. Ingrid Eckerman and Tom Børsen portray the Bhopal accident as misuse because of slack safety precautions, insufficient training, poor management, and so on. This case illustrates relationships among the chemical industry, public regulation of the industry, independent non-governmental organizations (NGOs), and critical journalism.

Other examples of chemical incidents include the Seveso accident (Moser & Dondi, 2016), the Deepwater Horizon oil spill (Beever & Hess, 2016), the Beirut explosion in 2020 (Al Daia & Yaacoub, 2021; Al-Hajj & Kazzi, 2021), and many others. Given the severity and frequency of such incidents, central issues in teaching ethics to chemical engineers are safety maintenance and accident prevention through chemical engineers’ proactive engagement in multi-stakeholder discourses (see next section).

A third type of misuse of ChE has to do with the right to patent chemical substances and processes. In short, the discussion revolves around materials that could potentially solve important problems but are limited in their widespread application because of patenting for commercial interests. Patenting is important in business models and corporate innovation practices (Grubb, 2016). Chemical startups, especially, secure their intellectual assets by patenting substances or chemical transformation processes. Without patents, monetizing their innovation activities is jeopardized. Patents can prevent more widespread use of the products or processes, which has many effects. When the use potential involves tackling an ethical (e.g., societal or environmental) challenge, deliberate patenting may be interpreted as a misuse of ChE competence and expertise for profit – when considering the missed chance of doing something good for people, society, or the planet. Ethical issues like equality, global justice, and transparency are at stake. The conflict is amplified when the patenting leads to hype (and, perhaps, 'patent racing') around a technological branch such as synthetic biology or genetics (Schummer, 2016).

Another ethical issue in the patenting context arises when the patenting of a synthesis of a natural chemical substance disregards the interests of those who – perhaps unknowingly – have lived or worked for centuries with this substance (McGonigle, 2016). Gerber and co-authors (2021) describe the case of psilocin and psilocybin, compounds from the psilocybin mushroom, successfully patented by Swiss pharmaceutical company Sandoz after their R&D team found a way to synthesize these molecules. While the company benefitted from a huge market for psychedelics, the Mazatec tribe, indigenous to Mexico, who discovered the psychedelic and other health effects of the mushroom, were left without any mention of their innovation or knowledge around its use. Despite the UN Convention for Biological Diversity (1992) that urges to "respect, preserve and maintain knowledge, innovations and practices of indigenous and local communities" (p. 1), Gerber et al. (2021) assert:

> There are still a series of important issues pending and ongoing debates on how to best achieve just and fair consultations and agreements, respecting ethical, epistemological, and ecological concerns, and how to properly share the benefits in the case of private properties of pharmaceutical corporations based on cumulative collective knowledge of indigenous peoples.
>
> *(p. 574)*

Lessons to be learned from cases where ChE is misused are that chemical engineers have a responsibility to reflect on the intentions and ethical legitimacy of their endeavors. Often it is difficult for engineers to formulate and make explicit a coherent ethical orientation system. Knowledge of different normative frameworks can assist in the formation of a feeling of what is right and wrong. Such formulation also requires toxicological and ecological knowledge to know whether a substance is harmful to humans or the environment. Hence, misuse of ChE requires a focus on the actual effects of ChE that can be compared to the intended ones. The harmful consequences for fellow humans, the environment, or even humanity that often accompany the – obviously morally wrong – misuse of ChE may also lead to a loss of public credibility of ChE in its entirety. This is more pertinent when misuse is not admitted and spun in secrecy and half-truths, making it ever more difficult to decide whom or what to trust.

Efforts to prevent misuse and manage uncertainty through interdisciplinarity

While the previous two subsections address ethical dimensions of ChE practices from the 'problem' perspective, this subsection intends to highlight the role of ChE in finding solutions. It often does so as an element of a network of many actors rather than regarding its own core competencies. The following examples illustrate how ChE practice can be supported by other disciplines and expertise realms, and how ChE can support multi-stakeholder efforts to tackle ethical challenges.

Concepts of sustainable and green chemistry as strategies to address environmental issues in chemistry and ChE are sometimes co-shaped by chemical and non-chemical experts. Green chemistry (GC) is an attempt to operationalize value-sensitive design in a chemical engineering context. GC was proposed by Paul Anastas and John Warner (1998) to translate ethical concerns into chemical engineering. It is defined as "the utilization of a set of principles that reduces or eliminates the use or generation of hazardous substances in the design, manufacture and application of chemical products" (Anastas & Warner, 1998, p. 2). It builds on 12 principles related to human and environmental health, safety, and security. Other ethical concerns can be added (Friedman & Hendry, 2019).

Illes and co-authors (2021) provide an example of how GC can be enacted in polyvinyl chloride (PVC) production that, according to them, poses three ethical challenges: *How can industry introduce safer technology? How can they be motivated? How can the complexity and uncertainty of new pathways be addressed?* GC deals with several stages in addressing these issues: extraction and conversion of raw materials, manufacturing, end-uses, end-of-life disposal, recycling, and reuse. Ethical values can be reflected in all steps. The idea is that chemical engineers can design their way out of ethical challenges – such as pollution and human health threats. Ethical dilemmas can be mitigated through altered synthesis design, the development of new reaction pathways, or the creation of novel chemicals that pose no threat to human or environmental health. GC is an interdisciplinary endeavor combining chemical engineering, ethics, toxicology, and ecology. It also involves different branches, including industry (management and employees), academia, legislative levels, and civil society. Hence, a potential obstacle arises from conflicts with other interests, including commercial ones and those associated with novel practices. Additionally, the challenge of uncertainty persists.

Carbon dioxide (CO_2) capture, as a form of climate engineering or geoengineering, aligns with the principles of green chemistry. Addressing climate change can involve either the chemical capture or removal of CO_2 or the chemical shading of sun radiation. Ethical concerns extend beyond the uncertainties surrounding climate engineering effects; the so-called 'moral hazard argument' is also pertinent. It states that attention given to unrealistic or uncertain technological solutions to climate change will make the world population less willing to change behavior and, in that way, worsen climate change: *Why bother when there is always an engineered technological solution at hand?* To prevent technocratic decision-making and overly optimistic reliance on technical solutions for addressing global challenges, chemical engineers must approach their interdisciplinary and public communication of possibilities and promises with care and caution.

A movement in the same direction as GC, but with significant dual-use potential, is the recent development of artificial intelligence (AI) technology applied to chemical contexts, such as identifying possibly harmful substances. Such application may support GC motifs by avoiding toxic or hazardous substances, for example, in drug discovery (Urbina et al., 2022). Yet, responsible use by chemists and chemical engineers is paramount given concerns regarding foreseeability, risk management, and accountability.

Ethical codes of conduct are often seen as a possible solution to the misuse of chemical technology. When a researcher or engineer pledges to serve the interests of clients, colleagues, humanity, and the environment and not intentionally misuse chemical technologies, the rationale is that the susceptibility to misconduct and malicious intentions is significantly reduced. Many companies and national or international ChE organizations have crafted codes of conduct for chemical engineers, for example, the American Institute of Chemical Engineers (AIChE), the Institution of Chemical Engineers (IChemE), the European Council of Engineers Chambers (ECEC), or the Philippine Board of Chemical Engineering (see links in the References section). The Organization for the Prevention of Chemical Weapons (OPCW) has compiled an overview of more than 120 codes of conduct for chemists and chemical engineers from around the world (link in the References section).

In the private-sector chemical industry, ChE activities are firmly entangled with corporate social responsibility (CSR) entrepreneurial strategies (see Chapter 6). Although some CSR efforts can be debunked as greenwashing or ethical whitewashing – the false claim of acting sustainably or philanthropically for marketing or publicity purposes – there are examples of positive effects of truthful CSR commitments on ChE practice in corporate settings. When German chemical company Bayer bought the infamous US-American company Monsanto, it saw itself exposed to criticism and loss of reputation. In response to this criticism that targeted disputable business and innovation practices at Monsanto, Bayer installed an internal ethics and integrity board tasked with overseeing R&D practices and giving competent advice on societal and environmental issues. Whether the board will have any visible effect remains to be seen. In any case, however, credible CSR approaches take discourses on the social impacts of ChE straight into R&D departments. This initiative is not unique. The chemical industry has institutionalized voluntary CSR activities in its 'Responsible Care' program (Belanger et al., 2014); the framework has been enacted in many countries around the world (Evangelinos et al., 2010; Gamper-Rabindran & Finger, 2013; Givel, 2007).

A case of successful collaboration of chemical experts with other experts involves the EU's development of regulations for the 'Registration, Evaluation, Authorization, and Restriction of Chemicals' (called REACH) and their enactment into law on 1 June 2007. REACH has been extended and developed iteratively over time. The European Chemical Agency (ECHA) oversees the enactment of regulations and supervises their execution and fulfillment. A core element of REACH is providing one coherent framework for new and existing chemicals. It seeks a balance between a reasonable workload of updating the database and the flood of new compounds and substances (considering, e.g., the countless nano-scaled compounds with different properties at different particle sizes). The idea is that the registry prioritizes the reporting of those substances that are relevant for industry and applications (sorted by various steps of production/consumption in tons per year). This form of organization only works with effective communication along the supply chain, from manufacturers, importers, and distributors to downstream users. Because of this, there is a shift away from public authorities towards industry. What at first may look like a burden for industrial stakeholders (and analytical chemistry) is, in fact, an increased efficiency in regulation. Given the vast number of new materials from chemistry and chemical engineering, no agency or other public authority would be able to manage the testing and regulation of all these chemicals in a feasible way. Through shared competencies and a clever combination of them, REACH has become one of the most sophisticated and helpful chemical registries worldwide. It is considered to be an institutionalization of the precautionary principle (Llored, 2017).

Responsibility domains

When discussing cases of ChE outputs that have ethical implications, the focus is usually on the professional competencies of the involved people. This means seeking to understand and clarify the specific responsibility that chemical engineers have regarding their professional competencies (in contrast to them being members of society). The attribution of professional responsibility hinges on the legitimacy of expectations on skill, knowledge, and the ability to comprehend and follow rules and guidelines (Mitcham, 2005; Williams, 2012). In the context of chemical engineering, it means that the decision-making and action of chemical engineers are the ethical focus only insofar as it concerns their chemical knowledge and their professional institutional or organizational context.

The principle of 'backward-oriented accountability' means that a chemical engineer should have known (or may be expected to know) about the specific effects of a design choice. Regarding the principle of 'forward-oriented responsibility,' a chemical engineer may be expected to anticipate or proactively assess a variety of effects and impacts. For a more nuanced understanding, it is helpful to distinguish among (a) internal and external domains of responsibility; (b) moral, legal, and institutional responsibility; and (c) individual and collective responsibility (Nihlén Fahlquist, 2017). Note that responsibility is an intrinsically ethical concept in such a way that taking responsibility always concerns ethical and other normative values.

Internal and external responsibility

While codes of professional conduct and guidelines of good engineering practice constitute the internal domain of responsibility, the external domain concerns the impact of engineering activity on the environment and human lifeworld (see also Mehlich et al., 2017). Analogously, Herkert (2005) distinguishes *micro-ethics* from *macro-ethics*. While many voices (especially from the engineering community) wish to see ethical discussion focusing on (or limiting itself to) the internal domain of good engineering practice, others point to the significant impact of engineering activity on the environment and human lifeworld.

In the internal (or micro-ethical) domain, responsibilities are often codified through professional guidelines and codes of conduct that are institutionally or organizationally enacted. Academic institutions have ethics boards, and companies encourage employees to read and sign their commitments to professional conduct. Some companies identify explicit corporate values. In such cases, the call for ethical behavior is often modeled as a virtue approach – the 'good chemical engineer' is committed to virtues such as objectivity, truthfulness, fairness, or skepticism. Merton's communalism, universalism, disinterestedness, and organized skepticism as the central virtues of science (CUDOS) approach, and related lists have been refined for specific chemical engineering contexts and purposes (Merton, 1973/1942). This domain is especially relevant in the context of cases related to the unpredictable nature of chemicals – chemical engineers are expected to use their professional skills and knowledge to identify and mitigate risks. The atrocities caused by Thalidomide might have been prevented had the chemical experts conducted and communicated their toxicological studies more thoroughly without technophile or commercial interests in mind. The Love Canal pollution might have been avoided by rigorously following life-cycle and risk-assessment protocols or hazard foresight strategies.

The external or macro-ethical domain gets less attention in the scientific literature on the topic. There are far fewer operational and methodological instructions and codified guidelines. However, technology development, design, and innovation tools such as value-sensitive engineering, ethical design thinking, or ethics-by-design (for more on these, see Chapter 22) aim at providing practicable frameworks for chemical engineers to systematically consider normative dimensions in their work. This domain mainly concerns the innovation entanglement of ChE in which chemical engineers defend and explain their choices in front of other actors and stakeholders such as corporate executive boards and management, regulatory bodies, or the public and civil society. An illustrative case is the development of chemical weapons. Responsibility, here, is not about the safe handling of the chemicals or the scientific rigidity of the study of their properties but about the known contribution to an unethical act (killing civilians with chemical weapons).

Disasters like the Bhopal and Seveso accidents illustrate how both domains can be interwoven. On one hand, scientifically sound study and communication regarding chemical properties is part of the internal responsibility domain of chemical engineers. Without reliable ChE knowledge, the safety of industrial facilities cannot be maintained. On the other hand, it is precisely that knowledge that empowers chemical engineers to supervise R&D practices and to communicate concerns about safety and risks, if necessary, in firm opposition to economic, commercial, or political interests. This is ChE's 'external responsibility.' External responsibility is sometimes called 'social responsibility.'

The unpredictable nature of chemicals touches upon the interplay of (or tensions between) internal and external responsibilities. Chemical engineers and their disciplinary societies need to address the unpredictable nature of chemicals as part of good professional practice. Doing so is also an expression of caring for societal and environmental impacts. In that way, chemical engineers are responsible for upholding the reputation and good practice of chemical engineering (internal responsibility) and for protecting society and the environment (external responsibility). Chemical engineers hold a responsibility to engage with different perspectives and manage uncertainty when trying to assess the sometimes unpredictable implications of chemical products.

Ethical issues related to the misuse of chemical engineering also transgress the distinction between the internal and external responsibilities of chemical engineers. If chemical engineers consciously violate good engineering practice, it has implications for citizens and the environment, as we have seen in several cases above, regarding accidents, war crimes, and unjust distributions of risks and benefits arising from chemical engineering.

Thus, we suggest including reflections and deliberations on both the purposes and the broader implications of chemical engineering projects as part of good practice in chemical engineering (for more on how to do this, see Chapters 6 and 25). Good engineering practices are intertwined with the societal role of engineering expertise. Luckily, we need not start from scratch when we try to bridge chemical engineers' internal and external responsibilities. Existing mechanisms to prevent misuse and manage uncertainty are ways of bridging the internal and external responsibilities of ChE. They include science-based legislation, formulating engineering programs to explicitly address ethical challenges (including green chemistry projects), and the many initiatives to formulate ethical codes of conduct in academia and industry. Almost all cases presented in this chapter can be made sense of by considering these two dimensions of responsibility. In the authors' teaching activities, students are happy to learn about this distinction because they find it clarifying.

Moral, legal, and institutional responsibility

Responsibility is attributed with different intentions by different stakeholders in view of different aspects of decision-making (rules, knowledge, competence, authority, etc.). It is useful for our purpose to distinguish the following dimensions of professional responsibility:

- Moral responsibility: to act in accordance with moral codes and to respect ethical values and principles
- Legal responsibility: to know and obey relevant laws and codified sets of rules
- Institutional responsibility: to follow institutional, for example an employer's, rules and binding contracts

Ethical behavior, here understood as an activity driven by a commitment to norms of any kind (morals, laws, culture, etc.), is not limited to one of the dimensions defined above but usually requires a navigation of choices that avoids conflicts between the different domains. For example, chemical engineers are, by contract, embedded into an institutional (e.g., firm-internal) network of rules and codes that they are expected to follow loyally. Sometimes, it may be necessary to challenge these rules in the face of moral dilemmas that surface during the development of a technological item, a prototype, or during the real-world implementation of something new (a chemical compound, a process, etc.). Imagine a chemical firm interested in the market implementation of one of its innovative products. When chemical engineers involved in the environmental impact assessment process identify safety risks beyond required assessment standards, they may face a conflict between their moral responsibility to point out these risks and their institutional obligation to follow executive board directives concerning the public communication of internal R&D information. The Love Canal and Thalidomide cases and historical chemical accidents are illustrative examples of chemical engineers under such tension.

In the cases described in this chapter, we see examples of ethical values such as no-harm and human dignity, international legal regimes such as the Chemical Weapons Convention and the Stockholm convention regulating POPs, and institutional rules such as codes of conducts, green chemistry principles, and corporate social responsibility arrangements like Responsible Care, that define different normative arrangements that professional responsibility can be defined by.

In view of this distinction, chemical engineers cannot solely rely on the codified terms of accountability when doing their job; they need to be aware of the ethical dimensions of their work that require proactive engagement and have the readiness to face discourses with stakeholders who pull into other (perhaps unethical) directions. Competent professional agency, in this respect, exceeds the realms of compliance (with codes of conduct) and standard protocols but may enter supererogatory areas in which ethical behavior is not a matter of top-down duty but of a personal commitment to '*doing more than one's duty*' (as, e.g., whistle-blowing even in the face of negative personal consequences).

Individual and collective responsibility

To fully comprehend one's professional responsibility ascription and act accordingly, it is necessary to understand the difference between attribution of individual responsibility (for example, for professional agency) and collective responsibility (e.g., to firms or social spheres such as *science* or *innovation*). The latter can never mean delegating responsibility and duties away from an individual and towards an intangible, non-personalized entity. In many of the cases described above, it would be possible for chemical engineers to argue that they didn't do wrong because their contribution to creating a problem was only marginal or because the chain of decisions that led to the conflict didn't allow identification of them personally as the causal origin of the problem. For instance, Thalidomide is a marketed product, so some could argue that disseminators rather than developers are to blame. The waste-disposal site disaster was created due to a community decision. The Bhopal disaster resulted from a long series of questionable economic decisions meeting unfortunate local working conditions, and so on. Yet, as a counter-argument, many of the decisions were based on factual information that was contributed by the chemical engineers. Thus, rather than denying any responsibility, it would be prudent for engineers to clarify their own contribution to the collective and to engage in co-operative discourse that constitutes the collective responsibility of a multi-agent entity such as a firm or a public service organization.

Responsibility in action: chemical engineers in discourses

Chemical engineers usually work in highly diverse teams, so their specific role in research, development, design, and innovation processes is described in this section to highlight the normative dimensions of interdisciplinary collaborative agency. This issue predominantly concerns the external responsibility domain: the elicitation, comprehension, and anticipation *by design* of future impacts of engineering, development, and innovation choices. In both academic and corporate realms, engineers make the normatively most significant decisions after connecting their factual contribution (research results, functionality and utility, risk and impact assessments, etc.) with values and norms that guide the outcome-oriented mission statements and purposes of the undertaken efforts.

For example, in their research about chemical engineers' experiences related to ethics in engineering in the health products industry, Kim and Kerr (2021) showed that when chemical engineers design products, such as drugs, they consider both benefits and risks to the patient, their customer, so that their products can ultimately improve patients' health outcomes. When engineers make a design decision, they ensure their arguments are based on data and scientific rigor so that their products can ultimately be safe and efficacious for patients. Kim and Kerr also showed that compliance with established technical standards and regulations is also an important consideration among engineers, and as engineers work in teams, acknowledging the contributions of different team members in the project is also very important. Likewise, chemical engineers consider multiple factors, including making data-driven and scientifically rigorous arguments, complying with technical standards and regulations, and cultivating a healthy teamwork environment to achieve their ultimate goal of serving the needs of their customers safely and efficaciously.

In these discourses, engineers are never purely knowledge exponents or providers of the necessary insights for the one right solution to a problem. Their input can potentially modify normative commitments and respective development paths. In this view, an ethical engineer is not an expert on ethics, but a responsible agent who stirs a constructive discourse among stakeholders to set factual knowledge and value claims into a plausible and scrutiny-withstanding relation. Pielke (2007) calls this the *honest broker*.

This practical discourse skill is not inborn but can and should be trained. Therefore, the education of chemical engineers in the ethical dimensions of ChE should be understood as practical coursework and not as a lecture on theories and principles. The case overview in the previous section indicates that the solution for problems arising from ChE undertakings, or the prevention of such problems, is not formal training in ethics and ethical principles but better development of responsibility, communication, critical thinking, teamwork, and interdisciplinary competencies. Chemical engineering education at most higher education institutions follows a highly practical curriculum. ChE students are used to approaching problems *at hand*. Ethical and other normative challenges are not different – they require pragmatic, practicable approaches that are best learned in a challenge- or problem-based fashion (Mehlich, 2022).

Conclusion and outlook

This chapter provided content for the ethics education of chemical engineers. Educators may consider broadly covering topics, including ethical issues arising from the unpredictable nature of chemicals, their potential misuse, and prevention efforts, employing specific examples we provided in this chapter. When teaching those issues, educators can consider the various responsibility domains we suggested. To support the design of teaching items in ChE, we suggest linking the presented cases with the different forms of responsibility discussed above. Since chemical engineering practice and its impact are interdisciplinary, educators may also need to consider how to help chemical engineers develop practical skills for navigating such environments.

Contemporary theoretical work on the philosophy and sociology of chemical engineering has laid out the foundations for understanding the practical agency of chemical engineers. This means the conceptual and methodological bridge towards practice and education has been built.

With appropriate and scientifically validated concepts of sustainability at hand, it is now possible to craft effective and efficient impact assessment procedures that not only study toxicity or life cycles but also address ethico-environmental issues such as biodiversity or environmental justice (Thompson, 2020), trust, or diversity. Thus, future chemical engineers will be empowered to anticipate these ethical aspects via their professional competence.

References

Al Daia, R., & Khayr Yaacoub, H. (2021). Beirut port blast: an escapable disaster in more than one way. *Emerald Emerging Markets Case Studies*, *11*(4), 1–16.

Al-Hajj, S., Mokdad, A. H., & Kazzi, A. (2021). Beirut explosion aftermath: Lessons and guidelines. *Emergency Medicine Journal*, *38*(12), 938–939.

American Institute of Chemical Engineers. *Code of ethics*. Retrieved May 2, 2023, from https://www.aiche.org/about/governance/policies/code-ethics

Anastas, P. T., & Warner, J. C. (1998). *Green chemistry: Theory and practice*. Oxford University Press.

Barry, B. E., & Herkert, J.R. (2015). Engineering ethics. In A. Johri & B. M. Olds (Eds.), *Cambridge handbook of engineering education research* (pp. 673–692). Cambridge University Press.

Beever, J., & Hess, J. L. (2016). *Deepwater Horizon oil spill: An ethics case study in environmental engineering*. American Society for Engineering Education.

Belanger, J., Topalovic, P., Krantzberg, G., & West, J. (2014). Responsible care: History & development. In P. Topalovic & G. Krantzberg (Eds.), *Responsible care: A case study* (pp. 1–20). De Gruyter.

Bielefeldt, A. R., Polmear, M., Knight, D., & Canney, N. (2019). Disciplinary variations in ethics and societal impact topics taught in courses for engineering students. *Journal of Professional Issues in Engineering Education and Practice*, *145*(4), 04019007.

Birkholm, K. (2016). The ethical judgment: Chemical psychotropics. *Hyle-International Journal for Philosophy of Chemistry*, *22*(1), 127–148.

Børsen, T., & Nielsen, S. N. (2017). Applying an ethical judgment model to the case of DDT. *HYLE–International Journal for Philosophy of Chemistry*, *23*, 5–27.

Børsen, T., Serreau, Y., Reifschneider, K., Baier, A., Pinkelman, R., Smetanina, T., & Zandvoort, H. (2020). Initiatives, experiences and best practices for teaching social and ecological responsibility in ethics education for science and engineering students. *European Journal of Engineering Education*, *46*(2), 186–209.

Bouwman, H., Bornman, R., van den Berg, H., & Kylin, H. (2013). DDT: Fifty years since silent spring. In David Gee (Eds.), *Late lessons from early warnings: Science, precaution, innovation* (p. 21). European Environment Agency.

Bowen, W. R. (2014). *Engineering ethics: Challenges and opportunities*. Springer.

Charles, D. (2011). *Between genius and genocide: The tragedy of Fritz Haber, Father of chemical warfare*. Random House.

Denn, M. M. (2011). *Chemical engineering: An introduction*. Cambridge University Press.

Eckerman, I., & Børsen, T. (2018). Corporate and governmental responsibilities for preventing chemical disasters: Lessons from Bhopal. *HYLE–International Journal for Philosophy of Chemistry*, *24*, 29–53.

Edwards, V., & Shelley, S. (2018). *Careers in chemical and biomolecular engineering*. CRC Press.

European Council of Engineers Chambers. *Code of conduct for European chartered engineers*. Retrieved May 2, 2023, from https://www.ecec.net/fileadmin/pdf/ECEC-Code-of-Conduct.pdf

Evangelinos, K. I., Nikolaou, I. E., & Karagiannis, A. (2010). Implementation of responsible care in the chemical industry: Evidence from Greece. *Journal of Hazardous Materials*, 177(1–3), 822–828.

Fjelland, R. (2016). When laypeople are right and experts are wrong: Lessons from love canal. *HYLE–International Journal for Philosophy of Chemistry*, *22*, 105–125.

Frank, H., Forman, J.E., & Cole-Hamilton, D. (2018). Chemical weapons: What is the purpose? The Hague ethical guidelines. *Toxicological & Environmental Chemistry*, *100*, 1.

Friedman, B., & Hendry, D. G. (2019). *Value sensitive design: Shaping technology with moral imagination*. MIT Press.

Fukuyama, F. (2003). *Our posthuman future: Consequences of the biotechnology revolution*. Farrar, Straus and Giroux.

Galston, A. (1972). *Science and social responsibility: A case history*. Annals New York Academy of Sciences.

Gamper-Rabindran, S., & Finger, S. R. (2013). Does industry self-regulation reduce pollution? Responsible Care in the chemical industry. *Journal of Regulatory Economics*, *43*, 1–30.
Gerber, K., García Flores, I., Ruiz, A. C., Ali, I., Ginsberg, N. L., & Schenberg, E. E. (2021). Ethical concerns about psilocybin intellectual property. *ACS Pharmacology & Translational Science*, *4*(2), 573–577.
Givel, M. (2007). Motivation of chemical industry social responsibility through responsible care. *Health Policy*, *81*(1), 85–92.
Godduhn, A., & Duffy, L. K. (2003). Multi-generation health risks of persistent organic pollution in the far north: use of the precautionary approach in the Stockholm Convention. *Environmental Science & Policy*, *6*(4), 341–353.
Green, D., & Southard, M. Z. (2018). *Perry's chemical engineers' handbook* (9th ed.). McGraw-Hill Education.
Grubb, P. W., Thomsen, P. R., Hoxie, T., & Wright, G. (2016). *Patents for chemicals, pharmaceuticals, and biotechnology* (6th ed.). Oxford University Press.
Harris, C. E., Pritchard, M. S. James, R. W., Englehardt, E. F., & Rabins, M. J. (2018). *Engineering ethics: Concepts and cases* (6th ed.). Cengage Learning.
Herkert, J. R. (2005). Ways of thinking about and teaching ethical problem solving: Microethics and macroethics in engineering. *Science and Engineering Ethics*, *11*(3), 373–385.
Huxley, A. (2010). *Brave new world* (11th ed.). Vintage.
Iles, A., Martin, A., & Rosen, C. M. (2021). Undoing chemical industry Lock-ins: Polyvinyl chloride and green chemistry. In J. Schummer and T.Børsen (Eds.), *Ethics of chemistry: From poison gas to climate engineering* (pp. 279–316). World Scientific.
Institution of Chemical Engineers. *Code of professional conduct*. Retrieved May 2, 2023, from https://www.icheme.org/about-us/governance/code-of-professional-conduct/
Jacob, C., & Walters, A. (2005). Risk and responsibility in chemical research: The case of agent orange. *HYLE–International Journal for Philosophy of Chemistry, 11*, 147.
Johnson, D. G. (2020). *Engineering ethics: Contemporary and enduring debates*. Yale University Press.
Kim, D., & Kerr, A. J. (2021). *Chemical engineers' experiences of ethics in the health products industry*. Paper presented at 2021 ASEE Virtual Annual Conference.
Kwiatkowski, C. F., Andrews, D. Q., Birnbaum, L. S., Bruton, T. A., DeWitt, J. C., Knappe, D. R., ... Blum, A. (2020). Scientific basis for managing PFAS as a chemical class. *Environmental Science & Technology Letters*, *7*(8), 532–543.
Levine, A. G. (1982). *Love canal: Science, politics, and people*. Lexington Books.
Llored, J. (2017). Ethics and chemical regulation: The case of REACH. *Hyle: International Journal for Philosophy of Chemistry*, *23*(1), 81–104.
Martin, A., Iles, A., & Rosen, C. (2021). Applying utilitarianism and deontology in managing bisphenol-A risks in the United States. In J. Schummer and T.Børsen (Eds.), *Ethics of chemistry: From poison gas to climate engineering* (pp. 249–278). World Scientific.
McGonigle, I. V. (2016). Patenting nature or protecting culture? Ethnopharmacology and indigenous intellectual property rights. *Journal of Law and the Biosciences, 3*(1), 217–226.
Mehlich, J. (2022). Technology assessment in the STEM curriculum: Teaching responsible research and innovation skills to future innovators. *TA Theory and Practice, 31*(1), 22–27.
Mehlich, J., Moser, F., Van Tiggelen, B., Campanella, L., & Hopf, H. (2017). On the ethical and social dimensions of chemistry: Reflections, considerations, and clarifications. *Chemistry: A European Journal*, *23*(6), 1210–1218.
Merton, R. K. (1973 [1942]). The normative structure of science. In R. K. Merton (Ed.), *The sociology of science: Theoretical and empirical investigations*. University of Chicago Press.
Mitcham, C. (2005). Responsibility. Overview. In C. Mitcham (Ed.), *Encyclopedia of science, technology, and ethics* (Vol. 3). Thomson Gale, Famington Hills.
Moser, F., & Dondi, F. (2016). Environmental protection between chemical practice and applied ethics: A critical review. *Toxicological & Environmental Chemistry*, *98*(9), 1026–1042.
Nihlén Fahlquist, J. (2017). Responsibility analysis. In S. O. Hansson (Ed.), *The ethics of technology*. Rowman & Littlefield.
Organisation for the Prevention of Chemical Weapons. *Compilation of codes of ethics and conduct*. Retrieved May 2, 2023, from https://www.opcw.org/sites/default/files/documents/SAB/en/2015_Compilation_of_Chemistry_Codes.pdf
Philippine Board of Chemical Engineering. *Code of ethics*. Retrieved May 2, 2023, from https://www.prc.gov.ph/sites/default/files/Board%20of%20Chemical%20Engineering%20-%20Code%20of%20Ethics_0.pdf

Pielke, R. A. (2007). *The honest broker: Making sense of science in policy and politics*. Cambridge University Press.

Resnik, D. B., & Elliott, K. C. (2015). Bisphenol a and risk management ethics. *Bioethics*, *29*(3), 182–189.

Ruthenberg, K. (2016). About the futile dream of an entirely riskless and fully effective remedy: Thalidomide. *HYLE–International Journal for Philosophy of Chemistry*, *22*, 55–77.

Schummer, J. (2016). 'Are you playing god?': Synthetic biology and the chemical ambition to create artificial life. *HYLE–International Journal for Philosophy of Chemistry*, *22*, 149–172.

Schummer, J. (2018). Ethics of chemical weapons research: Poison gas in world war one. *HYLE–International Journal for Philosophy of Chemistry*, *24*, 5–28.

Schummer, J., & Børsen, T. (Eds.). (2021). *Ethics of chemistry: From poison gas to climate engineering*. World Scientific.

Shallcross, D. C. (2010). Teaching ethics to chemical engineers: 2. Further class room scenarios. *Education for Chemical Engineers*, *5*(1), e13–e21.

Shallcross, D. C. (2017). *Chemical engineering explained: Basic concepts for novices*. Royal Society of Chemistry.

Shallcross, D. C., & Parkinson, M. J. (2006). Teaching ethics to chemical engineers—some class room scenarios. *Education for Chemical Engineers*, *1*(1), 49–54.

Stoltzenberg, D. (2004). *Fritz Haber: Chemist, Nobel Laureate, German, Jew*. Chemical Heritage Foundation.

Thompson, P. B. (2020). Ethics and environmental risk assessment. In P. B. Thompson (Ed.), *Food and agricultural biotechnology in ethical perspective* (pp. 137–165). Springer.

Tucker, J. B. (Ed.). (2012). *Innovation, dual use, and security: Managing the risks of emerging biological and chemical technologies*. MIT Press.

UN Convention on Biological Diversity. (1992). *Article 8(j) - traditional knowledge, innovations and practices*. Retrieved February 17, 2024, from https://www.cbd.int/traditional/default.shtml

Urbina, F., Lentzos, F., Invernizzi, C., & Ekins, S. (2022). Dual use of artificial-intelligence-powered drug discovery. *Nature Machine Intelligence, 4*(3), 189–191.

Whitbeck, C. (2011). *Ethics in engineering practice and research* (2nd ed.). Cambridge University Press.

Williams, G. (2012). Responsibility. In R. Chadwick (Ed. in Chief), *Encyclopedia of applied ethics* (2nd ed.). Elsevier.

World Health Organization. (2023). *Chemical incidents*. Retrieved May 2, 2023, from https://www.who.int/health-topics/chemical-incidents#tab=tab_1

18

ETHICAL ISSUES IN SOFTWARE ENGINEERING

Stephanie J. Lunn, Isis Hazewindus, Prajish Prasad, and Vivek Ramachandran

Introduction

From the first home computers in the 1970s to the use of algorithms in application engineering, technology has expansively progressed over the last several decades (O'Regan & O'Regan, 2008). The resulting hardware, software, and data generated have become ubiquitous in our lives, permeating numerous industries such as healthcare, finance, agriculture, manufacturing, and education. Creating and utilizing these technologies can wield tremendous influence, offering solutions across domains and society at large. Meanwhile, the conversation regarding ethical issues and implications surrounding these technologies expands.

Despite early twentieth-century critics of modern technology (Heidegger, 1977; Ellul, 1964; Mumford, 1934, 1967) warning about humans becoming slaves to technology, the prevailing view on technology has remained primarily utopian. Initially, technology and its artifacts were considered neutral, holding no intrinsic morality, with the user assessing its ethical aspects. Such approaches shift moral judgment from creators to users, subject to their end goals, thereby overlooking any underlying ethical concerns in the technologies themselves. Consequently, computers and software are perceived as objective entities, neglecting the influence of the software engineers behind the technologies, who bring their own viewpoints, beliefs, and biases into the products they and their teams create.

Nevertheless, the extensive integration of computing into society, the rise of 'Big Data,' and the impacts of created tools have necessitated questioning technology's purported neutral status. Although technology can enhance the quality of living, it can be harmful as well. Various scandals concerning algorithmic applications, data misuse, and security breaches make it increasingly evident that technology is dependent on the (conscious or subconscious) biases and decisions of the people and organizations building and maintaining them (Brown et al., 2021; Hysa et al., 2023). Computing solutions are applied to a range of topics and can enable access and dissemination of useful information, but can also exacerbate inequity, amplify stereotypes, and/or lead to issues around privacy and security – for example, through hacking into personal accounts (Xu & Tang, 2020). Thus, the influence of humans and the development and use of technology are inextricably linked, and decisions can have consequences not linked to the designers' intent.

DOI: 10.4324/9781003464259-22

To help software engineers become aware of their responsibilities in the creation, dissemination, and maintenance of software, as well as their consideration of data use, educators must provide environments that promote not only learning how to code, but also how to recognize ethical issues and accurately respond to them. This is where ethics education can play a crucial role; scholars have emphasized the need to focus on *how* rather than *what* to think (Horton et al., 2022). Integrating ethics and encouraging social justice early on can promote students' agency and encourage them to consider the implications of decisions (Moore, 2020; Ferreira & Vardi, 2021). It is critical for ethics to be conceptualized as part of practice, and to transition disciplinary mindsets away from 'avoiding traps' toward anticipating 'sociotechnical risks' (Andrus et al., 2020, p. 77).

Incorporating ethics into education is not novel but constitutes an ongoing effort. Numerous organizations, governmental funding agencies, and institutions of higher learning have previously acknowledged the need for concentrated and comprehensive ethics education. A joint task force comprised of representatives from the Association for Computing Machinery (ACM) and Institute of Electrical and Electronics Engineers (IEEE) – two major international computing societies – worked together to define a guiding 'Software Engineering Code of Ethics' to encourage standards for teaching and practice (Gotterbarn et al., 1997, 1999). Although this code can provide a beneficial foundation for framing ethical issues, decision-making in the real world can include any number of variables and requires all individuals to take ownership and responsibility for choices. Towards this goal, 'IEEE 7000' is the first standard that describes principles surrounding unintended risks and encourages 'responsible innovation' with consideration of human and social values (Spiekermann, 2021). Recognizing the importance of personal responsibility in software design and use, IEEE 7000 centers around combining material value ethics and moral philosophy to drive value-based engineering (Spiekermann & Winkler, 2022).

Despite such endeavors, attempts to foster ethical mindsets are not always widespread (Connolly, 2020). Ethics are often taught either as standalone courses or are incorporated within existing courses by engineering instructors (Fiesler et al., 2020). Yet, integrating ethical awareness and decision-making cannot be undertaken in isolation; ethics instruction must intertwine with technical understanding (Martin et al., 1996). Presenting ethical concepts sporadically may be insufficient to alter students' thinking. Instead, a widespread approach is required, with lessons embedded throughout curricula and recurring periodically throughout a student's matriculation.

As societies evolve and along with them the definitions of technology, the field of computing likewise advances, and dynamic subfields – such as artificial intelligence (AI), computer vision (CV), human-–computer interaction (HCI), and machine learning (ML) – emerge and expand. Continually considering the ways in which technology can impact accountability and transparency is essential. Moreover, we must reflect on how ethics can be applied to advocate for social justice and encourage diverse voices (Cheong et al., 2021; Ferreira & Vardi, 2021). With this chapter, we hope to contribute to a long-term and sustained approach to molding ethically minded software engineers by providing didactic tools to establish educational and professional environments where inclusion and morality are at the forefront of the discussion. We encourage conversations on and integration of ethics throughout computing curricula, touching upon themes like responsibility and social impact.

In the section that follows, we will introduce ourselves as authors of this chapter and describe our positionality to frame the chapter. Then, we will then elaborate on ethics and its place in software engineering (SE), before diving into practical approaches to integrate ethics in SE pedagogy. Finally, we outline case studies that instructors can use to convey important ethical issues in SE. Our tools can be applied within different courses with distinct objectives and assignments, with reinforcement of ethical thinking as a final goal, while providing the opportunity for students to practice ethics and empower them to "effectively actualize their (own) values in future professional decision making" (Cohen et al., 2021, p. 861).

Positionality

Given the role that personal values, experiences, and backgrounds can have in how topics are approached, interpreted, and defined (Martin et al., 2022), we want to be transparent as authors. We aim to disclose how we have influenced what is presented. In this section, we share information about who we are and the role that plays in the perspectives and approaches we use.

The first author, Stephanie, identifies as a white woman with a computer science (CS), computing education, and engineering education background. She is an assistant professor at a large, research-focused university in the southeastern United States. Her experiences with sexism and imposter phenomena in computing have driven her efforts to aid in students' technical and professional development and to cultivate more inclusive mindsets.

The second author, Isis, identifies as a white cis woman with a background in philosophy, data ethics, and the development of educational programs. Her role as a data ethics consultant and educator shaped her view on careful decision-making and ethics as a practice of continuous reflection and conscientiousness. Being the first in her family to receive training at university, she is conscious of the importance of introducing matters like ethics in ways explicitly tailored to the target group – to encourage enthusiasm and curiosity rather than alienation or indifference.

The third author, Prajish, identifies as an Indian man with a CS and educational technology background. He has experience working for an EdTech company as a software developer. He works as an assistant professor at a liberal arts university in India and teaches CS subjects like SE and introductory programming. His experiences in the software industry, as well as his research experience in educational technology, have made him realize that a more holistic instructional approach is required to help students engage with engineering problems equitably and effectively.

The fourth author, Vivek, is a non-binary Indian with a robotics, mechanical engineering, and engineering education background. His experiences in learning and teaching in Asia, Europe, and North America – where he was at times a member of the ethnic and/or racial minority – have shaped his worldview. His desire to see ethics and societal responsibility being prioritized in research and teaching in robotics and engineering domains drives his pedagogical practice.

Ethics and morality in software engineering

In philosophy, "if we are interested … in what our guiding ideals should be, in what sort of life is worth living, in how we should treat one another" (Shafer-Landau, 2018, p. 1), ethics is the domain that we turn to. This domain is a broad and can be categorized into three subdomains under which different types of looking at ethics can be grouped (Shafer-Landau, 2018):

1. *Normative ethics*: focuses on standards and what we ought to do. Irrespective of the circumstances, it considers *What are our moral duties?* and *Which actions are morally right, and which actions are morally wrong?* Normative ethics encompasses approaches like utilitarianism, consequentialism, deontology, feminist ethics, and virtue ethics.
2. *Meta-ethics*: focuses on the status of our moral claims and values. *Can we say moral theories or statements are 'true'?* We might even wonder about the existence of something like a universal moral viewpoint.

3. *Value theory*: focuses on questions concerning the nature of defining goodness and the good life: *What is a 'good life' and how do we pursue it?* It balances moral and natural goods.

More elaborate discussion of these ethical frameworks is provided in Chapter 2 of this handbook. Here, in this chapter, when we speak of ethics, we refer to normative ethics. We want to provide educators and students with the tools to help them grasp ethics and reflect on what is good in terms of actions and decision-making while keeping in mind that defining what is 'good' is complicated. Issues can be highly complex, contextual, and subjective, involving multiple actors with their own values, expectations, and worldviews. As such, we need a way to consider those issues cautiously and form as broad of a perspective as possible. With ethics, there may not be a simple solution offering clear-cut answers; rather, an ongoing process can guide and navigate conversations about these complex issues.

Cultivating ethical mindsets throughout the software development and data life cycles

We strive to instil moral values in computing students to empower them to make ethical decisions throughout the software development life cycle (SDLC) and the data life cycle (DLC). The term 'software development life cycle' describes the formal or informal process, or methodology, employed for the design, creation, and maintenance of software (Sommerville, 2016). It has been defined using multiple models (e.g., Waterfall, Spiral) (Ruparelia, 2010; Mall, 2018), and while the names for each of the phases and specifics vary, the core concepts remain more or less fixed. Linked to software design, the 'data life cycle' comprises distinct phases (Wing, 2019), and the term describes the process of generating, using, and managing data (e.g., datasets, databases, code). Once the data life cycle is complete, knowledge gained can result in feedback that can inform/improve additional data collection and creation. Throughout the SDLC and DLC, an 'Ethics by Design' approach can be employed to emphasize fairness, transparency, accountability, social and environmental well-being, privacy, data governance, and human agency (Dainow & Brey, 2021).

Teaching students to embrace 'Ethics by Design' involves tools, discussions, and assignments that promote ethical reflection and can aid in problem-solving. The approach can enhance choice consideration and inform planning, thinking, and actions in the life cycles. In Figure 18.1, we provide a conceptual model that unites the SDLC, DLC, and ethics (Ashok et al., 2022; Janeja, 2019; Karim et al., 2017; Wing, 2019).

Although each life cycle is presented separately, the arrows leading into each other are meant to emphasize that the two cycles are frequently interrelated. Data generation can follow software development, and data results can spark further software development. Ethical principles must be stressed across the process. We propose this new conceptualization, which we label the ETHOS model, as a way for educators to reconcile the components it encompasses:

Ethical (ethics)
Technological processes (software engineering life cycle),
Handling, and
Organizing data (data life cycle) for
Social impact (ethics)

These components appear in Figure 18.1 to make their connectivity explicit and highlight that consideration of ethics should encapsulate all decisions, choices, and actions taken – and all with

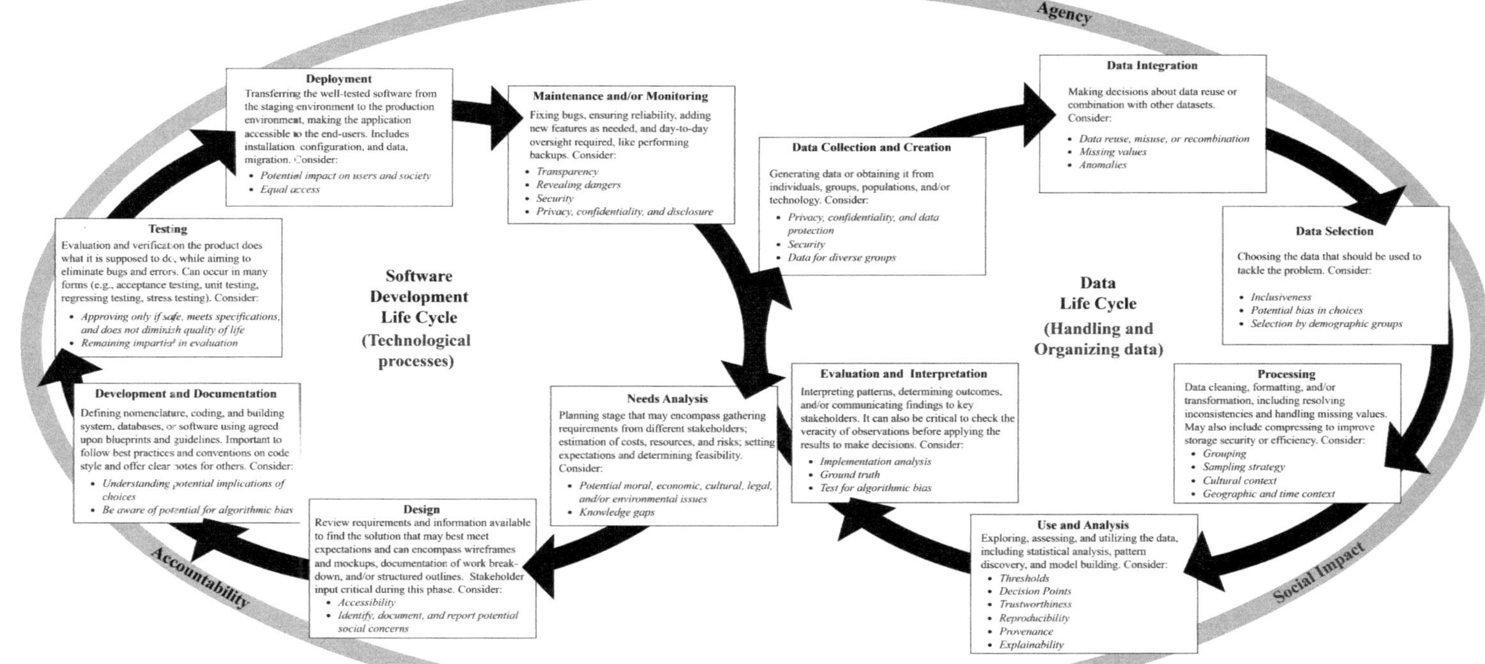

Figure 18.1 ETHOS conceptual model applying ethics throughout the SDLC and DLC (drawing from Ashok et al., 2022; Janeja, 2019; Karim et al., 2017; Wing, 2019).

regard for social impact. An 'Ethics by Design' perspective, as described by the ETHOS model, highlights the value of being mindful of and addressing ethical issues as early as possible. It also lends itself to following up on these issues and evaluating them throughout the process. As identified in the outer ring of this model, this necessitates consideration of additional concepts, including:

1. *Accountability and transparency*: To address the challenge posed by the 'many-hands problem' of attributing outcomes to a single individual in software projects that involve multiple people (van de Poel, 2015), the ETHOS model fosters a culture of individual ownership for actions, choices, and mistakes. In addition to complying with regulations, this involves collaboration, understanding expectations, ensuring accuracy, considering consequences, valuing quality, and being honest about decisions. Transparency involves communication, traceability (documenting or tracking goals, history, and assumptions), and intelligibility (comprehensibility and monitoring programs and systems to achieve outcomes).
2. *Agency*: Computing professionals make ethical decisions shaping the creation, maintenance, and application of products, programs, and data. They actively influence the ethical landscape through decision-making rather than passively following established norms. Thus, empowering individuals to address possible issues and propose solutions as they arise is valuable. Those involved ought to have autonomy over ethical deliberations, strategies, and actions that support the freedom and diversity of users. They should also reconcile differing interests and values while balancing project and product specifications.
3. *Ethical awareness*: Recognizing, analyzing, and navigating moral situations is crucial for individuals and organizations as it adds a human element of control. This goes beyond understanding the impact on various actors, including the environment. Hence, it is essential to prioritize the sustainability of products, programs, or data and anticipate potential repercussions at the conclusion of the item's life cycle. Proactively establishing safeguards in the SDLC and DLC to detect and mitigate problems is part of this, but ongoing appraisal and reflection are equally important for mindful decision-making.
4. *Professional ethics*: Conduct standards should be an integral part of the SDLC and the DLC. Individuals and organizations must uphold an ethical code – and values – centered on respect, transparency, honesty, fairness, privacy, and human-centric well-being. Individuals should understand diverse definitions and manifestations of these concepts, enabling them to make ethical decisions in various situations. Such ethical conduct may appear through fidelity, beneficence (i.e., promoting doing good or the well-being of others), non-maleficence (i.e., avoiding harming or injuring others), and justice (i.e., distributing benefits and burdens fairly).
5. *Social impact*: Ethical lapses in the SDLC and DLC can have serious consequences in real-world situations. Accordingly, it is important to extend beyond the technical aspects of software engineering to consider social, economic, cultural, and environmental impacts on humanity. Such thinking and behaviors demand global awareness, cultural competence, attention to accessibility, and the inclusion of diverse perspectives – encouraging proactive contributions to technology in alignment with varying ethical values. It necessitates addressing societal challenges, promoting social good, and identifying, seeking to prevent, and resolving potential negative impacts of programs, applications, and/or data.

We advise using the ETHOS model to emphasize human values and students' personal responsibility and encourage students to reflect on their own roles, biases, and decisions. Although we present bullet points in Figure 18.1 to provide instructors with ideas for further discussion points, a simplified version may make the content more digestible in the classroom. Introducing individuals to existing codes like the ACM's Code of Ethics or having students create their own codes fosters awareness of their professional responsibilities. Everyone taking moral responsibility and striving for continuous improvement and just behaviors can contribute to a positive shift in our interdependent society.

The following sections describe several ways educators can approach these topics. We encourage the exploration of different (ethical) viewpoints. Instructors can initiate meaningful conversations in classrooms using several 'dialogical approaches.' Students can also engage with various 'reflection tools' to help them ruminate on ethical issues and concerns in projects. These dialogue approaches and reflection tools can be considered to further students' engagement with ethical concepts, issues, and thought processes. The sections below focus on software engineering, and, again, for a more general discussion on using reflective and dialogical approaches in engineering ethics education, we refer readers to Chapter 25.

Dialogical approaches

Reflecting on choices and engaging in open discussions about intercultural responsibility, tolerance, and consequences aid ethical decision-making (Ferri, 2014). Integrating ethics into the classroom can be approached in various ways to help students develop ethical mindsets and address any issue they may encounter.

The purpose of ethical conversations in software engineering

Issues can arise in any project or system involving software or data, and three key aspects must be considered (van Veen & Visser-Knijff, 2022). The first involves the project's technological possibilities and feasibility, leading back to the SDLC. The second regards the legality of data storage and application (i.e., the DLC): *Is data's intended use and processing compliant with regulations?* These two aspects can easily be solved by personnel like privacy officers and can be assisted with a control check like a data protection impact assessment (DPIA).[1] However, even when all technical or legal questions are sufficiently answered in any project, a third aspect remains – concerning desirability. This aspect demands careful reflection regarding software and data projects at different points in time to ensure we consider future ethical implications a project might have in addition to complying with the rule.

One way to empower students is by incorporating ethical conversations into educational practice. Gaining experience in having ethical conversations about real-world case studies can help students gain what Aristotle calls *phrónēsis*, or practical wisdom, which to him means knowing how to act in a specific situation (Kraught, 2022). Gaining *phrónēsis* is a process of training to cultivate one's virtues or good traits. By applying principles of action to real-world situations, we will train ourselves to assess each situation carefully and then decide on our actions, which should align with a virtue. This way of looking at ethics is crucial because 'doing' ethics is not a matter of ticking boxes or choosing the right action from a predefined set, but rather requires us to assess each situation and its context on its own and decide our best course of action from there.

In addition to teaching students about ethical implications, conversations can aid decision-making. Facing ethical issues can potentially pressure students to make the right decision. At the same time, many students lack formal decision-making training and practice (Bond et al., 2008; Keeney, 2020) while also underestimating how their own biases affect decision quality. Discussing projects in a group may reveal diverse perspectives, offering invaluable input on project decisions.

Conversations may yield new insights, promoting ethically minded decision-making and risk mitigation. Below, we will summarize three conversation methods and how they can be used:

1. *Socratic dialogue*: The Socratic dialogue is a conversation method which, according to Knezic et al. (2010, p. 1105), is defined as:

   > a philosophical group dialogue in which the participants guided by a facilitator and a number of ground rules strive to reach a consensus in answering a fundamental question on the basis of a real-life example or incident with the purpose of achieving new insights.

For those in computing roles, the Socratic dialogue may appeal for cultivating several skills:

a) Asking the right questions – as the dialogue is centered around one core question, students are pushed to carefully consider what questions should be asked in different stages of the ETHOS model.
b) Inquiring into underlying assumptions or ideas – at its core, the Socratic dialogue is focused on inquiring into one's ideas, beliefs, assumptions, and dispositions. It is thus pre-eminently suited to encourage students to examine their own attitudes and identify any bias they might unconsciously hold that could impact projects.
c) Promoting mutual understanding using the Socratic dialogue is a collective effort, pushing participants to reach a unanimous conclusion. This encourages students to adapt their viewpoints based on new information and sound arguments, fostering better teamwork and openness to diverse perspectives that can alter project directions.

2. *Moral case deliberation*: Moral case deliberation is a conversation method commonly used in medical and business ethics. It involves discussing a real-life case to move towards a concrete approach or action.

   As moral case deliberation aims to obtain concrete project actions, it proves valuable with case studies or as group intervention for students dealing with ethical challenges in internships. Discussing problems and extracting actionable suggestions boosts students' confidence to address such issues in projects, enhancing their ethical agency. As per the description in a 2012 article by Karssing (2012), students will also be encouraged to explore the case and thus increase their ethical awareness and sense of personal responsibility.
3. *Thought experiments and scenario thinking*: The last method to initiate a conversation is to use thought experiments or scenario thinking. Thought experiments (e.g., Philippa Foot's 'Trolley Problem') are widely used in philosophy. Philosophers often use thought experiments to visualize scenarios that cannot be carried out in the real world due to physical, financial, technological, or ethical limitations. The *Stanford Encyclopedia of Philosophy* (Brown & Fehige, 2019) describes engaging in a thought experiment as follows: "we visualize some situation that we have set up in the imagination; we let it run or we carry out an operation; we see what happens; finally, we draw a conclusion."

Using thought experiments and scenario thinking lends itself especially well to software engineering, as it focuses on trying to gain insight by:

a) Identifying issues by describing possible outcomes of deployment (by sketching utopian and dystopian scenarios).
b) Reflecting on the nature of the problem that serves as the basis for a project (by asking: *Is this solution the means or the goal?*).

c) Reflecting on the necessity of using specific data-driven or technological solutions and exploring alternative options (by asking: *What problem(s) does this solution solve? What problem(s) doesn't it solve? What new problem(s) does it create?*).

Reflection tools

Ethical reflection tools can provide further guidelines and resources to help identify and mitigate potential ethical issues that may arise. Although the ETHOS model can function as one such reflection tool, others have been developed which may offer additional resources and guidance. They adopt a variety of formats, that we broadly classify as: (1) ethics-by-design tools, (2) process tools, and (3) ethical impact assessments. What they all have in common is that they offer a way to start ruminating on a project in a structured way. They can help individuals in computing make more informed decisions, consider how to address ethical issues in their projects, and weigh the potential benefits and harms of different options. Some tools that may be helpful to consider are:

- The *Data Ethics Canvas*[2] is a reflection tool created to help identify and manage ethical issues in the DLC. Designed for project-wide use, it poses predefined questions to provide an overview of problems encountered in data-based projects. It can help acquaint students with using ethical tools. Alongside the canvas (i.e., matrix), the tool offers a comprehensive user guide that promotes using the tool for enhanced project planning and encourages a human-centered perspective towards the design from the project's outset.
- *Consequence scanning*[3] involves a set of activities that can be incorporated into the SDLC, to examine the potential impact on people and/or society. It consists of a manual, printed headings, a proposed structure to follow, and two sets of prompts focused on consequence and context. It provides opportunities for software teams to have conversations about the potential implications of what they have built, what they will build, and how they can mitigate potential harms before they happen. It can also help students cultivate awareness of ethical and societal problems arising when one fails to reflect on potential consequences of their designs.
- The *Data Ethics Decision Aid* (DEDA)[4] is a toolkit designed to be used by teams involved in data-based projects. With a worksheet designed like a board game and with a questionnaire and handbook, the DEDA offers an interactive experience. Using the DEDA is a structured process in which each team member has their place, with the worksheet serving as the foundation of a conversational session. Applying the DEDA in educational settings can help students gain experience with identifying values *and* value conflicts, and approaching software products using an ethics-by-design approach, while also learning to work as a team.
- The *Ethical OS Toolkit*[5] serves as a guide for predicting the future consequences of current technology. It aims to encourage various stakeholders involved in its creation, dissemination, and use to foresee potential issues. The tool includes 14 scenarios to start conversations, a risk mitigation manual, and a checklist for considering ethical questions and issues. Educators can use the toolkit in computing or design courses in multiple ways, for example,having students read about the proposed 'Risk Zones' and tasking them with collecting examples in the real world.

Case studies

Using case studies to promote ethical thinking

As educators, facilitating critical thinking via case studies can help students understand moral issues in real-world situations and prompt them to connect ethics to their practice. Analyzing case studies can stimulate ethical awareness and a sense of personal and professional responsibility. Such activity can help students explore the values and interests of the various actors (Lim et al., 2011). Working with case studies can serve as problem-based learning, which may aid decision-making (Chowdhury, 2018) and improve critical thinking. By pairing reflection tools with the cases, teachers can provide students with further opportunities to internalize concepts and meta-cognition (Begley & Stefkovich, 2007). For specific advice on facilitating this type of discussion/activity, please see Chapter 20.

We note that 'flipped classrooms' can provide a viable alternative for educators concerned about finding time to integrate case studies into existing lessons. 'Flipped classrooms' is an instructional approach where traditional teaching methods are reversed, with students learning new material independently (outside of class) and engaging in activities, discussions, or case studies during class time. This approach can ensure material is covered while enhancing engagement and academic outcomes (Fuchs, 2021). Moreover, providing students with guidance as they solve problems (such as those described through hypothetical scenarios) can be aided via interactions with peers and/or instructors (Herreid & Schiller, 2013).

We now showcase three case studies to provide educators with possible starting points to engage students on topics related to current ethical concerns in software engineering. These case studies are hypothetical but are based on real-world examples. Although there are many ways to incorporate these cases into lessons, we encourage instructors to consider having ethical conversations (using the conversational methods described in the previous section) and providing students with reflection tools to tackle ethical issues arising from these scenarios. Alternatively, each could be integrated into small-group role-play (for more on that, see Chapter 20).

Case study 1: Algorithmic bias – fairness

Scenario

You head the Responsible AI division at SynGenAI, a software company based in Silicon Valley. SynGenAI is steadily gaining attention because it recently released Talk8ive, a free-to-access general-purpose chatbot that tries to answer users when they pose questions. Talk8ive is built on a large language model (LLM), a machine learning algorithm that can recognize, predict, and/or generate text based on large datasets of content, primarily text that the algorithm is fed as input training data. Talk8ive was released with much fanfare but within hours of its release, a number of users started flagging the chatbot for the content it was generating. Since Talk8ive is designed to answer any kind of question, all types of queries are treated the same way. Therefore, while queries with more deterministic responses (coding-related) generate less disputable answers, queries about predicting social outcomes (recidivism) produce untrue and racially biased responses. The behaviors displayed by the chatbot are attracting a lot of negative attention to the product and the company, and there are calls to revoke free access to the bot.

Discussion prompts:

1. *Did SynGenAI make the right decision to release Talk8ive? Why were the issues not flagged in the development stage itself?*
2. *Should the chatbot have been created to address all kinds of queries in the same manner? How did the algorithmic bias enter into the generated context?*
3. *As head of the Responsible AI division, how do you prioritize your responsibilities to the company, to the public, and to your team? What course of action should you adopt?*

Context:

Generative AI media, such as OpenAI's DALL-E or DALL·E2[6] and Speechify,[7] have gained popularity due to their ability to create realistic digital images or clone voices using machine learning algorithms (Chen & Lin, 2023; Pavlik, 2023). These algorithms are trained on vast amounts of data, allowing them to produce content that can be difficult to distinguish from human-created content. ChatGPT, developed by OpenAI, is one such example of a chatbot derived from the generative pretrained transformer (GPT) (Rudolph et al., 2023), that utilizes LLMs to generate responses in a conversational manner. Generative AI tools can be classified into three based on the kind of tasks they perform (Narayanan, 2019):

1. *Perception* tasks involve interpreting sensory data, such as images, audio, or text, to extract meaningful information.
2. *Judgment automation* involves making decisions or providing recommendations based on data analysis and predefined criteria.
3. *Social outcome prediction* involves predicting future events or results based on historical data and patterns.

An over-reliance and belief in generative AI's ability to solve problems can lead to complacency and additional problems. Although generative AI may positively contribute to society, it has limitations, because the underlying algorithms may be trained on datasets that reflect societal biases. If ignored, AI systems can 'learn' in ways that amplify incorrect points or replicate discriminatory patterns. For example, in areas such as hiring or loan approval, AI algorithms can reinforce existing prejudices and disadvantage certain groups, perpetuating social inequities rather than addressing them (Su & Yang, 2023). Broadly speaking, algorithmic bias can result in the following types of problems:

1. *Misrepresentation of minoritized communities*: Minoritized individuals in the tech industry often lack the influence to shape software development and data usage. Consequently, content related to these groups tends to be biased and reflective of societal stereotypes or exclusionary practices.
2. *Misinformation*: The presence of pseudoscientific content in the training data may result in its regurgitation by the AI when presented to users. Misinformation is compounded by automation bias because humans tend to trust the confident tone of interactive AI like chatbots.
3. *Automation and intellectual property theft*: Generative AI platforms are affecting the occupations of artists and journalists, whose occupations are being automated by large corporations because it is cited as being cheaper. Often, these automated generations neglect factors such as copyright or licensing in their training data – and effectively commit digital theft.

Educators can discuss algorithmic accountability using the ETHOS model to raise awareness of the ethical implications of creating and using AI. By ensuring that data sources are properly curated, algorithms are tested rigorously, and AI-generated outcomes are accounted for by developers, students will learn to develop a practice of developing ethical AI systems. Educators can use this case study in the classroom by integrating the dialogue approaches and reflection tools mentioned earlier. For example, moral case deliberation could be used to discuss algorithmic bias. Issues with unwanted output are a common problem with chatbots but often only seem to become apparent following release. Accordingly, we suggest holding a moral case deliberation on the question: *Should SynGenAI take Talk8ive offline completely?* We also suggest using the consequence scanning tool in combination with this case study to encourage reflecting on more careful implementation. The main ethical issues in this case revolve around the (unforeseen or unintended) consequences of generative AI.

Case study 2: User interface design – empowerment

Scenario

Your current phone is falling apart – it is running out of storage, it has fallen to the ground multiple times – and as a result, it runs very slowly. So, you decide to buy a new phone. Although you do not have a clear idea of what phone you want to purchase, you do have a set budget. By querying a search engine, you find a list of shopping websites selling phones within your target range. You browse some of these websites, looking up different phones and their specifications. After some time, you pick a phone offered at a 30% discount. You read the reviews and are satisfied with the performance and other aspects of the phone. You decide to wait a few days to buy it since there is no real sense of urgency to purchase it immediately. However, as you try to navigate away from this page, you encounter a pop-up message, with a limited time offer for an additional 10% discount that is valid only until the end of the day. A countdown timer is also shown, ticking down on the hours left to buy the phone at this price. Since this seems like a good deal – and you do not want to regret buying it later at a higher price – you add it to your shopping cart. As you check out, you look at the final price and realize that the total is much higher than what was shown earlier. Upon closer inspection, you find that the final breakdown includes a screen guard and a one-year phone protection plan that you did not intend to purchase, and did not add yourself.

Discussion prompts:

1. *How does this experience make you, as a user, feel? What specific design aspects of the shopping website contributed to this feeling?*
2. *How does using interface design like those mentioned above affect your ability (as a user) to make a well-informed, autonomous choice?*
3. *Do you note any ethical issues in how the shopping experience has been designed? If so, what are they and what would your feedback be for the user-interface designer?*
4. *What might the different motivations be that govern the design choices for the website user interface? How might web designers redesign the interface to improve the user experience while addressing potential ethical issues?*

Context:

The scenario describes a not-uncommon shopping experience. Many users encounter similar situations where choices and decisions are manipulated during their interactions with a given website. Moreover, these types of manipulation are not restricted to e-commerce. Companies have tried to force certain actions on users engaged in unrelated pursuits. For example, at one point, users were automatically opted into eBay's marketing emails when they signed in with their Google accounts.[8] Some companies design their user interfaces to make it difficult to perform certain actions, particularly those that can harm the company's revenue. For example, users from Europe have complained about Amazon's confusing cancellation process regarding Prime service, which tries to distract users by issuing several warnings to deter them from canceling their subscription.

Some designers use knowledge of human behavior to implement deceptive functionality that is not in the user's best interest (Gray et al., 2018) to manipulate them (Haselton et al., 2015). The design interfaces and choices described in the scenarios mentioned above are commonly known as 'dark patterns' (also known as 'deceptive designs'). Dark patterns are user interface design choices that nudge, manipulate, or deceive users into making unintended and potentially harmful decisions while using an online service (Mathur et al., 2019). The key issue in the above scenarios is that these services may sacrifice providing users with a positive experience for achieving the company's business goals. Dark patterns are asymmetric (i.e., available choices are presented differently to users), covert (i.e., the effect of the design choice is hidden from users), deceptive (i.e., they portray a misleading/false belief); and restrictive (i.e., they hide, obscure, or delay presentation of information from users) (Mathur et al., 2019). Tactics of dark patterns include sneaking (i.e., concealing or delaying information users might object to if overtly displayed), creating false urgency, misdirecting (i.e., using words or visuals to steer users), providing false 'social proof' (i.e., displaying fictitious user experiences to coerce a purchase), implying scarcity (i.e., falsely indicating product unavailability), causing obstruction (i.e., making it hard for users to exit situations), and forcing action (i.e., requiring tasks unrelated to current activities) (Mathur et al., 2019). Concerning the ETHOS model, such dark patterns can be associated with both the SDLC and the DLC, especially those regarding the design and development of software systems. Moreover, data from user interactions can be applied to design dark patterns, violating the professional and ethical codes of conduct that necessitate considering the well-being of the software's end-users.

Considering the prevalence of dark patterns, students are likely to have encountered such scenarios online. Hence, drawing from students' experiences can be an excellent way to introduce the ethics involved in user interface design. Instructors can describe examples of deceptive designs and ask students to classify designs into different categories. Researchers have suggested that interventions such as 'spot the dark pattern' and 'design bright patterns' can help create awareness regarding design patterns (Bongard-Blanchy et al., 2021) and hence can be used in classrooms. Instructors can also leverage software engineering projects to help students think of and design ethical user interfaces. Using dialogical approaches, such as the Socratic dialogue technique with a prompt like *What is autonomy?* can bring out diverse issues surrounding ethical user interfaces. DEDA can also be used as a reflection tool. Students can identify conflicting values between (a) companies that use dark patterns and (b) users confronted with dark-pattern user interfaces. As the DEDA encourages identifying (conflicting) values and actors, it can prompt helpful reflection on the implementation of dark patterns.

Although creating awareness of deceptive patterns is needed, a more crucial goal for educators is to help students integrate ethical reasoning into all aspects of the user interface design process. Elements of the ETHOS model, such as ethical awareness and accountability can prepare students to engage in critical debate about where to draw the line, based on the company's values and their own sense of ethics. Students should realize the power that user interfaces have over people's behavior and online interactions, and the social responsibility they as software designers have towards developing ethical user interfaces.

Case study 3: Privacy and surveillance – security

Scenario

You are a machine learning engineer in the computer vision department at Protos. This tech company manufactures virtual reality (VR) headsets and a social media space called Protospace for VR users' avatars to meet for work and entertainment. To enhance the experience, Protos has developed a headset called VISOR with inward-oriented cameras to obtain real-time information about users' gaze and non-verbal facial expressions. In addition, external, outward-oriented cameras emulate users' physical body movements in the virtual world. Protos states these sensors allow for deploying customized, photo-realistic avatars and adaptive content based on the users' motion and reactions. Users must accept Protos' terms to use VISOR. Critics of the company are skeptical about this justification – they believe that Protos will store and apply the data captured by the sensors as training for future projects. Moreover, the purported immersive nature of Protospace will grant Protos access to users' behavioral patterns, which can then be monetized by manipulating advertising and/or selling the data to third-party enterprises. Furthermore, there are growing concerns about the potential use of such a technology by autocratic governments that seek to surveil their citizens to quell civic unrest, repress all forms of opposition, and establish political dominance.

Discussion prompts:

1. *What are the potential risks of collecting this type of sensitive data from users? What measures can be taken to mitigate these risks?*
2. *What should be included as part of the informed consent terms? Are there other ways to help users recognize what they are agreeing to? How can users be made aware of future uses of the data collected?*
3. *As a machine learning engineer whose work involves processing the data gathered by the cameras, how could you improve the design's privacy and/or security?*

Context:

Ongoing attempts to improve facial recognition, non-verbal expressions, and gestures have resulted in data collection beyond what users have agreed upon or consented to. The example scenario mirrors real-life concerns around Meta's Quest Pro, a device collecting multi-dimensional data (Johnson, 2022). Although the technology was described as valuable for deploying customized, photo-realistic avatars, concerns arose around the detailed information collected about individuals and how it might subsequently be used for surveillance or otherwise infringe upon their privacy. Furthermore, Meta's privacy policy was vague regarding how the data would be used and shared with outside services.

Despite multiple calls to regulate the collection, storage, and application of user data, it appears that companies frequently collect more data than necessary. Personal information, preferences, and behaviors are often captured without explicit consent, leading to a loss of control over one's own data. This can result in unintended consequences for privacy, such as data breaches, identity theft, or the misuse of sensitive information. Scholars have highlighted the need for enhanced consideration of the interactions between technology and the law (Bernes, 2022). They have encouraged those involved to consider the purposes, implications, and uses of data collected; the transparency of tools developed; and ways to limit what is gathered, recorded, and disseminated for purposes more strictly aligned with goals. The European Union's General Data Protection Regulation (GDPR) has specific guidelines on this that students should be made aware of; readers are encouraged to see Chapter 13 on law in engineering ethics education.

There are several approaches educators can take to deliver lessons on privacy and security. For instance, they can introduce concepts, such as the benefits of sharing; privacy protection and perceptions; codes of conduct; cryptographic protocols; and international laws and regulations. The ETHOS model could help align discussion around design, deployment, development and documentation, or maintenance or monitoring when referencing components of the SDLC. The scenario above can also fit into discussions about the DLC throughout its phases. Concrete examples and discussion scenarios can raise awareness of possible issues and promote the need for greater accountability. For example, students could be asked to consider the trade-offs of retail loyalty cards, which can offer discounts but may also track shoppers' habits and sell data to third parties. Another example could be related to concerns regarding healthcare and genetic mapping and/or ancestry websites. By spanning multiple fields and topics, for example, from electronic payment systems to educational records, instructors can help illustrate the necessity of awareness, protection, and action to rectify issues.

Instructors can use dialogical approaches such as Socratic dialogue and scenario thinking in conjunction with this case study. For the Socratic dialogue, we suggest evaluating the concept of 'privacy by design' as a realistic development approach. For scenario thinking, imagining idealistic and problematic scenarios for collecting sensitive user data may shed light on data ethics and values. The Data Ethics Canvas can be an appropriate reflection tool introduced along with this case study. One of the central issues here is the collection and use of personal user-data to further the business interests of the company doing the collection. By utilizing the Data Ethics Canvas, students can engage with questions that address multiple facets of this case, including the legal implications of collecting and utilizing data on this scale; they can reflect on reasons for using this data and the need for transparency.

Conclusions

As technology evolves, so do concepts and definitions; engineering educators must continually update the definitions, pedagogical content, and activities we use. Although today's society is increasingly recognizing the urgent need to address ethical issues and the need for responsible software development, this recognition brings its own problems – problems that have an impact beyond the classroom. We face the very real danger of 'ethics' becoming a buzzword and another trend to be monetized by companies (the way corporate 'greenwashing' undermines sustainability initiatives, see Chapter 6). Ready-made courses or tools that seem able to solve our ethical issues rather than focus on imparting awareness and ethical thinking could exacerbate that potential problem. A criticism that has been raised regarding the Sustainable Development Goals (SDGs), which some organizations tout without achieving results in specific target areas, as mentioned in Chapter 6. Another possible concern involves the increasing reliance on tools; while ethics assessment tools can be very helpful in guiding conversations and steering us toward relevant questions, they can be used in more nefarious ways. They may be used to help companies identify or define what is easiest or cheapest rather than what is morally desirable (and desirability can also vary depending on who is asked). Using ethics assessment tools could devolve into merely another box-ticking exercise, rather than being used to prompt software developers to reflect on the possible implications of their decisions and the products they help create. Ongoing efforts must thus be made to provide meaningful examples, to articulate where potential issues may arise, and to help students make relevant and meaningful connections – to help ensure future generations of engineering practitioners are committed to ethical principles and that software engineers develop products and manage issues as equitably and justly as possible.

Notes

1 https://gdpr.eu/data-protection-impact-assessment-template/
2 https://www.theodi.org/article/the-data-ethics-canvas-2021/
3 https://www.tech-transformed.com/product-development/
4 https://dataschool.nl/en/deda/
5 https://oecd-opsi.org/toolkits/ethical-os-toolkit/
6 https://openai.com/dall-e-2
7 https://speechify.com/
8 https://twitter.com/darkpatterns/status/1470399874147438594

References

Andrus, M., Dean, S., Gilbert, T. K., Lambert, N., & Zick, T. (2020). AI development for the public interest: From abstraction traps to sociotechnical risks. In *2020 IEEE International Symposium on Technology and Society (ISTAS)* (pp. 72–79). IEEE.

Ashok, M., Madan, R., Joha, A., & Sivarajah, U. (2022). Ethical framework for artificial intelligence and digital technologies. *International Journal of Information Management, 62,* 102433.

Begley, P. T., & Stefkovich, J. (2007). Integrating values and ethics into post secondary teaching for leadership development: Principles, concepts, and strategies. *Journal of Educational Administration, 45*(4), 398–412.

Bernes, A. (2022). Enhancing transparency of data processing and data subject's rights through technical tools: The PIMS and PDS solution. In: Senigaglia, R., Irti, C., Bernes, A. (Eds.), *Privacy and Data Protection in Software Services. S*ervices and Business Process Reengineering. Springer, Singapore. https://doi.org/10.1007/978-981-16-3049-1_17

Bond, S. D., Carlson, K. A., & Keeney, R. L. (2008). Generating objectives: Can decision makers articulate what they want? *Management Science, 54*(1), 56–70.

Bongard-Blanchy, K., Rossi, A., Rivas, S., Doublet, S., Koenig, V., & Lenzini, G. (2021). "I am Definitely Manipulated, Even When I am Aware of it. It's Ridiculous!" – Dark Patterns from the End-User Perspective. In *Designing Interactive Systems Conference* (pp. 763–776).

Brown, J. R. & Fehige, Y. (2023). Thought experiments. In Zalta, E. N. & Nodelman, U. (Eds.), *The Stanford encyclopedia of philosophy*. Metaphysics Research Lab, Stanford University. https://plato.stanford.edu/entries/thought-experiment/

Brown, S., Davidovic, J., & Hasan, A. (2021). The algorithm audit: Scoring the algorithms that score us. *Big Data & Society, 8*(1), 2053951720983865.

Chen, J. J., & Lin, J. C. (2023). Artificial intelligence as a double-edged sword: Wielding the power principles to maximize its positive effects and minimize its negative effects. *Contemporary Issues in Early Childhood*, 14639491231169813. https://doi.org/10.1177/14639491231169813

Cheong, M., Leins, K., & Coghlan, S. (2021). *Computer science communities: Who is speaking, and who is listening to the women? Using an ethics of care to promote diverse voices*. In Proceedings of the *2021 ACM Conference on Fairness, Accountability, and Transparency* (pp. 106–115).

Chowdhury, M. (2018). Emphasizing morals, values, ethics, and character education in science education and science teaching. *MOJES: Malaysian Online Journal of Educational Sciences, 4*(2), 1–16.

Cohen, L., Precel, H., Triedman, H., & Fisler, K. (2021). A new model for weaving responsible computing into courses across the cs curriculum. In *Proceedings of the 52nd ACM Technical Symposium on Computer Science Education* (pp. 858–864).

Connolly, R. (2020). Why computing belongs within the social sciences. *Communications of the ACM, 63*(8), 54–59.

Dainow, B., & Brey, P. (2021). *Ethics by design and ethics of use approaches for artificial intelligence*. European Commission.

Ellul, J. (1964). *The technological society* (J.Wilkinson, Trans.). Vintage Books. (1954.

Ferreira, R., & Vardi, M. Y. (2021). Deep tech ethics: An approach to teaching social justice in computer science. In *Proceedings of the 52nd ACM Technical Symposium on Computer Science Education* (pp. 1041–1047).

Ferri, G. (2014). Ethical communication and intercultural responsibility: A philosophical perspective. *Language and Intercultural Communication, 14*(1), 7–23.

Fiesler, C., Garrett, N., & Beard, N. (2020). What do we teach when we teach tech ethics? A syllabi analysis. In *Proceedings of the 51st ACM Technical Symposium on Computer Science Education* (pp. 289–295).

Fuchs, K. (2021). Innovative teaching: A qualitative review of flipped classrooms. *International Journal of Learning, Teaching and Educational Research, 20*(3), 18–32.

Gotterbarn, D., Miller, K., & Rogerson, S. (1997). Software engineering code of ethics. *Communications of the ACM, 40*(11), 110–118.

Gotterbarn, D., Miller, K., & Rogerson, S. (1999). Computer society and ACM approve software engineering code of ethics. *Computer, 32*(10), 84–88.

Gray, C. M., Kou, Y., Battles, B., Hoggatt, J., & Toombs, A. L. (2018). The dark (patterns) side of UX design. In *2018 CHI Conference on Human Factors in Computing Systems* (pp. 1–14).

Haselton, M., Nettle, D., & Murray, D. (2015). The evolution of cognitive bias. In Buss, D. M. (Ed.), *The handbook of evolutionary psychology* (pp. 1–20). John Wiley & Sons, Inc. https://doi.org/10.1002/9780470939376.ch25

Heidegger, M. (1977). *The question concerning technology* (W. Lovitt, Trans.). Garland Publishing, Inc.

Herreid, C. F., & Schiller, N. A. (2013). Case studies and the flipped classroom. *Journal of College Science Teaching, 42*(5), 62–66.

Horton, D., McIlraith, S. A., Wang, N., Majedi, M., McClure, E., & Wald, B. (2022). Embedding ethics in computer science courses: Does it work? In *Proceedings of the 53rd ACM Technical Symposium on Computer Science Education* (pp. 481–487).

Hysa, X., D'Arco, M., & Kostaqi, J. (2023). Misuse of personal data: Exploring the privacy paradox in the age of big data analytics. In *Big data and decision-making: Applications and uses in the public and private sector* (A. Visvizi, O. Troisi, & M. Grimaldi, Eds.), pp. 43–57. Emerald Publishing Limited. https://doi.org/10.1108/978-1-80382-551-920231004

Janeja, V. (2019). *Do no harm: An ethical data life cycle.*American Association for the Advancement of Science (AAAS). https://www.aaaspolicyfellowships.org/blog/do-no-harm-ethical-data-life-cycle

Johnson, K. (2022). *Meta's new quest pro vr headset harvests personal data right off your face*. https://www.wired.com/story/metas-vr-headset-quest-pro-personal-data-face/

Karim, N. S. A., Ammar, F. A., & Aziz, R. (2017). Ethical software: Integrating code of ethics into software development life cycle. In *2017 International Conference on Computer and Applications (ICCA)* (pp. 290–298).

Karssing, E. D. (2012). Uit de boekenkast van de bedrijfsethiek 44: Stapvoets door een dilemma. opzet en achtergronden van een bewerkingsmodel en rob van es: Professionele ethiek: Morele besluitvorming in organisaties en professies. *Tijdschrift voor Compliance, 12,* 133–138.

Keeney, R. (2020). *Give yourself a nudge: Practical procedures for making better personal and business decisions.* Cambridge University Press.

Knezic, D., Wubbels, T., Elbers, E., & Hajer, M. (2010). The socratic dialogue and teacher education. *Teaching and teacher education, 26*(4), 1104–1111.

Kraught, R. (2022). *Aristotle's ethics.* In Zalta, E. N. & Nodelman, U. (Eds.), *The Stanford encyclopedia of philosophy.* Metaphysics Research Lab, Stanford University. https://plato.stanford.edu/entries/aristotle-ethics/

Lim, M. Y., Leichtenstern, K., Kriegel, M., Enz, S., Aylett, R., Vannini, N., Hall, L., & Rizzo, P. (2011). Technology-enhanced role-play for social and emotional learning context–intercultural empathy. *Entertainment Computing, 2*(4), 223–231.

Mall, R. (2018). *Fundamentals of software engineering (5th ed.).* PHI Learning.

Martin, C. D., Huff, C., Gotterbarn, D., & Miller, K. (1996). Implementing a tenth strand in the CS curriculum. *Communications of the ACM, 39*(12), 75–84.

Martin, J. P., Desing, R., & Borrego, M. (2022). Positionality statements are just the tip of the iceberg: Moving towards a reflexive process. *Journal of Women and Minorities in Science and Engineering, 28*(4), v–vii.

Mathur, A., Acar, G., Friedman, M. J., Lucherini, E., Mayer, J., Chetty, M., & Narayanan, A. (2019). Dark patterns at scale: Findings from a crawl of 11k shopping websites. In *Proceedings of the ACM on Human-Computer Interaction, 3*(CSCW), 1–32.

Moore, J. (2020). Towards a more representative politics in the ethics of computer science. In *Proceedings of the 2020 Conference on fairness, accountability, and transparency* (pp. 414–424).

Mumford, L. (1934). *Technics and civilisation.* Harcourt, Brace and Company, Inc.

Mumford, L. (1967). *The myth of the machine: Technics and human development* (Vol. 1). Harcourt, Brace Jovanovich.

Narayanan, A. (2019, November 18). *How to recognize AI snake oil.* Arthur Miller Lecture on Science and Ethics, Massachusetts Institute of Technology.

O'Regan, G. (2008). *A brief history of computing.* Springer.

Pavlik, J. V. (2023). Collaborating with chatgpt: Considering the implications of generative artificial intelligence for journalism and media education. *Journalism & Mass Communication Educator, 78*(1), 84–93.

Rudolph, J., Tan, S., & Tan, S. (2023). ChatGPT: Bullshit spewer or the end of traditional assessments in higher education? *Journal of Applied Learning and Teaching, 6*(1), 342–363.

Ruparelia, N. B. (2010). Software development lifecycle models. *ACM SIGSOFT Software Engineering Notes, 35*(3), 8–13.

Shafer-Landau, R. (2018). *The fundamentals of ethics* (4th ed.). Oxford University Press.
Sommerville, I. (2016). *Software engineering* (10th ed.). Pearson.
Spiekermann, S. (2021). What to expect from ieee 7000: The first standard for building ethical systems. *IEEE Technology and Society Magazine, 40*(3), 99–100.
Spiekermann, S., & Winkler, T. (2022). Value-Based Engineering With IEEE 7000. *IEEE Technology and Society Magazine, 41*(3), 71–80.
Su, J., & Yang, W. (2023). Unlocking the power of ChatGPT: A framework for applying generative AI in education. In *ECNU review of education* (pp. 355–366).
van de Poel, I. (2015). The problem of many hands. In *Moral responsibility and the problem of many hands* (pp. 50–92). Routledge.
van Veen, M., & Visser-Knijff, P. (2022). 8. Ethics in local politics: A case study of the city of eindhoven. In B. Wernaart (Ed.), *Moral design and technology* (pp. 149–168). Wageningen Academic Publishers. https://doi.org/10.3920/978-90-8686-922-0_8
Wing, J. M. (2019). The data life cycle. *Harvard Data Science Review, 1*(1), 6.
Xu, C. K., & Tang, T. (2020). Closing the gap or widening the divide: The impacts of technology-enabled coproduction on equity in public service delivery. *Public Administration Review, 80*(6), 962–975.

SECTION 4

Teaching methods in engineering ethics education

Diana Adela Martin

Engineering ethics education has undergone significant evolution over the past half-century. This evolution has included the deepening of methods such as case studies, proposed by pioneering handbooks on engineering ethics education and scholarly works in the field (see Harris et al., 1995; Herkert, 2005; Martin & Schinzinger, 1983; Whitbeck, 1995), alongside the emergence of innovative approaches that integrate arts and real-life partners into the classroom. Engineering ethics teaching today benefits from the groundwork laid by esteemed predecessors such as Charles Harris Jr., Michael Pritchard, Michael J. Rabins, Elaine Englehardt, Caroline Whitbeck, Vivien Weil, Michael Loui, Joe Herkert, Carl Mitcham, Michael Davis, Mike Martin, Ronald Schinzinger, and numerous other scholars and educators. This section provides a state-of-the-art perspective of diverse approaches to teaching ethics, aiming to inspire educators to enrich and expand these methods further. We take pride in inviting our readers to contemplate how ethics can be integrated into teaching, whether as a standalone subject or seamlessly woven throughout the curriculum.

Six of the seven chapters in this section offer detailed insights into specific methods utilized in contemporary engineering ethics education (EEE), while the opening chapter provides a comprehensive overview to contextualize these approaches within the literature. It is essential to recognize that the teaching methods discussed are not isolated entities with rigid boundaries; rather, they often intersect and complement each other. Educators exploring the applicability of EEE methods may find that multiple approaches described by our authors hold relevance. Indeed, there are many overlapping themes. Our goal in featuring these various teaching methodologies – case studies, problem-based learning, value-sensitive design, service learning, humanitarian engineering, arts-based methods, and reflective and dialogical approaches – is to equip educators with frameworks for conceptualizing how to effectively engage students in meaningful and memorable ethics education.

Chapter topics

The section opens with Chapter 19, ‘Literature Review of Teaching Methods: Trends and Ways Forward to Support Engineering Ethics Instruction,’ by Madeline Polmear, Tom Børsen, Heather A. Love, and Amir Hedayati. The chapter consolidates existing research on traditional and emerging teaching methodologies, crafting a heuristic framework to assist educators in selecting an

DOI: 10.4324/9781003464259-23

approach that aligns with their instructional goals and underlying principles. The authors map the landscape of engineering ethics teaching – describing classic approaches and identifying emerging ones. Noting the traditional dominance of case studies and coverage of ethics codes, they highlight more novel ways to teach ethics – like student-developed activities, co-curricular activities, role-plays, and gamification. They consider how engineering ethics can be taught in person and online. Additionally, to aid teachers, the alignment among objectives, underlying assumptions, teaching methods, and assessment strategies is emphasized to facilitate effective instruction. The chapter identifies future research opportunities for researchers, particularly focusing on non-cognitive aspects, behavioral ethics, student needs, and robust assessment methods for evaluating interventions and outcomes.

Case studies typically share three characteristics: they depict real or hypothetical scenarios requiring decision-making, include relevant contextual and technical details, and prompt learners to engage with diverse perspectives. The second chapter, Chapter 20 on 'Teaching Ethics Using Case Studies' by Christian Herzog, Aditya Johri, and Roland Tormey, presents a balanced discussion, providing both supporting and opposing arguments for employing case studies in ethics instruction. The chapter expands on Polmear et al.'s recognition of case studies as a predominant approach in teaching engineering ethics by systematically evaluating their purpose, pedagogical outcomes, and integration within the course or curriculum. The authors explore the literature specific to case studies in EEE. They emphasize the effectiveness of case studies for exploring ethical dilemmas, contextualizing situations, and providing detailed technical specifications. They point readers toward existing repositories of engineering ethics case studies, organizing them based on thematic categories, types, and depth of exposition.

Herzog, Johri, and Tormey distinguish "highly complex and highly indeterminate case studies" from "simpler vignettes with less veracity" and note that combining case studies with other methods can provide a robust approach. They deem case-based engineering ethics education to be highly beneficial because it directs attention to the intricate details influencing ethical assessments and encourages a methodical and ideally participative approach to analyzing complex situations. Role-playing activities, the authors explain, can enhance students' ability to consider different viewpoints. Including emotional content is highlighted as a way to convey the complexities of engineering decision-making and foster empathy among students toward stakeholders. In presenting a nuanced perspective, Herzog, Johri, and Tormey emphasize that the objective of engineering ethics education is not to transform "engineering students into ethicists" but to equip them with the skills to communicate effectively about "moral hazards and the values" inherent in their work. They underscore the importance of fostering the ability to openly debate and reflect on one's ethical agency, allowing for potential challenges or support.

In Chapter 21, titled 'Embedded Ethics in Problem Design: The Case of Problem-based Learning in Engineering and Science,' Henrik Work Routhe, Jette Egelund Holgaard, and Anette Kolmos underscore the significance of integrating sociotechnical content into educational settings through real-life projects. Within problem- and project-based learning (PBL) frameworks, students engage with authentic problems, which necessitates identifying, characterizing, analyzing, and formulating solutions. Ethical considerations may arise as students reflect on the contexts and stakeholders involved. The chapter probes the intricacies of ethics integration within a PBL environment, exploring various theoretical aspects such as the problem design process, the types of problems encountered, and their relationship to engineering ethics.

Additionally, the authors examine the Intended Learning Outcomes outlined in study regulations for engineering and science programs, aiming to pinpoint factors that facilitate ethical engagement. They reveal that although ethics topics are ingrained within PBL curricula, they are often implicit rather than explicit. The authors raise concerns regarding the effectiveness of this approach, questioning whether students fully grasp the ethical dimensions of their work or can apply ethical theories to practical scenarios. The chapter critically assesses PBL dimensions, offering insights into how these learning environments can subtly incorporate ethics. Moreover, it provides practical recommendations for educators, outlining steps to design problems and identify factors that promote ethical considerations within a PBL framework.

In Chapter 22, titled 'Teaching Responsible Engineering and Design through Value-Sensitive Design,' Andrea R. Gammon, Annuska Zolyomi, Richmond Y. Wong, Eva Eriksson, Camilla Gyldendahl Jensen, and Rikke Toft Nørgård explore value-sensitive design (VSD) as a pedagogical approach for instilling ethics in engineering education. VSD and other values-centered design methodologies explicitly acknowledge the significance of values (such as privacy, dignity, and sustainability) in the context of technological development. The chapter combines the theory behind VSD with practical examples and recommendations that will be extremely useful for those who may be intimidated to integrate values systematically in their teaching. The chapter introduces the key ideas before providing guidance on how to teach VSD via conceptual, technical, and empirical investigations, as well as strategies for assessing it. The authors introduce readers to various instructional activities, including stakeholder analysis, value hierarchies, and envisioning cards, providing opportunities for further exploration through additional resources such as recommended readings or access to the Value Sensitive Design in Higher Education (VASE) program. Additionally, the authors discuss extensions of VSD that address issues of power dynamics, cultural norms, and multicultural perspectives, broadening the scope of ethical considerations within engineering and design education. This chapter synergizes well with Chapter 12, on 'Ethics and Engineering Design Foundations' by Bairaktarova, Wint, and Nweke.

The next chapter, Chapter 23 on 'Ethics in Service-Learning and Humanitarian Engineering Education' by Scott Daniel, Adetoun Yeaman, and William (Bill) Oakes, continues the focus on real-life learning environments, this time via humanitarian and service learning. Humanitarian engineering involves engineering practices aimed at improving the well-being of underserved populations or collaborating with vulnerable communities and individuals. It spans a wide range of activities, from assistive technologies to disaster relief and peace-building efforts, across sectors such as water, sanitation, energy, and health. Service-learning is an educational approach where students participate in organized service activities that address community needs, reflect on their experiences, and gain a deeper understanding of course content and civic responsibility.

Daniel, Yeaman, and Oakes share their experiences developing humanitarian and service-oriented courses and programs. For the author team, this way of teaching ethics starts by challenging Western assumptions of what counts as knowledge and what it means to be an engineer. The chapter discusses the types of *deliverables* developed in such programs, the process of becoming more reciprocal and involving partners in more meaningful ways, and the growing focus on connecting the engagement activities more explicitly with 'ethics.' This should include intentional preparation before students engage with communities, introduction to ethics frameworks applicable within the context of the community and the needs being addressed, and the inclusion of dedicated periods of reflection at multiple points (beginning, during, and after the program).

The authors note the need for further research on how humanitarian and service-oriented programs influence students' engineering skills and ethical development and benefit participating communities and to identify or refine instruments for assessing development. They suggest that educators have ample opportunities to involve students in local and global projects to explore ethical dilemmas and foster ethical decision-making, yet they stress the importance of acknowledging power dynamics and advocating for sustained engagement and relationship cultivation.

In Chapter 24, titled 'Arts-Based Methods in Engineering Ethics Education,' Sarah Jayne Hitt, David D. Gillette, and Lauren E. Shumaker explore intersections between arts and engineering, particularly within educational contexts. They differentiate arts-based content (which encompasses artistic artifacts such as literary texts, film, music, visual arts, poetry, dance, and theatre) from arts-based methods (i.e., the creative problem-solving processes involved in generating aesthetic objects, environments, or experiences that can be shared with others). Arts-based methods include pedagogical approaches like Socratic-style seminar discussions, creative writing exercises, visual depictions, and role-play scenarios. The authors stress the effectiveness of integrating arts-based methods into engineering ethics education, highlighting their potential to enhance learning outcomes and engage students more deeply with ethical issues.

The authors further explore the perceived division between arts and humanities and the technical sciences within academia, noting successful examples of integration in disciplines such as business and medicine, as well as specific engineering programs like those at the Colorado School of Mines and California Polytechnic State University. They acknowledge the challenges associated with introducing arts-based activities into traditional engineering courses, emphasizing the importance of careful integration to avoid student resistance. Moreover, the chapter argues that incorporating arts into engineering education can foster a comprehensive understanding of system interactions and address shortcomings in engineering education, particularly in parts of the world that do not use a liberal arts or common core strategy to underpin students' broader education.

The goal of advocating for arts-based methods in engineering ethics education is to provide students with additional problem-solving tools and perspectives (as opposed to transforming them into artists). By occasionally prompting students to think from the viewpoint of artists while maintaining their identity as engineers, educators can facilitate deeper self-reflection and help students recognize the intrinsic connections between their professional roles and their personal identities. This approach underscores the transformative potential of integrating arts into engineering ethics instruction, catalyzing holistic development and ethical engagement among engineering students.

In Chapter 25, 'Reflective and Dialogical Approaches in Engineering Ethics Education,' authors Lavinia Marin, Yousef Jalali, Alexandra Morrison, and Cristina Voinea underscore the significance of reflection and dialogue in engineering ethics instruction. Drawing on Dewey's insights, the authors outline four key criteria for reflection: deriving meaning, systematic and disciplined thinking, communal interaction, and attitudes valuing personal and intellectual development. They delineate four levels of reflection: reporting/responding, relating, reasoning, and reconstructing, and suggest explicit prompts for ethical reflection by teachers. The authors advocate for using individual and collective contemplation methods, incorporating spoken communication alongside silent and written individual thinking (which the authors define as dialogical and monological reflection), to deepen students' understanding, enhance problem-solving abilities, and foster ethical growth. These techniques can be implemented synchronously and/or asynchronously.

Furthermore, Marin, Jalali, Morrison, and Voinea highlight the parallels between reflection and moral deliberation, emphasizing their shared elements of engaging with ethical dilemmas and considering potential actions. They propose that in education, reflection should not aim to yield definitive answers but rather encourage students to confront ambiguity and embrace complexity, thereby becoming more comfortable with uncertainty. Building upon the phenomenological tradition, reflection is portrayed as a dialogical and mediated encounter with various perspectives, aiding students in becoming aware of their experiences and situating themselves within their phenomenal world. The authors conclude by providing practical recommendations for educators about fostering ethical reflection within classrooms, emphasizing the role of reflection in nurturing students' ethical awareness and critical thinking skills.

Trends and implications

This section adopts a pragmatic approach compared to the earlier philosophical discussions, highlighting the ongoing potential for advancement and growth in EEE. A recurring theme in many chapters is the significance of reflection and exploring new perspectives to enrich engineering thinking and to connect ethics with everyday engineering practices. Case studies and emotions emerge in several chapters, and ways of engaging students in applying, internalizing, and making new meanings are evident throughout the section. Creating artifacts – essays, artworks, journals, community designs, or humanitarian proposals – as an outcome or deliverable of the teaching methods used constitutes another theme.

Engineering ethics education is global and local – necessitating *cross-cultural understanding and collaboration*. We encourage educators to consider diverse cultural perspectives and contexts in their teaching methods to foster an engineering community that is inclusive and globally aware. We highlight the value of *community engagement* in EEE and encourage educators to involve industry professionals, community stakeholders, and other relevant parties in teaching to provide students with real-world insights and experiences. We encourage educators to provide students with *the tools and skills to address real-world ethical dilemmas effectively*. We acknowledge that engineering ethics education prepares students for ethical and sociotechnical decision-making and equips them to navigate the ethical complexities they may encounter in their professional careers. Teaching inspired by real-world elements and features of professional practice may help cultivate students' abilities in nuanced decision-making and awareness of the complexities (brought by pluralist values and multi-stakeholder relations) that play a role in decision-making in the engineering profession.

The chapters of this section highlight the importance of *adaptability and flexibility* in teaching methods, acknowledging that the landscape of engineering ethics education is continuously evolving. The editors and authors encourage educators to remain open to exploring and incorporating innovative approaches to address emerging challenges and opportunities. We stress the importance of *continuous improvement and evaluation in teaching methods*. Participation in engineering ethics special interest groups, conferences, or workshops dedicated to different teaching methods provides an excellent opportunity for educators to discover novel teaching methods or more creative approaches to the methods already used. We also encourage educators to solicit feedback from students and colleagues, reflect on their teaching practices, adjust these based on feedback, and enhance the effectiveness of their pedagogical approaches. This may mean finding allies across different institutional departments representing varied disciplines – with the aim of connecting the teaching methods used with emerging disciplinary topics of high importance – as well as collaborating with educational researchers to examine the effectiveness of such innovations and experimentations using the classroom as a research site for collecting pedagogical data.

Overall, the chapters of this section provide practical advice to help teachers implement new methods – methods that may be beyond their own experience or training yet nonetheless valuable to try. The detailed guidance provided by the authors of Chapters 19–25 should prove valuable, but reaching out to others to gain first-hand exposure and experience delivering these methods is also essential. Team teaching, fellowships, and student and faculty/staff exchanges can be helpful for building capacity and developing skills and fluency.

Positionality

As editor of the section on teaching methods, Diana Martin strived to ensure a collaborative and dialogical process between the chapter authors. This meant brainstorming the author composition of the chapter teams to reflect diversity, inclusivity, and expertise (either established or in the course of development), as well as several hours spent in meetings bringing together author leads to discuss the plans for each chapter and share feedback with each other at the pre-draft stage. It has not been the easiest or least time-consuming approach for Diana or the authors. Still, she believes it aligns with her positionality as a member of the engineering education community. In this community, her main aim is to see its members building each other up rather than competing, and while doing so, also developing the discipline of engineering ethics education more broadly. She is glad to call many of the authors friends now and to see friendships developing throughout the collaborations of this handbook. The key values that drove her editorial approach are collaboration, inclusivity, fair play, ensuring fair acknowledgment of effort, and the ambition to continue to learn and develop the field further. These values are tremendously important given her background. Having been born during communism in Romania and growing up during the transition years to a post-Communist society (democratic, yet not really so; embracing Western values, yet not really embraced by the West), Diana is not a stranger to the broad range of socio-economic challenges that can shape negatively someone's professional opportunities and output. She attributes much of her overall success to her mother's efforts, as well as the support, kindness, trust, and opportunities provided by mentors, some of whom may not even recognize themselves as such or fully grasp the impact they've had. Her goal now is to open up possibilities for others and also oppose gatekeeping practices that may originate from hierarchical and elitist stances or as displays of power and privilege linked with any demographic characteristics. This concern was present at many points during the editing of this section and was also raised in the editorial process of the handbook, where she felt lucky to be part of a team sharing similar values.

Conclusions from the editor of this section

Diana thanks you for reading the contributions in this section, saying it has been an immense honor to bring together such fantastic authors at different levels of their careers and from different parts of the world to reflect on what it means to teach ethics *ethically* and as part of an international community of educators and researchers. Through the synergies of the diverse author teams reflecting on the methods to teach engineering ethics, Diana hopes the section will inspire educators and researchers alike to collaborate rather than compete in these endeavors of further developing the field. The community of voices reverberating across these chapters is open and eager to welcome any newcomer and continue, via other events or projects, the dialogue on how to teach ethics.

References

Harris, C. E., Pritchard, M. S. & Rabins, M.R. (1995). Ethics in Engineering: Concepts and Cases, Wadsworth.

Herkert, J. R. (2005). Ways of thinking about and teaching ethical problem solving: Microethics and macroethics in engineering. Science and Engineering Ethics. 11, 373–385. https://doi.org/10.1007/s11948-005-0006-3

Martin M. W. & Schinzinger R. (1983). Ethics in engineering. McGraw-Hill.

Whitbeck, C. (1995). Teaching ethics to scientists and engineers: Moral agents and moral problems. Science and Engineering Ethics, 1, 299–308.

19

LITERATURE REVIEW OF TEACHING METHODS

Trends and ways forward to support engineering ethics instruction

Madeline Polmear, Tom Børsen, Heather A. Love, and Amir Hedayati

Introduction

One of the enduring conversations in engineering ethics is what should be taught and how. The first question regarding what should be taught is discussed in Chapters 1–18 of this handbook. This section, Teaching Methods in Engineering Ethics Education (Chapters 19–25), delves into *how* engineering ethics is taught. Many approaches exist and there is limited consensus around which are most effective and which objectives they should achieve (Hess & Fore, 2018). Furthermore, the literature demonstrates that engineering ethics education is marked by a lack of attention to alignment between learning objectives, teaching methods, and assessment strategies that calls into question their effective design (Keefer et al., 2014). This is, in part, due to much of the engineering ethics literature focusing on specific examples of pedagogical strategies in practice (often presented in short-form conference proceedings) without learning theory to ground them or methodologies to evaluate them.

This chapter aims to provide a narrative literature review (Baumeister & Leary, 1997) to map teaching methods in engineering ethics education. This chapter does not detail the teaching methods because they are covered in standalone chapters within this handbook (Chapters 20–25). Instead, it attends to a broad view of the literature. The first part introduces and compares two 'classic' approaches, case studies and rules/codes of ethics. The literature on these teaching methods often focuses on their strengths and limitations (Colby & Sullivan, 2008; Conlon & Zandvoort, 2011).

In addition to these classic methods, we introduce and describe several emerging approaches being used to teach engineering ethics, such as student-developed activities (Alpay, 2013), role plays (Dempsey et al., 2017; Doorn & Kroesen; 2013; Hunger, 2013), co-curricular activities (Bielefeldt et al., 2020; Lee, 2021), gamification (Bekir et al., 2001; Briggle et al., 2016), and online education (Barak & Green, 2020). The chapter synthesizes the literature on classic and emerging teaching methods to develop a heuristic to guide educators in choosing a method aligned with their objectives and underlying assumptions.

Despite the variation in teaching methods, the literature reflects the prevalence of using case studies and ethics codes (Hess & Fore, 2018), which stems in part from the use of these approaches

DOI: 10.4324/9781003464259-24

in philosophy, business, and medicine, all of which have longer histories of teaching ethics in university education. Despite the dominance of case studies and ethics codes, engineering education reflects a recent critical examination and a shift toward new methods. Momentum toward experiential and active learning has opened new pathways for teaching engineering ethics.

Because the overall body of literature on this topic can be described as fragmented since it draws in large part on examples of practices in individual courses and programs, later sections of this chapter distill the scholarship, aiming to provide a broad-level view of the patterns that characterize how engineering ethics is being taught. Such trends and gaps in the literature can illuminate effective practices, contextual considerations, and areas where more research is needed.

This chapter is intended to be a resource for educators and researchers. By describing classic and emerging teaching methods and providing examples from research, the chapter can support educators who are teaching ethics to engineering students (and, in particular, those who are new to engineering education as postgraduate students, early-career faculty members, and faculty members who are based in technical engineering disciplines). The importance of alignment between objectives, underlying assumptions, teaching methods, and assessment strategies is discussed to aid in developing effective instruction. Finally, the chapter aims to support researchers in engineering ethics education by illuminating opportunities for future inquiry.

Learning objectives in engineering ethics education

The variation in teaching methods reflects the different courses and disciplines in which ethics is taught and the many objectives that ethics education is designed to achieve. A review of learning objectives (detailed in Chapter 1, with further discussion in Chapters 32–36 on accreditation) provides context for understanding teaching methods due to the interconnection between the two. Past reviews of engineering ethics education have detailed learning objectives. Davis (2006) described the learning outcomes of engineering ethics education as ethical sensitivity, knowledge, judgment, and willpower. Haws (2001) reviewed American Society of Engineering Education (ASEE) conference papers from 1996 to 1999 and identified objectives of improving divergent thinking, evaluating outcomes from the non-engineering perspective, and learning ethics-related vocabulary. Herkert (2000) similarly reviewed engineering ethics education in the United States and suggested the importance of professional responsibility while drawing a distinction between micro- and macro-ethics. More recently, Hess and Fore (2018) reviewed engineering ethics interventions in the United States and distilled three learning goals: sensitivity or awareness; judgement, decision-making, or imagination; and courage, confidence, or commitment. A common thread across these reviews is the spectrum on which the objectives fall, from being able to recognize an ethical issue to making an ethical decision to embodying ethical conviction. There are also broader objectives for engineering ethics education, such as supporting sustainable development and social responsibility (Børsen Hansen, 2005), which align with global priorities for addressing cascading humanitarian and environmental crises (United Nations, 2022).

Narrative review approach

This chapter presents a narrative literature review, since this approach can fulfill the aim of surveying the state of knowledge (Baumeister & Leary, 1997). The approach was selected because (1) it aligns with the chapter aim of developing an overview and integration of various teaching methods and (2) it complements existing, more formal systematic literature reviews (e.g., Hess & Fore, 2018; Martin et al., 2023). Our review focused on English language publications between

2010 and 2023. The decision was made to focus on contemporary scholarship, but we guide readers to foundational studies from before this period, such as Herkert (2000), Harris (2004), Haws (2001), and Colby & Sullivan (2008). The following databases and journals were searched: ERIC, Google Scholar, Scopus, EBSCOhost, Web of Science, Engineering Village, *Journal of Engineering Education, European Journal of Engineering Education, Australasian Journal of Engineering Education, Science and Engineering Ethics*, and *International Journal of Ethics Education*. Search terms broadly included ‘engineering ethics’ and ‘teaching’ and focused on the specific methods covered in this section (e.g., engineering ethics and case study) and snowballed to emergent methods (e.g., engineering ethics and games). The articles were recorded in a spreadsheet and Zotero database and critically analyzed to identify classic teaching methods, emerging teaching methods, trends, and gaps. These findings were then collaboratively synthesized to distill considerations for educators and researchers. In addition, research team members drew on in-progress literature review projects to identify broader trends and add nuance to the discussion of the evolution of engineering ethics teaching strategies.

Positionality

Our author team represents a range of national and disciplinary backgrounds, briefly summarized here, that influence our approach to the narrative review and our interpretation of the findings.

Madeline is an engineering education researcher and lecturer who explored educators’ teaching methods and perspectives related to engineering ethics through a mixed methods approach during her Ph.D. During a postdoctoral fellowship, she also studied macro-ethical development among civil and architectural engineering students. These projects and her research experience in the United Kingdom, Belgium, and the United States inform her qualitative and cross-cultural approach to understanding how ethics is taught and how curriculum can support students’ societal responsibility.

Tom holds a cross-disciplinary chemistry and philosophy degree and a Ph.D. in university STEM education. He is now an associate professor at Aalborg University in Denmark, where he does interdisciplinary research in the area between social sciences and humanities (SSH) and science, technology, engineering, and mathematics (STEM). Tom perceives himself as an educational activist as he strongly promotes the inclusion of ethical elements and participatory aspects in university STEM education.

Heather’s approach to engineering ethics education is informed by her background in interdisciplinary humanities research, which bridges literary studies, cultural studies, health humanities, composition pedagogy (she has taught writing at several American and Canadian universities), and service with the IEEE’s Society on Social Implications of Technology. She is part of research groups that study and develop arts/humanities approaches for cultivating ethical thinking within engineering programs, and she has led a cross-disciplinary team on a scoping review of pedagogical initiatives designed to foster attentiveness to technology’s socio-environmental impacts in engineering curriculum.

Amir is an educational researcher with a computing engineering, management, and human resource development background. His MBA thesis focused on analyzing organizations’ codes of ethics, and his Ph.D. was a grounded theory focused on how computer science students make decisions when they face ethical dilemmas. He recently conducted a literature review on ethics training programs in the workplace. He has served as a teaching assistant for two undergraduate courses on computing ethics and currently teaches a course he designed on ethics and diversity training in the workplace at the University of New Mexico.

Classic teaching methods

Recent literature reviews have demonstrated the relative prevalence of teaching methods in engineering ethics education (Hess & Fore, 2018; Martin et al., 2021). This work points to the preeminence of case studies and ethical codes or rules. Case studies have long been a staple of ethics education, with roots in philosophy, medicine, and business – all of which formalized ethics education before engineering did (Davis & Yadav, 2014). The underlying premise of this approach is that case-based knowledge is facilitated through information and context-rich scenarios that serve as prototypes and experiences on which students can reflect to inform their understanding and decision-making (Thiel et al., 2013). The learning goals of case studies include understanding professional conduct, integrating stakeholder perspectives, raising awareness of international issues, and supporting decision-making (Martin et al., 2021).

The use of professional codes of ethics is a common approach in the United States, where a recent literature review found that 85% of engineering ethics interventions employed codes as a pedagogical strategy (Hess & Fore, 2018). This approach stems from the emphasis on engineering as a profession accountable to ethical codes, the first of which dates to the American Society of Civil Engineers in 1914 (*Ethics | ASCE*, n.d.). Because the United States has a long history of formal ethics education in engineering relative to other places and often provides an example to other areas regarding engineering ethics education (see, e.g., Chapter 32), this teaching method has been widely adopted across the globe (Zandvoort et al., 2000). The underlying assumption is that the professional practice of engineering comes with expectations and obligations, which outline the key values of engineering such as protection of safety and welfare, sustainability, honesty, and professional competence (Colby & Sullivan, 2008).

Cases and codes have been criticized for their individualist approach, in which the actions of a single engineer are emphasized with minimal attention to context (Conlon & Zandvoort, 2011). The case-study approach has been criticized for portraying ethical dilemmas as having an obvious right and wrong answer in which the consequences are clear, when a clear answer is often only due to the benefit of hindsight (Colby & Sullivan, 2008). Some scholars have critiqued the focus of engineering ethics education on extreme and disaster-related case studies, arguing the negative effect of such an approach by promoting the misperception of morality as an infrequent concern (Pierrakos et al., 2019). Alongside this criticism, there are calls to evaluate the attributes of cases that make them effective while situating them more closely in the ambiguity of engineering practice (Martin et al., 2021). In response to these limitations, there is growing research around evaluating the content and structure of case-based methods. Such work has included integrating emotional content in case studies (Kotluk & Tormey, 2023; Thiel et al., 2013), taking a constructivist approach (Martin et al., 2018), and using wicked problems such as sustainability to broaden the macro-ethical scope (Byrne, 2012). Regarding codes of ethics, cultural context can affect the relevance of professional standards and organizational context can create competing priorities that codes cannot resolve (Harris, 2004). These critical perspectives and the recent growth in broader engineering ethics education research have opened the door to novel and emerging teaching methods.

Emerging teaching methods

This section describes and provides examples for emerging teaching methods: student-developed activities, role play, co-curricular activities, gamification, and online education. It complements this handbook's other chapters on emergent teaching methods (problem-based learning, value-sensitive design, field learning, arts-based, and reflective teaching methods).

Student-developed teaching activities

Engaging students in the design of engineering ethics education "helps to relocate ethics from the periphery of the curriculum to its core by empowering students to investigate ethics in the ways that are most meaningful to them" (Sunderland, 2019, p. 1781). The teaching approach is underpinned by the assumption that student engagement – students taking an active role in their learning – supports agency, which in turn, contributes to self-awareness and knowledge acquisition (Sunderland, 2019). Our literature search identified examples in which engineering students developed ethics-related teaching material. At Imperial College London (Alpay, 2013), students of aeronautics, bioengineering, chemical engineering, and computing were introduced to ethical theories and discipline-specific content and then asked to develop proposals to introduce engineering students to ethics in a fun and meaningful way. Proposals were collected in a database for possible future utilization and assessed by faculty members. As another example, engineering students at the Technical University of Berlin (Beier, 2013; Børsen et al., 2021) also developed self-contained teaching units for a course in engineering ethics. The units typically covered a complex topic and used moderators rather than expert lecturers. Two examples of teaching activities were titled 'Greenwashing or Decision Aids – Labels, Certificates and the Like' and 'Technology as a Drama – Technology in Drama.' The course enacting these teaching activities was itself developed and driven by students without interference from staff.

Role-plays

Role-play is a form of active learning that can be used to address macro-ethical issues or wicked problems (Doorn & Kroesen, 2013; Hunger, 2013; Dempsey et al., 2017; Carlson & Wong, 2020) as well as micro-ethical issues related to professional conduct and responsible research (Brummel et al., 2010), commercial dilemmas (White, 2020), and expert witnessing (Brummel & Daily, 2014). Role-playing involves interpersonal interaction and is typically conducted in small groups around a real-life or realistic hypothetical problem. Students undertake the role of an involved stakeholder and defend and promote the role's position while working together to negotiate a solution. Role-play activities are usually well received by students, although their success depends on how relevant the scenarios are to students' future professions and the skills of the facilitating teacher (Brummel et al., 2010).

Several outcomes can be embedded within role play, such as seeing multiple perspectives, developing frameworks for ethical conduct, and supporting ethical imagination (Birch & Lennerfors, 2020). Relatively limited understanding exists regarding the effectiveness of this teaching method compared to others (Hunger, 2013), and there is a need for additional assessment of its effectiveness (Dempsey et al., 2017). Chapter 18 of this volume addresses role-play discussions structured around case studies and argues that they can improve perspectival thinking and help link micro-meso-macro contexts.

Co-curricular activities

Co-curricular activities complement learning in the formal curriculum, and students engage voluntarily. Co-curricular is a broad category under which different teaching methods can be employed to introduce both micro- and macro-ethical topics (Bielefeldt et al., 2017; Knight et al., 2016). Co-curricular activities that engineering educators perceive as connecting to ethics include professional societies (e.g., IEEE), honor societies, design competitions, and research. In these contexts, teaching methods for integrating ethics include informal discussions, online training, study circles, guest speakers from industry, participation in conferences where students present their own work, self-assessment application essays, school (e.g., K-12) outreach, internships, community service, and field trips (Bielefeldt et al., 2020). Another example is the 'ethics bowl,' a co-curricular pedagogical tool to develop students' skills to "identify ethical dimensions of challenging problems, to elucidate and articulate the various ethical perspectives of affected communities, and to productively deliberate through to a practical solution" (Lee, 2020, p. 146). These skills are thought to lead to better ethical decision-making. In this approach, groups of students, typically from different disciplines, prepare all year for participation in a competition in which they present and discuss ethical issues from a case study.

The self-selected nature of co-curricular activities is both a strength and limitation for teaching engineering ethics. Student motivation and engagement are high in informal learning since participation is driven by interest and learning can be contextualized in application (National Research Council, 2009). However, not all students have equal access to co-curricular activities, so methods outside the classroom should not be relied on entirely for ethics education.

Gamification

Games can help translate ethical theories and concepts into everyday practices that challenge teacher-centered approaches. The underlying assumption is that effective engineering ethics teaching involves student *interaction* to address complex ethical issues. From this perspective, games can make ethical issues appear 'real' to students. Briggle et al. (2016) provide an example of how a card game can be used in engineering ethics teaching to complement ethical theories in a way that helps students develop higher-order skills, such as ethical problem-solving and decision-making, through an open-ended environment. Games should attend to considerations such as the balance between learning and fun, the treatment of unethical behavior in a game, and the assessment of the impact (Briggle et al., 2016).

Online engineering ethics education

The most common online engineering ethics education consists of individual work with literature and videos presenting ethical principles combined with a computer-graded multiple-choice test (Egilmez et al., 2019; Razavinia & Mydlarski, 2020). The advantage of online engineering ethics education is flexibility in time and place; it allows collaboration with peers far away (Hess et al., 2016; Leitch & Dittfurth, 2012). The disadvantages are that conventional standalone online ethics education has little effect on students' practices, does not include interaction between students, and does not allow supervision. To mitigate these limitations, online education can also utilize interactive elements (Lumgair, 2018) and be combined with additional teaching methods.

The outcomes of online ethics education depend heavily on how it is enacted and the specific methods it is employed in conjunction with (Canary et al., 2014; Plouff & Barakat, 2012). For example, videos are most effective when based on true stories (Itani, 2013). Other recommendations for online engineering ethics education include designing collaborative, case-based, and contextual learning (Barak & Green, 2020) and encompass three elements: (1) online lectures or videos presenting concrete, real-life examples of ethical dilemmas and how they were managed; (2) online forums where students can discuss ethical dilemmas; and (3) written assignments where individual dilemmas are analyzed.

Todd et al. (2017) compared online to face-to-face education and argue for a hybrid approach through a meta-analytic literature review. The study concluded that face-to-face ethics education is most efficient in delivering complex and process-oriented learning objectives. In contrast, online education efficiently delivers instructions and skills related to applying guidelines. They suggest hybrid educational formats, such as blended learning, for ethics education concerning responsible conduct of research. It is possible to digitalize tools used in analog engineering ethics teaching. As an example, Hoffmann and Borenstein (2014) present how software can help prepare digital argument maps. Reeves and Nadolny (2013) developed and tested virtual role-play, which takes place in a 3D environment. Students assessed the virtual role-play positively in terms of scientific learning, and there was some evidence that students' ethical awareness increased. Online ethics education can also enable students to reflect on their lived experiences with information and communication technology (ICT) and thus connect personal and professional experiences (Voss, 2013).

Trends

The following section provides a broad overview to elucidate trends in engineering ethics education. Such trends include the prevalence of particular teaching methods over time, their place in the curriculum, and their use within different engineering disciplines and across the globe. Despite the international scope, one acknowledged limitation of the review is that we only include literature published in English.[1]

Trends over time in teaching methods

The literature reveals that examples of engineering ethics pedagogy from a decade or more ago tend to place greater emphasis on lecture formats and case study-based curriculum, although exceptions are certainly present. Example exceptions include multidisciplinary capstone courses (Allenstein et al., 2013), community- and service-based projects (Canney & Bielefeldt, 2012; Croft et al., 2013; Krishnan & Nilsson, 2012; Wittig, 2013;), co-curricular training (Plouff & Barakat, 2012, 2014), game-based and storytelling assignments (Olwi, 2014; Sadowski et al., 2013), and role-playing (Reeves & Nadolny, 2013). Moving closer to the present, publication trends indicate that instructors are employing a wider variety of methods, with a greater emphasis on active learning. Examples include role-playing and game-based learning (Carlson & Wong, 2020; Dodson, 2017; Fan et al., 2015; Kumar & Kremer-Herman, 2019; Martin et al., 2019; Putko & Rooney-Varga, 2016; White, 2020; Xenos & Velli, 2020), narrative methods (Halada & Khost, 2017), community engagement (Catalano, 2016), metacognitive reflection often through writing-based activities (Badenhorst et al., 2020; Campbell et al., 2020; Chen & Orjuela-Laverde, 2018; Mogul & Tomblin, 2019; Robinson, 2019; Zain et al., 2017), and diversity, equity, and inclusion (DEI). Example DEI-focused strategies include culturally responsive pedagogy (Gomez & Svihla, 2018; Quigley et al., 2016), service learning (Hinds et al., 2020; Jung et al., 2016; Winkelman et al., 2016), accessibility-focused projects (Molina-Carmona et al., 2017), and decolonial approaches (Cruz, 2021).

These observations are primarily grounded in the literature about undergraduate-level teaching. Similar patterns are present in the literature on (post)graduate training; however, because fewer publications focus on this demographic, it is harder to draw firm conclusions. Additional research into the strategies to train master's and doctoral students in ethics and ethics-related issues would enrich our collective knowledge, particularly since these are the students most likely to become future teachers themselves and therefore shape the ongoing directions of undergraduate education.

Trends in engineering ethics across the curriculum

Engineering ethics education curriculum development occurs at a range of scales – from smaller standalone modules, workshops, and transferrable course components to revised or newly developed complete courses, through to multi-course sequences and program-level initiatives. Accounts of work done within single courses are most frequently published, and these shorter interventions can provide students with initial exposure to ethics topics; they can offer pathways for faculty members teaching non-ethics-focused courses to integrate ethical issues into that curriculum; and they can make possible various co- and extra-curricular involvement. On the other hand, programmatic changes that foreground ethics and ethical thinking represent efforts to cultivate wider-scale cultural change; they signal an emerging impulse to foster and valorize a deeper, more fully integrated ethical mindset within engineering programs and the discipline as a whole. Examples of programmatic efforts include Tampere University of Technology (Koskinen, 2015), James Madison University (Pierrakos et al., 2019), and Tecnologico de Monterrey (Ruiz-Soto et al., 2014)

A trend towards multi-pronged pedagogical approaches has emerged in recent years. For example, individual modules or courses might use case studies *in tandem with* group projects, game-based activities, or reflection tasks; they might *pair* traditional lectures with more active discussion- or project-based learning; or they might *blend* several of these strategies. If we segment for curricular level, additional patterns become apparent. First- and second-year undergraduate pedagogy has been moving toward emphasizing experiential learning, group activities, and design projects, implying a growing sense of the need to actively engage students with ethics-related topics early in their university studies. Third- and fourth-year undergraduate initiatives are frequently linked to capstone or other project-based courses, allowing students opportunities to explore more complex aspects of ethical issues within practical and applied engineering contexts. (Post)graduate examples often adopt more advanced frameworks (Llopis-Albert et al., 2020; Sekiguchi & Hori, 2019), operate through the lens of professionalization (Berdanier et al., 2018), and provide opportunities for student to learn from one another (Brey et al., 2019; Celik et al., 2020) or from direct interactions with experienced professionals (Bernstein et al., 2017).

Discipline-specific trends

Engineering ethics education is not always discipline-specific, with most initiatives (both course-based and programmatic) offering general ethics instruction and/or simultaneously engaging students from multiple engineering subfields. However, more targeted pedagogy exists that can signal educators' and curriculum designers' commitment to exposing students to the ethical issues most relevant for their specific, future professional contexts and offering them early opportunities to practice navigating those often-challenging questions and scenarios.

Furthermore, the past decade has seen an uptick in the number of interdisciplinary pedagogical initiatives working towards more holistic approaches to engineering ethics education (e.g., Birch & Lennerfors, 2020; Campbell et al., 2020; Di Biasio et al., 2018; Kallergi & Zwijnenberg, 2019). Several adjacent STEM fields are regularly represented among the partner disciplines (e.g., applied sciences, biology, chemistry, computer sciences, mathematics, physics). However, examples from a broader range of fields also appear with increasing frequency; these encompass the social sciences (anthropology, political science, psychology, sociology), arts and humanities (communication and rhetoric, creative/visual/performing arts, liberal arts, philosophy), and other professional or interdisciplinary fields (business, design, human–computer interaction, law). These interdisciplinary partnerships represent a trend for engineering ethics education more broadly, as they suggest a move away from the often-siloed work. Additional research can help us better understand the challenges, successes, institutional contexts, and potential future questions related to cross-disciplinary pedagogical methods as they become more common components of engineering programs.

Global trends in engineering ethics education

Publications in English related to engineering ethics education are heavily skewed towards initiatives based in the United States.[2] Nonetheless, relatively strong representation in the literature exists from Canada, several European countries (the United Kingdom, Spain, the Netherlands, and Germany in particular), and the Asia-Pacific region (including Malaysia, India, China/Hong Kong, and Australia); furthermore, smaller clusters of papers on the topic have come out of Central and South America, Africa and the Middle East, and international collaborations. Across the globe, we see all levels of engineering education curricula in development. Future collaborative research into how ethics instruction takes place in non-English programs and contexts – and targeted translation projects designed to foster more widespread knowledge and appreciation of the diverse approaches currently in use – would greatly enrich our collective understanding of *global* and *comparative* cultural trends in engineering ethics pedagogy. The systematic literature review by Martin et al. (2023) on global engineering ethics education and research further details trends across the world.

Gaps in the literature

Three main gaps in approaches to teaching engineering ethics are identified in the literature. First, research predominantly focuses on cognitive aspects of ethical development. Second, research often has limited engagement with the needs of learners to inform the teaching approaches. Third, evaluation of engineering ethics education efforts has focused mainly on a narrow set of outcomes.

Gap 1: The limited attention to non-cognitive aspects of ethics education

Engineering ethics education has focused on cognitive aspects of ethical decision-making. As suggested by some scholars in the field, undergraduate students do not tend to emotionally engage in the study of ethics (Troesch, 2014, 2015). This reflects a broader, historical belief that emotions are problematic in ethical decision-making (Lönngren et al., 2023). Nonetheless, emotions can provide insights for making more desirable ethical decisions (Roeser, 2012). This suggests a disconnect between desired student learning outcomes and the pedagogical approaches we use to teach engineering ethics (Hitt & Lennerfors, 2022). Whereas the dominant approaches are cognitive – such as case studies and ethical theories (Troesch, 2014) – supporting an individualist and analytical approach, experiential and emotional learning is often neglected. This calls for new approaches to engineering ethics education that go beyond students' cognitive engagement.

Arts-based instructional strategies (detailed in Chapter 24) are among the teaching approaches that can potentially engage students emotionally. In traditional approaches to teaching, opportunities for dramatic dialogue with the potential to "show how specific circumstances and actions can lead to trauma" have been neglected (Monk, 2009, p. 113). For example, the literature suggests that the field has not utilized fictional films to teach ethics at the level one can see in other fields such as management and medicine (Hitt & Lennerfors, 2022). Films can evoke "emotions, imagination, and a connection to personal lived experiences … and help achieve" student learning outcomes (Hitt & Lennerfors, 2022, p. 44). Our review suggests there is a need for conducting evaluative studies on the effectiveness of different arts-based pedagogies because, with some exceptions (e.g., Birch & Lennerfors, 2020; Mullin et al., 2006), there are not many articles reporting the outcomes of using these approaches to teach ethics in engineering. The existing limited publications on the subject have focused mainly on proposing specific new approaches and discussing their potential contributions to engineering ethics education (e.g., Hitt & Lennerfors, 2022; Monk, 2009; Troesch, 2014).

Another approach to engaging students emotionally in the process of ethical decision-making is through developing empathy, as several recent studies have suggested (Hess & Fila, 2016; Walther et al., 2017). A related strategy is using reflective approaches (see Chapter 25 for more on this) that can improve student empathy (Sochaka et al., 2020) and bring ethical concerns to the core of engineering (Valentine et al., 2020). As stated by Badenhorst et al. (2020), reflective practice is not mainstream in engineering education since the dominance of cognitive approaches and a technically oriented mindset means that "reflective thinking and practice is not easy to implement in engineering classes" (p. 3).

Gap 2. Lack of sufficient attention to behavioral ethics and student needs

In general, research on engineering ethics education has not focused on moral psychology and behavioral ethics that emphasize why people act unethically. Likewise, it can be argued that the research does not focus significantly on why people act unethically without meaning to (Bazerman & Tenbrunsel, 2011) and on developing strategies to increase the likelihood of ethical behavior among engineering students (High et al., 2011). Bazerman and Tenbrunsel (2011) stated that solutions to reduce unethical decisions generally do not consider the limitations of the human mind. As Drumwright et al. (2015) suggested, "the philosophically based traditional approach to teaching business ethics should be significantly supplemented with the psychologically and sociologically based learning of behavioral ethics" (p. 433). The main argument of behavioral ethics is that the situation matters and people with good character or skilled in moral reasoning can still make unethical decisions. Studies attentive to how engineering students make decisions in ethical situations and why they might engage in unethical behaviors are limited. In one of the exceptions, High et al. (2011) proposed an approach to integrate moral psychology into a course on research ethics. In another example, Gelfand (2016) proposed an approach to teach "students about situational factors that may affect ethical judgment and behavior … [and] how their own personality traits or organizational structures might affect their judgment and behavior" (p. 1532). Finally, another study examined how computing students made decisions when faced with ethical scenarios and provided suggestions on improving ethics education courses, including both identifying students' misconceptions that may hinder them from making ethical decisions and improving students' self-confidence in addressing ethical issues (Hedayati Mehdiabadi, 2022). These are topics that future research can seek to address.

Gap 3. Limitations in evaluation of engineering ethics education efforts

Engineering ethics education research has a narrow focus when it comes to quality and evaluating the outcomes of designed interventions (Bairaktarova & Woodcock, 2015; Bombaerts et al., 2019; Clancy & Gammon, 2021). One example of a quality framework has four elements (relevance, consistency, practicality, and effectiveness), but only relevance and effectiveness have received robust attention in engineering education research (Bombaerts et al., 2019). For the most part, evaluation has focused on the effectiveness of educational efforts based on outcomes such as ethical awareness (Bairaktarova & Woodcock, 2015), student self-reports (e.g., Davis, 2006), and ethical reasoning skills (e.g., Drake et al., 2005). Moreover, other outcomes such as moral efficacy and moral courage, with a few exceptions (e.g., Douglas et al., 2022; May & Luth, 2013), are rarely examined.

Clancy and Gammon (2021) argued that long-term ethical behaviors should be the ultimate goal of ethics education. They suggested using empirical moral psychology to evaluate ethics education's effectiveness in improving ethical behaviors. From this perspective, ethical behaviors are not exclusively affected by moral reasoning – the outcome that most engineering courses attempt to assess (Clancy & Gammon, 2021). Other factors, including moral disengagement mechanisms (Bandura, 1999), biases, and situational variables, have often been overlooked in evaluating engineering ethics education, and thus, future research could examine how to integrate such factors in an assessment heuristic.

Considerations for educators

Engineering ethics can be integrated into various courses and activities with the literature providing many examples. However, it can be daunting for educators who are new to teaching ethics to make sense of topics and pedagogies that might be outside their technical expertise. With the ever-evolving societal and technological context of engineering and the emergence of novel teaching methods, even experienced engineering ethics instructors may find it challenging to keep informed of effective practices. Findings from our review suggest several considerations for educators in choosing and implementing teaching methods. First, it is crucial to consider the objective/purpose and context. Objectives such as raising awareness of ethical issues, reasoning through an ethical dilemma, and behaving ethically will have different instructional implications – as will teaching a single intervention in an engineering course, teaching an ethics-focused course, developing a program-wide initiative, and teaching a seminar of 15 students versus a group of 200. In response to these variations and to address alignment across instruction design, frameworks for quality would be welcome future research topics.

Table 19.1 Heuristic for instructional alignment

Objective/purpose[3]	*Teaching method*	*Underlying assumption*
Knowledge of ethical issues	Online education, ethics codes and rules, problem-based learning (PBL)	Knowledge of ethical issues is a pre-requisite to ethical sensitivity and decision-making
Ethical awareness/sensitivity	Case study, role play, gamification, value-sensitive design, PBL	Without recognition of ethical implications in practice, one cannot reason through and address them
Ethical reasoning/decision-making/judgment	Case study, role play, reflection, co-curricular, PBL	Ethical reasoning/moral judgment is a good predictor of moral behavior
Ethical behavior (inclusive of moral courage/confidence/commitment)	Case study, role play, co-curricular, PBL	Ethics education can affect future behavior
Ethical culture	Role play, co-curricular, service learning, PBL	An organization's culture or environment can affect individuals' decisions and actions

The heuristic below has been developed to guide instructional design-making and point educators to other chapters in the handbook for more detail on the respective methods. It is important to note that multiple teaching methods can be oriented toward each objective, and each teaching method can fulfill more than one objective. As examples, Bairaktarova and Woodcock (2017) discussed the importance of integrating ethical awareness and behavior; Wittig (2013) described the use of project-based learning through a co-curricular organization to support engineering students' ethical sensitivity and decision-making; and Herkert and colleagues (2020) used the case study of Boeing crashes to highlight moral courage and its effect on ethical culture. The varying ways in which the methods are employed also impact the objectives and contexts for which they are appropriate; see, for example, the taxonomy of case studies developed by Martin and colleagues (2021). The alignment and impact depend as much on the facilitation as on the method itself, as detailed in the other chapters of this section.

Summary of areas for future research

Gaps identified in the literature illuminate opportunities for future research. There is a need for further research on emotional engagement in engineering ethics pedagogy. It is first essential to recognize emotion's role in education and ethics to better understand how teaching methods can engage students emotionally. Future work can also explore the efficacy of emotion-oriented approaches, such as reflection (Chapters 6 and 25) and art-based (Chapter 24) methods, in achieving desired objectives.

Given the lack of studies identifying engineering students' needs (e.g., Zhu et al., 2022), research in this realm can significantly improve how we teach engineering ethics. Engineering ethics has been traditionally taught using a top-down approach, but recent trends point to increasing consideration of student-driven approaches to support students' engagement.

Another opportunity for future research involves the impact of teaching approaches on ethical behavior. Although often considered a short and long-term objective in ethics education, there is limited research exploring if/how ethics instruction affects students' and professional engineers' decisions and behaviors. Engagement with other disciplines, such as moral psychology and behavioral ethics, and longitudinal research can support this line of inquiry.

A final recommendation is expanding research into evaluation of teaching methods. There is a breadth of examples of teaching practices in the literature, but there is less engagement with the evaluation of their efficacy. Future research can compare and evaluate the relative efficacy of different teaching approaches for various objectives and instructional contexts. There is a multitude of factors that affect instruction and assessment, which can be a challenge for making sense of different approaches and selecting an appropriate one for a given context. Developing flexible assessment strategies that can be adapted by educators and researchers would be a contribution to the field.

Conclusion

There is ongoing growth in research and scholarship on teaching and learning related to engineering ethics education, providing an array of instructional approaches. This chapter took a narrative review approach to highlight classic and emerging teaching methods – as well as trends and gaps in the literature. Although case studies and rules/codes of ethics are most prevalent in the literature, many methods offer novel ways to support student engagement and active learning. Given the limited consensus around which approaches are most effective, future research is needed to evaluate and compare the teaching methods while accounting for the contexts in which they are employed.

Notes

1 Content in this section draws from and synthesizes findings from an ongoing scoping review of pedagogical initiatives designed to foster ethical thinking within the engineering community, preliminary results from which are available here: https://uwspace.uwaterloo.ca/handle/10012/17764. For pre-publication access to the full dataset, please contact Heather A. Love.
2 Content in this section draws from and synthesizes findings from an ongoing scoping review of pedagogical initiatives, please see note 1 above.
3 Adapted from Tkachenko & Hedayati Mehdiabadi (2022) and Hess & Fore (2018).

References

Allenstein, J.T., Rhoads, B., Rogers, P., & Whitfield, C. A. (2013). Examining the impacts of a multidisciplinary engineering capstone design program. *ASEE*. https://doi.org/10.18260/1-2--19574
Alpay, E. (2013). Student-inspired activities for the teaching and learning of engineering ethics. *Science and Engineering Ethics*, *19*, 1455–1468.
Baumeister, R. F., & Leary, M. R. (1997). Writing narrative literature reviews. Review of General Psychology, 1(3), 311–320. https://doi.org/10.1037/1089-2680.1.3.311
Badenhorst, C. M., Moloney, C., & Rosales, J. (2020). New literacies for engineering students: Critical reflective-writing practice. *Canadian Journal for the Scholarship of Teaching and Learning*, *11*(1), n1.
Baier, A. (2013). Student-driven courses on the social and ecological responsibilities of engineers: Commentary on "student-inspired activities for the teaching and learning of engineering ethics". *Science and Engineering Ethics*, *19*, 1469–1472.
Bandura, A. (1999). Moral disengagement in the perpetration of inhumanities. *Personality and Social Psychology Review*, *3*(3), 193–209.
Barak, M., & Green, G. (2020). Novice researchers' views about online ethics education and the instructional design components that may foster ethical practice. *Science and Engineering Ethics*, *26*, 1403–1421.
Bazerman, M. H., & Gino, F. (2012). Behavioral ethics: Toward a deeper understanding of moral judgment and dishonesty. *Annual Review of Law and Social Science*, *8*, 85–104.
Bairaktarova, D., & Woodcock, A. (2015). Engineering ethics education: Aligning practice and outcomes. *IEEE Communications Magazine*, *53*(11), 18–22.
Bairaktarova, D., & Woodcock, A. (2017). Engineering student's ethical awareness and behavior: A new motivational model. *Science and Engineering Ethics*, *23*, 1129–1157.
Bazerman, M. H., & Tenbrunsel, A. E. (2011). *Blind spots: Why we fail to do what's right and what to do about it*. Princeton University Press.
Bekir, N., Cable, V., Hashimoto, I., & Katz, S. (2001, October). Teaching engineering ethics: a new approach. In *31st Annual frontiers in education conference. Impact on engineering and science education. Conference proceedings* (Cat. No. 01CH37193) (Vol. 1, pp. T2G-1). IEEE.
Berdanier, C. G. P., Tang, X., & Cox, M. F. (2018). Ethics and sustainability in global contexts: Studying engineering student perspectives through photoelicitation. *Journal of Engineering Education*, *107*(2), 238–262. https://doi.org/10.1002/jee.20198
Bernstein, M. J., Reifschneider, K., Bennett, I., & Wetmore, J. M. (2017). Science outside the lab: Helping graduate students in science and engineering understand the complexities of science policy. *Science and Engineering Ethics*, *23*(3), 861–882. https://doi.org/10.1007/s11948-016-9818-6
Bielefeldt, A. R., Lewis, J., Polmear, M., Knight, D., Canney, N., & Swan, C. (2020). Educating civil engineering students about ethics and societal impacts via cocurricular activities. *Journal of Civil Engineering Education*, *146*(4). https://doi.org/10.1061/(asce)ei.2643-9115.0000021
Bielefeldt, A. R., Polmear, M., Swan, C., Knight, D., & Canney, N. (2017, October). An overview of the microethics and macroethics education of computing students in the United States. In *2017 IEEE Frontiers in Education Conference (FIE)* (pp. 1–9). IEEE.
Birch, P., & Lennerfors, T. (2020, October). Teaching engineering ethics with drama. In *2020 IEEE Frontiers in Education Conference (FIE)* (pp. 1–5). IEEE.

Bombaerts, G., Doulougeri, K., & Nieveen, N. (2019). Quality of ethics education in engineering programs using Goodlad's curriculum typology. In *47th SEFI Annual Conference: Varietas delectat: Complexity is the new normality*.

Børsen, T., Serreau, Y., Reifschneider, K., Baier, A., Pinkelman, R., Smetanina, T., & Zandvoort, H. (2021). Initiatives, experiences and best practices for teaching social and ecological responsibility in ethics education for science and engineering students. *European Journal of Engineering Education, 46*(2), 186–209.

Børsen Hansen, T. (2005). *Teaching ethics to science and engineering students*. Report From a Follow-Up Symposium to the 1999 World Conference on Science. Copenhagen, April 15–16, 2005. University of Copenhagen, Copenhagen.

Brey, E.M., Hildt, E., Laas, K., Miller, C., & Taylor, S. (2019). Empowering graduate students to address ethics in research environments. *Cambridge Quarterly of Healthcare Ethics*, *28*(3), 542–550. https://doi.org/10.1017/S096318011900046X

Briggle, A., Holbrook, J. B., Oppong, J., Hoffmann, J., Larsen, E. K., & Pluscht, P. (2016). Research ethics education in the STEM disciplines: The promises and challenges of a gaming approach. *Science and Engineering Ethics*, *22*, 237–250.

Brummel, B. J., & Daily, J. S. (2014, June). Developing engineering ethics through expert witness role plays. In *2014 ASEE Annual Conference & Exposition* (pp. 24–400).

Brummel, B. J., Gunsalus, C. K., Anderson, K. L., & Loui, M. C. (2010). Development of role-play scenarios for teaching responsible conduct of research. *Science and Engineering Ethics*, *16*(3), 573–589. https://doi.org/10.1007/s11948-010-9221-7

Byrne, E. P. (2012). Teaching engineering ethics with sustainability as context. *International Journal of Sustainability in Higher Education*, *13*(3), 232–248. https://doi.org/10.1108/14676371211242553

Campbell, R. C., Reible, D. D., Taraban, R., Kim, J.-H., & Na, C. (2020). *Fostering reflective habits and skills in graduate engineering education via the arts and humanities*. ASEE. https://doi.org/10.18260/1-2--34685

Canary, H. E., Taylor, J. L., Herkert, J. R., Ellison, K., Wetmore, J. M., & Tarin, C. A. (2014). Engaging students in integrated ethics education: A communication in the disciplines study of pedagogy and students' roles in society. *Communication Education*, *63*(2), 83–104.

Canney, N. E., & Bielefeldt, A. R. (2012). *Engineering students' views of the role of engineering in society*. ASEE. https://doi.org/10.18260/1-2--21315

Carlson, C. H., & Wong, C. W. (2020). *If engineers solve problems, why are there still so many problems to solve? Getting beyond technical solutions in the classroom*. ASEE.

Catalano, G. D. (2016). *Integrating compassion into an engineering ethics course*. ASEE.

Celik, S., Kirjavainen, S., & Björklund, T. A. (2020). *Educating future engineers-student perceptions of the societal linkages of innovation opportunities*. ASEE. https://doi.org/10.18260/1-2--34490

Chen, L. R., & Orjuela-Laverde, M. (2018). *Implementing reflective writing in large non-technical engineering courses*. CEEA. https://doi.org/10.24908/pceea.v0i0.12995

Clancy, R. F., & Gammon, A. (2021, July). The ultimate goal of ethics education should be more ethical behaviors. In *Proceedings of the 2021 ASEE Virtual Annual Conference*.

Colby, A., & Sullivan, W. M. (2008). Ethics teaching in undergraduate engineering education. *Journal of Engineering Education*, *97*(3), 327–338. https://doi.org/10.1002/j.2168-9830.2008.tb00982.x

Conlon, E., & Zandvoort, H. (2011). Broadening ethics teaching in engineering: Beyond the individualistic approach. *Science and Engineering Ethics*, *17*(2), 217–232. https://doi.org/10.1007/s11948-010-9205-7

Croft, E. A., Winkelman, P., Boisvert, A., & Patten, K. (2013). *Global engineering leadership design and implementation of local and international service learning curriculum for senior engineering students*. CEEA. https://doi.org/10.24908/pceea.v0i0.4868

Cruz, C.C. (2021) Brazilian grassroots engineering: A decolonial approach to engineering education. *European Journal of Engineering Education, 46*(5), 690–706. https://doi.org/10.1080/03043797.2021.1878346

Davis, C., & Yadav, A. (2014). Case Studies in Engineering. In A. Johri & B. M. Olds (Eds.), *Cambridge handbook of engineering education research* (pp. 161–180). Cambridge University Press. https://doi.org/10.1017/CBO9781139013451.013

Davis, M. (2006). Integrating ethics into technical courses: Micro-insertion. *Science and Engineering Ethics 12*, 717–730, 726–727.

Dempsey, J., Stamets, J., & Eggleson, K. (2017). Stakeholder views of nanosilver linings: Macroethics education and automated text analysis through participatory governance role play in a workshop format. *Science and Engineering Ethics*, *23*(3), 913–939.

Di Blasio, D., et al. (2018). *Many hands on the elephant: How a transdisciplinary team assesses an integrative course*. ASEE. https://doi.org/10.18260/1-2--30787

Dodson, L., et al. (2017). *How role-playing builds empathy and concern for social justice*. ASEE.
Doorn, N., & Kroesen, J. O. (2013). Using and developing role plays in teaching aimed at preparing for social responsibility. *Science and Engineering Ethics*, *19*, 1513–1527.
Douglas, E. P., Holbrook, J. B., & Corbo-Ferreira, F. (2022, October). Engineering ethics education for social justice: Implementation and preliminary data. In 2022 IEEE Frontiers in Education Conference (FIE) (pp. 1–4). IEEE.
Drake, M. J., Griffin, P. M., Kirkman, R., & Swann, J. L. (2005). Engineering ethical curricula: Assessment and comparison of two approaches. *Journal of Engineering Education, 94*(2), 223–231.
Drumwright, M., Prentice, R., & Biasucci, C. (2015). Behavioral ethics and teaching ethical decision making. *Decision Sciences Journal of Innovative Education, 13*(3), 431–458.
Egilmez, G., Viscomi, P., & Carnasciali, M. I. (2019). Assessing an online engineering ethics module from experiential learning perspective.
Ethics | ASCE. (n.d.). Retrieved January 19, 2023, from https://www.asce.org/career-growth/ethics
Fan, Y., Zhang, X., & Xie, X. (2015). Design and development of a course in professionalism and ethics for CDIO curriculum in China. *Science and Engineering Ethics*, *21*(5), 1381–1389. https://doi.org/10.1007/s11948-014-9592-2
Gelfand, S. D. (2016). Using insights from applied moral psychology to promote ethical behavior among engineering students and professional engineers. *Science and Engineering Ethics, 22*(5), 1513–1534.
Gomez, J., & Svihla, V. (2018). Rurality as an asset for inclusive teaching in chemical engineering. *Chemical Engineering Education*, *52*(2), 99–106.
Halada, G. P., & Khost, P. H. (2017). The use of narrative in undergraduate engineering education. *ASEE*. https://doi.org/10.18260/1-2--29018
Harris, C. E. (2004). Internationalizing professional codes in engineering. *Science and Engineering Ethics*, *10*(3), 503–521. https://doi.org/10.1007/s11948-004-0008-6
Haws, D. R. (2001). Ethics instruction in engineering education: A (mini) meta-analysis. *Journal of Engineering Education*, *90*(2), 223–229. https://doi.org/10.1002/j.2168-9830.2001.tb00596.x
Hedayati-Mehdiabadi, A. (2022). How do computer science students make decisions in ethical situations? Implications for teaching computing ethics based on a grounded theory study. *ACM Transactions on Computing Education, 22*(3), 1–24.
Herkert, J. R. (2000). Engineering ethics education in the USA: Content, pedagogy and curriculum. *European Journal of Engineering Education*, *25*(4), 303–313. https://doi.org/10.1080/03043790050200340
Herkert, J. R., Borenstein, J., & Miller, K. (2020). The Boeing 737 MAX: Lessons for engineering ethics. *Science and Engineering Ethics*, *26*, 2957–2974.
Hess, J. L., & Fila, N. D. (2016, June). *The development and growth of empathy among engineering students*. Paper presented at 2016 ASEE Annual Conference & Exposition, New Orleans, Louisiana.
Hess, J. L., & Fore, G. (2018). A systematic literature review of US engineering ethics interventions. *Science and Engineering Ethics, 24*, 551–583.
High, M. S., Harrist, S., & Gelfand, S. D. (2011, June). Tools to craft ethical behavior. In *2011 ASEE Annual Conference & Exposition* (pp. 22–1534).
Hinds, T., Buch, N., Delgado, V., & Morgan, J. (2020). Development of a peace engineering initiative within a first-year engineering program. *Journal of Engineering Education Transformations, 33*, 112–117. https://doi.org/10.16920/jeet/2020/v33i0/150077
Hitt, S. J., & Lennerfors, T. T. (2022). Fictional Film in engineering ethics education: With Miyazaki's the wind rises as exemplar. *Science and Engineering Ethics*, *28*(5), 44.
Hoffmann, M., & Borenstein, J. (2014). Understanding ill-structured engineering ethics problems through a collaborative learning and argument visualization approach. *Science and Engineering Ethics, 20*, 261–276.
Hunger, I. (2013). Some personal notes on role plays as an excellent teaching tool: Commentary on "using and developing role plays in teaching aimed at preparing for social responsibility". *Science and Engineering Ethics*, *19*, 1529–1531.
Itani, M. (2013, June). The effectiveness of videos as a learning tool in an engineering ethics course: A students' perspective. In 2013 ASEE Annual Conference & Exposition (pp. 23–1193).
Jung, H. Y., Zhou, Z., & Ni, L. (2016). *Growing together with the community through service learning*. ASEE.
Kallergi, A., & Zwijnenberg, R. (2019). Educating Responsible Innovators-to-Be: Hands-on Participation with Biotechnology. *Lecture Notes in Electrical Engineering. 532*, 79–94. https://doi.org/10.1007/978-3-030-02242-6_7

Keefer, M. W., Wilson, S. E., Dankowicz, H., & Loui, M. C. (2014). The importance of formative assessment in science and engineering ethics education: Some evidence and practical advice. *Science and engineering ethics, 20*, 249–260.
Knight, D. W., Canney, N. E., Bielefeldt, A. R., & Swan, C. (2016, October). Macroethics instruction in co-curricular settings:The development and results of a national survey. In *2016 IEEE Frontiers in Education Conference* (FIE) (pp. 1–4). IEEE.
Koskinen, J. A. (2015). E-learning of ethics, awareness, hacking and research. 43rd Annual SEFI Conference. Orleans, France.
Krishnan, S., & Nilsson, T. L. (2012). *Engineering service learning case study on preparing students for the global community*. ASEE. https://doi.org/10.18260/1-2--21311
Kotluk, N., & Tormey, R. (2023). Compassion and engineering students' moral reasoning: The emotional experience of engineering ethics cases. *Journal of Engineering Education, 112*(3), 719–740.
Kumar, S., & Kremer-Herman, N. (2019). *Integrating ethics across computing: An experience report of three computing courses engaging ethics and societal impact through roleplaying, case studies, and service learning*. IEEE FIE. https://doi.org/10.1109/FIE43999.2019.9028568
Lee, L. M. (2021). The growth of ethics bowls: A pedagogical tool to develop moral reasoning in a complex world. *International Journal of Ethics Education, 6*, 141–148.
Leitch, K. R., & Dittfurth, R. B. (2012). Online and In-Seat Engineering Ethics Instruction: The View from Both Sides. In *American Society for Engineering Education*. American Society for Engineering Education.
Llopis-Albert, C., Rubio, F., Valle-Falcones, L. M., & Grima-Olmedo, C. (2020). Use of technical computing systems in the context of engineering problems. *Multidisciplinary Journal for Education Social and Technological Sciences, 7*(2), 84–99. https://doi.org/10.4995/muse.2020.14283
Lönngren, J., Direito, I., Tormey, R., & Huff, J. L. (2023). Emotions in engineering education. In A. Johri (Ed.), *International handbook of engineering education research* (1st ed., pp. 156–182). Routledge. https://doi.org/10.4324/9781003287483
Lumgair, B. (2018, June). The effectiveness of webinars in professional skills and engineering ethics education in large online classes. In *2018 ASEE Annual Conference & Exposition.*
Martin, D. A., Conlon, E., & Bowe, B. (2019). The role of role-play in student awareness of the social dimension of the engineering profession. *European Journal of Engineering Education, 44*(6), 882–905. https://doi.org/10.1080/03043797.2019.1624691
Martin, D. A., Conlon, E., & Bowe, B. (2018). A constructivist approach to the use of case studies in teaching engineering ethics. In M. E. Auer, D. Guralnick, & I. Simonics (Eds.), *Teaching and learning in a digital world* (Vol. 715, pp. 193–201). Springer International Publishing. https://doi.org/10.1007/978-3-319-73210-7_23
Martin, D. A., Conlon, E., & Bowe, B. (2021). Using case studies in engineering ethics education: The case for immersive scenarios through stakeholder engagement and real life data. *Australasian Journal of Engineering Education, 26*(1), 47–63.
Martin, D. A., Gwynne-Evans, A., Kazakova, A. A., & Zhu, Q. (2023). Developing a global and culturally inclusive vision of engineering ethics education and research. In *International handbook of engineering education research* (pp. 87–114). Routledge.
May, D. R., & Luth, M. T. (2013). The effectiveness of ethics education: A quasi-experimental field study. *Science and Engineering Ethics, 19*, 545–568.
Mogul, N. F., & Tomblin, D. (2019). *Just add context? Analyzing student perceptions of decontextualized and contextualized engineering problems and their use of storytelling to create context*. ASEE. https://doi.org/10.18260/1-2--33035
Molina-Carmona, R., Satorre-Cuerda, R., Villagrá-Arnedo, C., & Compañ-Rosique, P. (2017). Training socially responsible engineers by developing accessible video games. In P. Zaphiris & A. Ioannou (Eds.), *Learning and collaboration technologies. Technology in education*. LCT 2017. *Lecture Notes in Computer Science, 10296*. Springer. https://doi.org/10.1007/978-3-319-58515-4_15
Monk, J. (2009). Ethics, engineers and drama. *Science and Engineering Ethics, 15*(1), 111–123.
Mullin, J., Lohani, V. K., & Lo, J. (2006, October). Work in progress: Introduction to engineering ethics through student skits in the freshman engineering program at Virginia Tech. In *Proceedings. Frontiers in education. 36th Annual Conference* (pp. 21–22). IEEE.
National Research Council. (2009). Learning science in informal environments: People, places, and pursuits.
Olwi, I. A. (2014). *Story telling as an effective mean for stimulating students passion in engineering classes*. ASEE.

Pierrakos, O., Prentice, M., Silverglate, C., Lamb, M., Demaske, A., & Smout, R. (2019, October). Reimagining engineering ethics: From ethics education to character education. In *2019 IEEE Frontiers in Education Conference (FIE)* (pp. 1–9). IEEE.

Plouff, C., & Barakat, N. (2012). *Infusion of ABET - Specified professional and academic content into off-campus work experiences via distance learning modules*. IEEE FIE. https://doi.org/10.1109/FIE.2012.6462309

Plouff, C., & Barakat, N. (2014). *A model for engineering ethics education through a co-op program*. ASEE.

Putko, M., & Rooney-Varga, J. N. (2016). *World energy in engineering design*. ASEE.

Quigley, D., Sonnenfeld, D., Brown, P., Silka, L., He, L., & Tian, Q. (2016). Research ethics training on place-based communities and cultural groups. *Journal of Environmental Studies and Sciences*, *6*(3), 479–489. https://doi.org/10.1007/s13412-015-0236-x

Razavinia, N., & Mydlarski, L. (2020). A short online course targeting professionalism, the impact of engineering on society and the environment, and ethics and equity. Proceedings of the Canadian Engineering Education Association (CEEA).

Reeves, J., & Nadolny, L. (2013, June). Ethics in engineering education using virtual worlds. In *2013 ASEE Annual Conference & Exposition* (pp. 23–547).

Roeser, S. (2012). Emotional engineers: Toward morally responsible design. *Science and Engineering Ethics, 18*, 103–115.

Ruiz-Soto, G., Montesinos, L., Santos-Díaz, A., & de Paz-Arroyo, S. (2015). Ethics and Citizenship Education across the Biomedical Engineering Curriculum. In VI Latin American Congress on Biomedical Engineering CLAIB 2014, Paraná, Argentina 29, 30 & 31 October 2014 (pp. 968–971). Springer International Publishing.

Sadowski, J., Seager, T. P., Selinger, E., Spierre, S. G., & Whyte, K. P. (2013). An experiential, game-theoretic pedagogy for sustainability ethics. *Science and Engineering Ethics*, *19*(3), 1323–1339. https://doi.org/10.1007/s11948-012-9385-4

Sekiguchi, K., & Hori, K. (2019). Can ethics enhance creative design activity? *ICED, 19*, 3181–3190. https://doi.org/10.1017/dsi.2019.325

Sochacka, N., Walther, J., & Miller, S. E. (2018). Fostering empathy in engineering education. *Scientia, 119*, 110–113.

Sochacka, N. W., Youngblood, K. M., Walther, J., & Miller, S. E. (2020). A qualitative study of how mental models impact engineering students' engagement with empathic communication exercises. *Australasian Journal of Engineering Education*, *25*(2), 121–132.

Sunderland, M. E. (2019). Using student engagement to relocate ethics to the core of the engineering curriculum. *Science and Engineering Ethics*, *25*(6), 1771–1788. https://doi.org/10.1007/s11948-013-9444-5

Thiel, C. E., Connelly, S., Harkrider, L., Devenport, L. D., Bagdasarov, Z., Johnson, J. F., & Mumford, M. D. (2013). Case-based knowledge and ethics education: Improving learning and transfer through emotionally rich cases. *Science and Engineering Ethics*, *19*(1), 265–286. https://doi.org/10.1007/s11948-011-9318-7

Tkachenko, O., & Hedayati, A. (2022). Employee ethics training: Literature review and integrative perspective. In *Academy of management proceedings* (Vol. 2022, no. 1, p. 17746). https://doi.org/10.5465/AMBPP.2022.17746abstract

Todd, E. M., Watts, L. L., Mulhearn, T. J., Torrence, B. S., Turner, M. R., Connelly, S., & Mumford, M. D. (2017). A meta-analytic comparison of face-to-face and online delivery in ethics instruction: The case for a hybrid approach. *Science and Engineering Ethics*, *23*, 1719–1754.

Troesch, V. (2014, May). A phenomenological approach to teaching engineering ethics. In *2014 IEEE International Symposium on ethics in science, technology and engineering* (pp. 1–9). IEEE.

Troesch, V. (2015). Teaching engineering ethics: A phenomenological approach. *IEEE Technology and Society Magazine, 34*(2), 56–63.

United Nations Department for Economic and Social Affairs. (2022). Sustainable Development Goals Report 2022. United Nations. https://unstats.un.org/sdgs/report/2022/The-Sustainable-Development-Goals-Report-2022.pdf

Valentine, A., Lowenhoff, S., Marinelli, M., Male, S., & Hassan, G. M. (2020). Building students' nascent understanding of ethics in engineering practice. *European Journal of Engineering Education*, *45*(6), 957–970.

Voss, G. (2013). Gaming, texting, learning? Teaching engineering ethics through students' lived experiences with technology. *Science and Engineering Ethics*, *19*(3), 1375–1393.

Walther, J., Miller, S. E., & Sochacka, N. W. (2017). A model of empathy in engineering as a core skill, practice orientation, and professional way of being. *Journal of Engineering Education, 106*(1), 123–148.

White, A. R. (2020). *A simulation system for exploring ethical situations that arise from conflicting engineering team goals*. ASEE. https://doi.org/10.18260/1-2--34052

Winkelman, P. M., Penner, J., & Beittoei, A. (2016). Sustainable development for engineers through a thematic restructuring of experiential learning. In W. Leal Filho & S. Nesbit (Eds.), *New developments in engineering education for sustainable development*. World Sustainability Series. Springer. https://doi.org/10.1007/978-3-319-32933-8_26

Wittig, A. (2013). Implementing problem based learning through engineers without borders student projects. *Advances in Engineering Education, 3*(4), 1–20.

Xenos, M., & Velli, V. (2020). A serious game for introducing software engineering ethics to university students. *The Challenges of the Digital Transformation in Education,* 579–88. https://doi.org/10.1007/978-3-030-11932-4_55

Zain, S. M., Basri, N. E. A., Mamat, L., Ghee, T. K., Syaril, S., & Rahman, N. S. A. (2017). Community service as platform for environmental ethics in CITRA education. *Journal of Engineering Science and Technology*, *12*(12), 67–79.

Zandvoort, H., VanDePoel, I., & Brumsen, M. (2000). Ethics in the engineering curricula: Topics, trends and challenges for the future. *European Journal of Engineering Education*, *25*(4), 291–302. https://doi.org/10.1080/03043790050200331

Zhu, Q., Clancy, R. F., Streiner, S., Gammon, A., Thorper, R., & Angeli, A. (2022). Exploring the ethical perceptions of first year engineering students: Public welfare beliefs, ethical behavior, and professional values. In *American Society for .engineering education* Zone IV Vancouver.

20
TEACHING ETHICS USING CASE STUDIES

Christian Herzog, Aditya Johri, and Roland Tormey

Introduction

This chapter provides a critical overview of case studies in engineering ethics education. The chapter aims to support the adoption of case studies by engineering ethics educators through a systematic evaluation of their purpose, pedagogical outcomes, and overall fit within the course or curriculum.

We will employ a broad definition of the concept, where case studies provide a context-driven approach to teaching ethics that allows learners to think through ethical issues and debates using grounded information and make decisions that would – or even can – potentially have real consequences. Case studies for teaching engineering ethics range from short narratives that encapsulate a real-world problem or dilemma (called vignettes) to longer forms, including many cases that are over ten pages long. They can be narrowly focused on a problem in the workplace, such as deciding whether to use a specific chemical or process, or they can be used to examine a large-scale project or disaster, such as an aircraft accident.

For this chapter, we will first outline the arguments for and against teaching engineering ethics using case studies in order to aid educators in assessing what case studies can bring to the classroom. After a broader discussion of what case studies are and their role within engineering ethics education, we will examine how the use of case studies furthers specific learning goals through their integration into teaching, drawing on a Neo-Kohlbergian view (dubbed the 'four-component model'), and relating this to different styles of choosing and integrating case studies in higher education. We will briefly and pre-emptively refer to challenge-based learning (CBL) as a particular way of integrating real-world, non-prepared, and non-academic case studies, contrasting this with the more classical way of discussing historical or fictional case studies in engineering ethics. We will also discuss role-play discussions structured around case studies and their advantages for improving perspectival thinking and linking micro-meso-macro contexts. We will argue that there are no emotionally neutral cases and that it is a matter of the learning goals to be achieved whether and how one should explicitly constructively deal with this emotionality.

A separate section will briefly showcase various available repositories of engineering ethics case studies, categorized into available themes, types, and scope of expositions. Finally, we will

DOI: 10.4324/9781003464259-25

summarise key takeaway points for those interested in designing successful engineering ethics course syllabi based on case studies.

Our approach to this chapter reflects our own different backgrounds and intellectual trajectories. We are all interdisciplinary researchers, with two of the writing team being trained as engineers and combining a focus on both technical aspects of computing and engineering research and teaching, with a focus on ethical and social dimensions. The third member of the writing team is a sociologist who also teaches and researches both engineering ethics and learning sciences. Our interdisciplinary, evidence-informed, and teaching-focused approach is reflected in this chapter.

Case studies in engineering ethics education – a critical discussion

In professional education, case studies are generally identified as first introduced in legal education at Harvard in the late 1800s; they were widely used in that field by the early 1900s. The method spread to business education before being adopted in other professional domains, including medical, teacher, and engineering education. The use of cases as a teaching method has been the mainstay of business education for over a century since the method was introduced to this domain at Harvard Business School in the 1920s. Although the use of case studies for teaching has come a long way since then, many of the core components of using the case method remain the same. Case studies come with three main characteristics. Most cases

1) describe an actual or hypothetical situation where a decision needs to be made;
2) include contextual and technical information that the reader can use; and
3) involve a decision or solution that requires the learner to engage with and develop various perspectives

(D. A. Martin et al., 2021b; Merseth, 1994)

The learner is also often placed in a specific role or has to approach the problem from the viewpoint of the role – e.g., an engineer or a manager. Furthermore, the case can be used in a group setting, where the entire class or smaller groups sometimes discuss the situation and decision-making.

Case studies provide a context-driven approach to teaching ethics that allows the instructors and learners to think through ethical issues and debates using grounded information and make decisions that can potentially have real consequences. Ethics case studies are implemented through standalone courses, co-curricular activities, immersion programs, and modules within courses. The materials used to teach cases include articles, books, videos, audio, and other curricular material. They can be used as part of the curriculum to teach how to apply professional codes or even broader societal considerations of engineering.

Given the intersection of business and engineering, it may not be surprising that the use of cases has found application in teaching across various engineering topics. Even though lectures remain the dominant form of teaching in higher education, cases have achieved significant application in engineering ethics and are one of the most prevalent techniques (Davis, 1997). Hess and Fore (2017) identified case study–based instruction as the most frequently employed method in engineering ethics. Cases are a powerful teaching strategy, especially for courses and problems that deal with 'real-world' applications of engineering knowledge. It has been contended that "there is widespread agreement that the best way to teach professional ethics is by using cases" (Harris et al., 1996, p. 94).

Case-based instruction is effective as a component of ethics teaching. Several meta-analyses of ethics education in both science and business, for example, have found that case-based learning has more significant positive effects than many other teaching approaches (Antes et al., 2009; Waples et al., 2009; Watts, Todd, et al., 2017). For example, the meta-analysis by Logan Watts and colleagues (2017) found that programs with a stronger focus on case-based instruction also tended to have an above-average impact on ethics learning.

From this perspective, it may even seem that using case studies in ethics education – and elsewhere – has become a tradition, a signature style for scholars from particular disciplines that is applied in a rather unreflected manner. One of the purposes of this chapter is to raise awareness of why and when the case-study approach may be helpful and, more importantly, which form it may take to align with the course learning goals.

For instance, for ethics education, Martin et al. (2019) describe case-based role-playing activity as being situated "within a microethical frame, focused on describing individual dilemmas set in scenarios of crisis that can be solved through the application of ethical heuristics and by appealing to the precepts of professional codes and ethical theories" (p. 1). However, cases often resist a specific or narrow answer, leading to sustained discussion, which may not necessarily end in finding a consensual 'solution.' Furthermore, cases need not necessarily be solely embedded within a micro-ethical frame but can be designed to take on macro-ethical subjects or extend toward them.

This is where tensions exist between different views of how and whether case-based instruction of applied ethics should be adopted. For instance, Lawlor (2021) objects to too heavily relying on case-based ethics instruction when its mode is confined to a single teaching session per case, focused on individuals, and lacking context. Lawlor contends that, instead, instruction – case-based or otherwise – should allow for considering the ethics of the macro-level aspects as well, thus taking into account more wide-arching circumstances that may prevent individuals from doing the right thing. Consequently – and according to Lawlor – in case-based teaching, room must be given to address and discuss precisely those conditions that prevent specific or narrow answers from being applicable. Furthermore, for Lawlor (2021), content may often even be more relevant than the actual ethics. Accordingly, it can be argued that case-based instruction must make space to not just focus on the ethical dilemma itself but allow for conveying – and unearthing – the relevant circumstances and the technical and scientific details accompanying the case.

Rottmann and Reeve (2020), reviewing prior work in the use of case studies in engineering ethics instruction, have identified two key dimensions along which case-study application in engineering varies: (1) analytical strategy or a deductive-inductive dimension and (2) level of analysis or a micro–macro dimension (p. 149). In a deductive approach, instructors encourage students to apply a specific theory or viewpoint to analyze an event. In an inductive approach, students are asked to draw various ethical lessons from a given case. Regarding the level of analysis, cases can be used for micro-ethical scenarios that usually depict individual practitioners facing difficult situations or highlight the socio-political consequences of engineering. The micro approach is a powerful pedagogical technique because it asks a learner to decide from the viewpoint of someone in that situation. The macro approach is helpful as it forces learners to connect with different stakeholders and larger organizational and societal concerns.

In addition, approaches to case studies can also differ in terms of (3) the timeframe or historicity, (4) the veracity or the amount of hypothetical versus factual elements, and (5) the duration in which the case is considered and the student's role. A taxonomy informed by these elements has been developed by Diana Martin and colleagues (D. A. Martin et al., 2021a). This taxonomy certainly has its appeal, even though it operates on vastly different ontological domains, ranging from the highly descriptive (e.g., the length of the case) to the reflective-subjective (e.g., the micro- vs. macro-ethical scope). This is why, for the remainder of the chapter, we would like to propose a simplified distinction between two general types of case studies, reflecting a combination of the five dimensions proposed by Martin et al. (2022):

A. Highly complex and highly indeterminate case studies
B. Simpler vignettes with less veracity

Our streamlined taxonomy is, in fact, complementary to Martin et al. as, for example, highly complex cases or simpler vignettes can both be approached inductively or deductively, concern macro- or micro-level issues, and be current or historical, factual or hypothetical. Although there may be tendencies in engineering education practice towards specific characteristics in types A and B, parsing that out is beyond the scope of this chapter. Instead – and perhaps quite generally – case studies of type A may be better in helping people to decide in the real world, while case studies of type B tend to help people learn to apply principles. As stated above, scholars should be able to choose the type of case studies appropriate to the learning situations.

The meta-analysis by Logan Watts and colleagues (2017), already mentioned above, found that longer cases seem to have more impact than shorter ones, those with moderate complexity have more impact than both simple and complex cases, and cases with low to moderate realism have more impact than realistic cases which may involve highly emotive content such as multiple deaths or family tragedies. This is an important finding given the extent to which realistic cases of big news stories (e.g., case studies of the Columbia and Challenger space shuttle explosions in engineering ethics) are ubiquitous in many ethics textbooks. The use of well-known and even spectacular cases that have resulted in considerable negative and emotive media echoes may appear so distant from the experience of most students that they generate emotional distance and closure. Even moderately sensitized students may quickly dismiss these cases as depicting situations in which they will never find themselves. At the same time, knowing how cases turned out can limit students' opportunities to develop reasoning and perspective-taking skills. Before discussing practical choices in employing cases for ethics education, we will commence by going deeper into the subject of learning goals that may be pursued.

What is the purpose of using case studies in engineering ethics education?

Overall, *what can meaningfully be said about the goals of professional ethics education?* As Chapter 10 (on the psychological foundations of engineering ethics education) has explored, a cognitive, rationalist, and individualist perspective largely dominated thinking on moral reasoning in the late twentieth century. This was (and to some extent still is) reflected in professionals' ethics education goals. The last 20 years have seen an increasing focus on supplementing the focus on moral reasoning with a broader concern for other components of moral action, such as ethical sensitivity, motivation, agency, and imagination. Having developed these capacities, students then need to learn to transfer them into professional practice situations. *So what, then, can be identified as essential goals of using case studies in engineering ethics education?*

Neo-Kohlbergian learning goals of ethics education

The set of possible goals pursued by case study–based engineering ethics instructors, according to an interview-based study by Martin et al. (2021a), comprise (1) epistemic and (2) value- and virtue-driven aims, as well as (3) awareness and (4) agency forming. Epistemic goals try to elucidate the complexity behind the cases, while value- and virtue-driven goals relate to fostering engineering habits such as accepting moral responsibility. Awareness is intended to be brought to the broader context, while ethical agency is encouraged to be taken seriously within the range or even at the fringes of legal frameworks.

Martin et al.'s (2021a) empirical study seems to suggest an awareness among educators of the fact that focusing on moral reasoning is not enough to improve subsequent moral behavior. Empirical research has since indicated that a narrow focus on cognitive abilities in the moral domain is insufficient. As Chapter 10 has explored, one way of doing this is the Neo-Kohlbergian four-component model of moral behavior (Bebeau et al., 1999; Narvaez & Rest, 1995). These components are:

- *Moral sensitivity* as the ability to recognize ethically salient conditions
- *Moral judgment* as the outcome of the process of moral reasoning that involves identifying relevant values or principles
- *Moral motivation* as the ability to prioritize moral over other values
- *Moral character* (or agency) as the ability to persevere in following through with one's moral judgment.

The critical difference to the traditional Kohlbergian model (Kohlberg, 1974) lies in recognizing moral judgment (or reasoning) as necessary but insufficient to result in ethical behavior. Instead, Bebeau et al. (1999) argue that all four components interact and contribute to producing moral behavior. Moral sensitivity integrates both the ability to recognize vulnerabilities – to (emotionally) empathize and to take others' perspectives – and to integrate professional knowledge (e.g., facts about working conditions or ecologically relevant material properties). Moral judgment denotes the traditional Kohlbergian notion of 'post-conventional moral reasoning' as the highest form of morality but is considered insufficient. Moral motivation – or a lack thereof – can explain the dissonance between recognizing a moral issue and acting accordingly. In contrast, moral character has to do with mustering the courage to overcome obstacles that impede the action demanded by one's moral judgment.

What is evident from this brief sketch of the 'four-component model' is that it recognizes the intertwinement of cognitive and emotional factors. It marks a clear departure from the Kohlbergian rejection of irrational or emotive factors influencing moral development (Haidt, 2001) and a turn towards learning from a feminist ethics of care that understands the moral significance of conditions in terms of the vulnerability of and the situated relationships between people (Gilligan, 1993; Noddings, 1988). For more on this see Chapter 4 on reason and emotion in engineering ethics education.

The problem of transfer

It is worth noting that if the goal of ethics education is that students learn to make ethical decisions as engineering practitioners, then developing moral sensitivity, reasoning, motivation, and agency in university-based education programs is unlikely to be enough. This is because one of the perennial problems in professional education is that learning things in university classrooms does not automatically mean that people use that knowledge and skill in settings outside the classroom. To learning scientists, this is known as the problem of transfer. In professional education, it can also be seen as a facet of the theory–practice divide and has been called 'the problem of enactment' (Darling-Hammond & Baratz-Snowden, 2007).

As John Bransford, Ann Brown, and Rodney Cocking have noted (2000, p. 235), many approaches to teaching that appear equivalent when the ability to recall class-based learning is measured turn out to be quite different in their effects when transfer of learning to new settings is assessed. Learning science researchers have identified several features of teaching and learning that facilitate transfer. One of the features of learning is that people do not simply learn an idea that is the target of learning but also encode aspects of the context within which the memory is encoded. In this sense, knowledge is said to be conditionalized (Bransford et al., 2000, p. 43). This makes it easier for people to remember things in contexts similar to those in which they were learned. This 'conditionalization' of learning can be used in teaching by providing students with learning experiences closer to the situations in which they will practice their profession. Experiential learning, such as project work, challenge-based learning, and fieldwork or internships, may help close the gap between university learning and professional practice (Tormey et al., 2021; for more, see Chapters 21 and 23). However, these experiential learning situations give rise to challenges. Students can rapidly become overloaded with new information, making it difficult to focus on essential and relevant parts of the experience. Such experiences are often, therefore, useful when introduced progressively into the learning process; once students have gained some understanding of key ideas and practices, richer and more complex settings can provide them with opportunities to apply these ideas in more 'real-world' settings (Tormey et al., 2021, pp. 200–204).

What do possible practical implementations of engineering ethics courses using case studies look like? The following section delves into three specific topics: one focuses on introducing real-world case studies via the challenge-based learning approach, another deals with case-based role-playing, and a third discusses ways of making dealings with emotions explicit.

From the above discussion, it should have become clear that very often, a fundamental trade-off may appear when choosing between a highly unstructured and complex case as a means to train, for example, moral sensitivity and judgment, and a more structured case with clear potential outcomes as a means to focus on training moral motivation and character. Hence, it may appear as if much rests on the choice of cases, and indeed, the pre-processing endowed upon a case can be used to emphasize the learning goals to be achieved.

However, the nature of the pedagogy built around a case may even be more significant, meaning that it is less about the case *per se* and more about what students are led to do with it that determines which learning goals are targeted. Plainly speaking, whatever the general method utilized to present an ethics case study to the learners, it will not be a sufficient approach for instructors to throw a case at the class and then stand by and watch. Instead, one way of going about careful considerations on employing case studies in engineering ethics courses can be structured as per the curriculum typology of Goodlad, Klein, and Tye (1979), that is, along the lines of the five aspects of the ideal, formal, perceived, operational, and experienced curriculum. In this vein, Herzog et al. (2022a) considered a simplified version of Goodlad's typology applied to CBL. By iterating through the temporal dimension and traversing the duties and goals of instructors and learners (and potentially others), Herzog et al. provided a matrix-like approach to laying down and planning necessary preparatory steps to deal with a particular case.

Practical ways of integrating engineering ethics case studies

Challenge-based learning as a way of using highly complex, indeterminate case studies

One method of introducing work with cases in engineering ethics on a collaborative level is challenge-based learning (CBL) (Nichols & Cator, 2008). In CBL for engineering ethics, (typically groups of) learners are tasked with identifying, analyzing, and trying to solve their own challenges based on real-life problems. Here, the focus is on thinking about the challenge as a case: CBL gives ample opportunities but also imposes the necessity to tailor the pedagogy of the coursework on a case towards particular learning goals (Herzog et al., 2022a). Both instructors and external stakeholders must provide constant guidance and exchange. The external stakeholders mainly provide the general case (business, technological, or otherwise) to allow learners to inquire, reflect, and develop potential solutions. Even though cases are real or based on real-world developments, instructors can and should find ways to reduce (or, though less likely, enrich) complexity. For instance, views on a case based on, say, a health companion app can be restricted to the particular ethically salient field of data protection, privacy, and confidentiality or enriched by letting learners envision and extrapolate an early-stage solution toward deployment. While the former may amount to tailoring the case study towards the learning goals more directed towards heightening moral motivation and agency, the latter – arguably more complex – case would tend to promote moral sensitivity and judgment. In CBL, this tailoring towards a specific emphasis on particular learning goals is both a matter of choice for the instructors and subject to negotiations with the external stakeholders. Exemplars typically do not provide a ready-made case but rather a setting where a case can be embedded or emerge.

This flexibility means it may be possible to tailor the approach to the skill levels in terms of the four components (moral sensitivity, judgment, motivation, and agency) present at an individual or group level. This yields the potential to align CBL with highly individual learning trajectories. Such complexity by diversity can be regarded as a potential asset. By letting learners decide which group to join and which case to work on, they can also effectively choose their emphasis on specific learning goals. One could hope this would work out entirely tacitly and implicitly, assuming that learners are drawn to the case setting that is most appealing from a learning gain point of view. This assumption, however, may be naïve, as a multitude of other factors – interest in the technological content of the case at the very least – will probably dominate the decision. This suggests that some facility of conveying the significance of the choice of the case in terms of, for example, if the Neo-Kohlbergian learning goals should be deployed to guide learners. Potentially pre-processed real-world cases could be analyzed in co-operation with the external stakeholders to assign them according to the tendency to promote any learning goal. With a sufficient mix of cases provided, this strategy could accommodate the breadth of moral skillsets students bring with them into what may typically be the only ethics course within an engineering curriculum.

Alternatively, it is imaginable, albeit elaborate, that instruments such as the Defining Issues Test (DIT) or the DIT2 (Rest et al., 1999; for more on DIT see Chapter 10) could be employed to conduct a quantitative assessment of the learners' patterns of moral reasoning, provide feedback, and suggest case assignments. From a moral skillset perspective, CBL appears to be a suitable candidate for case-based engineering ethics education despite its instructional and organizational complexity (cf. Bombaerts, 2020; Herzog et al., 2022b).

However, there are other reasons why CBL fares well as a framework for case-based instruction. We will briefly provide – potentially non-exhaustive – arguments here. Challenge-based learning alleviates one of Lawlor's (2021) main criticisms in that it typically confronts learners with a single case on which to develop and identify ethical challenges, research and inquire about background information (technical and otherwise), and consider, propose, and evaluate individual and organizational conduct. Furthermore, CBL is not necessarily confined to the micro-ethical frame. Based on the experience of this chapter's first author, some external stakeholders even seem to gravitate towards proposing macro-ethical issues as their case of interest. In addition, perhaps one of the most obvious – and so far neglected – reasons to engage in the extra effort CBL presents is that learners will work on real-world cases. Challenge-based learning generates an experience of transfer, thus aiming for sustained learning that translates into the engineers' professional work-life.

Role-play case studies for teaching perspectives

Role-playing uses cases or scenarios to discuss a problem or an issue where each participant is assigned a specific role (Loui, 2009). In broader terms, role-plays are one kind of simulation-based learning exercise (Hertel & Millis, 2011). Role-playing is often related to cases that are simpler vignettes with complex background conditions, but the role-playing will be confined to a particular snap-shot situation. However, there are also role-play-based courses which go on for the entire semester and keep adding elements to the case to increase complexity over time. In any role-play case-study discussion, students take on the role of a person involved with or impacted by the case, and they have to participate in a discussion that reflects the viewpoint or perspective of their role. Role-playing is used to emphasize the real-world side of things and provide students an opportunity to engage deeply in a close approximation to experience since 'being' a character in a role-play introduces a social component that cannot be achieved with a mere case discussion (Hertel & Millis, 2011). Role-play supports perspective-taking (Pusateri et al., 2009), emotional engagement (Heyward, 2010), critical thinking (Poling & Hupp, 2009), and communication skills (Nestel & Tierney, 2007). Role-plays are also helpful in teaching and consolidating student knowledge (DeNeve & Heppner, 1997; McCarthy & Anderson, 1999; Poling & Hupp, 2009) and for teaching students how to link micro- and macro-level issues, as well as issues in between (Johri & Hingle, 2022).

Role-play is challenging as there is usually the need to reach consensus. Reaching consensus can be argued to contribute to all four Neo-Kohlbergian learning goals: all participants must develop enough *moral sensitivity* to recognize and agree on the ethically salient conditions worth arguing about; participants must arrive at a *moral judgment* and formulate arguments based on relevant values or principles. While reaching consensus, students must also exhibit *moral motivation* by prioritizing and agreeing on moral values. Finally, *moral character* can be displayed by swaying from one's resolve only when one's moral judgment is questioned for good reasons. All learning goals are effectively addressed, especially when participants must assume roles that do not reflect their moral standpoint. Accordingly, prior research on the use of role-plays has shown that, initially, learners find them uncomfortable. This discomfort is often due to the experience of "disjuncture" (Jarvis, 2012, pp. 79–84) between the role-play experience and the student's previous conceptualizations of a given topic. Studies have shown that disjuncture can prime students to engage vigorously in efforts to develop knowledge and skills to bring "equilibrium" back to their conceptualization (Jarvis, 2012, p. 80).

Furthermore, students also learn to adopt different perspectives (Van De Sande & Greeno, 2012). Role-playing is a student-centered approach to learning, and by being actively involved in their roles, students experience tension and conflict and feel more attached to the issues. Scholars have referred to role-playing as an unstructured drama. While the instructor provides the setting and the characters, students/participants must decide their characters' arguments and the direction of the discussion.

To participate fully in a role-play, students need to research the case and their role to make informative decisions that represent their character's perspective. The amount of research needed depends on the characters or roles they are playing and the complexity of the case study. In some instances, the case used can draw on students' personal experience – for example, a discussion of research ethics – and require almost no preparation, whereas, in others, they might need from a few days to a week to prepare well. Learning outcomes for role-play discussions are further supported through pre-discussion and post-discussion assignments, including essay-based questions or concept maps, either individual-level or group-based (Hingle et al., 2021; Johri & Hingle, 2022).

Role-plays have been used extensively in professional and applied disciplines (Rao & Stupans, 2012; Shaw, 2004), including medicine (Lane & Rollnick, 2007; Nestel & Tierney, 2007), mental health fields (Schwitzer et al., 2001), teacher training (Kilgour et al., 2015), leadership roles (Brown, 1994; Shapiro & Leopold, 2012), research ethics (Brummel et al., 2010), and engineering (e.g., Herkert, 1997; Hingle et al., 2022). Role-play scenarios are a popular engineering ethics intervention because they allow students to engage in empathy and perspective-building exercises in the classroom. This participation is fundamental to early-career engineers, who Loui (2005) argues are likely to face ethical dilemmas that include conflicting viewpoints during everyday work. It is essential to recognize that "students come to a course with various backgrounds and developmental stages, and different students internalize different ideas" (Loui, 2005, p. 383). As there is certainty that students will face these ethical dilemmas but have different readiness levels, researchers argue that the focus is on readying students to empathize. Hess and Fila (2016) argue that empathy and perspective-taking exercises should be incorporated within engineering curricula, which are fundamental to holistic engagement efforts. As Doorn and Kroesen (2013) describe, role-play scenarios serve as an instrument to guide students to engage with, debate, and evaluate decisions from the perspective of different roles. Doorn and Kroesen also acknowledge that by making students aware of other perspectives, they better understand pressures and influences that would otherwise have been hidden.

As with most pedagogical interventions, role-plays have their limitations. They are resource-intensive in developing adequate role-play case studies, role descriptions, and implementation. They require designing a prompt script to facilitate the discussion. They also need a relatively high amount of time within a class or lecture to be effective and require the instructor or a trained person to lead the discussion, at least initially. Finally, assessing learning requires more time as most of the information or data is qualitative. Despite these limitations, the benefits of using role-plays make them an effective case-based instruction method.

Dealing with emotive content

A further question of practical significance – one already touched upon in the previous section – relates to how to deal with the emotive content that case-based ethics instruction contains. Chapter 4 discusses the relationship between emotion and reason in engineering ethics education, and Chapter 10 probes the psychological aspects of this. The focus in this section, therefore, is on the emotionality of ethics cases. As mentioned above, such content may distress a student to the point that it hinders the achievement of learning goals (Watts, Medeiros, et al., 2017). However, some evidence (e.g., Thiel et al., 2013) indicates that emotionally rich case studies can support learning and even learning transfer. Higgs et al. (2020) argue that the traditional striving for emotionless ethical evaluation can "result in cognitive processes that could lead to less ethical decisions" (Higgs et al., 2020, p. 53).

Noting that including a low/modest level of emotional content seems to improve students' learning, Kotluk and Tormey (2022) have found that including mildly emotive content does not introduce biases to the moral reasoning process. Their findings invite instructors to actively introduce some emotional content to improve learning. Kotluk and Tormey also note that engineering ethics cases – even those seen as emotionally neutral – already include emotional content. Kotluk and Tormey asked students to quantify the degree to which they thought the leading actor of a case would have experienced emotions. Students identified moderate levels of guilt, embarrassment, compassion, and anger, even when the emotional content (e.g., actual feelings or significant adverse outcomes) was not explicitly mentioned in the case description, revealing the idea of an emotionless ethics case and emotionless deliberation to be a myth.

Other researchers have also contributed to advancing the idea that processing emotions is integral to assessing and working on ethics cases. For instance, Justin Hess and others (2016, 2017, 2019) documented statistically significant increases in empathic perspective-taking and emotion-regulation skills when instructing students within the Scaffolded, Interactive, and Reflective Analysis (SIRA) framework (Kisselburgh et al., 2016), an elaborate framework for case-based ethics instruction that integrates high levels of student interaction and e-learning platform–supported reflective exercises. Watts et al. (2017) even found that explicit consideration for processing emotions more effectively impacted learning than cognitive or values-oriented processes.

However, empathy can also limit prosocial responses towards dissimilar people, for example, regarding social status, place, or time. So-called 'empathetic overarousal' can also be distressing to the point that a person's focus wanders inwards, potentially resulting in moral or social disengagement, for example, in ecological matters (Stanley et al., 2021). A combination of teaching learners to take perspective, empathize with others, and regulate emotion can, therefore, be conceived as an additional important goal of ethics education.

These insights ask instructors to consider emotive content carefully in their case-based engineering ethics education designs. Role-play-based instruction is well suited for constructively addressing this. However, especially in CBL settings with external stakeholders, emotive content may appear in even more multifaceted ways. Apart from the typical subjects affected, students could also empathize with the external stakeholders (e.g., companies) when getting to understand economic, and perhaps even personal, struggles that kept them from adequately addressing ethical challenges. Such sentiments could also develop into anger. Clearly, such emotions need proper management by the instructors in order to detect, address, and mitigate unwanted effects. This adds to the complexity of this type of case-based instruction, which indeed requires further research.

Repositories of engineering ethics case studies

Having discussed various theoretical aspects of employing case studies for engineering ethics education, challenge-based learning, role-playing, and making emotional processes explicit as three highly practical approaches, we follow up with a short but lightly annotated list of repositories and resources for getting started. The list is not exhaustive and does not present a consistent resource from which one could simply mix and match cases to build one's syllabus. However, we perceive the value of referencing these repositories as showcasing the breadth of styles and content to which case-based instruction could adhere.

- The 'Engineering Ethics Toolkit' of the Engineering Professors Council (https://epc.ac.uk/resources/toolkit/ethics-toolkit/ethics-toolkit-case-studies/): The case studies section features a range of topics to which respective disciplines, overarching ethical issues, and student levels are assigned. Disciplines range from mechanical, electrical, energy, and nuclear to civil engineering, while issues are listed as sustainability, honesty, integrity, corporate social responsibility, accountability, justice, and public health, among others. Each case is accompanied by learning and teaching notes, further resources, and academic and non-academic literature. Cases are presented as brief dilemmas, split into parts, and are accompanied by suggested questions and activities. An enhancement is available for some cases, increasing the respective case's complexity. All work is licensed under Creative Commons Attribution-ShareAlike 4.0 International Licenses.

- The 4TU Centre for Ethics and Technology's 'Engineering Practise Cases' hosted on the 'edusources' platform (https://edusources.nl/en/collections/bc0c75e4-0bcd-4e26-ad8e-78498643d868): Currently 20 cases are available for download, with topics ranging from autonomous driving to offshore drilling and ethical issues ranging from transparency and professional responsibility to, for example, bias. The materials range from longer cases with elaborate descriptions and task suggestions to quite brief vignettes.
- The Online Ethics Center for Engineering and Science, particularly its 'Resources' subsection on 'Collections' (https://onlineethics.org/): This diverse list of resources includes collections from specific problems in research ethics to biographies of noteworthy role models in engineering and science. This is not an ethics case repository per se, but rather a meta-repository linking to further, mostly specialized, repositories.
- The Markkula Center for Applied Ethics at Santa Clara University's 'Ethics Cases' repository (https://www.scu.edu/ethics/ethics-resources/ethics-cases/): The repository contains cases from various disciplines beyond technology often associated with innovation. Case descriptions are brief, linking a few articles on the referenced real-world event. Discussion questions usually follow the case.
- The Princeton 'Dialogues on AI and Ethics Case Studies' (https://aiethics.princeton.edu/case-studies/): The Princeton 'Dialogues on AI and Ethics Case Studies' adhere to five guiding principles (empirical foundations, broad accessibility, interactivity, presenting/requiring multiple viewpoints, and depth). Six case studies are available with comparatively elaborate dossiers, successive discussion questions, explicit potential ethical objections, and a separate section on reflection and discussion questions aligned with ethical principles/issues.
- The Mason Technology Ethics 'Role-Play Case' repository (http://www.ist.gmu.edu/eecl/techethics/rps-cases.html): Four cases are compiled and tailored specifically to let students engage with the matter in a role-playing style.
- The 'Role-Playing Exercises' from the Science Education Resource Center at Carleton College (https://serc.carleton.edu/introgeo/roleplaying/index.html): This webpage provides role-playing exercises, information for instructors, and a collection of currently 28 role-playing scenarios. The resources and role-playing exercises are all centered around topics related to the Earth system, ranging from the atmosphere to oceans and Earth history. Roles, suggested ways of preparing the role-playing, and assessments are tailored to the specific cases.

Conclusion and further discussion

We have provided an introduction to teaching engineering ethics by using case studies, employing a wide-arching interpretation of the concept of a case. From a Neo-Kohlbergian point of view, the main point lies in guiding students through enhancing their moral sensitivity (as their ability to recognize ethically salient conditions), their moral judgment (as the ability to reason morally based on relevant values or principles), their moral motivation (as the ability to prioritize moral over other values), and their moral character (as the ability to follow through). Cases are employed via tasking students to make decisions in real or hypothetical situations, researching background information, and engaging with various perspectives. At this point, a different emphasis on learning goals is typically determined by choosing between highly complex and indeterminate case studies or simpler vignettes with less veracity. Beyond the many ways in which case-based instruction contributes to the above learning goals, utilizing cases from domains students are likely to experience in their later professional work will also help in tackling the problem of transfer, that is,the difficulty of endowing students with the ability to apply knowledge from the classroom to the outside world.

We have discussed challenge-based learning as a promising candidate for achieving this by bringing in external stakeholders who offer their technological or business approaches as real-world cases to be analyzed and discussed. We have presented a range of aspects to consider when tailoring a challenge-based-learning-flavored pedagogy towards the Neo-Kohlbergian learning goals. Additionally, we have presented the basic modes and advantages of an alternative way to introduce cases by using role-playing. Such exercises in perspective-taking – among other things – require learners to empathize. Accordingly, as a third practically important aspect to consider when using case-based ethics instruction, we have constructively elaborated on the emotional content of cases. Making the emotional side of ethics explicit means contributing to learners' perspective-taking and emotional regulation skills. We have also provided a brief and lightly annotated list of current ethics case repositories to get instructors started with designing or refining their own case-based engineering ethics courses.

We feel that the issue is highly pressing, as engineering ethics education must focus on substantiating the relevance of ethics in engineering practice in what is, perhaps, a period of unprecedented urgency. Engineers should take their power to bring ethical quandaries to others and perhaps even the general public's serious attention. Similarly, engineers should be aware of and carefully handle their power to make ethically salient and relevant decisions. Engineering students need a chance to learn and practice these skills. Choosing highly relatable topics or even letting students have a choice within a selection of case-study subjects represent worthy first steps.

Portrayed from a different angle, we purport that it is not the aim to turn engineering students into ethicists. On the other hand, it is of utmost importance to demand that engineers be willing and able to communicate about moral hazards and the values that are implicitly or explicitly pursued when working towards some particular remedy, that is, innovation. Correspondingly, engineering ethics education must work towards endowing the skills that make developers able communicators and reflective personalities about their own ethical agency, such that their intent and methods can be openly debated and – possibly – challenged or supported. Case-based engineering ethics education appears to be highly useful, as it hones attention to the details that can influence ethical assessments and cultivate a methodical – and perhaps, ideally, participative – approach to analyzing a particular, and potentially highly complex, situation.

Acknowledgments

We would like to acknowledge the valuable discussions and the peer-review process managed by the editors. Johri's research was partly supported by U.S. NSF Awards # 1937950, 2335636, 1954556 and USDA/NIFA Award # 2021-67021-35329. Any opinions, findings, conclusions, or recommendations expressed in this material are those of the authors and do not necessarily reflect the views of the funding agencies.

References

Antes, A. L., Murphy, S. T., Waples, E. P., Mumford, M. D., Brown, R. P., Connelly, S., & Devenport, L. D. (2009). A meta-analysis of ethics instruction effectiveness in the sciences. *Ethics & Behavior, 19*(5), 379–402. https://doi.org/10.1080/10508420903035380

Bebeau, M. J., Rest, J. R., & Narvaez, D. (1999). Beyond the promise: A perspective on research in moral education. *Educational Researcher, 28*(4), 18–26.

Bombaerts, G. (2020). Upscaling Challenge-Based Learning for humanities in engineering education. *SEFI 48th Annual Conference*, 11.

Bransford, J. D., Brown, A. L., & Cocking, R. R. (2000). *How people learn: Brain, mind, experience, and school: Expanded edition.* National Academies Press. https://doi.org/10.17226/9853

Brown, K. M. (1994). Using role play to integrate ethics into the business curriculum a financial management example. *Journal of Business Ethics, 13*(2), 105–110. https://doi.org/10.1007/BF00881579

Brummel, B. J., Gunsalus, C. K., Anderson, K. L., & Loui, M. C. (2010). Development of role-play scenarios for teaching responsible conduct of research. *Science and Engineering Ethics, 16*(3), 573–589. https://doi .org/10.1007/s11948-010-9221-7

Davis, M. (1997). Developing and using cases to teach practical ethics. *Teaching Philosophy, 20*(4), 353–385. https://doi.org/10.5840/teachphil199720445

Darling-Hammond, L., & Baratz-Snowden, J. (2007). A good teacher in every classroom: Preparing the highly qualified teachers our children deserve. *Educational Horizons, 85*(2), 111–132.

DeNeve, K. M., & Heppner, M. J. (1997). Role play simulations: The assessment of an active learning technique and comparisons with traditional lectures. *Innovative Higher Education, 21*(3), 231–246. https://doi .org/10.1007/BF01243718

Doorn, N., & Kroesen, J. O. (2013). Using and developing role plays in teaching aimed at preparing for social responsibility. *Science and Engineering Ethics, 19*(4), 1513–1527. https://doi.org/10.1007/s11948 -011-9335-6

Gilligan, C. (1993). *In a different voice: Psychological theory and women's development.* Harvard University Press.

Goodlad, J., Klein, M., & Tye, K. (1979). The domains of curriculum and their study. In J. Goodlad (Ed.), *Curriculum inquiry* (pp. 43–76). McGraw-Hill.

Haidt, J. (2001). The emotional dog and its rational tail: A social intuitionist approach to moral judgment. *Psychological Review, 108*(4), 814–834. https://doi.org/10.1037/0033-295X.108.4.814

Harris, C. E., Davis, M., Pritchard, M. S., & Rabins, M. J. (1996). Engineering ethics: What? Why? How? And when? *Journal of Engineering Education, 85*(2), 93–96. https://doi.org/10.1002/j.2168-9830.1996 .tb00216.x

Herkert, J. R. (1997). Collaborative learning in engineering ethics. *Science and Engineering Ethics, 3*(4), 447–462. https://doi.org/10.1007/s11948-997-0047-x

Hertel, J. P., & Millis, B. (2011). *Using simulations to promote learning in higher education.* Stylus Publishing.

Herzog, C., Breyer, S., Leinweber, N.-A., Preiß, R., Sonar, A., & Bombaerts, G. (2022a). *Everything you want to know and never dared to ask – A practical approach to employing challenge-based learning in engineering ethics.* SEFI Annual Conference, Barcelona, Spain.

Herzog, C., Leinweber, N.-A., Engelhard, S., & Engelhard, L. (2022b). *Autonomous Ferries and Cargo ships: Discovering ethical issues via a Challenge-Based Learning approach in higher education.* IEEE International Symposium on Technology and Society, Hong Kong.

Hess, J., & Fila, N. (2016). The development and growth of empathy among engineering students. *2016 ASEE Annual Conference & Exposition Proceedings, 26120.* https://doi.org/10.18260/p.26120

Hess, J. L., Beever, J., Zoltowski, C. B., Kisselburgh, L., & Brightman, A. O. (2019). Enhancing engineering students' ethical reasoning: Situating reflexive principlism within the SIRA framework. *Journal of Engineering Education, 108*(1), 82–102. https://doi.org/10.1002/jee.20249

Hess, J. L., & Fore, G. (2017). A systematic literature review of US engineering ethics interventions. *Science and Engineering Ethics, 24*, 551–583. https://doi.org/10.1007/s11948-017-9910-6

Hess, J. L., Strobel, J., & Brightman, A. O. (2017). The development of empathic perspective-taking in an engineering ethics course. *Journal of Engineering Education, 106*(4), 534–563. https://doi.org/10.1002/ jee.20175

Heyward, P. (2010). Emotional engagement through drama: Strategies to assist learning through role play. *International Journal of Teaching and Learning in Higher Education, 22*(2), 197–205. http://www.isetl .org/ijtlhe/

Higgs, C., McIntosh, T., Connelly, S., & Mumford, M. (2020). Self-focused emotions and ethical decision-making: Comparing the effects of regulated and unregulated guilt, shame, and embarrassment. *Science and Engineering Ethics, 26*(1), 27–63. https://doi.org/10.1007/s11948-018-00082-z

Hingle, A., Johri, A., & Brozina, C. (2022). *Instructing first-year engineering students on the ethics of algorithms through a role-play*. 2022 ASEE Annual Conference & Exposition, Minneapolis, MN. https://peer.asee.org/41199

Hingle, A., Rangwala, H., Johri, A., & Monea, A. (2021). Using role-plays to improve ethical understanding of algorithms among computing students. *2021 IEEE Frontiers in Education Conference*, 1–7. https://doi.org/10.1109/FIE49875.2021.9637418

Jarvis, P. (2012). *Adult learning in the social context*. Routledge. https://doi.org/10.4324/9780203802724

Johri, A., & Hingle, A. (2022). Learning to link micro, meso, and macro ethical concerns through role-play discussions. *IEEE Frontiers in Education Conference*, 1–8. https://doi.org/10.1109/FIE56618.2022.9962560

Kilgour, P. W., Reynaud, D., Northcote, M. T., & Shields, M. (2015). Role-playing as a tool to facilitate learning, self reflection and social awareness in teacher education. *International Journal of Innovative Interdisciplinary Research, 2*(4), 8–20.

Kisselburgh, L., Hess, J., Zoltowski, C., Beever, J., & Brightman, A. (2016). Assessing a scaffolded, interactive, and reflective analysis framework for developing ethical reasoning in engineering students. *2016 ASEE Annual Conference & Exposition Proceedings*, 26288. https://doi.org/10.18260/p.26288

Kohlberg, L. (1974). Education, moral development and faith. *Journal of Moral Education, 4*(1), 5–16. https://doi.org/10.1080/0305724740040102

Kotluk, N., & Tormey, R. (2022). Emotional empathy and engineering students' moral reasoning. *SEFI 50th Annual conference of The European Society for Engineering Education. "Towards a new future in engineering education, new scenarios that european alliances of tech universities open up"*. Barcelona: Universitat Politècnica de Catalunya, 2022, pp. 458–467. https://doi.org/10.5821/conference-9788412322262.1124

Lane, C., & Rollnick, S. (2007). The use of simulated patients and role-play in communication skills training: A review of the literature to August 2005. *Patient Education and Counseling, 67*(1–2), 13–20. https://doi.org/10.1016/j.pec.2007.02.011

Lawlor, R. (2021). Teaching engineering ethics: A dissenting voice. *Australasian Journal of Engineering Education, 26*(1), 38–46. https://doi.org/10.1080/22054952.2021.1925404

Loui, M. C. (2005). Ethics and the -development of professional identities of engineering students. *Journal of Engineering Education, 94*(4), 383–390. https://doi.org/10.1002/j.2168-9830.2005.tb00866.x

Loui, M. C. (2009). What can students learn in an extended role-play simulation on technology and society? Bulletin of science, *Technology & Society, 29*(1), 37–47. https://doi.org/10.1177/0270467608328710

Martin, D. A., Conlon, E., & Bowe, B. (2019). The role of role-play in student awareness of the social dimension of the engineering profession. *European Journal of Engineering Education, 44*(6), 882–905. https://doi.org/10.1080/03043797.2019.1624691

Martin, D. A., Conlon, E., & Bowe, B. (2021a). Using case studies in engineering ethics education: The case for immersive scenarios through stakeholder engagement and real life data. *Australasian Journal of Engineering Education, 26*(1), 47–63. https://doi.org/10.1080/22054952.2021.1914297

Martin, D. A., Conlon, E., & Bowe, B. (2021b). A multi-level review of engineering ethics education: Towards a socio-technical orientation of engineering education for ethics. *Science and Engineering Ethics, 27*(5), 60. https://doi.org/10.1007/s11948-021-00333-6

Martin, D., Herzog, C., Papageorgiou, K., & Bombaerts, G. (2022). Three European experiences of co-creating ethical solutions to real-world problems through Challenge Based Learning. In E. Vilalta-Perdomo, J. Membrillo-Hernández, R. Michel-Villarreal, G. Lakshmi, & M. Martínez-Acosta (Eds.), *Emerald Handbook on Challenge-based Learning* (pp. 251–279). Emerald Publishing Limited. https://doi.org/10.1108/978-1-80117-490-920221011

McCarthy, J. P., & Anderson, L. (1999). Active learning techniques versus traditional teaching styles: Two experiments from history and political science. *Innovative Higher Education, 24*(4), 279–294. https://doi.org/10.1023/B:IHIE.0000047415.48495.05

Merseth, K. K. (1994). Cases, case methods, and the professional development of educators (No. ED401272; ERIC Digests). *ERIC Clearinghouse on Teaching and Teacher Education*. www.eric.ed.gov

Narvaez, D., & Rest, J. (1995). The four components of acting morally. Moral behavior and moral development: An introduction. In W. Kurtines & J. Gewirtz (Eds.), *Moral development: An introduction* (pp. 385–400). Allyn & Bacon.

Nestel, D., & Tierney, T. (2007). Role-play for medical students learning about communication: Guidelines for maximising benefits. *BMC Medical Education, 7*(1), 3. https://doi.org/10.1186/1472-6920-7-3

Nichols, M. H., & Cator, K. (2008). *Challenge Based Learning White paper*. Apple.

Noddings, N. (1988). An ethic of caring and its implications for instructional arrangements. *American Journal of Education, 96*(2), 215–230.

Poling, D. A., & Hupp, J. M. (2009). Active learning through role playing: Virtual babies in a child development course. *College Teaching, 57*(4), 221–228. https://doi.org/10.1080/87567550903218703

Pusateri, T., Halonen, J. S., Hill, B., & McCarthy, M. (Eds.). (2009). *The assessment cyberguide for learning goals and outcomes*. American Psychological Association.

Rao, D., & Stupans, I. (2012). Exploring the potential of role play in higher education: Development of a typology and teacher guidelines. *Innovations in Education and Teaching International, 49*(4), 427–436. https://doi.org/10.1080/14703297.2012.728879

Rest, J. R., Narvaez, D., Thoma, S. J., & Bebeau, M. J. (1999). DIT2: Devising and testing a revised instrument of moral judgment. *Journal of Educational Psychology, 91*(4), 644–659.

Rottmann, C., & Reeve, D. (2020). Equity as rebar: Bridging the micro/macro divide in engineering ethics education. *Canadian Journal of Science, Mathematics and Technology Education, 20*(1), 146–165. https://doi.org/10.1007/s42330-019-00073-7

Schwitzer, A. M., Gonzalez, T., & Curl, J. (2001). Preparing students for professional roles by simulating work settings in counselor education courses. *Counselor Education and Supervision, 40*(4), 308–319. https://doi.org/10.1002/j.1556-6978.2001.tb01262.x

Shapiro, S., & Leopold, L. (2012). A critical role for role-playing pedagogy. *TESL Canada Journal, 29*(2), 120. https://doi.org/10.18806/tesl.v29i2.1104

Shaw, C. M. (2004). Using role-play scenarios in the IR classroom: An examination of exercises on peacekeeping operations and foreign policy decision making. *International Studies Perspectives, 5*(1), 1–22. https://doi.org/10.1111/j.1528-3577.2004.00151.x

Stanley, S. K., Hogg, T. L., Leviston, Z., & Walker, I. (2021). From anger to action: Differential impacts of eco-anxiety, eco-depression, and eco-anger on climate action and wellbeing. *The Journal of Climate Change and Health, 1*, 100003. https://doi.org/10.1016/j.joclim.2021.100003

Thiel, C. E., Connelly, S., Harkrider, L., Devenport, L. D., Bagdasarov, Z., Johnson, J. F., & Mumford, M. D. (2013). Case-based knowledge and ethics education: Improving learning and transfer through emotionally rich cases. *Science and Engineering Ethics, 19*(1), 265–286. https://doi.org/10.1007/s11948-011-9318-7

Tormey, R., Isaac, S., Hardebolle, C., & Duc, I. L. (2021). *Facilitating experiential learning in higher education: Teaching and supervising in labs, fieldwork, studios, and projects* (1st ed.). Routledge. https://doi.org/10.4324/9781003107606

Van De Sande, C. C., & Greeno, J. G. (2012). Achieving alignment of perspectival framings in problem-solving discourse. *Journal of the Learning Sciences, 21*(1), 1–44. https://doi.org/10.1080/10508406.2011.639000

Waples, E. P., Antes, A. L., Murphy, S. T., Connelly, S., & Mumford, M. D. (2009). A meta-analytic investigation of business ethics instruction. *Journal of Business Ethics, 87*(1), 133–151. https://doi.org/10.1007/s10551-008-9875-0

Watts, L. L., Medeiros, K. E., Mulhearn, T. J., Steele, L. M., Connelly, S., & Mumford, M. D. (2017). Are ethics training programs improving? A meta-analytic review of past and present ethics instruction in the sciences. *Ethics & Behavior, 27*(5), 351–384. https://doi.org/10.1080/10508422.2016.1182025

Watts, L. L., Todd, E. M., Mulhearn, T. J., Medeiros, K. E., Mumford, M. D., & Connelly, S. (2017). Qualitative evaluation methods in ethics education: A systematic review and analysis of best practices. *Accountability in Research, 24*(4), 225–242. https://doi.org/10.1080/08989621.2016.1274975

21

EMBEDDED ETHICS IN PROBLEM DESIGN

The case of problem-based learning in engineering and science

Henrik Worm Routhe, Jette Egelund Holgaard, and Anette Kolmos

Introduction

The importance of ethics in engineering has emerged in codes of conduct at national and international levels, including codes of the National Society of Professional Engineers (NSPE), the European Federation of National Engineering Associations (FEANI), the Institute for Electrical and Electronic Engineering (IEEE), and the Accreditation Board for Engineering and Technology (ABET). When such codes of conduct and standards are incorporated into engineering education practices, they inevitably differ in emphasis. Thus, there is a need for a collective and global vision of ethics in engineering education and research (Martin et al., 2023). Moreover, the ethical codes used by organizations like the NSPE are often expressed negatively, using terms such as 'not' and 'only' and might be referred to as 'preventive ethics' (Harris, 2008, pp. 153–154).

Osbeck et al. (2018) have highlighted the risk of neglecting the contextual, situational, and knowledge-related aspects of ethical competence. The contextual aspects relate to what Herkert (2001) has termed macro-ethics, which apply "to both the collective social responsibility of the engineering profession and to societal decisions about technology" (Herkert, 2001, p. 404). Conversely, micro-ethics relate to individuals' ethical decisions and the engineering profession's internal relationships (Herkert, 2001). Combining these two definitions provides a comprehensive approach to ethics in engineering education.

Bringing such a comprehensive understanding of ethics into teaching will increase the complexity of integrating ethics into engineering education. Adding to this complexity is the constantly changing nature of the engineering field's ethical concerns. Although there has been an increasing focus on including context in engineering education, this is not a typical consideration in traditional engineering ethics (Barry & Herkert, 2014). According to Elliott and June (2018, p. 32), an open question is whether ethics education meets students' needs considering the ethical dilemmas of a changing world. Meeting these needs requires dynamic educational models that embrace contextual and situational requirements, as well as a flexible curriculum that allows students to reflect on and react to current societal challenges; for example, exploring the use and prospects of emergent technologies like artificial intelligence (AI) or the grand challenge of fostering more sustainable societies.

DOI: 10.4324/9781003464259-26

There are several dimensions of problem- and project-based learning (PBL). One dimension considers how much influence students have on their own learning processes. In a teacher-driven PBL environment, the problem is designed for students, whereas in a student-driven environment, students identify the problems their work will address within a given framework (Kolmos & de Graaff, 2014; Kolmos et al., 2009). According to Barry and Herkert (2014), PBL can be viewed as an alternative to case-based instruction, the most common pedagogical method in engineering education. Yet, these two approaches can be mutually supportive rather than contradictory. Indeed, case studies can help to provide a nuanced understanding of what a problem is before addressing a problem as a part of PBL (Børsen et al., 2021). Within its various implementations, PBL can include a student- or learning-centered approach, where problems are ill-defined; lectures may support but do not determine student projects (Kolmos & de Graaff, 2014). Together with the intended learning outcomes of the curriculum, the perceived relevance for society also has implications for the types of problems addressed in PBL (Habbal et al., 2024).

Some of the barriers in engineering education include what Newberry (2004) refers to as the 'technical gravity' of the curriculum. In contrast to such technical gravity, a curriculum can extend the technical aspects of education by integrating contextual aspects and issues of student responsibility. However, an open question involves how this integration occurs and becomes visible through the formal curriculum.

In this chapter, we seek inspiration from a PBL environment to characterize the intended learning outcomes of formal curricula. We assume that ethics in a PBL approach is not necessarily explicitly stated but indirectly embedded in the problem design and the problem types that students work with. We point at possible enablers of ethics in PBL, and with this outset, the objective is to exemplify ethics embedded in a PBL curriculum.

Positionality

All three authors have a background at Aalborg University, though following different trajectories, from a background in social science to engineering and engineering practice. However, as engineering education researchers we all share a common interest in adapting engineering curricula to meet the challenges of the twenty-first century, including interdisciplinary competences and understanding the variation related to different contextual situations. As researchers at Aalborg University, a university established in 1974 with a problem-oriented and project-organized approach to teaching and learning, the scene for researching forms an excellent base for interdisciplinarity in combination with engineering education. From the beginning in 1974, the university's pedagogy was based on German critical theory developed in the late 1960s and 1970s. During this period, several reformist universities were established with the idea that universities should develop a socially oriented perspective and integrate societal problems in their curricula. This created entirely new challenges for academia, as it represented a shift from a theoretical academic approach to a more societal approach in terms of market orientation and critical societal discourse. These early developments toward a new university culture would go on to serve as examples for the development of many other universities, in terms of the emphasis on competencies and skills, from the 1990s onward. In the 2010s, the university began working to integrate the United Nations (UN) Sustainable Development Goals (SDGs) in students' learning. The case of Aalborg University is special, as the critical pedagogy was grounded and practiced before integrating ethics into engineering education became mainstream.

As researchers we work from a pragmatic view following a problem-based approach integrating the theories and methods needed to address the specific problem at hand and to point to appropriate solutions. In other words, we carry out problem-based research. In this chapter, we have set out to address the problem of ethics being implicit in a PBL curriculum.

Ethics from a PBL perspective

Beyond the explicit mention of ethics in engineering education curricula, the theoretical framework presented in this section attempts to conceptualize potential enablers of ethics. The specificity of PBL is that the problem is the *point of departure*, whereas the *problem design* and the *problem type* are the primary focus of the following. This is not to say that ethical considerations do not happen in the problem-solving phase, but problem-solving is seen as a continuous interaction and contribution to the problem design, which is iteratively altered through the PBL process.

Problem design in a PBL environment

De Graaf and Kolmos (2003, 2007) have presented three approaches embedded in a PBL framework: the learning, the content, and the social. The learning approach emphasizes that learning *starts from* and is *organized around* problems, which are exemplary for societal practices and change. The content approach concerns the disciplinary, interdisciplinary, and exemplary content, meaning that the problem can call for the combination of different knowledge combinations and is exemplary for the intended learning outcomes. The social approach embraces team-based, participant-directed aspects and considers learning a social act. Concerning participant-directed learning, it is important to note, though, that a PBL curriculum design must allow some space and freedom for the students to have the possibility to identify and analyze problems (Habbal et al., 2024). Thus, the problem is grounded in a careful problem design process, which ensures that the problem and the way that the students address the problem matches the conditions of the above-mentioned PBL approaches.

In this regard, Hung (2006) has developed the '3C3R model,' a comprehensive model of problem design components for faculty to use. The three core components of this model include content, context, and connection (the 'three Cs'). Connections "interweave (1) the concepts and information within the conceptual framework, and (2) content into context" (Hung, 2006, p. 61). The processing components include researching, reasoning, and reflecting (the 'three Rs'), which concern the learners' processes and problem-solving skills. Based on this framework, Hung (2009) presents a nine-step PBL problem design model intended to "help instructional designers and educators use the 3C3R PBL problem design model" (Hung, 2009, p. 123). The nine steps are as follows (Hung, 2009, p. 123):

1. Set goals and objectives
2. Conduct content/task analysis
3. Analyze context specification
4. Select/generate PBL problem
5. Conduct PBL affordance analysis
6. Conduct correspondence analysis
7. Conduct calibration processes
8. Construct reflection component
9. Examine intersupporting relationships of 3C3R components

In his later work, Hung (2019) highlighted the affective aspects of problem design and further developed the second generation of the 3C3R PBL problem design model. This added a third class of components related to the affective and social aspects of learning (Hung, 2019); namely, problem difficulty, teamwork, and affect. Hung's comprehensive work – based on the interaction of core, processual, affective, and social components – seeks to inspire educational designers to design problems.

Holgaard et al. (2017) have also presented a framework to support students in designing their own problems as part of the PBL process. Based on Hung (2006) and others, five steps of problem design were defined (Holgaard et al., 2017, p. 1077):

1. Relating to a theme to clarify boundaries (e.g., provided by the intended learning outcomes)
2. Mapping the problem field to screen for opportunities, challenges, and unknowns
3. Narrowing down the problems to select one problem for further analysis
4. Analyzing the problem and contextualizing to pinpoint specific motivations for action
5. Formulating the problem to create the bridge between the problem analysis and problem-solving process

It should be noted that the analysis of the problem context is emphasized in both process models described above for the design of problems. In this regard, Holgaard et al. (2017) further relate problem analysis to other types of analyses, including analysis within the fields of sustainability and ethics, stakeholder analysis, actor analysis, and constructive technology assessment. Furthermore, Holgaard et al. (2017) conclude, based on an empirical study of students' experiences as problem designers, that a conceptual model for students' problem design activities should draw attention to "the process of moving from a broad theme to an initiating problem and starting up a problem analysis" (p. 1083).

Regardless of the steps taken in the problem design process, and whether the students 'own and direct' the problem design process, the type of problem that emerges through the design has implications for the complexity of the problem design process.

Problem types and their implications for problem design

Holgaard et al. (2017) describe problem design as an exploratory process that considers the existing situation as well as arguments and possibilities for change. From this perspective, a problem can be understood as a discrepancy between what is and what could be. Jonassen (2011) characterizes problems based on the extent to which they are structured, complex, contextual, dynamic, and domain-specific. In the following section, we elaborate on these problem types as dimensions, whereas ill-structuredness and dynamicity are merged under the term 'integrated aspects of complexity.'

The first dimension represents the simple versus the complex. Complex problems are ill-structured. Ill-structured problems provide multiple potential solutions and solution paths and unknown problem elements. An example of an ill-structured problem is the problem of self-medication, in which considerable effort must be given to the problem analysis to outline user needs. The problem design process is rather simple for structured problems, whereas it is multi-directional and time-consuming for ill-structured problems. As a result, numerous assumptions and limitations are typically involved in the problem design process of ill-structured problems to reach the problem-solving stage.

Complexity also relates to the level of emergence and number of relationships embedded in a problem. As noted by Kurtz and Snowden (2003) in their view of complex problems:

> This is the domain of complexity theory, which studies how patterns emerge through the interaction of many agents. There are cause and effect relationships between the agents, but both the number of agents and the number of relationships defy categorization or analytic techniques. Emergent patterns can be perceived but not predicted.
>
> *(p. 469)*

From this, it follows that a complex problem must be addressed dynamically through emergent practices if students are to perceive emergent patterns. It is not sufficient, or even possible, to analyze existing practices from a distance. Indeed, practices must include lived experiences interacting with the field, and engaging with knowledge providers and stakeholders. In other words, complex problems require students to be *in* the problem context, which adds an enactment dimension to the problem analysis as part of the problem design process. Complex problems are 'wicked,' where 'wicked' refers to a state where it is simply not clear what the problem is, and likewise even less clear how to effectively intervene (Rittel & Webber, 1973). For example, a complex design problem for potential assistive technology for elder care might require students to experience daily practices (of, e.g., being at a nursing home) to enable them to understand how the technology can be of assistance. Another example of a complex problem is an emergent and unexpected biodiversity loss, where it is not clear why this is happening or how to intervene most effectively.

On the other hand, simple problems are structured. They have prescriptive and known elements and a fixed or expected solution. A typical textbook math problem is an example, as the problem is clearly stated and includes all necessary information that must be employed in clear problem-solving procedures to arrive at a fixed solution. Whereas structuredness describes problems in terms of the predictability of the problem space, complexity highlights interrelatedness in the problem space (Hung, 2016). Simple problems do not exhibit interrelationships between many elements, and "the objectivity is such that any reasonable person would accept the constraints of best practice" (Kurtz & Snowden, 2003, p. 468).

The second dimension, 'from text to context,' relates to the situation in which the problem is embedded. This dimension relates to the relative nature between *text* and *context*. In an educational curriculum, the 'text' relates to the delimited problem of what the specific engineering discipline can contribute to. In contrast, the context relates to what must be addressed in order to qualify the use of disciplinary knowledge. In other words, what is used as the *text* for a student in architecture and design might instead be the *context* for a student in civil engineering – and vice versa.

In engineering and science, the concept of context is also a way of acknowledging that technological artifacts are socially constructed and, therefore, closely interrelated with societal problems and social groups, as noted by Bijker et al. (1989). Societal problems also relate to the analysis of known and potential consequences, for example, through a constructive technology assessment (CTA), as elaborated by Rip et al. (1995). The problem analysis can be further expanded based on theories in Science and Technology Studies (STS), but this also increases the complexity and methodological span of a problem analysis.

The third and final dimension concerns domain specificity, which concerns problem-solving strategies specific to particular domains (Jonassen, 2011). To underscore the increasing interdisciplinary aspects of problem-solving, we position problem-solving strategies that span multiple domains (interdomains) aligned with complex and contextual problems. For example, architecture and design students and civil engineering students may work together to solve a problem by combining problem-solving strategies within their respective domains. The interdomain approach, reflecting an academic context of interdisciplinarity, aims to interrelate different types of epistemologies in the problem-solving process. Thus the problem design is open to much broader problem formulations. Wenger (1998) has used the notion of 'communities of practice' (CoP) as a central aspect of social learning theory, presenting boundary-crossing as an essential element to explain the interactions between different CoPs.

To work in the inter-domain sphere, students – and educators – must remain open-minded and do considerable boundary work. It is not enough to 'borrow' from other knowledge fields to design a problem and then narrow it down to accommodate problem-solving processes within disciplinary bounds. Instead, the problem-solving process itself must become interdisciplinary. In an educational context, an interdisciplinary problem expands to encompass a combination of disciplines before the problem can be solved meaningfully. For example, in the case of self-medication above, the context of study might end up with limited insight into the psychological aspects of self-medication. In a disciplinary project, students will view this as a project delimitation. In contrast, from an interdisciplinary problem perspective, they will view it as an opportunity to collaborate in a meaningful interaction with students from other disciplines.

Ethics in problem design and potential enablers of ethics

In the previous sections, we outlined that problems that are ill-structured, complex, and highly contextually dependent require assumptions, limitations, enactment, interventions, and, most likely, a move across established domains. We have also argued that this complicates problem design, as it becomes a multi-directional process governed by an overarching question: *Who determines the problem, when and where, in what direction, and with which arguments?* Overall, problem designers must consider the 'right' way to proceed, with the right arguments, and making certain assumptions.

The range of ethical concerns we see as related to problem design includes:

- Some problems are initially addressed; others are not.
- Some stakeholder interests are considered; others are not.
- Some success criteria are selected; others are not.
- Some solutions are addressed; some are not.
- Some impacts are assessed; others are not.
- Some trade-offs are accepted; others are not.
- Some project types are seen as appropriate responses to the problem; others are not.
- Some team members might agree on the chosen decisions; others might not.

These concerns not only activate moral values but also reflect the moral itself; thus, they become ethical questions. Furthermore, more explicit ethical questions might also be put forward, such as consideration of the ethical consequences of a specific emerging technology by carrying out an ethical constructive technology assessment (eCTA) as elaborated by Kiran et al. (2015).

Van de Poel and Royakkers (2011) have presented six categories of morals in the domain of engineering ethics, which are aligned with different goals for engineering ethics education: sensibility, analysis, creativity, judgment, decision-making, and argumentation. Based on an extensive literature review, Martin et al. (2021) further elaborate on this list, adding categories related to knowledge, design, agency and action, character and virtuous development, emotional development, and situatedness. Each category implies specific kinds of responsibility for students in a self-directed PBL environment, as exemplified in Table 21.1, while the responsibility of curriculum designers and teaching staff is to support students in taking on such responsibilities.

Table 21.1 Examples of student responsibility in a PBL environment, based on the categorization made in van de Poel and Royakkers (2011)* and Martin et al. (2021)**

Moral categories	*Examples of student responsibility in a PBL environment*
Sensibility*	Recognize and acknowledge ethical issues in problem design and problem-solving.
Analysis*	Incorporate an analysis of the underlying moral problems embedded in the problems addressed in relation to technology.
Creativity*	Be open to alternative solutions and taking time to explore them from an ethical point of view.
Judgement*	Work to understand all viewpoints and make an informed judgement based on transparent criteria.
Decision-making*	Participate in team discussions, negotiate, and compromise toward a shared decision, taking into consideration different views of team members.
Argumentation*	Justify one's own actions regarding the project outcome, learning objectives, and professional identity.
Knowledge**	Obtain knowledge of ethical theories, codes, and language and use these to inform moral arguments, judgments, and decisions.
Design**	See moral considerations as an integrated part of the design and use of technological artifacts.
Agency and action**	Engage with and contribute to the reshaping of the society for common good.
Character and virtuous development**	Be able to articulate one's own virtues and use them to define the virtues of a team.
Emotional development**	Endeavour to understand one's own and others' emotions, and how they can impact the project outcome and process.
Situatedness**	Connect both technology and engineering practice to relevant contextual settings.

One way that educational designers can support students is to introduce a systematic method of integrating moral problems into the problem design. Van de Poel and Royakkers (2011) have presented such a method which they term the 'ethical cycle,' which takes its point of departure in a case. In many ways, the ethical cycle recapitulates the problem design process described above, with its steps of problem identification, analysis, statement, and solution. However, the ethical cycle more explicitly requires ethical problem statements, evaluations as well as action. On the other hand, the PBL approach can initiate a series of built-in cases for ethical consideration, contemplating these moral requirements.

Another way for educational designers to support and motivate students to engage with ethical considerations is to clarify that ethical responsibility and moral problems are among the intended learning outcomes in the curriculum. Ethics can be an explicit part of the curriculum delivered through explicit statements, or they can be explored indirectly via the establishment of an obligation for students to take ethical responsibility upon themselves (as exemplified in Table 21.1). In keeping with the desire to support and motivate students to engage with ethics, the authors of Chapter 26 present a framework for assessing the 'ethical competencies and affective dispositions' of students based upon the same 'moral categories' listed in Table 21.1.

Furthermore, we argue that PBL makes ethical considerations a team concern related to care ethics instead of a purely individual matter. In this way, the team structure becomes another factor enabling students to address ethical considerations; indeed, students might have the same duties

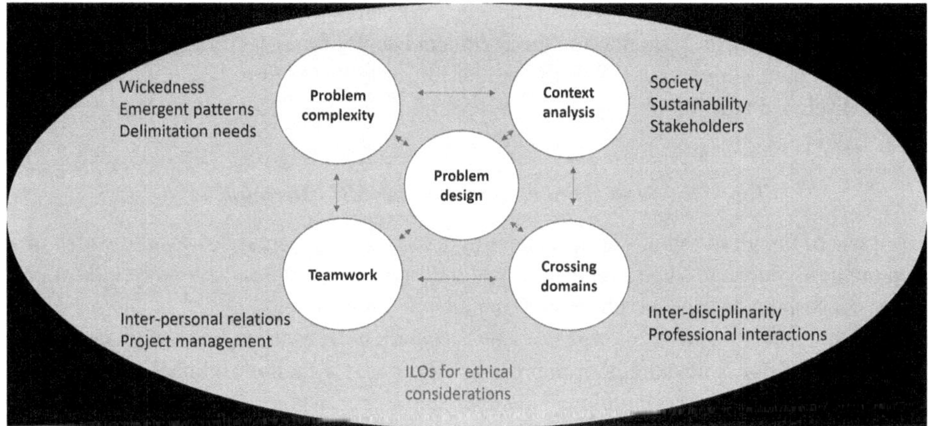

Figure 21.1 Ethical enablers in a PBL curriculum.

in the context of a curriculum, but the negotiated virtues and values of the team are of high importance for coordinated action in the design and solving of a problem. To provide a complete picture of the factors enabling students' ethical engagement, Figure 21.1 presents a theoretical framework for the empirical study to be discussed in this section. Four factors influence the problem design: problem complexity, context analysis, crossing domains, and teamwork. Moreover, curricula considerations to the problem design itself will be considered.

In the following sections, we relate the factors enabling engagement with ethical concerns to examples of explicit and implicit ways to integrate moral and ethical considerations into the curriculum. Our point of departure is the case of a university implementing PBL systemically at the institutional level, namely, Aalborg University in Denmark. The examples have been extracted from a comprehensive analysis of Aalborg University's engineering and science education curricula.

PBL at Aalborg University

Engineering education programs at Aalborg University have included a course on Technology and Society, usually in the first year of study. Students learn to analyze problems from a societal perspective and, if possible, apply sociological or more action-driven methodologies. Over the last 12 years, the integration of Sustainable Development Goals (SDGs) has been emphasized by focusing on problems in the area of sustainability. Students learn to analyze problems from a sustainability point of view and to identify related dilemmas in society. They are encouraged to recognize that there are moral issues related to each problem in society, including how we understand and analyze the problem, who the stakeholders are, and the values and worldviews underlying the chosen solutions.

Beyond PBL, Aalborg University is distinguished by its close collaboration with industry and its efforts to balance business and academic needs. Since 2023, the university has aimed to become a mission-driven institution, developing interdisciplinary competences through initiatives such as the formation of interdisciplinary teams consisting of groups or individual students from both STEM (science technology, engineering, and mathematics) and SSH (social sciences and humanities) fields. In turn, this impacts the problems that students are given to work on. In addition, it has a vital role as an essential element in the cultivation of technical and non-technical excellence, as discussed by Harris (2008, p. 163).

The following section starts by exploring the study regulations and the curricular contexts of the Faculty of Engineering and Science and the Technical Faculty for IT [Information Technology] and Design at Aalborg University. Using the model illustrated in Figure 21.1, the curricula are analyzed and discussed accordingly.

Ethics discourses and enablers in the PBL curricula

The explicit use of the word 'ethics' is infrequent in the studied curricula; however, the activities mentioned implicitly indicate a comprehensive exploration of ethics. The curricula include ethical perspectives concerning energy, AI, chemistry, and techno-anthropology.

One example is that students exploring authentic technological case studies formulate technological dilemmas and use ethical methods to propose solutions, for example, ethical, technological assessment, and ethical scenario-building. This approach is consistent with the advice given by Jonassen et al. (2009, p. 235), suggesting that rather than explicitly teaching students about ethics, providing students with experiences of solving authentic 'everyday' engineering ethics problems may be more impactful.

Another approach is to include the notion of ethics together with other contextual components, for example, mentioning ethics alongside a societal and theory-of-science perspective. Other elements directly related to the notion of ethics include engineering science, technological development and use, responsibility, value-sensibility, future scenarios, privacy, trust, fairness, research implications, and change agency. Students are asked to understand, analyze, apply, evaluate, argue, discuss, construct visions and interventions, and further reflect on and contextualize ethical considerations, problems, and representations. Other requests are more descriptive, for example, asking students to describe the traditions of engineering, the engineer's role in society, or ethical issues in engineering science. The descriptive nature of such requests implies that a given set of norms is followed.

Notably, there is a lack of attention to character, virtues, and emotional development regarding moral categories, as Martin et al. (2021) emphasized (which Chapters 26–31 of this handbook seek to address). The term 'ethics' describes a responsibility or orientation toward something or someone – not necessarily in the sense of positioning oneself in terms of morals and virtues. Furthermore, the domain of ethics is typically treated as comprehensive, as seen in the framing of courses such as 'Technology and Ethics,' 'Ethics and Technological Intervention Processes,' and 'Privacy and Ethics in Computer Systems' – in contrast, ethical considerations are not explicitly connected to project modules. At the project level, ethics is more implicitly integrated by including opportunities to engage with ethical considerations.

Van de Poel and Royakkers (2011) noted that engineering is an inherently morally motivated activity. In contrast, ethics involves systematically reflecting upon morality and dialogically expressing what is perceived as 'right' and 'wrong.' In the previous sections, we have explored how moral opinions, decisions, and actions of students in a participant-directed learning environment can open up discussion of ethical considerations (for more on reflective and dialogical approaches in engineering ethics education, see Chapter 25). The objective of the following section is to provide a richer description of the factors enabling such integration of ethical thinking without the explicit mention of ethics – and discuss implications of these factors.

Problem design processes and problem complexity as ethical enablers

Some of Aalborg University's curricula explicitly call for students to design the problems they will work with, for example, by setting as a learning goal the ability to provide a problem analysis, a problem statement, and perspectives relating to the context in which the problem is defined. Furthermore, some indications are made in the curricula of the types of actions expected from students in the problem design process – including the ability to justify, critically evaluate, and argue. The evaluation aspect implies that students should be able to assess the possible solutions they have identified to decide which is optimal. This decision-making competence is just as crucial as any problem-solving competence and is highly complex due to the influence of numerous technical, environmental, social, economic, and ethical constraints (El-Zein et al., 2008, p. 170).

Other curricula emphasize the importance of students being able to argue for the chosen solutions, explain how they have narrowed an open problem space to something possible to complete within the given timeframe, and describe the potential of disciplinary contributions. It is explicitly stated that students must open the solution space themselves; this is an integrated part of the problem design process, as noted by Holgaard et al. (2017). In one example, the curriculum explicitly calls for a minimization of the proposed solution, which, inadvertently, might lead to a reductionist approach. Others direct students to argue for their choice of problem bounds – the chosen delimitations. Byrne and Mullally (2014) highlight the necessity of challenging reductionist thinking and suggest that a broader and more contingent view of the engineer's professional role is required. In either case, however, it remains somewhat unclear how the 'narrowing down' process is happening, and whether students are reflecting on the value propositions they make in this process.

Problem complexity is reflected in how students combine the various of aspects of a problem during the problem design process, particularly in relation to actors, organizational conditions, and institutional framings. Such a combination of aspects implies the 'wickedness' of the problem. Complexity can also be understood from a system perspective, in which the challenge is to narrow down a technological system. The system perspective includes a call for students to understand the relationships and interdependencies in a system. Complexity is also seen in the emergent patterns considered in relation to future systems, within a comprehensive approach to emergent technologies. Some curricula explicitly state that students should be able to solve complex problems, although some make it clear that this is within disciplinary borders. Other curricula address the need to 'futureproof' solutions using methodologies such as scenario-building, life-cycle assessment, and cost–benefit analyses based on a set of pre-defined criteria. The ethical question – which may, in fact, be a part of practice – is for the students to consider the ways in which these criteria are defined, and by whom. As stated by Bucciarelli (2008), however, learning about the social, organizational, and even political complexities of practice may be a more fundamental prerequisite for students but without neglecting ethics in engineering education.

Ethics enabled by teamwork and boundary-crossing activities

Interpersonal reflections provide a way to nurture students' awareness of personal virtues. One example is a request for students to participate actively, collaboratively, constructively, and critically in order to develop communicative solutions with a specific focus on culture and values. Some curricular requirements highlight students' ability to reflect on their own role in a team, consider group dynamics, identify their own and others' competencies, and reflect on their individual and collaborative learning processes. The process of developing this awareness is a component of developing team norms within a group work context, which the curricula explicitly name as an objective of group projects. In some cases, the study regulation (i.e., syllabus or project brief) calls for students to demonstrate specific virtues in the context of teamwork (e.g., to be tolerant and resilient). Concerning project management, students are encouraged to be analytical and forward-looking. For example, students may be instructed to analyze how their team organizes its work in order to identify strengths and weaknesses in their approach and, based on this analysis, provide recommendations for enhancing teamwork in the future. This implicitly calls for the team to set specific analytical parameters and evaluation criteria. These are examples of what Conlon (2008) refers to as generic competencies, which are non-technical competencies (like communication, project management, leadership, and teamwork) that help make engineers more effective and engineering students more prepared for future management tasks.

The importance of a boundary-crossing perspective is highlighted to students via activities that involve collaboration across teams. The recognition of different disciplinary languages is also emphasized, most often by requesting students use the language of a particular discipline. Other activities ask students to recognize specific academic norms within their field of study, which implies that someone (e.g., a facilitator) is actively defining those specific norms. From a broader perspective, there is an explicit discussion of interdisciplinary work in the curricula, but the level of interdisciplinarity called for is often left open to interpretation. This overall situation reflects Nair and Bulleit's (2020, p. 71) argument that engineering ethics should be taught in a way that embraces interdisciplinary thinking, including the recognition and use of disciplinary knowledge from beyond engineering within the practice of engineering ethics.

In addition, connections to the professional sphere are made in the curricula by emphasizing the importance of concepts, models, methods, and techniques that are relevant to professional teamwork. There are examples of discussions of organizational cultures, structures, and decision-making, and in some cases, the interdisciplinary and cross-departmental perspective is put on the same footing. A question that emerges here involves how the difference between the two types of boundary-crossings, in terms of disciplines and professions alike, can ease the transition from engineering education to the workplace. Another open question involves what students expect of their future workplaces and whether integrating such reflections and awareness into the curriculum will benefit students professionally. Discussing engineering virtues and what characterizes a 'good engineer' (Harris, 2008) could be a point of departure to connect the educational domain with the professional domain.

Contextual analysis as ethical enablers

A contextual analysis moves technological considerations to the societal level and emphasizes grand challenges such as sustainable development. As Aalborg University has embedded contextual learning and exemplarity in its PBL model, the curricula are especially rich in this aspect related to the problem design and problem-solving processes. Overall, however, there are two broad types of curricula: one is focused on integrating specific technologies into contexts, while the other (examples of which include curricula related to design) takes an inherited and integrated approach to contextual factors. In other words, they are part of the 'text.' Although historical, political, cultural, and philosophical contexts are mentioned, along with considerations to the theory of science, the most frequent reference to context is societal – and, more implicitly, a business context. As a part of the societal context, the consideration of various actors and sustainability are recurring themes.

Discussions of sustainability and ethics are often connected to and embedded within each other (Chance et al., 2021, p. 94). Sustainability is also an integrated part of most of the engineering curricula at Aalborg University, using a variety of approaches. Some intended learning outcomes focus on the calculation of environmental impacts; others focus on sustainability standards; still others focus on designing and re-designing for sustainability. Reference to the United Nations' SDGs is also present, either with explicit reference to specific goals or as a broad guideline for alignment with the goals as a whole. Keywords are 'work environment,' 'ecology,' 'eco-systems,' 'safety,' 'circular economy,' 'life-cycle assessment,' 'principles for sustainability,' and, more broadly, the 'interplay between humans and nature.' Students are expected to analyze, assess, discuss, design for, and reflect on various aspects of sustainability. For example, students are asked to evaluate trade-offs between environmental, social, and economic sustainability. Further, globalization is often discussed as an essential societal consideration in decision-making regarding options for development, with attention to both local and global consequences. This approach reflects the importance of global awareness for engineers, as discussed by Nair and Bulleit (2020), who recommend a focus on how human well-being in the local context may be affected, not only from a market perspective.

Market-driven discourse, as described by Jamison et al. (2014) is also present, and issues such as competitiveness and socio-economics on the micro-, meso- and macro-levels are mentioned. Another discourse related to the business context involves entrepreneurship and students' ability to work in a commercial value-oriented approach and to address business cases and models, which include value propositions. For example, students should be able to understand and create a business case for a given technological system and must evaluate the effectiveness and applicability of certain technologies. In this case, the importance of establishing criteria is implicit.

Students are also asked in some curricula to identify and engage relevant actors in the assessment of technological consequences, which gives the students the responsibility of evaluating what and who is relevant in the given context. Students' ability to assess conflicts of interest is also mentioned in the curricula. In some cases, students are expected to make actor analyses using specific approaches like actor network theory (ANT). Engagement with users is specifically highlighted through principles of design, for example, co-design or participatory design. Other actors are mentioned in relation to societal responsibility, including researchers, experts, professionals, and companies. The goal of these requirements is to help students recognize that the problem-solving process is a community activity that must involve input from all involved parties (Nair & Bulleit, 2020, p. 71). Research indicates that engagement with users and stakeholders can enhance students' ethical sensitivity and reflexivity while also stimulating ethical decision-making in the design process (Corple et al., 2020).

The contextual factors related to sustainability, market orientation, and the cast of actors involved are numerous, as are the interdependencies between them; a significant amount of decision-making is thus necessary in the problem-solving process. This might present a challenge to students' critical and holistic thinking and stimulate their engagement with the ethical considerations embedded in the sustainability and market discourse. Furthermore, disciplinary framings, which are continuously referenced throughout many of the curricula, might also challenge students' motivation to work in an interdisciplinary context.

Curricula frequently refer to society as a context, as a framework condition, and/or as an object of technological implications. At the same time, however, it remains open for students to define the societal aspects of a problem and proposed solution, and the 'relevance' of the problem is often used as a criterion. This suggests that students are left with the challenge of deciding what is relevant, and although implicit, this carries considerable learning potential for ethics in engineering education.

Final remarks

Even if there is no explicit learning outcome regarding ethics in a given engineering curriculum, students can implicitly learn to analyze, understand, and resolve a range of ethical issues inherited from the involvement of stakeholders, collaborative behavior, and the impact of developed technologies. However, although the curriculum may open doors for such ethical considerations, the implicitness of ethics might impede students' transition from the problem design cycle in a technological context to the ethical cycle. This transition would enhance students' specific attention to moral considerations and actions. The question is how much emphasis on ethical theory, methods, and mindset is needed for educators and students to integrate ethics throughout engineering education. This chapter intends to argue for exploring the depths of possible ethical considerations implicitly present in a PBL environment. The currently implicit opportunities for ethical thinking must be studied further to fully examine their associated learning potential, and it might not be sufficient to expect students to enter these openings independently. Rather, ethics in engineering education must be scaffolded and deliberately nurtured.

Acknowledgments

This work is part of InterPBL, a research project aimed at developing innovative educational models to educate engineers to work proactively and interactively in an interdisciplinary work environment. It has been funded by The Poul Due Jensen Foundation.

References

Barry, B. E., & Herkert, J. R. (2014). Engineering ethics. In A. Johri & B. Olds. (Eds.), *Cambridge handbook of engineering education research* (pp. 673–692). Cambridge University Press.

Bijker, W. E., Hughes, T. P., & Pinch, T. (1989). The social construction of technological systems: New directions in the sociology and history of technology. *Technology and Culture*. https://doi.org/10.2307/3105993

Bucciarelli, L. L. (2008). Ethics and engineering education. *European Journal of Engineering Education, 33*(2), 141–149. https://doi.org/10.1080/03043790801979856

Børsen, T., Serreau, Y., Reifschneider, K., Baier, A., Pinkelman, R., Smetanina, T., & Zandvoort, H. (2021). Initiatives, experiences and best practices for teaching social and ecological responsibility in ethics education for science and engineering students. *European Journal of Engineering Education, 46*(2), 186–209. https://doi.org/10.1080/03043797.2019.1701632

Byrne, E. P., & Mullally, G. (2014). Educating engineers to embrace complexity and context. *Proceedings of the Institution of Civil Engineers-Engineering Sustainability, 167*, 241–248.

Chance, S., Lawlor, R., Direito, I., & Mitchell, J. (2021). Above and beyond: Ethics and responsibility in civil engineering. *Australasian Journal of Engineering Education, 26*(1), 93–116. https://doi.org/10.1080/22054952.2021.1942767

Conlon, E. (2008). The new engineer: Between employability and social responsibility. *European Journal of Engineering Education, 33*(2), 151–159. https://doi.org/10.1080/03043790801996371

Corple, D. J., Zoltowski, C. B., Kenny Feister, M., & Buzzanell, P. M. (2020). Understanding ethical decision-making in design. *Journal of Engineering Education, 109*(2), 262–280. https://doi.org/10.1002/jee.20312

De Graaf, E., & Kolmos, A. (2003). Characteristics of problem-based learning. *International Journal of Engineering Education, 19*(5), 657–662.

De Graaff, E., & Kolmos, A. (2007). History of problem-based and project-based learning. In E. d. G. a. A. Kolmos (Ed.), *Management of change. Implementation of problem-based and project-based learning in engineering* (pp. 1–8). Sense Publishers.

El-Zein, A., Airey, D., Bowden, P., & Clarkeburn, H. (2008). Sustainability and ethics as decision-making paradigms in engineering curricula. *International Journal of Sustainability in Higher Education, 9*(2), 170–182.

Elliott, D., & June, K. (2018). The ēvolution of ethics education 1980–2015. In E. E. Englehardt & M. S. Pritchard (Eds.), *Ethics across the curriculum—pedagogical perspectives* (pp. 11–37). Springer International Publishing. https://doi.org/10.1007/978-3-319-78939-2_2

Habbal, F., Kolmos, A., Hadgraft, R. G., Holgaard, J. E., & Reda, K. (2024). *Reshaping engineering education. Addressing complex human challenges*. Springer. https://doi.org/https://doi.org/10.1007/978-981-99-5873-3

Harris, C. E. (2008). The good engineer: Giving virtue its due in engineering ethics. *Science and Engineering Ethics*, *14*(2), 153–164. https://doi.org/10.1007/s11948-008-9068-3

Herkert, J. R. (2001). Future directions in engineering ethics research: Microethics, macroethics and the role of professional societies. *Science and Engineering Ethics*, *7*(3), 403–414. https://doi.org/10.1007/s11948-001-0062-2

Holgaard, J. E., Guerra, A., Kolmos, A., & Petersen, L. S. (2017). Getting a hold on the problem in a problem-based learning environment*. *International Journal of Engineering Education*, *33*(3), 1070–1085.

Hung, W. (2006). The 3C3R model: A conceptual framework for designing problems in PBL. *Interdisciplinary Journal of Problem-Based Learning*, *1*(1), 6.

Hung, W. (2009). The 9-step problem design process for problem-based learning: Application of the 3C3R model. *Educational Research Review*, *4*(2), 118–141. https://doi.org/https://doi.org/10.1016/j.edurev.2008.12.001

Hung, W. (2016). All PBL starts here: The problem. *Interdisciplinary Journal of Problem-Based Learning*, *10*(2). https://doi.org/10.7771/1541-5015.1604

Hung, W. (2019). Problem design in PBL. In W. Hung, M. Moallem, & N. Dabbagh (Eds.), *The Wiley handbook of problem-based learning* (pp. 249–272). Wiley.

Jamison, A., Kolmos, A., & Holgaard, J. E. (2014). Hybrid learning: An integrative approach to engineering education. *Journal of Engineering Education*, *103*(2), 253–273. https://doi.org/10.1002/jee.20041

Jonassen, D. H. (2011). *Learning to solve problems—A handbook for designing problem-solving learning environments*. Routledge.

Jonassen, D. H., Shen, D., Marra, R. M., Cho, Y. H., Lo, J. L., & Lohani, V. K. (2009). Engaging and supporting problem solving in engineering ethics. *Journal of Engineering Education*, *98*(3), 235–254. https://doi.org/10.1002/j.2168-9830.2009.tb01022.x

Kiran, A. H., Oudshoorn, N., & Verbeek, P.-P. (2015). Beyond checklists: Toward an ethical-constructive technology assessment. *Journal of Responsible Innovation*, *2*(1), 5–19.

Kolmos, A., & de Graaff, E. (2014). Problem-based and project-based learning in engineering education: Merging models. In A. Johri & B. M. Olds (Eds.), *Cambridge handbook of engineering education research* (pp. 141–160). Cambridge University Press. https://doi.org/10.1017/CBO9781139013451.012

Kolmos, A., de Graaff, E., & Du, X. (2009). Diversity of PBL: - PBL learning principles and models. In X. Du, E. de Graaff, & A. Kolmos (Eds.), *Research on PBL practice in engineering education* (pp. 9–21). Sense Publishers.

Kurtz, C. F., & Snowden, D. (2003). The new dynamics of strategy: Sense-making in a complex and complicated world. *IBM Systems Journal*, *42*, 462–483. https://doi.org/10.1147/sj.423.0462

Martin, D. A., Conlon, E., & Bowe, B. (2021). A multi-level review of engineering ethics education: Towards a socio-technical orientation of engineering education for ethics. *Science and Engineering Ethics*, *27*(5), 60. https://doi.org/10.1007/s11948-021-00333-6

Martin, D. A., Gwynne-Evans, A., Kazakova, A. A., & Zhu, Q. (2023). Developing a global and culturally Inclusive vision of engineering ethics education and research. In *International handbook of engineering education research* (pp. 87–114). https://doi.org/10.4324/9781003287483-6

Nair, I., & Bulleit, W. M. (2020). Pragmatism and care in engineering ethics. *Science and Engineering Ethics*, *26*(1), 65–87. https://doi.org/10.1007/s11948-018-0080-y

Newberry, B. (2004). The dilemma of ethics in engineering education. *Science and Engineering Ethics*, *10*(2), 343–351. https://doi.org/10.1007/s11948-004-0030-8

Osbeck, C., Franck, O., Lilja, A., & Sporre, K. (2018). Possible competences to be aimed at in ethics education – Ethical competences highlighted in educational research journals. *Journal of Beliefs & Values*, *39*(2), 195–208. https://doi.org/10.1080/13617672.2018.1450807

Rip, A., Misa, T. J., & Schot, J. (1995). *Managing technology in society*. Pinter Publishers.

Rittel, H. W. J., & Webber, M. M. (1973). Dilemmas in a general theory of planning. *Policy Sciences*, *4*(2), 155–169. https://doi.org/10.1007/BF01405730

van de Poel, I., & Royakkers, L. (2011). *Ethics, technology and engineering: An introduction*. Wiley-Blackwell.

Wenger, E. (1998). *Communities of practice: Learning, meaning, and identity*. Cambridge University Press.

22

TEACHING RESPONSIBLE ENGINEERING AND DESIGN THROUGH VALUE-SENSITIVE DESIGN

Andrea R. Gammon, Annuska Zolyomi, Richmond Y. Wong, Eva Eriksson, Camilla Gyldendahl Jensen, and Rikke Toft Nørgård

Introduction

Value-sensitive design (VSD) and other values-based design approaches are advanced as means for creating better technologies, that is, technologies that support human values (Friedman & Hendry, 2019; van den Hoven et al., 2015). But so too do these approaches have considerable educational potential which, while observed by others (Cummings, 2006; Eriksson et al., 2021; Rocco et al., 2022), has received far less attention. In this chapter, we propose VSD as an effective approach for teaching ethics in design and engineering education and thus a way to cultivate designers and engineers who are socially and ethically responsive and responsible in their (future) work. VSD and other values-based design approaches explicitly consider values (e.g., privacy, dignity, sustainability) expressed using technology. Through teaching these methods, we teach designers and engineers to think about values in the various stages of technology development so as to create better technologies, and in so doing, to interrogate what 'better' means and for whom, and how to achieve this. We mean to emphasize VSD as an approach to learning by doing, a formalized but flexible process for the continuous work of designing and engineering better technologies. So, too, can VSD offer educators a highly flexible and open-ended approach for equipping students to think more carefully, comprehensively, and inclusively about how the things they make, and the ways they make them, impact others.

This chapter invites teachers to explore VSD's educational offerings and possibilities: we aim to introduce VSD to those unfamiliar and provide ideas and perspectives to deepen the knowledge of those with more experience. We consider the VSD literature an excellent resource for teaching: methods, critiques, and a wide array of examples and applications can be found in various fields and domains, much of which is accessible to engineering and design students. For this reason, this chapter draws on research and developments in VSD, values in design, and education-specific research, questions, and issues. We begin by introducing the key ideas and methods of VSD and introducing an example we drew from the literature and developed by working it through our chapter together. The example functions in two main ways: First, we use it to demonstrate key strategies for teach-

DOI: 10.4324/9781003464259-27

ing VSD we take to be useful for teachers. Second, the example shows the generative possibilities of teaching VSD, as it is through developing this example for teaching VSD that new possibilities open up. Through the example, we consider the treatment of values in engineering and design classrooms, cover specific strategies for teaching VSD, and suggest methods for assessing VSD and the teaching aims we have suggested – of cultivating responsible engineering and design students. We conclude the chapter by turning to some critiques of VSD to highlight emerging work that also pushes VSD and its educational potential in new directions. Given the broad applicability of VSD to various classroom styles and settings, we have aimed at a level of generality in writing this chapter. We encourage instructors to maximize VSD's flexibility and open-ended nature by experimenting and adapting the methods and strategies presented here to best suit their classrooms and students.

We, the authors of this chapter, all have experience teaching values-based design to engineering and design students in the United States or Europe, coming to this from different academic backgrounds – philosophy, user-centered design, information management, education, and human–computer interaction.

Where to start with VSD?

Friedman et al.'s "Value sensitive design and information systems" (2008) is a natural starting place as a key paper that provides an introduction and overview useful for newcomers and accessible for students. The authors introduce VSD as a "grounded theoretical approach to values in design" (Friedman et al., 2008, p. 69), explain its methods, and propose 13 values with ethical relevance for design systems.[1] Although this paper, like most of the VSD literature, does not position itself in terms of teaching or education, it provides an excellent starting resource for introducing students to VSD. The list of values and accompanying definitions and descriptions acquaint students or other readers less familiar with values with what they are, and three detailed examples illustrate how these values take shape through technologies. This is also a starting place for this chapter. After briefly introducing VSD's main ideas and methods in the next section, we return to an example from the paper by Friedman et al., which will then reappear throughout the chapter, demonstrating how examples can be put to use for educational purposes in encouraging the development of responsible engineers and designers.

Main ideas and methods

Value-sensitive design understands technologies as value-laden: whether software, bridges, or the screen or page you read this on, technologies are designed with particular uses and aims, and so are invariably shaped by values. This means that designers and engineers, whether they mean to or not, embed values into the things they make. VSD is a leading approach for acquainting – or sensitizing – engineers and designers with values, and further, guiding how they engage values more deliberatively, comprehensively, and with the involvement of stakeholders in engineering and design processes (Davis & Nathan, 2015; Friedman & Hendry, 2019).[2] The insight that values are expressed in technologies, whether or not their designers or engineers gave any thought to these values ahead of time, is important for design and engineering students who often view technology as neutral with respect to values. That VSD as an approach helps students recognize and understand *how* values can be embedded in technologies, rather than just that they are, makes it especially useful in educational contexts for challenging the pervasive idea of technologies as neutral tools.

Methodologically, VSD takes the form of (1) conceptual, (2) technical, and (3) empirical investigations, which are complementary and should be mutually reinforcing, but can be done separately. That they can come apart makes VSD an attractive approach for teaching, as even partial methodological efforts can help students appreciate the value-ladenness of technologies and the manifold challenges and opportunities this introduces. *Conceptual investigations* explore what values are at play, how values might be impacted by a specific technology, and for whom. Conceptual investigations will often draw on theoretical or normative frameworks to determine relevant values, understand the meanings of these values, and identify the ways in which their meanings have changed. This can be done in a typical classroom setting, through examples, brainstorming, and discussion, using existing lists of values in VSD toolkits or in codes of conduct, for instance, and drawing from existing VSD literature for specific domains, technologies, or values. *Empirical investigations* use social science research to interrogate how, in practice, stakeholders experience values in a specific technology or design. Methods can be qualitative or quantitative, determined by what best suits the stage and needs of the project; focus groups, surveys, interviews, and behavioral studies are typical. Empirical investigations can bring to light how stakeholders respond when conflicts arise, and such investigations are needed to determine if a design, in the real world, with real stakeholders, supports intended values. Because of this, empirical investigations are difficult to achieve in classroom settings where opportunities to see a design in practice, or to survey real stakeholders, may be extremely limited. Nevertheless, partial and modified empirical investigations are possible.[3] *Technical investigations* focus on designing or adapting a technological artifact to be responsive to values and stakeholder contexts drawn from the conceptual and empirical investigations. Technical investigations turn the attention to the technology itself to see how it is or isn't supporting intended values and what technological re-designs could constitute improvements.[4]

Conducting all stages of VSD's methods isn't possible in most educational settings. However, using examples – whether from the extensive VSD research literature or based on the specific educational context – is an essential and effective way to introduce VSD and involve students in thinking through its processes. We demonstrate this strategy by developing an existing VSD case to illustrate VSD concepts and teaching activities.

Example: the Augmented Window

We find the use of examples especially important in teaching VSD. Examples provide focal points for discussion and can bring to light for students how designs afford certain uses and values and close off others.[5] Examples can be presented to students in a course simply via a case from existing literature others have researched (as we will demonstrate below), or might involve a physical or digital object for students to tinker with or re-design with specific values in mind (van Grunsven et al., 2023). Asking students to prepare and share (additional) examples of their own helps ensure that students' diverse personal interests and study backgrounds are reflected in the classroom. The 'Augmented Window' example could be presented in class by asking students to read (parts of) Friedman et al. (2006 & 2008) or by introducing the salient points of the example to students in class, leading into discussion or group work. In this chapter, the example does double work. We refer to it repeatedly throughout the chapter to illustrate several methods for teaching and assessing VSD, but additionally, and more generally, it shows VSD as amenable to iteration and development over time. The flexibility and dynamism of VSD make it especially useful for teaching in various contexts.

We will refer to an example originally presented in "Value sensitive design and information systems" (Friedman et al., 2008): the 'Augmented Window' example, which was discussed again in "The watcher and the watched," Friedman et al. (2006). The Augmented Window comes from VSD research exploring social judgments about privacy related to surveillance and sensing technologies. In this case, Friedman et al. (2008) studied a scenario in which an office uses plasma screens to "continuously display the local real-time outdoor scene" (p. 77) of a nearby public plaza located on a university campus and frequented by the general public. Employees in offices with no view to the outside would effectively have a view to the outside through the 'augmented window' of the plasma screen in their office. As normal, passersby on the plaza would be visible to surrounding office workers with windows onto the plaza, but through this technology, so too were their images captured and broadcast by HDTV cameras onto the plasma screens for office workers whom they could not see. Friedman et al. (2008) initially investigated this case with productivity and creativity for the office workers in mind. The additional paper, "The watcher and the watched," involves conceptual and empirical investigations into privacy that develop this example in more detail and serve as a very useful resource for teaching this case or similar ones in the classroom. Different notions of privacy as a socio-technological construct are presented, followed by an empirical analysis of stakeholder views around privacy in the case of the augmented window technology. The difference in position and power between direct and indirect stakeholders is discussed, as are other dimensions: gender, cultural norms about privacy, when violations of privacy are more permissible (for instance, for security), when consent is needed, etc. As Friedman et al. (2008) suggest, this example opens discussions of indirect versus direct stakeholders, value conflicts, and how different data sources can inform empirical investigations.[6]

This example will help us illustrate how the methods of VSD can be adapted for the classroom. In returning to it throughout the chapter, we additionally hope to show how using a classic VSD example in an educational context breathes new life into it by opening it up to novel questions, approaches, and demands.

Teaching VSD

VSD is well-poised to serve as a formalized, active learning approach to ethics, especially since VSD is designed to be applied through an iterative, reflexive design process (Cummings, 2006). VSD offers a conceptual framework for practical investigations that can teach design and engineering students to grapple with the real-world complexities of technology. However, VSD was developed in research and design contexts, and the development of education-specific resources for teaching VSD has lagged behind (Eriksson et al., 2021). In this section, we focus on *teaching* VSD. We discuss issues with introducing students to values, and then turn to three foundational VSD approaches, which we translate into an educational context.

Teaching 'values'

Typically, in VSD, 'values' are defined as "what a person or group of people consider to be important in life" (Friedman et al., 2008, p. 70).[7] Even this broad definition can be challenging for students – especially those encountering human or moral values for the first time and finding values discussions abstract and vague. Nevertheless, we suggest that VSD is an excellent approach for acquainting students with values. VSD doesn't remain at a level of high abstraction but always uses concrete technologies to bring values into focus. Offering an example – of a technology where a value (or potential conflict) is salient – is a good way to begin. Turning again to the Augmented Window example, the value of privacy is likely to arise, and indeed, Friedman et al. (2006) addressed individual privacy in "The watcher and the watched." Privacy, the authors claim, is "an enduring human value" (Friedman et al., 2006, p. 237) and is one students will have personal experience and working knowledge of.[8] We suggest that in this and all cases, thinking about the value through the technology at issue, in this case, through the Augmented Window, helps to make otherwise abstract ideas and concepts more concrete for students. Privacy is thus, in this case, understood *through* the "technological capture and display of people's images" (Friedman et al. 2006, p. 237) shown to others remotely.

Another strategy for teaching values is to consider in what ways design and engineering students may have encountered the notion of values in their education, lived experiences, and training for professional practice. Ethics in engineering has been legitimized through education accreditation organizations and software engineering professional organizations, which introduce values and ethical principles in their codes of conduct or other guiding documents. Such codes describe how professionals should behave, approach creating systems, and strive to design and implement high-quality systems. For example, the Accreditation Board for Engineering and Technology (ABET) requires a student learning outcome of "an ability to recognize ethical and professional responsibilities in engineering situations and make informed judgments, which must consider the impact of engineering solutions in global, economic, environmental, and societal contexts" (Accreditation Board for Engineering and Technology, 2022). The Association for Computing Machinery's 'Code of Ethics and Professional Conduct' mentions several values that resonate with VSD, including privacy, security, and confidentiality (Association for Computing Machinery, 2018).

Students may also engage with values when learning to elicit user requirements and assess software quality or create design briefs. User requirements often encompass ways a system or product is meaningful to the user regarding functionality and value or refer to fundamental engineering quality requirements, such as robustness and security. In software engineering, the paradigm of value-based software engineering (VBSE) argues for the importance of considering values throughout software engineering principles and practices (Association for Computing Machinery, 2018) and frames 'value' as an ultimate benefit of the project – as perceived by project stakeholders – "whether tangible or intangible, economic or social, monetary or utilitarian, or even aesthetic or ethical" (Biffl et al., 2006, p. X). VBSE takes a practical approach to values, considering them as drivers in software engineering decisions that occur in management-oriented and software lifecycle activities. Working from this framework and highlighting human or moral values is another approach for improving students' proficiency with values.

Design students educated in the paradigm of human-centered design (HCD) focus their practice on designing for people and society (Meyer & Norman, 2020). However, they may assume or be taught that technology is value neutral. Meyer and Norman argue for design schools to teach students not just on the skills for creating polished design, but how to embrace and design for the complexity, frustrations, and tensions of the real world. Modern design students "must meet new ethical challenges that go along with an expansion into different global territories with different sustainability issues, different cultures, and different value systems" (Meyer & Norman, 2020, p. 26).

Stakeholder analysis

A critical step in any VSD conceptual investigation is an analysis of those interested in or impacted by the technology and the relevant values they hold: the *stakeholder analysis*. Stakeholders include direct stakeholders, who will use the product, and indirect stakeholders, including project decision-makers and bystanders impacted by product use. The list of values can be based on empirical investigation, conceptual work by the design team, or values identified in existing VSD literature.

Table 22.1 Initial stakeholder analysis – Augmented Window

Type	*Stakeholder*	*Key values*	*Benefits*	*Harms*
Direct	Office employees (watchers)	Productivity, health	Connection, workplace comfort	Unease, distractions
Indirect	Passersby through plaza (the watched)	Privacy, safety, autonomy	Witnesses if there is an incident	Loss of anonymity and privacy; feeling surveilled; lack of consent
Indirect	Research team	Ethical research, responsibility, reflection	Empirical research, rich understanding of values in technology	Probing sensitive issues related to safety, surveillance, consent

Table 22.1 shows the beginnings of a stakeholder analysis of the Augmented Window case. In the classroom, an instructor might introduce the case, asking students to create or complete a similar table using their knowledge, related literature, and possibly short interviews with university students and personnel. Being able to distinguish between, and identify, direct and indirect stakeholders is an important learning outcome for students who may have previously focused their design attention on a narrowly scoped set of end-users and may not have considered how indirect stakeholders influence the adoption and impact of technology. Commonalities and differences in the students' stakeholder analyses are useful to discuss together in the classroom.

As stakeholder values are identified and more deeply understood, value tensions within and across stakeholder groups become evident. For example, the Augmented Window may give employees an increased sense of community or gains in productivity, which all stakeholders likely value broadly, but not at the expense of the privacy lost when public spaces are surveilled. Therefore, it is essential for stakeholder analysis to be considered a living document that VSD researchers can enrich as they integrate their deepening knowledge through empirical investigations, group reflection, and class discussions.

We also suggest increasing student reflection on students' own responsibilities and values by asking them to reflect on and include the research team's values, as seen in Table 22.1. This aligns with Borning and Muller's call for VSD researchers and designers to be more transparent and explicit about their values, methodological choices, and analysis (Borning & Muller, 2012). A shared and prioritized set of values is challenging to craft, yet is crucial because design action is guided by the character and responsibilities of the designer (Gray & Boling, 2016). The process of students considering, sharing, and negotiating project values greatly benefits from instructor guidance, as teachers can draw from well-established value frameworks, such as Schwartz's Theory of Basic Values (Schwartz, 2012).

Values hierarchies

Building upon stakeholder analysis, students benefit from a formalized approach to thinking about how values connect to or are translated into design. Such a translation is aided by the values hierarchy technique introduced by van de Poel (2013). A values hierarchy is a strategy for visualizing the relations between values, norms, and design requirements. It can be read from values downwards, with values *specified* into relevant norms, which are further *specified* into design requirements, or from design requirements upwards, where design requirements are *for the sake* of the specific norm, which is *for the sake* of the key value.

Thinking again of the Augmented Window case, building a values hierarchy would work from the key values already identified in the stakeholder analysis (Table 22.1). While any of the named values could be used, productivity, the original value motivating the Augmented Window case, is a good candidate for illustration. For this value, norms for the work environment are indicated. Norms can be brainstormed based on the example (as in Figure 22.1) or involve research into health and safety requirements or empirical findings on productivity in the workplace. Relevant norms are then further translated, or specified, into design requirements for the Augmented Window. The values hierarchy in Figure 22.1 provides an initial sketch; much more detail could be elaborated, especially concerning specific design requirements per norm.

Students can create 'values hierarchies' for any value. Even constructing cursory values hierarchies can help students get a better grip on values, as this technique works by turning values, which may be vague and abstract, into concrete, actionable requirements by connecting them to norms and then to design requirements. Including multiple values or stakeholders can also illuminate conflicts between values as they are translated into design requirements that may be inconsistent. (Already in Figure 22.1 we see a conflict between an imagined design requirement of having openable windows for fresh air and the use of a plasma display, substituting a real window. If expanded to include additional values, a values hierarchy would show conflicts between values of privacy and productivity.) Distinguishing between values, norms, and design requirements, and understanding the relationships between these, are key VSD intended learning outcomes that the values hierarchy technique helps students achieve.

Envisioning cards

Envisioning cards are another example of a VSD method useful for teaching. Nathan et al. (2008) proposed criteria – stakeholders, time, value, and pervasiveness – for envisioning systemic effects on persons and society. These were then formalized into 32 themed envisioning cards and accompanying toolkit (Friedman & Hendry, 2012). Each theme is accompanied with a tailored activity spurring designers to engage in actions such as think, identify, sketch, or act. The cards and activities together can engage students in directed and more purposeful ethics-based brainstorming than other ideation techniques.[9]

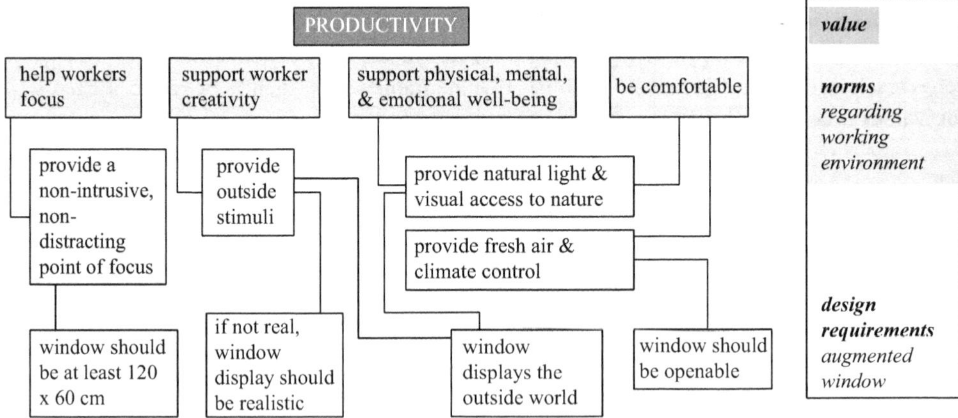

Figure 22.1 Productivity values hierarchy – Augmented Window.

In the Augmented Window example, students could pick an envisioning card they find relevant and thought-provoking to discuss and share. A possibility could be "The Long Now," which invites designers to think of long-term use of the technology, across generations and for people who would grow up with this technology. The power of the envisioning cards lies in their provocative design prompts. "The Long Now" prompts designers to sketch the interactions occurring over 5-year intervals as the technology and stakeholders shift over time. Another interesting card could be "Choosing Not to Use," presenting the possibility of a deliberate choice not to use the technology. This card could raise important questions of who actually is in a position to choose not to engage, as indirect stakeholders caught on camera may not be able to make this choice for themselves. Student engagement with envisioning cards can lead to vibrant discussions alongside probing design work.

Additional resources

In addition to the teaching-focused suggestions already elaborated, many VSD resources and toolkits are available to educators to guide or scaffold teaching. The most extensive and programmatic set of resources is the Value Sensitive Design in Higher Education (VASE) program, which provides a pedagogical framework and teaching resources targeted towards university-level design and development programs. VASE identifies three core competency pillars: (1) Ethics and Values, (2) Designers and Stakeholders, and (3) Technology and Design, which the creators use to anchor design phases, defined as value theory, research, synthesis, ideation, and evaluation. In the VASE pedagogical framework, each pillar is mapped to 28 teaching activities with corresponding assessment activities to support teachers in VSD teaching and evaluating students' learning outcomes in value-sensitive ways (Value Sensitive Design in Higher Education, 2021). Resources can also be found through the University of Washington's VSD Lab Cooperative[10] and the Delft Design for Values Institute.[11]

Finally, we suggest that the teachers make use of the broad, multidisciplinary, and ever-growing research literature on VSD. A key resource is the *Handbook of Ethics, Values, Technological Design* (van den Hoven et al., 2015) for its range of chapters on designing for specific values (from 'Human Wellbeing' to 'Presence'), as well as domains (from Economics and Fashion to Water Management). Conference proceedings and technical literature (especially from conferences affiliated with the Association for Computation Machinery, e.g., the CHI conference on Human Factors in Computing Systems; Conference on Computer Supported Cooperative Work; Conference on Computers and Accessibility) are excellent places to look for VSD literature that is especially accessible for engineering students and that offer some of the most recent applications of VSD.

Assessment

When applying VSD to educate responsible designers, there is a need to not only teach students about values in design but also assess their learning (Eriksson et al., 2022). Using different assessment forms that align with VSD teaching activities and support VSD-based intended learning outcomes (ILOs) can create a more substantiated VSD teaching and learning culture. The assessment forms and activities thus need to provide students involved in design and technology with an awareness of the role values play in design. This section connects VSD teaching and assessment activities to show how assessment *of*, *for*, and *as* learning creates different opportunities for assessing values in design. Four different assessment forms (summative, formative, ipsative, and authentic) are introduced, and teaching activities from the above section are connected with assessment activities developed specifically for teaching values in design, continuing to use the Augmented Window example for illustration.

Values-based teaching in education encompasses cognitive understanding as well as affective and behavioral components, requiring more than measuring knowledge or skills.[12] This makes it challenging to develop standardized assessment criteria (Dann, 2014; William & Thompson, 2008). Creating assessment methods that accurately gauge a student's ethical decision-making process is challenging and requires assessment activities that focus on this process – which is not easily quantifiable or measurable. In addition, design is a practical discipline. Assessing how students translate their understanding of values into practical design processes and solutions is something that requires methods sensitive to the methodological characteristics and specific teaching activities of VSD. Assessments focusing solely on, for example, theoretical knowledge may not effectively assess students' ability to incorporate values into their design work (Dann, 2014; Friedman & Hendry, 2019; William & Thompson, 2008).

Overall, assessing students' learning about values in design is complex due to the subjective nature of values, the abstract concepts involved, the multifaceted nature of assessment, and the ethical considerations of VSD. It requires thoughtful consideration and appropriate assessment strategies to ensure fair and meaningful evaluation (Frauenberger et al., 2016; Friedman & Hendry, 2012; William & Thompson, 2008). To address these challenges in teaching, a combination of assessment methods can be used, including open-ended assignments, portfolio reviews, case studies, and peer feedback. This is to attain alignment between *how teachers teach* VSD and *how students' VSD learning is evaluated*, as well as for teachers to gain a more comprehensive understanding of students' learning about values in design. It is crucial to provide VSD-focused and operational assessment criteria that can foster open discussions, critical thinking, and self-reflection among students (Hughes et al., 2014).

Below, using the VASE resource, we illustrate how concrete VSD teaching and assessment activities can be aligned to evaluate students' learning. Here, the VASE teaching activities of *listing stakeholders and their values* (using stakeholder analysis), *understanding value tensions* (working with values hierarchies), and *envisioning future scenarios* (using envisioning cards) make use of the four complementary assessment types.

Summative assessment/activities assess students' knowledge and comprehension of teaching material. Summative assessment refers to evaluating the student's learning, knowledge proficiency, or success at the end of the process in terms of their understanding, application, or critical analysis of design principles and ethical considerations. A VSD-relevant and sensitive summative assessment activity related to envisioning cards could be *case-based assessment for responsible designers*,[13] where students are asked to focus on imaging potential consequences, long-term effects, and societal impacts through a value scenario that goes beyond what would normally be described as intended use. Students can then analyze the potential consequences of the case, using relevant envisioning criteria (including values) through a value scenario and provide suggestions for how to mitigate negative consequences (e.g., regarding re-design, further stakeholder dialogue, possible tensions) through re-thinking the design.

Formative assessment/activities support students and teachers in identifying and closing knowledge gaps. Formative assessment promotes reflection about learning and teaching and charts the development of these processes over time. It is utilized by teachers to gain an understanding of their students' current knowledge and skills to guide future and formative learning. Concerning the above example of working with values hierarchies, teachers can formatively evaluate students' understanding of value tensions by having them write a *reflective values report*[14] about perceived value tensions within the case. Value tensions show how a product, system, or service, like the Augmented Window, can contain elements that compromise or undermine some stakeholders' values. Students reflect on how identified value tensions relate to or emerge from the design's values, how the values of different stakeholders may be at odds with each other, and why value tensions are important to consider.[15] Formative assessment and activities are ideal for guiding how this knowledge affects students' future thinking and practice.

Ipsative assessment/activities enable teachers to compare a student's current work with previous work – either in the same field through time or in comparison with other fields. Ipsative assessment is a highly personalized form of assessment where progress is measured against the needs and goals of the individual, not in comparison to external standards or peers' performance. As such, ipsative assessment activities focus on the student's ability to describe changes within their acquisition of knowledge, concepts, ideas, beliefs, and facts related to working with values in design. A *personal values-reflection video*[16] is a useful method for ipsative assessment. Here, teachers can ask students to record a group video about stakeholder values with a focus on how the VSD analysis shifted students' views on what being a responsible designer requires and how they might now approach design differently. To ensure alignment between the teaching and assessment activity, students should focus on how they now understand the diversity of possible stakeholders in new ways and how they can reflect on the possible consequences of considering diverse stakeholders in ways they were not able to before.

Authentic assessment/activities emphasize the importance of contextualized design activities and assignments focusing on a problem identification that points to a 'real-life practice' wherein students must present their ability to translate and integrate their knowledge. The participation of stakeholders or people from an external community of practice often constitutes an essential premise for conducting authentic assessment. However, this might not always be possible. Here, students can use their emerging understanding and identity of what it means to be a responsible designer to engage in authentic assessment dialogues with each other. Teachers can facilitate authentic assessment through *peer feedback for responsible designers*[17] by asking students to take on the role of VSD teachers and designers in relation to specific VSD assessment criteria applied to each other's stakeholder analysis. In this assessment activity, it is important for the teacher to focus on the students' abilities to capture and address the visible signs of learning through peer feedback: to think, act, and behave like a responsible designer.

In conclusion, using different assessment forms can create a more varied teaching and learning culture as well as enhance alignment between teaching and evaluating VSD. It is crucial to align assessment methods with teaching approaches and ILOs to effectively teach values in design. Even then, however, assessing students' understanding of values in design presents unique challenges. Values are subjective and contextual and often involve complex and abstract concepts. Evaluating the students' comprehension of ethical values requires more than measuring knowledge or skills – it requires assessing their attitudes, decision-making processes, and practical application in real-world scenarios. By adopting the principles of assessment *for* learning and assessment *as* learning, teachers can create a more holistic approach to assessing students' understanding of values in design. This demands an intentional and comprehensive assessment strategy that combines different assessment types and activities that are aligned with VSD teaching activities and ILOs, and that encourages student engagement, reflection, and critical thinking. By aligning assessment practices and criteria with teaching practices and ILOs, teachers can support and promote students' understanding of values in design and their ability to apply them effectively in real-world scenarios and future professional practice.

Extending VSD

The classroom setting provides an opportunity to adapt and refine VSD's methods for teaching. We, as educators of future engineers and designers, have the responsibility to ensure that what and how we teach responds to the changing contexts and demands of their future work. We conclude the chapter by highlighting three critiques of VSD that, in our view, represent essential developments in the VSD literature and that pose rich educational opportunities. Here again, we use the Augmented Window example to show how established ideas and cases can be adapted and updated based on developments in the field.

Power

Accounting for power is not explicitly addressed in the VSD framework, beyond legitimizing direct and indirect stakeholders (Friedman & Hendry, 2019). This has been a point of recent critique (Jacobs et al., 2021). Even the act of deciding how to identify relevant values involves deciding *whose* values to elicit. But this can also be limited, which the Augmented Window example aptly illustrates. Friedman et al. (2006) explicitly considered power dynamics in their research, inquiring about and contrasting the privacy beliefs of both the direct and indirect stakeholders, all of whom were people working at or present on a university campus. While directly relevant in this case and methodologically sound, choosing to engage people physically present on a university campus is a choice that may overrepresent socially privileged populations and fail to include populations whose perspectives are often not considered in technology design. This is especially relevant in the United States, where communities of specific racial or socio-economic backgrounds are over-surveilled (Billies, 2015). Regarding power dimensions, this reveals a limitation of the VSD framework which makes the concrete technology the focal point: by homing in on this specific window operating in this specific location – a space of privilege – the researchers gained a robust understanding of privacy vis-à-vis the Augmented Window, but missed understanding how power dynamics determine who even features in this research and how conceptions of privacy may vary more broadly.

One approach that can address this shortcoming in VSD is to give voice to communities historically underrepresented in design and technology initiatives. Research on the experiences of autistic students in higher education, a setting in which autistic students experience lower graduation rates than neurotypical peers (White et al., 2016), demonstrates this approach. Rather than interviewing faculty and neurotypical students, Zolyomi et al. (2017, 2018) interviewed autistic students and disability services staff in higher education to understand the dynamics of class-based teamwork from the perspective of neurodivergent students and those who directly support them with disability services. Based on interview insights, key values of autistic students were identified as freedom from stigma, individual comfort, social comfort, social connection, and team cohesion. Investigating the more dominant stakeholders in higher education would elicit different values and value tensions.

The decisions regarding how to account for power relations highlight the responsibility of researchers to practice reflexivity. VSD researchers can describe their own positionality, reflecting on the influence of their lived experiences and potential bias on their methodological choices and interpretation of data – as demonstrated in Alsheikh et al. (2011) and delineated as a researcher stance by Yoo et al. (2013). Bringing these considerations into the classroom trains engineers and designers from an early stage to be sensitive to who is represented, who has agency, and how power dimensions shape these and other aspects of design.

Instructors can also prepare students to navigate issues of *organizational power* when using VSD in practice beyond educational environments. Many challenges to implementing VSD in practice are organizational. In other organizational and institutional contexts, students may work with others unfamiliar with VSD or may not have the social power to implement VSD on their own. There can be tension between an individual's knowledge and desire to practice VSD, co-workers' awareness of values as considerations in design, and ongoing organizational practices (Chivukula et al., 2020) – particularly in private industry when taking the time and resources to address values may be seen as conflicting with a company's profit motive. Although the Augmented Window example project was conducted by a research lab explicitly interested in using and developing VSD principles, a team in a different research lab or private company could easily develop a similar Augmented Window focusing on its technical development or its potential as a profitable product. In those situations, a VSD-minded engineer would need to navigate the social dynamics of that team in order to convince others to follow a VSD approach.

This suggests that students should also learn how to navigate organizational contexts when attempting to implement VSD. Strategies might include having students view themselves as a 'values advocate' to help educate others on a team or in an organization (Shilton, 2013); attending to the emotional labor often required to advocate for values within organizations (Su et al., 2021; Wong, 2021); considering when they might choose to resist or conform with organizational norms when advocating for addressing values issues (Wong, 2021); understanding potential allies within an organization such as 'ethics owners' who may be more empowered decision-makers (Metcalf et al., 2019); or seeking support through community and collective action (Pillai et al., 2022). Introducing these strategies in a classroom environment might use Authentic Assessment techniques and involve group discussions or role-playing, e.g. (Shilton et al., 2020).

Norms and multiculturalism

In addition to considering whose values to elicit, designers should also consider whose cultural norms are reflected in design processes. Much VSD literature suggests human values are or could be universal, aligning, for instance, with the United Nations' Declaration of Human Rights. However, this perspective has been critiqued. JafariNaimi et al. (2015) argue that values are situated in people's lived experiences and practices, and the expression of the same value may look very different in different times and places. For instance, conceptions and practices of privacy differ across regions and cultural backgrounds and are based on the use of online tools (Abokhodair & Hodges, 2019). Designers can attune themselves to these issues by working with people from diverse communities and backgrounds, considering multiple conceptions and dimensions of the same value – for example, Mulligan et al. (2016)'s 'Analytic' tool for privacy – by using tools that foreground consideration of multiculturalism (e.g., the multicultural envisioning card). Here, we suggest that education has a key role to play. Already pointing out cultural differences and the possibilities for bias as part of value conceptualization helps students reflect on their own positions and assumptions. Raising values as hypotheses (JafariNaimi et al., 2015) to be explored and investigated from various perspectives opens up VSD to more diverse and participatory practices.

Alternatively, Martin et al. (2023) propose *norm*-sensitive design, arguing that norms better capture behaviors and avoid Western biases encoded in values-based frameworks. However, norms can also reinforce existing inequalities or arrangements of powers within a community (McDonald & Forte, 2020). The Augmented Window example, developed in the early 2000s, showcases these issues. Norms about recording in public spaces have changed such that people's responses would likely differ (informed perhaps by more stringent data regulations or the proliferation of doorbell cameras). Discussing this case with students could illustrate how norms and values change even over relatively short timescales and how values are interpreted differently in different cultural contexts.

Multi-species VSD

The human-centered perspective in design and engineering privileges humans over all other species but becomes inadequate in the era of the Anthropocene (Crutzen, 2006; Haraway, 2015; see also Chapter 6). We need new approaches in engineering education capable of engaging and caring for multiple species and environments, and where non-human beings are also considered users, designers, participants, and stakeholders of technologies. In addition to this being a ripe area for research, we suggest educators can contribute to opening up VSD to multi-species frameworks.

Speculative design presents a common approach for considering values and multi-species perspectives (e.g., Nijs et al., 2020; Smith & Qaurooni, 2020), and works well in educational contexts. Speculative design relies on speculation and proposition, aims to enact change, and can be useful for understanding future consequences and implications of the entangled relationship between multiple species, technology, and humans (Auger, 2013; Dunne & Raby, 2013). A hands-on approach is to train students to adapt VSD methods to include multi-species perspectives in the design process, for example, by creating a set of non-human personas (Tomitsch et al., 2021) representing various species to be included as direct or indirect stakeholders. Non-human personas can be used to inform the initial concept design, to evaluate potential solutions, or to critique a design solution through the perspective of non-human stakeholders.

Like most other VSD applications, the Augmented Window is anthropocentric for considering only human stakeholders. But more deeply, it might exemplify VSD's limited engagement with the non-human. VSD rarely considers non-human stakeholders in its processes. And this example shows that if non-human nature does feature, it is as something to behold, from indoors, something that might be instrumentalized (in this case, to increase worker productivity), not something to engage, or to design for or with. A multi-species interpretation of VSD would challenge this relation, urging that non-human perspectives and agencies are not only considered in stakeholder analyses but that new forms of engagement are developed, bringing multi-species actors into the design process. We believe this broadening of design and engineering is needed and that a multi-species expansion of VSD in our classrooms can cultivate engineers and designers with the requisite attention and responsiveness for addressing mounting environmental crises.

Conclusion

We have advanced VSD as an approach for helping students develop sensitivity to values and stakeholders in the design and engineering practices needed to be responsive and responsible designers and engineers. VSD provides an excellent framework for teaching values through technologies in development. We have highlighted VSD's open-ended and iterative nature, showing how existing examples from the VSD literature can be further developed and how existing VSD tools and methods are fruitfully repurposed for teaching. We have also prompted teachers to respond to challenges posed to VSD to make the framework more responsive to power dimensions, cultural differences, and non-human perspectives. We conclude with a final point of encouragement for teachers: while we have highlighted the educational potential of VSD for students, especially because of its flexibility and openness, VSD is also rewarding for teachers. As teachers using VSD in our classrooms, we have found it to be highly generative, allowing for experimentation, iteration, connections with research, and rich discussions with our students, who often come to see technologies and their role in creating them in new ways. We hope this chapter motivates more educators to adopt and adapt VSD for their students.

Acknowledgments

The authors thank Ibo van de Poel, Esther Matemba, Christine Boshuijzen-van Burken, the editors, and the other authors in the Teaching Ethics subsection for their critical and constructive feedback, and Nassim Parvin for her contributions to the ideation of this chapter.

Notes

1 Human welfare, ownership and property, privacy, freedom from bias, universal usability, trust, autonomy, informed consent, accountability, courtesy, identity, calmness, and environmental sustainability.
2 Again, we direct readers and students to Friedman et al. 2008 for a more thorough introduction to VSD than we can provide here.
3 Possible strategies from our institutions include the use of online testimonials in healthcare technologies (van Grunsven et al., 2023) and Amazon reviews of voice assistants to analyze user experiences (Olya Kudina).
4 Andersen and Cawthorne (2021) use technical investigations to address educational challenges in value-sensitive drone design.
5 Langdon Winner's "Do Artifacts have Politics?" (1980) shows this through the famous example of Robert Moses's low-hanging bridges and the far less famous example of the mechanical tomato harvester.
6 As well as their limitations: see Friedman et al. (2006).
7 This definition has been criticized: e.g., Manders-Huits (2011). Van De Poel (2015) provides a helpful orientation to values for engineers.
8 The extended stakeholder research and discussion in Friedman et al. (2006) can be used to illustrate the full extent of empirical investigations in VSD, or to supplement when conducting empirical investigations not possible in class.
9 Card decks can be downloaded and printed for free: https://vsdesign.org/toolkits/
10 https://vsdcoop.ischool.uw.edu/index.php/VSD_Coop
11 https://www.delftdesignforvalues.nl/fundamentals/
12 See Chapter 27 for a detailed discussion of assessment and common learning objectives in engineering ethics.
13 See https://teachingforvaluesindesign.eu/A9_casebasedassessment.html for a step-by-step description.
14 See https://teachingforvaluesindesign.eu/A2_reflectivevaluesreport.html.
15 See https://teachingforvaluesindesign.eu/20_understandingvaluetensions.html.
16 See https://teachingforvaluesindesign.eu/A3_personalvaluesreflectionvideo.html.
17 See https://teachingforvaluesindesign.eu/A8_peerfeedback.html.

References

Abokhodair, N., & Hodges, A. (2019). Toward a transnational model of social media privacy: How young Saudi transnationals do privacy on Facebook. *New Media & Society*, *21*(5), 1105–1120. https://doi.org/10.1177/1461444818821363

Accreditation Board for Engineering and Technology. (2022). *Criteria for accrediting engineering programs, 2022 – 2023 | ABET*. https://www.abet.org/accreditation/accreditation-criteria/criteria-for-accrediting-engineering-programs-2022-2023/

Alsheikh, T., Rode, J. A., & Lindley, S. E. (2011). (Whose) value-sensitive design: A study of long-distance relationships in an Arabic cultural context. *Proceedings of the ACM 2011 Conference on Computer Supported Cooperative Work*, 75–84. http://dl.acm.org/citation.cfm?id=1958836

Andersen, T., & Cawthorne, D. (2021). Value-sensitive design of unmanned aerial systems: Using action research to bridge the theory-practice gap. *International Journal of Technoethics*, *12*(2), 17–34. https://doi.org/10.4018/IJT.2021070102

Association for Computer Machinery. (2018). *ACM code of ethics & professional conduct*. https://www.acm.org/code-of-ethics

Auger, J. (2013). Speculative design: Crafting the speculation. *Digital Creativity*, *24*(1), 11–35. https://doi.org/10.1080/14626268.2013.767276

Biffl, S., Aurum, A., Boehm, B., Erdogmus, H., & Grünbacher, P. (Eds.). (2006). *Value-based software engineering*. Springer.

Billies, M. (2015). Surveillance threat as embodied psychological dilemma. *Peace and Conflict: Journal of Peace Psychology*, *21*(2), 168–186. https://doi.org/10.1037/pac0000070

Borning, A., & Muller, M. (2012). Next steps for value sensitive design. *Proceedings of the SIGCHI Conference on Human Factors in Computing Systems*, 1125–1134. http://dl.acm.org/citation.cfm?id=2208560

Chivukula, S. S., Watkins, C. R., Manocha, R., Chen, J., & Gray, C. M. (2020). Dimensions of UX practice that shape ethical awareness. *Proceedings of the 2020 CHI Conference on Human Factors in Computing Systems*, 1–13. https://doi.org/10.1145/3313831.3376459

Crutzen, P. J. (2006). The "anthropocene". In E. Ehlers & T. Krafft (Eds.), *Earth system science in the anthropocene* (pp. 13–18). Springer. https://doi.org/10.1007/3-540-26590-2_3

Cummings, M. L. (2006). Integrating ethics in design through the value-sensitive design approach. *Science and Engineering Ethics*, *12*(4), 701–715. https://doi.org/10.1007/s11948-006-0065-0

Dann, R. (2014). Assessment as learning: Blurring the boundaries of assessment and learning for theory, policy and practice. *Assessment in Education: Principles, Policy & Practice*, *21*(2), 149–166. https://doi.org/10.1080/0969594X.2014.898128

Davis, J., & Nathan, L. P. (2015). Value sensitive design: Applications, adaptations, and critiques. In J. Van Den Hoven, P. E. Vermaas, & I. Van De Poel (Eds.), *Handbook of ethics, values, and technological design* (pp. 11–40). Springer Netherlands. https://doi.org/10.1007/978-94-007-6970-0_3

Dunne, A., & Raby, F. (2013). *Speculative everything: Design, fiction, and social dreaming*. The MIT Press.

Eriksson, E., Barendregt, W., Nilsson, E. M., & Nørgård, R. T. (2021). Teaching values in design in higher education - Towards the new normal. *Conference on the Ethical and Social Impacts of ICT – Ethicomp, 2021*.

Eriksson, E., Kok, A. L., Barendregt, W., Gyldendahl Jensen, C., & Nilsson, E. M. (2022). Teaching for values in interaction design: A discussion about assessment. *Interaction Design and Architecture(s)*, *52*, 221–233. https://doi.org/10.55612/s-5002-052-012

Frauenberger, C., Rauhala, M., & Fitzpatrick, G. (2016). In-action ethics. *Interacting with Computers*. iwc;iww024v1. https://doi.org/10.1093/iwc/iww024

Friedman, B., & Hendry, D. (2012). The envisioning cards: A toolkit for catalyzing humanistic and technical imaginations. *Proceedings of the SIGCHI Conference on Human Factors in Computing Systems*, 1145–1148. https://doi.org/10.1145/2207676.2208562

Friedman, B., & Hendry, D. G. (2019). *Value sensitive design*. MIT Press.

Friedman, B., Kahn Jr, P. H. K., & Borning, A. (2008). Value sensitive design and information systems. In K. E. Himma and H. T. Tavani (Eds.), *The handbook of information and computer ethics* (pp. 69–101). John Wiley & Sons, Inc.

Friedman, B., Kahn Jr., P. H., Hagman, J., Severson, R. L., & Gill, B. (2006). The watcher and the watched: Social judgments about privacy in a public place. *Human–Computer Interaction*, *21*(2), 235–272. https://doi.org/10.1207/s15327051hci2102_3

Gray, C. M., & Boling, E. (2016). Inscribing ethics and values in designs for learning: A problematic. *Educational Technology Research and Development*, *64*(5), 969–1001. https://doi.org/10.1007/s11423-016-9478-x

Haraway, D. (2015). Anthropocene, capitalocene, plantationocene, chthulucene: Making Kin. *Environmental Humanities*, *6*(1), 159–165. https://doi.org/10.1215/22011919-3615934

Hughes, G., Wood, E., & Kitagawa, K. (2014). Use of self-referential (ipsative) feedback to motivate and guide distance learners. *Open Learning: The Journal of Open, Distance and e-Learning*, *29*(1).

Jacobs, M., Kurtz, C., Simon, J., & Böhmann, T. (2021). Value sensitive design and power in socio-technical ecosystems. *Internet Policy Review*, *10*(3). https://doi.org/10.14763/2021.3.1580

JafariNaimi, N., Nathan, L., & Hargraves, I. (2015). Values as hypotheses: Design, inquiry, and the service of values. *Design Issues*, *31*, 91–104. https://doi.org/10.1162/DESI_a_00354

Manders-Huits, N. (2011). What values in design? The challenge of incorporating moral values into design. *Science and Engineering Ethics*, *17*(2), 271–287. https://doi.org/10.1007/s11948-010-9198-2

Martin, D. A., Clancy, R. F., Zhu, Q., & Bombaerts, G. (2023). Why do we need norm sensitive design? A WEIRD critique of value sensitive approaches to design. *Global Philosophy*, *33*(4), 40. https://doi.org/10.1007/s10516-023-09689-9

McDonald, N., & Forte, A. (2020). The politics of privacy theories: Moving from norms to vulnerabilities. *Proceedings of the 2020 CHI Conference on Human Factors in Computing Systems*, 1–14. https://doi.org/10.1145/3313831.3376167

Metcalf, J., Moss, E., & boyd, d. (2019). Owning ethics: Corporate logics, silicon valley, and the institutionalization of ethics. *Social Research: An International Quarterly*, *86*(2), 449–476. https://doi.org/10.1353/sor.2019.0022

Meyer, M. W., & Norman, D. (2020). Changing design education for the 21st century. *She Ji: The Journal of Design, Economics, and Innovation*, *6*(1), 13–49. https://doi.org/10.1016/j.sheji.2019.12.002

Mulligan, D. K., Koopman, C., & Doty, N. (2016). Privacy is an essentially contested concept: A multi-dimensional analytic for mapping privacy. *Philosophical Transactions: Mathematical, Physical and Engineering Sciences*, *374*(2083), 1–17.

Nathan, L. P., Friedman, B., Klasnja, P., Kane, S. K., & Miller, J. K. (2008). Envisioning systemic effects on persons and society throughout interactive system design. *Proceedings of the 7th ACM Conference on Designing Interactive Systems*, 1–10. http://dl.acm.org/citation.cfm?id=1394446

Nijs, G., Laki, G., Houlstan, R., Slizewicz, G., & Laureyssens, T. (2020). Fostering more-than-human imaginaries: Introducing DIY speculative fabulation in civic HCI. *Proceedings of the 11th Nordic Conference on Human-Computer Interaction: Shaping Experiences, Shaping Society*, 1–12. https://doi.org/10.1145/3419249.3420147

Pillai, A. G., Sachathep, T., & Ahmadpour, N. (2022). Exploring the experience of ethical tensions and the role of community in UX practice. *Nordic Human-Computer Interaction Conference*, 1–13. https://doi.org/10.1145/3546155.3546683

Rocco, R, Thomas, A., & Novas-Ferradás, M. (Eds.). (2022). *Teaching design for values*. TU Delft OPEN Books. https://doi.org/10.34641/mg.54

Schwartz, S. H. (2012). An overview of the Schwartz theory of basic values. *Online Readings in Psychology and Culture*, *2*(1). https://doi.org/10.9707/2307-0919.1116

Shilton, K. (2013). Values levers: Building ethics into design. *Science, Technology, & Human Values*, *38*(3), 374–397.

Shilton, K., Heidenblad, D., Porter, A., Winter, S., & Kendig, M. (2020). Role-playing computer ethics: Designing and evaluating the privacy by design (PbD) simulation. *Science and Engineering Ethics*, *26*(6), 2911–2926. https://doi.org/10.1007/s11948-020-00250-0

Smith, N., & Qaurooni, D. (2020). The inclusion zone: Grounded speculations in Chernobyl. *Proceedings of the 2020 ACM Designing Interactive Systems Conference*, 1929–1942. https://doi.org/10.1145/3357236.3395574

Su, N. M., Lazar, A., & Irani, L. (2021). Critical affects: Tech work emotions amidst the techlash. *Proceedings of the ACM on Human-Computer Interaction*, *5*(CSCW1), 1–27. https://doi.org/10.1145/3449253

Tomitsch, M., Fredericks, J., Vo, D., Frawley, J., & Foth, M. (2021). *Non-human personas: Including nature in the participatory design of smart cities*. http://dx.doi.org/10.55612/s-5002-050-006

Value Sensitive Design in Higher Education (VASE). (2021). *Teaching for values in design: Creating conditions for students to grow into responsible designs and developers of future technologies*. https://teachingforvaluesindesign.eu/assets/materials/VASEBooklet_25082021.pdf

van de Poel, I. (2013). Translating values into design requirements. In D. P. Michelfelder, N. McCarthy, & D. E. Goldberg (Eds.), *Philosophy and engineering: Reflections on practice, principles and process* (Vol. 15, pp. 253–266). Springer Netherlands. https://doi.org/10.1007/978-94-007-7762-0_20

van de Poel, I. (2015). Design for values in engineering. In J. Van Den Hoven, P. E. Vermaas, & I. Van De Poel (Eds.), *Handbook of ethics, values, and technological design* (pp. 667–690). Springer Netherlands. https://doi.org/10.1007/978-94-007-6970-0_25

van den Hoven, J., Vermaas, P. E., & van de Poel, I. (Eds.). (2015). *Handbook of ethics, values, and technological design: Sources, theory, values and application domains*. Springer Netherlands. https://doi.org/10.1007/978-94-007-6970-0

van Grunsven, J., Gammon, A. R., Franssen, T., & Marin, L. (2023). Tinkering with technology: An exercise in inclusive experimental engineering ethics. In E. Hildt, K. Laas, & E. Brey (Eds.), *Building inclusive ethical cultures in STEM* (pp. 289–311). Springer.

White, S. W., Elias, R., Salinas, C. E., Capriola, N., Conner, C. M., Asselin, S. B., Miyazaki, Y., Mazefsky, C. A., Howlin, P., & Getzel, E. E. (2016). Students with autism spectrum disorder in college: Results from a preliminary mixed methods. *Research in Developmental Disabilities*, *56*, 29–40. https://doi.org/10.1016/j.ridd.2016.05.010

William, D., & Thompson, M. (2008). Integrating assessment with learning: What will it take to make it work? In C. A. Dwyer (Ed.), *The future of assessment*. Routledge.

Winner, L. (1980). Do artifacts have politics? *Daedelus*, *109*(1), 121–136.

Wong, R. Y. (2021). Tactics of soft resistance in user experience professionals' values work. *Proceedings of the ACM on Human-Computer Interaction*, *5*(CSCW2), 355:1–355:28. https://doi.org/10.1145/3479499

Yoo, D., Lake, M., Nilsen, T., Utter, M. E., Alsdorf, R., Bizimana, T., Nathan, L. P., Ring, M., Utter, E. J., Utter, R. F., & Others. (2013). Envisioning across generations: A multi-lifespan information system for international justice in Rwanda. *Proceedings of the SIGCHI Conference on Human Factors in Computing Systems*, 2527–2536. http://dl.acm.org/citation.cfm?id=2481349

Zolyomi, A., Ross, A. S., Bhattacharya, A., Milne, L., & Munson, S. (2017). Value sensitive design for neurodiverse teams in higher education. *Proceedings of the 19th International ACM SIGACCESS Conference on Computers and Accessibility*, (ASSETS '17), 353–354. https://dl.acm.org/doi/10.1145/3132525.3134787

Zolyomi, A., Ross, A. S., Bhattacharya, A., Milne, L., & Munson, S. A. (2018). Values, identity, and social translucence: Neurodiverse student teams in higher education. *Proceedings of the 2018 CHI Conference on Human Factors in Computing Systems*, 1–13. https://doi.org/10.1145/3173574.3174073

23

ETHICS IN SERVICE-LEARNING AND HUMANITARIAN ENGINEERING EDUCATION

Scott Daniel, Adetoun Yeaman, and William Oakes

Introduction

This chapter reviews different ways ethics is operationalized in educational experiences in various field contexts in service-learning, humanitarian engineering, and more. After describing our positionality as authors and then defining and problematizing humanitarian engineering and service-learning, we discuss different ways in which ethics education can be integrated, and sometimes targeted, in these pedagogies. We close the chapter by considering classroom approaches, such as scenario-based teaching, which can prime students to consider ethical dimensions in these field experiences, before concluding with some overall insights and recommendations for educators.

We recognize that this type of learning experience can challenge traditional Western ideas of what counts as engineering knowledge, what it is to be an engineer, and how engineering relates to international development – see for example Cruz (2021) or Kleba & Reina-Rozo (2021) for some perspectives from the Global South about decolonizing engineering education or rethinking development. In this chapter, we will not address these grander epistemological and ontological questions but focus instead on our own understanding of ethics education within these fields, while acknowledging that service-learning and humanitarian learning experiences can be transformative for students and lead them to raise many of these questions for themselves.

Positionality statements

In the following paragraphs, each of us has shared our positionality regarding this chapter's content.

Scott Daniel's pathway to humanitarian engineering has been circuitous. Having come from a privileged, culturally near-homogenous white WEIRD (Western, educated, industrialized, rich, democratic) Sydney upbringing, he has become aware of how abnormal, globally, his background is. This came first through teaching in a low socio-economic, multicultural high school, then doing science outreach in remote, under-resourced schools around Australia, to now having worked and traveled in almost 30 developing countries, helping Scott develop an acute sense of his privilege and the importance of cultural humility.

DOI: 10.4324/9781003464259-28

Scott's career has explored STEM, education, and social justice, with them all combining in his recent work with Engineers without Borders Australia and his current role in developing humanitarian engineering education at the University of Technology Sydney. Although his technical training was in mathematics and physics, he is drawn to humanitarian engineering education in part because of its inherent ambiguity and complexity. He sees the ultimate purpose of ethics education as developing ethical behaviors in his students, with the more immediate goals of ensuring students engage respectfully with community stakeholders, normalizing that ethical decision-making is a part of engineering, and equipping students with tools and frameworks to support ethical reasoning.

Adetoun Yeaman grew up in Nigeria, getting early exposure to humanitarian work through her mother, a research scientist who engages in action research on socioeconomic issues. Beyond this informal connection to the topic of this chapter, she has engaged more formally through taking an undergraduate service-learning course and getting involved with a humanitarian engineering organization (Engineering World Health) as a Master's student. Subsequently, she worked in a hospital maintenance unit in Nigeria as part of her service in the National Youth Service Corps, which increased her consciousness about humanitarian donation issues and medical equipment management. These experiences motivated her interest in understanding empathy, especially as it relates to engineering design and service-oriented projects and programs. Adetoun's Ph.D. dissertation research on empathy within service-learning contexts positioned her to better understand various service-learning models, how they operate, the challenges they experience, the role that empathy plays in such experiences (while also increasing her curiosity about community engagement more broadly), and some of the reasons for 'successes' and 'failures' of such endeavors. Lastly, she recently concluded a postdoctoral fellowship at Wake Forest University, supporting curriculum development in engineering ethics education through a character/virtue ethics lens. Adetoun continues to value and engage in activities at the intersection of engineering, community engagement, and societal impact.

William (Bill) Oakes was born and raised in the Midwest of the United States. Bill started his career as a mechanical engineer in the aerospace industry and shifted to an academic career, where he was assigned to teach in a community-engaged design course, EPICS (Engineering Projects in Community Service). That experience flipped his view of engineering education upside down, and the work connected to the social activism that was part of his family history. Bill has worked for 25 years in community-engaged learning, building a large program at his own university and working with colleagues from within the United States and 17 other countries. His scholarship now is focused on engineering education research and in particular how to enhance design-based community-engaged learning.

Defining and contrasting humanitarian engineering and service-learning

Humanitarian engineering

Humanitarian engineering has been defined as "design under constraints to directly improve the wellbeing of underserved populations" (Mitcham & Muñoz, 2009, p. 191). However, the term remains contested, with humanitarian engineering being conceptualized or defined in numerous ways (Turner et al., 2015). One overarching definition recently put forward is that "Humanitarian Engineering involves engineering practice to support and work with vulnerable communities and/ or individuals" (Brown et al., 2022, p. 2). Such a broad umbrella term has meant that humanitarian engineering is sometimes seen to encompass a spectrum from assistive technologies through to development assistance, humanitarian disaster relief, and peace-building (Brown et al., 2022), covering sectors including water, sanitation, energy, agriculture, health, and more (Hazeltine &

Bull, 1999). With such diversity, it is often considered an approach (or perhaps a set of approaches) that can be applied to all engineering, rather than a specific engineering discipline in its own right (Brown et al., 2022).

Perhaps because of this complexity, and different practitioners highlighting different aspects or nuances, a number of terms have been (and are being) used internationally, thwarting comparisons of education and practice in the area (Smith, Tran, et al., 2019). Some of these terms include 'engineering for developing communities,' 'development engineering,' 'engineering for social justice,' 'peace engineering,' 'global engineering,' and more. This morass of terminology has led to some concerted efforts to define terms and reach consensus, which remains an ongoing challenge in progressing the field (Brown et al., 2022; Burleson et al., 2023; Turner et al., 2015).

What is certain is the growing interest and recognition of the area, with, for example, the first Australian humanitarian engineering subject being offered at the Australian National University in 2015. Now, more than a dozen Australian universities offer subjects, minors, or majors. This growth has been paralleled internationally, with several reports mapping the increase in 'Engineering for Global Development' education programs in different global regions being produced by 'Engineering for Change' in recent years, see for example Kunwar (2020), Peiffer (2019), and Rojas (2020).

Although acknowledging the diversity of terms used almost synonymously with humanitarian engineering, we will use the one term from here on. This is in part for simplicity, and in part because, at least in the Australian context, the term has gained official sanction. The peak body Engineers Australia (EA) declared 2011 the 'Year of Humanitarian Engineering,' and then in 2019 the Humanitarian Engineering Community of Practice was established within the EA College of Leadership and Management (Smith, Lynch, et al., 2019). At the time of writing, the community of practice engages its 800+ members with regular newsletters and events, all under the banner of humanitarian engineering.

In the academic sector, the Australian and New Zealand Standard Research Classification tracks research using a catalog of numerical codes. Until 2020, humanitarian engineering research was categorized anonymously under *Interdisciplinary Engineering not elsewhere classified*, if at all. However, in 2020, a 'Field of Research' code was assigned specifically to Humanitarian engineering (Australian Bureau of Statistics, 2020). This ratification meant that for the first time, the increasing number of academics with named positions in humanitarian engineering could have their research officially recognized.

This official endorsement of the term means that, at least in the Australian context, the term is taken as given, which has meant that the research conversation is starting to move on to other questions. These include investigating what skills and competencies are associated with (or required in) humanitarian engineering, how it relates to other disciplines and to engineering in general, and if and to what extent it should be professionalized (Brown et al., 2022) in light of critiques of the practice of humanitarian engineering being associated with, for example, neo-colonialism (Sagoe, 2012) or voluntourism (Birzer & Hamilton, 2019).

Service-learning

One widely used definition of service-learning is a "course-based, credit-bearing educational experience in which students participate in an organised service activity that meets identified community needs, and reflect on the service activity in such a way as to gain further understanding of course content, a broader appreciation of the discipline, and an enhanced sense of civic responsibility" (Bringle & Hatcher, 1995, p. 112). Service-learning has also been described as a "form of experiential education whose pedagogy rests on principles established by Dewey … where learning occurs through a cycle of action and reflection and not simply through being able to recount what was learned through reading and lecture" (Eyler & Giles, 1999, p. 7). These definitions highlight three important components where students are:

a) engaged in a credit-bearing course.
b) engaged with a community and working towards meeting a need or solving a problem within that community.
c) reflecting on their experiences.

Bringle and Hatcher highlight how service-learning can promote "understanding of course content" (1995, p. 112). Eyler and Giles (1999) also point to this opportunity for learning, though not necessarily of course content, through action and reflection. Service-learning experiences present an opportunity for students to deeply understand people and situations by virtue of authentic engagement and reflection, with the potential to cultivate character strengths such as empathy and compassion. Bringle and Hatcher's definition also addresses 'civic responsibility,' which, we argue, supports achieving real objectives for the communities students in service-learning engage with. These definitions point to the relevance of competencies beyond the technical areas of these curricular efforts with implicit touchpoints to ethics in the curriculum.

In the last couple of decades, the term 'service-learning' has often been replaced by terms that include 'community-engaged learning,' 'community-based learning,' and 'civically engaged learning.' The word 'service' can imply a vertical and power differential approach to partnerships that is not in alignment with the goals of the approach. These alternative terms attempt to more accurately reflect a more reciprocal approach to partnerships. The highest faculty award in the United States for this type of work is awarded by the Campus Compact, and was renamed from the Thomas Ehrlich Faculty Award for Service-Learning to the Thomas Ehrlich Civically Engaged Faculty Award in 2009. The term 'service-learning' is still used and will be for this chapter, while recognizing that the ideals of the pedagogy move beyond the traditional use of the words. The naming of the approach itself provides opportunities to engage students in early ethical discussions about the implications of the work they are embarking on.

Most models and definitions of service-learning arise from social sciences and humanities where the engagement is a placement of students into a community setting for a specified time. Within engineering, the engagement is often a design project. The value that the faculty and students bring is expertise that can be applied to a project deliverable. A model for a project approach to service-learning that aligns with engineering was developed by Leidig and Oakes (2021) and breaks out the engagement into two parts – a deliverable and the process. Most of the engineering engagement activities involve developing a design, plan, or other deliverable. The interactions and the processes of developing the deliverable are also very important and tap resources from our partners and potentially add value to them. All stakeholders can derive benefits and be impacted by the deliverable, as well as the process. Within ethics education in service-learning, there are opportunities for ethical learning and exploration of the projects or deliverables that are developed, as well as the process of engagement.

Ethical aspects and considerations in service-learning

This section addresses components of service-learning expanding on deliverable development, engagement in the design process, and connecting the engagement with ethics through reflective practice.

Deliverable development

Often, in engineering service-learning, the deliverable is a design, but it can also be an analysis, process, or plan. The development process often follows standard engineering processes that could be done in a wide range of courses. Ethical considerations follow traditional engineering practices regarding honesty, following laws and codes, and ensuring safety for people and the environment. Service-learning adds authentic contexts because the deliverable has real use for real people. Ideally, the students will meet the real users or recipients of the deliverable, and that humanizes their work and adds to the purpose of the activities. Situating this work with partners who are, or who work with, underserved communities provides additional avenues to employ and explore ethical frameworks. A utilitarian approach would leave out the stakeholders, often the very partners underserved by traditional societal practices. The ethics of care, virtue, and justice can play important roles within the context of the partnerships.

Process

As mentioned earlier, the terminology for pedagogy in some areas has moved away from the term 'service,' which implies doing something *for* someone else. Too often in engineering and higher education, we do things *to* others. The goal of service-learning is to engage *with* others in a manner that is reciprocal and involves the partners in the design and development process. This methodology aligns with the pillars of human-centered design and is used within programs that include EWB (Engineers without Borders) and EPICS (Zoltowski et al., 2012). Achieving the 'with' is vital to being effective and an opportunity to explore social constructs, biases, and preconceived ideas that influence ethical decision-making. The ethics of the engagement is an area ripe with opportunities for ethical development.

The Lakota are an indigenous people in North America. They have a name for outsiders who engage with them poorly. They refer to them as 'white-van people,' which means people of any race who show up driving white vans, a typical rental vehicle in their region. These white-van people view themselves as helping a group who are poor and need help while operating on their own timeline, arriving when convenient, and leaving when it fits their schedule. They are more interested in feeling good than creating meaningful partnerships with mutual value. Within engineering and higher education, there is a pull to operate in this mode as we fit engagement within our academic timelines and structures. Engagement with others requires considerations that include the ethics of the engagement itself and how we may need to bend our comfortable ways of thinking to be the most effective and ethical. Relationships often take time to develop, and partnerships can be the same. *How do we engage so there is mutual respect and value for each other over time?*

Within international engagement, there are a lot of discussions and efforts toward moving into a postcolonial mode. Decolonializing our approaches to engagement offers challenges and opportunities, especially within the area of ethics. We have all grown up within a colonial-influenced society, so operating outside of that takes intentionality. The opportunity is to engage students in exploring questions like: *How can one use their own engineering expertise without displacing or inhibiting local engineers or other professionals? Is the work one does taking away from work that could be done locally? How do we operate in true partnerships that offer mutual value and respect? How do we create mechanisms to hear and appreciate the expertise of our partners? What are we learning from our partners?* Viewing the work to be *with* our partners is central to the engagement mindset.

Intentionally connecting engagement and ethics

Students who are engaged in work with community partners address many ethical issues and considerations but often do not connect their experiences to ethics. In a design-based engagement program, for instance, students were interviewed and rarely discussed ethics explicitly but nevertheless described things related to or within the realm of ethics (Feister et al., 2016). Similarly, a study of engineering graduates who participated in EWB-USA programs reported a lower impact on ethics than on professional skills that included communication, teamwork, and leadership (Oakes et al., 2023). That finding contrasted with how these graduates talked in subsequent interviews where they described ethical situations but did not label them as ethics *per se*. Students experience issues and considerations that we would classify as ethical reasoning, but they do not connect them to ethics. This study reveals that we as educators can make the mistake of assuming that implicit learning is taking place when, instead, there is the opportunity to make it explicit through reflective practices. Thus, locating where to explicitly make ethical dimensions visible to the participants within the service-learning curriculum is an area for future exploration.

The need for further research on how service-learning can influence ethics education has also been pointed out by Corple et al. (2020). They examined students' ethical decision-making within the context of an undergraduate engineering service-learning design course that incorporated human-centered design. This approach places people at the center of the design process. While the paper did not mention that students were taught ethics directly, the authors discovered specific ways that students practiced ethical decision-making at various phases of the design process within the described service-learning context. Of particular focus were the ethical principles of autonomy, justice, and non-maleficence (i.e., to 'do no harm'). It also examined the aspects of the students' experiences that triggered ethical reflection. For example, in the project identification and specification development phase, interactions with community partners and with professors triggered reflections on autonomy. The paper suggested that service-learning can facilitate authentic reflections on students' ethical decision-making.

Additionally, some instruments have been suggested (Bringle et al., 2004) to assess ethics-related learning outcomes in service-learning. Such instruments include the Defining Issues Test (Bebeau, 2002; Rest, 1990; Rest & Narvaez, 1998), the Sociomoral Reflection Objective Measure (Gibbs et al., 1984) and the Ethics Position Questionnaire (Forsyth, 1980; O'Boyle & Forsyth, 2021). The suggestion by Bringle et al. (2004) of using these assessment tools within a service-learning context seems to indicate that there is an opportunity for students to experience growth or reflect on the ethical outcomes that these scales were created to measure (such as moral reasoning, moral judgment, and moral thought) within service-learning programs.

While many publications do not explicitly address the *how* of bringing ethics into the curriculum within service-learning contexts, there are many references to where the opportunities are. It is interesting that several articles and conference papers that describe service-learning courses and programs have ethics in their keywords but do not include descriptions of ethics-focused activities in the body of such literature (Chang et al., 2011; 2014; Kang & Chang, 2019; Verharen, 2014). In some cases, the authors subtly refer to ethics-related learning activities such as discussion sessions that confront the issue of homelessness to try to help students develop empathy for homeless people in the process of engaging with the complexities of litter and waste disposal (Wolfand et al., 2022). Sometimes ethical frameworks are mentioned in the context of a service-learning course curriculum, but it is not necessarily discussed how those frameworks are used (Sanchez & Lasso, 2014). However, much of the literature with discussions at the intersection of ethics and service-learning, especially in engineering education, offers some guidance on how we might intentionally integrate ethics in service-learning contexts looking at three aspects of the pedagogy.

1. Preparation. Before students work with communities, it is important to engage in some intentional preparation. Delaine and colleagues (2021) suggest that some implicit beliefs that students tend to have could be problematic to the objectives of service-learning programs. Such beliefs include "1) engineering is predominantly technical, 2) engineering requires deliverables and tangibles and 3) engineers are the best problem solvers" (p. 21). The paper also discusses how communities can be exploited at the expense of student learning, especially in resource-constrained communities where proper preparation and intention are not incorporated into the service-learning experience. To disrupt such beliefs that could result in unintended problems, the authors suggest including materials that introduce students to the underlying social and political contexts that influenced the need for the service in the community, broadening the focus of the course beyond the students' role in product delivery to include supporting the community to develop a solution collaboratively, and encouraging students to consider the community context alongside engineering design as well as the expertise and experiences the community partners bring to the solution.
2. Introducing ethics frameworks within the context of the community and the needs being addressed. Many students have learned ethics frameworks in more traditional ways that appear far removed from their contexts. We can weave those frameworks into authentic situations where they are most relevant to better motivate students in engaging with ethics. For example, a service-learning course described by Wolfand et al. (2022), which was designed for computing students and focused on data analysis using MATLAB, implicitly integrated ethics. Students collected litter, identified research questions related to litter, analyzed their research questions, and proposed data-driven solutions to the problem. These students worked in teams and the course incorporated discussion sessions to engage students in confronting the complexities of waste disposal, including the topic of homelessness. The authors mentioned that these discussions helped students develop empathy for the homeless. While not explicitly mentioned, virtue ethics could be incorporated in these kinds of discussions as a way of highlighting the virtue of empathy as a needed trait for contemplating societal issues like homelessness (for more on virtue ethics, see Chapter 22). Another example is a social inclusion project involving undergraduate engineering students collaborating with trainee coaches within a non-governmental organization (NGO) that supported people with mental illnesses to work within paid jobs (Chang et al., 2011). This project brought into focus the ethics of human and technology interactions.
3. Reflection at multiple phases of the project, especially in the beginning and at the end.

Reflection is a major component of the pedagogy of service-learning and vital to preparing students for an experience, processing their experience, and connecting it to desired learning outcomes (Eyler, 2002). Reflection and reflexivity are discussed more in Chapters 6 and 25.

There are many ways to prompt students' reflection including written assignments, readings, discussions, and combinations thereof. Research has shown that formal and informal discussions promote learning and in particular the broader learning of civic mindedness (Richard et al., 2016).

Models for reflection include the work of Ash and Clayton (2004), who identified three components for reflection:

1. Reflective observation of the experience during the week to spark this reflection (Consider identifying an experience and its context and/or impact.)
2. Conceptualizing and connecting your observation to a broader concept in one of the reflection themes (Consider: *What did I learn? How did I learn it? Why does the learning matter?*)
3. Connecting how you will use your experience and learning in the future, within and beyond the current course or experience (Consider: *What will/could I or others do in light of this learning?*)

A simplified version that can be used to elicit individuals' thoughts and feelings is the 'What? So what? Now what?' scaffolded questions proposed by John Driscoll (1994):

1. What: *What did they do and/or learn?*
2. So What: *Why is it important?*
3. Now What: *What can they do with that learning in the future?*

Reflection can and should be integrated across multiple aspects of the engagement experience. As mentioned above, the first place is as part of students' preparation, before any engagement has taken place. During the process of engagement, regular reflections allow students to connect the work with broader learning outcomes. Reflections offer ways for an instructor to evaluate how the process is progressing and identify any issues that need to be addressed or content that should be added to the experience. Summative reflections that include guided questions to connect the learning are the final step. These reflections also offer insights to the instructors and provide artifacts that can be graded and used to demonstrate learning.

For example, Schaad et al. (2008) described an innovative project that involved students in a service-learning project focused on disaster relief related to Hurricane Katrina. The project engaged students in deep and multifaceted reflection that addressed social, personal, and professional aspects of their experiences. The reflections happened at various phases of the project including before and after a visit to the site in New Orleans. Additionally, students watched lectures and presented research papers on their projects at the end of the semester. This example illustrates how to incorporate multiple opportunities for students to reflect. Some of this reflection can be facilitated by lecture topics on pertinent ethical issues.

In summary, reflection can involve readings, videos, discussions, and writing. Reflection is a crucial component of service-learning and must be integrated throughout the phases of projects with targeted prompts that bring ethics to the forefront of student experiences.

Humanitarian engineering education

Humanitarian engineering education programs are diverse, growing, and becoming increasingly integrated into mainstream engineering curricula (Kunwar, 2020; Peiffer, 2019; Rojas, 2020; Smith, Tran, et al., 2019). Humanitarian engineering principles come from the sustainable development sector, and the mainstreaming of humanitarian engineering is evident in how engineering competency frameworks are increasingly referencing the UN Sustainable Development Goals (e.g., International Engineering Alliance, 2021). As all engineering has the potential to contribute to (or detract from) the sustainable development agenda, humanitarian engineering principles apply to all engineering work.

Engineers without Borders' education programs exemplify the growth and diversity of humanitarian engineering education. For example, Engineers without Borders Australia (EWBA) has been offering the EWB Challenge for over 15 years, where first-year students respond to a design brief developed by EWBA and their community partners to address development priorities. Annually, more than 10,000 students around Australia and overseas participate, typically as part of their first subject addressing professional engineering practice (Jolly et al., 2011). For later-year students, since 2015, EWBA has been running Design Summits, which are immersive study tours to developing countries. So far, more than 1,000 students have learned about human-centered design in context, working cross-culturally, and have experienced how engineering can affect positive change (Daniel & Brown, 2018). Lastly, for final-year students, EWBA runs a Research Challenge supporting capstone projects in humanitarian engineering.

EWBA has made its methods explicit, publishing its Technology Development Approach (TDA), which builds on human-centered design approaches combined with principles from the community and international development sectors. EWBA's TDA guides their engineering work and is used with students as a teaching resource. Without necessarily labeling factors as ethical considerations, it addresses key tenets of best practice in humanitarian engineering (with implicit ethical underpinnings), such as an ethics of care founded on empathy and compassion for stakeholders, inclusion for all, 'do no harm,' and strengths-based approaches. These tenets are operationalized to varying degrees in EWBA's education programs. For example, strengths-based approaches feature in the immersive study-tours, where there is a push to shift students away from the typical default of deficit-thinking to having them recognize the strengths of the host community. This can be through framing 'how might we' questions as design *opportunities* rather than design *problems* and encouraging a question-style with stakeholders of *Why do you do things that way?* rather than *Why* don't *you do things like we do?* Another mechanism where strengths-based approaches are evident is in the reverse project brief that capstone students are asked to complete, where they re-frame the design brief in their own words to demonstrate understanding, and are explicitly asked to identify multiple community strengths.

Classroom exercises to prepare students for field experiences

This section discusses scenario-based teaching and assessment, with some relevant examples to help prime students for field experiences. We also briefly discuss other teaching approaches to sensitize students to some implicit and explicit ethical considerations in such contexts.

Introduction to scenario-based teaching and assessment

Scenarios are one strategy for teaching and assessing the many skills required in engineering professional practice. Other assessment strategies include self-reporting against Likert-scale surveys or observational studies of behaviors in authentic scenarios, for example, in assessment centers. Survey instruments are easy to administer and analyze with large numbers of students and offer insights into students' self-concept and confidence. However, the extent to which they predict the application of these skills in practice is questionable (Mazzurco & Murzi, 2017). Conversely, assessment centers (Ilgen et al., 2015) or simulation-based assessments (Hoffman et al., 2015), in which participants' actions are evaluated in authentic situations, typically lead to more reliable and valid assessments but are much more resource- and time-intensive and so are difficult to scale.

Scenario-based assessment offers a compromise between (a) the scalability but questionable predictive validity of self-report survey instruments and (b) the authentic but time-consuming nature of simulation-based assessments and other approaches involving observations of behaviors in context. Scenario-based instruments have been developed to assess a range of engineering skills, including global competency (Jesiek et al., 2020), design skills (Atman et al., 2007), interdisciplinary problem-solving (Adams et al., 2010), and more.

Although often used solely for assessment, scenario-based approaches can easily be adapted as a teaching and learning approach in ethics and other areas. They can prime students for particular considerations in fieldwork or other professional experiences, offering a safe space to explore and discuss different ethical dimensions without repercussion. Beyond using the scenario and student responses as discussion prompts, comparing responses between students or within the same student over time can be an effective device to guide reflection.

Although scenario-based approaches are diverse, such tools share three typical components: a scenario, some related questions, and some scoring criteria. The scenario should authentically depict an open-ended situation the participant could conceivably expect to encounter in their career, and the scenario may vary in length from one sentence (Kilgore et al., 2007) up to multiple paragraphs (McMartin et al., 2000). Scenarios are developed from practitioner interviews, case studies of practice, or some combination thereof (Jesiek et al., 2020).

Questions should prompt some reflections or responses to the scenario and can be recorded using a transcript or think-aloud protocol or written in a paper-based or online form. Questions can be either closed- or open-ended and scored quantitatively or with some interpretive scheme. The scheme can involve, for example, coding responses into different categories (Kilgore et al., 2007) or using some rubric to score responses across different dimensions (McMartin et al., 2000).

The Energy Conversion Playground task

One example relevant to service-learning and humanitarian engineering is the Energy Conversion Playground (ECP) task, originally developed by Mazzurco et al. (2014) and then elaborated to address sociotechnical thinking and co-design (Daniel & Mazzurco, 2020; Mazzurco & Daniel, 2020). This scenario involves designing a power supply for a remote primary school in a developing country, with one suggestion being a merry-go-round-powered dynamo, and asks students to identify relevant considerations and outline the steps they would take in designing a solution:

> *Scenario: In developing countries, energy production is one of the most critical problems. Resources or technologies to produce energy are often not available. Thus, human power conversion systems might be used to power small appliances. Imagine that you and your team are assigned to a design project in partnership with a Non-Governmental Organization (NGO) of a developing country. The NGO needs a low-cost power system that can generate enough energy for the lights of a primary school. One of the members of your team suggests using merry-go-round, seesaw, and swing to produce energy that can be converted to electricity for the lights.*
>
> 1. *What considerations do you need to take into account to solve the problem described in the scenario? List and describe all constraints and justify their inclusion.*
> 2. *How would you proceed to solve the problem described in the scenario? List and describe concisely all the steps you would take to solve the problem described in the scenario.*

The scenario was developed by combining a paper that explored, from a purely technical perspective, how playground equipment could be used to generate electricity (Pandian, 2004) with the story of the PlayPump, a much-heralded project in which water pumps were designed to be powered by children playing on merry-go-rounds, that – with its failure – became a fable about the importance of understanding the social contexts of technology (Case, 2010).

The ECP task allows bringing up ethical dimensions in a realistic context, in advance of the situations students may encounter in service-learning and/or humanitarian engineering experiences. Both implicit (e.g., making decisions on behalf of marginalized communities, with no stakeholder involvement) and explicit (e.g., using child labor to produce energy) ethical dimensions arise naturally in discussing this scenario. It has been used to draw out features of co-design, highlight key sociotechnical design considerations of context and ethics, and as a prompt for reflection.

Rubrics were developed for both the first question (on sociotechnical considerations) and the second question (on co-design). The sociotechnical thinking rubric categorized considerations into three domains: *technical*, *people*, and *broader context*. The *technical* domain was further broken into considerations around *inputs* (e.g., required power), *functionality* (e.g., efficiency, friction), and *long-term* (e.g., maintenance), while *broader context* was broken into *local norms* (e.g., culture, gender norms), *laws and ethics*, and *other socio-material contexts* (e.g., the local economy). The *people* domain was scored by whether respondents merely *mentioned* people's needs or expertise, or whether they described *listening* (e.g., consulting with) or *collaborating* (e.g., co-constructing) with stakeholders. An expert sociotechnical response included considerations from all sub-domains of *technology* and *broader context*, and also described *collaboration* with local stakeholders.

The co-design rubric had two dimensions, the first categorizing respondents' design steps into three main phases of the design process (i.e., scope, develop, deliver). The second dimension characterized the extent to which stakeholders were involved in these different phases. An expert co-design response described collaboration across all three design phases.

Teaching approaches using the ECP instrument

The ECP instrument can be used in different ways in the classroom. One approach could involve the following steps (distilled from different ways the lead author has used the instrument in his own teaching):

1. Students respond to the prompts individually by compiling individual considerations (or design steps) on separate sticky notes.
2. Students compare and consolidate their responses, first in pairs and then in groups of four students (i.e., Think-Pair-Square). An alternative could be to have students discuss and respond to the prompts in a group from the start.
3. In whole-class discussion, groups compare their responses by, for example, each group calling out a consideration or a design step and continuing around the room until all considerations have been listed (or all design steps put in order).
4. The rubric is introduced, and student groups self-assess their responses. Alternatively, groups could assess their neighbors' responses.
5. This self-assessment is used as a discussion prompt, for example, about which areas were over-subscribed (typically the technical considerations and the scoping phase) and which under-subscribed (typically the broader contextual considerations and the delivery phase).

An alternative approach could be collecting students' responses at the start of some teaching block, and then at the end of the teaching block having students respond to the scenario a second time. The next step would be returning students' original submissions and inviting them to compare their responses – both to the rubric and from before and after the teaching session – and using this comparison as a reflective prompt. In either approach, the key steps are responding to the scenario, evaluating the response against the rubric, and using this evaluation to prompt reflection on what has, and has not, been identified.

Having been introduced to the two rubrics as frameworks for sociotechnical thinking and co-design, the students could apply these frameworks to future scenarios or situations. Note that there are other frameworks for unpacking context (such as PESTEL – political, economic, social, technological, environmental, legal), but these do not always highlight ethical considerations.

Other scenario-based tools

Other scenario-based approaches that could be used with engineering students include the:

- 'Midwest Floods design task' to elicit sociotechnical and co-design considerations (Kilgore et al., 2007).
- 'Engineering and Science Issues Test' to consider ethical dilemmas in science and engineering (Borenstein et al., 2010).
- suite of scenarios from the Global Engineering Competency (GEC) project to explore cross-cultural skills (*some of which* focus on ethical issues).

Although these skills are relevant to humanitarian engineering and service-learning, these instruments are not focused explicitly on such contexts, unlike the Energy Conversion Playground task.

Other collections of ethical scenarios include those curated by the Markkula Center for Applied Ethics or the Engineering Professors Council. These are mainly focused on Western contexts; however, some involve situations appropriate to this chapter, such as a water supply project in Ghana, a community solar project in East Africa, a water filtration project in Bolivia, or a service program in Belize. One classroom approach could be to have multiple groups consider each scenario and how they would respond, and then discuss with other groups about whether they have reached the same conclusions. Connections to ethics could be made explicit by asking students to consider which ethical principles or theories their response represents.

Additional classroom resources

Although not explicitly addressing ethics, other classroom activities that touch on ethical considerations could be used to prepare students for fieldwork. Such activities include conducting background context research (using resources like GapMinder to contrast diverse measures of different countries' development and status and to challenge misconceptions about international development); exercises to build empathy (e.g., journey mapping, empathy maps, personas); or activities combining the two, such as reverse project briefs. More generally, activities like the Privilege Walk (Silverman, 2013) can foster discussion on power dynamics and connect to bigger topics like colonialism or 'white savior-ism,' as comically explored by Radi-Aid. Radi-Aid's *Africa for Norway* and *Who wants to be a volunteer?* parodies offer a gateway to challenging stereotypes about poverty, development, and how narratives around aid and social responsibility are communicated. Role-playing games like Engineering with People can foster empathy for different stakeholders and an appreciation of ethical complexity. Finally, Leydens and Lucena (2009) have argued for the importance of listening in engineering and offered suggestions on how this might be addressed in the curriculum. One fantastic resource to open discussion on this issue is the TedTalk "Want to help someone? Shut up and listen!" Although not situated in an engineering context, this TedTalk powerfully demonstrates the importance of respectful community engagement.

There are some overlaps between service-learning and humanitarian engineering, especially in how educators address them. One major overlap is that both focus on meeting the needs of communities, especially vulnerable ones. Another overlap exists structurally in some models. For instance, the humanitarian engineering minor at the Colorado School of Mines has a service-learning component (Mitcham & Englehardt, 2019). In it, service-learning is a practical pedagogical approach to fulfill the requirements of the broader humanitarian engineering program. Consequently, the exercises and resources we suggest for humanitarian engineering could be applied to service-learning, and vice versa.

Conclusion and recommendations

In this concluding section, we offer some recommendations for exploring this space for researchers and educators, individually and at a programmatic level.

Education researchers

In the service-learning and humanitarian space, there are many research opportunities that connect to ethics. We can not only explore the impact on student ethical learning outcomes but also the impact of our projects, processes, and partnerships on the communities involved. We can explore the development of students' ability to manage complex and ambiguous contexts. There are also ethical considerations for conducting research with specific communities, where we should weigh the benefits of research outcomes against the impact on stakeholders. While there are many questions research can help us answer, the very nature of community work – especially the sensitivity of building trust with communities – means that we may need to forego studying some elements of the engagement. However, some research areas include:

- Evaluating the impact of interventions on students, communities, and other stakeholders;
- Developing and validating research instruments to assess learning in these contexts (see Chapters 26–31 in the assessment theme for a deeper exploration of such questions)
- Exploring different teaching approaches to support students in dealing with the complexity and ambiguity of these social contexts (versus oversimplifying sociocultural differences or ethical complexities)

Classroom educators/subject teachers

For educators, there are numerous opportunities to engage students locally and globally while exploring ethical issues and developing ethical reasoning. It is imperative that we examine the ethical considerations of the engagement, our project(s), and the processes by which they are developed. While there is potential for positive impact on students and communities, there is also potential for unintended consequences that must be considered. Educators need to familiarize themselves with the cultural and social contexts while supporting students to step beyond ethnocentrism to examine their own social and cultural background and challenge any implicit assumptions. Particular attention should be paid to power dynamics between the institution, community, and other stakeholders. The relationships with community partners should be structured so that the partners have a significant voice in the process and partnership, for example, by using participatory development methods. Preparing students for the context before they are engaged and exploring ethical frameworks relevant to the context are very important. Reflection before, during, and after engagement is critical and offers opportunities to make learning explicit and provide insights into the student experiences. Finally, seek out colleagues, including those from outside of engineering, who may have expertise that can contribute to the engagement as partners, mentors, or resources.

Programmatic

From a programmatic standpoint, community needs require long-term commitments and relationship building. Programs such as EWB or the EPICS Program at Purdue University (Zoltowski & Oakes, 2014) typically commit to at least 5-year engagements. This length allows for relationships and projects to be developed and supported. Curricular structures need to be created to support long-term engagement. Assessing the value brought to the partnership and the time and resources invested can be pivotal in building long-term and high-impact partnerships. Community engagement can place faculty in new or uncomfortable settings. Robust support systems for faculty can benefit the program tremendously. Some of this support can come from colleagues in other disciplines, such as international development or intercultural communication. They can also come from institutional centers such as offices of community engagement. Assessment and curricular structures should be aligned with developing reciprocal partnerships. Finally, program leaders should seek ways to connect to institutional goals to institutionalize programs to be sustained into the future. These programs can contribute to accreditation by addressing connections to broader societal, environmental, and human issues. The active engagement, hands-on aspects, and societal contexts of the approaches can help to attract more and more diverse students to our programs as well as retain them at higher rates.

References

Adams, R. S., Beltz, N., Mann, L., & Wilson, D. (2010). Exploring student differences in formulating cross-disciplinary sustainability problems. *International Journal of Engineering Education*, *26*(2), 324.

Ash, S. L., & Clayton, P. H. (2004). The articulated learning: An approach to guided reflection and assessment. *Innovative Higher Education*, *29*(2), 137–154.

Atman, C. J., Adams, R. S., Cardella, M. E., Turns, J., Mosborg, S., & Saleem, J. (2007). Engineering design processes: A comparison of students and expert practitioners. *Journal of Engineering Education*, *96*(4), 359–379. https://doi.org/10.1002/j.2168-9830.2007.tb00945.x

Australian Bureau of Statistics. (2020). *Australian and New Zealand Standard Research Classification (ANZSRC)*. Commonwealth of Australia. Retrieved November, 11, from https://www.abs.gov.au/statistics/classifications/australian-and-new-zealand-standard-research-classification-anzsrc/latest-release

Bebeau, M. J. (2002). The defining issues test and four component model: Contributions to professional education. *Journal of Moral Education*, *31*(3), 271–295.

Birzer, C. H., & Hamilton, J. (2019). Humanitarian engineering education fieldwork and the risk of doing more harm than good. *Australasian Journal of Engineering Education*, *24*(2), 51–60.

Borenstein, J., Drake, M. J., Kirkman, R., & Swann, J. L. (2010). The Engineering and Science Issues Test (ESIT): A discipline-specific approach to assessing moral judgment. *Science and Engineering Ethics*, *16*(2), 387–407. https://doi.org/10.1007/s11948-009-9148-z

Bringle, R. G., & Hatcher, J. A. (1995). A service-learning curriculum for faculty. *Michigan Journal of Community Service-Learning*, *2*(1), 112–122.

Bringle, R. G., Phillips, M. A., & Hudson, M. (2004). *The measure of service-learning: Research scales to assess student experiences*. American Psychological Association. https://doi.org/10.1037/10677-000

Brown, N. J., Smith, J., Daniel, S., & Birzer, C. (2022). *Exploring the position of humanitarian engineering in Australia*. Paper presented at the 33rd Australasian Association for Engineering Education Conference (AAEE 2022): Future of Engineering Education: Future of Engineering Education.

Burleson, G., Lajoie, J., Mabey, C., Sours, P., Ventrella, J., Peiffer, E., ... Aranda, I. (2023). Advancing sustainable development: Emerging factors and futures for the engineering field. *Sustainability*, *15*(10), 7869. https://doi:10.3390/su15107869

Case, J. (2010). *The painful acknowledgement of coming up short*. The Case Foundation. https://casefoundation.org/blog/painful-acknowledgment-coming-short/

Chang, Y. J., Chen, Y. R., Wang, F. T. Y., Chen, S. F., & Liao, R. H. (2014). Enriching service-learning by its diversity: Combining university service-learning and corporate social responsibility to help the NGOs adapt technology to their needs. *Systemic Practice and Action Research*, *27*(2), 185–193. https://doi.org/10.1007/s11213-013-9278-8

Chang, Y. J., Wang, T. Y., Chen, S. F., & Liao, R. H. (2011). Student engineers as agents of change: Combining social inclusion in the professional development of electrical and computer engineering students. *Systemic Practice and Action Research*, *24*(3), 237–245. https://doi.org/10.1007/s11213-010-9183-3

Corple, D. J., Zoltowski, C. B., Kenny Feister, M., & Buzzanell, P. M. (2020). Understanding ethical decision-making in design. *Journal of Engineering Education*, *109*(2), 262–280. https://doi.org/10.1002/jee.20312

Cruz, C. C. (2021). Brazilian grassroots engineering: A decolonial approach to engineering education. *European Journal of Engineering Education*, *46*(5), 690–706. https://doi.org/10.1080/03043797.2021.1878346

Daniel, S. A., & Brown, N. J. (2018). *The impact of the EWB design summit on the professional social responsibility attitudes of participants*. 2018 ASEE Annual Conference & Exposition, Salt Lake City.

Daniel, S., & Mazzurco, A. (2020). Development of a scenario-based instrument to assess co-design expertise in humanitarian engineering. *European Journal of Engineering Education*, *45*(5), 654–674. https://doi.org/10.1080/03043797.2019.1704689

Delaine, D. A., Desing, R., Wang, L., Dringenberg, E., & Walther, J. (2021). Identifying and disrupting problematic implicit beliefs about engineering held by students in service-learning. *International Journal for Service-learning in Engineering, Humanitarian Engineering and Social Entrepreneurship*, *16*(2), 14–38. https://doi.org/10.24908/ijsle.v16i2.14747.

Driscoll, J. (1994). Reflective practice for practice—a framework of structure reflection for clinical areas. *Senior Nurse*, *14*(1), 47–50.

Eyler, J. (2002). Reflection: Linking service and learning—linking students and communities. *Journal of social issues*, *58*(3), 517–534.

Eyler, J., & Giles, D. (1999). *Where's the learning in service-learning?* (1st ed.). Jossey-Bass.

Feister, M. K., Zoltowski, C. B., Buzzanell, P. M., & Torres, D. H. (2016). *Integrating ethical considerations in design*. 2016 ASEE Annual Conference & Exposition.

Forsyth, D. R. (1980). A taxonomy of ethical ideologies. *Journal of Personality and Social Psychology*, *39*, 175–184.

Gibbs, J. C., Arnold, K. D., Morgan, R. L., Schwartz, E. S., Gavaghan, M. P., & Tappan, M. B. (1984). Construction and validation of a multiple-choice measure of moral reasoning. *Child Development*, *55*(2), 527–536. https://doi.org/10.2307/1129963

Hazeltine, B., & Bull, C. (1999). *Appropriate technology; Tools, choices, and implications*. Academic Press, Inc.

Hoffman, B. J., Kennedy, C. L., LoPilato, A. C., Monahan, E. L., & Lance, C. E. (2015). A review of the content, criterion-related, and construct-related validity of assessment center exercises. *Journal of Applied Psychology*, *100*(4), 1143.

Ilgen, J. S., Ma, I. W., Hatala, R., & Cook, D. A. (2015). A systematic review of validity evidence for checklists versus global rating scales in simulation-based assessment. *Medical Education*, *49*(2), 161–173. https://doi.org/10.1111/medu.12621

International Engineering Alliance. (2021). *In International Engineering Alliance graduate attributes & professional competencies*.

Jesiek, B. K., Woo, S. E., Parrigon, S., & Porter, C. M. (2020). Development of a situational judgment test for global engineering competency. *Journal of Engineering Education*, *109*(3), 470–490.

Jolly, L., Crosthwaite, C., Brodie, L., Kavanagh, L., & Buys, L. (2011, December 5–7). *The impact of curriculum content in fostering inclusive engineering: data from a national evaluation of the use of EWB projects in first year engineering*. Australasian Association for Engineering Education Conference 2011: Developing engineers for social justice: Community involvement, ethics & sustainability, Fremantle, Western Australia.

Kang, Y. S., & Chang, Y. J. (2019). Sharing the voice and experience of our community members with significant disabilities in the development of rehabilitation games. *Systemic Practice and Action Research, 32*(1), 1–12. https://doi.org/10.1007/s11213-018-9449-8
Kilgore, D., Atman, C. J., Yasuhara, K., Barker, T. J., & Morozov, A. (2007). Considering context: A study of first-year engineering students. *Journal of Engineering Education, 96*(4), 321–334.
Kleba, J. B., & Reina-Rozo, J. D. (2021). Fostering peace engineering and rethinking development: A Latin American view. *Technological Forecasting and Social Change, 167*, 120711. https://doi.org/10.1016/j.techfore.2021.120711
Kolb, D. A., Rubin, I. M., & McIntyre, J. M. (1984). *Organizational psychology: Readings on human behavior in organizations*. Prentice-Hall.
Kunwar, P. (2020). *State of engineering for global development Asia*. https://www.engineeringforchange.org/wp-content/uploads/2020/12/E4C_State-of-EGD-Asia_December-2020.pdf
Leidig, P., & Oakes. W. (2021). Model for project-based community engagement. *International Journal for Service-learning in Engineering, Humanitarian Engineering and Social Entrepreneurship, 16*(2), 1–13.
Leydens, J. A., & Lucena, J. C. (2009). Listening as a missing dimension in engineering education: Implications for sustainable community development efforts. *IEEE Transactions on Professional Communication, 52*(4), 359–376.
Leydens, J. A., & Lucena, J. (2016). *Making the invisible visible: Integrating engineering-for-social-justice criteria in humanities and social science courses*. Proceedings for the American Society for Engineering Education Annual Conference and Exposition.
Mazzurco, A., & Daniel, S. (2020). Socio-technical thinking of students and practitioners in the context of humanitarian engineering. *Journal of Engineering Education, 109*(2), 243–261. https://doi.org/10.1002/jee.20307
Mazzurco, A., Huff, J. L., & Jesiek, B. K. (2014). The Energy Conversion Playground (ECP) design task: Assessing how students think about technical and non-technical considerations in sustainable community development. *International Journal for Service-learning in Engineering, 9*(2). https://doi.org/10.24908/ijsle.v9i2.5585
Mazzurco, A., & Murzi, H. (2017). *Evaluating humanitarian engineering education initiatives: A scoping review of literature*. Annual Conference of the Australasian Association for Engineering Education, 28 2017: Sydney.
McMartin, F., McKenna, A., & Youssefi, K. (2000). Scenario assignments as assessment tools for undergraduate engineering education. *IEEE Transactions on Education, 43*(2), 111–119.
Mitcham, C., & Englehardt, E. E. (2019). Ethics Across the -curriculum: Prospects for broader (and deeper) teaching and learning in research and engineering ethics. *Science and Engineering Ethics, 25*(6), 1735–1762. https://doi.org/10.1007/s11948-016-9797-7
Mitcham, C., & Muñoz, D. R. (2009). The humanitarian context. In S. H. Christensen, B. Delahousse, & M. Meganck (Eds.), *Engineering in context* (pp. 183–195). Academica.
Oakes, W., Leidig, P. A., & Holloway, E. (2023). *Impact of engineers without borders USA experiences on professional preparation*. Proceedings of the 2023 ASEE Annual Conference, Baltimore, Maryland, USA.
O'Boyle, E. H., & Forsyth, D. R. (2021). Individual differences in ethics positions: The EPQ-5. *PLoS One, 16*(6), e0251989.
Pandian, S. R. (2004, June 17–19). *A human power conversion system based on children's play*. 2004 International Symposium on Technology and Society, Worcester, MA, USA.
Passino, K. M. (2009). Educating the humanitarian engineer. *Science and Engineering Ethics, 15*(4), 577–600. https://doi.org/10.1007/s11948-009-9184-8
Peiffer, E. (2019). State of engineering for global development: United States and Canada. https://www.engineeringforchange.org/wp-content/uploads/2019/10/State-of-EGD-North-America-Full-Report-Google-Docs.pdf
Rest, J. R. (1990). *DIT: Manual for the defining issues test*. University of Minnesota Center for the Study of Ethical Development.
Rest, J., & Narvaez, D. (1998). *Guide for DIT-2*. Center for the Study of Ethical Development, University of Minnesota.
Richard, D., Keen, C., Hatcher, J. A., & Pease, H. A. (2016). Pathways to adult civic engagement: Benefits of reflection and dialogue across difference in higher education service-learning programs. *Michigan Journal of Community Service-learning, 23*(1), 60–74.

Rojas, C. (2020). State of engineering for global development - Latin America. https://www.engineeringforchange.org/wp-content/uploads/2020/11/State-of-EGD-Latin-America-Full-Report-ENG-October-2020.pdf

Sagoe, C. (2012). The Neo-colonialism of development programs. https://www.e-ir.info/2012/08/12/the-neo-colonialism-of-development-programs/

Sanchez, D. A. M., & Laso, A. G. (2014). Experiences in social innovation: A platform for ethics through a school of engineering studies. *Journal of Cases on Information Technology*, *16*(3), 4–17.

Schaad, D. E., Franzoni, L. P., Paul, C., Bauer, A., & Morgan, K. (2008). A perfect storm: Examining natural disasters by combining traditional teaching methods with service-learning and innovative technology. *International Journal of Engineering Education*, *24*(3), 450–465.

Silverman, K. (2013). Lessons in injustice: Privilege walks. *The Intellectual Standard*, *2*(2), 3.

Smith, J., Lynch, E., Care, R., Greet, N., & Mitchell, R. (2019). *Building a humanitarian engineering community of practice*. 30th Annual Conference for the Australasian Association for Engineering Education (AAEE 2019): Educators Becoming Agents of Change: Innovate, Integrate, Motivate, Brisbane.

Smith, J., Tran, A. L., & Compston, P. (2019). Review of humanitarian action and development engineering education programmes. *European Journal of Engineering Education*, 1–24.

Turner, J., Brown, N., & Smith, J. (2015). *Humanitarian engineering-what does it all mean?* Paper presented at the AAEE 2015: Blended Design and Project Based Learning: a future for engineering education, Geelong.

Verharen, C., Tharakan, J., Bugarin, F., Fortunak, J., Kadoda, G., & Middendorf, G. (2014). Survival ethics in the real world: The research university and sustainable development. *Science and Engineering Ethics*, *20*(1), 135–154. https://doi.org/10.1007/s11948-013-9441-8

Wolfand, J. M., Bieryla, K. A., Ivler, C. M., & Symons, J. E. (2022). Exploring an engineer's role in society: Service-learning in a first-year computing course. *IEEE Transactions on Education*, *65*(4), 568–574. https://doi.org/10.1109/TE.2022.3148698

Zoltowski, C. B., Oakes, W. C., & Cardella, M. E. (2012). Students' ways of experiencing human-centered design. *Journal of Engineering Education*, *101*(1), 28–59.

Zoltowski, C. B., & Oakes, W. C. (2014). Learning by doing: Reflections of the EPICS program. *International Journal for Service Learning in Engineering, Humanitarian Engineering and Social Entrepreneurship*, 1–32. https://doi.org/10.24908/ijsle.v0i0.5540

24

ARTS-BASED METHODS IN ENGINEERING ETHICS EDUCATION

Sarah Jayne Hitt, David D. Gillette, and Lauren E. Shumaker

Introduction

Educational institutions, professional organizations, and accreditors now recognize that engineers must be prepared to face the complex challenges of the future in a globally responsible and responsive way. This has led to the growth and development of engineering ethics as a discipline over the last few decades and its acceptance as an essential component of engineering education. During the same time, many scholars and thinkers have urged a coming together of science, engineering, and the arts, resulting in global initiatives like STEAM (Science, Technology, Engineering, Arts, and Math) and the PISA Creative Thinking assessment (Belbase et al., 2021; OECD, 2022). These, in turn, have spurred new engineering programs and institutions around the world that privilege a more integrated and expansive educational model that often includes non-technical components (Graham, 2018). Students and accrediting bodies too are demanding more of a focus on the human and environmental contexts of engineering that can be revealed through the inclusion of liberal arts content (ABET, 2020; ENAEE, 2021; Engineers Australia, 2008; Engineering Council, 2020), while educators have recognized the value of problem- and community-based learning pedagogies that naturally include and reveal the ethical and cultural implications of engineering practice (Walton et al., 2022; Rogers et al., 2021). Teaching and learning methods that encourage and enable collaboration, creativity, communication, and reflection are now seen as essential to cultivating the habits of mind required of today's engineers (Lucas & Hanson, 2014). In this evolving educational context, arts-based methods in engineering ethics can respond to these changes and, therefore, have great potential.

However, while there is undoubtedly growing interest in this area, the link between the arts and engineering ethics specifically has not been as clearly described. We first aim to define what practitioners mean by arts-based methods and then describe how scholars conceptualize their relationship to engineering ethics. We next look to both historical and contemporary precedents for the connections inherent in these disciplines. This review will show that despite the long-standing associations between approaches to engineering and the arts, structural and cultural constructs have erected barriers against linking them in contemporary academic environments. We address these and use recent examples of research and practice indicating these barriers can be overcome. For instance, pedagogy integrating the arts into ethics education has yielded intriguing and prom-

DOI: 10.4324/9781003464259-29

ising results in other fields such as business and medicine. In pulling from adjacent disciplines to lend specificity and further understanding to this field, we also highlight the absence of analogous research within engineering education. Throughout, we suggest many opportunities in teaching and research that have the potential to bring benefits to larger engineering and ethics contexts.

Definitions and precedents

When we discuss 'the arts' in this chapter, we adhere to the *Encyclopedia Britannica*'s definition of this general term, which comprises fields and practices in the categories of literature, the visual arts, the graphic arts, the plastic arts, the decorative arts, the performing arts, music, and architecture. This definition may be broader than that typically used by universities to delineate where different disciplines 'belong' or are organized institutionally. Still, we believe these categories, with their various technical, theoretical, and creative components, can lend much-needed expertise to teaching methods within the engineering context (as explained below). This broader definition is especially important for the European engineering education context, where a 'core curriculum' rooted in the liberal arts that is so common to higher education in the United States is much less prevalent. Lang (1999) explores the 300-year history of this 'distinctively American' approach that was as much focused on developing students' character and civic responsibility as preparing them for a profession. This educational philosophy has left a legacy whereby even students at US universities that specialize in engineering are often still required to take liberal arts modules (ABET, 2020). Despite being the birthplace of the concept of 'liberal arts' as originally articulated by Greek philosophers and exemplified in the curricula of its earliest universities, Europe experienced a decline of this educational tradition due to a variety of social, political, and economic factors beginning in the nineteenth century (van der Wende, 2011). This, combined with a system where students must choose a specialization earlier, means that European engineering students may be much less likely than their US counterparts to encounter liberal arts content within their degree programs.

We must, therefore, also distinguish between arts-based content and arts-based methods. Arts-based content refers to artifacts such as literary texts, film, music, visual arts, poetry, dance, theater, and so forth, which may be used or referenced in an exercise, assignment, lecture, discussion, or other educational activity. In referring to arts methods, we mean the artistic invention and problem-solving processes that relate to "the creation of aesthetic objects, environments, or experiences that can be shared with others" (*Encyclopedia Brittanica*). Arts-based methods, then, comprise the learning and teaching activities common to or rooted in 'the arts' as broadly defined above such as Socratic-style seminar discussions, creative writing exercises, visual or graphical depictions of ideas, and role-play scenarios. Students might, therefore, engage with arts-based content without educators using an arts-based method; for instance, they could view a film that focuses on an engineering ethics issue, such as *Oppenheimer* or *Gattaca*, without analyzing the musical score, acting out a scene, or re-writing the ending. However, we believe that when educators use pedagogies that allow students to learn from and through the arts, this content becomes most effective. Indeed, the use of arts-based methods in engineering ethics education can range from one-time adoption of an arts-inspired exercise fulfilling a specific educational purpose to full integration of the arts across a technical curriculum. We describe examples of approaches across this spectrum later in this chapter.

Our goal in advocating for the use of arts-based methods in engineering ethics is not to turn engineers into artists but instead to offer engineering students additional methods for problem-solving and production, asking them to occasionally think from the point of view of artists while still working as engineers. Adopting and using arts-based methods in an engineering ethics classroom does not require engineering faculty to become art critics or ask them to assess the resulting engineering products as forms of artistic expression. As Haidet et al. (2016) explain, the sustainable adoption of arts-based methods into an engineering program or course requires that the approaches not rely on an individual's specialized arts expertise or interests for success. Moreover, engineering educators already possess some skills that intersect with the arts but which may be framed in different terms that obscure the similarities. For instance, usability study (UI/UX) creates a linguistic and conceptual bridge between human experience and engineering construction, with an arts-based, human-focused methodology for systems design and evaluation. These bridges also connect arts-based components to the engineered system – the use of color for verification of identity, the use of animation to entice interaction, visual ranking order to create hierarchies of importance and relevance, algorithmic selection and presentation of topics based on prior conscious human choice. Engaging with these design choices as artistic considerations rather than just technical ones may allow their implicit ethical dimensions to be more easily revealed. In the case of UI/UX, linking aesthetics with usability makes some ethical concerns about accessibility unavoidable (e.g., who is excluded from seeing or engaging with the interface due to color blindness). Arts-focused areas of engineering design can, therefore, place the human at the forefront of decisions rather than at the mercy of technical criteria. Even though ethical aspects of engineering are always present, they can be obscured by a limited focus on technical problem-solving. Embedding the arts in engineering ethics thus allows for a more human-centered approach.

Educators can be reassured that, once learned, many arts-based methods can be effectively adapted for diverse settings, audiences, and curricular needs without specialized forms of arts critique and assessment which may be unfamiliar or intimidating to educators from a primarily technical background. Still, it is important to avoid being dismissive of the scholarly expertise garnered through years of studying and working with arts-based content and methods and to recognize that initial resistance to the use of arts-based methods in engineering ethics education may be based on more than a basic discomfort with using alternative methods for problem-solving and may also be rooted in the common but false narrative that engineering and arts are so different from each other that they cannot be effectively combined (Snow, 1959). This narrative is, unfortunately, already deeply ingrained in many of our discussions about the connection between engineered and human systems, which position engineering as distinctly separate from human processes (Booker et al., 2021). For example, even when engineering creates products intimately connected to the human form such as a prosthetic arm, artery stent, or 3D-printed heart valve, the conversation about the invention and use of those devices often celebrates the mechanical aspects of the product and how it improves the life of the user in a mechanistic fashion, excluding considerations of the value that it brings to the way humans live and interact with the world (Hewa & Hetherington, 1995). The focus is on one machine connecting to another kind of machine (when viewing the body from a mechanistic lens) instead of the engineered solution being an inherently value-laden process for managing human well-being.

We believe that by better integrating arts-based methods into the process of problem framing, problem solving, and reflection within engineering ethics education and practice, we can help switch the standard narrative for discussing the integration of human and engineered systems to a more expansive and holistic understanding of system interactions and their essential interdependencies. For instance, one of the key ideological precepts that drives usability study (UI/UX) is privileging human experience over the machine and making machines better adapted to and more closely aligned with (and possibly amplifying and improving) human experience. This human-centered design process inevitably raises questions of *which* humans and how to accommodate conflicting needs among a diverse user base. Decision-making that requires trade-offs in which some groups are disadvantaged relative to others must be supported by a robust ethical foundation. In other words, the arts can allow values to drive decision-making.

Barriers to overcome

As a discipline, ethics has always been interwoven with the arts, with philosophers having debated for millennia the extent of this connection and whether morality is the root of aesthetics or vice versa (Eaton, 1997; Haney, 1999). However, disconnects and limitations have also always existed when considering the application to technical learning. This becomes especially relevant to engineering through disagreements on what Aristotle meant by *techne*, posed for example by Aquinas in the thirteenth century, who "ascribes ethical value to doing, aesthetic value to making" (Eaton, 1997, p. 355) all the way through to Mitcham's (2022) twenty-first-century assertion that "technology is more akin to art" because its making is conditioned by social needs and philosophical ideas (p. 134). Either way, the liberal arts heritage of ethics may, in part, enable connections between educational methods traditionally associated with the arts, such as those that require emotion and empathy, imagination, self-discovery, reflection, and even creative thinking (Pizarro et al., 2006). Ethics teaching is often characterized by methods found within the liberal arts such as Socratic seminars or small group discussions or debates (Avci, 2017). Thus, when applied ethics (which includes the areas of business, medical, and engineering ethics) emerged in the mid-twentieth century as a distinct field, it is not surprising that scholars saw it in part as an attempt "to carry the banner of the humanities beyond traditional disciplinary boundaries" (Caplan, 1980, p. 24). Indeed, Kline (2001) has shown that engineering ethics "pedagogical methods come from moral philosophy, history, and sociology" (p. 14).

However, engineering ethics teaching often has, as Newberry (2004) describes, "the tendency to force square pegs of non-technical knowledge into round holes of technical learning, in order to accommodate the thinking preferences of engineering faculty and students" (p. 350). This results in more individualist and cognitivist pedagogies such as teaching that is limited to codes of practice or focused narrowly on knowledge around legal compliance like health and safety regulations as a proxy for ethics education (Walling, 2015) or positioning students as 'instrumentalists' whose only concern is solving a problem adequately and efficiently (Snieder & Zhu, 2020). Rethorst (2019) outlines the many ways this approach creates a 'lack' that disconnects the aesthetic from the ethical and limits the moral perception critical to moral education. He argues that "algorithms for action implied by moral principles may fall short of sufficient guidance" in addressing an ethical dilemma (p. 156). This amounts to a situation where the methods and content of engineering ethics education limit the aims and outcomes that are possible. A quiz that asks students to identify elements of a professional code of ethics may not be as effective as having students write a response to a hypothetical client seeking an engineered solution that breaches that code. Walling (2015) acknowledges this by pointing out that even where active learning methods are embraced in the teaching of engineering ethics, their purpose is "largely to further cognitive competence" rather than a more expansive emotional, social, or cultural understanding that we might see through an arts-based approach (p. 1641).

What drives this disconnect? Educators who are accustomed to process-based inquiry may fall prey to the common misperception that the arts lack a method or structure that can be analyzed, reviewed, and taught. For instance, a lecture that explains the difference between ethical theories might feel more manageable for an engineering educator to conduct than a Socratic discussion that explores and questions those theories. Génova and Gónzalez (2016) describe how in their discussion-based engineering ethics class, they find they need "to show that the real is not only what can be measured, and to teach how to reason about realities that are not corporeal or empirically verifiable" (p. 571). This relates to the acknowledged challenge of assessment within engineering education, which traditionally hews to exams with right/wrong answers instead of authentic assessments that support deeper, more contextualized learning (Villaroel et al., 2020). Skepticism about qualitative assessment can raise concerns and resistance from students as well; the clarity of a grading system based on problem sets and exam scores aligns with the mindsets and skillsets that draw many students to engineering in the first place. In contrast, assessment that targets the demonstration of critical thinking, reflection, and creativity may feel at once obscure and alien to some students. Taken together, this culture of 'traditional' methods of engineering learning might prevent educators from attempting to teach or assess ethics through arts-based methods. However, assessment methods used within active learning approaches and other emerging best practices in education can also be applied to arts-based methods, providing a bridge of familiarity for some educators.

Another barrier to using arts-based methods in engineering ethics is the historically siloed structure of academic institutions and disciplines, which is not a problem unique to this field. It is well acknowledged that the academy tends to maintain a conservative infrastructure and instructional system that perpetuates itself rather than embracing innovation (Selznick et al., 2021; Winebrake, 2015). Kazerounian and Foley (2007) argue that a pervasive emphasis on accuracy and risk reduction impedes engineering programs from embracing opportunities for creativity that lend themselves to the inclusion of the arts. Where the arts and engineering may be connected, concern remains that one discipline may get short shrift at the expense of the other. For instance, in a team-taught interdisciplinary module that used the arts in nursing education, the educators reflected on the need to avoid "reducing art to a means toward an end" and their own challenges of grappling with new ways of teaching and learning (McKie et al., 2008).

Lessons learned that can unite research and practice

Despite these barriers, much evidence exists of educators and even institutions breaking through them. The literature contains many descriptions of activities, projects, or modules that explicitly bring the arts to ethics education and to engineering education more broadly. From these examples, we can elicit application within the specific area of engineering ethics education.

Lessons from other disciplines: Business and medicine

As indicated above, business and medical education scholars have studied the use of the arts in ethics education for many years. Educational researchers within medicine have formally defined different types of arts-based approaches (as in Rodenhauser et al., 2004), and Kinsella and Bidinosti (2016) reviewed over two dozen studies of arts-informed approaches to healthcare education generally. Learning outcomes for arts-informed medical curricula have been described (as in Kinsella & Bidinosti, 2016), while Haidet et al. (2016) have taken this work still further by creating a "conceptual model to guide design, evaluation and research of the use of the arts in medical education" (p. 328). Engineering education can learn much from these examples, and several studies on arts-based methods of ethics teaching in healthcare programs could be replicated in an engineering context.

For instance, role play, skits, and theater have been studied as a method of ethics education for both doctors and nurses. Coleman and Dick (2016) researched a collaboration between nursing and theater departments where situations that might be encountered in healthcare contexts were simulated by theater students to elicit a process of ethical decision-making by the nursing students. While practicing dealing with uncertain situations, students developed not only their ethical decision-making skills but also skills in collaboration and in "soliciting and appreciating varied viewpoints" (Coleman & Dick, 2016, p. 265). This has obvious relevance to an engineering context, where students could use role play to 'try out' possible approaches to a professional ethical situation, such as a conflict with a client or manager. De la Croix et al. (2011) analyzed student reflections on three years of a 'Performing Medicine' program created by the Clod Ensemble theater company in collaboration with the London School of Medicine and Dentistry and the Department of Drama at Queen Mary University. Student feedback revealed that learning acting techniques helped alleviate professional anxiety, and "development of skills used by artists, such as those of detailed observation and interpretation, as well as those required to identify preconceptions and prejudice, was considered to be transferable to clinical practice for application in, for instance, medical diagnostics" (De la Croix et al., 2011, p. 1094). Efforts to highlight issues of equality, diversity, and inclusion within engineering (all ethical concerns) might benefit from a similar approach. Role play has already been used in a Theatrical Technology Assessment activity to allow students to learn about and practice stakeholder engagement (Visscher, 2023), and an interesting result was found by Baliga et al. (2017) as their study showed that medical students who were taught bioethics were more interested in skits as a teaching method than those who were not taught bioethics, suggesting that exposure to ethics might also increase enthusiasm for the arts.

In the field of business, Freeman et al. (2015) used storytelling and narratives as methods to help achieve outcomes in leadership and business ethics. This is because they wanted to situate ethics as "a conversation about how we describe and re-describe self, other, and communities to live together and collaborate in making a better world" (p. 526). Koehn and Elm's 2014 edited collection *Aesthetics and Business Ethics* considers many ways that the fine and liberal arts can – and in their view should – be incorporated in business schools "to enhance our aesthetic sensibility and to improve our ethical judgments in order to live better lives" (p. 5). The substantial research outlined in this book, both philosophical and practical, provides many lessons that could be learned from engineering ethics educators and suggests many openings for research in this field.

Lessons from experience: Colorado School of Mines and California Polytechnic State University

Moving beyond descriptions of practice within the literature, we can learn from the experience garnered by educators working in programs that have deliberately adopted arts-based methods to teach engineering ethics. In a later section of this chapter, these methods are described more thoroughly to serve as a guide for others.

The interdisciplinary design, communication, and ethics course 'IDEAS: Innovation and Discovery in Engineering, Arts, and Sciences' at Colorado School of Mines (Mines) applies a humanities lens to design problem-solving in complex socio-technical contexts. Arts-based methods are fundamental to the course learning outcomes (including those related to ethics), and first-year students self-select into the year-long program in part out of a desire to integrate humanities with their STEM coursework. Despite the students' genuine interest in humanities topics, the educator team has noticed that they struggle with (real or perceived) ambiguity in assignments and assessments. In response, the instructor team has developed assessment strategies that alleviate these concerns and simplify the grading process for arts-based and project-based content in the course.

Assignments that allow students to express themselves creatively in a medium they already enjoy (or choose to explore) invite them to connect personally with the material and push them into a mental state that favors the critical thinking and reflection required in many arts-based approaches. (For more on reflection and reflexivity, see Chapters 6 and 25.) For example, a self-portrait assignment asks students to creatively depict who they are as a team member in any medium (visual, written, musical, or performing arts); this first requires critical reflection on their team experiences and their own contributions to those outcomes, and then requires development of a written or visual metaphor to communicate abstractions such as values and mindsets. This in-depth consideration of team membership reveals the ethical components of working collaboratively that are often obscured when outputs are the primary measure of team effectiveness in engineering projects. In developing a fair assessment for an assignment such as this, it is critical to keep the goals and intended outcomes of the work in mind. For instance, both the level of artistic skill a student demonstrates and the aesthetic value of the final deliverable are largely irrelevant to the assignment, which is designed to evoke critical and creative thinking. A simple written 'artist's statement' accompanying the creative work prompts the students to explain their intended message and draw explicit connections between their communicative goals and the concrete elements of their expressive work. Noting in simple, straightforward terms the significance of certain colors or turns of phrase, or why their musical composition speeds up or changes key, prompts the students again to reflect on process and engage metacognitively with their own work. Notably, the artist's statement is the linchpin of the grade; while students earn credit for essentially following instructions (using a creative medium, exploring a specific topic, and so on), the 'subjective' portion of their grade that requires careful evaluation by the instructor is focused on the student's ability to connect the abstract to the concrete. This approach gives students the confidence to take risks with their creative expression, knowing that its quality will have no bearing on their grade, and relieves instructors of the need to become art critics in order to evaluate their students' work. While this is just one example, the model opens the door to assignments that provide immense creative freedom and a tangible connection to the arts, while also providing future engineers much-needed practice in communicating complex ideas to a broader audience.

Similarly, in project-based coursework, the bulk of the project grade falls on the student's critical reflection of their process work, which must meet criteria designed to prompt depth and specificity in the response. Existing frameworks, such as those described in the University of Edinburgh Reflection Toolkit (McCabe & Thejll-Madsen, 2018), provide a solid foundation for developing assessment criteria. Reflection on the process helps students center the learning experience over the tangible outcomes of the project. Valuing process over product is a central mindset in arts education, where students are encouraged to experiment, take risks, and prioritize revision in crafting quality work. This mindset is also critical for ethics education, in which students must feel obligated and empowered to analyze their own thinking and decision-making through myriad lenses and societal contexts, and to re-think their positions when encountering new information or perspectives.

The Liberal Arts and Engineering Studies program (LAES) at California Polytechnic State University, San Luis Obispo (Cal Poly) offers a hybrid engineering and humanities Bachelor of Science degree that was created to help retain engineering students who were seeking more diverse approaches to engineering design and training, and to allow them more individualized control over their studies. The program requires the same core modules in math, sciences, and introductory engineering as the first 2 years of the engineering degree, but during the third year, students divide their studies equally between upper-level modules from the engineering college and the liberal arts college. The four core required LAES modules focus on project-based learning, team collaboration, project management techniques, and integration of practices and design approaches used in engineering and the liberal arts. LAES students focus on how language shapes distinctions of disciplinary practice between engineering and humanities and distinctions of professional persona and purpose. Students work on interdisciplinary, community-centered projects that are developed for, and delivered to, real-world clients for immediate use in a public, community and/or industry context. Authentic applications and communication with clients help students see the importance and complexity of ethical considerations in their work.

As students work in small scrum teams, operating within an academically adapted version of Agile project management, they are asked to constantly integrate the language and design methodologies they have learned in their engineering and liberal arts modules into a user-focused critique of their work-in-progress. The language that students use to describe this integrative design practice often arises from the language of usability study, especially when their project deliverable is a software-based system. But students also adopt some of the design and development language used in architecture, civic management, community support, and performing arts that aligns with their project and professional identities of their clients, who may be community organizers, theater directors, product managers, systems engineers, public education administrators, and museum directors. Students learn how to speak about their work in terms that make sense to their client but that also align metaphorically and structurally with their team's integrated arts and engineering practices.

LAES students who are fresh to this process worry about how their individual work will be assessed. Because they are required to repeat the core project-based-learning modules twice, students enrolled in the module for the second time take on leadership roles to guide new students through the anxieties and ambiguities of the integrative design process, encouraging the new students to trust the Agile and LAES systems, and making them more comfortable, peer-to-peer, with learning how to translate between design and work languages and practices. Ultimately, the students must demonstrate how they are putting sound engineering and arts-based methods to use in the iterative improvement process of creating a useful, professional-quality final product for their client in a responsive and responsible way.

Further opportunities

The lessons that can be learned from other disciplines as well as from innovators in engineering institutions point to many opportunities for meaningfully using arts methods in engineering ethics education despite the deeply rooted and complex barriers outlined above. These opportunities can be characterized as small shifts, such as the words we use to describe our work, or as more significant changes such as the adoption of different activities and pedagogies, or as broader structural efforts to align research and practice across a variety of areas of engineering education.

What we can say

To help guide future efforts of combining art with engineering ethics, we propose that curriculum designers use definitions and explanations of key concepts that arise from the arts while also being directly connected to the practice of engineering. We recommend a fresh use of language in engineering education that considers engineering as art and art as engineering, therefore providing a linguistic framework for a productive interdisciplinary integration of art practice and conception throughout every level of engineering pedagogy. The key to the reconnection of art with engineering is the human form and how it – how we – interact with, learn from, and communicate to the world through our manifested creations.

Above all else, the most effective method for teaching about how to combine art and engineering is the actual practice of it. Being directly engaged with producing actual deliverables for real-world users is a far more successful and lasting form of teaching than simply discussing these processes and approaches in a lecture and classroom setting. In many ways, learning the language and process of integrative design and industrial practice is similar to learning a foreign language and culture – language and culture are living, evolving contexts that are best experienced first-hand with full immersion to provide for the most fluent and lasting understanding of how they work.

What we can do

The language used in public critiques of, and legislative efforts to manage, social media (one of the most recognizable forms of use of engineering technology), often suggests that Big Tech (engineering) needs to be more human and therefore more responsive to the complexities of how these engineered systems are used in a complex social/human context (Milmo, 2021). These critiques can then influence the calls for improving engineering practice by improving engineering education by integrating more humanistic (such as arts-based) forms of design, problem-solving, and ethical training and practice throughout engineering curricula (Osgood, 2017; Schwartz, 2007). We argue that these recommended improvements in engineering education cannot be effectively addressed by just adding a few arts and ethics courses to the curriculum; instead, arts-based methods and ethics need to be integrated throughout the engineering curriculum as an acknowledgment that engineering is an important, interrelated aspect of being human, and that arts and ethics have always been important, interrelated aspects of being an engineer. This section outlines a range of approaches to integration of engineering ethics and the arts.

Perry et al. (2011) outline two overarching ways in which the arts can be used in healthcare education; we expand on these for an engineering ethics context and add a third possibility, all of which could allow engineering education to embrace a fuller, deeper, more comprehensive approach to ethics learning. First, students could experience art first-hand, including reading literature, attending an art gallery or play, or watching a film. Watching films is one option to deepen and broaden engineering ethics instruction; Hitt and Lennerfors (2022) describe how this could manifest through the use of Miyazaki's *The Wind Rises*. In our experience using science fiction to teach ethical concepts in the IDEAS course at Mines described by Burgess (2019), we found that students displayed some of the highest levels of engagement in the discussions and activities during the class sessions devoted to the short stories. The stories addressed topics such as cultural norms in engineering, personal ethical decisions and their effects on society, and the impacts of emerging technologies. In the context of medical education, Delany and Gaunt (2018) used educational sessions based on students viewing art in a gallery to foster critical thinking about ethics in clinical practice. In both these examples, educators collaborated with experts who had disciplinary expertise to implement the activity (i.e., a professor with a Ph.D. in literature and an art curator, respectively). Key to adopting these methods, therefore, is a willingness for interdisciplinary collaboration, something that educators are not necessarily trained in or have comfort with. Mentorship and reaching out to exemplars can help alleviate this discomfort.

Second, students could be engaged in creating art of their own. Creativity has been increasingly cited as essential to engineering practice (Bruhl & Bruhl, 2020); however, it is less discussed in relation to ethics. Some scholars have explicitly linked the use of arts-based methods to creativity development in engineering design (Laduca et al., 2017; Spuzic et al., 2016), but we believe engineering ethics education is also an excellent avenue for fostering creativity. Kazerounian and Foley (2007) set forth ten maxims of creativity in education, of which many relate directly not only to the process of creating art but also to the practice of ethical deliberation: for example, 'keep an open mind,' 'ambiguity is good,' and 'search for multiple answers.' Baliga et al. (2017) found that using skits to teach components of bioethics fostered "critical thinking and creative brainstorming when dealing with controversial topics" (p. 78) and Freeman et al. (2015) show how "the exercise of the kind of creative muscles involved in … theater" (p. 522) develops empathy, encourages self-reflection, and connects to the complexities of professional issues. Additionally, our teaching experience has demonstrated how engaging in a creative project with a group necessitates communication and collaboration, other transferable skills required for twenty-first-century engineers.

Finally, we suggest that students could be engaged in a deliberate process of making meaning through arts-informed experiences. This interpretive process is vital because it is possible to observe and create art (and reflect on those experiences) without considering what those experiences mean for, or how they can apply to, professional engineering practice (Haidet et al., 2016). This is a challenging process to engage in, due to the pervasive 'two cultures' biases. However, it is essential for strengthening the relationship between engineers and the public that they serve; as Seedhouse (1988) explains, ethical actions stem from a person's awareness that what she/he does is socially important. Meaning-making is also concerned with revealing how engineering is a human activity and therefore necessarily encompasses the 'whole human' range of experiences – values, emotions, complexities, and relationships (Snieder & Zhu, 2020). Practice in interpreting an arts-informed experience can serve as a proxy for interpreting a professional situation or ethical dilemma and can allow for the incorporation of students' "own emerging professional identity into the ethical analysis and reflection" (Delany & Gaunt, 2018, p. 522). Prompting students to reflect on specific questions can guide them to make meaning from arts-informed experiences and give them the opportunity to find direct connections to their own technical knowledge and interests.

Each of the three approaches illustrated above could be adopted on a spectrum from relatively simple 'micro-insertions' into existing curricula (see Davis, 2006) up to foundational integration across an entire unit, module, or program. Educators seeking to try out arts-based methods may naturally be inclined to begin with small interventions in order to observe benefits and barriers within the context of their specific courses. The instructor must first consider their own skills and comfort levels with the range of arts-based approaches they might choose from. For example, the challenges of moderating a live in-class discussion are distinct from those of assessing or responding to students' artistic work. Possible micro-insertions that do not require significant on-the-spot improvisation or synthesis from the instructor include activities such as (1) viewing a video, listening to music, or reading a short story followed by small-group discussions or individual reflections on a prewritten set of questions or prompts; (2) working in small groups to conduct brief 'vox pop' style video interviews, which can then be screened in class or assigned as homework viewing, to gain a broad set of perspectives on an ethical question or topic; (3) individually maintaining a learning journal or other artifact (blog, vlog, sketchbook) that prompts periodic reflection on course topics, perhaps supported by an existing reflection framework such as the University of Edinburgh Toolkit (McCabe & Thejll-Madsen, 2018). These examples are also likely to be successful (in producing the desired learning outcomes) in classrooms where a culture of trust and open discourse has not yet been developed.

Perhaps counter-intuitively, there is some risk when introducing a single arts-based activity or lesson plan into an otherwise traditional course that students will be particularly attuned to its 'oddness' and, therefore, be more skeptical or resistant. Taking time to contextualize the activity and transparently state the motivations for assigning it can help assuage students' concerns. More broadly, integrating arts-based perspectives and practices across multiple facets of the course can normalize these approaches and earn students' trust over time. Such efforts can still be small, such as prompting written reflection after class activities or assignments, as advocated outside the context of arts-based methods (e.g., Harding et al., 2015).

Arts-based adoptions that require more commitment in terms of class time and instructor preparation include activities that may span multiple class sessions or require scaffolding across an entire unit. For example, York and Conley (2020) detail an approach they term 'Creative Anticipatory Ethical Reasoning,' which engages students in the ethics of fictional future techno-social scenarios such as the widespread adoption of autonomous vehicles. The case study occupied 2.5 weeks of class time in their implementations and was preceded by introductions to key concepts through various readings. Such efforts may stretch the expertise of engineering educators working independently but offer rich opportunities for collaboration across disciplines, which can mitigate the concerns around imposter syndrome and mismatched professional expertise.

Embedding the arts in engineering beyond ethics

We believe that the necessary expansion of this field of study can begin with an understanding of how arts-based methods in engineering ethics link to and build from three more well-established research fields which are useful entry points for future scholars. These are active learning (including the service learning and challenge-based learning discussed, e.g., in Chapter 23), the STEAM movement widely adopted in primary and secondary education, and the integration of liberal arts and engineering more generally (Hitt et al., 2023). These three existing movements can be built upon to bolster further research and practice in using the arts to extend and improve engineering ethics education.

Additionally, addressing the attitudes and attributes that professionals should cultivate can be used to leverage research on the inclusion of the arts, such as the curiosity, creativity, and reflection articulated as 'Engineering Habits of Mind' (Lucas & Hanson, 2014). Professional, industrial, civic, and commercial contexts demand the transferable skills that arts-based methods promote. For instance, there is wide consensus on the sustainability competencies that engineering practitioners must cultivate (ABET, 2020; AdvanceHE, 2021; Arizona State University, 2018; Engineering for One Planet, 2022; European Commission, 2022). These typically include systems thinking, values thinking, strategic thinking, collaboration, and futures thinking. Besides values thinking and collaboration, which we have already shown to link neatly with the arts-based methods outlined above, Kinsella and Bidinosti (2016) have shown that the arts can enable 'anticipatory competence,' another term for futures thinking: that is, the ability for students to "imaginatively [project] themselves into their future practice in terms of the values they intend to bring, the actions they wish to take, and the type of practitioner they'd like to become" (p. 314). This could be extended to emphasize engineering ethics: thinking about the kind of future we want to create. Further, the arts have been shown to foster eco-centric views in engineering students, as described by Paek and Kim in their 2021 study of a workshop used to promote empathy and ecological sensitivity. We encourage researchers to systematically study the connection between arts-based methods and professional competencies in engineering education.

Conclusion

Successful adoption of arts-based methods in engineering ethics requires a complicated and critical scholarly effort that can demonstrate the value of the arts in ethics education and the effectiveness of arts-based methods in engineering education more generally. Practitioners reconnecting these disciplinary lenses must also analyze real boundary constraints and learn when these approaches work most effectively and when they might be likely to fail. Additionally, scholarship needs to move beyond accounts of practice and calls to action – and more towards research on implementation and systematic studies. The good news is that there are many entry points for educators and researchers from all disciplines to contribute to this exciting field that can make arts and practical ethics a more comprehensive and interconnected part of the engineering curricula and thus create a broader educational experience for tomorrow's engineers.

Finally, we believe that education is fundamentally a human process of becoming. Engineering students are often young adults who are still discovering and questioning their values as they connect to new perspectives and communities through their peers and mentors. Using the arts in teaching engineering ethics can help students reveal themselves to themselves on a personal as well as a professional level, serving as a mirror for self-reflection and re-establishing the fundamental connections between being a human and being an engineer.

Acknowledgments

The authors would like to thank Toni Lefton for her invaluable feedback on each stage of the writing process from proposal to final draft, as well as to our reviewers, who provided insightful suggestions for improvement.

References

ABET. (2020). *Criteria for accrediting engineering programs.* ABET Accreditation. https://www.abet.org/accreditation/accreditation-criteria/criteria-for-accrediting-engineering-programs-2020-2021/

AdvanceHE and Quality Assurance Agency for Higher Education. (2021). *Education for sustainable development guidance.* QAA. https://www.qaa.ac.uk/the-quality-code/education-for-sustainable-development

Arizona State University. (2018). *Key competencies in sustainability.* School of Sustainability. https://static.sustainability.asu.edu/schoolMS/sites/4/2018/04/Key_Competencies_Overview_Final.pdf

Avci, E. (2017). Learning from experiences to determine quality in ethics education. *International Journal of Ethics Education, 2*, 3–16. https://doi.org/10.1007/s40889-016-0027-6

Baliga, M. S., Palatty, P. L., D'Souza, O., Naik, T. S., Adnan, M., Boloor, R., & Shivashankara, A. R. (2017). Medical undergraduate students opinion on use of visual arts (skit) in teaching-learning bioethics: A questionnaire based study. *Global Bioethics Enquiry Journal, 5*(2), 78. https://doi.org/10.38020/GBE.5.2.2017.78-82

Belbase, S., Mainali, B. R., Kasemsukpipat, W., Tairab, H., Gochoo, M., & Jarrah, A. (2022). At the dawn of Science, Technology, Engineering, Arts, and Mathematics (STEAM) education: Prospects, priorities, processes, and problems. *International Journal of Mathematical Education in Science and Technology, 53*(11), 2919–2955. https://doi.org/10.1080/0020739X.2021.1922943

Booker, N., Gates, J. D., & Knights, P. (2021). Cognitive biases and the cultural disconnect between engineers and decision-makers. *Technium Social Sciences Journal, 35*(17), 35–62.

Bruhl, J., & Bruhl, W. (2020). *Engineering creativity: Ideas from the visual arts for engineering programs.* ASEE Virtual Conference. https://doi org/10.18260/1-2 34550

Burgess, O. (2019). Stand where you stand on Omelas. *Teaching Ethics, 19*(1), 63–70.

Caplan, A. L. (1980). Ethical engineers need not apply: The State of applied ethics today. *Science, Technology, & Human Values, 6*(33), 24–32.

Coleman, J. J., & Dick, T. K. (2016). Nursing and theater: Teaching ethics through the arts. *Nurse Educator, 41*(5), 262–265. https://doi.org/10.1097/NNE.0000000000000271

Davis, M. (2006) Integrating ethics into technical courses: Micro-insertion. *Science and Engineering Ethics, 12*(4), 717–730.

De la Croix, A., Rose, C., Wildig, E., & Willson, S. (2011). Arts-based learning in medical education: The students' perspective. *Medical Education, 45*(11), 1090–1100. https://doi.org/10.1111/j.1365-2923.2011.04060.x

Delany, C., & Gaunt, H. (2018). "I left the museum somewhat changed": Visual arts and health ethics education. *Cambridge Quarterly of Healthcare Ethics, 27*(3), 511–524. https://doi.org/10.1017/S0963180117000913

Eaton, M. M. (1997). Aesthetics: The mother of ethics? *The Journal of Aesthetics and Art Criticism, 55*(4), 355–364. https://doi.org/10.2307/430923

ENAEE. (2021). *EUR-ACE® framework standards and guidelines.* European Accreditation of Engineering Programs. https://www.enaee.eu/wp-content/uploads/2022/03/EAFSG-04112021-English-1-1.pdf

Encyclopedia Britannica. (n.d.). https://www.britannica.com/topic/Britannica-Online

Engineering Council. (2020). *The Accreditation of Higher Education Programmes (AHEP)* (4th ed.). https://www.engc.org.uk/media/3464/ahep-fourth-edition.pdf

Engineering for One Planet. (2022). *The engineering for one planet framework: Essential sustainability-focused learning outcomes for engineering education.* https://engineeringforoneplanet.org/wp-content/uploads/2022_EOP_Framework_110922.pdf

Engineers Australia. (2008). *Accreditation management system: Education programs at the level of professional engineer document no. S02: Accreditation criteria summary.* https://www.engineersaustralia.org.au/sites/default/files/content-files/2016-12/S02_Accreditation_Criteria_Summary.pdf

European Commission Joint Research Centre. (2022). *GreenComp: The European sustainability competence framework.* https://publications.jrc.ec.europa.eu/repository/handle/JRC128040

Freeman, R. E., Dunham, L., Fairchild, G., & Parmar, B. (2015). Leveraging the creative arts in business ethics teaching. *Journal of Business Ethics, 131*(3), 519–526. https://doi.org/10.1007/s10551-014-2479-y

Génova, G., & Gónzalez, M. R. (2016). Teaching ethics to engineers: A socratic experience. *Science and Engineering Ethics, 22*(2), 567–580. https://doi.org/10.1007/ s11948-015-9661-1

Graham, R. (2018). *The global state of the art in engineering education.* Massachusetts.

Haidet, P., Jarecke, J., Adams, N. E., Stuckey, H. L., Green, M. J., Shapiro, D., Teal, C. R., & Wolpaw, D. R. (2016). A guiding framework to maximize the power of the arts in medical education: A systematic review and metasynthesis. *Medical Education, 50*(3), 320–331. https://doi.org/10.1111/medu.12925

Haney, D. (1999). Aesthetics and ethics in Gadamer, Levinas, and Romanticism: Problems of phronesis and techne. *PMLA, 114*(1), 32–45. https://doi.org/10.2307/463425

Harding, T. S. et al., (2015). The role of collaborative inquiry in transforming faculty perspectives on use of reflection in engineering education. *2015 IEEE Frontiers in Education Conference* (FIE), El Paso, TX, USA, pp. 1–7. doi: 10.1109/FIE.2015.7344183.

Hewa, S., & Hetherington, R. W. (1995). Specialists without spirit: Limitations of the mechanistic biomedical model. *Theoretical Medicine 16*, 129–139. https://doi.org/10.1007/BF00998540

Hitt, S. J., & Lennerfors, T. T. (2022). Fictional film in engineering ethics education: With Miyazaki's *The Wind Rises* as exemplar. *Science and Engineering Ethics, 28*(5). https://doi.org/10.1007/s11948-022-00399-w

Hitt, S. J., Banzaert, A., & Pierrakos, O. (2023). Educating the whole engineer by integrating engineering and the liberal arts. In A. Johri (Ed.), *International handbook of engineering education research*. Routledge. https://doi.org/10.4324/9781003287483-25

Graham, R. (2018). The global state of the art in engineering education. Massachusetts Institute of Technology. https://www.cti-commission.fr/wp-content/uploads/2017/10/ Phase-1-engineering-education-benchmarking-study-2017.pdf

Kazerounian, K., & Foley, S. (2007). Barriers to creativity in engineering education: A Study of instructors and students perceptions. *Journal of Mechanical Design, 129*(7), 761–768. https://doi.org/10.1115/1.2739569

Kinsella, E. A., & Bidinosti, S. (2016). "I now have a visual image in my mind and it is something I will never forget": An analysis of an arts-informed approach to health professions ethics education. *Advances in Health Sciences Education : Theory and Practice, 21*(2), 303–322. https://doi.org/10.1007/s10459-015-9628-7

Kline, R. R. (2001). Using history and sociology to teach engineering ethics. *IEEE Technology and Society Magazine, 20*(4), 13–20. https://doi.org/10.1109/44.974503

Kochn, D., & Elm, D. (2014). *Aesthetics and Business Ethics* (D. Koehn & D. Elm, Eds.). Springer Netherlands. https://doi.org/10.1007/978-94-007-7070-6

Laduca, B., Ausdenmoore, A., Katz-Buonincontro, J., Hallinan, K. P., & Marshall, K. (2017). An arts-based instructional model for student creativity in engineering design. *International Journal of Engineering Pedagogy, 7*(1), 34–57. https://doi.org/10.3991/ijep.v7i1.6335

Lang, E. M. (1999). Distinctively American: The Liberal Arts College. *Daedalus*, *128*(1), 133–150.

Lucas, B., & Hanson, J. (2014). *Thinking like an engineer: Using engineering habits of mind to redesign engineering education for global competitiveness.* SEFI 2014 Annual Conference, University of Birmingham, Birmingham, United Kingdom.
McCabe, G., & Thejll-Madsen, T. (2018). *The reflection toolkit.* University of Edinburgh. https://www.ed.ac.uk/reflection
McKie, A., Adams, V., Gass, J. P., & Macduff, C. (2008). Windows and mirrors: Reflections of a module team teaching the arts in nurse education. *Nurse Education in Practice, 8*(3), 156–164. https://doi.org/10.1016/j.nepr.2007.03.002
Milmo, D. (2021, October 24). Frances Haugen: "I never wanted to be a whistleblower, but lives were in danger." *The Guardian.* https://www.theguardian.com/technology/2021/oct/24/frances-haugen-i-never-wanted-to-be-a-whistleblower-but-lives-were-in-danger
Mitcham, C. (2022). *Thinking through technology: The path between engineering and philosophy.* University of Chicago Press.
Newberry, B. (2004). The dilemma of ethics in engineering education. *Science and Engineering Ethics, 10*(2), 343–351. https://doi.org/10.1007/s11948-004-0030-8
OECD. (2022). PISA 2022 creative thinking. Program for international student assessment. https://www.oecd.org/pisa/innovation/creative-thinking/
Osgood, K. (2017, May 1). *Engineers need the liberal arts too.* The Chronicle of Higher Education. https://www.chronicle.com/article/engineers-need-the-liberal-arts-too/
Paek, K.-M., & Kim, H. (2021). Promoting a pro-ecological view: The effects of art on engineering students' perceptions of the environment. *Sustainability, 13*(24). https://doi.org/10.3390/su132414072
Perry, M., Maffulli, N., Willson, S., & Morrissey, D. (2011). The effectiveness of arts-based interventions in medical education: A literature review. *Medical Education, 45*(2), 141–148. https://doi.org/10.1111/j.1365-2923.2010.03848.x
Pizarro, D. A., Detweiler-Bedell, B., & Bloom, P. (2006). The creativity of everyday moral reasoning. In J. Kaufman & J. Baer (Eds.), *Creativity and reason in cognitive development* (pp. 81-98). Cambridge University Press. https://doi.org/10.1017/CBO9780511606915
Rethorst, J. (2019). Moral density: Why teaching art is teaching ethics. *Philosophy and Literature, 43*(1), 155–172. https://doi.org/10.1353/phl.2019.0009
Rodenhauser P., Strickland M. A., & Gambala, C. T. (2004). Arts-related activities across U.S. medical schools: A follow-up study. *Teaching and Learning in Medicine, 16*(3), 233–239. https://doi.org/10.1207/s15328015tlm1603_2
Rogers, H. L., Hitt, S. J., & Allan, D. G. (2021). Problem-based learning in institutional and curricular design at the New Model Institute for Technology and Engineering (NMITE). *Journal of Problem-Based Learning in Higher Education, 9*(2). https://doi.org/10.5278/ojs.jpblhe.v9i2.6814
Schwartz, J. (2007, September 30). Re-engineering engineering. *The New York Times Magazine*, 94(L).
Seedhouse, D. (1988). *Ethics: The Heart of health care.* Wiley.
Selznick, B. S., Mayhew, M. J., Zhang, L., & McChesney, E. T. (2021). Creating an organizational culture in support of innovation education: A Canadian case study. *Journal of College Student Development, 62*(2), 219–235. https://doi.org/10.1353/csd.2021.0018
Snieder, R., & Zhu, Q. (2020). Connecting to the heart: Teaching value-based professional ethics. *Science and Engineering Ethics, 26*(4), 2235–2254. https://doi.org/10.1007/s11948-020-00216-2
Snow, C. P. (1959). *The two cultures and the scientific revolution.* Cambridge University Press.
Spuzic, S., Narayanan, R., Abhary, K., Adriansen, H. K., Pignata, S., Uzunovic, F., & Guang, X. (2016). The synergy of creativity and critical thinking in engineering design: The role of interdisciplinary augmentation and the fine arts. *Technology in Society, 45*, 1–7. https://doi.org/10.1016/j.techsoc.2015.11.005
Villarroel, V., Boud, D., Bloxham, S., Bruna, D., & Bruna, C. (2020). Using principles of authentic assessment to redesign written examinations and tests. *Innovations in Education and Teaching International, 57*(1), 38–49. https://doi.org/10.1080/14703297.2018.1564882
Visscher, K. (2023). Experiencing complex stakeholder dynamics around emerging technologies: A role-play simulation. *European Journal of Engineering Education.* https//doi.org/10.1080/03043797.2023.2196940
Walling, O. (2015). Beyond ethical frameworks: Using moral experimentation in the engineering ethics classroom. *Science and Engineering Ethics, 21*(6), 1637–1656. https://doi.org/10.1007/s11948-014-9614-0
Walton, M., Baker-Olson, A., Luckes, H., Blaber, I., Walker, E., Folkes-Lallo, B.,Becton, M., & Thomas, E. (2022). The integration of classroom and community learning in narrative accounts of co-curricular service. *Michigan Journal of Community Service Learning, 28*(1). https://doi.org/10.3998/mjcsl.434

Wende, M. C. van der. (2011). The emergence of liberal arts and sciences education in Europe: A comparative perspective. *Higher Education Policy*, *24*(2), 233–253. https://doi.org/10.1057/hep.2011.3

Winebrake, J. J. (2015). The integrative liberal arts and engineering – The 'grand challenge' of curricular implementation. *Engineering Studies, 7*(2–3), 193–195. https://doi.org/10.1080/19378629.2015.1062486

York, E., & Conley, S. N. (2020). Creative anticipatory ethical reasoning with scenario analysis and design fiction. *Science and Engineering Ethics*, *26*(6), 2985–3016. https://doi.org/10.1007/s11948-020-00253-x

25

REFLECTIVE AND DIALOGICAL APPROACHES IN ENGINEERING ETHICS EDUCATION

Lavinia Marin, Yousef Jalali, Alexandra Morrison, and Cristina Voinea

Reflection in higher education

Reflection is a central competency in higher education and one of the highest forms of cognitive achievement in Bloom's taxonomy (Bloom et al., 2020), meaning that it is difficult to achieve but worthwhile to pursue. In this section, we aim to describe what reflection is and what its benefits are for our context of application, namely engineering ethics education (EEE). We cannot offer an overview of the existing models or the history of the concept of reflection since this would require a chapter in itself. Prominent scholars like John Dewey (1933), Donald Schön (1984), and David Kolb (1984) proposed detailed accounts of reflection that have been used as conceptual underpinnings for developing structured processes of reflection and reflective practice. Dewey's description of reflection as a general mode of thought and cognitive process that highlights the interactions between experience and self provides a valuable lens to describe what reflection is and how it can be incorporated into general educational settings. Dewey defines reflective thought as "Active, persistent, and careful consideration of any belief or supposed form of knowledge in the light of the grounds that support it and the further conclusions to which it tends" (1933/2008, p. 118). Dewey's account was philosophical and guided by existential and phenomenological principles, and much of the ensuing research on reflection tried to operationalize Dewey's insights into more applicable principles for education (English, 2023).

Rodgers (2002, p. 845) summarizes Dewey's view, distilling four main criteria for defining reflection:

- It is "a meaning-making process"
- It is a "systematic, rigorous, disciplined way of thinking"
- It takes place "in community, in interaction with others"
- It would require "attitudes that value the personal and intellectual growth of oneself and of others."

Briefly, as Rodgers (2002) explains, the first criterion points out the primary function of reflection, to grapple with the various interpretive possibilities of ethical situations; the need to reconstruct the experience to understand the problem initially obscured in layers of complexity. The second

DOI: 10.4324/9781003464259-30

criterion addresses the process of reflection as a way of conscious and deliberate thinking. In this process, one draws on the meaning of experience, develops possible alternatives and hypotheses for a given situation, and then subjects these to testing and experimentation. The third criterion highlights that reflection is not merely a solitary action. While it is plausible that moments of pause and engaging in solitary research may serve as a valuable exercise, it is through dialogue that one can see the experience from a different lens and further expand one's understanding. This is also true for those who teach praxis and facilitate reflection practices; relationality is the essence of reflective thought (Buber, 1958; Freire, 2005). The fourth criterion points to a set of attitudes needed for an individual to engage in reflective practice, mainly awareness of one's own limited perspective, open-mindedness and willingness to seek counter-evidence, and being responsive to the particularities of the unique situation and the needs of others.

Focusing on the practice of reflection within *higher education*, Ryan (2013) elucidated this concept by delineating two key elements and four levels. The two elements of reflection are "making sense of experience in relation to self, others, and contextual conditions" and "reimagining and/or planning future experience for personal and social benefit" (Ryan, 2013, p. 145). The two elements capture the core of the process of reflection illustrated by Dewey: experience and interpretation of the experience, and developing and experimenting with potential alternatives (Rodgers, 2002). Further, Ryan illustrates *four levels of reflection*, which provide direction to both the teacher/facilitator and students. In educational practice, the four levels of reflection are:

- reporting/responding
- relating
- reasoning
- reconstructing

These point to identifying and reporting key issues, relating issues to background and experience, analyzing situations considering different perspectives, and alternatively, reframing and experimenting with the course of action. In a similar vein, some scholars of service-learning provide practical advice for incorporating reflection into educational settings, highlighting the need for understanding the meaning of experience, surfacing and challenging assumptions, and creating opportunities for sharing perspectives to develop more complex views of situations or problems (Eyler, 2002; Hatcher & Bringle, 1997).

In this chapter, we are particularly interested in how reflection emerges in the context of EEE and what methods exist for systematically fostering ethical reflection in formal engineering education.

Reflection in engineering ethics: ethical reflection

In engineering education, reflection has been recognized as facilitating students' learning and skill development (e.g., Turns et al., 2014; Woods et al., 2000). Specifically, in the context of engineering ethics, scholars have emphasized the benefits of incorporating reflective practices, such as ethical reasoning, awareness of experience, the meaning-making process, fostering openness to new possibilities, and developing ethical sensitivity and commitment (e.g., Beever and Brightman, 2016; Bielefeldt et al., 2020; Bombaerts et al., 2022; Bucciarelli, 2008; Corple et al., 2020; Kim et al., 2019; Lönngren, 2021). Ethical reflection is considered foundational for most ethics classes in most professional fields (Chadwick, 2012, p. 718), beyond the cognitive reasons that make reflection a worthwhile process to pursue. In engineering ethics, the goal is not only to learn something (i.e., an epistemic goal) but is also existential, that is, to transform how one views the profession as a whole in the context of larger political, economic, and social structures. Self-transformation without reflection is hard to imagine (Mezirow, 2006), which is why engineering ethics pedagogy, if it is to succeed, must also resist the tendency toward instrumentalization. Dewey theorized that reflection only 'gets off the ground' when students are given the space to question and experience ambiguity. Benefits for students are often described concerning the 'process' of ethical reasoning, considering broader non-technical factors and being more critical in decision-making. Reflection as a mode of thinking enables us to continuously monitor our assumptions and values and bridge experiences, self, and situation. While reflection is a general pedagogical practice that can be deployed in almost any curriculum, ethical reflection is a competency more specific to EEE scholarship (Bielefeldt et al., 2020; Bucciarelli, 2008; Colby & Sullivan, 2008; Marin, 2020; Royakkers & van de Poel, 2011). In this chapter, we are concerned with the process of ethical reflection, what it can borrow from reflective practice, and what is unique about ethical reflection *qua* reflection. If reflection is "a careful examination and bringing together of ideas to create new insight through ongoing cycles of expression and re/evaluation" (Marshall, 2019, p. 411), *how is ethical reflection distinctive?*

To define ethical reflection, we turn to a model put forth by van de Poel and van Gorp (2006). They take ethical reflection to be a form of moral deliberation in which:

> engineers should take into account all relevant moral values. Designing engineers should, for example, reflect on the choices they make regarding the relative importance of safety, economic, and sustainability considerations ... Typical for ethical reflection is that the actual existing way of dealing with moral issues is not taken for granted.
>
> *(van de Poel & van Gorp, 2006, p. 335)*

Thus, for ethical reflection, students and practitioners first recognize that there is a normative issue at stake that existing ethical frameworks or codes of conduct cannot solve straightforwardly (Grunwald, 2000). If the need for a non-trivial solution is recognized, then they need to launch into a process of ethical reflection. Ethical reflection shares with the wider concept of reflection its four-component model (of cognitive assessment, active, iterative, and integrative aspects), but all these are applied to the ethical theory realm. What this realm contains is up for debate, though. While van den Poel and van Gorp (2006) confine the realm of ethical reflection to ethical values and theories, others, such as Erin Czech, also identify political and social values as legitimate ethical concerns, hence worthy of ethical reflection (see Morrison, 2020).

The process of reflection shares some of the elements of moral deliberation – from our engagement with an ethical problem to experiencing perception and action in imagining possible courses of action and transforming the situation and the self. Within educational praxis, the goal of the process is not necessarily to arrive at a particular answer but to provide opportunities for students to grapple with the perplexity of a given situation, envisioning various courses of action and critically evaluating their relative merits, thereby enhancing their understanding.

Ethical reflection can take many forms in educational practice, which we will delve deeper into in the third section. But for a quick insight into how it might look, let us consider the 'Revenge Test' (Jalali et al., 2021), a scenario in which students imagine taking revenge in a situation. The facilitator asks students to think about why they would take revenge. Group discussion provides an opportunity for communicating different perspectives and understanding alternative meanings of experience. Students can increase their awareness of their own values, question their assumptions, and see new emerging questions and ideas. Next, the facilitator presents a challenge: while we often imagine someone else's future experiences in a negative or cruel manner, we may lack insight into envisioning positive future experiences for others and fostering meaningful relationships. The facilitator begins by asking students to describe a given story/scenario to encourage participation. Then, the discussion can move to identify the main issue, inviting students to consider 'what' and 'why' questions (Jalali et al., 2022). The facilitator can assist in uncovering (i.e., making explicit) the students' values, backgrounds, and experiences during this process, fostering an environment where students are encouraged to reframe their perspectives and adopt new lenses to examine the issue. Consistent with embodied perspectives of reasoning, pedagogical methods, and the design and reflection on intervention outcomes, these rely on students' lived experiences (Civjan & Jalali, 2022). Sharing, feedback, and reflection on students' perspectives provide opportunities for experimentation, out of which more questions may be raised. This example showcases important constituents of reflection – connecting experience with a given situation, questioning values and assumptions, discussing alternative perspectives, and stretching reasoning in considering different possibilities (Eyler, 2002; Rodgers, 2002; Ryan, 2013).

Based on research in phenomenology and cognitive science, we emphasize that addressing an ethical dilemma requires understanding the problem and simulating potential scenarios through imagination (Johnson, 1993). This process cannot be isolated from who we uniquely are and what experiences, values, and emotions bring to our sense and interpretation of the given situation (Marin & Steinert, 2022). Suppose moral deliberation is not about applying habitual patterns of thought and fixed rules. In that case, there needs to be an ongoing interplay between thinking and experience where we can continually expand our boundaries and reorient and adjust our thought patterns (Johnson, 1993). There is a clear connection between reflection and ethical awareness in engineering. A deeper understanding of the ethical implications of professional activities can foster an ethically informed community of tech and engineering students.

In this chapter, we propose that adopting reflective and dialogical approaches can familiarize (and habituate) engineering students with the process of ethical reflection. This may, over time, cultivate a professional culture that prioritizes ethics in technology development and implementation. We argue that all reflective approaches in ethics education are grounded in dialogical encounters with oneself, others, and texts. We show that reflection is fundamentally dialogical and that successful dialogical methods will stir reflection. We aim to examine the existing reflective methods used in EEE in order to reveal their modes of dialogical engagement, based on this theoretical premise. It is important to note that every reflective method has its own set of advantages and drawbacks, which we will briefly describe. We end the chapter with practical recommendations for instructors aiming to instill ethical reflection in their classrooms.

Some theory: the dialogical nature of ethical reflection

Reflection as a dialogical and mediated encounter

In this section, we will discuss how the phenomenological philosophical tradition informs how we think about reflection. Briefly, phenomenology is a major current in European philosophy that emerged at the turn of the twentieth century with Edmund Husserl's meticulous studies of lived experience. Following him came a procession of philosophers like Martin Heidegger, Jean-Paul Sartre, Simone de Beauvoir, and Maurice Merleau-Ponty. While these philosophers have distinctive and not-always-compatible philosophical views, they nevertheless share a methodological commitment to rigorous descriptions of concrete lived experiences.

As this is an expanding area of research that affects different fields such as cognitive and neurosciences, philosophy of technology, healthcare pedagogy, and the social sciences, it has become evident that our experiences, precisely because we are bodies, are not trapped in a realm of personal mental representations, but rather that our embodied selves are always already caught up in and shaped by a historical and sociocultural milieu. This highlights the numerous ways in which our experiences are shaped by our physical bodies and the environment in which we live. When objects appear to us, they always do so within a particular horizon of implicit meanings, a tacit interpretive framework that structures the modes of appearance and the possibilities for our involvement. Thus, the field of our experience always has a social and 'intersubjective' character.

What does this mean specifically for reflection? First, we must pause and ask where reflection is happening and what its object is. The term 'reflection' itself might lead us to think that what is at issue here is an inquiry directed towards the self, toward one's own inner life. There is a long philosophical heritage going back to René Descartes' famous 'cogito' argument ('I think therefore I am') behind this idea of reflection as self-directed introspection. As the Cartesian tradition exemplifies, this approach tends to lead to a kind of 'mind–body dualism' insofar as it treats the mind as something 'interior' and detached from the world. This presupposition of detachment has profound implications for how we think about agency, ethics, and our involvement with others and with technologies. Phenomenology, on the other hand, when it uses the term 'reflection,' has in mind a kind of attentive directedness toward the field of lived experience itself. Its methodological aim is to avoid presuppositions and begin with a description of how experience happens. This brings to light certain features of experience that can be taken as guides for philosophical inquiry into ethical life.

First, in attending to the happening of experience, I notice that most of the time, *I* am not the object of my experience. My attention is instead directed toward taking care of projects in the world. For example, I am frequently absorbed in tasks like buying groceries and traveling between home and work, as well as attending to the larger projects of my career and family life. Many phenomenologists draw our attention to the way in which our experience is *seamless* – that we are, first of all, and for the most part, *absorbed* in meaningful tasks and contexts of action. For instance, utilizing my car to drive to the store, taking out my wallet while at the checkout counter, typing on my laptop, sharing announcements on the learning management software I use to interact with students, and employing various other technological devices – these are all continuously shaping the form that my experience takes. And yet, I am not reflecting on my use of those tools and devices but rather on the sense and the overall aim of my engagement. Again, I find that my lived experience is most often not explicitly self-aware.

Once we attend to the goal-directed character of our experience, we notice that these tasks are always undertaken within a coherent and contextualized whole. I never encounter raw 'data points' or feel 'bare' sensations. I am absorbed in situations that are always already sense laden. The inherent meaningfulness of these contexts lets me be absorbed and attentive to them. Driving to the store this morning, I did not have to explicitly cognize, step by step, how to shift the gears to slow down or signal left into the parking lot. At that very moment, all of my attention was on two pedestrians, a mother and child, who were motioning to cross the road in front of my car. Yet I can vaguely recall, many years ago, when I first learned to drive, that driving was an 'alienated' and self-conscious experience that is emphatically *not* how it is now. Because my body and my consciousness are inextricably intertwined, after practice and eventual habituation, my car functions as a seamless extension of my bodily intentions. I am a skillful driver precisely because I 'forget' the explicit details of driving, allowing me to pay attention to pedestrians. This example is not at all extraordinary – most of our experiences take this form. However, it does mean that we often take the intelligence of our embodiment for granted because, much of the time, we are busy enacting meaning in the world through our projects. We are only able to make sense of the world through our actions. We are embodied because we are 'gearing into' the world, both literally and figuratively.

This brings us back to that essential component of lived experience that we mentioned at the beginning of this section – which is that it is always involved with others – and the essentially 'intersubjective' character of this involvement. The world appears to me as a 'context of significance' open to me and others, which confers on my experiential contexts their latent sense of 'objectivity.' I am immediately aware of the significance of others' actions, and I am aware that they are aware of mine. In our seamless bodily involvement with the world, our actions are also expressive. We are geared into a shared cultural horizon of meaning. This means that, fundamentally, others are *inside* of my experience. For example, while slowing down to make way for the pedestrians crossing the road, I made eye contact with the mother. In a split second, she read my intention – just as I felt her concern. My glance conveyed that I had seen them and that they were safe to cross.

Again, this is not extraordinary. Because my experience is constitutively intersubjective, my awareness of myself as a moral agent is dialogically mediated through others. All of our experiences take this general form: We are attentive to the meanings of our embodied actions with respect to their 'interrelatedness' or interpretability by others. It is my recognition of the 'gaze' of the other (whether literal or imaginatively anticipated) interpreting the meaning of my actions in particular contexts that directs my own 'gaze' back to myself. That my experience is always open to others calls me to respond and to be responsible. Recall that the first feature we noticed about our lived experience was that we are not explicitly self-aware most of the time. It takes others to get us there.

Before we conclude this section, it is worth noting that being attentive to the 'intersubjective' character of experience is further complicated by the fact that more and more of our relations with others are technologically mediated. Reflection on the social and shared character of meaning-making necessarily includes grappling with the material contexts of our relations with others – because the meaning of those relations is transformed when mediated through technological artifacts and systems. American sociologist Sherry Turkle has reflected deeply on the contemporary digitally mediated social world, and she describes the particular and peculiar phenomenon of 'being alone together' (Turkle, 2011). According to Turkle, our phones and other communication technologies create a false sense of connection with others by disconnecting us from the meaningful contexts in which we first encounter them. Instead, they connect us in ways that are abstract and do not account for the shared meaning and understanding that comes with face-to-face communication. Here, we find a danger that certain technical mediations of our experience can thwart our capacity to attend to the real sources of meaning, including ethical meaning, in our experience.

To put it briefly, phenomenology is a technique that helps us understand and explore our experiences in their own context, with the goal of uncovering the ways in which significance arises. Taking seriously these phenomenological insights (that experience is embodied, intersubjective, and technically mediated) poses a challenge to traditional approaches to professional ethics that rely on abstract rules or codes. This approach often views ethical living as analogous to using tools, where the focus is solely on determining the appropriate rules for using the tools available to us. Phenomenology, by highlighting the way ethical meaning arises in our experiences, redirects our attention towards exploring how our tools (and collections of tools) influence our relationships with the world and others. It prompts us to examine how they bring certain things into focus while obscuring others, and how they shape our perception of what (and who) is significant. Ethics then becomes less 'a simple matter of correct tool use' and more a question of 'design and responsible agency.' Such phenomenological reflection, in the context of engineering ethics pedagogy, enables students to see for themselves the emergence of ethical meaning and responsibility in their experience and in the professional context for which they are becoming prepared.

Dialogical education and dialogism

Dialogical education has been a growing trend in educational theory in the last decades (Mercer et al., 2020). Dialogism as an educational movement started from and inspired by Mikhail Bakhtin's theoretical work : "Dialogism is a philosophy of language which places central importance on the reality of socio-verbal interaction in understanding the kind of phenomenon that language is" (Skidmore, 2020, p. 27). A constitutive principle of dialogism is that "Truth is not born nor is it to be found inside the head of an individual person; it is born between people collectively searching for truth, in the process of their dialogic interaction" (Bakhtin, 1929/1984, p. 110, cited from Wegerif, 2020). Dialogue is then defined as a method in which "students learn through being called out by others into active engagement in ongoing dialogues" (Wegerif, 2020, p. 23), where this other can be another human being, a generalized other (e.g., society, a body of knowledge), or a non-human other (e.g., nature, a technological artifact). The fundamental principle of dialogism is ethical and epistemic, as it entails that epistemic values and achievements are always found in encounters with another. Moreover, seeking opportunities for encounters is something valuable that one should seek systematically if one wants to develop oneself.

Dialogue is not a mere conversation – talking about something in front of another; it is also affected by how the other responds (verbally or non-verbally). There are many educational formats centered around dialogue in EEE: interacting with stakeholders (e.g., interviews), having discussions with peers about a case study, interacting and deliberating via online platforms, role-playing, mock trials, and so on. Yet, not all such interactions are dialogical; the possibility of being affected by others varies based on the specific configuration. There are also other practices that one could call *monological* (based on a simple distinction of how many voices one finds in practice), such as writing reflective journals or essays, which are used in engineering ethics instruction to promote reflection.

Our central claim is that any pedagogical activity aiming to instill an experience of ethical reflection in the ethics classroom needs to be infused with dialogism at some level. This claim is based on discriminating between superficial dialogical exchanges and genuine dialogical exchanges. A superficial dialogical exchange is one where we merely enact a dialogue as an exchange of replies: A says this, B replies, and A then takes their turn, and so on. We call this 'superficial exchange' because taking turns while speaking does not ensure a dialogue between those involved. One can see such a non-dialogical exchange in formal debates or in the 'Ethics Bowls,' where students can respond to each other's arguments for the sake of winning the debate without letting the debate change their opinions on the matter at hand. In a genuine dialogical exchange, by contrast, the other – be this human, non-human, or a generalized other – can challenge and change the interlocutors, who are vulnerable and open to listening. This means that dialogical experiences are not necessarily about encountering others; one can encounter oneself through technological mediation or when writing a text. Even 'classical' monological practices, such as lecturing, journaling, or watching a movie, can be injected with dialogical elements (formally) and serve the same purposes depending on how open and engaged the participants are.

Drawing from the theory of dialogism with its ethos of being attentive and vulnerable to the voices of others as potentially changing ourselves, and the phenomenological nature of reflection – as a transformative experience, mediated, happening 'in between' – we will now examine activities and methods that promote dialogical encounters and that thus seem promising as sites for ethical reflection.

Some praxis for fostering ethical reflection – methods and approaches

Before we review the existing educational methods for fostering ethical reflection, we need to emphasize that there are complexities in instilling reflection, and there is not one single bullet-proof method for this endeavor. Here, we single out two main difficulties to be expected and planned for when fostering reflection systematically.

The first difficulty concerns the effort required for reflection and the self-transformation entailed. These may come as unpleasant surprises for many students and, perhaps, for instructors as well. Regarding fostering reflection in classroom practices, some commonly used methods include reflective notebook writing and in-class discussions (Walker, 2013). These are valuable methods when used systematically, yet they do not work by themselves without being tweaked and adapted to the specific cohort of students. Teachers will not trigger reflection by merely assigning a journal entry or leading a class discussion on ethical issues, because reflection is not idea generation. Reflection is not merely 'thinking about' something, a brainstorming session, or jotting down strings of opinions about a controversial case. When we, as educators, ask for reflection from our students in the ethics classroom, we ask for more than simple assignments. We ask for an effort that is uncomfortable emotionally (Mikalayeva, 2020); we ask for vulnerability and self-disclosure. For this, we need to showcase what reflection is and provide examples of it. We can start with simple models and move toward more complex ones.

The second significant difficulty lies in the open-ended nature of reflection. Reflection, as we construe it here (drawing as we have from Dewey), is an experience of thinking that the subject undergoes once they encounter resistance from the world. Engineering students are well-versed in problem-solving. Even when confronted with an ethical dilemma, their first approach is to treat it as a problem with only one correct solution. The problem-solving mindset (sometimes called the 'techno-fix mindset,' see Huesemann & Huesemann, 2011[1]) conceptualizes ethical concerns as something ultimately solvable through the power of reason and knowledge in a rationalistic vein (Warford, 2022). This attitude focuses more on finding a solution to what is perceived as a problem rather than dwelling on the problem itself and exploring its complexity. A central goal of reflection is not merely to 'solve' the problem as such – although, based on ethical reflection, arriving at new designs is encouraged (van den Poel & van Dorp, 2006) – but rather to make the student aware of their situated thinking and how their assumptions play a role in what they perceive as viable solutions. The techno-fix mindset clashes with the ethos of reflection, which treats problems as open-ended, complex, and as a source for self-knowledge. Due to its prevalence in engineering education, the techno-fix mindset often stands in opposition to the practice of ethical reflection. Moreover, the so-called 'hidden curriculum' in engineering (Tormey et al., 2015) makes it seem that ethics and ethical reflection are not necessary for doing solid engineering work and are somewhat at odds with engineering. Engineering students are, by and large, trained not to care about ethical issues and to avoid ethical reflection on the issues emerging in everyday engineering practices.

Nonetheless, the methods advanced throughout this chapter are meant to encourage collaborative learning, divergent thinking, and critical, constructive in-class debates, which might open students to reflection. More importantly, all these methods are egalitarian, as they stress the need to listen and respond to others and to build on their inputs while focusing not on the interlocutor's social position but on their arguments and grounds. No matter how sophisticated our methods may be, engaging in reflection is a task that is effortful and emotionally vulnerable – especially when we ask students to reflect in front of others. As educators, it's crucial to delineate our intentions behind incorporating reflection into a course. We must identify when, during the course or learning process, we aim to promote reflection, whether it involves individuals, classmates, or even inanimate objects. Subsequently, we should tailor our pedagogical methods to effectively foster this reflective practice.

How can educators create opportunities for students to participate in critical inquiry processes that prioritize essential aspects of moral deliberation? Ones that place emphasis on the 'qualitative unity' of a situation, individual values, backgrounds, and experiences, as well as encouraging the imagination and evaluation of various alternatives? This section explores the main existing methods to incorporate ethical reflection. We have divided these methods into four main categories – dialogical and monological, synchronous and asynchronous – based on the temporality of the method. Via monological/dialogical polarity, we aim to stress that the dialogical experience will vary depending on whether the main challenge of the method is encountering others (e.g., colleagues with diverging ethical intuitions and arguments or the unseen stakeholders for whom one is designing) or encountering oneself (e.g., one's beliefs, attitudes, and biases).

Four main types of activities for instilling ethical reflection

We have identified four main categories of methods for teaching reflection:

A. Monological and synchronous
  - Writing prompts in the classroom for individual students
  - Exam with essay-type answers (e.g., argue for … explain why … analyze this case …)

B. Dialogical and synchronous
  - Case studies with complex iterative deliberation (ethical cycle)
  - Role-plays
  - Mock-trials with deliberation
  - Tinkering with artifacts (design, redesign, optimization)
  - Group design of educational activities
  - Group essays written collaboratively, simultaneously

C. Monological and asynchronous
  - Essay as homework
  - Reflective notebooks

D. Dialogical and asynchronous
  - Online deliberation (forum-like debates with threads of nested messages)
  - Commenting on another's written reflection (peer feedback)
  - Group essays written sequentially
  - Below, we analyze a token from each of the four categories of methods for instilling reflection.

A. Dialogical and synchronous

Case studies

One common dialogical approach used, especially in engineering and business ethics, is the case study that presents students with various morally problematic situations and invites them to find solutions or to imagine new ways of tackling the issues presented (Colby & Sullivan, 2008; see also Chapter 20). The case-study method can help students familiarize themselves with moral judgment processes and acquaint them with the ethical standards for their profession (Davis, 1997). Despite its centrality in engineering ethics classes, the case-study method has been criticized for being exclusively individualistic in its scope, leaving aside the complexities of the context (Bucciarelli, 2008) and the broader macro-ethical context, meaning 'the profession's collective social responsibility … to societal decisions about technology' (Herkert, 2005, p. 374). Moreover, the overuse of dramatic disaster case studies tacitly suggests to students that ethical decision-making is an exceptional occurrence rather than a day-to-day demand (Morrison, 2019). One way to remedy these deficiencies of the case-study method is the role-play strategy, as proposed by Martin et al. (2019). In role play–based case studies, students are asked to form groups representing the stakeholders involved (Doorn & Kroesen, 2013). This encourages students to take a more active stance when trying to find a solution to the problem presented in the case study. It also familiarizes students with the different interests of the parties involved. Assuming a role is about adopting a situated position in the world, with its epistemic limitations and values, is helpful if we want students to reflect on the situatedness of their own position and help them contextualize their thinking. There are caveats to role-playing, however. The role can be assumed superficially, played based on stereotypes about the profession, or overtly focused on performance rather than reflection – and ultimately fail to highlight ethical positions.

Deliberation on case studies can help – and can be done in various ways, some more sophisticated than others. We have identified several effective ways to facilitate deliberation based on our experience as educators.

Iterative and complex ethical cycle

A method that deploys case-study deliberation is the 'ethical cycle,' created to help students grapple with and embrace the ambiguous, non-linear character of ethical judgment (van de Poel & Royakkers, 2011, 2007). With the ethical cycle, students can make well-considered judgments on real-life situations through a series of iterative steps. Typically, the first 'walk through' of the cycle is done individually, and then the students get together to compare and discuss the divergences in their interpretations and evaluations. For the individual 'walk through,' students write down their reflections on each stage of the cycle; this solitary work also has dialogical elements. As we argued above, it is more accurate to describe the meaning-making realm as being 'in-between' us and others rather than as residing in some incorrigible and inscrutable 'interiority.' Writing is always a process that involves circling repeatedly, and the ethical cycle emphasizes this. Ethical reflection takes time, and although it requires knowing facts about a situation or 'case,' it also requires self-awareness. Often, we only come to understand our earlier motivations for making particular choices long after we've chosen. Our desire for expediency and to see ourselves in a certain way often hinders honest self-assessment. If implemented thoughtfully, the ethical cycle can help to habituate these reflective behaviors.

The five basic steps of the ethical cycle involve moral problem identification, problem analysis, options for action, ethical evaluation, and reflection. At each stage, there are opportunities to expand and increase the theoretical and contextual considerations that could deepen or even fundamentally change the students' initial interpretations of the earlier stages, prompting them to return to an earlier stage and rearticulate, for example, their initial moral problem statement. Depending on the complexity of the case and the depth of the critical inquiry engaged by the students, the cycle may take several iterations over subsequent weeks. Groups come together to select their ethical scenario or 'case study' and reconvene periodically throughout the term as they are exposed to additional frameworks for analysis. Since many of their ethical scenarios involve technologies, they may need to consider the agentive character of a particular technology. Sometimes, they need time to gather more relevant information, such as the issue's history in the communities involved in their chosen scenario. Other times, if there is too much group consensus, they might need to engage more deeply with different normative frameworks to uproot a deeply embedded culturally hegemonic way of seeing. Experimenting with the ethical cycle group process over the term underscores the importance of pragmatic social contexts and the necessity of time and care for robust ethical reflection. Furthermore, this kind of dialogue-based approach stresses the importance of the meaning-generating nature of concrete experience. Not only are students engaging in ethical reflection about an imagined professional scenario, but together, they are simultaneously enacting the process of ethical community building. Suppose their instructor underscores the value of difference rather than consensus throughout the process – that can free the students to gain awareness and respect for the uniqueness of the varied lived experiences of other persons. This iterative approach starts from the presupposition that ethics is not about individuals simply applying principles but rather that group dialogue is about building moral and emotional relationships of mutual trust and respect for difference.[2]

Design your own ethics curriculum

Another method for stimulating engineering students' engagement with ethics has been advanced by Alpay (2013). Instead of offering a predefined task that students must solve in class or at home, instructors can ask students to develop, in groups, resources, methods, or activities that are meant to familiarize their colleagues with ethics meaningfully. In this way, roles are reversed, and students instruct. To avoid over-burdening students who might not be acquainted with the ethics of their profession, a series of lectures prepares them for the task by introducing the main concepts, issues, and applications of moral philosophy relevant to the students' profession (Alpay, 2013, p. 1457). After these introductory lectures provide a baseline understanding, the students work collectively in groups to develop proposals for their peers about how ethics should be taught. Proposing educational resources and activities prompts students to reflect on the importance of ethics for their profession and fosters "a culture of shared responsibility in learning and development" (Alpay, 2013, p. 1466). Moreover, students might devise interesting approaches that can be enacted subsequently to stimulate reflection and critical thinking among engineering students. Each group should present its proposal to the class and receive feedback that can further be integrated into the advanced activities. In this way, everybody participates in the other groups' work, which can stimulate reflection – and a sense of community and shared responsibility. This method aims to make students think of the relevant ethical topics that could be useful for their profession. In this way, they see beyond the immediate technical aspects of what they are learning and think about the implications of what they are doing. As we mentioned in the previous section, reflection is dialogical; it can be stimulated by engaging in a conversation with others, which is precisely what the collaborative dimension of this method aims at.

The emotional deliberation approach

The group deliberative methods described above can complement the emotional deliberation approach (Roeser & Pesch, 2016). Emotions should be taken into consideration in attempts to foster reflection in EEE. Creating a symmetric setup for discussion, where the students and instructors are placed on an equal footing (i.e., in a circle), helps everyone feel freer to express their analysis and emotions regarding what is being discussed. According to Roeser and Pesch, the main idea is to convey respect to every participant so that they can feel safe talking and critically reflect on their emotions and thoughts. This ties back to the idea mentioned in previous sections: reflection is not solely a rational process. It involves an emotional component. Ignoring the emotions stirred by discussions will not make them disappear. Instead, Roeser and Pesch argue that it is more fruitful to start with emotions in mind and analyze these emotional reactions as indications of the values and norms one endorses. This can prompt reflecting on what one takes for granted about the normative fabric of the world. This method's effectiveness is dependent on classroom size; large groups struggle with emotional deliberation, whereas tutorial groups find it easier.

B. Monological and asynchronous methods

Reflective journaling/notebooks or essays

Another interesting method of fostering ethical reflection in engineering education is to invite students to reflect on their own values and reasons for studying engineering – by prompting them to write an auto-biographical essay or to keep a reflective notebook in which, given a specific ethical situation and learning activity, students log the development of their opinions. Although these methods might seem ill-fitted for science-oriented education, they provide essential methods to explore one's development as an individual and a professional (Kim et al., 2019). The auto-biographical essay puts students in the position to think in a structured way about their own lives and experiences. Thus, it promotes self-understanding, reflection, and critical examination of one's choices. Kim et al. (2019) present some interesting questions that help guide students in approaching such an assignment. The auto-biographical essay starts with questions regarding one's personal life, such as *What experience has contributed to the person I am becoming?* or *What were or are the challenges in my life, and how do I make sense of them?* It moves to questions touching upon professional life, like *What kind of an engineer (or other professional) do I want to become?* and *What is it that I want to do with a degree in engineering (or another field)?* The journal method asks students to reflect, for a whole semester, on a particular technology (be it smartphones, cars, artificial intelligence systems, etc.) or a moral issue raised by technologies (privacy in the case of Internet apps, pollution in the case of cars, fragmentation of attention in the case of social media). By writing a weekly entry in the journal, students are encouraged to reflect freely about how design choices influence their interaction with different technologies and how they shape their lives.

Despite their appeal, both the auto-biographical essay and the journal appear to be monological pedagogy techniques. One way to add interactivity to these methods is to ask students to discuss their entries in class, with the instructor and other colleagues. Infusing the auto-biographical essay and the journal with the benefits of dialogical approaches can allow students to find affinities and common interests with their colleagues and also to critically filter their thoughts and reflections through the perspectives and worldviews of others.

C. Monological and synchronous methods

Monological and synchronous methods ask students to reflect on their own during classroom time. For example, the instructor may tell students to take 5 minutes to think about problem X before students discuss it in groups or individually write brief responses concerning the ethical issue. Such methods are suitable for generating material to think about further in groups or pairs, and these exercises are helpful as pre-reflection by asking students to make up their minds concerning an ethical issue – such that this initial opinion can be challenged and further refined through subsequent activities. The value of such exercises is that they are not confrontational, specifically because what the student reflects is kept private. A teacher could assign such exercises at the beginning of the semester and again at the end of the semester so that students can assess on their own how far they have arrived in refining their reflective capacities.

D. Dialogical and asynchronous methods

In asynchronous methods, the student's reflection is mediated by an online collaboration platform. Such methods entail, for example, asking students to comment on a paper online, annotate a text online, or build a mind map on a collaborative online platform. What students get to see from each other are only digital traces in the form of comments and, perhaps, some images. These methods are more akin to brainstorming, but when students edit an existing text by adding questions or suggestions, the collective reflection can be quite deep.

It may seem then that the main difference in synchronous versus asynchronous methods lies in the mediation aspect. However, the kind of dialogism entailed by mediation concerns us, rather than the mediation itself. This is because all pedagogical methods are mediated to some extent. The phenomenology of intersubjectivity recognizes mediated access – our sense-making activities are, at the same time, expressing themselves through behavior and speech. Given this mediated access to our own thoughts, the difference made by digital or paper-based platforms should not be radically different. There is a mediation of technology when we ask students to collaborate on a paper and comment on each other's responses to a text. This mediation does something other than the mediation of speech and body when students are in a room. When our methods require that students engage digitally with one another in an asynchronous way (i.e., not at the same time and not seeing each other instantly as would be the case with a video call), the resistance posed by others to the thinking process becomes less tangible and less immediate, and one could choose to ignore it. Reflection is still possible in asynchronous digital methods. However, it hinges on how seriously one engages with others' textual traces; it may be easier for students to engage in self-reflection rather than reflecting with others.

Assessment of reflection in EEE

Assessing the success of the educational methods in instilling reflection depends on the kind of classroom and the format where we find ourselves. Ethics in engineering education is taught either in standalone classes or integrated into learning pathways, where it is incorporated throughout engineering courses (van Grunsven et al., 2021).

If ethics is taught in a standalone class, we can take several steps, spaced through time, to foster reflection and iteratively revisit the results of reflective practice. Asynchronous methods, where the students keep a log or a notebook, will work effectively since the students will have a reference point to return to and re-evaluate. These asynchronous methods also facilitate students' self-assessment. Teachers can assign reflective notebooks at the beginning of the class, asking students to jot down their thoughts and insights throughout the semester. At the end of the course, students can be asked to reflect on their reflective processes and what they learned. Educators can assess this meta-reflection while the logbooks stay private to protect the students' fledging reflective processes. Rubrics for assessing reflection should involve the four previously described categories – integration with previous knowledge, interaction, systematicity, and active engagement (Rodgers, 2002). When assessing the dialogical activities, the instructor should also assess the group dynamics: *Did only one student engage in reflection, or was the activity constructed with insights from most group members?* This can be achieved by observing the interactions or, when this is not possible, by asking students to log their discussions in class and provide graphic emphasis to signify when they changed their minds or arrived at a new conclusion.

Reflection assessment is included in the general assessment methods of engineering courses when ethics is taught through modules embedded in the curriculum. Either way, reflection assessment should usually be linked with other learning goals' assessment in EEE or engineering education. Most engineering ethics classes do not prioritize reflection as their primary learning objective. Rather, reflection is a process to be fostered that enhances other ethics learning goals such as ethical awareness, ethical judgment, and deliberation. Hence, it makes sense to assess these other learning goals primarily – and then to have reflection as a sub-category of these. For example, when we assess ethical awareness/sensitivity, we can add a rubric on whether this ethical awareness improved through reflection or was showcased in a non-reflective way. While reflection is a high-level learning goal (Bloom et al., 2020), it should be assessed alongside other contributing goals in EEE. With ethical reflection, we can see assessment more as feedback rather than grading. As ethics instructors, we need to create opportunities for formative assessment throughout the semester by facilitating the self-assessment of students – peer assessment, feedback on journals, and even group presentations and discussions should receive feedback regarding how reflective these were.

Some practical take-away points for teaching ethical reflection

Integrating regular reflection into engineering ethics curricula and practices is necessary, as reflection is the primary component of ethical reasoning and moral judgment. In creating opportunities for reflection, it is critical to pay attention to the choice of situations and, in general, the cases and scenarios used in instruction. Students should be guided to see and engage in ethical situations considering (i) the situation's contextual reality; (ii) their own assumptions, values, and experiences; and (iii) dialogical practices. For instance, in writing reflective journals or essays, students can be prompted to redefine the problem; address their values, feelings, and assumptions; raise potential questions; and analyze the situation considering the aforementioned factors and a given model or text. Further, reflection can be operationalized through engagement with stakeholders in real ethical cases – as well as class presentations and discussions. For educators, it is important to address the complexities involved in developing the competencies required for reflection and to establish clear criteria for evaluating the reflective process.

We offer several practical takeaways for instructors aiming to instill the experience of reflection in the ethics classroom.

- Integrate reflection as a learning experience with other, more easily measurable learning goals:
  - When assessing these other goals, such as ethical deliberation, ethical sensitivity, and so on, provide a separate sub-rubric regarding how reflective the process and the outcomes were.
- When ethical reflection is embedded in another engineering course, use reflection as a sub-goal for the other learning goals (e.g., when assessing the design of an artifact, one can add the reflective component to the design evaluation).
- Create a safe space for reflective engagement by providing clear guidelines at the beginning, recognizing that dialogical exercises are spontaneous; people can easily hurt each other when they speak their minds without considering the effect on others.
  - Provide a set of clear rules and expectations so that all students can feel included in this process.
  - Start by announcing the rules of respectful engagement at the start of the class, reminding students of these rules and enforcing them.

- Make sure that all students feel heard and seen.
- Acknowledge students' contributions.
- Showcase examples of reflection, for example:
  - Engage in reflection yourself concerning a sample case study, or comment on a role-play acted by students in front of the classroom.
  - The examples teachers provide can be personal and should model spontaneity and vulnerability. For example, when teaching, you can explain how you changed your mind about issue X, mention the emotions entailed in that experience, and thus show students that emotions are to be expected and that nobody is a perfect epistemic agent, having the 'correct' answer from the start.
- Start with simple models of reflection and increase their complexity as the semester continues.
- Try to use a mix of dialogical and monological methods and have these interact, for instance:
  - You can promote dialogical methods during class time and then ask students to reflect privately in their notebooks on what they learned through the interactions.
  - Do not rely solely on monological or dialogical methods since these do not target the same kinds of reflective experience, and you'll want to create a variety of experiences for the students.
- Use reflection beyond the fleeting experience created in class:
  - The more students think and reflect about their reflections, the easier it becomes for them to perform.
  - You can ask students to refer back to their classroom or online discussions and use these insights or be critical about them in their individual assignments such as essays.

Conclusions

This chapter addressed the ambiguities and challenges in understanding and implementing reflective thinking in EEE. We argue that EEE instructors should pursue ethical reflection in a context-specific manner, as a worthwhile goal. We conceptualized reflection drawing from existing literature in pragmatism and phenomenology and argued that it is a sophisticated experience that can be nicely captured by experiential, first-person accounts. First-person experience, however, is what makes reflection tricky to assess and notice in the classroom. Whereas for ethical reasoning, instructors can look at the quality and complexity of the propositions advanced by students and thus use early responses as benchmarks for evaluating students' later proposals for dealing with an ethical problem, assessment of student reflections is trickier. In reflection pedagogy, teachers should not evaluate as such the propositions or design outcomes, but the process itself. (We do, however, recognize that reflection is often combined with ethical reasoning, and thus these do come as a package.) The process that teachers assess should encompass students' self-awareness, transformation, and spontaneity in interaction.

In this chapter, we used a phenomenological lens to argue why dialogism and emotional engagement are foundational for engaging in genuine, spontaneous reflection. While dialogism is a tool in the reflection toolbox, used alongside other monological tools, it has often been overlooked. We think dialogism shows a lot of potential when used correctly.

In the final section of the chapter, we presented several methods for instilling reflection as well as some ideas for assessment. Then, we provided some practical tips for educators who want to instill ethical reflection in engineering classes. Although we argued for the potential of the dialogical dimension of reflection, we also encourage instructors to use a mix of dialogical and monological methods – to introduce variation and provide periodical moments of feedback – and to provide time for students to think. Ethical reflection is a transformative experience and, as such, works well for formative assessments, for enriching the quality of the moral deliberation judgment, and for fostering ethical awareness.

Funding and acknowledgments

Lavinia Marin's work has been supported by the 4TU Centre for Engineering Education (4TU. CEE), grant number TPM.23.25.4TU.CEE.TUD, and by the research program Ethics of Socially Disruptive Technologies (ESDIT), NWO grant number 024.004.031. Cristina Voinea's work was supported by the European Commission [grant number 101102749] and UK Research and Innovation (UKRI) [grant number EP/Y027973/1]. Yousef Jalali would like to thank Christian Matheis and Vinod Lohani for inspiring theoretical exploration and innovative pedagogical practices. Alexandra Morrison would like to thank Professor Charles Wallace in the College of Computing at Michigan Tech for his ongoing collaboration in STEM ethics pedagogy.

Notes

1 The term "techno-fix" points to "a variety of technologies employed to respond to intractable societal problems, which have proven to be difficult or insoluble through political, legal and cultural reform" (Sand et al., 2023).

2 From one author's experience, comments like the following have not been uncommon regarding the updated version of the ethical cycle process: "I appreciated being pushed to rethink my initial formulation of the problem statement. My classmates' point that by using the utilitarian normative framework I wasn't able to see the real moral problem was eye-opening" (MTU EE student testimony, 2021).

References

Alpay, E. (2013). Student-inspired activities for the teaching and learning of engineering ethics. *Science and Engineering Ethics, 19*(4), 1455–1468. https://doi.org/10.1007/s11948-011-9297-8

Beever, J., & Brightman, A. O. (2016). Reflexive principlism as an effective approach for developing ethical reasoning in engineering. *Science and Engineering Ethics*, *22*, 275–291.

Bielefeldt, A. R., Polmear, M., Swan, C., Knight, D., & Canney, N. E. (2020, June). Variations in reflections as a method for teaching and assessment of engineering ethics. In *2020 ASEE Virtual Annual Conference Content Access*.

Bloom, B. S., & Krathwohl, D. R. (2020). *Taxonomy of educational objectives: The classification of educational goals. Book 1, Cognitive domain*. Longman.

Buber, M. (1958). *I and Thou* (R. G. Smith, Trans.). Charles Scribner's Sons.

Bombaerts, G., Martin, D., Watkins, A., & Doulougeri, K. (2022, March). Reflection to support ethics learning in an interdisciplinary challenge-based learning course. In *2022 IEEE Global Engineering Education Conference (EDUCON)* (pp. 1393–1400). IEEE.

Bucciarelli, L. L. (2008). Ethics and engineering education. *European journal of engineering education, 33*(2), 141–149.

Civjan, S., & Jalali, Y. (2022, August). Can you feel it? A case for reflexive response and imagination in ethics discussions [Theory Paper]. In *2022 ASEE Annual Conference & Exposition*.

Chadwick, R. (2012). *Encyclopedia of applied ethics*. Academic Press.

Colby, A., & Sullivan, W. M. (2008). Ethics teaching in undergraduate engineering education. *Journal of Engineering Education, 97*(3), 327–338. https://doi.org/10.1002/j.2168-9830.2008.tb00982.x

Corple, D. J., Zoltowski, C. B., Kenny Feister, M., & Buzzanell, P. M. (2020). Understanding ethical decision-making in design. *Journal of Engineering Education*, *109*(2), 262–280.

Davis, M. (1997). Developing and using cases to teach practical ethics. *Teaching Philosophy, 20*(4), 353–85. https://doi.org/10.5840/teachphil199720445

Dewey, J. (2008). *The Later Works of John Dewey, Volume 8, 1925—1953: 1933, Essays and How We Think, Revised Edition* (J. A. Boydston, Ed.). Southern Illinois University Press.

Doorn, N., & Kroesen, J. O. (2013). Using and developing role plays in teaching aimed at preparing for social responsibility. *Science and Engineering Ethics*, *19*(4), 1513–1527. https://doi.org/10.1007/s11948-011-9335-6

English, A. R. (2023). Dewey, existential uncertainty and non-affirmative democratic education. In M. Uljens (Eds.), *Non-affirmative theory of education and bildung. Educational governance research* (Vol. 20). Springer.

Eyler, J. (2002). Reflection: Linking service and learning—Linking students and communities. *Journal of social issues*, *58*(3), 517–534.

Freire, P. (2005). *Pedagogy of the oppressed* (M. B. Ramos, Trans.). Continuum International Publishing Group.

Grunwald, A. (2000). Against over-estimating the role of ethics in technology development. *Science and Engineering Ethics, 6*(2), 181–196.

Hatcher, J. A., & Bringle, R. G. (1997). Reflection: Bridging the gap between service and learning. *College Teaching*, *45*(4), 153–158.

Herkert, J. R. (2005). Ways of thinking about and teaching ethical problem solving. Microethics and macro-ethics in engineering. *Science and Engineering Ethics, 11*(3), 373–85. https://doi.org/10.1007/s11948-005-0006-3

Huesemann, M., & Huesemann, J. (2011). *Techno-fix: Why technology won't save us or the environment*. New Society Publishers.

Jalali, Y., Matheis, C., & Edwards, M. (2021, July). A graduate-level engineering ethics course: An initial attempt to provoke moral imagination. In *2021 ASEE Virtual Annual Conference Content Access*.

Jalali, Y., Matheis, C., & Lohani, V. K. (2022). Imagination and moral deliberation: A case study of an ethics discussion session. *International Journal of Engineering Education*, *38*(3), 709–718.

Johnson, M. (1993). *Moral imagination: Implications of cognitive science for ethics*. University of Chicago Press.

Kim, J. H., Campbell, R., Nguyen, N., Taraban, R., Reible, D., & Na, C. (2019, July). Exploring ways to develop reflective engineers: Toward phronesis-centered engineering education. In *Proceedings of the American Society for Engineering Education (ASEE) Annual Conference* (Vol. 21).

Kolb, D. A. (1984). *Experiential learning: Experience as the source of learning and development.* Prentice-Hall.

Lönngren, J. (2021). Exploring the discursive construction of ethics in an introductory engineering course. *Journal of Engineering Education*, *110*(1), 44–69.

Marin, L. (2020). Ethical reflection or critical thinking? Overlapping competencies in engineering ethics education. In van der Veen, J., van Hattum-Janssen, N., Järvinen, H.-M., de Laet, T., & ten Dam, I. (Eds.), *48th SEFI Annual Conference (Online): Engaging Engineering Education* (pp. 1354–1358).

Marin, L., & Steinert, S. (2022). Twisted thinking: Technology, values and critical thinking. *Prometheus*, *38*(1). https://doi.org/10.13169/prometheus.38.1.0124

Marshall, T. (2019). The concept of reflection: A systematic review and thematic synthesis across professional contexts. *Reflective Practice*, *20*(3), 396–415. https://doi.org/10.1080/14623943.2019.1622520

Martin, D. A., Conlon, E., & Bowe, B. (2019). The role of role-play in student awareness of the social dimension of the engineering profession. *European Journal of Engineering Education*, *44*(6), 882–905. https://doi.org/10.1080/03043797.2019.1624691

Merleau-Ponty, M. (2012). *Phenomenology of perception* (D. A. Landes, Trans.). Routledge.

Mezirow, J. (2006). An overview of transformative learning. In P. Sutherland & J. Crowther (Eds.), *Lifelong learning.* Routledge.

Mikalayeva, L. (2020). Introduction: Encouraging student reflection—approaches to teaching and assessment. *European Political Science*, *19*(1), 1–8. https://doi.org/10.1057/s41304-018-0182-7

Morrison, A. (2019). The ethics of authenticity: Heidegger on the struggle to be what one is. *Anekaant: A Journal of Polysemic Thought*, *8*(Autumn), 17–26.

Morrison, L. A. (2020). Situating moral agency: How postphenomenology can benefit engineering ethics. *Science and Engineering Ethics*, *26*(3), 1377–1401.

Rodgers, C. (2002). Defining reflection: Another look at John Dewey and reflective thinking. *Teachers -College Record*, *104*(4), 842–866.

Roeser, S., & Pesch, U. (2016). An emotional deliberation approach to risk. *Science, Technology, & Human Values, 41*, 2, 274–297. https://doi.org/10.1177/0162243915596231

Ryan, M. (2013). The pedagogical balancing act: Teaching reflection in higher education. *Teaching in Higher Education, 18*(2), 144–155.
Sand, M., Hofbauer, B. P., & Alleblas, J. (2023). Techno-fixing non-compliance—Geoengineering, ideal theory and residual responsibility. *Technology in Society, 73*, 102236. https://doi.org/10.1016/j.techsoc.2023.102236
Schön, D. A. (1984). *The reflective practitioner: How professionals think in action* (Vol. 5126). Basic Books.
Skidmore, D. (2019). Dialogism and education. In Mercer, N., Wegerif, R., & Major, L. (Eds.), *The Routledge international handbook of research on dialogic education*, (pp. 27–37), Routledge.
Tormey, R., LeDuc, I., Isaac, S., & Hardebolle, C. (2015). *The formal and hidden curricula of ethics in engineering education*. SEFI Proceedings, SEFI, Orléans, France.
Turns, J. A., Sattler, B., Yasuhara, K., Borgford-Parnell, J. L., & Atman, C. J. (2014, June). Integrating reflection into engineering education. In *2014 ASEE Annual Conference & Exposition* (pp. 24–776).
Turkle, S. (2011). *Being alone together: Why we expect more from technology and less from each other*. Basic Books
van de Poel, I., & Royakkers, L. (2007). The ethical cycle. *Journal of Business Ethics, 71*, 1–13. https:// doi.org/10.1007/s10551-006-9121-6
van de Poel, I., & Royakkers, L. (2011). *Ethics, technology, and engineering: An introduction*. Wiley-Blackwell.
Van De Poel, I., & Van Gorp, A. C. (2006). The need for ethical reflection in engineering design: The relevance of type of design and design hierarchy. *Science, Technology, & Human Values, 31*(3), 333–360. https://doi.org/10.1177/0162243905285846
van Grunsven, J. V., Marin, L., Stone, T., Roeser, S., & Doorn, N. (2021). *How to teach engineering ethics? A retrospective and prospective sketch of TU Delft's approach to engineering ethics education*. AEE.
Walker, D. (2013, October). Writing and reflection. In *Reflection* (pp. 52–68). Routledge.
Warford, E. (2022). Engineering students' affective response to climate change: Toward a pedagogy of "critical hope" and praxis. *Teaching Ethics, 22*(1), 1–15. https://doi.org/10.5840/tej2022829117
Wegerif, R. (2020). Towards a dialogic theory of education for the internet age. In *The Routledge international handbook of research on dialogic education* (pp. 14–26). Routledge.
Woods, D. R., Felder, R. M., Rugarcia, A., & Stice, J. E. (2000). The future of engineering education: Part 3. Developing critical skills. *Chemical Engineering Education, 34*(2), 108–117.

SECTION 5

Assessment in engineering ethics education

Gunter Bombaerts

Introduction

Assessing students regarding ethics aspects of engineering education is far from straightforward. This section explores assessment challenges by considering the tension between two extremes. On the one hand, high expectations are placed on engineering ethics education, as many engineering universities claim to develop their students into individuals with solid character, critical professionals, or socially responsible citizens. On the other hand, engineering ethics, as part of the curriculum, is required to have reliable (comparative-fair, neutral) and valid (they should measure the high goals universities put forward) assessment methods.

While writing and editing this section, two dimensions were discussed among chapter authors and handbook editors to probe this gap. First, we considered which aspects of engineering ethics can and should be assessed; here, the focus was on competencies, attitudes and character, epistemic cognition, and behavior. Second, we considered how engineering ethics assessment is systemically embedded in the university as an organization, the university ecosystem, and engineering globally.

Chapter topics and trends

Moral subjectivation

When reading across the six contributions of this section, Foucault's (2020) view on ethics comes to mind. By no means do we want to argue that this Foucauldian approach should be a central way to interpret morality in engineering ethics education research, yet we hope to show that it helps to picture the complexity of assessment in engineering ethics education.

Let us start with Foucault's (2020) notion of power. For him, power is a hermeneutical interaction (e.g., focused on existential understanding) between the agencies of the individuals and the structures. Considered as such, students, teachers, educational support and management, general management, and external university partners are all involved in assessment. They are thus individuals whose agencies enact structures, and these structures enable the agency of individuals.

Foucault (2020) describes four aspects of the constitution of moral subjectivity. First, the determination of the ethical substance (*détermination de la substance éthique*) is "the way in which the

DOI: 10.4324/9781003464259-31

individual has to constitute this or that part of himself as the prime material of his moral conduct" (p. 26). These substances can include soul, desire, emotion, passions, will, pleasure, suffering, and the like. For students in an engineering ethics course, these can be parts of themselves relating to intended learning objectives, such as knowledge of ethical theories, an ethical argumentation skill, ethical awareness, mastery in dealing with ambiguity, efficiency, and so on.

Second, there is the mode of subjectivation (*mode d'assujetissement*), that is, "the way in which the individual establishes his relation to the rule and recognizes himself as obliged to put it into practice" (Foucault, 2020, p. 27). Modes can be many things, such as individuals considering themselves members of a specific group; relating themselves to a spiritual tradition, reason, a cosmological order, or a divine law; or having personal solid goals. For individuals in engineering ethics assessment, the mode refers to what makes a good student, such as passing an exam, genuinely engaging in a project, developing a commitment to become a socially responsible engineer in the future, and so forth.

Third, there is self-practice or ethical work (*élaboration, travail éthique*). These involve activities that one decides to do to shape oneself or to "transform oneself into the ethical subjects of one's behavior" (Foucault, 2020, p. 27). Examples are learning, memorizing, assimilation of a systematic ensemble of precepts, renunciation of pleasures, deciphering desires, or relentless combat. In engineering ethics education, individuals have to relate to the assessment as encouraging rather than inhibiting, achieved by such things as feedback, reflection, or group discussions on how to engage in an evaluation.

Last, teleology (*téléologie*) refers to whether "an action is moral … by virtue of the place it occupies in a pattern of conduct" (Foucault, 2020, p. 28). This can involve increasingly complete mastery of the self, detachment, tranquility of soul, insensitivity to the agitations of the passions, purification for salvation. For assessment, this could mean becoming an individual of strong character, a critical professional, or a socially responsible citizen.

Assessment of ethics in engineering education

These four aspects of the constitution of subjectivity are part of the hermeneutical interaction between individual agencies and the structures they are part of. Assessment, thus, can be seen as disciplinary power that enables the subjects to form themselves. Assessment, thus, is not about a kind of dialectic of oppression of the free student. It is about an inevitable and continuous becoming an individual subject in a structured system embedded in this disciplinary power (Table S5.1).

Table S5.1 Four aspects of subject constitution and potential substantiations in the interaction between individual agency and structure for assessment in engineering ethics

Aspects of subject constitution	*Individual agency – Structure*
Determination of the ethical substance	Competencies, attitudes and character, epistemic cognition, behavior, [knowledge production, emotions, attention, soul, suffering, etc.]
Mode of subjectivation	University and its curriculum, eco-system, global context, [grading, etc.]
Self-practice or ethical work	Pedagogical activities, [individual self-practices]
Teleology	Vision, mission, [individual ideals]

"[]" designates aspects that are less discussed in this section.

Determination of the ethical substance

Using the view on ethics described above, four chapters of this section focus strongly on determining the ethical substance. They explore the importance of assessing engineering ethics education beyond just the knowledge of ethical theories (cf., Section 1, where learning outcomes of engineering ethics education are discussed).

Chapter 26, 'A Framework for the Assessment of Ethical Competencies and Affective Dispositions' by Elena Mäkiö, Tijn Borghuis, Juho Mäkiö, and Jolanta Kowal, studies the alignment of learning objectives, learning and teaching activities, and assessment methods for moral competencies. The literature's ethical objectives and assessment forms are categorized using Bloom's taxonomy of cognitive and affective domains (Anderson & Krathwohl, 2001). On this basis, a heuristic framework is proposed to support educators in (re)designing and broadening engineering ethics education courses/modules.

Determining the ethical substance can also refer to moral attitudes and character. Chapter 27, 'Assessing Attitudes and Character in Engineering Ethics Education: Current State and Future Directions' by Adetoun Yeaman, Balamuralithara Balakrishnan, Olga Pierrakos, and Elise M. Dykhuis, shows that most assessment methods of moral attitudes and character are quantitative, self-reported, *ad hoc*, done at the course level, and serve to help evaluate the course rather than to provide the students with formative feedback. However, the authors do discuss cases where the assessment of attitudes and character is situational, and the feedback is mainly formative.

In Chapter 28, 'Employing Epistemic Micropractices to Assess Progress and Barriers in Engineering Students' Ethics Development,' authors Siara Isaac and Ashley Shew note that to assess ethics also means to assess epistemic cognition, "the practical application of our ideas about the nature of knowledge – that is, how we identify correct answers and the roles we allow ourselves in generating or validating knowledge." Isaac and Shew state that ethics educators should provide pedagogical activities that engage students in more sophisticated epistemic micropractices. These methods should challenge the students to choose between competing solutions and relate to more uncertainty, as such enacting their moral subjectivation.

Chapter 31, 'Two Criticisms of Engineering Ethics Assessment: The Importance of Behaviors and Culture' by Rockwell Clancy, Xin Luo, Chunping Fan, and Fumihiko Tochinai, argues that the ultimate goal of engineering ethics education should be moral behaviors because they are often seen to follow naturally and unproblematically from moral cognition, and adopting ethical behaviors as the goal of engineering ethics education and assessment faces important validity and reliability difficulties.

Mode of subjectivation

Assessment can be seen as a mode of subjectivation in which individuals establish their relations to the rules and recognize themselves as obliged to put them into practice. However, this hermeneutic relation of the individual student with the structure is, in practice, a very complex dynamic at the university, university eco-system, and global levels.

Chapter 29, 'Aspirations for Ethical Education in Engineering Curricula Envisioned through the Quality Lens of Goodlad's typology' by Emanuela Tilley, Nienke Nieveen, Christine Boshuijzen-van Burken, and Folashade Akinmolayan Taiwo, uses Goodlad's (1979) curriculum typology to picture the engineering ethics education dynamics of university ideologies and formalizations, the perceptions and operationalizations of teachers, and the experiences and learnings of students. Quality from a university perspective implies how the students' individual agency is sufficiently influenced in the direction of the university's own vision and mission.

Chapter 30, 'Evaluating Stakeholder Engagement Opportunities Toward Building Practical Ethics Education into Engineering Programs' by Alison Gwynne-Evans, Irene Magara, Esther Matemba, and Sarah Junaid, posits that the focus on embedding ethics education within an engineering program needs to extend beyond the university to include opportunities for practical engagement with industry and community stakeholders at the university eco-system level. These engagements provide critical opportunities for students to reflect on how engineering responsibility is connected to and practiced within communities rather than in isolation. Stakeholder engagement thus needs to be recognized as playing a potentially important role in the setting, achievement, and assessment of the objectives of engineering programs explicitly relating to ethics education and engineering responsibility.

Chapter 31, 'Two Criticisms of Engineering Ethics Assessment: The Importance of Behaviors and Culture' by Clancy, Lou, Fan, and Tochinai, zooms further out and analyzes if assessments in engineering ethics education are reliable globally. Engineering is a global profession, but measures of engineering ethics education have been developed by researchers in and with samples from mostly the United States. The United States is culturally WEIRD (Western, educated, industrialized, rich, and democratic) and, relative to global populations, individuals from WEIRD cultures are outliers on various psychological and social measures.

Conclusions from the section editor

The authoring teams worked on aspects of engineering ethics assessment and how the engineering ethics assessment is systemically embedded. Using Foucault's ethics model of subjectivation, we can, in hindsight, determine some limitations. First, ethics education should not be reduced to knowledge reproduction – but knowledge reproduction and good assessment should not be overlooked, especially in light of particular developments such as artificial intelligence (AI), upscaling education to large ethics classes, multiple-choice assessment, and so forth. Second, although a broad set of ethical substances have been discussed, gaps remain, such as emotions, attention, soul, or suffering. Third, the contributions do not elaborate in-depth on the role of grading. Fourth, the mode of subjectivation refers to establishing a relation to the rule. This relation can be consonant or dissonant. The assessment of students' critical thinking towards the organization is challenging and could have been further developed. Fifth, the chapters approached assessment more from a structure perspective than individual agency and how students experience the proposals mentioned. There are differences between how pedagogical activities and aims are meant by course designers and how they help students in their self-practice, ethical work, and teleology.

We hope this section can further support all the excellent work currently being done in and outside this section on the challenging and vital topic of assessment in engineering ethics education.

Positionality

The lead editor of this section, Gunter Bombaerts, considers himself more of an engineering ethics teacher than an engineering ethics education researcher. When he thinks of himself as a researcher, he would not say 'assessment' is a core expertise. Therefore, it was a great pleasure for him as editor to collaborate in this section with people who have more expertise regarding assessment. This is particularly poignant because he considers assessment as an essential challenge for the current and future practice of engineering ethics education – and because the consequences of AI for ethics education are immense – and because, with the growing urgency to answer global challenges, the societal role of universities will change. Gunter is convinced that *self-practice of ethical work* and *teleology* will have to become more central in the study of assessment in engineering ethics education. As an attentive reader will have noticed, the subtitles 'self-practice or ethical work' and 'teleology' are missing above, which indicates even in this section an existing gap. He sincerely hopes this gap will be addressed in the future. Overall, he is most thankful for the experience of being a co-editor of this handbook.

References

Anderson, L. W., & Krathwohl, D. R. (2001). *A taxonomy for learning, teaching, and assessing: A revision of Bloom's taxonomy of educational objectives*, abridged edition. Longman.

Foucault, M. (2020). *The History of Sexuality. Vol. 2: The Use of Pleasure*. Reprinted. London: Penguin Books.

Goodlad, J. I. (1979). *Curriculum inquiry. The study of curriculum practice*.

26

A FRAMEWORK FOR THE ASSESSMENT OF ETHICAL COMPETENCIES AND AFFECTIVE DISPOSITIONS

Elena Mäkiö, Tijn Borghuis, Juho Mäkiö, and Jolanta Kowal

Introduction

The Cambridge online dictionary defines engineering as "the study of using scientific principles to design and build machines, structures, and other things, including bridges, roads, vehicles, and buildings." In the past, until the 1970s, the ethical aspects of engineering were not explicitly addressed, as the results of engineering were technical objects and, as such, considered ethically neutral. This view changed when the societal consequences of using these technical objects were recognized. This led to engineering ethics, that is, the application of ethical theories and moral principles in the field of engineering, and the need for ethical education of engineering students (Barry & Herkert, 2014).

Ethics education in engineering is considered as the development of students' higher-level moral reasoning skills by solving realistic ethical cases and dilemmas (Goldin et al., 2015) and making ethical decisions (Bonde et al., 2016). However, the ability to morally reason does not automatically lead to students' ethical awareness, moral development, or future ethical behavior (Bairaktarova & Woodcock, 2015). Therefore, ethics education in engineering must address the development of both moral cognitive skills and ethical behavior.

In this chapter, we conceptualize ethical thinking and action in terms of two dimensions: competencies and affective dispositions. We refer to competence as "combinations of those cognitive, motivational, moral, and social skills available to (or potentially learnable by) a person" (Weinert, 2001, p. 2433). In this sense, competence is always more than just knowledge or just experience. We refer to affective dispositions as to the personal traits, habits of mind, or attitudes which characterize a personality – see for example critical thinking affective dispositions in (Facione, 1990). The cognitive and affective "domains interact significantly in instruction and learning" (section in Miller, 2005).

Assessment of students' ethical competencies and dispositions is considered difficult and is done in a variety of ways (Goldin et al., 2015; Martin et al., 2021). To make teaching and learning effective, three central educational aspects – the intended learning outcomes (ILOs), the learning activities, and the assessment tasks – need to be constructively aligned (Biggs & Tang, 2007),

DOI: 10.4324/9781003464259-32

which is also true for engineering ethics education (Bonde et al., 2016; Keefer et al., 2014; Martin et al., 2021; Miñano et al., 2017; Shuman et al., 2005). A granular alignment of these three aspects can be made by categorizing them according to the Bloom's revised taxonomy, which Spivey (2007) suggested for the Cognitive Domain. In this chapter, we extend Spivey's proposal and use Bloom's taxonomies of the Cognitive and Affective Domains to align ethical outcomes and forms of assessment. Based on this alignment, we propose a heuristic framework that can guide educators in developing assessment methods according to the ILOs defined for modules in engineering ethics, and we demonstrate the framework's application.

The structure of the chapter is as follows: we first review the literature on the ethical competencies required of engineers (they serve as ILOs) and existing forms of assessment of ethical competences and dispositions. In subsequent sections, we describe the methodology used to develop a framework and then provide an example application. In the final section, we conclude and outline avenues for future research.

This chapter was written by authors from Eastern and Western European countries with different professional backgrounds – computer science, philosophy, and psychology. The authors base this chapter on research conducted and published in Western industrial societies and their experience teaching ethics to engineering students in this context.

Ethical competencies and affective dispositions required of engineers

Engineering ethics is an important element in professional practice and engineering education. This is reflected by (1) the inclusion of ethical codes in regulations and recommendations of professional associations (organizations (1) to (20) (in Center, 2023)); (2) the increasing manifestation of ethical principles in the accreditation criteria for higher education worldwide (Lucena et al., 2008); and (3) addressing engineering ethics in engineering education research (Martin et al., 2021).

Professional engineering associations, such as the Institute of Electrical and Electronics Engineers (IEEE), Association for Computer Machinery (ACM), the Engineering Council (EC) and the Royal Academy of Engineering (RAE), have established ethical codes that focus primarily on concrete aspects of ethical behavior to guide engineers in the practice of their profession. These aspects include the philosophical understanding of engineering practice, honesty and integrity, ethical aspects of products and their use, respect for life, law, the environment and public good, and the professional conduct and responsibility of the individual engineer (ACM, n.d.; Engineering Council, n.d.; IEEE, 1963; Ingenia, 2023).

The International Engineering Alliance (IEA) has defined a professional competence profile for engineers called the 'Washington Accord' (Alliance 2014). This profile defines bench-marked standards for engineering education and expected competencies for engineering practice. It includes personal skills such as communication, ethical practice, judgment, taking responsibility, and protecting society. These skills align with the concept of the 'social contract' (Snoeyenbos et al., 1983), which expects professionals to act according to their professional code of ethics, among other things. It is worth noting that the ethical aspects mentioned in the competence profile focus mainly on character dispositions and less on ethical competencies. The ethical requirements formulated by engineering professional associations and accreditation organizations represent the competencies associated with general ethical professional acting and conduct and character building.

Accreditation organizations, similar to professional associations, formulate accreditation criteria that emphasize primarily concrete aspects of ethical behavior such as professional and ethical responsibility (ABET, 2021) and understanding of professional ethics and codes of conduct in relation to professional practice. For example, the Accreditation Agency for Study Programs in Engineering, Informatics, Natural Sciences and Mathematics (ASIIN) formulates the requirements for Electrical Engineering/Information Technology as follows: "Graduates possess social and professional ethical competences and are able to shape social processes critically, reflectively and with a sense of responsibility and in a democratic spirit" (ASIIN, 2022a). (For more on accreditation, please see Chapters 32–36 of this handbook.)

The accreditation requirements related to ethical aspects are specific to a field of engineering and often formulated in general terms associated with general ethical professional acting and conduct and character building. The curricular content is not uniform (Durst et al., 2021; Grohman et al., 2020; Katz et al., 2020; Li & Fu, 2012), and curricular objectives depend on various factors, such as existing cultural values, worldviews, and national professional codes of ethics (Downey et al., 2007). This means that the overarching accreditation criteria are named or given by the accreditation organizations, but their concrete design and implementation are generally a matter for the teaching institutions. Thus, the sociocultural context triggers the curricular change, as Lattuca and Stark (2009) note. Additionally, the curricular content varies over time (Lattuca & Stark, 2009) and location as shown in alZahir and Kombo (2014), where the authors compared international codes of ethics of engineering societies in more than 30 countries with the IEEE code of ethics.

Engineering education research has identified goals posited for engineering ethics education and grouped them into 12 major categories: moral sensibility, analysis, creativity, judgment, decision-making, argumentation, moral knowledge, design, agency, character and virtuous development, emotional development, and situatedness (Martin et al., 2021). The academic literature formulates the goals – that is, ethical competencies and affective dispositions – sought in engineering ethics education in a more precise and differentiated way than professional engineering societies or accreditation organizations, which emphasize the behavioral and action-oriented aspects of ethics.

These goals identified in the academic literature by Martin et al. (2021) are listed in Table 26.1 along with the goal 'Moral conduct,' which was added based on the requirements of professional engineering associations and accreditation organizations. The goals in Table 26.1 are interpreted in this chapter as ILOs for engineering ethics education. The table is structured as follows: (1) the 'Categories' column contains 13 major categories of ILOs, (2) the 'Sub-Cat.' column enumerates ILOs within each category, (3) the 'Goals' column lists ILOs, (4) the 'Literature' column contains references from the academic literature, and (5) the 'A/P' column contains references from accreditation organizations and professional associations.

Existing forms of assessment of ethical competencies and dispositions

To prepare a proposal for assessment opportunities of ethical competencies, a literature review was conducted investigating how ethical competencies and dispositions are currently assessed in engineering education. The examined literature relied on the core collection of the Web of Science for assessing ethical competencies in engineering education. To retrieve sources the following search string was used to search in all fields of publications during the period 2013–2022:

ALL=(ethic*" AND "Engineering" AND "education*" AND "assess*")
ONLY journals and 2013–2022

To ensure a more comprehensive analysis, the process of retrieving sources based on this search string was followed by an overview of the references mentioned by the publications found.

Table 26.1 Intended learning outcomes for teaching ethics in engineering education from scientific literature and documents of accreditation (A) and professional (P) organizations. Source: The content of the Goals and Literature columns below, and the table's structure, comes from Martin et al. (2021)[1]. The names of categories 1–6 were posed initially by van de Poel and Royakkers (2011)[2] and categories 7–12 by Martin et al. (2021)[1]. Category 13 and the A/P column are new contributions by this chapter's authors.

Categories	*Sub-Cat.*	*Goals (ethical competencies and affective dispositions)*[1]	*Literature*[1]	*A/P*
1. Moral sensibility[2]	1	Developing proficiency in recognizing social and ethical issues in engineering	(Harris Jr et al., 1996) (Pritchard, 2005) (Poel & Royakkers, 2011) (Martin & Schinzinger, 2013)	(ASIIN, 2022a) (ASIIN, 2022b)
	2	Increasing students' sensitivity to ethical issues	(Davis, 1999) (Harris Jr et al., 1996)	
	3	Encouraging students to take ethics seriously	(Harris Jr et al., 1996)	
2. Moral analysis[2]	1	Analyzing moral problems in terms of facts, values, stakeholders, and their interests	(Poel & Royakkers, 2011)	
	2	Comprehending, clarifying, and assessing arguments on opposing sides of moral issues	(Martin & Schinzinger, 2013)	
	3	Facilitating the analysis of key ethical principles	(Harris Jr et al., 1996)	
	4	Exploring the perspective of those in other positions	(Lynch & Kline, 2000) (Martin et al., 2019)	
3. Moral creativity[2]	1	Considering different options for action in the light of (conflicting) moral values and relevant facts	(Poel & Royakkers, 2011)	
	2	Stimulating ethical imagination	(Coeckelbergh, 2006) (Harris Jr, et al. 1996) (Martin & Schinzinger, 2013) (Pritchard, 2005)	
	3	Creatively exploring solutions rather than choosing a dilemma horn	(Lynch & Kline, 2000)	
	4	Enhancing divergent thinking	(Haws, 2001)	
4. Moral judgment[2]	1	Making moral judgments based on different ethical theories or frameworks, including professional ethics and common-sense morality	(Poel & Royakkers, 2011)	
	2	Improving ethical judgement	(Davis, 1999) (Harris Jr et al., 1996)	
	3	Ability to reliably respond to any situation with a course of action that makes life better	(Pritchard, 2005) (Davis, 2012)	
	4	Forming consistent and comprehensive viewpoints based on consideration of relevant facts	(Martin & Schinzinger, 2013)	

5. Moral decision-making[2]	1	Enabling students to make decisions based on different ethical theories and frameworks	(Poel & Royakkers, 2011)	
	2	Providing conceptual tools for reflecting on how organizational practices can potentially threaten public safety and welfare and how to counter the normalization of deviance	(Lynch & Kline, 2000)	
	3	Helping students deal with ambiguity in decision-making situations	(Harris Jr et al., 1996)	
6. Moral argumentation[2]	1	Developing the ability to morally justify one's actions and to discuss and evaluate them	(Poel & Royakkers, 2011)	
7. Moral knowledge[1]	1	Gaining knowledge of professional standards, codes, and principles	(Davis, 1999) (Harris Jr et al., 1996) (Pritchard, 2005)	
	2	Giving students access to the language of ethics to express and support one's moral views adequately to others	(Haws, 2001) (Martin & Schinzinger, 2013)	
	3	Grounding one's views and decisions in moral theory	(Lynch & Kline, 2000)	
8. Moral design[1]	1	Considering how values, as well as modes of use and interaction, can be implicitly or explicitly inscribed into engineering artifacts at the design stage	(Poel & Verbeek, 2006) (Verbeek, 2008)	
9. Moral agency and action[1]	1	Empowering students to reshape the social, economic, and legal context of practice	(Conlon & Zandvoort, 2011)	(ASIIN, 2022a) (ASIIN, 2022b)
	2	Responding wisely and responsibly to situations in a way that satisfies as many potentially competing constraints as possible	(Whitbeck, 1995)	
	3	Encouraging students to take an activist stance 'for what is right, good and just'	(Hodson, 1999)	
	4	Inspire the engineers of the future to challenge the status quo and to strengthen the profession	(Lawlor, 2021)	
10. Moral character and virtuous development[1]	1	Fairness		(IEEE, 1963) (Engineers Canada, 2016) (ACM, n.d.)
	2	Honesty and trustworthiness		(ACM, n.d.)
	3	Increasing students' ethical willpower	(Davis, 1999) (Harris Jr et al., 1996)	
	4	Cultivating students' sense of professional identity	(Loui, 2005) (Miller, 2018)	

(*Continued*)

Table 26.1 (Continued)

Categories	*Sub-Cat.*	*Goals (ethical competencies and affective dispositions)1*	*Literature1*	**A/P**
	5	Cultivating virtues, such as respect for nature, to support engagement in environment-friendly engineering	(Harris Jr et al., 2019)	
	6	Cultivating virtues for objectivity, care, and honesty	(Moriarty, 2009) (Nair & Bulleit, 2020)	
	7	Cultivating virtues for identifying certain decision situations and actions as ethically relevant	(Frigo et al., 2021)	
11. Moral emotional development[1]	1	Reflecting on the role of emotions in the development and acceptability of risky technologies	(Roeser, 2012)	
	2	Reflecting on the role of emotions in the effects of climate change	(Lönngren et al., 2020)	
	3	Engaging learners in their emotional life so as to develop a sense of empathy with people across physical, social, and cultural distances and a language for emotions	(Tormey, 2005) (Hess & Fila, 2016) (Hess et al., 2017)	
12. Moral situatedness[1]	1	Understanding the social relations of expertise in connection with technology management and decision-making	(Devon, 1999)	(ASIIN, 2022a) (ASIIN, 2022b)
	2	Helping students situate their work in its contribution to their community	(Haws, 2001)	(ASIIN, 2022a) (ACM, n.d.)
	3	Acknowledging the social dimension of engineering practice	(Martin et al., 2019)	(ASIIN, 2022a)
13. Moral conduct	1	Commitment to appropriately act according to professional ethics, accountability, and norms set by the technical-scientific practice	(Clancy & Zhu, 2023)	(ASIIN, 2022b)
	2	Upholding the professional code	(Clancy & Zhu, 2023)	(IEEE, 1963) (Engineers Canada, 2016)
	3	Upholding responsible behavior and ethical conduct in professional activities	(Clancy & Zhu, 2023)	(IEEE, 1963) (Alliance, 2014)

[1] Column or Category taken from Martin et al. (2021), please see Table 2 available at https://link.springer.com/article/10.1007/s11948-021-00333-6/tables/2
[2] Category title from van de Poel and Royakkers (2011)

Assessing students' ethical competencies and especially their dispositions is difficult. It is done in various ways (Goldin et al., 2015; Martin et al., 2021), with an assessment of dispositions rarely conducted, and when it is, it is informal rather than formal (Shiveley & Misco, 2010). Both locally developed tailored instruments and standardized instruments are used to assess students' ethical competencies and dispositions. Tailored instruments are used for both formative and summative assessments. Formative assessments provide immediate and meaningful feedback to the students on their progress and help them to improve; they can also inform instruction (Keefer et al., 2014). Summative assessments at the end of a course are used to formally measure student achievement against ILOs.

The following tailored instruments and methods developed to assess specific ethical learning outcomes of modules have been used in the engineering ethics education Field (Martin et al., 2021): reflective essays, individual assignments graded with a rubric, presentations, group projects (including an artifact), and portfolio. However, ethical outcomes often remain unassessed or are subjected to a binary assessment of pass/fail (Keefer et al., 2014). Although reflective essay writing is a popular form of assessment, it often does not accurately reflect the level of ethical competence of, for example, science students, as essay writing is not practiced in science education (Johnson, 2010). Some authors have introduced instruments in the form of rubrics that (1) provide students with practical guidance in essential components of realistic ethical problem-solving and (2) assess students' work on ethical cases and dilemmas (see Table 26.2).

In contrast to tailored instruments that assess whether students have achieved specific ethical learning outcomes, standardized instruments assess a student's general level of moral development, reasoning, or judgment compared to a baseline and, therefore, can provide an independent and comparable assessment (Keefer et al., 2014). These instruments were developed and validated in Western industrial societies' cultural and moral environments. Most standardized instruments used in engineering ethics education were derived from Kohlberg's (1976) approach to morality (see Table 26.2, and more on this topic in Chapter 9). Based on a cognitive developmental perspective, this approach defines morality as rational reasoning about justice or fairness (Graham et al., 2013). The rubrics and standardized instruments are listed in Table 26.2 together with the literature and their brief descriptions and the competencies and dispositions assessed.

As Kohlberg's (1976) theory has been criticized by researchers (e.g., in Lind, 2002), the Moral Foundations Theory (MFT) by Graham et al. (2013) can be considered as an alternative in this chapter. MFT defines moral functioning as an unconscious process or moral intuition, a type of cognition that is not based on reasoning. It includes five moral foundations instead of Kohlberg's moral reasoning one: (1) Harm/Care, (2) Fairness/Reciprocity, (3) Ingroup/Loyalty, (4) Authority/Respect, and (5) Purity/Sanctity. MFT has its own instrument to measure these five foundations, called the Moral Foundations Questionnaire (MFQ) (Graham et al., 2011). Glover et al. (2014) examined the relationship between Kohlberg's Defining Issues Test DIT-2 (see Table 26.2) and MFQ and suggested that MFQ may not be an appropriate measure for capturing advanced moral functioning, as moral judgments assessed by DIT2 may not reflect the MFQ foundations. MFQ has sporadically been used in engineering ethics education, without the research results being published.

Discussion of methodological approach

Three aspects – ILOs, learning activities, and assessments – need to be constructively aligned in well-designed modules. In this chapter, we have decided to align the assessment methods used in engineering ethics education with the ethical learning outcomes, as Gil-Jaurena and Kucina Softic (2016) have done for education in general, and omitted the alignment with learning activities. The

Table 26.2 The rubrics and standardized instruments to assess ethical competencies and dispositions (Authors' elaboration based on the literature. Sources are included in column 2: Instrument.)

Code	*Instrument*	*Assessed competencies/dispositions. Specificity of the instrument*
Tailored rubrics		
Rub_1	The Decision Procedure Checklist (**DPC**) (Keefer et al., 2014).	Used to conduct a formative assessment of students' written commentary on a case study involving moral issues in bioengineering and mechanical engineering. Four key learning outcomes are assessed that relate to complex ethical problems: the ability to (1) identify ethical issues and professional responsibilities, (2) identify additional important information when investigating the problem, (3) consider alternative courses of action in response to the case, and (4) consider the long and short-term consequences of proposed solutions.
Rub_2	**Decision Procedure Scoring Guide (DPSG)** (Keefer et al., 2014).	Used to conduct a summative assessment based on the student's answers to **DPC**.
Rub_3	The **Assessment Instrument of Analytical Components of Methods of Moral Reasoning** (Goldin et al., 2015) (Pinkus et al., 2015).	Assesses students' higher-level moral reasoning skills when working on moral dilemmas and cases posed in bioengineering ethics. Students (1) created their own ethical case based on their technical area of engineering research, (2) presented the case to the class for comment, and (3) wrote a paper by analyzing the case using the methods taught in the class. Five key learning outcomes are assessed that relate to complex ethical problems: the ability to (1) employ professional engineering knowledge to frame issues, (2) view the problem from the perspectives of different stakeholders, (3) link multiple perspectives of different stakeholders throughout the analyses, (4) identify analogous cases and find similarity to past cases by using problem structure, (5) employ a method of moral reasoning in conducting the analysis.
Rub_4	A **measurement tool** (rubric) (Sindelar et al., 2003).	Assesses students' abilities to recognize and resolve ethical dilemmas and cases in engineering ethics. Students wrote essays to ethical dilemmas contained in the given cases. Five key learning outcomes are assessed: the ability to (1) recognize an ethical dilemma, (2) morally argue an issue, (3) morally analyze a dilemma, (4) use multiple perspectives, (5) resolve an ethical dilemma.

Standardized instruments		
DIT	Defining Issues Test (DIT) (Rest, 1979).	DIT, derived from Kohlberg's (1976) theory of moral development, measures general students' moral reasoning ability and the maturity of students' reflection on ethical issues. DIT contains a series of moral dilemmas that need to be judged. It is not tailored to ethical situations in engineering practice.
DIT-2	Defining Issues Test (DIT-2) (Rest et al., 1999).	DIT-2, a revised version of DIT, is a more reliable and valid measure of general moral reasoning ability. It includes updated moral dilemmas that address such contemporary ethical issues as diversity, social justice, and environmental ethics.
TESSE	Test of Ethical Sensitivity in Science and Engineering (TESSE) (Borenstein et al., 2008) (based on DIT-2).	Unlike other instruments, TESSE examines ethical sensitivity and the ability to identify and recognize relevant ethical issues with scenarios related to engineering. It measures students' ability to identify and recognize ethical issues emerging from a situation.
ESIT	Engineering and Science Issues Test (Borenstein et al. 2010; based on DIT2 test).	ESIT, derived from Kohlberg's (1976) theory of moral development, measures students' moral reasoning ability, like DIT-2, but using technical dilemmas from the fields of science and engineering.
EERI	Engineering Ethics Reasoning Instrument (Zhu et al., 2014).	EERI, derived from Kohlberg's (1976) theory of moral development, measures students' moral decision-making ability. It consists of a series of ethical dilemmas that are commonly encountered in engineering practice. EERI also measures a student's ability to recognize ethical issues, understand the consequences of different courses of action, and make ethical decisions based on sound ethical principles.
EDM	Ethical Decision Making Measure (Mumford et al., 2006).	EDM, based on the taxonomy of ethical behavior developed by Helton-Fauth et al. (2003), by analyzing codes of conduct, measures the moral decision-making ability of researchers in different disciplines, including engineering. In a series of discipline-specific scenarios, an individual's ability to (1) recognize ethical issues, (2) analyze the situation, and (3) identify possible courses of action is assessed.
MFQ	Moral Foundations Questionnaire (MFQ) (Graham et al., 2011).	MFQ is the instrument based on the Moral Foundations Theory (Graham et al., 2013) and measures five moral foundations: (1) Harm/Care, (2) Fairness/Reciprocity, (3) Ingroup/Loyalty, (4) Authority/Respect, and (5) Purity/Sanctity. It is used to measure overall moral judgment.

alignment was accomplished within the framework of Bloom's revised taxonomies of Cognitive and Affective Domains (Anderson & Krathwohl 2001), which we briefly introduce here.

Bloom's revised taxonomy of the Cognitive Domain has two dimensions: (1) the cognitive process in the development of intellectual competencies and (2) the knowledge dimension (Anderson & Krathwohl, 2001). The cognitive process dimension includes six cognitive levels (labeled C1–C6, ordered from the lowest to the highest one):

(C1) *remember* – retrieve relevant knowledge from long-term memory;
(C2) *understand* – construct meaning from instructional messages, including oral, written, and graphic communication;
(C3) *apply* – carry out or use a procedure in a given situation;
(C4) *analyze* – break material into its constituent parts and determine how the parts relate to one another and to an overall structure or purpose;
(C5) *evaluate* – make judgments based on criteria and standards; and
(C6) *create* – put elements together to form a coherent or functional whole; reorganize elements into a new pattern or structure.

The knowledge dimension consists of four types of knowledge (labeled K1-K4):

(K1) *factual knowledge* of terminology and specific details and elements;
(K2) *conceptual knowledge* of classifications, principles, generalizations, theories, and models;
(K3) *procedural knowledge* of subject-specific skills, algorithms, techniques, and methods; and
(K4) *meta-cognitive knowledge* – strategic understanding; cognitive awareness, which is the ability to recognize one's own thought processes; and cognitive control, which facilitates the regulation and adaptation of thought to improve efficiency.

Conceptual knowledge is traditionally the mainstay of ethics education, but factual knowledge also plays a role, particularly in case-based education activities.

Procedural knowledge comes in the form of knowledge of processes for ethical analysis and decision-making, such as the ethical cycle (van de Poel & Royakkers, 2011), but also concerning formalized ethical reasoning. Specific to engineering ethics is knowledge regarding procedures that plug into the engineering design process, such as methods for elucidating and operationalizing values as ethical requirements for a design (Vermaas et al. 2015).

Meta-cognitive knowledge comes into play in engineering ethics in the form of self-knowledge, which is about how individuals think about their internal ways of thinking in relation to their ethical values. This also involves the ability to recognize one's own thought processes in relation to ethical reasoning and to regulate and adapt these thought processes to improve efficiency.

Combined, the cognitive process and knowledge dimensions produce the 24 categories in the revised Bloom's taxonomy of the Cognitive Domain (Anderson & Krathwohl, 2001).

Bloom's taxonomy of the Affective Domain (Anderson & Krathwohl, 2001) used to categorize dispositional (affective) outcomes, has five categories (labeled A1–A5) that describe changes in behavior as values or attitudes are learned (Miller, 2005):

(A1) *receiving* – awareness, willingness to hear, selected attention;
(A2) *responding* – appreciating or internalizing;
(A3) *valuing* – accepting, preferring, becoming committed to;
(A4) *conceptualizing/organizing* – incorporating into a value system; and
(A5) *characterizing by value* – orientation toward / identification with.

We found that mapping the ethical ILOs and assessment methods to the dimensions of Bloom's revised taxonomies was not a straightforward task, as it involved interpretation. It became clear early on that the mapping would not be one-to-one because, for instance, a single ILO or assessment form could relate to multiple cognitive processing levels, knowledge types, or affective levels of Bloom's taxonomies. In the absence of a foundation in the literature, the following procedure was followed to achieve the best possible result in mapping the ILOs:

1. Two authors of this chapter independently categorized the ethical ILOs described in Table 26.1 to the extended Bloom's taxonomy of the Cognitive Domain. ILOs referring to students' emotional and attitudinal development were mapped to the five levels of the taxonomy of the Affective Domain.
2. Differences in the mappings of the two authors were marked.
3. The two authors revised their mappings consulting the literature underlying Table 26.1, and explained in writing why they placed the marked items in a particular category.
4. All authors of the chapter discussed the placement of all ethical ILOs, especially the marked ones, to reach an agreement.

The assessment instruments listed in Table 26.2 were first mapped to six levels of the cognitive process dimension of Bloom's taxonomy based on a mapping made by Maffei et al. (2022). This mapping was then extended to the four types of the knowledge dimension. The tailored rubrics and standardized instruments as proprietary self-report measures (Table 26.2) were placed in the corresponding categories according to their descriptions given in the literature and the learning outcomes they tested. Tailored instruments and rubrics do not assess students' emotional and attitudinal development, and therefore they have not been mapped to the categories of Bloom's taxonomy of Affective Domain.

The standardized instruments derived from Kohlberg's (1976) theory of moral development (DIT, DIT-2, ESIT, EERI) were mapped to the affective categories of Bloom's taxonomy based on the following considerations. Kohlberg's (1976) theory defines six stages of cognitive moral reasoning. It suggests that cognitive moral reasoning and development are closely tied to forming and prioritizing personal values (see Newton, 1978). As individuals progress through stages of moral development, their values become more complex and aligned with moral principles, impacting their attitudes, behaviors, and decision-making processes. Kohlberg's stages 3 and 4, where individuals develop a sense of societal values and norms, can be associated with the category (A3) *Valuing* in Bloom's taxonomy of Affective Domain. Stages 5 and 6, where individuals transcend societal norms and develop their own ethical principles, can be associated with the category (A4) *Conceptualizing/Organizing.*

The standardized tools TESSE, EDM, and MFQ have not been included in the mapping to categories of Bloom's taxonomy of Affective Domain for the following reasons: (1) TESSE is not validated and is used to a limited extent; (2) mapping of the Helton-Fauth taxonomy of ethical behavior, on which EDM is based, to Bloom's taxonomy of Affective Domain is beyond the scope of this chapter; and (3) MFQ is rarely used in engineering education.

Framework for assessment of ethical learning outcomes

The framework proposed in this section includes (1) a mapping of assessment methods and ethical learning outcomes to the relevant dimensions of Bloom's revised taxonomies of Cognitive and Affective Domains and (2) a procedure for matching ILOs with assessment forms to help educators develop or improve their engineering ethics teaching. For this mapping, we classified the ethical competencies and dispositions required of engineers (Table 26.1) and the existing forms of their assessment (Table 26.2) in relation to Bloom's taxonomies of Cognitive and Affective Domains.

Table 26.3 presents the mapping of the ethical ILOs listed in Table 26.1 and the tailored and standardized assessment forms from Table 26.2 to the cognitive categories and knowledge dimension of the Bloom's revised taxonomy of Cognitive Domain. In contrast, Table 26.4 shows the mapping of these ILOs and assessment forms to the affective categories of the Bloom's taxonomy of Affective Domain.

Table 26.3 demonstrates that most ethical cognitive competencies are in the (K2) *Conceptual* knowledge column at the cognitive levels (C2) *Understand,* (C4) *Analyze*, (C5) *Evaluate*, and (C6) *Create*. This means that the cognitive, ethical learning goals are mainly related to understanding ethical concepts and higher-order thinking in the conceptual categories. Some selected ethical competencies are only in the columns for (K1) *Factual* and (K3) *Procedural* knowledge, while only one competence is in the column for (K4) *Meta-cognitive* knowledge. In contrast, most forms of assessment are distributed across all cognitive levels in the (K2) *Conceptual* knowledge column, while the other knowledge types contain no or only a very limited selection of assessment methods. This means that the forms of assessment mainly test students' conceptual knowledge at different levels of thinking rather than at the factual, procedural, and meta-cognitive levels on which the range of assessment forms used is limited. It is worth noting that the competencies in the (K1) *Factual* and the (K3) *Procedural* columns extend to higher cognitive levels than the assessment forms, which means that there are no appropriate assessment forms for the cognitive levels (C4) *Analyze*, (C5) *Evaluate*, and (C6) *Create* for the factual and procedural knowledge types.

Table 26.4 demonstrates three important aspects: (1) standardized instruments (in this case derived from Kohlberg's theory of moral development) rather than tailored instruments were identified as appropriate for assessing ethical affective dispositions; (2) most ethical affective dispositions fall into the highest category (A5) *Characterizing;* and (3) there are suitable tools in the (A3) *Valuing* and (A4) *Conceptualizing/Organizing* categories, but not in the category (A5) *Characterizing*. This means that the achievement of the ambitious learning goals set by professional engineering organizations and ethical engineering education – namely that engineering students change their values and even their professional behaviors based on these values as a result of engineering ethics education – cannot be assessed with the existing tools.

Despite this discrepancy, Table 26.3 and Table 26.4 show substantial overlap in the occupied categories for competencies and assessment forms in the taxonomies. This provides an opportunity for systematic thinking about constructive alignment. To be constructively aligned, the mapping must classify the three facets of a module – ILOs, learning activities, and assessment tasks – in the same categories in Bloom's taxonomies. If not, the ILO, learning activity, and assessment task could each be addressing a different cognitive process level, knowledge type, or affective level. However, mapping to the same category is a necessary but insufficient condition for alignment. It is possible, for instance, that an assessment task is not suitable for a teaching activity, even though they are mapped to the same category. Based on this observation, educators can use Table 26.3 and Table 26.4 as a heuristic framework to find assessment forms that align with the ethical competencies targeted in a module when designing or revising their modules. Another way of using these tables is to analyze which ILOs and assessment forms are not included in existing or newly developed modules.

In the process-based approach to ethics engineering teaching, every step in an assignment has a characteristic learning activity, which aims to contribute to one (or more) learning outcome(s). For instance, the ethical cycle of van de Poel and Royakkers (2011) has the following five steps linked to ILOs from Table 26.1:

Table 26.3 Mapping of ethical competencies and assessment forms to the dimensions of cognitive process and knowledge (Bloom's revised taxonomy of Cognitive Domain)

Category/ Knowledge dimensions	*K1 Factual (of terminology and specific details and elements)*		*K2 Conceptual (of classifications, principles, generalizations, theories, and model)*		*K3 Procedural (of subject-specific skills, algorithms, techniques, and methods)*		*K4 Meta-cognitive (strategic knowledge, knowledge about cognitive tasks, and self-knowledge)*	
	Ethical competences (Table 26.1)	*Assessment forms (Table 26.2)*	*Ethical competences (Table 26.1)*	*Assessment forms (Table 26.2)*	*Ethical competences (Table 26.1)*	*Assessment forms (Table 26.2)*	*Ethical competences (Table 26.1)*	*Assessment forms (Table 26.2)*
C1 Remember (recognizing, recalling)	Moral knowledge[1] 1[2] Moral knowledge[1] 2[2]	Individual assignments graded with a rubric	Moral knowledge[1] 1[2] Moral knowledge[1] 2[2]	Individual assignments graded with a rubric		Individual assignments graded with a rubric		
C2 Understand (interpreting, exemplifying, classifying, summarizing, inferring, comparing, explaining)			Moral knowledge[1] 1[2] Moral knowledge[1] 2[2] Moral decision making[1] 2[2] Moral sensibility[1] 1[2] Moral sensibility[1] 3[2] Moral situatedness[1] 1[2]	• individual assignments graded with a rubric • presentation • **Rub_1, Rub_2, Rub_3, Rub_4** • **DIT, DIT-2** • **TESSE, ESIT, EERI** • **EDM**				
C3 Apply (executing, implementing)			Moral judgment[1] 1[2]	• individual assignments graded with a rubric • group project • **Rub_1, Rub_2, Rub_3, Rub_4**	Moral analysis[1] 1[2] Moral analysis[1] 4[2] Moral decision-making[1] 2[2]	• individual assignments graded with a rubric • group project • **Rub_1, Rub_2, Rub_3, Rub_4**		

Table 26.3 (Continued)

Category/ Knowledge dimensions	*K1* *Factual* *(of terminology and specific details and elements)*		*K2* *Conceptual* *(of classifications, principles, generalizations, theories, and model)*		*K3* *Procedural* *(of subject-specific skills, algorithms, techniques, and methods)*		*K4* *Meta-cognitive* *(strategic knowledge, knowledge about cognitive tasks, and self-knowledge)*	
	Ethical competences (Table 26.1)	*Assessment forms (Table 26.2)*	*Ethical competences (Table 26.1)*	*Assessment forms (Table 26.2)*	*Ethical competences (Table 26.1)*	*Assessment forms (Table 26.2)*	*Ethical competences (Table 26.1)*	*Assessment forms (Table 26.2)*
C4 Analyze (differentiating, organizing, attributing)	Moral analysis[1] 1[2] Moral judgment[1] 3[2]		Moral analysis[1] 1[2] Moral analysis[1] 2[2] Moral analysis[1] 3[2] Moral analysis[1] 4[2] Moral judgment[1] 3[2] Moral decision making[1] 2[2] Moral decision making[1] 3[2] Moral emotion[1] 1[2]	• Individual assignments graded with a rubric • Portfolio • group project • **Rub_1, Rub_2, Rub_3, Rub_4** • **DIT, DIT-2** • **TESSE, ESIT, EERI** • **EDM**				
C5 Evaluate (checking, critiquing)	Moral judgmen[1]t 2[2] Moral agency[1] 1[2]		Moral judgment[1] 2[2] Moral argumentation[1] 1[2] Moral decision making[1] 3[2] Moral agency[1] 1[2] Moral judgment[1] 3[2]	• individual assignments graded with a rubric • Portfolio • group project • **Rub_4** • **DIT, DIT-2** • **ESIT, EERI** • **EDM**	Moral judgment[1] 2 Moral argumentation[1] 1[2]			Reflective essay
C6 Create (generating, planning, producing)	Moral creativity[1] 1[2] Moral design[1] 1[2]		Moral creativity[1] 1[2] Moral creativity[1] 2[2] Moral creativity[1] 3[2] Moral design[1] 1[2] Moral agency[1] 2[2] Moral decision-making[1] 1[2] Moral decision-making[1] 3[2] Moral situatedness[1] 2[2]	• Individual assignments graded with a rubric • Portfolio • Group project • **Rub_3**	Moral design[1] 1[2] Moral agency[1] 2[2]		Moral creativity[1] 4[2]	Reflective essay

[1] – The "Categories" column in Table 26.1

[2] – The "Sub-Cat." column in Table 26.1

- (creating a) *Moral Problem Statement* – Moral sensibility 1.
- (performing) *Problem Analysis* – Moral analysis 1.
- (generating) *Options for Action* – Moral creativity 1.
- *Ethical Evaluation* (of options in different frameworks) – Moral judgment 1.
- *Reflection* (to find an equilibrium solution) – Moral decision-making 1.
- In all steps: Moral argumentation 1.

Using Table 26.3 and Table 26.4, the ILOs can be matched with the assessment forms. For example, in categories C6-K1 C6-K2, 'Moral creativity 1' matches with 'individual assignment graded with rubric, portfolio, group project' and 'Rub_2.'

Table 26.4 Mapping of ethical affective dispositions and assessment forms to the affective categories (Bloom's taxonomy of Affective Domain)

Categories	*Ethical affective dispositions*	
	Ethical ILOs	*Assessment forms*
A1 Receiving (awareness, willingness to hear, selected attention)		
A2 Responding (appreciating or internalizing)		
A3 Valuing (accepting, preferring, becoming committed to)	Moral sensibility[1] 2[2] Moral situatedness[1] 3[2]	DIT, DIT-2, ESIT, EERI
A4 Conceptualizing/Organizing (incorporating into a value system)	Moral knowledge[1] 3[2] Moral character[1] 1[2] Moral character[1] 2[2] Moral character[1] 3[2] Moral emotional development[1] 2[2] Moral agency and action[1] 1[2]	DIT, DIT-2, ESIT, EERI
A5 Characterizing (orientation toward / identification with)	Moral creativity[1] 2[2] Moral creativity[1] 4[2] Moral character[1] 1[2] Moral character[1] 2[2] Moral character[1] 3[2] Moral character[1] 4[2] Moral character[1] 5[2] Moral emotional development[1] 2[2] Moral agency and action[1] 2[2] Moral agency and action[1] 3[2] Moral agency and action[1] 4[2] Moral conduct[1] 1[2] Moral conduct[1] 2[2] Moral conduct[1] 3[2]	

[1] – The "Categories" column in Table 26.1
[2] – The "Sub-Cat." column in Table 26.1

Example: applying the framework to an ethical case study

To make the application of the matching procedure more concrete, we look at the case-based exercise "Apple vs. FBI – the encrypted phone" by Udo Pesch.[1] It is based on an actual event in which the FBI requested Apple to unlock an iPhone 5C of one of the suspects in a major terrorist attack that happened in California in 2015. The FBI believed it would find valuable data about planning the attack on the phone. Since Apple had improved its encryption earlier to the point where it could no longer unlock iPhones, Apple refused to adjust its technology to allow for investigation in this specific case, arguing that this could affect the security of all iPhones and provide the government with undue power over its products.

In individual, group, and in-class work, students need to analyze, argue, discuss, and debate this dilemma using ethical frameworks, and come up with regulatory and technological measures to solve it. The ILOs of this exercise are specified by the author as moral sensitivity and moral argumentation. In the more granular conceptualization of Table 26.1, we can discern the following ILOs in the exercise's seven steps:

- *Home assignment* – moral knowledge 1+2, moral sensibility 1+3, moral analysis 1
- *In group value selection* – moral argumentation 1, moral analysis 1
- *In group argument building* – moral argumentation 1
- *In-class debate* – moral argumentation 1
- *Individual work and aggregation* – moral judgment 3
- *Entire class interactive discussion* – moral creativity 1
- *Interactive feedback session and concluding* – moral sensibility 1+2+3

One way of using the framework would be to look at each of the steps separately, use the ILOs of the step and Tables 26.3 and 26.4 to determine the relevant categories, and find possible aligned assessment forms in these categories.

However, since most of the ILOs pertain to multiple steps and one may not want to assess every step separately, one can first aggregate over the steps to obtain the total set of 9 ILOs for the exercise: moral knowledge 1+2; moral analysis 1; moral argumentation 1; moral judgment 3; moral creativity 1; moral sensibility 1+2+3. When mapping these to the two tables, we find categories: (C1-K1) *Remember-Factual*; (C1-K2) *Remember-Conceptual*; (C2-K2) *Understand-Conceptual*; (C4-K1) *Analyze-Factual*; (C4-K2) *Analyze-Conceptual*; (C5-K2) *Evaluate-Conceptual*; (C5-K3) *Evaluate-Procedural*; (C6-K1) *Create-Factual*; (C6-K2) *Create-Conceptual* in Table 26.3 and (A3) *Valuing* in Table 26.4. For these categories, the tables suggest the following possible forms of assessment:

- Individual assignment graded with a rubric (Table 26.3: C1-K1; C1-K2; C2-K2; C4-K2; C5-K2; C6-K2)
- Portfolio (Table 26.3: C4-K2; C5-K2; C6-K2)
- Group project (Table 26.3: C4-K2; C5-K2; C6-K2)
- Rub 1 (Table 26.3: C2-K2; C4-K2)
- Rub 2 (Table 26.3: C2-K2; C4-K2)
- Rub 3 (Table 26.3: C4-K2)
- Rub 4 (Table 26.3:C4-K2; C5-K2)
- TESSE (Table 26.3: C2-K2)
- EDM (Table 26.3: C4-K2; C5-K2)
- DIT, DIT-2, ESIT, EERI (Table 26.3: C5-K2; Table 26.4: A3)

Although in principle the standardized instruments (Rub. 1 and following) could be used to assess some of the ILOs, tailored instruments would be more appropriate here. As the students work in the particular context of the case, it would be preferable to also assess them in this context.

Of the tailored instruments, 'Individual assignment graded with a rubric' covers all relevant categories[2] with one exception (A3), which makes it a good candidate for aligned assessment. The exercise has steps that are performed individually as well as steps performed in groups. Still, we can easily imagine that a student's written answers and performance in group activities are scored individually according to a set of rubrics. Another reason to go with a single form of assessment here is that the exercise is designed to be rather short, involving 2 hours of home preparation and 2 hours in class.

The one category not covered by the proposed assessment form is A3, for moral sensitivity 2. This is a dispositional ILO, so assessing requires measuring a change in behavior, by administering a standardized test (DIT, DIT-2, ESIT, EERI, see Table 26.4) before and after the exercise. Although possible in principle, in practice, this is typically done for longer-form activities, like an entire unit or series of units.

In the example, we applied the framework to a single short exercise. But the same way of thinking applies to more complex and extensive exercises (it is not uncommon for case-based exercises to fill an entire 3 or 5 credit-hour course) and can be used in designing units with multiple educational elements. Starting from the ethical ILOs, aligned possible forms of assessment can be determined from Tables 26.3 and 26.4. From these candidate assessment forms, those that best match the educational activities developed for the unit (which themselves must be aligned with the ILOs) can then be selected and implemented. Conversely, the framework could be used to check for an existing course if the used assessment forms align with the ILOs.

Conclusions and future work

This chapter has presented a methodology for aligning ethical competencies and affective dispositions with forms of assessment in engineering ethics education. To elaborate this methodology, the authors examined current academic literature and documents from professional and accreditation organizations to identify ethical competencies and dispositions that need to be addressed in engineering ethics education and explored tailored and standardized forms used to assess ethical competencies and dispositions. Following the principle of constructive alignment, the authors matched ethical competencies and affective dispositions to assessment forms by mapping both to Bloom's revised taxonomies of Cognitive and Affective Domains. As a result, a heuristic framework consisting of Tables 26.3 and 26.4 has been developed to assist educators in developing new and improving existing engineering education modules.

This framework can be used in four ways: (1) to see how the ethical ILOs are distributed over the cognitive categories and knowledge levels of the Bloom's Cognitive Domain and the categories of the Bloom's Affective Domain: which cells are filled and which are empty, pointing to potential gaps in literature and practice; (2) to identify the assessment methods that best match the ILOs of a newly developed module; (3) to check whether the assessment methods used in existing modules actually test the ILOs in those modules; and (4) to survey the tables of the framework to see which ILOs and assessment forms are not addressed or included in existing or newly developed modules. The heuristic framework reflects the authors' current understanding and interpretation of the theoretical and conceptual base and, therefore, requires further validation.

Based on the results obtained, several observations can be made. Examination of tables 26.3 and 26.4 shows that these tables are not evenly filled out with ILOs and assessment methods and that some cells are empty. It is worth noting that the 'Assessment forms' columns in these tables have more empty cells than the corresponding 'Ethical competencies' columns. The empty cells could point to gaps both in ILOs and assessment forms for ethics engineering teaching, especially in the knowledge types (K3) *Procedural* and (K4) *Meta-cognitive*, which are significant for the development and transfer of ethical thinking (Billing, 2007). Our findings show that methods for assessing affective dispositions, including (a) the ethical learning goals of higher education and (b) the codes and requirements of professional and accreditation organizations, are underdeveloped.

Comparisons within Table 26.3 and Table 26.4 show that the ethical learning goals outside of the conceptual knowledge column (K2) are generally located at higher cognitive and affective levels than the assessment methods and are therefore, overall, more challenging. There are several possible explanations for this: (1) the ethical ILOs have broadened and deepened and have become more complex in recent decades due to new societal demands (see the second section of this chapter), and developments in the academic field of engineering ethics while assessment methods have not kept pace with these changes; or (2) there are appropriate assessment methods for all ILOs of ethics modules, but engineering ethics teachers are not aware of them and therefore do not fully deploy them; or (3) engineering faculties are content to assess ethical competencies at a lower level than they are taught. At this point, these explanations are speculative. Determining whether one or more of them is true requires a study beyond the scope of this chapter.

To further develop the proposed framework and address the gaps identified, future work is proposed on several fronts:

- Conduct a literature review of learning activities used in engineering ethics education and map them onto the Revised Bloom's Taxonomy and Bloom's Affective Taxonomy. This will provide commensurable overviews of all three educational facets, enabling teachers to consider the complete alignment of ILOs, learning activities, and assessment forms in ethics engineering teaching.
- Examine the 'empty cells' in Tables 26.3 and 26.4 to see if ILOs can be defined or if forms of assessment useful for ethics engineering teaching should be constructed. Of particular interest in this respect are the empty cells of the knowledge types (K3) *Procedural* and (K4) *Metacognitive* and the cell for assessment forms in (A5) *Characterizing*.

Notes

1 See https://edusources.nl/en/materials/a506f5e6-6078-472a-aec7-a56042b88ce6/

2 Note that not all relevant categories in Tables 26.3 and 26.4 are occupied; C2-K1, C4-K1, C5-K3, C6-K1 do not have entries for forms of assessment.

References

ABET, Accreditation Board for Engineering and Technology. (2021). *Criteria for Accrediting Engineering Programs*. Engineering Accreditation Commission. https://www.abet.org/accreditation/accreditation-criteria/criteria-for-accrediting-engineering-programs-2022-2023/

ACM. (n.d.). *ACM Code of Ethics*. Retrieved April 21, 2023, from https://www.acm.org/code-of-ethics

Alliance, I. E. (2014). *25 years Washington Accord*. https://www.ieagreements.org/assets/Uploads/Documents/History/25YearsWashingtonAccord-A5booklet-FINAL.pdf

alZahir, S., & Kombo, L. (2014). *Towards a global code of ethics for engineers*. 2014 IEEE International Symposium on Ethics in Science, Technology and Engineering, Chicago, 1–5.

Anderson, L. W., & Krathwohl, D. R. (2001). *A taxonomy for learning, teaching, and assessing: A revision of Bloom"s taxonomy of educational objectives, abridged edition*. Longman.

ASIIN. (2022a). *Subject-specific criteria of the Technical Committee 02 – Electrical Engineering/Information Technology*. https://www.asiin.de/files/content/kriterien/ASIIN_SSC_06_Industrial_Engineering_2019-09-20.pdf

ASIIN. (2022b). *Subject-Specific Criteria of the Technical Committee 05 – Materials Science, Physical Technologies*. https://www.asiin.de/files/content/kriterien/ASIIN_SSC_05_Materials_Science_Physical_Technologies_2022-03-18.pdf

Bairaktarova, D., & Woodcock, A. (2015). Engineering ethics education: Aligning practice and outcomes. *IEEE Communications Magazine*, *53*(11), 18–22.

Barry, B. E., & Herkert, J. R. (2014). Engineering Ethics. In A. Johri & B. M. E. Olds (Eds.), *Cambridge handbook of engineering education research* (pp. 673–692). Cambridge University Press. https://doi.org/10.1017/CBO9781139013451.041

Biggs, J. B., & Tang, C. (2007). *Teaching for quality learning at university: What the student does*. McGraw-Hill/Society for Research into Higher Education/Open University Press.

Billing, D. (2007). Teaching for transfer of core/key skills in higher education: Cognitive skills. *Higher education*, *53*(4), 483–516.

Bonde, S., Briant, C., Firenze, P., Hanavan, J., Huang, A., Li, M., Narayanan, N., Parthasarathy, D., & Zhao, H. (2016). Making choices: ethical decisions in a global context. *Science Engineering Ethics*, *22*(2), 343–366.

Borenstein, J., Drake, M. J., Kirkman, R., & Swann, J. L. (2010). The engineering and science issues test (ESIT): A discipline-specific approach to assessing moral judgment. *Science and Engineering Ethics*, *16*(2), 387–407.

Borenstein, J., Drake, M., Kirkman, R., & Swann, J. (2008). The test of ethical sensitivity in science and engineering (tesse): A discipline specific assessment tool for awareness of ethical issues. *2008 Annual Conference & Exposition*, 13–1270. https://doi.org/10.18260/1-2--3253

Center, O. O. E. (2023). *Chapter 1: Ethical codes in physics and related fields*. https://onlineethics.org/cases/chapter-1-ethical-codes-physics-and-related-fields

Clancy, R. F., & Zhu, Q. (2023). Why should ethical behaviors be the ultimate goal of engineering ethics education? *Business and Professional Ethics Journal*, *42*(1), 33–53.

Coeckelbergh, M. (2006). Regulation or responsibility? Autonomy, moral imagination, and engineering. *Science, Technology, & Human Values*, *31*(3), 237–260.

Conlon, E., & Zandvoort, H. (2011). Broadening ethics teaching in engineering: Beyond the individualistic approach. *Science and Engineering Ethics*, *17*(2), 217–232.

Davis, M. (1999). *Ethics and the university*. Routledge.

Davis, M. (2012). A plea for judgment. *Science and Engineering Ethics*, *18*, 789–808.

Devon, R. (1999). Towards a social ethics of engineering: The norms of engagement. *Journal of Engineering Education*, *88*(1), 87–92.

Downey, G. L., Lucena, J. C., & Mitcham, C. (2007). Engineering ethics and identity: Emerging initiatives in comparative perspective. *Science and Engineering Ethics*, *13*, 463–487.

Durst, S., Pevkur, A., & Parts, V. (2021). Ethical attitudes among engineering students: Some preliminary insights. In *Educating Engineers for Future Industrial Revolutions: Proceedings of the 23rd International Conference on Interactive Collaborative Learning* (ICL2020), Volume 2 (pp. 780–788). Springer International Publishing.

Engineers_Canada. (2016). *Public Guideline on the code of ethics*. Retrieved April 21, 2023, from https://engineerscanada.ca/publications/public-guideline-on-the-code-of-ethics#-the-code-of-ethics

Engineering Council. (n.d.). Statement of ethical principles. Retrieved from https://www.engc.org.uk/media/2337/statement-of-ethical-principles-2014.pdf

Facione, P. (1990). *Critical thinking: A statement of expert consensus for purposes of educational assessment and instruction (The Delphi Report)*. California Academic Press.

Frigo, G., Marthaler, F., Albers, A., Ott, S., & Hillerbrand, R. (2021). Training responsible engineers. Phronesis and the role of virtues in teaching engineering ethics. *Australasian Journal of Engineering Education*, *26*(1), 25–37.

Gil-Jaurena, I., & Kucina Softic, S. (2016). Aligning learning outcomes and assessment methods: A web tool for e-learning courses. *International Journal of Educational Technology in Higher Education*, *13*(1), 1–16.

Glover, R. J., Natesan, P., Wang, J., Rohr, D., McAfee-Etheridge, L., Booker, D. D., Bishop, J., Lee, D., Kildare, C., & Wu, M. (2014). Moral rationality and intuition: An exploration of relationships between the defining issues test and the moral foundations questionnaire. *Journal of Moral Education*, *43*(4), 395–412.

Goldin, I. M., Pinkus, R. L., & Ashley, K. (2015). Validity and reliability of an instrument for assessing case analyses in bioengineering ethics education. *Science and Engineering Ethics*, *21*(3), 789–807.

Graham, J., Haidt, J., Koleva, S., Motyl, M., Iyer, R., Wojcik, S. P., & Ditto, P. H. (2013). Moral foundations theory: The pragmatic validity of moral pluralism. In P. Devine & A. Plant (Eds.), *Advances in experimental social psychology* (Vol. 47, pp. 55–130). Elsevier.

Graham, J., Nosek, B. A., Haidt, J., Iyer, R., Koleva, S., & Ditto, P. H. (2011). Mapping the moral domain. *Journal of personality and social psychology*, *101*(2), 366.

Harris Jr., C. E., Davis, M., Pritchard, M. S., & Rabins, M. J. (1996). Engineering ethics: What? why? how? and when? *Journal of Engineering Education*, *85*(2), 93–96.

Harris Jr., C. E., Pritchard, M. S., Rabins, M. J., James, R., & Englehardt, E. (2019). *Engineering ethics: Concepts and cases*. Cengage Learning.

Haws, D. R. (2001). Ethics instruction in engineering education: A (mini) meta-analysis. *Journal of Engineering Education*, *90*(2), 223–229.

Helton-Fauth, W., Gaddis, B., Scott, G., Mumford, M., Devenport, L., Connelly, S., & Brown, R. (2003). A new approach to assessing ethical conduct in scientific work. *Accountability in Research: Policies Quality Assurance*, *10*(4), 205–228.

Hess, J. L., Beever, J., Strobel, J., & Brightman, A. O. (2017). Empathic perspective-taking and ethical decision-making in engineering ethics education. In D. P. Michelfelder, B. Newberry, & Q. Zhu (Eds.), *Philosophy and engineering: Exploring boundaries, expanding connections* (pp. 163–179). Springer.

Hess, J. L., & Fila, N. D. (2016). *The development and growth of empathy among engineering students*. Paper presented at 2016 ASEE Annual Conference & Exposition, New Orleans, Louisiana.

Hodson, D. (1999). Going beyond cultural pluralism: Science education for sociopolitical action. *Science Education*, *83*(6), 775–796.

IEEE. (1963). *IEEE code of ethics*. Retrieved April 21, 2023, from https://www.ieee.org/about/corporate/governance/p7-8.html

Ingenia, 2023. Ethics and the Engineer. Retrieved from https://www.ingenia.org.uk/ingenia/issue-25/ethics-and-the-engineer.

Johnson, J. (2010). Teaching ethics to science students: Challenges and a strategy. In B. Rappert (Ed.), *Education and ethics in the life sciences: Strengthening the prohibition of biological weapons* (pp. 197–214). ANU Press. http://www.jstor.org/stable/j.ctt24hc5p.17

Katz, A., Reid, K., Riley, D., & Van Tyne, N. (2020). Overcoming challenges to enhance a first year engineering ethics curriculum. *Advances in Engineering Education, 8*(3), 10.

Keefer, M. W., Wilson, S. E., Dankowicz, H., & Loui, M. C. (2014). The importance of formative assessment in science and engineering ethics education: Some evidence and practical advice. *Science and Engineering Ethics*, *20*(1), 249–260.

Kohlberg, L. (1976). Moral stages and moralization: The cognitive-development approach. In T. Lickona (Ed.), *Moral development and behavior: Theory research and social issues* (pp. 31–53). Holt, Rienhart, and Winston.

Lattuca, L. R., & Stark, J. S. (2009). *Shaping the college curriculum: Academic plans in context*. John Wiley & Sons.

Lawlor, R. (2021). Teaching engineering ethics: A dissenting voice. *Australasian Journal of Engineering Education*, *26*(1), 38–46.

Lee, E. A., Gans, N., Grohman, M., Tacca, M., & Brown, M. J. (2020). Guiding engineering student teams' ethics discussions with peer advising. Science and Engineering Ethics, *26*, 1743–1769.

Li, J., & Fu, S. (2012). A systematic approach to engineering ethics education. *Science and Engineering Ethics*, *18*, 339–349.

Lind, G. (2002). *Ist Moral lehrbar?: Ergebnisse der modernen moralpsychologischen Forschung*. Logos-Verlag.

Lönngren, J., Adawi, T., Berge, M., Huff, J., Murzi, H., Direito, I., Tormey, R., & Sultan, U. (2020). Emotions in engineering education: Towards a research agenda. *2020 IEEE Frontiers in Education Conference (FIE)*, 1–5. doi: 10.1109/FIE44824.2020.927395

Loui, M. C. (2005). Ethics and the development of professional identities of engineering students. *Journal of Engineering Education*, *94*(4), 383–390.

Lucena, J., Downey, G., Jesiek, B., & Elber, S. (2008). Competencies beyond countries: The re-organization of engineering education in the United States, Europe, and Latin America. *Journal of Engineering Education*, *97*(4), 433–447.

Lynch, W. T., & Kline, R. (2000). Engineering practice and engineering ethics. *Science, Technology, & Human Values*, *25*(2), 195–225.

Maffei, A., Boffa, E., Lupi, F., & Lanzetta, M. (2022). On the design of constructively aligned educational unit. *Education Sciences*, *12*(7), 438.

Martin, D. A., Conlon, E., & Bowe, B. (2019). The role of role-play in student awareness of the social dimension of the engineering profession. *European Journal of Engineering Education*, *44*(6), 882–905.

Martin, D. A., Conlon, E., & Bowe, B. (2021). A Multi-level review of engineering ethics education: Towards a socio-technical orientation of engineering education for ethics. *Science and Engineering Ethics*, *27*(5), 1–38.

Martin, M. W., & Schinzinger, R. (2013). *Ethics in engineering*. Mcgraw-Hill Book Co.

Miller, G. (2018). Aiming professional ethics courses toward identity development. In E. E. Englehardt & M. Pritchard (Eds.), *Ethics across the curriculum: Pedagogical perspectives* (pp. 89–105). Springer.

Miller, M. (2005). Teaching and learning in affective domain. In M. Orey (Ed.), *Emerging perspectives on learning, teaching, and technology*. Retrieved March 6, 2008, from http://projects.coe.uga.edu/epltt/

Miñano, R., Uruburu, Á., Moreno-Romero, A., & Pérez-López, D. (2017). Strategies for teaching professional ethics to IT engineering degree students and evaluating the result. *Science and Engineering Ethics*, *23*, 263–286.

Moriarty, G. (2009). *The engineering project: Its nature, ethics, and promise*. The Pennsylvania State University Press.

Mumford, M. D., Devenport, L. D., Brown, R. P., Connelly, S., Murphy, S. T., Hill, J. H., & Antes, A. L. (2006). Validation of ethical decision making measures: Evidence for a new set of measures. *Ethics & Behavior*, *16*(4), 319–345.

Nair, I., & Bulleit, W. M. (2020). Pragmatism and care in engineering ethics. *Science and Engineering Ethics*, *26*, 65–87.

Newton, R. R. (1978). Kohlberg: Implications for high school programs. *Living Light*, *15*(2), 231–239. http://hdl.handle.net/2345/4397

Pinkus, R. L., Gloeckner, C., & Fortunato, A. (2015). The role of professional knowledge in case-based reasoning in practical ethics. *Science and Engineering Ethics*, *21*(3), 767–787.

Pritchard, M. (2005). Perception and imagination in engineering ethics. *International Journal of Engineering Education*, *21*(3), 415–423.

Rest, J. R. (1979). *Development in judging moral issues*. University of Minnesota Press.

Rest, J. R., Narvaez, D., Thoma, S. J., & Bebeau, M. J. (1999). DIT2: Devising and testing a revised instrument of moral judgment. *Journal of Educational Psychology*, *91*(4), 644–659.

Roeser, S. (2012). Emotional engineers: Toward morally responsible design. *Science and Engineering Ethics*, *18*, 103–115.

Shiveley, J., & Misco, T. (2010). But how do I know about their attitudes and beliefs? A four-step process for integrating and assessing dispositions in teacher education. *The clearing House A Journal of Educational Strategies, Issues and Ideas*, *83*(1), 9–14.

Shuman, L. J., Besterfield-Sacre, M., & McGourty, J. (2005). The ABET professional skills—Can they be taught? Can they be assessed? *Journal of Engineering Education*, *94*(1), 41–56.

Sindelar, M., Shuman, L., Besterfield-Sacre, M., Miller, R., Mitcham, C., Olds, B., Pinkus, R., & Wolfe, H. (2003). Assessing engineering students' abilities to resolve ethical dilemmas. *33 rd Annual Frontiers in Education, 2003, FIE 2003, 3*, S2A–25.

Snoeyenbos, M., Almeder, R. F., & Humber, J. M. (1983). *Business ethics: Corporate values and Society*. Prometheus Books.

Spivey, G. (2007). *A taxonomy for learning, teaching, and assessing digital logic design*. 2007 37th Annual Frontiers in Education Conference Global Engineering Knowledge without Borders, Opportunities without Passports, F4G–9.

Tormey, R. (2005). The costs of values: questioning the application of the term in development education. *Development Education Journal*, *11*(2), 9.

van de Poel, I., & Verbeek, P.-P. (2006). Ethics and engineering design. *Science, Technology, & Human Values*, *31*(3), 223–236.

van de Poel, I., & Royakkers, L. (2011). *Ethics, technology, and engineering: An introduction*. John Wiley & Sons.

Verbeek, P. P. (2008). Morality in design: Design ethics and the morality of technological artifacts. In P. Kroes, P. E. Vermaas, A. Light, S. A. Moore, & P.-P. Verbeek (Eds.), *Philosophy and design: From engineering to architecture* (pp. 91–103). Springer.

Vermaas, P. E., Hekkert, P., Manders-Huits, N., & Tromp, N. (2015). Design methods in design for values. In J. van den Hoven, I. Pieter, and van de Poel Vermaas (Eds.), *Handbook of ethics, values, and technological design: Sources, theory, values and application domains* (pp. 179–201). Springer Netherlands. https://doi.org/10.1007/978-94-007-6970-0_10

Weinert, F. E. (2001). Competencies and key competencies: Educational perspective. *International Encyclopedia Social Behavioral Sciences, 4*, 2433–2436.

Whitbeck, C. (1995). Teaching ethics to scientists and engineers: Moral agents and moral problems. *Science and Engineering Ethics*, *1*, 299–308.

Zhu, Q., Zoltowski, C. B., Feister, M. K., Buzzanell, P. M., Oakes, W. C., & Mead, A. D. (2014). *The development of an instrument for assessing individual ethical decisionmaking in project-based design teams: Integrating quantitative and qualitative methods*. 2014 ASEE Annual Conference & Exposition, 24–1197.

27

ASSESSING ATTITUDES AND CHARACTER IN ENGINEERING ETHICS EDUCATION

Current state and future directions

Adetoun Yeaman, Balamuralithara Balakrishnan, Olga Pierrakos, and Elise M. Dykhuis

Introduction

In this chapter, drawing from case studies, we discuss the current state of assessing engineering ethics through the lens of attitudes and character development. Evaluating engineering students' character and ethical development requires assessing their knowledge, skills, and attitudes (KSA) related to moral, intellectual, and performance dimensions (Seider et al., 2017). Systematically and meaningfully measuring the dimensions of ethics and character can provide insights into what processes and experiences of learning environments make a positive difference. As engineering ethics educators, refining our assessment practices can help us (as individuals and as a community) teach ethics better and more effectively and define pedagogical methods that develop our students holistically.

Assessing KSAs related to character and ethical development poses noteworthy challenges. Confusion is evident regarding conceptualization (definitions) and operationalization (i.e., how the concrete measures will look) (Card, 2016). There is additional confusion evident in the literature regarding what experiences to assess and how to provide meaningful feedback on developmental character-related dimensions without being punitive or assuming there is some 'right' amount or type of character necessary to be successful and ethical (Davis & Feinerman, 2012; Martin et al., 2021). The myriad approaches to framing character and ethical development in engineering stem from various nuanced contexts that concern character-relevant attitudes in the engineering curriculum. Before we delve further into the assessment of character development, it is important to define a few things. Thus, in the following section, we provide definitions for character and attitudes based on our desk and field research. We provide an overview and then describe eight cases where pedagogies and research methods have been implemented to develop and test various aspects of development in this realm. We use these case examples – drawn from the literature and our own research work – to discuss how character and attitudes have been represented and assessed in the engineering curriculum. Subsequently, we identify and reflect on existing gaps and pose recommendations for developing and assessing character and ethical attitudes in engineering

DOI: 10.4324/9781003464259-33

ethics education. We end with concluding thoughts to summarize key takeaways. Our first step is to introduce ourselves as authors so that readers can better understand the experiences, assumptions, and underlying beliefs that inform our work.

Positionality

In her Ph.D. dissertation, this chapter's lead author, Adetoun Yeaman, studied empathy, including its role and how it can be supported within undergraduate engineering students' experiences in service-learning programs. This research sparked Adetoun's interest in socio-ethical competencies more broadly. After completing her Ph.D., she spent three years as an engineering education postdoctoral fellow in the Department of Engineering at Wake Forest University. In this role, she collaborated with an interdisciplinary team to reimagine ethics education in engineering and design curricular modules that approached ethics from a virtue ethics lens (as reported in cases 4 and 8 below, e.g.). Now, as an assistant teaching professor of first-year Engineering at Northeastern University, she continues exploring ways to integrate ethics and engineering teaching. She strives to help students see ethics and engineering as related and to build students' confidence in practicing engineering with the well-being of humanity and the environment in mind.

Balamuralithara Balakrishnan, the second author, lives and works in Malaysia. He is a proponent of holistic development for future engineers. He promotes educational techniques that go beyond typical classroom teaching and learning – to emphasize developing ethical character and promote a positive professional attitude. As a transdisciplinary researcher in ethics, sustainability, creativity, and STEM education, Balamuralithara has prioritized research on character and attitude. His research investigating students' perceptions of and attitudes toward engineering ethics is featured in cases 6 and 7 below. Balamuralithara teaches engineers to meet technical requirements *and* constructively contribute to society via ethical decision-making by emphasizing integrity, responsibility, and a collaborative approach.

The third author, Olga Pierrakos, was born in Greece and moved to the United States at the age of 10. She identifies as a naturalized citizen, first-generation college student, first-generation engineer (with BS, MS, and Ph.D. degrees), woman, engineering educator, engineering education researcher, biomedical engineering and mechanical engineering researcher, inclusive and innovative higher education leader, and interdisciplinary scholar. As a founder of two new engineering programs in the United States, she strongly believes ethics education, leadership and character development, and innovation can push the boundaries of engineering education and help reimagine the culture in engineering. As a program director at the National Science Foundation, she witnessed the power of policy and research to advance knowledge and positive societal impact through high ethical and equitable standards.

The fourth author, Elise Dykhuis, is a developmental psychologist trained in North American contexts, using a paradigm that takes into account the complexity of individual development with regard to social, cultural, and biological diversity. Elise's expertise is in character development and measurement. She looks at how interventions can be developed for population-level integration, particularly in university contexts.

As an authoring team, our diverse perspectives guide us to remain open and learn from the expertise and diverse experiences of others. Together, we believe that ethical engineering extends beyond technical proficiency to include a strong character and a good attitude – both required for navigating the engineering profession's complicated ethical landscape. We believe engineering ethics education should be transformative, developing cognitive components of ethical reasoning, character characteristics, and attitudes contributing to ethical behaviors. We believe engineers have an ethical responsibility to society – a bigger purpose to better humanity – and that we need to educate the next generation to understand this bigger purpose. Yet we also believe that few researchers are involved in this subject area because character and attitude toward ethics are extremely complex to measure and assess. The four of us are all involved in producing research in this realm as well as teaching future engineers. Writing this chapter has provided the opportunity to consider and assess our work in new ways.

Background

This section defines *attitudes* and *character* and situates the work done in engineering education (both teaching and research) within the broader literature.

Attitudes

Eagly and Chaiken (1993, 2007) defined attitude as "a psychological tendency that is expressed by evaluating a particular entity with some degree of favor or disfavor" (pp. 1, 582). Lamprianou and Athanasou (2009) described attitudes as "a system of beliefs, values or tendencies that predispose a person to act in certain ways" (p. 254). Although several definitions for attitude(s) exist, Eagly and Chaiken (1993, 2007) emphasized that at the core of attitude are evaluation, the object of the attitude (what the attitude is directed towards), and tendency. Individuals can have attitudes towards myriad categories, including people, places, opinions, values, and even themselves. Attitudes can be positive, negative, or ambivalent. Attitudes can be expressed in responses like evaluative judgments, emotions, or even behaviors – but these responses do not comprise attitudes (Eagly & Chaiken, 2007).

Character

For this chapter, we focus on character, as Berkowitz (2012) defined it: the constellation of positive psychological attributes that allow individuals to live self-transcendent, flourishing lives. Our conceptualization of character is widely accepted and detailed by many, including Baehr (2017). It is comprised of various strengths, or virtues, that are emblematic of self-transcendent, flourishing lives, and with major domains identified as (1) *moral* – including empathy, compassion, humility, honesty; (2) *intellectual* – including curiosity, creativity, and intellectual humility; (3) *performance* – including resilience, persistence, and tenacity; and (4) *civic* (which often overlaps moral) – including fairness and respect. Specific models in this realm vary depending on their theoretical approach.

Attitudes, character, and ethics education assessment

Connections between character and attitudes have important implications for ethics education and assessment. A major goal of ethics education is to promote ethical decision-making. At the core, engineering ethics educators hope to promote ethical decisions and behavior in their students for future engineering practice. Although we cannot fully predict that an individual will make ethical decisions in the future, we can educate for and measure attitudinal changes which are a precursor to behavioral (among others) changes (Ajzen, 1985; Ajzen & Fishbein, 2004).

Character development is associated with understanding attitudes – or thoughts, feelings, and/or beliefs –associated with a particular subject or idea (Eagly & Chaiken, 1993). A person who embodies a certain character trait or virtue is more likely to behave in a way that aligns with such a character trait. For example, an empathetic person will likely show empathy towards others in different situations. A kind person is likely to be kind in various contexts. Yet, virtues like empathy and kindness are hard to measure. It could be presumptuous to conclude that an individual is empathetic or kind without having long-term familiarity with the individual. We might go about such a quest by measuring attitudes emblematic of such virtues instead.

Research on character is, in part, attitudinal research, in the sense that we can have attitudes toward oneself, others, and the world – and these attitudes partly constitute one's character. Additionally, a person's character can dictate their attitude toward any object (Eagly & Chaiken, 2007). Character and attitudes have a rich history in being assessed in education settings given the formative requirements of educational systems (Berkowitz & Bier, 2005), but have a much shorter history in the specific context of assessment in engineering ethics education.

Character and attitudes are represented in various ways in the engineering curriculum. Character has been discussed as a curricular goal to be developed through engaging with specific issues like sustainability (de la Riva de la Riva et al., 2015). Sometimes, engineering educators discuss character and virtues as ancillary to other efforts such as entrepreneurship education (Jen et al., 2012) or leadership development (Hawks, 2009), or in the context of introducing virtue ethics as one of several frameworks for engaging students in ethics discussions (Gomez, 2013; Van Tyne, 2020) often centered around resolving dilemmas (see Chapter 22 for more on virtue ethics). Sometimes, educators focus on building specific character traits such as 'grit' in students (Direito et al., 2021; Golding et al, 2018); 'grit' is defined as the perseverance and passion an individual exhibits over time to overcome challenges and setbacks (Duckworth, 2016). Attitudes have also been discussed in engineering ethics curricula in diverse ways. In the literature, attitudes are sometimes discussed in exploring student attitudes towards ethics, ethics curricular implementations (Sethy, 2017), sustainability, or diversity. 'Ethical attitude' is one of the ways attitudes are discussed in engineering education publications (Lee, 2018). Importantly, within the engineering education literature, the terminology regarding 'attitudes' has been more connected to assessment than character has been. As the teaching and research cases described later in this chapter will show, several *attitudinal assessments* have been used to measure ethics-related learning objectives.

Assessing *character* has proven to be more difficult than assessing *attitude*: (1) incorporating character and virtue in engineering education is still in its infancy; (2) many character education efforts in higher education are grappling with how to conduct assessment; (3) in terms of evidence, we need to use caution when attempting to assess character traits (i.e., it is challenging to gauge that a person has attained better character or is more kind, just, empathetic, etc. as a result of an intervention). Consequently, in the next section, we will use the term 'character education' to include all attempts to incorporate character or support character development in the engineering curriculum. We acknowledge that character education is a holistic process informed by virtue ethics and concerned with cultivating character in individuals to promote human flourishing (Pierrakos et al., 2019). For simplicity, we will also use the term 'ethical attitudes' in this chapter – to mean attitudes related to ethics education outcomes.

Assessing attitudes and character education efforts in engineering ethics

Because there is no singular way that attitudes and character are taught or assessed in engineering ethics curricula, this section features a variety of examples, including some that relate to either (a) attitudes or (b) character and (c) those that connect to both. We present these examples as a set of eight cases that involve teaching (i.e., pedagogical/educational interventions/treatments) and formal assessment measures. We present the cases based on the three assessment types evident in the literature: (1) assessment of cognition and skill, (2) assessment of student attitudes, and (3) assessment of pedagogical methods. These examples are summarized in Table 27.1.

We selected a range of intervention/teaching methods and assessment/measurement types and identified one published case of each type to share here. Some of the eight cases featured below (specifically cases 4, 6, 7, and 8) represent projects we ourselves were involved in. The process we used to select the cases featured below includes a selection of papers from literature informed by results of a prior scoping literature review that helped reveal engineering education interventions related to character development (see Yeaman, 2022). We focused on only the ones with descriptions of the assessment approaches that were applied. We extended that work with additional searches to account for new scholarly works that may have and cases we knew about because of our involvement. Our inclusion criteria were the following: published research, ethics, engineering education, and either character development assessment or attitude assessment.

Most of the assessment efforts identified in the research studies (listed in Table 27.1) were implemented at the course level, and the researchers endeavored to measure character traits, character-focused outcomes, or attitudes within students. In a few instances, the treatment intervention or assessment method expanded beyond the boundaries of a single course.

The measures used in the featured research cases draw from quantitative and qualitative approaches. Most of the cases used quantitative measures like self-reported survey questionnaires. Some used standardized measures (like quizzes and exams) or *ad hoc* measures tailored to the needs of the respective program.

Our intention is to provide examples of what has been done to date so that our community can learn from and build upon existing work. Our aim was not to make evaluative judgments about the efficacy of the intervention/programs or assessments. Nevertheless, we provide some preliminary analysis based on supporting literature and our own experience as researchers in this realm. This realm of inquiry is nascent and emerging; published research is thus, at this point, exploratory in nature. The cases show approaches used in published research; yet the results are not as robust as in more established realms of research on engineering education, and the method of reporting findings doesn't provide the reader with a very high level of confidence. Nevertheless, the development of this nascent line of inquiry rests on those who pioneer tools, methods, and methodologies – those who develop and seek to test their hypotheses and pedagogical designs empirically.

Table 27.1 Summary of cases to exemplify attitudes and character education assessments in engineering education

Assessment outcomes and objectives	*Types of assessment measures/methods (researchers reporting results)*
Assessment of student learning and skill	Case 1 – Standardized ethics questions (Davis & Butkus, 2008) Case 2 – Case-based action plan (Dillon et al., 2020) Case 3 – Self and peer performance evaluation (Hawks & Terry, 2009)
Assessment of student attitudes	Case 4 – Self and team reflections in project work (Gross et al., 2021) Case 5 – Self-assessment of grit (Golding et al., 2018) Case 6 – Self-assessment of ethical attitudes, course level (Balakrishnan, 2015) Case 7 – Self-assessment of learning, program level (Balakrishnan et al., 2019)
Assessment of pedagogical methods	Case 8 – Assessment of pedagogies and character cultivation, program level (Koehler et al., 2023)

Assessment of student learning and skill

The cases in this category comprise engineering courses or programs that incorporated character education (to any degree) and applied assessments regarding whether students learned something (e.g., ethics content knowledge), gained a skill (e.g., action plan development), or developed new abilities (e.g., leadership). The assessments in these cases involved standardized tests, focus groups, self and peer evaluations, and self-report questionnaires.

Case 1: Standardized exam questions to assess impact regarding ethics and character

In this case, Davis and Butkus (2008) used standardized, multiple-choice questions to assess students' learning of ethical topics and character development. We use this case as an example of how standardized tests have been used in the engineering ethics education literature. It is important to note that despite the age of this example, it has been very difficult to find examples of standardized tests used in higher education for the purposes of assessing character. More recently, there has been some work to explore the use of situational judgment tests to assess character in higher education (Kuncel et al., 2020).

Assessment measures – Davis and Butkus (2008) used a set of standardized, multiple-choice questions from the NCEES Fundamentals of Engineering (FE) exam. Questions came from the Fundamentals of Engineering (FE) exam ethics portion (NCEES, 2023). Some questions were based on ethics terminology and definitions, while others were behavioral questions regarding some written scenarios.

Context – Davis and Butkus (2008) of the United States Military Academy engaged students in an ethical education and training program to support cadets in moral character development. The program includes a series of ethical education activities and training events spread through a 47-month experience including one philosophy course/module and a leadership course. Within this program, engineering students engaged in case-based learning which incorporated cases reflecting ethical situations that they could face on and off the battlefield. Student cadets also analyzed their ethical views regarding various contemporary issues – probing their own perspectives of medical, social, and engineering ethics – in small group discussions facilitated by faculty. The researchers wanted to know if students' performance on the exam would improve due to the training program. Thus, they were attempting to gauge their character program's impact on students' learning by assessing students' performance on ethics questions. They conducted an experiment by comparing scores of voluntary participants from a sample of entering students who had no training with the scores of students who had experienced some of the institution's training program. None of the participants had experienced formal ethics training previously (in contrast to character training).

Findings – Davis and Butkus (2008) conducted a t-test of the scores. The data did not show statistically significant differences between the scores of the two groups.

Primary relevance to this chapter – The authors acknowledged a need for better ways to carry out their assessments in the future. The faculty at this institution has continued working to improve the character program and assessment since the time of the (Davis & Butkus, 2008) publication. This case provides an example of work underway to help inspire readers and encourage others to take a similarly iterative approach to developing research in this realm.

Case 2: Case-based action plan

This case's intervention and assessment were reported by Dillon et al. (2020).

Assessment measures – Dillon et al. (2020) focused on assessing a specific skill in students – the ability to develop an action plan for ethics. The tools used for research purposes were observations as well as a self-report survey with students that included Likert-scale and open-ended questions. The results of the observation and open-ended questions were not discussed in Dillon et al. (2020), so we can only summarize the survey results here.

Context – Drawing insights from Mary Gentile (2010) that oftentimes people know what is right but lack the preparation to practice it when faced with an issue that challenges their ethics or character in a professional setting, Dillon and colleagues (2020) designed a course-based module to support students in this preparation. The three objectives of their module were for students to (1) develop an action plan to modify or address an ethics or character issue, (2) explore multiple solution paths, and (3) identify the needs and motivations of various stakeholders. The module was implemented in a heat transfer course with junior-level (i.e., third year) students and embedded into an existing class project. The class comprised around 40 students with two or three students in each project group. The module presented students with an ethical dilemma related to vaccine transportation. The simulated dilemma involved a conflict between an engineering design team and their management team. More details on the dilemma and how the project was implemented can be found in the paper (Dillon et al., 2020).

To assess this module, the research team focused on students' ability to develop an action plan for ethics as the main skill to be assessed. They assessed this ability via a student self-reported survey. The survey included Likert questions that asked students to self-assess specific learning objectives and free-response questions that asked for examples to illustrate their growth in the learning objective. The researchers (Dillon et al., 2020) presented statistical data from the closed-ended survey results. They used natural language processing to analyze the open-ended responses.

Findings – According to Dillon et al. (2020), students indicated the educational activities helped build ethics-action-planning skills, including the objective of exploring multiple solution paths. Some of the notable themes that came up related to the objective of developing "an action plan to modify or address an ethics or character issues" are that the project "helped them consider the impact of their own decisions as an engineer" and "helped them consider multiple perspectives" (p. 8). The researchers claimed that the kind of module they implemented may support character development in engineers.

Relevance to this chapter – This case's teaching and research methods seem worthy of additional study. The reported survey results (Dillon et al., 2020) are interesting and may be useful to readers.

Case 3: Self and peer performance evaluation

This case was published by Hawks and Terry (2009).

Assessment measures – The case's assessment of a character and leadership program's outcomes involved focus groups, a '360-degree self-evaluation instrument,' and a post-course assignment survey. We note that 360-degree self-evaluation instruments have gained traction in organizational settings for performance evaluation that considers feedback from peers, superiors, and subordinates in addition to an individual's ratings of themselves (Craig & Hannum, 2006).

Context – Hawks and Terry (2009) discussed a plan developed by the the Ira A. Fulton College of Engineering and Technology at Brigham Young University to prepare and develop students as leaders for success in a changing global environment. According to the authors, one of the five focus areas of this plan was character development, which included ethics. The other areas included technical excellence, systems emphasis, leadership, global awareness, and innovation. A course was designed to help students develop these five attributes within the college. Engineering instructors taught the course, and non-engineering students across the university were welcome to take it as a general education elective. An important part of the course emphasized leadership and the development of leadership attributes and skills. These attributes included 'absolute integrity' and 'sound moral values.' Hawks and Terry (2009) alluded to the notion that character and leadership are intertwined in the sense that there are aspects of one's character that support good leadership. The course's leadership component involved several opportunities for experiential learning via in-class team-based activities, case studies, and out-of-class team projects. There was also an emphasis on self-awareness – viewed as a vital part of leadership development. To promote self-awareness in students, the educators delivered instruction on team development and dynamics using the Kolb Learning Style Model, the Myers-Briggs Type Index, and the Tuckman Model of team development. Students completed inventories to identify their own learning styles and personality types (even though the report was unclear whether students took all three inventories).

The class used team projects to explore topics like corruption in the technology industry in a country of their choice. One of the team projects engaged students in investigating a major societal issue. Students were asked to create a survey to better understand the issue and write an executive summary of the investigation to a hypothetical media outlet's chief executive officer (CEO). The students were introduced to leadership theories (e.g., trait, character, contingency theories) and various leadership models.

The assessments for the course were focused on three questions. First, *Do students see themselves as leaders?* This question was not formally addressed in the course but pursued through a focus group sponsored by the college administration. Some of the seven prompts for the focus group included the following: *Before this class did you see yourself as a leader?* and *What does it mean to be a leader, in your minds?*

The second question was the following: *How did the course learning activities help inform the students about their leadership potential?* This question was assessed via self-reports corroborated by assessments from other people using a 360-degree self-evaluation instrument. In this context, students took the assessment and gave it to three people who knew them.

The third and final was the following: *Have students thought much about leadership theories and principles and what leadership principles are important to the students?* Addressing this question involved an assignment in which students formulated their own personal leadership theory and defended it in a short paper. Following the assignment, students were asked to fill out a survey of Likert-type questions on a scale of 1 (do not agree) to 5 (strongly agree).

Findings – Hawks and Terry (2009) presented the focus-group results via bullet points addressing each question (the researchers defined seven questions overall). The paper provides greater detail, but the results most relevant to this chapter were that some students saw that they were put in leadership positions continually but did not see themselves as leaders. Going through the class helped them realize that they had been leaders before the course. Some students' perspectives on leadership shifted from seeing it as 'managing others' to being more about 'personal character.'

The survey regarding leadership had four items. One of the survey items regarding the assignment was, "I intend to use this as a basis for my leadership activities in the future." The average rating of student responses for this item was 4.36 (with 5 being the upper limit of the scale).

The focus group and survey data provided a more overarching view of what students had gained from the course, helping the researchers ascertain how experiencing the course could be improved.

Relevance to this chapter – It was interesting and novel to us to see how this study applied several forms of assessment, including formative and summative assessments that corroborated evidence regarding students' growth. Inventories allowed students to get to know themselves better. Students also had the opportunity to learn about leadership and character and practice their learning through team-based projects, which were assessed formatively. Notably, this case provided formative feedback to students that they could implement and improve upon. Summative assessments were made of the students' personal leadership theories and reflections on them.

We were also intrigued that there were mixed results in relation to the usefulness of the 360-degree instrument. Some students indicated it was not helpful yet but might be helpful for them later in the future. Others indicated it helped them see areas of leadership they needed to improve; others said it helped them to focus on their strengths.

As researchers reading the report by Hawks and Terry (2009), we would have liked to find a better description of how the project assignments were assessed. We encourage future research to apply more structured, reliable, and transparent approaches to analyzing the survey and focus group data.

On the plus side, Hawks and Terry (2009) documented a rare example of character-based interventions at the course level that have applied several assessment types, particularly ones that corroborate various types of evidence regarding an individual's growth/performance. Although the leadership intervention discussed in this case was delivered within one course, it was part of a larger institutional effort that afforded methods like focus groups.

The resource-intensive nature of the model presents a challenge to adopting a similar model at other institutions. Nevertheless, we believe the instruction methods, student assignments, and assessment rubrics may prove helpful in other contexts. Techniques to promote self-awareness using the 360-degree instrument (or similar tools) can be adapted to other settings. We encourage future researchers to consult the paper by Terry and Hawks (2009) and consider using these assessment approaches based on the resources available within their respective institutional contexts.

Assessment of student attitudes

Case 4: Self and team reflections regarding character in project work

Authors 1 and 3 worked on this study, published in conference proceedings (Gross et al., 2021).

Assessment measures – Generating reflections on personal character and team dynamics during collaborative projects facilitates individual and collective learning. Self and peer reflections during team-based project work can take different forms; an example of one is provided in this case.

Context – Gross et al. (2021) described a module dedicated to fostering teamwork as a virtue in undergraduate engineering students. The module focused on teamwork embedded in a project-based course for first-year engineering students. The course was divided into two halves; each engaged students in a project. The module was introduced in the middle of the two projects as students were transitioning from one project to the next, but it was completed at the end of the second project. One of the projects involved using sensors to collect and analyze data to determine which method is better for transferring patients with potential spinal injuries. The other project required designing a water filtration device. The module engaged students in hands-on activities such as brainstorming exercises, discussions, and reflections. It applied the seven strategies of character development (Lamb et al., 2021) as a framework for guiding the activities: habituation through practice, reflection on personal experiences, dialogue that increases virtue literacy, awareness of situational variables, engagement with virtuous exemplars, moral reminders, and friendships of mutual accountability. The assessments used in this intervention involved both self and peer assessments guided by specific prompts provided by the instructor.

In the beginning, students produced their own lists of qualities they believed described a good team member. The instructor introduced the definition of 'a virtue,' three categories of virtues (Jubilee Framework, 2022), and examples of virtues for each category. The instructor described teamwork as a performance virtue. In a class-wide brainstorming activity, students categorized the qualities of a good team worker using a Venn diagram. They collectively developed a 'top-15 list' of attributes of a good team worker, which became a metric for assessment in the module. Students rated which attributes they thought they exhibited during the first project and identified tangible examples. Each student rated, and was also rated by, their teammates. Then each student chose one attribute they would like to work on during the semester. The class repeated the rating exercise at the end of the second project. The rating process was supported with discussions and reflections about how students saw themselves versus how they were rated by others. Gross and colleagues (2021) provide details regarding how the module was executed.

Findings – Students, over the course of the module, refined their ideas of what it means to embody teamwork. Their ratings, involving both individual and peer assessments, provided an opportunity for students to reflect on their character, how they wanted to grow, and any changes they saw in themselves over the course of working in project teams. This assessment was formative: students had initial thoughts on what it meant to be a good team worker and the qualities they saw in themselves, and then they received feedback from peers and had the opportunity to refine their thoughts.

Relevance to this chapter – This case is an example that illustrates a curricular intervention geared towards cultivating a specific virtue or trait, in this case, teamwork. It serves as a helpful example of providing formative feedback to students to help them improve over time. These assessments also exemplify the use of *ad hoc* approaches.

Case 5: Self-assessment of grit

This case was published by Golding et al. (2018).

Assessment measures – In this case, Golding et al. (2018) used multiple approaches to address different outcomes. The investigators assessed grit using a validated self-assessment instrument – conducting pre- and post-tests with a treatment and a control group. They also created an ad-hoc survey that addressed students' perceived growth from the reported intervention. Lastly, they collected qualitative responses to an open-ended question.

Context – Golding et al. (2018) described a technology-supported learning program called 'STEMGrow' implemented at two interconnected institutions (El Paso Community College and The University of Texas at El Paso). The program involved first-year students entering science, technology, engineering, and mathematics (STEM), although the students in this report specifically majored in engineering and biology. Theoretical assumptions underpinning the research included grit and self-control as important determinants for success and the idea that grit can be self-developed. A cloud-based app – EduGuide for STEM online learning activities, coaching, and mentoring – was used. STEMGrow focused on developing the character strength of grit in students, among other outcomes of the intervention. The EduGuide online toolkit increased students' autonomy by providing an avenue for self-paced learning and helped students and their mentors maintain communication.

The assessment of grit levels utilized a validated instrument, the 'Short Grit Scale' or 'Grit-S,' to assess passion and perseverance (Duckworth & Quinn, 2009). The scale consists of items like 'new ideas and projects distract me from previous ones' and 'I am a hard worker' rated from 1 (not like me at all) to 5 (very much like me) or vice versa (for reverse scored items). The researchers reported results of 38 students in the experimental group and 12 students in the control group as these were the consistent participants in both the pre- and post-test. This assessment aimed to determine the program's impact on students via the development of grit, science motivation, and perceived growth attributed to the STEMGrow program.

To assess students' perceived growth due to the intervention, an ad-hoc survey was used with the question "So far, how have you grown through the work you've done with EduGuide?" The response items included a list of 14 impact areas (e.g., I'm more self-motivated, enjoy learning more, manage stress better, and get over setbacks quicker). For each impact area, students could select no growth, little/slight growth, moderate growth, considerable growth, or significant/very considerable growth.

The researchers also collected qualitative responses to an open-ended question: *In your own words, how has your work with EduGuide helped you so far this year?* Students responded to this question as they saw fit.

Findings – Golding et al. (2018) presented the results of Short Grit Scale scores, which indicated improvement in average grit levels within students' first semester. In contrast, the control group experienced a decline in grit scores. Regarding student perceptions of impact, the researchers combined the percentages of students who responded with considerable and very considerable growth in all impact areas. The results showed that the highest percentage of students reported considerable' to 'very considerable' growth on 13 of the 14 items including 'get over setbacks quicker' (an item important to the grit framework that guided the study). Moreover, the open-ended question elicited many positive responses from students including "It has helped me realize that anyone can do or become anything, all we need is more commitment" and "I have improved by not focusing on setbacks."

Relevance to this chapter – This case provides one example of how a specific character trait can be assessed in an engineering context using both quantitative (self-report surveys) and qualitative approaches (open-ended responses). Nevertheless, it would have been more helpful to see the researchers code the free responses based on the grit framework they used. For researchers wanting to conduct research on grit, we recommend consulting the systematic literature review on grit in engineering education by Direito et al. (2021), which provides recommendations for reporting.

Case 6: Assessing Malaysian engineering students' attitudes towards socio-ethical issues using survey questionnaires and interviews.

Author 2 led this research study, which was published in a book chapter (Balakrishnan, 2015).

Assessment measures – Balakrishnan (2015) used a mixed-methods approach involving a survey questionnaire and an interview process to assess students' attitudes towards socio-ethical issues.

Context – Balakrishnan (2015) focused on the students' attitudes regarding socio-ethical issues in engineering in general, rather than the understanding of the content of the subject-related socio-ethical education. A survey questionnaire was given to 43 final-year engineering students at University C in Malaysia. The questionnaire contained statements designed to evaluate the students' attitudes towards socio-ethical engineering issues. The statements were adopted from studies by Lathem et al. (2011) and Balakrishnan et al. (2013) but were modified to align with the needs of this study; the modified questionnaire can be found via Balakrishnan (2015). The study participants came from various engineering disciplines. All were in their final semesters before graduation and had completed a compulsory engineering ethics course called 'Engineers in Society.' That course covers socio-ethical issues, environmental impacts, and economic implications of the engineering profession, and its goal is to introduce students to the history of science and technology, issues concerning the impact of technology on economic development and the environment, engineering issues in the Malaysian context, the engineering profession, the code of ethics, and professionalism.

A five-point Likert-type scale (5 for *strongly agree* and 1 for *strongly disagree*) was used in the questionnaire to rank the level of respondents' agreement and disagreement with given statements. Three experts in engineering ethics, who each had more than 10 years of experience, helped validate the questionnaire. The data collected were normally distributed and the reliability value for the collected data (Cronbach's Alpha) was calculated to be a = 0.824. Thus, the data collected for this study is considered reliable and within the accepted range.

Of the 43 survey participants, 20 were randomly chosen to interview regarding their perspectives on (a) engineering ethics education and (b) their roles and responsibilities as future engineers; the students' interviews were used to support and confirm the study's findings.

Findings – The mean values recorded for the questionnaire ranged from 1.95 to 3.12, with a standard deviation ranging from 0.109 to 0.391. According to the mean value scores, the respondents had low mean scores in response to all the statements in the questionnaire that assessed their attitude towards socio-ethical issues in engineering. The statement that measured belief towards sustainability issues in engineering design or projects had the lowest mean score (1.95), indicating that awareness and knowledge of sustainable development in engineering were not emphasized in the program. According to the interviews, most participants took engineering ethics without realizing the importance of socio-ethical issues in the engineering profession. Moreover, there was a lack of socio-ethical-related activities conducted throughout their program of study. Some interviewees commented on the lecturers' lack of expertise in teaching engineering ethics, particularly in relation to socio-ethical issues and recent technological developments in engineering.

Relevance to this chapter – Assessing attitude can be done quantitatively and qualitatively – it largely depends on the context of attitudes one would like to assess. Measuring attitudes based on knowledge is appropriate with a quantitative method, while attitudes based on skills and values are better determined via a qualitative approach (Thurstone, 1929).

In this case, both the questionnaire and interview questions were used to measure respondents' attitudes toward ethics. The method enabled the researchers to understand how (in terms of pedagogy and content) the ethics course shaped students' attitudes. Balakrishnan's (2015) study indicated that the knowledge and skills students acquire in the ethics classroom mold their attitude. The report by Balakrishnan (2015) provides further details about the methodology and statistical analysis procedures used, and reviewing these can be helpful to readers wanting to assess the credibility of the findings, conduct follow-up or replication studies, and conduct systematic reviews or meta-analyses once the body of knowledge on this topic has grown.

Case 7: Engineering student perceptions (self-assessment) of ethics-focused learning goal attainment

Author 2 led this research study, which was published in a journal (Balakrishnan et al., 2019).

Assessment measures – Perception can influence an individual's character development by shaping their beliefs, behaviors, and values (Berry & McArthur, 1986). This case's researchers, Balakrishnan et al. (2019), iteratively developed and implemented a self-report questionnaire to assess students' perceptions of ethics instruction, based on King's (1981) theory of goal attainment and Newberry's (2004) nine core engineering ethical objectives. The instrument is available for those wishing to review or use it (see Balakrishnan et al., 2019).

King (1981) proposed that "the theory of goal attainment explains how individuals grow and develop throughout their lifespan and experience changes in structure and function of their bodies over time, which influence their perception of self through gaining knowledge via learning" (p. 19). King's theory can be used to analyze students' perceptions of attainment of major objectives – as this study aimed to do within engineering ethics education.

The nine core objectives of engineering ethics education, identified by Newberry (2004), address emotional engagement, intellectual engagement, and specific knowledge. *Emotional involvement* helps meet students' affective needs while developing their abilities to handle ethical concerns. *Intellectual engagement* strengthens students' critical thinking skills for confronting difficult ethical concerns. *Specific knowledge* deals with students' understanding of ethical norms, principles, and cases of ethical standards. The nine core objectives nest within these broader categories. (Newberry's objectives have also been used in Chapter 33 to map ethics education and accreditation contextually, nationally, and internationally.)

In developing the questionnaire to use for this study, Balakrishnan et al. (2019) drew from and adapted perception statements about pertinent ethical issues from Balakrishnan et al. (2013), which were based on Newberry's (2004) specified aims. They used the updated instrument to assess students' perceptions of achieving the goals; respondents were asked to rank each questionnaire statement using a five-point Likert scale ranging from 5 (*strongly agree*) to 1 (*strongly disagree*).

Context – The researchers aimed to assess the effectiveness of engineering ethics education by querying undergraduate students regarding their perceptions of their own attainment of educational objectives related to engineering ethics. The researchers focused on Japanese and Malaysian engineering students' evaluations of attaining the objectives and their attitudes toward socio-ethical topics. They used the questionnaire they had developed (published in Balakrishnan et al., 2019) to collect data at two institutions using two different educational approaches and draw comparisons. A random sampling procedure was used to choose two groups of respondents: 163 engineering students from a university in Japan and 108 engineering students from a university in Malaysia. All the respondents were in their fourth year and had completed all the relevant engineering ethics courses. Leading support to the accuracy and appropriateness of the measures, three engineering ethics education experts verified the questionnaire's validity. Moreover, the Cronbach Alpha value was identified as 0.874, indicating the instrument can be considered reliable.

Findings – In the study, the mean scores and standard deviations (related to perceptions of attaining nine key ethics education objectives) of the Japanese institution's undergraduate students were high. In contrast, the mean scores of data collected at the Malaysian institution reflected medium mean scores. Thus, the researchers reported that the impact of engineering ethics education on undergraduates' positive perceptions of attainment was significantly greater at the university in Japan than at the university in Malaysia.

The authors attributed the disparity in students' perceptions of attainment, in part, to the different teaching mechanisms or approaches to ethics education used in the respective institutions. The data suggested that experiences at the university in Japan – where the overall pedagogical mechanism combines both traditional and interactive case-based instructions – positively impacted students' perception of the attainment of key objectives of ethics education and attitude toward engineering socio-ethical issues. In contrast, the university providing data in Malaysia utilizes only traditional (i.e., teacher-centered) approaches in engineering ethics education, and the measured impact of students was notably lower. Other plausible explanations for the Japanese undergraduates reporting higher attainment of objectives and positive attitude include that their university systematically employs ethics education that spans all years of engineering study (from the first year to the final year). A longer learning span can provide many opportunities for undergraduate students to learn and practice ethics throughout their studies, the authors noted.

Balakrishnan et al. (2019) asserted that their findings were consistent with Barak and Green (2020) that identified interactive case-based learning as an opportunity for meaningful learning that can help students appreciate the importance of ethics in the engineering profession. It is important to recognize that the attainment of engineering ethics education must be built on a personal level, integrating the knowledge that students have received in the classroom with their personal convictions about engineering ethics (Balakrishnan et al., 2019).

Relevance to this chapter – Gauging how well engineering programs are meeting ethics education objectives and developing students' positive attitudes toward socio-ethical issues is important. Such assessments can provide educators in program development, helping them understand the degree to which their ethics teaching efforts are informing students' ethical values. This case illustrates a way to gauge such and, importantly, describes an approach to assessing student attitudes beyond a single instructional experience. By sampling fourth-year students, the investigation explored the impact of systematic integration of ethics engineering across multiple courses in a program and compared effects with an institution not using an integrated approach.

Assessment of pedagogical methods supporting ethical attitudes and character cultivation

Case 8: Mixed-method assessment of pedagogies and character cultivation (program level)

Authors 1 and 3 worked on this study. The text below explains the program, which we feel will provide a helpful example to readers, and the empirical research done to support it. The description draws from a report published in conference proceedings (Koehler et al., 2023) and our own personal experience with the program.

Assessment measure – In this case, Koehler et al. (2023) developed a virtue-guided questionnaire (as reported in their 2023 paper) to assess the impact of program-wide pedagogical approaches to cultivating character, and administered it across an entire cohort of engineering students (from first year students to senior, i.e. fourth-year/final-year, students). The survey included a list of 15 virtues with simple definitions and asked students to identify the character virtues they developed in the course and what aspects of the course led to that growth.

Context of the educational program – Wake Forest University launched a new engineering program in 2017, which (1) provided an interdisciplinary Bachelor of Science engineering degree, (2) leveraged aspects of liberal arts education, and (3) reimagined engineering education (Pierrakos, 2024). The vision and mission of Wake Forest Engineering was to "educate the whole engineer with a commitment to humanity" by integrating entrepreneurial mindset and ethical character within the fundamental engineering education competencies. Activities geared toward student development were infused across the curriculum with buy-in from nearly all engineering faculty members (permanent, visiting, and adjuncts). The Wake Forest Program for Character and Leadership introduced the engineering faculty to seven strategies for character development (Lamb et al., 2021). Under the leadership of the third author (who was serving as Founding Chair of Wake Forest Engineering and Principal Investigator of a Kern Family Foundation KEEN grant to infuse character and entrepreneurial mindset across the curriculum), ethical theories from the lens of philosophy, psychology, and professional education were interwoven within the Wake Forest Engineering curriculum. The Wake Forest Engineering faculty were given autonomy to create modules around virtues that resonated with them for reasons such as connection to the overall subject or specific topics relevant to their course, relevance to the modalities of their course (e.g., working in teams), and personal interests.

The curriculum design team was guided by a framework developed by the Jubilee Center for Character and Virtues (2022) in an effort to integrate a diverse set of virtues across the curriculum. The framework targets specific and desirable forms of development – intellectual (e.g., critical thinking, creativity, curiosity), performance (e.g., resilience, teamwork, zest), moral (e.g., honesty, courage, humility), and civic (e.g., empathy, service, purpose). Some of the engineering faculty members collaborated with experts from other disciplines (e.g., religious studies, history, psychology, anthropology) to support the implementation of their character-based modules (Gross et al., 2021; Henslee et al., 2021a, 2021b; Hitt et al., 2023; Kenny et al., 2021; Koehler et al., 2023; Pierrakos et al., 2023, 2024). More than 12 virtue modules were developed and integrated across 14 engineering courses that students experience across 4 years.

Context of the research/assessment – To assess the impact of these modules on students' development and understand the aspects of the curriculum that contributed to those impacts, the research team (Koehler et al., 2023) conducted a mixed-methods study focused on measuring students' perceptions of their learning gains. They distributed a survey with fixed and open-ended questions to all students across all four years of the curriculum. To maximize participation, the survey items were added to the regular end-of-semester course evaluation surveys. In all, 161 students (nearly all students enrolled in the program) completed the survey.

Findings – As reported by Koehler et al. (2023), student responses provided insight into specific pedagogies that were effective in character cultivation. Students attributed perceived character growth in courses where virtue modules were embedded as well as courses where virtue modules were not embedded (based on, e.g., engineering instructors serving as role models of character). Pedagogies like project-, team-, and mastery-based learning were highly rated by students as supportive of their perceived growth in ethical attitudes and character cultivation. The most prominent virtues cultivated were performance and intellectual virtues (e.g., teamwork, critical thinking, resilience, creativity, curiosity, humility), followed by moral and civic virtues (e.g., purpose, honesty, courage, empathy, service, and justice). The authors reported that context, faculty role models, and pedagogy matter in cultivating character.

Relevance to this chapter – Assessment of pedagogy is as important as assessment of student learning and student growth. The educational program and the assessment process used can serve as a model for others engaged in integrated ethics teaching.

Reflections: Strengths, gaps, and future directions

Assessment of (a) engineering ethics education models and pedagogical strategies and (b) individual character and attitudes related to ethics is challenging because it is complex and multifaceted. Multiple factors need to be considered. Considering research/assessment models, no broadly applicable approach exists yet to assess character and ethical attitudes in engineering education. Usually, researchers attempt to address a specific dimension of character or assess attitudes related to an ethics outcome or pedagogical approach.

Although *ad hoc* measures to assess students' development within specific courses/environments (as featured in cases 3 and 4) offer the advantage of ensuring that assessments are tailored to specific pedagogical interventions, designing them to a high level of quality can be time-consuming, and their context-specific nature makes them hardly generalizable. In many examples discussed above, there was no clear evidence that the researchers' *ad hoc* questionnaires had been validated. One solution could be that engineering educators try to use existing instruments, where validity and reliability have been previously established. Another could be inter-institutional collaboration to develop validated measures to increase their quality and generalizability.

For efficiency and feasibility, standardized tests and questionnaire instruments are often used, as featured in many of the cases above. Yet the use of standardized testing has received much criticism in the past decade or two. Their accuracy in measuring student learning and abilities has come into question.

In terms of assessing character traits, one popular validated measure is the 'Values in Action' (VIA) assessment instrument (Peterson & Seligman, 2004), created to be diagnostic and cross-sectional. The VIA was not designed to assess whether students are growing within an educational context, but it is nonetheless an available measurement tool worthy of consideration. There are shorter versions of the VIA available that are more user-friendly than the full version for academic settings, like the VIA 'Global Assessment of Character Strengths' (VIA GACS-72) (McGrath, 2019), which have been used in higher education character development programs (e.g., Lamb et al., 2021).

Although self-report measures (e.g., used in cases 6, 7, and 8) are the most common approach to attitudinal assessments, they have inherent limitations for understanding student learning gains and growth. One limitation involves the subjective nature of the subject. One limitation involves the subjective nature of the subject. Self-report measures can only gauge certain aspects of development and are not necessarily accurate. For instance, there is potential for a perceived social desirability to influence results (i.e., a 'halo effect') which could lead to a majority of participants reporting high scores (producing a 'ceiling effect') (Vogt & Johnson, 2011).

Alternatives to self-report measures exist, such as observations, focus groups, or student responses on written assignments. These can be coded for meaningful themes, but the process is very time-consuming. There is, therefore, a need to develop effective *and* efficient models for assessing development in the realm of ethics attitudes and character development.

We believe that situational judgment tests could present more authentic tools for assessing the character traits (e.g., resilience) needed to address real-life challenges (Teng et al., 2020) and might be worth exploring in the engineering ethics education space. A situational judgment test (SJT) is an assessment method used to evaluate an individual's ability to make appropriate decisions in work-related scenarios. SJTs present candidates with realistic workplace situations and ask them to choose the most effective course of action from a set of response options. These tests are designed to assess an individual's judgment, problem-solving skills, and decision-making abilities within the context of specific job roles or situations. Some researchers in engineering education have developed and are applying SJTs to assess global competency (the ability to understand and interact effectively in diverse cultural contexts) and meta-cognitive behavior (the awareness and control of one's own thought processes) (Carthy, 2021; Davis et al., 2023; Jesiek et al., 2020). This work could be extended to understand students' responses in situations that challenge various character strengths or virtues as well as attitudes towards others or specific ethical situations.

Few efforts have been reported regarding *program-level* interventions and assessments related to character and attitudes in higher education ethics education (and so it is noteworthy that all eight cases above looked at program-level interventions). Since character and attitudes take time to develop (Gal et al., 1997; Lamb et al., 2021), *module-level* experiences encompassed within a semester (a few months) might not be enough to support character growth in students. We, the authors of this chapter, encourage future studies on the longitudinal impacts of multiple ethics education interventions within a 4-year engineering curriculum.

Gauging program-level impacts will, we believe, require measuring and evaluating students' development of positive ethical attitudes and character, and the levels of motivation within students to be ethical engineers. We think the interventions and assessment models should focus on improving outcomes regarding character and positive attitudes. Research in this realm should use change-sensitive measuring tools, although these are not yet common in the literature.

Meindl and Dykhuis (2022) proposed that the following three domains of existing change-sensitive metrics might translate to measuring character change: growth mindset and self-efficacy to change character (e.g., 'I know my character can be developed'; see Dweck, 2009; Han et al., 2017); motivation to be a person of character ('It is important for me to embody character'; e.g., Aquino & Reed, 2002); and knowledge of the mechanisms or skills that would promote character (e.g., knowledge checks by the student regarding their goal-setting strategies). Each of these domains has been shown to be relatively malleable, and there exist relatively good, validated measures for each (e.g., the examples provided above are examples of those metrics). Programs such as the United States Military Academy are attempting to shape their institutional character assessments around these domains and to determine the assessment instruments' validity in the context of character formation (Erbe et al., 2023). The type of tools discussed by Meindl and Dykhuis provide relatively new ways of conceptualizing character assessment but seem to help address short- and long-term change-sensitivity to ethics programming, tedium of assessment, and validity and reliability of measures. Such models and measures that do not necessarily exist in the realm of *assessment of character* but do exist in the realm of *attitude evaluation* and other psychological constructs.

Conclusions

As a community of educators, we need effective approaches to assess character development and ethical attitudes in the engineering classroom setting. Situational judgment tests, assessments involving change-sensitive measures, and 360-degree performance evaluations are promising directions, but they require further research. We would like to see more mixed-method and multi-method approaches that promote more corroborated evidence of student development in ethics and character-based outcomes.

Crucially, we argue it is vital that assessments keep students at the center. Many of the cases we discussed were focused on determining whether the reported interventions 'worked', but only a few – notably case 3 (Hawks & Terry, 2009) and case 4 (Gross et al., 2021) – provided formative feedback to students that they could implement and improve upon. If the goal of ethics education is to support students' holistic development and ethical decision-making processes, we need to prioritize how we might use assessments for ongoing learning and growth, especially in engineering course contexts.

At present, there are many ways in which attitudes and character are being measured in engineering ethics education, although some methods are much more popular and simpler to implement than others. It is much more common to see methods, especially self-reports utilized to determine successful engineering ethics education implementation, but there is much to be desired when looking at the internalization of ethical attitudes and character by students. As noted above, only a few institutions take a program-level approach to assessing character and attitudes – but character and attitudes take time to develop, which the confines of single-course or even classroom (i.e., formal in-class) interventions may not be sufficient to address and assess. Nevertheless, we assert that growth within students as a result of formal class interventions really must be assessed, and well-designed character-based pedagogies and findings of prior research can help inform new interventions and situated assessments.

Acknowledgment

The work reported regarding Wake Forest University was supported by a Kern Family Foundation (KFF) Kern Engineering Entrepreneurial Network (KEEN) grant.

References

Ajzen, I. (1985). From intentions to actions: A theory of planned behavior. In J. Kuhl & J. Beckmann (Eds), *Action control. SSSP Springer series in social psychology*. Springer. https://doi.org/10.1007/978-3-642-69746-3_2

Ajzen, I., & Fishbein, M. (2004). The influence of attitudes on behavior. In D. Albarracín, B. T. Johnson, & M. P. Zanna (Eds.), *The handbook of attitudes* (pp. 173–221). Lawrence Erlbaum Associates.

Aquino, K., & Americus, R. (2002). The self-importance of moral identity. *Journal of Personality and Social Psychology*, *83*(6), 1423–1440. https://doi.org/10.1037/0022-3514.83.6.1423

Baehr, J. (2017). The varieties of character and some implications for character education. *Journal of Youth and Adolescence*, *46*(6), 1153–1161. https://doi.org/10.1007/s10964-017-0654-z

Balakrishnan, B. (2015). Engineering ethics education: Issues and student attitudes. In *Contemporary ethical issues in engineering* (pp. 133–143). IGI Global. https://doi.org/10.4018/978-1-4666-9624-2.ch061

Balakrishnan, B., Er, P. H., & Visvanathan, P. (2013). Socio-ethical education in nanotechnology engineering programmes: A case study in Malaysia. *Science and Engineering Ethics*, *19*(3), 1341–1355.

Balakrishnan, B., Tochinai, F., & Kanemitsu, H. (2019). Engineering ethics education: A comparative study of Japan and Malaysia. *Science and Engineering Ethics*, *25*(4), 1069–1083.

Barak, M., & Green, G. (2020). Novice researchers' views about online ethics education and the instructional design components that may foster ethical practice. *Science and Engineering Ethics*, *26*, 1403–1421.

Berkowitz, M. W. (2012). *Understanding effective character education.* The Literacy and Numeracy Secretariat Capacity Building Series.

Berkowitz, M. W., & Bier, M. C. (2005). *What works in character education: A research-driven guide for educators*. Character Education Partnership.

Berry, D. S., & McArthur, L. Z. (1986). Perceiving character in faces: the impact of age-related craniofacial changes on social perception. *Psychological Bulletin*, *100*(1), 3–18.

Card, N. A. (2016). Methodological issues in measuring the development of character. *Journal of Moral Education*, *13*(2), 30–47.

Carthy, D. (2021). *Development of a situational judgement test and an assessment of its efficacy as a stimulus of metacognitive behaviour in engineering students* [Doctoral thesis, Technological University Dublin]. https://doi.org/10.21427/2DVM-FB16

Craig, S., & Hannum, K. (2006). Research update: 360-degree performance assessment consulting. *Psychology Journal: Practice and Research*, *58*(2), 117–122.

Davis, G., & Butkus, M. (2008). *A study in engineering and military ethics*. ASEE Annual Conference and Exposition, Conference Proceedings. https://doi.org/10.18260/1-2--3655

Davis, K. A., Jesiek, B. K., & Knight, D. B. (2023). Exploring scenario-based assessment of students' global engineering competency: Building evidence of validity of a China-based situational judgment test. *Journal of Engineering Education*, *112*(4), 1032–1055. https://doi.org/10.1002/jee.20552

Davis, M., & Feinerman, A. (2012). Assessing graduate student progress in engineering ethics. *Science and Engineering Ethics*, *18*(2), 351–367. https://doi.org/10.1007/s11948-010-9250-2

de la Riva de la Riva, G. A., Espinosa Fajardo, C. C., & Juárez Nájera, M. (2015). Sustainability in engineering education: An approach to reach significant learning and character skills. *Sustainability in Higher Education*, 97–125. https://doi.org/10.1016/b978-0-08-100367-1.00005-6

Dillon, H. E., Welch, J. M., Ralston, N., & Levison, R. D. (2020). *Students taking action on engineering ethics*. ASEE's Virtual Conference.

Direito, I., Chance, S., & Malik, M. (2021). The study of grit in engineering education research: A systematic literature review. *European Journal of Engineering Education*, *46*(2), 161–185.

Duckworth, A. L. (2016). *Grit: The power of passion and perseverance*. Scribner.

Duckworth, A. L., & Quinn, P. D. (2009). Development and validation of the Short Grit Scale (GRIT–S). *Journal of Personality Assessment*, *91*(2), 166–174. https://doi.org/10.1080/00223890802634290

Dweck, C. (2009) Mindset: Developing talent through a growth mindset. *Olympic Coach*, *21*(1), 4–7.

Eagly, A. H., & Chaiken, S. (1993). *The psychology of attitudes*. Harcourt Brace Jovanovich College Publishers.

Eagly, A. H., & Chaiken, S. (2007). The advantages of an inclusive definition of attitude. *Social Cognition, 25*(5), 582–602.

Erbe, R. G., Schnack, D., & Meindl, P. (2023). Creating a culture of character growth: Developing faculty character and competence at the United States Military Academy. *Journal of Character and Leadership Development, 10*(3), 43–51. https://doi.org/10.58315/jcld.v10.275

Fundamental of Engineering (FE) Exam. (2023). National Council of Examiners for Engineering and Surveying (NCEES). Retrieved June 2, 2023, from https://ncees.org/exams/fe-exam/

Gal, I., Ginsburg, L., & Schau, C. (1997). Monitoring attitudes and beliefs in statistics education. In I. Gal & J. B. Garfield (Eds.), *The assessment challenge in statistics education* (pp. 37–51). International Statistical Institute.

Gentile, M. C. (2010). *Giving voice to values: How to speak your mind when you know what's right*. Yale University Press.

Golding, P., Arreola, C., Pitcher, M. T., Fernandez-Pena, C., Geller, H. E., Andrade, G., Golding, D. E., Nevarez, H. E. L., Espinoza, P. A., Gomez, H., Hemmitt, H., & Stearns, M. (2018). *Growing character strengths across boundaries*. American Society for Engineering Education Annual Conference and Exposition.

Gomez, C., Taboada, H. A., & Espiritu, J. F. (2013). *To be green or not to be green? Ethical tools for sustainability engineering*. ASEE Annual Conference and Exposition, Conference Proceedings. https://doi.org/10.18260/1-2--22636

Gross, M. D., Wiinikka-Lydon, J., Lamb, M., Pierrakos, O., & Yeaman, A. (2021). *The virtues of teamwork: A course module to cultivate the virtuous team worker*. ASEE Annual Conference and Exposition, Conference Proceedings.

Han, H., Kim, J., Jeong, C., & Cohen, G. L. (2017). Attainable and relevant moral exemplars are more effective than extraordinary exemplars in promoting voluntary service engagement. *Frontiers in Psychology, 8*(MAR). https://doi.org/10.3389/fpsyg.2017.00283

Hawks, V., & Terry, R. (2009). *Teaching leadership principles to undergraduate engineering and technology students*. ASEE Annual Conference and Exposition, Conference Proceedings. https://doi.org/10.18260/1-2--5361

Henslee, E. A., Luthy, K., Crowe, W. N., & Gray, L. J. (2021a). *Lab every day!! Lab every day? *&%#ing Lab every day!? Examining student attitudes in a core engineering course using hands-on learning every day of class*. ASEE Annual Conference and Exposition, Conference Proceedings.

Henslee, E., Yeaman, A., Wiinikka-Lydon, J., Vasquez, M. A., & Pike, E. (2021b). *ETHICS-2021 special session 7: Integrating virtue ethics into STEM courses: ETHICS-2021 Session, Sunday October 31, 2021, 3:00 AM-4:30 PM ET*. Proceedings - 2021 IEEE International Symposium on Ethics in Engineering, Science and Technology: Engineering and Corporate Social Responsibility, ETHICS 2021. https://doi.org/10.1109/ETHICS53270.2021.9632711

Hitt, S. J., Banzaert, A., & Pierrakos, O. (2023). Educating the whole engineer by integrating engineering and the liberal arts. In *International handbook of engineering education research* (p. 457). http://dx.doi.org/10.4324/9781003287483-25

Jen, C. C., Helmus, T. S., & VanderLeest, S. H. (2012). *Teaching entrepreneurship through Virtues*. American Society for Engineering Education Annual Conference & Exposition.

Jesiek, B. K., Woo, S. E., Parrigon, S., & Porter, C. M. (2020). Development of a situational judgment test for global engineering competency. *Journal of Engineering Education, 109*(3), 470–490.

Jubilee Centre for Character and Virtues. (2022). *A framework for character education in schools*. 3rd ed. University of Birmingham, Research Report 978-0- 244-91301-4.

Kenny, M. C., Pierrakos, O., & O'Connell, M. (2021). *Infusing the liberal arts in first-year engineering: A module on history, professional identity, and courage*. ASEE Annual Conference.

King, I. M. (1981). *A theory for nursing: Systems, concepts, process*. Delmar.

Koehler, J., Pierrakos, O., & Yeaman, A. (2023, June). *Character development in the engineering classroom: An exploratory, mixed-methods investigation of student perspectives on cultivating character*. In 2023 ASEE Annual Conference & Exposition.

Kuncel, N., Tran, K., Zhang, S. H. (2020). Measuring student character: Modernizing predictors of academic success. In M. E. Oliveri & C. Wendler (Eds.), *Higher education admissions practices: An international perspective* (pp. 276–301). Cambridge University Press.

Lamb, M., Brant, J., & Brooks, E. (2021). How is virtue cultivated? Seven strategies for postgraduate character development. *Journal of Character Education, 17*(1), 81–108.

Lamprianou, I., & Athanasou, J. A. (2009). Assessment of attitude and behavior. In *A Teacher's guide to educational assessment* (pp. 253–265). Brill. https://doi.org/10.1163/9789087909147_015

Lathem, S. A., Neumann, M. D., & Hayden, N. (2011). The socially responsible engineer: Assessing student attitudes of roles and responsibilities. *Journal of Engineering Education*, *100*(3), 444–474.

Lee, J. E., Park, K. Y., & Kim, J. H. (2018). *Engineering ethics education: A meta-analysis*. Paper presented at the International Conference on Teaching, Assessment, and Learning for Engineering, Wollongong, Australia. https://doi.org/10.1109/TALE.2018.8615133

Martin, D. A., Conlon, E., & Bowe, B. (2021). A multi-level review of engineering ethics education: Towards a socio-technical orientation of engineering education for ethics. *Science and Engineering Ethics, 27*(5), Springer Netherlands. https://doi.org/10.1007/s11948-021-00333-6

McGrath, R. E. (2019). *Technical report: The VIA assessment suite for adults: Development and initial evaluation (rev. ed.)*. VIA Institute on Character.

Meindl, P., & Dykhuis, E. M. (2022, September). *Making character assessment far less terrible*. Jubilee Conference for Character and Virtues: Integrating Research: 10 Years of Impact. Oxford, UK. Paper.

Newberry, B. (2004). The dilemma of ethics in engineering education. *Science and Engineering Ethics*, *10*(2), 343–351.

Peterson, C., & Seligman, M. E. (2004). *Character strengths and virtues: A handbook and classification*. American Psychological Association.

Pierrakos, O. (2024). *Transforming engineering education is possible! A descriptive case study of reimagining engineering education and delivering a wake forest engineering student experience promoting inclusion, agency, holistic learning, and success*. 2024 Collaborative Network for Engineering and Computing Diversity (CoNECD) Conference, Crystal City, Virginia.

Pierrakos, O., Prentice, M., Silverglate, C., Lamb, M., Demaske, A., & Smout, R. (2019, October). Reimagining engineering ethics: From ethics education to character education. In *2019 IEEE Frontiers in Education Conference (FIE)* (pp. 1–9). IEEE, Covington, KY, USA.

Pierrakos, O., Yeaman, A., Luthy, K., & Gentile, M. (2023, October). Moral and civic virtues in engineering: Reimagining engineering ethics to cultivate virtuous engineers. In *2023 IEEE Frontiers in Education Conference* (FIE) (pp. 1–3). IEEE, Covington, KY, USA.

Seider, S., Jayawickreme, E., & Lerner, R. M. (2017). Theoretical and empirical bases of character development in adolescence: A view of the issues. *Journal of Youth and Adolescence*, *46*, 1149–1152.

Sethy, S. S. (2017). Undergraduate engineering students' attitudes and perceptions towards 'professional ethics' course: A case study of India. *European Journal of Engineering Education*, *42*(6), 987–999. https://doi.org/10.1080/03043797.2016.1243656

Teng, Y., Brannick, M. T., & Borman, W. C. (2020). Capturing resilience in context: Development and validation of a situational judgment test of resilience. *Human Performance, 33*(2–3), 74–103. https://doi.org/10.1080/08959285.2019.1709069

Thurstone, L. L. (1929). Theory of attitude measurement. *Psychological Review*, *36*(3), 222–241.

Van Tyne, N. C. T. (2020). *WIP: How should we decide? The application of ethical reasoning to decision making in difficult cases*. ASEE's Virtual Conference.

Vogt, W. P., & Johnson, B. (2011). *Dictionary of statistics & methodology: A nontechnical guide for the social sciences*. Sage.

Yeaman, A., Koehler, J., Pappas, J., & Pierrakos, O. (2022, October 8–11). *Scoping review of character development pedagogies in engineering education* [Paper presentation]. 2022 IEEE Frontiers in Education Conference (FIE), Uppsala, Sweden. https://doi.org/10.1109/FIE56618.2022.9962514

28

EMPLOYING *EPISTEMIC MICRO-PRACTICES* TO ASSESS PROGRESS AND BARRIERS IN ENGINEERING STUDENTS' ETHICS DEVELOPMENT

Siara Isaac and Ashley Shew

Introduction

Engineering is often characterized as problem-solving. When undergraduate engineering students conceptualize their role as 'problem solvers,' it creates an expectation that they should solve ethical problems encountered during their disciplinary problem-solving. Naïve epistemic beliefs, which are more present (Wise et al., 2004) and more persistent (Felder & Brent, 2004; King & Magun-Jackson, 2009; Marra et al., 2000) among engineering students than among students of many other disciplines, lead students to expect their problem-solving practices to identify single correct answers that hold 'true' across all contexts. Additionally, engineering students may react by separating ethics from engineering when ethical dilemmas challenge the assumed universality of a 'good' response (Lönngren, 2021). This chapter explores the interaction between ethical reasoning and epistemic cognition to encourage engineering educators and researchers to consider how epistemic cognition can hinder or accelerate students' ethical development in their courses and curricula. Specifically, we argue for an approach that attends to specific actions of problem-solving and reasoning, called *epistemic micro-practices*, that engineering students employ when considering ethical dilemmas.

The cognitive intuitions that people hold about knowledge are known as 'epistemic beliefs' and encompass our conceptions about how to identify correct answers, what sources of knowledge we determine to be reliable, and the role we perceive for ourselves in generating or validating knowledge. 'Epistemic cognition' describes the thinking that enacts these beliefs by focusing on people's decision-making activities, such as evaluating an answer's correctness. Epistemic beliefs are discipline- and context-dependent and usually more sophisticated in one's own specific fields of expertise (Hofer, 2000; Muis et al., 2006). Therefore, engineering students are highly likely to bring naïve epistemic approaches to their ethical decision-making, where they often have less expertise. Making ethically sound decisions may call for more sophisticated epistemic skills than some engineering students have available (Gainsburg, 2015; Isaac, 2021). This situ-

DOI: 10.4324/9781003464259-34

ation would negatively affect engineering graduates' ability to engage with ethical dilemmas in their work and is insufficient for the ethical leadership expected of them. This chapter explores the intriguing connection between ethical reasoning and epistemic cognition, with a particular focus on assessment.

Early models of ethical reasoning developed by Kohlberg (Colby et al., 1983) assumed that people made these crucial decisions intentionally and rationally. Models based on these approaches are well represented in engineering ethics education, such as moral argumentation skills, as described by van de Poel and Royakkers (2007). Later research has considered the role of emotion in ethics, including Jonathan Haidt's (2001) moral foundations work, which argued that we should consider how rapid emotional judgments influence logical problem-solving and decision-making. Although we, the authors of this chapter, do not seek to dismiss the importance of emotion in ethical decision-making, we argue that we should also consider how people's cognitive intuitions influence their ethical reasoning.

Engineers are currently designing technologies that will disrupt known ethical problems and create novel issues: new technologies introduce more nuance to existing ethical problems and generate new ethical problems. It is, therefore, imperative that engineering graduates are equipped with robust ethical reasoning skills and the epistemic sophistication to guide their own ethical judgments. Despite considerable efforts to improve the integration of ethics into engineering curricula (Watts et al., 2017), meta-analyses of the teaching of ethics have found low to moderate effect sizes for the development of the targeted skills (Antes et al., 2009; Watts et al., 2017). Although some studies have found engineering undergraduates' ethical perspectives develop during their university years (Clancy, 2020), others have found students' scores on measures of ethical reasoning and social responsibility either remain static or decline over the course of their education (Bielefeldt & Canney, 2016; Lönngren, 2021; Monzon et al., 2010; Tormey et al., 2015). This is a disappointing outcome because of the recognized importance of ethics for engineers and the effort invested in ethics education. Students in other fields have been observed to develop their ethical thinking more during their studies than engineering students (O'Flaherty & Gleeson, 2014). This chapter proposes that one overlooked obstacle to the development of engineering students' ethical reasoning may be the epistemic practices they employ.

This chapter reviews the assessment of students' ethical thinking from the perspective of epistemic practices by using the steps of moral argumentation described by van der Poel and Royakkers (2007). We also draw significantly from the work of the primary author, Siara Isaac, and her ongoing research around epistemic cognition (Isaac, 2021). Ashley Shew completes the author team and provides her expertise as a multiply-disabled scholar engaged in research about engineering approaches to disability technology, describing in her work how 'technoableism' features in many problem-solving approaches to disability (Shew, 2023). In this paper, the two of us explore how to assess students' ability to engage where solutionism fails. We begin by discussing epistemic beliefs and epistemic cognition in more detail, before returning to argue for the relevance of assessing ethical thinking from an *epistemic micro-practices* perspective.

From epistemic beliefs to epistemic cognition

Epistemology is a branch of philosophy concerned with the theory of knowledge, exploring questions such as: *Where does knowledge come from? How do we know something is true?* The field of personal epistemology launched by William Perry (1970) is now more commonly known by the term 'epistemic beliefs.' It examines people's personal conceptions of how we identify correct answers, how we determine if a source of knowledge is reliable, and our own roles in generating, parsing, and validating knowledge.

Perry (1970) conducted longitudinal interviews with male Harvard students about how they navigated decision-making in their lives, and he created a six-stage model. His model served as an entry point for teachers to a domain previously dominated by philosophical approaches; it continues to be the reference point for many despite significant limitations. Two major issues are (1) how to manage the apparent rigidity of the stages and (2) the fact that the empirical tools developed to identify and assess an individual's stage have not proven reliable. Perry's ideas, however, intrigued thousands of researchers and there has been significant research activity endeavoring to clarify concepts related to epistemic development. The many models produced by these efforts rarely offer simplicity or rival the overall coherence of Perry's model, while often introducing additional empirical inconsistencies. This section avoids documenting the labyrinthine development of this field. It provides only an overview sufficient to support our argument that models of epistemic cognition via *cognitive processes* are a more coherent approach than the *cognitive structures* approaches that focus on beliefs rather than actions.

Research efforts regarding epistemic beliefs have been dominated by *cognitive structures* approaches (Briell et al., 2011), which seek to characterize peoples' beliefs about knowledge and knowing. Stage-based models (i.e., Baxter Magolda, 1992; Perry, 1970) were superseded by more nuanced ones, such as Hofer and Pintrich's (1997) highly influential model. These nuanced models posit that individuals develop some aspects of their epistemic beliefs before others. Other disciplinary models have been developed, positing that people's approach to knowledge is discipline-specific (i.e., Muis, 2004; Palmer & Marra, 2004). *Cognitive processes* approaches focus on how people seek, apply, and evaluate knowledge and remain rare, despite excellent work by King and Kitchener (2004) and Kuhn and colleagues (2000). A fundamental distinction between *cognitive processes* and *cognitive structures* models is apparent in how they are assessed. Evaluation tools for *cognitive structures* models generally seek to directly access students' ideas about knowledge (i.e., their agreement with items like 'When I read it in an engineering book, then I know that it is true') based on the assumption students' responses will draw on a static, organized structure underpinning their beliefs. Alternately, *cognitive processes* models anticipate that students' responses will not be consistent across contexts because students respond to each specific knowledge claim or situation. Therefore, *cognitive processes* models employ more direct observations of students' thinking or reasoning activities using, for example, think-aloud protocols (Isaac, 2021). Although *cognitive processes* approaches do employ models to characterize students' epistemic perspectives (as explored in detail below), they are based on observations of students' thinking processes rather than asking students to report how they think. Both *cognitive structures* and *cognitive processes* models typically describe a developmental trajectory away from naïve approaches associated with absolutism (i.e., where knowledge is assumed to be either 'true' or 'false' and people rely on 'experts' to evaluate knowledge claims), and toward increased personal agency to make nuanced and contextual decisions. The relevance of more sophisticated epistemic approaches for ethical decision making is evident, where context and ambiguity reign, and absolute experts and absolute correct answers are scarce.

However, despite considerable and diverse efforts, developing a robust, empirically grounded model of epistemic beliefs long evaded researchers and resulted in a plethora of overlapping terminologies and models. Siara Isaac developed a new approach, arguing that we should avoid the hegemony of *cognitive structures* models in favor of *cognitive processes*. Although acknowledging the attractiveness of categorizing students in neat levels of epistemic development, Isaac argues that the significant effort expended in attempts to establish *cognitive structures* models of students' overarching beliefs about knowledge by inferring from a small number of their statements has not worked because these methods take an abstract, decontextualized, macro-level approach. Much like ethics, abstract ideas and 'should' statements about knowledge and problem-solving can differ significantly from actions and applications in context. Wanting to reconcile apparent contradictions in students' conceptions resulted in researchers attempting to assess discipline-specific epistemic beliefs, for example, in geology or history (e.g., Muis et al., 2006). Elby and Hammer (2001, 2010) critiqued these approaches as being insufficient and imprecise; they observed that while an epistemically sophisticated stance understands knowledge to be tentative and evolving, assuming this stance to each knowledge claim is not an appropriate measure of epistemic cognition. They argued instead that epistemic sophistication is demonstrated by selecting an appropriate or effective approach for the specific knowledge claim at issue. Their characterization of epistemic sophistication resolved persistent inconsistencies identified in multiple empirical studies that were often dismissed as inconsequential. For example, Zhu et al. (2019) dismiss their own observations of students who simultaneously expressed both epistemically naïve and advanced thinking as resulting from "transitional stages in their epistemological development" (p. 4) rather than taking the inconsistency as a prompt to reconsider their use of the Perry model. Given that the majority (about 75%) of their study participants exhibited this co-existence, this is a prime example of how persistent issues with prevailing *cognitive structures* models of epistemic beliefs have been ignored.

Isaac's empirically based *epistemic micro-practices* model indicates how *cognitive processes* approaches can capture the range of ways that people interact with knowledge; it avoids the apparent inconsistencies of *cognitive structures* models. Further, Isaac argues that the range of practices is a relevant observation about the quality of thinking and problem-solving. This diversity is lost in both *cognitive structures* and prior *cognitive processes* approaches when characterizations exclusively assess progress towards an advanced level of epistemic development.

Let's take a civil engineering example to show how inconsistencies a researcher might perceive in a student's epistemic approach can represent coherent disciplinary practice instead. In a hydroelectric project, it is likely appropriate (for a student or a practicing engineer) to simply apply a model to estimate or predict the corrosion of iron. Yet, they should also question and test that the models for continuous flow do apply. In this scenario, there is likely little to challenge the application of the molecular level corrosion model, but it is appropriate to verify that the assumptions of the continuous flow model are not violated. So, in the same situation, effective problem-solving could involve a student naïvely accepting one model as received (e.g., from the textbook or the teacher) while subjecting another model to *epistemic micro-practices* from more facets. This example of context-specificity explains measurement difficulties in *cognitive structure* approaches – using items such as "Engineering knowledge should be accepted as an unquestionable truth" (from Yu & Strobel, 2011, p. 100) – as we do not know the context the students had in mind when answering. A researcher would need to provide more context to derive accurate interpretations from a student's responses.

Elby and Hammer (2010) claimed that considering epistemic beliefs at the level of the discipline is insufficient and that researchers should take a fine-grained, contextual approach. They used a *cognitive processes* approach to demonstrate that the long-held characterization (i.e., epistemic sophistication as the consistent use of epistemically sophisticated actions) was false (Hammer & Elby, 2003). They proposed that the effectiveness of a given approach in context is a better way to assess students' epistemic skills.

Building on this, Isaac proposes taking a firmly *cognitive processes* approach in her model for classifying students' interactions with knowledge during problem-solving. Her approach considers four facets (as illustrated in Figure 28.1), and her term for specific actions that enact an epistemic perspective is *epistemic micro-practices*. Three complementary examples of *epistemic micro-practices*, each employing a different facet, related to checking answers are (1) asking a teacher, (2) using dimensional analysis, and (3) considering a real-life application. Isaac does not accord hierarchical importance to these different *epistemic micro-practices*, noting that although some *epistemic micro-practices* are more accessible or appropriate in a given context, the use of *micro-practices* from several different facets will generally provide a more robust result.

The first facet of Isaac's four-part model resembles naïve levels in previous models (i.e., Hofer & Pintrich, 1997; Perry, 1970), with the important distinction of being characterized by specific (micro)*practices* and not describing overarching *beliefs* about knowledge. *Epistemic micro-practices* belonging to Facet 1, Absolute, seek single, exact answers considering only the core objective or task. Facet 1 *epistemic micro-practices* function as though a 'right' answer will be 'true' across all contexts. Facet 2, Local Coherence, *micro-practices* seek to establish logical connections and explanations relying only on information internal to the specific context. The existence of different solutions is dismissed by Facet 2 *micro-practices* as the result of different but equally valid approaches, thereby avoiding the need to pass judgment. *Epistemic micro-practices* in Facet 3, Coherence, recognize that the situation, constraints, and criteria for this problem must also be considered in the broader context. This is the type of thinking that one would hope to stimulate with case studies, the most widely used way of teaching ethics (Bairaktarova & Woodcock, 2017; see Chapter 20). *Epistemic micro-practices* in Facet 4, Skeptical Reverence, involve exercising engineering judgment where the student assumes the authority to determine the best answer in the current context.

Although the specific examples of *micro-practices* regarding correct answers presented in Figure 28.1 were developed from observations of students' disciplinary problem-solving, they can be readily applied to ethical reasoning. For Facet 1, we can see that the expectation of the existence of a single correct answer known to experts is coherent with a reliance on ethical codes and rules or norms-based reasoning, as shown in Rest's (1994) model. The two approaches intersect again for Facet 4, where attention to the specific context and necessity of making judgments when applying principles to determine ethically appropriate actions are coherent with Rest's principle-based reasoning. In the following section, we apply an *epistemic micro-practices* model to consider how students may approach the cognitive activities of moral argumentation.

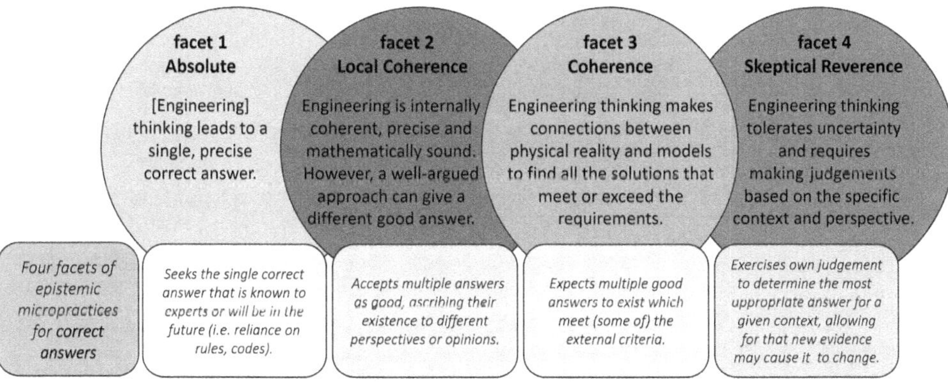

Figure 28.1 Definitions of the 4 facets of *epistemic micro-practices* with example *micro-practices* regarding correct answers.

Epistemic micro-practices *analysis of moral argumentation skills*

Van de Poel and Royakkers' (2007) model of moral argumentation skills is well suited to our purpose of illustrating how *epistemic micro-practices* inform ethical reasoning. Their model was developed to teach how to make good ethical decisions and not to describe how people actually make decisions. That is, the authors sought to elucidate the cognitive steps in a robust process for making ethical decisions. As such, their model allows us to show how students' cognitive intuitions (enacted via their *epistemic micro-practices*) influence how they approach and resolve the challenges of making ethical decisions. We have represented the five core steps of their model of moral reasoning in Figure 28.2 as a linear path for simplicity rather than to the iterative, recursive process proposed by van de Poel and Royakkers. For each step in their model, we applied Isaac's (2021) model to identify both Facet 1 (left side of the figure) and Facet 4 *epistemic micro-practices* (right side of the figure).

The formulation of a moral problem statement is the first step in van de Poel and Royakkers' (2007) model. The necessity of this initial step of formulating the ethical issue was highlighted by the update to the UK-based Engineering Council's Statement of Ethical Principles (Institution of Civil Engineers, 2021). The previous version simply stated "exercise responsibilities in an ethical manner" (Institution of Civil Engineers, 2021, p. 11), while the new text makes it the responsibility of the engineer to be attentive and aware to the potential ethical issues might arise by prefacing the previous text with "Understand the ethical issues that may arise in their role and ...." Taking an *epistemic micro-practices* perspective on this first step of problem formulation, we can see that multiple facets would be involved in a thorough exploration of moral sensitivity. Facet 1 *epistemic micro-practices* could involve formulating moral problem statements, when prompted, that reflect the issues identified by experts. This is an appropriate action. However, students could also use Facet 4 *epistemic micro-practices* in complementary and parallel ways to determine for themselves what ethical issues are present. These two *epistemic micro-practices* are distinguished by the agency of the engineer themselves: Do they consider it their role to identify ethical issues, or do they rely on experts to make this decision? Below, we illustrate these two approaches with the case of software design.

Ivan Selenkzy (Székely et al., 2011) found that while IT professionals typically complied with privacy or ethical standards, they did so because this requirement was explicitly stated in the project criteria provided by their employers and not due to the engineers' own agency. Cécile Hardebolle et al. (2023) seek to support a more sophisticated approach in terms of the use of edu-

Figure 28.2 Examples of *epistemic micro-practices* from 2 facets for each step of the 2007 moral argumentation skills model by van der Poel and Royakkers (labeled 'vdP+R').

cational software by teachers through the application of their template (or 'canvas' as mentioned in Chapters 7 and 18) to analyze potential risks. Assuming that the essential underpinning for an engineer to enact ethical sensitivity is the *awareness* of the potential for ethical issues overlooks the need for the engineer to have the *epistemic sophistication* to accord themselves the authority to judge that such issues are indeed potentially present. That an engineer needs to be able to act in epistemically sophisticated ways in interaction with their ethical sensitivity illustrates the central thesis of this chapter. If engineers do not conceive of identifying ethical issues to be within their (respons)abilities but rather expect that an expert will make such a judgment, then they will not be able to enact ethical sensitivity in practice. This epistemic cognition perspective should be investigated to assess potential obstacles to engineering students applying their ethical reasoning.

The second step in the model (Figure 28.2) is problem analysis, a step in which the relevant elements of the moral problem are described, including "the stakeholders and their interests, the moral values that are relevant in the situation and the relevant facts" (van de Poel & Royakkers, 2007, p. 4). Once again taking the extreme *epistemic micro-practices* to illustrate their influence, Facet 1 micro-practices could involve directly applying project criteria or a specific model in the analysis and considering such elements as constant and exact. In contrast, *epistemic micro-practices* from a broader range of facets would also leverage the students' own judgment to determine which criteria, models, and conditions are salient while accounting for available information necessarily being incomplete, subject to change, and potentially biased.

Fundamental physical sciences and engineering courses tend to present students with single-answer problems for which highly precise answers can be determined. The result is that engineering students encounter ill-structured problems insufficiently often during their studies (McNeill et al., 2016) and can form the idea that good engineering thinking should produce single exact solutions. Ill-structured problems make engineering students uncomfortable as they require sophisticated epistemic practices (Isaac, 2021). Research about how engineering students perceive and engage with ill-structured problems, in general, is salient here, as naïve epistemic practices collide with the uncertainty, open-mindedness, and judgment required for solving ill-structured problems, including ethical ones (Bendixen et al., 1998). Especially in contexts where political issues or the values of the community (and even implicit judgments determining who is acknowledged as a member of the community or stakeholder) are involved, students can be ill-prepared for the more complex discussion and nuance in considering what counts as a problem and what might count as a solution. Performance indicators for the technical aspects of an engineering product can typically be specified in the project assignment, directly measured in the solution, and, therefore, unambiguously assessed.

In contrast, ethical concerns often involve extended networks of stakeholders and longer-term impacts which take specifying and assessing relevant performance indicators beyond the skill set of many engineers. This means engineers may relegate ethical concerns to lower importance than technical concerns, employing *epistemic micro-practices* that prioritize immediately quantifiable outcomes of technical performance over the more nuanced outcomes of ethical considerations. Thus, *epistemic micro-practices* that seek absolute, specific correct answers reduce engineers' engagement with ethical concerns.

The third step in van de Poel and Royakkers' (2007) model involves exploring options for action. Facet 1 *epistemic micro-practices* are directed towards identifying a single correct solution, while other facets would also evaluate the advantages, disadvantages, and differences of multiple potential solutions. The anticipated number and 'absoluteness' of correct answers is a central pillar in models of epistemic cognition. Steps 4 and 5 of the moral reasoning model focus on evaluating options and assessing outcomes; these high-level cognitive activities sharpen the contrast between

Facet 1 *micro-practices* (that rely on external expertise) and Facet 4 *micro-practices*. Evaluation and judgment are actions associated with Facet 4 *epistemic micro-practices,* where engineers must take responsibility for exercising their own judgment based on the specific context. Although such practices are essential to professional engineering work (Gainsburg, 2007), they are infrequently observed among engineering students working on disciplinary problems (Gainsburg, 2015; Isaac, 2021). Further, in the context of education, teachers consistently hold the role of exercising judgment (as enacted by the assigning of grades) and, therefore, students often do not enact these sophisticated epistemic *micro-practices*. Thus, it is likely that many ethics courses do not challenge engineering students to employ the full range of *epistemic micro-practices* for moral judgment (Step 4) and moral decision-making (Step 5).

In the preceding analysis, we omitted intermediate *epistemic micro-practices* and provided only examples of Facet 1 and Facet 4 *epistemic micro-practices*. Well-documented intermediate *micro-practices* – between the dichotomous certainty of absolutism and the contextual application of principle-based reasoning – recognize multiple ways of reasoning and, therefore, multiple outcomes without applying a critical perspective to these approaches or outcomes. The result is that differences in, for example, ethical reasoning are ascribed to differences in opinion and taken to be equally valid. This phenomenon is coherent with Haws' (2001) review of the proceedings of American Society for Engineering Education, wherein she identified students' acceptance of 'ethical relativity' (i.e., belief that everyone's ethical 'opinion' is of equal value) as a major and persistent issue. An exclusively naïve approach is clearly ill-suited to real life engineering problems, and it likely also impedes students from developing their capacities as ethical practitioners.

This is an instance where approaches to teaching different learning outcomes stated by the Accreditation Board for Engineering and Technology (ABET) such as 'Criteria 3: Student Outcomes' exist in tension with one another. Teaching toward Outcome 1, "an ability to identify, formulate, and solve complex engineering problems by applying principles of engineering, science, and mathematics" (ABET, 2023, p. 6) looks very different from Outcome 4, "an ability to recognize ethical and professional responsibilities in engineering situations and make informed judgments, which must consider the impact of engineering solutions in global, economic, environmental, and societal contexts" (p. 6). Both outcomes are important to engineering practice, but the act of identifying and formulating a problem through engineering, science, and math will conflict with approaches to problem identification that start with community-based needs. Often, in the world of disability technology, for instance, every disabled person's body or mind is seen as calling out to the 'humanitarian' engineer for intervention (Shew, 2023). Engineering courses that aim to convey a societal edge or 'technology for good' approach often center their projects on a disability case study, asking students to prototype, for example, a prosthetic hand. This is often done in the absence of any engagement with hand or arm amputees, who may have more complicated reasons for their choices and desires. Their concerns include available technologies, financial access, access to repair, addressing stigma and people who stare, pain and functionality, and more (Shew, 2023). Simply devising a hand device that meets a professor's specs does not engage with people's lived experiences of disability – which may differ greatly between people and in different cultural environments. Here, the format of education matters. Even with case-based ethics teaching (a prevalent approach), students can remain in a naïve position where the expert/teacher assigns a case-based problem (like they might for a typical problem assignment to solve for x). 'The answer' in these cases is known to the expert/teacher who grades students' attempts to analyze and decide or offer ethical judgment on a case. Having solid answers that can be judged in the binary of right or wrong certainly lends itself to expedience in grading and assessment but does not necessarily change or expand students' universe of moral considerations or make them face deeply held cultural assumptions that may be inaccurate. When engineering students perceive engineering as problem-solving, it informs who they see as offering relevant and useful information, what they take as problems, and what they consider constitutes a good solution.

Epistemic micro-practices perspective on the assessment of ethics in engineering education

The epistemological approach of engineering as a discipline is a relevant consideration, as ethical reasoning must integrate and dialogue with the other types of reasoning occurring in engineering problem-solving. Further, many people teaching in engineering programs have been trained as engineers. As such, they may be unfamiliar with the social sciences' disciplinary traditions and epistemic practices regarding how knowledge is produced and evaluated and how students' work is assessed in social sciences. Another contextual issue, which has been raised in other chapters of this book, regards the influence of cultural factors on ethical reasoning both directly and, we argue, indirectly through epistemic practices.

Ethics is generally understood as a noun – something that can be described or absent. Ethical reasoning is the process of 'doing ethics,' of examining, formulating, and determining how to respond to a moral problem. Engineering programs aim to develop ethical reasoning skills in graduates to navigate future ethical issues as engineers and citizens. This poses an inherent challenge to the assessment approaches typically used in engineering programs, as these usually evaluate the product (i.e., the circuit designed) or the result (i.e., the value calculated) rather than the process employed to obtain the product or the result. While we would argue that technical courses should also attend more to *process* to foster higher order thinking skills and transfer, we will restrain ourselves to ethics education in this chapter. The attention afforded to *process* by *epistemic micro-practices*, in contrast to the prioritization of *epistemic structures* in belief models, further supports our argument in favor of *epistemic micro-practices*. In addition to the conceptual shift towards evaluating the thinking process, the logistical challenges of assessing a process are not insignificant, particularly at the large scale of engineering cohorts. This section examines different approaches to evaluating engineering students' ethical reasoning from an *epistemic micro-practices* perspective.

Statements from engineering schools and accreditation bodies indicate that engineering ethics education should ensure that graduates act as ethical professionals in their discipline. Making and assessing direct observations of recent graduates' *epistemic micro-practices* are beyond the scope of what engineering programs can do regularly. Indeed, given the high number of students studying in engineering programs, the constraints of large-scale administration, even within the university's walls, are considerable. These constraints apply particularly to assessments that require assigning grades/marks, as teachers must assume the workload yearly. In this section, we look first at assessments for grading ethical thinking and then at research tools from an *epistemic micro-practices* perspective.

Case studies are the most frequently used method for teaching ethics in engineering (Bairaktarova & Woodcock, 2017). Exploring situations where there is no obvious answer is a key element that motivates teachers to use case studies, but the same teachers report that their teaching with case studies does not align well with their ideas of how to best offer students a chance to explore ethical reasoning (Martin et al., 2021). Teachers often use 'thin' or simplified case studies due to logistical and organizational constraints, such as having large cohorts, which prevent them from creating more authentic or immersive learning experiences for their students. Tim Healy (1997) says that 'thin' ethics cases have pedagogical value for reducing the complexity that can overwhelm students new to the field. However, he cautions that such cases are over-used due to their ease in writing, discussing, and (although he doesn't explicitly say it) evaluating. It is a straightforward extrapolation to see how using case studies to assess students' ethical reasoning is subject to the same constraints identified for teaching with cases.

Monteiro et al. (2019) reviewed the teaching and assessment of ethics presented in the documents of Portugal's 33 degree programs; written exams were the most common assessment method described in all contexts. Yet, the documents studied did not provide information about when these written exams assessed ethical principles and codes and when the exams challenged students with ill-structured problems presented as complex cases. In apparent coherence with findings regarding 'thin' cases in engineering teaching, Monerio et al. estimated that engineering programs were roughly half as likely to assess the development of ethical arguments compared to other university programs and that engineering programs were highly unlikely to include debate and discussion in the assessment criteria.

In research contexts, the relevance and richness of the observations of qualitative approaches are well-suited to assessing the process of ethical reasoning. However, quantitative assessments are better suited to collecting hundreds of observations from large cohorts, and they have higher face value for faculty members who are more accustomed to quantitative methods. The ESIT (Engineering and Science version of the Defining Issues Test) is a popular tool used to evaluate students' ethical reasoning abilities. This test is similar to the Defining Issues Test (DIT), which is discussed in other chapters of the book and extensively in Chapter 10. However, the ESIT focuses on technical dilemmas in science and engineering, and is intended to measure moral judgment in this specific context (Borenstein et al., 2010). The instrument was developed via extensive qualitative studies (Borenstein et al., 2010), and it seeks to evaluate how students make their ethical decisions by the considerations or elements deemed germane to their decision-making. Indeed, while the cases of the ESIT ask students to make a dichotomous choice to take 'action A' or 'action B,' that decision is not considered in terms of evaluating students' *ethical reasoning*. That is, the *result* of the students' *ethical reasoning* (i.e., the choice of action A or B) is not the basis of the assessment. Rather, students' ethical reasoning is determined by which *elements* students identify as important to make their decision. This approach enables the quantitative ESIT instrument to access students' internal ethical reasoning processes. An effective quantitative instrument to assess epistemic cognition requires a similar capacity to access the reasoning process. Further, the short descriptions of the situations that make up the ESIT are classic 'thin' cases as they have little of the complexity found in real life. The current lack of a robust quantitative instrument for both epistemic beliefs and epistemic cognition models – including those assessing the ethics dimension – leads us to recommend using qualitative approaches.

Qualitative methodological approaches are well suited to explore the range of different perspectives and approaches students use to make their ethical decisions. The observations produced by these methodologies also provide a rich basis for theory generation toward resolving persistent issues identified in the second section of this chapter. Although interviews are a perennially popular qualitative method, the applied nature of engineering makes direct observations more salient. Interviews can provide information regarding students' thoughts about how they should or could make ethical decisions. However, humans' actions often deviate from the opinions or beliefs they claim to hold. Thus, methodologies that allow more direct observation of students' *epistemic micro-practices* are a better fit to advance the field. Games and think-aloud protocols are two types of experimental design that create relevant opportunities to observe contextualized reasoning and decision-making.

While games are a popular strategy to induce moral decision-making, their nature as synthetic environments can undermine their validity. That the consequences of the strategies employed by players in a game do not have real-world impact (i.e., actual environmental pollution) is visibly apparent to people participating in these studies (Reall et al., 1998). Both from the anecdotal experiences of the authors of this chapter and in studies that compared students' DIT scores to their game strategy (e.g., Reall et al., 1998), we suggest that students' actions in the patently artificial context of games are not coherent with their values.

A more effective way to evaluate students' ethical reasoning in practical situations is to observe them while they work on authentic tasks. A methodology well-suited to this type of observation is think-aloud protocols. Originating in the field of cognitive processing in psychology (Ericsson & Simon, 1998), this approach involves setting a task for participants to complete while narrating their actions and thoughts. Think-aloud protocols are highly effective for studying complex thinking processes (Olson et al., 2018), providing rich information about students' approaches to problem-solving and decision-making (Kuusela & Paul, 2000), and generating authentic, contextual observations about how students think. In their think-aloud study of a software design task, Isaac et al. (2023) found that computer science students' in-the-moment thinking tended to overlook privacy issues, minimize design criteria related to inclusion, and omit issues of sustainability. Isaac et al. interpret this not as students' lack of concern or knowledge about the ethical implications of these issues, but rather as students not thinking about incorporating these aspects into their disciplinary thinking. This illustrates the importance for engineering students to develop ethical reasoning and epistemic micro-practices to constructively with ill-structured and design problems.

Assessing how engineering students engage with ill-structured problems, ethical and otherwise

Robert C. Solomon wrote, "The aim of ethics … is not to teach the difference between right and wrong, but to make people comfortable facing moral complexity" (cited in McWilliams & Nahavandi, 2006, p. 421). Epistemic sophistication is exactly about engaging with complexity and conflicting claims, and therefore, well suited as a lens for evaluating engineering students' ethical reasoning. Indeed, Bendixen et al. (1998) found that epistemic factors play an essential role in solving ill-structured problems. Although this section should be a major component of this chapter (given its position in the handbook's assessment theme), robust methods are missing for evaluating epistemic cognition alone, let alone in conjunction with ethical reasoning. Yet, given our argument that teachers and researchers should be more attentive to *epistemic micro-practices* in engineering ethics education, this section will offer some suggestions.

The frustration of engineering students with uncertainty has been identified with both ill-structured disciplinary problems (McNeill et al., 2016) and ethical issues (Lönngren, 2021). It is important that students develop skills to engage with conflicting goals, interdisciplinarity, multiple solution methods, and unanticipated issues of ill-structured problems to prepare them for their professional lives (Jonassen et al., 2006). Discomfort with not having absolutely correct answers, as is typical of ill-structured problems, is a fundamental characteristic of low epistemic sophistication. Intermediate *epistemic micro-practices* that tolerate different answers or perspectives as simply 'matters of opinion' not subject to evaluation have been found to persist among engineering students (Felder & Brent, 2004; King & Magun-Jackson, 2009; Marra et al., 2000). Although the intermediate realm represents progress beyond the most epistemically naïve practices, the awareness of differences without the agency to evaluate different perspectives or solutions is inadequate for a professional engineer. The intermediate level is insufficient, and educators must lead students beyond this level of subjective thinking.

Engineering is often characterized as problem-solving, where technical and scientific knowledge is applied to create the best solution. Paired with training toward efficiency, students can feel immense pressure to produce the 'right' solution. The discomfort can lead students to avoid or dismiss ethics as a vital facet of engineering work. As Lönngren (2021) explains, dismissing ethics as part of the act of 'engineering' may be a self-protective mechanism to avoid the potential discomfort of examining the ethical implications of one's work. Therefore, developing students' *epistemic micro-practices* for working with uncertainty and exercising judgment could decrease this pressure and allow them to engage with ethical issues more productively. The cognitive focus of this chapter does not integrate the emotional dimensions of ethical reasoning identified in recent work (i.e., Haidt, 2001), work on ethics of care, and the themes of sustainability and equity – doing so would be a fascinating challenge that is, however, outside the scope of this chapter.

Teachers and researchers should consider the role of assessment methods and teaching strategies in perpetuating students' overuse of *epistemic micro-practices* from Facets 1 and 2. For example, case studies where the student is expected to identify the issues previously identified by the teacher in order to achieve a good grade offer little scope for students to employ *epistemic micro-practices* around managing uncertainty and imposing their judgment, which will be required for authentic, real-world problems. Counter-examples include teaching strategies and formative assessments that increase opportunities for teachers and students to discuss the process of ethical reasoning, thereby supporting students' adoption of a broader range of *epistemic micro-practices*. To borrow from the think-aloud observation protocols, instructors could ask students to explain how they accomplished a specific task and how they determined that it was a sufficient outcome. Depending on the objective, this could help evaluate the pedagogical potential of the task to elicit a desirable range of *epistemic micro-practices* or assess a student's reasoning skills. This is the approach of many oral examinations; however, student engineering projects are typically assessed via a report or presentation focusing on the result rather than the process.

In traditional format courses, assessment tasks with students working directly at the level of evaluating solutions may be a more direct way to promote sophisticated *epistemic micro-practices*. For example, students could be furnished with several solutions to an ethical problem (rather than generating the solutions) to prompt them to enact *epistemic micro-practices* from several facets to generate relevant criteria and apply their own judgment to assess the solutions. This would encourage students to become aware of recurrent criteria and underlying principles relevant to many situations. Of course, this does not address the fundamental conflict between exercising engineering judgment and having an expert/teacher assess your work to provide a specific numerical grade. It may be appropriate to remind ourselves of the level of epistemic cognition engineers typically have at graduation. Previous work (Gainsburg, 2015; Isaac, 2021; Pavelich & Moore, 1996; Wise et al., 2004) suggests that students rarely achieve an advanced level of epistemic cognition in their disciplinary thinking. Hence, it is likely not appropriate to expect them to be highly advanced in ethical reasoning as they start their professional journeys; they need openness in their mindset and appropriate *epistemic micro-practices* to continue their exploration beyond their time as undergraduates. Indeed, Baxter Magolda (1992) showed that although university students' epistemic cognition increased gradually through their studies, a more significant improvement occurred in the 2 years after graduation.

To encourage students to engage with ethical complexity, teachers should introduce activities and assessments that allow for multiple perspectives and solutions, starting from the first semester. Open-ended activities that engage creativity and increase students' experience with *epistemic micro-practices* to manage uncertainty and exercise their own judgment should be a goal of engineering ethics education. Problem-, project-, and challenge-based curricula, where students take a more active role in directing their own learning, offer excellent opportunities for such epistemic practices. Challenge-based learning (CBL) requires students to engage more with ambiguity and uncertainty through highly contextualized problems with multiple stakeholders (Bombaerts et al., 2021). Problem- and project-based learning (PBL) approaches are the subject of Chapter 21 of this handbook. The plurality of approaches and possible answers, the necessary contextualization and situational knowledge, and often the lack of the full scope of impact make ethical analysis not only hard to teach but also hard to evaluate in ways commensurate with typical assessment of student outcomes.

Conclusion

Engineering curricula should pay more attention to the role of sophisticated epistemic cognition in developing ethical reasoning skills. Ethics courses and units should seek to stimulate and measure students' epistemic cognition and to minimize the detrimental impact of teaching and grading approaches that do not encourage students to diversify their epistemic approaches. We are not proposing a wholesale replacement of models of ethical reasoning in engineering curricula but rather that greater attention should be afforded to epistemic cognition as a means of moral and professional growth.

Not assessing students' epistemic sophistication risks overlooking an essential blockage in students' ethical engagement and development. A succinct illustration is the potential explanatory power of epistemic naïveté for Haws' identification of engineering students' persistent stance of 'ethical relativity,' where students conclude that if there is not a single, definitely correct answer, then everyone's ethical 'opinion' must be equally valid. Engineering ethics teachers' intention for students to engage with authentic cases (Martin et al., 2021) can be reinforced by setting analysis tasks that have students engage in the more sophisticated epistemic micro-practices of exercising their own judgment – setting and applying criteria to evaluate competing solutions, and working with more ambiguity. This correlates to the idea of providing 'thick' cases, as are already used in ethics education and project work. However, we believe that teachers can provide better support to students in developing their skills by introducing contextual factors that enable Facet 3 *epistemic micro-practices* and allowing students to choose between competing solutions (Facet 4) in paper, calculation, or fundamental technical engineering assignments from their first year of studies.

The lack of empirical and theoretically robust quantitative tools available for measuring epistemic sophistication (Isaac, 2021) is unfortunate. This lack of quantitative instruments makes directly measuring epistemic cognition in large engineering cohorts unfeasible. The significant insights offered by epistemic cognition to ethical reasoning include greater attention to the process of ethical reasoning and, therefore, to assessing not the product or result as in common engineering disciplines. Being more explicit about this can also assist teachers to be more attentive to developing students as professionals. Given the comparatively weak epistemic sophistication of engineering students (Gainsburg, 2015; Marra et al., 2000; Wise et al., 2004), these epistemic issues can exert a significant negative force.

Engineers' capacity to contribute to solving big societal issues is much vaunted, but moral leadership is essential for resolving ill-structured, volatile, and interdisciplinary problems. The speed at which new technologies and technological capabilities emerge means that we also need engineers to be proactively formulating moral problem statements and not waiting for someone else to identify a potential issue. Epistemic cognition is a promising element for ensuring that the next generation of engineers is prepared for the epistemic reality of ethics in engineering practice.

References

ABET. (2023). *Criteria for accrediting engineering programs - Effective for reviews during the 2024–2025 accreditation cycle*. Retrieved November 3, 2023, from www.abet.org

Antes, A. L., Murphy, S. T., Waples, E. P., Mumford, M. D., Brown, R. P., Connelly, S., & Devenport, L. D. (2009). A meta-analysis of ethics instruction effectiveness in the sciences. *Ethics & Behavior*, *19*(5), 379–402. https://doi.org/10.1080/10508420903035380

Bairaktarova, D., & Woodcock, A. (2017). Engineering student's ethical awareness and behavior: A new motivational model. *Science and Engineering Ethics*, *23*(4), 1129–1157. https://doi.org/10.1007/s11948-016-9814-x

Baxter Magolda, M. B. (1992). *Knowing and reasoning in college: Gender-related patterns in students' intellectual development*. Jossey Bass.

Bendixen, L. D., Schraw, G., & Dunkle, M. E. (1998). Epistemic beliefs and moral reasoning. *The Journal of Psychology*, *132*(2), 187–200.

Bielefeldt, A. R., & Canney, N. E. (2016). Changes in the social responsibility attitudes of engineering students over time. *Science and Engineering Ethics*, *22*, 1535–1551. https://doi.org/10.1007/s11948-015-9706-5

Bombaerts, G., Doulougeri, K., Tsui, S., Laes, E., Spahn, A., & Martin, D. A. (2021). Engineering students as co-creators in an ethics of technology course. *Science and Engineering Ethics*, *27*(4), 48. https://doi.org/10.1007/s11948-021-00326-5

Borenstein, J., Drake, M. J., Kirkman, R., & Swann, J. L. (2010). The Engineering and Science Issues Test (ESIT): A discipline-specific approach to assessing moral judgment. *Science and Engineering Ethics*, *16*, 387–407.

Briell, J., Elen, J., Verschaffel, L., & Clarebout, G. (2011). Personal epistemology: Nomenclature, conceptualizations, and measurement. In J. Elen, E. Stahl, R. Bromme, & G. Clarebout (Eds.), *Links between beliefs and cognitive flexibility* (pp. 7–36). Springer Netherlands. https://doi.org/10.1007/978-94-007-1793-0_2

Clancy, R. F. (2020). The ethical education and perspectives of Chinese engineering students: A preliminary investigation and recommendations. *Science and Engineering Ethics*, *26*(4), 1935–1965.

Colby, A., Kohlberg, L., Gibbs, J., Lieberman, M., Fischer, K., & Saltzstein, H. D. (1983). A longitudinal study of moral judgment. *Monographs of the Society for Research in Child Development*, *48*, 1–124.

Elby, A., & Hammer, D. (2001). On the substance of a sophisticated epistemology. *Science Education*, *85*(5), 554–567. https://doi.org/10.1002/sce.1023

Elby, A., & Hammer, D. (2010). Epistemological resources and framing: A cognitive framework for helping teachers interpret and respond to their students' epistemologies. In L. D. Bendixen & F. C. Feucht (Eds.), *Personal epistemology in the classroom: Theory, research, and implications for practice* (pp. 409–434). Cambridge University Press.

Ericsson, K. A., & Simon, H. A. (1998). How to study thinking in everyday life: Contrasting think-aloud protocols with descriptions and explanations of thinking. *Mind, Culture, and Activity*, *5*(3), 178–186. https://doi.org/10.1207/s15327884mca0503_3

Felder, R. M., & Brent, R. (2004). The intellectual development of science and engineering students. Part 2: Teaching to promote growth. *Journal of Engineering Education*, *93*(4), 279–291. https://doi.org/10.1002/j.2168-9830.2004.tb00817.x

Gainsburg, J. (2007). The mathematical disposition of structural engineers. *Journal for Research in Mathematics Education*, *38*(5), 477–506. https://doi.org/10.2307/30034962

Gainsburg, J. (2015). Engineering students' epistemological views on mathematical methods in engineering. *Journal of Engineering Education*, *104*(2), 139–166. https://doi.org/10.1002/jee.20073

Haidt, J. (2001). The emotional dog and its rational tail: A social intuitionist approach to moral judgment. *Psychological Review*, *108*, 814–834. https://doi.org/10.1037/0033-295X.108.4.814

Hammer, D., & Elby, A. (2003). Tapping epistemological resources for learning physics. *Journal of the Learning Sciences*, *12*(1), 53–90. https://doi.org/10.1207/s15327809jls1201_3

Hardebolle, C., Macko, V., Ramachandran, V., Holzer, A., & Jermann, P. (2023). The digital ethics canvas: A guide for ethical risk assessment and mitigation In the digital domain. In *Proceedings of the 51st SEFI Annual Conference 2023*. Dublin, Ireland. https://doi.org/10.21427/9WA5-ZY95

Haws, D. R. (2001). Ethics instruction in engineering education: A (mini) meta-analysis. *Journal of Engineering Education*, *90*(2), 223–229. https://doi.org/10.1002/j.2168-9830.2001.tb00596.x

Healy, T. (1997, March 14). *Parallels between teaching ethics and teaching engineering*. Annual Meeting of the Pacific Southwest Section of the American Society for Engineering Education, San Luis Obispo, CA. https://www.scu.edu/ethics/focus-areas/technology-ethics/resources/teaching-ethics-and-teaching-engineering/

Hofer, B. K. (2000). Dimensionality and disciplinary differences in personal epistemology. *Contemporary Educational Psychology*, *25*(4), 378–405. https://doi.org/10.1006/ceps.1999.1026

Hofer, B. K., & Pintrich, P. R. (1997). The development of epistemological theories: Beliefs about knowledge and knowing and their relation to learning. *Review of Educational Research*, *67*(1), 88–140.

Institution of Civil Engineers. (2021). *ICE attributes updates: A guide for trainees*. Version 1, revision 1. https://www.ice.org.uk

Isaac, S. (2021). *Epistemic practices: A framework for characterising engineering students' epistemic cognition* [PhD thesis]. Lancaster University.

Isaac, S., Kothiyal, A., Borsò, P., & Ford, B. (2023). Sustainability and ethicality are peripheral to students' software design. *International Journal of Engineering Education*, *39*(3), 52–556.

Jonassen, D., Strobel, J., & Lee, C. B. (2006). Everyday problem solving in engineering: Lessons for engineering educators. *Journal of Engineering Education*, *95*(2), 139–151. https://doi.org/10.1002/j.2168-9830.2006.tb00885.x

King, B. A., & Magun-Jackson, S. (2009). Epistemological beliefs of engineering students. *The Journal of Technology Studies*, *35*(2), 56–64.

King, P. M., & Kitchener, K. S. (2004). The reflective judgment model: Twenty years of research on epistemic cognition. In B. K. Hofer & P. R. Pintrich (Eds.), *Personal epistemology: The psychology of beliefs about knowledge and knowing* (Vol. 37). Erlbaum.

Kuhn, D., Cheney, R., & Weinstock, M. (2000). The development of epistemological understanding. *Cognitive Development*, *15*(3), 309–328.

Kuusela, H., & Paul, P. (2000). A comparison of concurrent verbal protocol analysis retrospective. *American Journal of Psychology*, *113*(3), 387–404.

Lönngren, J. (2021). Exploring the discursive construction of ethics in an introductory engineering course. *Journal of Engineering Education*, *110*(1), 44–69. https://doi.org/10.1002/jee.20367

Marra, R. M., Palmer, B., & Litzinger, T. A. (2000). The effects of a first-year engineering design course on student intellectual development as measured by the Perry scheme. *Journal of Engineering Education (Washington)*, *89*(1), 39–46.

Martin, D. A., Conlon, E., & Bowe, B. (2021). Using case studies in engineering ethics education: The case for immersive scenarios through stakeholder engagement and real life data. *Australasian Journal of Engineering Education*, *26*(1), 47–63. https://doi.org/10.1080/22054952.2021.1914297

McNeill, N. J., Douglas, E. P., Koro-Ljungberg, M., Therriault, D. J., & Krause, I. (2016). Undergraduate students' beliefs about engineering problem solving. *Journal of Engineering Education*, *105*(4), 560–584. https://doi.org/10.1002/jee.20150

Monteiro, F., Leite, C., & Rocha, C. (2019). Ethical education as a pillar of the future role of higher education: Analysing its presence in the curricula of engineering courses. *Futures*, *111*, 168–180. https://doi.org/10.1016/j.futures.2018.02.004

Monzon, J. E., Ariasgago, O. L., & Monzon-Wyngaard, A. (2010). *Assessment of moral judgment of BME and other health sciences students*. 2010 Annual International Conference of the IEEE Engineering in Medicine and Biology, 2963–2966.

Muis, K. R. (2004). Personal epistemology and mathematics: A critical review and synthesis of research. *Review of Educational Research*, *74*(3), 317–377. https://doi.org/10.3102/00346543074003317

Muis, K. R., Bendixen, L. D., & Haerle, F. C. (2006). Domain-generality and domain-specificity in personal epistemology research: Philosophical and empirical reflections in the development of a theoretical framework. *Educational Psychology Review*, *18*(1), 3–54.

O'Flaherty, J., & Gleeson, J. (2014). Longitudinal study of levels of moral reasoning of undergraduate students in an Irish university: The influence of contextual factors. *Irish Educational Studies*, *33*(1), 57–74.

Olson, G. M., Duffy, S. A., & Mack, R. L. (2018). Thinking-out-loud as a method for studying 11 real-time comprehension processes. In D. E. Kieras & M. A. Just (Eds.), *New methods in reading comprehension research* (pp. 253–286). Routledge.

Palmer, B., & Marra, R. (2004). College student epistemological perspectives across knowledge domains: A proposed grounded theory. *Higher Education, 47*(3), 311–335. https://doi.org/10.1023/B:HIGH.0000016445.92289.f1

Pavelich, M. J., & Moore, W. S. (1996). Measuring the effect of experiential education using the Perry model. *Journal of Engineering Education, 85*(4), 287–292.

Perry, W. G. (1970). *Forms of intellectual and ethical development in the college years: A scheme*. Holt, Rinehart, and Winston.

Reall, M. J., Bailey, J. J., & Stoll, S. K. (1998). Moral reasoning 'on hold' during a competitive game. *Journal of Business Ethics, 17*(11), 1205–1210. https://doi.org/10.1023/A:1005832115497

Rest, J. R. (1994). *Moral development in the professions: Psychology and applied ethics*. Psychology Press.

Shew, A. (2023). *Against technoableism: Rethinking who needs improvement.* W.W. Norton & Company.

Székely, I., Fuchs, C., Boersma, K., Albrechtslund, A., & Sandoval, M. (2011). What do IT professionals think about surveillance? In *Internet and Surveillance* (Vol. 16, pp. 198–219). Routledge.

Tormey, R., Le Duc, I., Isaac, S., Hardebolle, C., & Cardia, I. V. (2015, July 29). *The formal and hidden curricula of ethics in engineering education.* Proceedings of the 43rd SEFI Annual Conference. Diversity in Engineering Education: An Opportunity to Face the New Trends of Engineering - SEFI 2015, Orléans, France.

van de Poel, I., & Royakkers, L. (2007). The ethical cycle. *Journal of Business Ethics, 71*(1), 1–13. https://doi.org/10.1007/s10551-006-9121-6

Watts, L. L., Medeiros, K. E., Mulhearn, T. J., Steele, L. M., Connelly, S., & Mumford, M. D. (2017). Are ethics training programs improving? A meta-analytic review of past and present ethics instruction in the sciences. *Ethics & Behavior, 27*(5), 351–384. https://doi.org/10.1080/10508422.2016.1182025

Wise, J. C., Lee, S. H., Litzinger, T., Marra, R. M., & Palmer, B. (2004). A report on a four-year longitudinal study of intellectual development of engineering undergraduates. *Journal of Adult Development, 11*(2), 103–110. https://doi.org/10.1023/B:JADE.0000024543.83578.59

Yu, J. H., & Strobel, J. (2011). Instrument development: Engineering-specific epistemological, epistemic and ontological beliefs. *Proceedings of the Research in Engineering Education Symposium*, Madrid, Spain, 95–105.

Zhu, J., Liu, R., Liu, Q., Zheng, T., & Zhang, Z. (2019). Engineering students' epistemological thinking in the context of Project-Based Learning. *IEEE Transactions on Education, 62*(3), 188–198. https://doi.org/10.1109/TE.2019.2909491

29

ASPIRATIONS FOR ETHICAL EDUCATION IN ENGINEERING CURRICULA ENVISIONED THROUGH THE QUALITY LENS OF GOODLAD'S TYPOLOGY

Emanuela Tilley, Nienke Nieveen, Christine Boshuijzen-van Burken, and Folashade Akinmolayan Taiwo

Introduction

Ethical consideration within engineering curricula represents a modern advancement of the core topics included in engineering degree programs within higher education. However, it is widely recognized that the development of teaching ethics to students in engineering programs since its establishment in the 1970s has been gradual (Martin et al., 2021; Mitcham, 2009; Weil, 1984; for more on this, see Chapter 32 on foundational perspectives on ethics in engineering accreditation). Consequently, it is unsurprising that research on evaluating the quality of ethics education included in engineering courses and programs is not yet mature and, thus, is variable and sparse. Nevertheless, this chapter provides a critical reflection of the research – in engineering education and the broader discipline of education – on evaluating the quality of curricula to provide an aspirational proposal for envisioning an ideal curriculum. *What should be strived for from an institutional perspective regarding quality assessment and enhancement practices of ethics education within engineering curricula?* This chapter was written to emphasize what quality ethics education looks like within engineering curricula, with essential connections to the other chapters of the assessment theme included within this handbook.

Curricula: the connection between a higher education institution and its students

Curricula constitute a fundamental element of universities and higher education, as the root of conventional student learning is found within them. A curriculum's purpose is galvanized to influence the development of a learner. In the broadest sense, the curriculum is defined as a plan for learning or, more specifically, an organized set of learning experiences to modify the behavior of learners in a desired and predetermined manner (Kropp, 1973). Moreover, Priestley, and Philippou (2019) describe curricula as "the multi-layered social practices, including infrastructure, pedagogy and

DOI: 10.4324/9781003464259-35

assessment, through which education is structured, enacted and evaluated" (p. 3). At this point in the chapter (and again in its following sections), it is helpful to look to John Goodlad (an educator and researcher who is widely known as an advocate of education) for advice on "preparing young people to be active and engaged citizens in a participatory democracy" (Britannica, 2023). The book *Fifty Modern Thinkers on Education: From Piaget to the Present* by Psychology Press (Palmer et al., 2001) lists him, indicating the significance of his ideas on curricula and curriculum development. Goodlad (1979) developed a curriculum typology (outlined below) as a valuable framework for delineating how curriculum, manifested through various yet interconnected representations, serves as an intermediary mechanism in students' progression toward achieving learning outcomes.

1. The *Intended* Curriculum includes *Ideological* and *Formal* domains – curricula that emerge within the context of higher education from idealistic planning processes led by institutional leaders and/or gain official approval by state, professional accreditation, and university approval boards and are adopted by choice or force, by teachers.
2. The *Implemented* Curriculum includes *Perceived* and *Operational* domains – curricula of the mind of the teacher (or user of the intended curriculum) and that which is put in place and goes on in the classroom hour after hour and day after day, led by the teacher (i.e., pedagogy).
3. The *Attained* Curriculum includes *Experiential* and *Learned* domains – the curriculum as experienced by students and that results in learning outcomes of the individual student.

Here, the premise is that a curriculum can be described by the intentions of a university (and its departments/schools) to set out a purpose for its education and a plan to incorporate associated knowledge, skills, values, and, in many ways, the metacognition (i.e., awareness and understanding of one's own thought processes) required to foster top-class graduates. There are two other significant parties whose perspectives are essential in understanding 'quality' within a curriculum: the teachers responsible for delivering the planned education and the students who are the recipients of it. According to the literature presented in this chapter, the goals established for curricula play a crucial role in determining the quality of ethics education in engineering. These goals heavily influence the intended outcomes of the learning process, and therefore it is essential to set clear and appropriate expectations.

Quality evaluation: a theoretical, practical, and political problem

Developing an effective and ideal ethics curriculum presents theoretical and practical challenges. It not only involves developing high-quality engineering programs to foster ethical graduates, but it also involves realizing or operationalizing such programs while achieving sustainable improvements and successfully developing capacity among the teachers who deliver ethics education within engineering. It is important to note that curriculum decision-making is a socio-political process that involves addressing questions such as: *Who determines what 'effective' and 'ideal' mean in this context? Who decides? What are the roles of university management, program leaders, industrial partners, employers, teachers, students, and others in this respect?*

Filippakou (2011) proposed a conceptual approach to assess the quality of higher education. This approach considers curriculum evaluation as a crucial part of educational evaluation. It involves making judgments on the value of the school curriculum, constantly improving it, and achieving value-added education. Filippakou argued that it includes the evaluation of three interlinked elements: the curriculum plan with consideration of its disciplinary standards and teaching materials, the curriculum implementation, and the curriculum's effect on students. Wu and Liu (2021) and Pavel et al. (2015) identified the main factors affecting curriculum quality (when introducing modern overarching themes and topics into existing curricula) as curriculum concept, resources, and management.

These authors identify factors that are well aligned, at least in part, with Goodlad's (1979) framework introduced in this section of our chapter and explored in further detail in our subsequent section on literature review: 'curriculum concept' can be linked to Goodlad's *intended* curriculum, while 'curriculum resources' can be associated with Goodlad's *delivered* curriculum, which includes what happens in the classroom. The factor of 'curriculum management,' however, has a less evident link to Goodlad's framework. This factor highlights the importance of having support from various levels of the institution – for example, the messages and actions of teachers in the classroom being aligned with institutional requirements and educational vision. It is related to the importance of 'practicability,' which Bombaerts et al. (2019) link to the *attained* curriculum (explored below).

Positionality

As authors of a chapter positioned within the assessment theme of this engineering ethics education handbook, we started by observing that the concept of 'quality' in the context of curricula is interpreted too narrowly in assessments when they refer only to the student achievements (i.e., learning outcomes). In this chapter, we call for a broader approach to the concept of quality ethics curricula within engineering programs by also looking at the quality of teaching practices and the relevance of those teaching practices for engineering students.

Our four-person authoring team includes three engineers and one specialist in science, technology, engineering, and mathematics (STEM) education and curriculum design. We all currently work in higher education. All four authors have experience with higher education within Europe. Emanuela and Nienke bring extensive experience in program and curriculum design and conducting training programs as well as leading various communities of practice for staff to help them deliver curricula (within and outside of engineering education). Christine has deep knowledge and interest in the ethics and philosophy of technology. Folashade brings research strengths in the investigation of innovative pedagogies for embedding graduate attributes. For all of us, studying and evaluating quality in ethics curricula at the program level within engineering education was a new and intriguing opportunity to collaborate on a relatively unexplored topic in literature. What is provided herein has been shaped by our interest in the future progress of engineering programs to embed ethics consistently, practically, and effectively – and with contextual relevance that is engaging to students and representative of institutional values as well as the continuous evolution of the profession.

Goodlad curriculum typology: a means of evaluating quality

A typology foregrounded by Goodlad (1979) is used in this chapter to categorize the discussion within ethics curriculum research. Examples wherein this approach has been used in education-based research include Stokking et al. (2003), who refer to Goodlad to characterize the implementation strategy of environmental education and conclude that only an *ideological* curriculum was presented in the case investigated; and Terwel et al. (2004), who described the innovation waves in the Dutch mathematics curriculum and concluded that much is known about the *ideal* and *formal* curriculum levels, but that still no-one knows what happens in the classroom at the *operational* level. A recent study that involved two authors of this chapter (Bombaerts et al., 2019) uses the Goodlad typology to propose four distinct quality criteria to evaluate and improve ethical education in engineering curricula: relevance, consistency, practicality, and effectiveness. Bombaerts et al. argued that a high-quality ethics curriculum should have virtues and components that serve all four quality criteria. However, their research shows that there are some disparities and unknowns.

Relevance and effectiveness as measures of quality

The work of Bombaerts et al. (2019) suggests that current engineering education research in the context of ethics education is focused mainly on the quality evaluation associated with the *relevance* and *effectiveness* of the provision, with little to no inclusion of studies examining the *consistency* and *practicality* of ethics education in engineering programs. In other words, plenty of publications describe the aim to guide the vision/rationale/basic philosophy underlying an engineering ethics curriculum (i.e., an aspirational aspect of the *intended* curriculum set out and controlled by institutional leaders). Moreover, a robust research foundation exists on initiatives aimed at assessing and informing educators about ethics-centered learning experiences as perceived by students. These initiatives result in individual learning outcomes achieved through assessments and self-reflection, reflecting the *attained* curriculum. Attention to the *relevance* of ethics education is prominent through, for instance, studying and discussing (1) the conceptual goals surrounding ethics education within engineering programs, such as honesty, integrity, and/or social responsibility; (2) the support of engineering concepts of complexity, risk, and security; and (3) the compliance of national, disciplinary, or institutional education standards, such as accreditation and policy. The *effectiveness* of ethical education within engineering education research is often studied and debated through themes of student attributes and competence development.

Lack of research on consistency and practicality as a measure of quality

To a lesser extent, attention is paid to issues related to *consistency*, which is described as the need for coherence among all course components and the entire curriculum, including the aspects of learning listed below that are associated with all three types of curricula distilled from Goodlad (1979).

- Rationale: *Why are students learning?*
- Aims and objectives: *Towards which goals are students learning?*
- Content: *What are students learning?*
- Learning activities: *How are students learning?*
- Teacher role: *How is the teacher facilitating the students' learning?*
- Materials and resources: *With what are students learning?*
- Grouping: *With whom are students learning?*
- Location: *Where are students learning?*
- Time: *When are students learning?*
- Assessment: *How is learning of students assessed?*

The aspects of learning listed above are taken from the work of the Netherlands Institute for Curriculum Development (SLO, 2009, p. 12) and van den Akker et al. (2003), who present them together as a so-called *curricular spider web* with the aim of clarifying the key aspects of learning within curricula and what coherence within curricula comprises. The curricular spider web is drawn in these referenced texts to visualize the relationship between all the key aspects of learning within a curriculum to support moving away from the notion that curriculum generally only concerns the aims and contents of learning. The *rationale* serves as a central link and supports all other aspects, which within the curricular spider web are connected, providing coherence and, ideally, the consistency that Bombaerts at al. (2019) sought to find in research studies on the quality of engineering ethics curricula.

Based on reviewing the engineering ethics education research literature, Bombaerts et al. (2019) claimed that consistency in ethics education is not given much importance beyond the ongoing debate on whether ethics should be taught as an embedded approach within degree programs or as a separate one-off or strand of ethics course(s). It is often argued that there is a shortage of engineering faculty members who possess the necessary expertise and comfort in teaching ethics. This is frequently cited as the reason for having a separate teacher specifically dedicated to engineering ethics. The idea behind this argument is to ensure consistency and provide quality education. This is because the considerable generational gap between modern curricula and the educational background of engineering academic staff can hinder their ability to teach ethics effectively. However, Farahani and Farahani (2014) pointed out that the 'practice-what-you-preach' principle impacts the quality of a student's ethical education; through teachers' consistent and daily obligation to show respect for the privacy, health, and safety of students, teachers can model trust, respect, tolerance, and openness.

Bombaerts et al. (2019) described *practicality* as being more explicitly linked to the *implemented* or *operational* curriculum through the perceptions and abilities of the teachers whose role it is to enact or operationalize the *intended* curriculum. Bombaerts et al. explained that teachers face practical considerations in their daily work of delivering the *intended* curriculum. Furthering from the work of Jansen et al. (2013), they believe that teachers should do this by considering the following questions when delivering the lesson associated with the ethics syllabus or curricula: (1) *Is it possible or helpful to teach it (i.e., instrumentality)?* (2) *Does it fit within their given circumstances (i.e., congruence)?* (3) *Is it feasible to teach within the available time and resources allotted (i.e., cost)?* (p. 1424).

In contrast, Bombaerts et al. (2019) found that research on the practical quality of engineering ethics education has focused on whether the assessment is valid and reliable – another aspect of the curricular spider web, alongside that highlighted above as the teacher's role. Bombaerts et al. (2019) found that practicality and consistency are two key aspects missing from ethics education in engineering programs. They stated that 'practical courses show consistency between both forms of the *implemented* curriculum, the perceived as well as the *operational*' (2019, p. 1427), which means that practical courses should align with both the *intended* curriculum and the actual implementation of the curriculum. Very little research has been done on the relationship between practicality and quality, especially regarding teachers' perceptions of how well they are implementing the curriculum. This suggests that more work needs to be done within the engineering education ethics research community to move beyond simply understanding "whether assessment is valid and reliable in ethics education" (2019, p. 1429).

By framing teachers' hesitation, Bardfod and Bentsen (2018) used the Goodlad framework to evaluate Education Outside the Classroom (EOtC) – a relatively common practice in Scandinavian schools. The research indicates the importance of examining the practical issues experienced and described by teachers and the issues' impact on delivering a high-quality curriculum. Bardfod and Bentsen asserted that quality may include broader aspects (such as confidence levels of teachers to teach a novel subject or the potentially increased workload affecting the effort teachers may wish to put into their delivery), as well as how some subjects are logically/chronologically interwoven with others (which has ramifications for the 'coherence' of the curriculum and each teacher's role within it).

Literature review: search for constituent elements of quality

The literature review below analyzes the ideal representation of ethics curricula in engineering education. It explores the various aspects of learning that are part of the curriculum's flexible yet vulnerable nature, as described in the previous section of this chapter using the metaphor of a spider web. The literature review presents a theoretical conceptualization of the *intended* curriculum for ethics education within engineering programs, using the Goodlad typology as a lens. This conceptualization is based on research on the evaluation of engineering ethics education and considers all aspects of the curricular spider web, including assessment. The goal is to create a clear and comprehensive representation of what ethics education in engineering programs should look like. Adopting this approach allows us to respond more to the fundamental query of what could constitute high-quality, ethical education in engineering curricula while acknowledging key issues because of the infancy of its evolution and development.

Our literature review utilized the Web of Science and Scopus databases to identify research about curriculum content, pedagogy, and assessment associated with undergraduate engineering ethics education. To retrieve sources related to the three representational aspects of curricula as described in Section 1, we used the following combination of key terms to search in the titles and abstracts of publications during the period 2000–2023: "ethic*" AND "engineer*" AND "education*" OR "course" OR "program*" OR "curricul*" OR "teach*" OR "assess*" OR "implement*" OR "pedagogy*". After the initial database searches, we retrieved additional sources by 'pearling' or 'snowballing' the reference lists of the collected, relevant publications.

Below, we discuss the results of our literature search. We have identified key components of ethical education that should be included in undergraduate engineering degree programs. These components are aligned with Goodlad's typology of curriculum and are integrated into the different aspects of learning found within the curricular spider web.

The intended engineering ethics curriculum

In this section, we look at the rationale, aims, and objectives of the intended engineering ethics curriculum, with the role of the teacher at its core. Goodlad (1979) argued that teachers' interpretations of the *intended curriculum* are a preface to its execution. There are debates and opinions on its best approach and ideological purpose in engineering ethics education. Additionally, the components or aspects of learning (e.g., the nine threads of the curricular spider web) of the *intended curriculum* will look different depending on the fundamental purpose (the rationale/vision at the center of the spider web) chosen. Our literature review found two main perspectives on the fundamental reasoning behind ethics curricula in engineering education. On the one hand, Mitcham (2009) and others argue that engineering, particularly its inclusion of ethical considerations, should be made part of the 'lifeworld' and endeavor "to advance a deeper human understanding of the Good" (p. 50). Arguments aligned with this stance posit that engineering curricula must include more authentic contexts and show how social and political interests contribute in important ways to the forms of technologies and technical solutions engineers produce (Bucciarelli, 2008). On the other hand, another group of scholars has foregrounded the crucial role of curriculum in motivating engineering students to learn ethics – highlighting the historical significance of engineering as a profession and practice on society and the planet while promoting the diverse future roles of responsible engineers (Bombaerts & Vaessen, 2022; Colby & Sullivan, 2008; Hess et al., 2017; Rayne et al., 2006;

Silvast et al., 2020). Within this argument, Pfatteicher (2001) asserted that students' motivation would be improved if more of the curriculum was focused on supporting students' understanding of the nature of engineering ethics – meaning that the curriculum should focus on why it is crucial to be an ethical engineer, and how to address and resolve moral problems.

There is a debate in the literature regarding whether ethical codes should be used as a baseline for incorporating ethics into engineering education (Li & Fu, 2012). One body of research suggests that codes can offer guidance and a shared understanding of a commitment to ethics that can uphold the professional image (Li & Fu, 2012). However, there is a growing interest in moving beyond the offering of these basics, which suggests that codes cannot substitute either for individual capabilities in solving ethical dilemmas (Bucciarelli, 2008; Martin & Schinzinger, 2005) or serve as the substantive knowledge base for ethics education. Voices in this debate posit that engineering ethics curricula should equip students with a broader knowledge base beyond the core technical aspects offered in traditional engineering programs, as well as a diverse set of skills necessary to address the current ethical challenges. From a curricular point of view, codes can be seen as aims/objectives and content (and thus can be placed at those two threads of the spider web); however, the need for codes would depend on the purpose of ethics education (see Chapter 5).

Who should teach ethics in engineering education? Instructors justify their curricular choices according to their vision of what engineering practice 'is' (Monteiro, 2021; Quinlan, 2002) and their understanding of an engineer's responsibilities (Downey et al., 2007) – thus the teacher's alignment to the curricular rationale is crucial for delivering coherent, unified messages. The literature suggests that newly hired engineering teachers must be expected/required to teach contemporary ethics alongside advancements in disciplinary content. Bucciarelli (2008) proposed the idea of recruiting/assembling a mixed faculty, where each member may not possess comprehensive knowledge of all dimensions but, akin to engineering practice, each can "articulate their own interests to others and to listen with full respect" (2008, p. 147). There are compelling reasons to incorporate ethics education into current engineering curricula. This means that engineering instructors should teach ethics education as part of their engineering courses, and from a personal perspective that reflects the profession (Li & Fu, 2012).

Challenges of consistency within the intended curriculum

This section investigates the challenges of achieving consistency within the intended curriculum, considering aspects of content, learning activities, materials and resources, grouping, location, and time. Within the literature, many works discuss the various and numerous challenges associated with implementing ethics into engineering education curricula with consistency. Authors who frame their studies around consistency consider the challenges related to the ethics curricula as part of the theoretical conceptualization of engineering programs. They touch on aspects of how ethics should be taught, what resources and institutional support are needed, and how learning can be grouped and situated in a time and space that makes sense and is easily accessible to the learners physically and cognitively. Some of the most significant challenges regard teachers' low familiarity with ethics and the lack of broader institutional support or resources (Boshuijzen-van Burken et al., 2022; Martin et al., 2021). There have been reports of unsystematic implementation of ethics in engineering (Colby & Sullivan, 2008; Flynn & Barry, 2010; Polmear et al., 2018), leading to inconsistency. In this section, we aim to bring together the various pedagogies used for teaching ethics in engineering curricula, which are heavily focused on technical aspects. We will explore how effective and coherent the different teaching and assessment methods are and how well they align with the goals and theoretical frameworks envisioned for engineering ethics education.

A significant and fundamental challenge with sufficiently and consistently implementing ethics into an existing curriculum stems from identifying which disciplinary or 'technical' elements of learning will be reduced to make room for the ethical components required (Polmear et al., 2018; Romkey, 2015). Below are the two most common methods utilized to include ethics education into degree programs/course curricula:

1. A freestanding course, often taught by the philosophy department or another entity external to the engineering department, can give the impression that this is not a 'core' skill required by engineers as it is not integrated within their studies (Wolverton & Wolverton, 2003).
2. Ethics are embedded across the degree program, and each teacher is expected to include an element in their course materials. These range from references to well-known cases of engineering failure to classroom or homework exercises in which students grapple with trade-offs between potentially conflicting values such as cost and safety (Herkert, 2005).

Bucciarelli, later supported by Bielefeldt et al. (2018), argues that there was a "need to open up the engineering classroom to discussion and debate" (Bucciarelli, 2008, p. 147) in favor of embedding ethics education into engineering curricula and suggests a possible way forward would be the inclusion of project-based learning to encourage the development of graduate attributes, such as critical analysis, communication, and respect. Also included in this conversation is the subdivision of engineering ethics into categorizations of (1) 'micro-ethics,' which considers individuals and internal relations of the engineering profession (Li & Fu, 2012); and (2) 'macro-ethics,' which Herkert (2005) identifies as concerning the collective social responsibility of the profession to make societal decisions about technology. Li and Fu (2012) argue that most contemporary engineering ethics teaching has focused on micro-ethics and presented oversimplified situations. Indeed, there is support within the literature indicating that this educational focus neglects the social nature of engineering practice (Bucciarelli, 2008; Herkert, 2005; Huff & Frey, 2005). When Bielefeldt et al. (2018) surveyed engineering and computing faculty across the United States, they found that teaching social justice as a macro-ethics issue has expanded. The literature suggests an educational trend toward adopting a broader focus on engineering ethics. This reflects calls to go beyond micro-ethical considerations to include topics of, for example, social responsibilities (Hollander & Arenberg, 2009).

Incorporating ethical decisions into the technical content covered in a course enables students to see how abstract concepts can be applied to real-world problems. Like technical problems, students need to practice solving ethical dilemmas first-hand. Many engineering students discuss ethical case studies like the Space Shuttle Challenger incident or the Ford Pinto gas-tank problem (Wolverton & Wolverton, 2003). However, this promotes an unrealistic simplification of events and fails to help students identify complexities without the benefit of hindsight (see Chapter 20). Most ethical issues in the real world appear subtly, and engineering students need more practice in typical ethical deliberation and decision-making.

Designing an ideal curriculum considering student engagement and reflection

Our review of recently published research on engineering ethics indicates that predominant methods for assessing student engagement and perspectives on engineering ethics within engineering programs include presentations, group projects, portfolios (Sunderland et al., 2013), reflective essays, and individual assignments evaluated using a rubric comprising multiple criteria and thresholds (Bielefeldt et al., 2016). Assessing such outputs – and commonly used case study assignments – can be challenging due to the ill-structured and open-ended nature of the problems they address (Goldin et al., 2015). Assessment is one of the nine curricular aspects of learning and is thus represented on the curricular spider web, so it needs to be discussed when designing the *ideal* curriculum. Students view the assessments that are part of the ethics curriculum in engineering programs as significantly distinct from the other core assessments required for the degree. This is because they need different skills and knowledge related to the topics covered. As a result, these assessments are often not aligned with the students' expectations of what constitutes professional engineering work.

Furthermore, due to the typically limited engagement with ethics in engineering curricula and some documented student resistance towards learning about ethics, students are likely to allocate minimal time during their formal education as engineers (i.e., progressing through their degree program/course) to understanding the relevance of ethics to their future roles. When students receive ethics instruction, there is a likelihood that many of them do not take it seriously. This can adversely affect the way engineering graduates view ethics and can lead to deficiencies in their capability to respond to ethical issues emerging in the workplace (Valentine et al., 2020), despite the intentions of academic leaders and curriculum designers to foster ethical graduates.

Considering this, Gwynne-Evans, Chetty, and Junaid (2021) explored how ethics can be more extensively incorporated into accreditation documents to validate engineering degree programs. They recommended that Washington Accord signatories reposition ethics so that it becomes the heart of the engineering graduate attributes, and they demonstrated how to do so. In alignment with the earlier discourse around the various rationales associated with engineering ethics curricula, Gwynne-Evans et al. evoked Davis' (1991) argument for developing a professional identity where engineering ethics is conceptualized as an essential part of thinking like an engineer. Thus, intentionally integrating opportunities into engineering ethics curricula to engage in reflection and having teachers provide reflective assessments (and therefore modeling reflection) can help students claim aspects of that identity in a personal and engaging way while simultaneously developing their own sense of professional responsibility (Gwynne-Evans, 2021).

Reflections and recommendations for practice and research

As authors of this chapter, we have attempted to provide a theoretical conceptualization of an idealized high-quality ethics curriculum within engineering education for third-level/postsecondary education through our investigation of different aspects of learning that appear promising within a recent body of engineering ethics education research. Throughout this investigation, we held the belief that central to the mission and vision of institutions representing higher education globally is an aim to provide education that fosters high-quality graduates who will make an impact in the world upon graduation. For engineering education, we also believe this is a shared goal of the accreditation and validation processes and criteria set out by national and international representative engineering bodies (as mentioned in previous chapters of this handbook). However, through our collective years of experience working within higher education, we also understand that educational change to embed contemporary themes, topics, and challenges (and the associated skills, values, and metacognitive aspects) – including ethics, sustainability, and diversity, equity, and inclusion (DEI) – can be very slow. An ideal curriculum, in many ways, reflects the aims and intentions of higher education institutions, academic departments/programs, and accrediting bodies. However, we understand from the literature on curriculum design and development and the research assessing the quality of the ethics curriculum within engineering education that it takes more than a *mandate* (i.e., the *intended* curriculum as per Goodlad typology) to achieve institutional goals like the one above – it requires a *movement*, one that holds the roles of the teacher and the student central to its delivery and attainment. This is described by curriculum and its spider web (SLO, 2009) as a fragile balance that requires consideration of its rationale and the ten aspects of learning, including assessment.

The evaluation or, perhaps more appropriately, the *assessment* of quality ethics education within engineering programs is unclear as there isn't a large body of work within engineering education research to draw from to inform the necessary curriculum development that is being called for to prepare engineering graduates for the challenges and complexities of our future world within which they will work and progress their professional careers. We utilized the four quality criteria within Goodlad's curriculum typology (relevance, consistency, practicality, and effectiveness) while looking at all aspects of learning in the context of the central rationale for an engineering ethics curriculum (SLO, 2009). What jumps out is the need for coherence of all aspects of learning, from the point of view of the teachers and the students, within a curriculum to effectively deliver upon its objectives. This means going beyond just emphasizing the aims and contents of learning and optimizing each aspect to balance all others. Assessment, the theme of this section of the handbook, the authors would argue, needs to have a better balance with all other aspects of learning and not be seen as an afterthought. We argue for understanding the balance across pedagogy (i.e., learning activities), the teacher role (i.e., experience, expertise, approach, enthusiasm, and confidence level), time and content (what to teach? when/at what level? e.g., year 1 versus year 3), and the core aims and objectives within engineering ethics curriculum design and development (as described in the chapters that follow in this section).

Considering the literature review that underpins this work, the paragraphs that make up the rest of this section include recommendations to curriculum developers and educational and institutional leaders in their work to establish an ideal ethics curriculum for engineering programs within higher education. Furthermore, we offer suggestions for research on engineering ethics education to enhance our comprehension of how engineering ethics curricula, across different modalities and encompassing all aspects of learning, including assessment, can be improved. The goal is to graduate ethical students well-prepared to make meaningful contributions as professionals in our dynamic and continually evolving world.

Fundamentally, academic departments and programs must ensure that their leadership aligns with and promotes the notion that an excellent modern engineering degree includes ethics education. Educational programs should manage student expectations before students start their studies and provide a coherent narrative in which engineering and ethics are consistently interwoven within their curricula. What constitutes an *ideal* ethics curriculum within the context of an engineering degree program needs to be informed by the multitude of stakeholders associated with the university and its engineering graduates and include a balanced consideration of all nine of the curricular aspects of learning, underpinned by a central, agreed, and prominent rationale.

When it comes to how ethics should be taught, we advise introducing various approaches to ethics and their associated relevance/rationale to help students understand the dynamic nature of ethics within a professional context and frame the potential impact of ethics in their future work. Many excellent examples have been presented and discussed above. Thus, we believe that students should be introduced to ethics as often and within as many places and learning opportunities (e.g., pre-arranged or authentic, inside and outside the curriculum) as possible to provide consistent access and a consistent sense of the centrality of ethics within engineering. Building a coherent ethics curriculum of quality takes time to design, implement, and develop. We recommend that the evaluation of such a curriculum be incorporated into its development plan (or perhaps more aptly, its management plan) to work towards addressing the challenges associated with the teacher's role and implementation and its effect on students.

Concerning who should teach ethics within engineering, we advocate for diversity in the expertise of staff responsible for teaching various aspects of ethics curricula. This approach aims to integrate ethical specialists alongside academic staff within engineering departments. The goal is to contextualize learning by incorporating themes associated with ethics (such as sustainability, DEI, risk/security, social responsibility, design, materials, and systems thinking) and to provide in-depth learning opportunities regarding professional responsibility. Practicality is an important aspect of curriculum quality and may include looking to literature to better understand the challenges – including the challenges within the *intended* curriculum. We urge academic departments to foster a better sense of community among the teachers who will implement the ideal engineering curriculum to address the practical aspects of implementation. Such a community of practice could then advocate for and implement more active learning pedagogies throughout the students' engagement with their engineering education to promote a natural sense of discussion and debate around ethics that is student-centered and student-led.

The literature presented in this chapter and other chapters of the handbook acknowledges that the conventional methods of teaching ethics in engineering education are too limited. These approaches are often added as an afterthought or an add-on to the core curriculum and rely too much on ethical codes to meet the accreditation requirements. We believe it is essential to broaden the traditional categorizations of engineering ethics to include macro-ethics, which will have students considering the collective social responsibility of the profession beyond the micro-ethics of their individual actions and professional relations. There is a growing belief that engineering must incorporate a more comprehensive understanding of ethics. This can be achieved by academic departments and programs purposefully integrating ethical decision-making and reflective practice into the assessment of their students. By doing so, they can encourage their future engineers to normalize ethical considerations and make this a part of their professional practice. This will help them become responsible and adaptable citizens who can thrive in a complex and ever-changing world.

Conclusions

Improving the quality of ethics education is an ongoing theme of higher education reform within universities and engineering departments. Evaluating curriculum quality is a challenge that has not yet matured within engineering ethics education practice. Similarly, there is a relatively small body of research looking to define quality ethics curricula within engineering programs, and thus, its overall impact on fostering ethical engineering graduates is nominal. Knowing that curricula generally comprise layers of social practice, as well as interconnected aspects of learning such as rationale, aims, content, resources, pedagogy, and assessment through which education is formalized, operationalized, and evaluated, here, the Goodlad framework has provided a lens to understand what form an ideal ethics curriculum could take. Our literature search revealed, through this lens, various ideas and approaches to address multiple aspects of learning that are currently being heralded and debated.

Quality in engineering education is frequently associated with the criteria incorporated into the accreditation of engineering degree programs. These criteria serve to define and validate the curriculum content requirements, extending beyond the technical disciplinary aspects of an engineering degree. More recently, accreditation standards have emphasized the integration of ethics and other contemporary themes or topics related to engineering ethics, including sustainability, risk/security, DEI, and more. This chapter presented literature associated with the quality evaluation of education in general so as to provide a set of lenses from which to envision an ideal curriculum for engineering ethics education. Essential themes emerged, such as the integration of ethics in practice, the role of the teacher, and the rationale for ethics curricula, including the need to bring students' attention to professional identity and responsibility at the core of their learning experience. Another contribution of this chapter was to highlight aspirations and potential for further developments in ethics curricula within engineering programs across the globe while serving the engineering education research community on ethics by tracing future lines of inquiry.

References

Barfod, K., & Bentsen, P. (2018). Don't ask how outdoor education can be integrated into the school curriculum; ask how the school curriculum can be taught outside the classroom. *Curriculum Perspectives, 38*, 151–156.

Bielefeldt, A. R., Canney, N. E., Swan, C., & Knight, D. (2016). *Efficacy of macroethics education in engineering*. In 2016 ASEE Annual Conference & Exposition.

Bielefeldt, A. R., Polmear, M., Knight, D., Swan, C., & Canney, N. (2018). Intersections between engineering ethics and diversity issues in engineering education. *Journal of Professional Issues in Engineering Education and Practice, 144*(2), 04017017.

Bombaerts, G., Doulougeri, K., & Nieveen, N. (2019). *Quality of ethics education in engineering programs using Goodlad's curriculum typology*. In 47th SEFI Annual Conference: Varietas Delectat: Complexity is the New Normality.

Bombaerts, G., & Vaessen, B. (2022). Motivational dynamics in basic needs profiles: Toward a person-centered motivation approach in engineering education. *Journal of Engineering Education, 111*(2), 357–375.

Boshuijzen-van Burken, C., Singh, S., & Baggiarini, B. (2022). A feasibility study for inclusion of ethics and social issues in engineering and design coursework in Australia. In H.-M. Järvinen, S. Silvestre, A. Llorens, & B. Nagy (Eds.), *Towards a new future in engineering education, new scenarios that European alliances of tech universities open up* (pp. 140–150). Universitat Politècnica de Catalunya.

Britannica. (2023). *John Goodlad: Canadian-born educator and author*. Retrieved September 23, 2023, from https://www.britannica.com/biography/John-Goodlad

Bucciarelli, L. L. (2008). Ethics and engineering education. *European Journal of Engineering Education, 33*(2), 141–149.

Colby, A., & Sullivan, W. M. (2008). Ethics teaching in undergraduate engineering education. *Journal of Engineering Education, 97*(3), 327–338.

Davis, M. (1991). Thinking like an engineer: The place of a code of ethics in the practice of a profession. *Philosophy & Public Affairs, 20*(2), 150–167.

Downey, G. L., Lucena, J. C., & Mitcham, C. (2007). Engineering ethics and identity: Emerging initiatives in comparative perspective. *Science and Engineering Ethics, 13*(4), 463–487.

Farahani, M. F., & Farahani, F. F. (2014). The study on professional ethics components among faculty members in the engineering. *Procedia-Social and Behavioral Sciences, 116*, 2085–2089.

Filippakou, O. (2011). The idea of quality in higher education: A conceptual approach. *Discourse: Studies in the Cultural Politics of Education, 32*(1), 15–28.

Flynn, R., & Barry, J. (2010). *Teaching ethics to engineers: Reflections on an interdisciplinary approach*. In 3rd International Symposium for Engineering Education.

Goldin, I. M., Pinkus, R. L., & Ashley, K. (2015). Validity and reliability of an instrument for assessing case analyses in bioengineering ethics education. *Science and Engineering Ethics, 21*, 789–807.

Goodlad, J. I. (1979). *Curriculum inquiry. The study of curriculum practice*. McGraw-Hill.

Gwynne-Evans, A. J. (2021). More than learning: Teaching and learning ethics within an electrical engineering undergraduate capstone course. *SAIEE Africa Research Journal, 112*(4), 171–180.

Gwynne-Evans, A. J., Chetty, M., & Junaid, S. (2021). Repositioning ethics at the heart of engineering graduate attributes. *Australasian Journal of Engineering Education, 26*(1), 7–24.

Herkert, J. R. (2005). Ways of thinking about and teaching ethical problem solving: Microethics and macroethics in engineering. *Science and Engineering Ethics, 11*, 373–385.

Hess, J. L., Strobel, J., & Brightman, A. O. (2017). The development of empathic perspective-taking in an engineering ethics course. *Journal of Engineering Education, 106*(4), 534–563.

Hollander, R. & Arenberg, C. R. (2009). Ethics education and scientific and engineering research: What's been learned? What should be done? Summary of a workshop. Retrieved September 21, 2023, from National Academy of Engineering: http://www.nap.edu/catalog.php?record_id=12695.

Huff, C. W., & Frey, W. (2005). Moral pedagogy and practical ethics. *Science and Engineering Ethics, 11*(3), 389–408.
Janssen, F., Westbroek, H., Doyle, W., & Van Driel, J. (2013). How to make innovations practical. *Teachers College Record, 115*(7), 1–43.
Kropp, R. P. (1973). Curriculum development: Principles and methods. *American Journal of Agricultural Economics, 55*(4), 735–739.
Li, J., & Fu, S. (2012). A systematic approach to engineering ethics education. *Science and Engineering Ethics, 18*, 339–349.
Martin, D. A., Conlon, E., & Bowe, B. (2021). A multi-level review of engineering ethics education: Towards a socio-technical orientation of engineering education for ethics. *Science and Engineering Ethics, 27*(5), 60.
Martin, M. W., & Schinzinger, R. (2005). *Ethics in engineering* (4th ed.). McGraw Hill.
Mitcham, C. (2009). A historico-ethical perspective on engineering education: From use and convenience to policy engagement. *Engineering -Studies, 1*(1), 35–53.
Monteiro, F. (2021). *Ethical skills needed by engineers to face future's challenges*. The 4th International Conference of the Portuguese Society for Engineering Education (CISPEE), Lisbon, Portugal, 1–6.
Palmer, J., Cooper, D. E., & Bresler, L. (Eds.). (2001). *Fifty modern thinkers on education: From Piaget to the present*. Psychology Press.
Pavel, A. P., Fruth, A., & Neacsu, M. N. (2015). ICT and e-learning–catalysts for innovation and quality in higher education. *Procedia Economics and Finance, 23*, 704–711.
Pfatteicher, S. K. (2001). Teaching vs. preaching: EC2000 and the engineering ethics dilemma. *Journal of Engineering Education, 90*(1), 137–142.
Polmear, M., Bielefeldt, A. R., Knight, D., Swan, C., & Canney, N. E. (2018, June). *Faculty perceptions of challenges to educating engineering and computing students about ethics and societal impacts*. In 2018 ASEE Annual Conference & Exposition.
Priestley, M., & Philippou, S. (2019). Curriculum is – or should be – at the heart of educational practice. *The Curriculum Journal, 30*(1), 1–7.
Quinlan, K. M. (2002). Scholarly dimensions of academics' beliefs about engineering education. *Teachers and Teaching, 8*(1), 41–64.
Rayne, K., Martin, T., Brophy, S., Kemp, N. J., Hart, J. D., & Diller, K. R. (2006). The development of adaptive expertise in biomedical engineering ethics. *Journal of Engineering Education, 95*(2), 165–173.
Romkey, L. (2015, June). Engineering, society, and the environment in the teaching goals and practices of engineering instructors. In *2015 ASEE Annual Conference & Exposition* (pp. 26–650).
Silvast, A., Laes, E. J. W., Abram, S., & Bombaerts, G. (2020). What do energy modellers know? An ethnography of epistemic values and knowledge models. *Energy Research and Social Science, 66*, 101495.
SLO – Netherlands Institute for Curriculum Development. (2009). *Curriculum in development*. Enschede, the Netherlands.
Sunderland, M. E., Ahn, J., Carson, C., & Kastenberg, W. E. (2013). Making ethics explicit: Relocating ethics to the core of engineering education. In *2013 ASEE Annual Conference & Exposition* (pp. 23–881).
Stokking, K., Leenders, F., De Jong, J., & Van Tartwijk, J. (2003). From student to teacher: reducing practice shock and early dropout in the teaching profession. *European Journal of Teacher Education, 26*(3), 329–350.
Terwel, J., Volman, M., & Wardekker, W. (2004). Substantive Trends in Curriculum Design and Implementation: An Analysis of Innovations in the Netherlands. In Curriculum Landscapes and Trends. Springer, Dordrecht. https://doi.org/10.1007/978-94-017-1205-7_9
Valentine, A., Lowenhoff, S., Marinelli, M., Male, S., & Hassan, G. M. (2020). Building students' nascent understanding of ethics in engineering practice. *European Journal of Engineering Education, 45*(6), 957–970.
van den Akker, J., Kuiper, W., Hameyer, U., Terwel, J., Volman, M., & Wardekker, W. (2003). Substantive trends in curriculum design and implementation: An analysis of innovations in the Netherlands. In J. van den Akker, W. A. J. M. Kuiper, & U. Hameyer (Eds.), *Curriculum landscapes and trends* (pp. 137–156). Kluwer Academic Publishers.
Wolverton, R., & Wolverton, J. (2003). Implementation of ethics education throughout an engineering college. In *2003 Annual Conference* (pp. 8–661).
Weil, V. M. (1984). The rise of engineering ethics. *Technology in Society, 6*, 341–345.
Wu, N., & Liu, Z. (2021). Higher education development, technological innovation and industrial structure upgrade. *Technological Forecasting and Social Change, 162*, 120400.

30

ASSESSING ENGINEERING ETHICS EDUCATION LEVERAGING STAKEHOLDER ENGAGEMENT IN ENGINEERING PROGRAMS

Alison Gwynne-Evans, Irene Magara, Esther Matemba, and Sarah Junaid

Introduction

Ethics is considered central to the professional practice of engineers (Harris et al., 2014), where engineers face a range of decisions relating to ethics daily (Kim et al., 2020). In this context, ethics education aims at "developing responsible and caring future engineers who can discharge their duty as professionals" (Balikrishnan et al., 2019) in diverse and changing contexts. Research has identified various options for incorporating ethics within engineering programs (Li & Fu, 2012; Martin & Polmear, 2023), the need to align engineering ethics education within university programs, and the ethical requirements required in the working environments where students are expected to practice. There have been multiple responses to this challenge, ranging from increasing measurability (Davis & Feinerman, 2012) and micro-assessments to preparing students for professional practice through broader experiences that foster reflection on identity and professional responsibility, both as individuals and as professionals (Hess & Fore, 2018). Engineering practice is positioned as requiring critical thinking, reflective action, and the exercise of ethical judgment (Riley, 2008/2022), thus requiring the redefinition of engineering responsibility beyond the technical field to include the context and culture in which engineering is practiced. Conlon and Zandvoort (2011) recommended expanding how ethical, professional, and social responsibility should be addressed within engineering programs. As such, ethics education needs to provide students with opportunities to develop the ability and confidence to exercise judgment in various contexts related to the practice of engineering (Basart & Serra, 2013).

Studies on ethics education within industry recommend that industry engagement may be effective in improving the quality of engineering education (Bucciarelli, 2008) and that improvements in ethics education require engagement with industry stakeholders. Academic–industry partnerships have been proposed and tested. Some efforts have been made to establish ethics advisory councils for industry practitioners to improve ethics education (Kim et al., 2020).

This chapter argues for the importance of broader stakeholder engagement, where this engagement provides opportunities for ethics education to connect with engineering practice. It will

DOI: 10.4324/9781003464259-36

examine four case studies wherein the assessment of ethics education can potentially be embedded in these stakeholder engagement opportunities.

This will require two interrelated processes: critical assessment of the process of establishing the stakeholder partnership and an examination of what can be learned about ethical practice in these interactions. Scaffolded stakeholder involvement in both these processes is seen as key to the learning experience, and a provisional framework is proposed, as indicated in Table 30.1. The table identifies elements of the stakeholder engagement process relating to the assessment of ethics education – highlighting contrasting elements of the stakeholder engagement process of each case presented in this chapter and suggesting opportunities for stakeholder engagement as potential sites for assessing ethics education.

The case studies profile a range of existing engagements among university and stakeholders, including industry engagement in program design, work placements, site visits, and community involvement in skills development. This recognizes that, although the formal ethics education of engineering students may take place within the boundaries of academia, it can usefully be supplemented by the strategic design of ethics assessments that build on stakeholder engagements.

The authors of this chapter are all practicing engineering educators situated across three continents; three of us teach ethics to engineering students, and one researches engineering education and co-leads the African Engineering Education Research Network. Our interests align on finding opportunities to enhance the learning experience of engineering students outside of the formal focus on disciplinary knowledge, recognizing the value of diverse contexts and backgrounds, and the necessity of building assessment practices that incorporate reflective engagement with practical experience. The case studies have been drawn from the universities where we work.

The need for ethics education relating to stakeholder engagement

Ethics education must permeate the engineering curriculum and anticipate engineering practice, incorporating the relational imperatives of ethics. The different types of relationships between a range of stakeholders (public or private, individual or group) demand a more open, participatory, and decentralized approach to engineering ethics (Basart & Serra, 2013).

How ethics is defined affects how ethics education is positioned in a program, where it is visible, and how it is assessed. Thus, it is strategically vital to draw on an expansive understanding of ethics that invites critical examination of the ethical practice rather than limiting attention to ethical knowledge. This requires a model of ethics education that engages with knowledge, concepts, skills, values, and attitudes and encourages examination of the interplay between theory, identity, and action (Gwynne-Evans et al., 2021). The concept of 'ethical becoming' proffered by Fore and Hess (2020) introduces dynamism to learning ethics and emphasizes the relational nature of the process.

This kind of ethics education requires students to develop practical strategies to make choices that affect action in complex situations, where options are not self-evident, and to see themselves as part of communities with intersecting interests. It looks at the interconnection of individual and group choices – to one another and to professional codes of conduct (Colby & Sullivan, 2008) – and examines the justification of these choices within a particular context. Stakeholder input can be necessary to position ethics education as a collaborative undertaking beyond the individual actor's control (Zhu & Clancy, 2023; Zhu & Jesiek, 2017). This provides a rationale for exploring stakeholder engagement as an opportunity to develop assessment strategies that measure various elements related to ethics education for engineers.

Table 30.1 Elements of the stakeholder engagement process relating to the assessment of ethics education

Location of project	*Stakeholder engagement*	*Key stakeholders*	*Point of entry*	*Form of assessment*	*Assessment criteria by university*	*Assessment criteria by external stakeholders*	*Explicit criteria for assessment of ethics education*
Mechanical; Biomedical and Design Engineering Aston University; United Kingdom	Industrial advisory board program review	Students; industry professionals; regulatory and professional bodies	N/A, part of the engagement is a voluntary activity for 2nd and 3rd years	Course review; in-person meetings and project pitch for students	N/A	N/A	Not yet explicit
Construction Management; University of Cape Town; South Africa	Service-learning project	Students; lecturers NGOs community members	1st year	Academic assessment and course evaluation	Yes; skills variety; task identification; task significance; autonomy and feedback	Yes, partially technical skills within a specified time period versus community response	Not explicit – possible to build in through collaboration
Faculty of Applied Sciences and Technology; Mbarara University; Uganda	Industrial internship program	Engineering and construction firms; engineering professionals; technical and administrative staff; university faculty; students	2nd and 3rd year	Surveys; feedback reports and in-person meetings	Yes, communication skills; teamwork; time keeping	Yes, work ethics; problem-solving; self-management and development	Yes, direct engagement of engineering students with industry supervisors for the two months' period where the values of work ethics are observed
Civil Engineering; University of Cape Town; South Africa	Professional practice site visit	Engineering and construction firms; engineering professionals; consultants; technical and administrative staff; students	4th year	Reflective essays relating to Graduate Attributes (GA7, GA8; GA10 & GA11)	Yes, professional skills such as punctuality; courtesy; teamwork; abiding by rules; communicating professionally	Yes, partial input from professionals within an academic setting but not formally incorporated in site visit or in assessment processes	Reflective essays by students on site experience and interactions on site relating to: GA7 Impact of engineering on the environment; GA8 Teamwork; GA10 Engineering professionalism; GA11 Engineering management

Definition of stakeholders within the engineering education ecosystem

Stakeholders are recognized as individuals, groups, and organizations that have an interest in the objectives, processes, and outcomes of an organization (Freeman, 1984). As such, stakeholders are not neutral actors but have vested interests that may or may not align with those of the educational institution or engineering program with which they interact (Martin et al., 2021).

In higher education, engineering program stakeholders may be divided into internal and external stakeholders. *Internal* stakeholders include the students, academics, administrators, and the academic institution. To some degree, parents and carers of students may also be included as internal stakeholders, but these are not a focus of this chapter. *External* stakeholders include industry, companies, and organizations that intersect with the academic institution and/or fund activities and people, as well as the communities – including non-profit organizations, engineering professional bodies, and social enterprises – that potentially benefit from the engineering program activities and engineering practice more widely. This chapter will focus on interactions with external stakeholders, other than the accreditation bodies, who play a significant role in determining the direction of engineering programs.

Stakeholders, including the educational institution itself, the students, and community and industry partners, can be seen to impact and be impacted by the engineering program's projects and interactions. The institution's implicit and explicit engagement processes with external parties potentially impact the scope of engineering ethics education and its assessment. This highlights a largely underexplored area of ethics education within engineering relating to evaluating stakeholder engagement – in terms of how its processes model both ethical engagement and student learning and assessment.

The opportunity to assess ethics education provided by stakeholder engagement

Stakeholder engagement is increasingly recognized as part of engineering practice. This requires students in engineering programs to anticipate and rehearse how to manage processes that involve the exercise of judgment and decision-making relating to stakeholder interaction. This relates to technical problem-solving and management decisions concerning people, where consequences may not be predicted with the same confidence as technical processes, yet where decisions must be made. Stakeholder partnerships introduce a range of possible interventions where the assessment of ethics education extends beyond individual student responses and scores to include evaluation of the opportunities for engagement with industry and community stakeholders. This emphasizes accountability as a collective responsibility.

Practical considerations concerning the engagement process need to be identified, as these can affect the quality of stakeholder engagement (O'Riordan & Fairbrass, 2014). These considerations affect power relationships, priorities, outputs, and reach. They include but are not limited to funding, affiliations of program contributors, and the process of determining who contributes to the program's benefits and objectives.

From the above, it is evident that the extent and nature of partnerships and stakeholder engagements influence priorities and ethical decision-making. Envisioning an undergraduate ethics education program that engages communities and industry stakeholders as partners requires critical reflection on the following:

The role and responsibilities of the university and its faculty

The university operates primarily at the management level to provide an environment for teaching, learning, and research by ensuring staff and student programs run smoothly.

Faculty members collectively facilitate teaching, learning, and assessment through classroom and extra-curricular activities.

The role and responsibilities of industry stakeholders

Industry stakeholders determine the trajectory of future employment and must position themselves strategically to further their commercial and strategic interests. They play a vital role as a partner in ensuring students and professionals acquire and maintain ethical practice. Industry engagement provides work experience through internships, graduate trainee positions, or full-time employment opportunities and thus influences priorities. Industry's vested interest requires careful consideration during stakeholder engagement.

The role and responsibilities of community stakeholders

The role of community stakeholders is equally important (to that of industry and the university) in identifying, promoting, and assessing ethical practice. This role includes the regulatory component required by engineering professional bodies. It involves highlighting social and environmental obligations in ethical practice profiled through the work of non-profit organizations and social enterprises with a strong focus on community development.

These considerations affect choices and synergies regarding program goals and learning objectives for ethics education and require careful planning and design to incorporate. There can be significant areas of synergy but also areas where interests conflict and cannot align. These differences need to be recognized and dealt with strategically.

Making decisions within contexts with unpredictable outcomes involves risk, and making judgments necessitates grappling with ethical theory and action alongside technical proficiency. Stakeholder engagements thus provide a fertile context for evaluating this intersection.

Donaldson and Preston (1995) distinguished three kinds of stakeholder engagement analysis: descriptive, instrumental, and normative. This chapter will translate these into three reflection levels regarding the stakeholder engagement process. Reflection is positioned as an activity that connects theoretical content to understanding (Correia & Bleicher, 2008).

As such, the first level of reflection requires a *descriptive engagement* where facts relating to the interaction are relayed, and where stakeholders such as industry and community groups may have little or no significant role in contributing to educational objectives and outcomes. This is distinguished from *instrumental engagement*, where the goal of the interaction may be identified in terms of a contractual benefit to the stakeholders. Here, ethics education aims to result in students acquiring "an understanding of professional and ethical responsibility" (Bucciarelli, 2008, p. 141). The third alternative is a *normative engagement* with the stakeholder engagement process, potentially influencing educational objectives and outcomes. These three levels of reflection correlate with determining the value of an intervention to the stakeholders and measuring their experience of and engagement with ethics education.

Four case studies, each selected to capture breadth in terms of stakeholders involved and across three disparate global contexts, explore the interaction of stakeholders across these three levels.

Assessment of ethics education through reflection at individual and program levels

Assessment of ethics education relates to students' outcomes and abilities and to the internal coherence and goals of the engineering program in which the assessment is embedded. At an ecosystem level, it is crucial to distinguish assessment concerning program or institutional goals from assessment of learning objectives. This process of evaluating program goals recognizes that the assessment process itself builds in values that guide and shape engagement with stakeholders and can, in turn, potentially be shaped by the stakeholders.

Assessment is a powerful tool in the learning process, highlighting both the content areas and skills that are important and those that require attention (Davis & Feinerman, 2012). Assessing ethics education within the curriculum can be enhanced by reflecting on real-life opportunities where students engage with stakeholders as role players in various contexts (Hess et al., 2023). Stakeholder engagement thus provides a valuable context to evaluate where this can be achieved. As such, an assessment of ethics education needs to be formulated to assess the students' grasp of the different elements of ethics, including concepts, knowledge, skills, values, and attitudes, occurring in stakeholder engagement. These areas relate to the intersection of knowledge, identity, and action within a context. Here, the value of the assessment is recognized as the quality of the students' reflection on their learning relating to ethics, which is distinct from the overt testing of content knowledge.

In the context of ethics education aligned with stakeholder engagement, students are learning to anticipate and justify a plan of action to deal with processes that are not under their direct control as actors. This draws attention to the way in which choice is justified within a context and that the justification of choice is dependent on principles, values, and processes of engagement rather than only on the outcomes of engagement. As such, it also draws attention to the role and profile of attitudes and values within the learning process and within professional practice.

Assessment in engineering ethics education in connection to curricula crucially affects what is taught and learned through practical engagements (Davis & Feinerman, 2012). Evaluation of the effectiveness of ethics education within an engineering program has traditionally focused on measuring students' knowledge and awareness in terms of case studies, the details of codes of conduct, and skills such as the ability to reference or develop an argument (McGinn, 2003). The engagement with stakeholders impacts students' learning in both explicit and implicit ways, potentially building and strengthening the ethical practice of various stakeholders, including the engineering students, faculty, staff, and the community and industry stakeholders they interact with.

Interactions and partnerships with industry and communities during the undergraduate degree potentially provide opportunities to enhance the ethical proficiency and judgement of students as well as to draw on and to enhance the established ethical processes of stakeholders.

Case studies exploring assessment of ethics education with stakeholder engagement

We selected the four case studies involving stakeholder engagement as examples of extending ethics education concerning what is taught and how learning is assessed within engineering programs. Each case study's analysis is structured in four sections: a contextual background, a description of the engagement with the respective stakeholder, an analysis of the instrumental value and benefit of the intervention, and a normative reflection with recommendations. The presentation is based on Donaldson and Preston's (1995) recommendations demonstrating the potential for engagement to enhance the learning and assessment of ethics in engineering education.

Case study 1: Aston University, United Kingdom: industrial panel program review

Contextual background

The Mechanical, Biomedical, and Design Engineering (MBDE) Department at Aston University sits within the College of Engineering and Physical Sciences. MBDE runs four programs:

Mechanical Engineering, Design Engineering, Biomedical Engineering, and Product Design. This case study explores the introduction of an industrial program review process to help monitor and enhance two of these programs: Mechanical Engineering and Design Engineering.

Aston University developed as a traditional technological university, working closely with industry, commerce, and the surrounding communities and pioneering research in metallurgy and other sciences in its early days. These roots have continued, and the MBDE department today strongly emphasizes developing students to be industry-ready – this involves equipping students with the interpersonal skills necessary to enter an interdisciplinary, team-based workforce. The UK Engineering Council accredits the university's engineering and product design programs through the Institute of Mechanical Engineers (IMechE) and the Institution of Engineering Designers (iED), respectively. An external panel of chartered engineers from academic and industrial backgrounds runs these accreditation evaluations.

'Chartered Engineer' is a professional status recognized by UK regulatory engineering bodies, such as the IMechE. Becoming a Chartered Engineer requires achieving the requisite academic qualification and documenting specified competencies (based on work-based learning) after graduation. Although being chartered is not a legal requirement to work as an engineer in the United Kingdom, the university's engineering community conveys the expectation that a graduate engineer and designer from an accredited program will continue their membership and build on their professional development postgraduation to chartered status after several years of work experience.

However, beyond the formal accreditation process in engineering programs in the United Kingdom, there are no formal methods to review the teaching programs where there is industrial involvement. Academic reviews occur internally and externally annually and cover operational elements of the program, such as curriculum structure, delivery, assessment, and processes for feedback and moderation. There is a clear gap in stakeholder involvement within program development. From the viewpoint of ethics, the challenge to address is that the ethical outcomes, skills, awareness, reflection, and practice required are not well-informed or well-influenced by industrial stakeholders.

One area where this challenge can be addressed is the growing emphasis on entrepreneurship within engineering programs. Entrepreneurship presents an opportunity to embed practical ethics education in the program. Developing entrepreneurial skills lends itself well to industry involvement and aligns well with existing engineering, commerce, and business programs. Entrepreneurship is a potentially powerful way to provide the experiential learning needed for ethical character traits to develop within the practical context of industry and business.

The Mechanical and Design Engineering programs at Aston University embed the CDIO (Conceive-Design-Implement-Operate) teaching framework (Worldwide CDIO Initiative, n.d.; Malmqvist et al., 2022) and the 12 Standards (The CDIO Standards v 2.0, 2010) into the curriculum structure, curriculum delivery, faculty development, and teaching spaces (Worldwide CDIO Initiative, n.d.). The curriculum structure puts active learning at the core. It brings to the heart of the programs the interpersonal skills taught within a professional teamwork setting to develop well-rounded, ethically responsible, and technically competent graduate engineers ready for the workplace. The curriculum at Aston University consists of cycles of 12-week team projects that address real-world problems within a team-based professional working environment. On completion of each project, a product or artifact is delivered and tested as part of the assessment. In parallel, technical modules or courses provide the technical know-how to support the team projects. Within this structure, one of the second-year team projects on 'designing for the user' requires student teams to develop a consumer product based on a need from a niche user group. The design-build-test team project was used in the case study to review the program with the industry stakeholders. This is where entrepreneurial skills are introduced, and an entrepreneurial pipeline from product ideas to incubator funding is introduced to students.

Description of stakeholder engagement

The MBDE group at Aston University has an industrial advisory board that meets quarterly to discuss program challenges and strategic developments, drawing on changes and trends observed by industrialists on the panel who represent a range of engineering and design sectors. This case study, therefore, explores the use of the industrial advisory board to formally review the program for Mechanical Engineering and Design Engineering at an operational level. The advisory board actively supports the entrepreneurial pipeline by running a voluntary panel event where students can pitch their ideas and products. The review process was designed to capture the ethical awareness and skills that the industrialists on the panel viewed as most important for graduates to know.

In this case, a review form was developed and used to capture reflections for the second-year team project on the following areas: assessments, innovation and good practice, learning and achievement opportunities, engineering ethics (ethics; sustainability; equality, diversity, and inclusion or EDI), and any other matters or suggestions. The review was conducted in person, with the panel visiting the teaching spaces and reviewing the project-based structure, delivery, and artifacts produced. The form was completed after the visit, and the academic lead incorporated the reviews into the program's subsequent development. The panel where students volunteered to pitch their ideas and products was held after all assessments in the course were completed. This is part of the entrepreneurial pipeline where students take on board the feedback and implement changes to apply for a local funding opportunity and start-up support available for students in the region.

Analysis of stakeholder engagement

The review highlighted the need to enhance commercial awareness among students in design projects. It indicated a need for more focus on scoping problems and setting realistic product design specifications. The evaluation panel highly valued the practical hands-on elements and team environment, validating the importance of working in teams with dissimilar skill sets in industry. Working with technicians with experience and know-how was valuable to students' learning and entrepreneurial skills development. There is scope to develop assessments and marking matrices around various industry practices, such as measuring against a product design specification. These detailed criteria assist students in taking ownership and responsibility for delivering on the product requirements they had set.

For the second-year team project, providing design briefs with different corporate objectives may also enhance commercial and entrepreneurial awareness, such as considering customization versus mass production. These approaches will have associated challenges, including implications for finance, human capital, ethical factors, and sustainability.

The panelists drew from their experience working with placement students in their companies. One observation was the need to train for external and disruptive factors that impact the critical pathway in product development, such as a machine going offline or a standard part no longer being available. Industry seeks experience in agility and adaptability to changes. Program development actions could include focusing on manufacturing processes, introducing a disruption point in the project, or setting hypothetical questions in oral assessments around these factors.

In reviewing the program's ethical considerations and equality, diversity, and inclusion (EDI), the industrial panel highlighted the importance of understanding the user, not second-guessing what the consumer experiences throughout the stages of drafting specifications, ensuring safety, inclusive access, and ergonomics. While these are challenging to implement due to the ethical evaluation needed for user-testing work, in-class exercises can be introduced to explore this.

The panel's recommendations largely concerned the need for additional focus on entrepreneurship training and support in defining business and market goals. The ideas pitch event was attended by just under 10% of the teams that completed the second-year team-based project. This aligns with the national picture, where the total early-stage entrepreneurial activity rate is approximately 10% over 2020–2021 (Hart et al., 2022). This correlation may be coincidental. However, the number of engineers who become entrepreneurs in their field is small compared to engineers working in industry and research. Yet entrepreneurial skills are still sought after within small to medium businesses and large corporations where innovation and enterprise activity are standard.

Reflection and recommendations

The aim of introducing an industrial program review was to evaluate the program at an *operational* level to bring more practical insights into program delivery from industry and strengthen the entrepreneurial pipeline for students with ideas and an interest in starting their own businesses. The review focused primarily on the nature of assessments, processes, standards, and ethical practices used in industry.

The panel's recommendations concerned additional focus on entrepreneurship training and support in developing agility and adaptability via the design process. On evaluating the review, the program committee sought to expand on the conceive phase (The CDIO Standards v 2.0, 2010), both from an ethical and commercial standpoint, designing assessments that are comparable to industry practice and that bring agility and adaptability to the learning experiences to cope with changes in a team-project setting. The action from this was to focus more on the early stages of the project, where teams are scoping the problem and identifying the users' needs more deeply, ensuring that critical review cycles occur to refine this phase. This brings a more sustainable and ethical approach to product development for the academic team and student learning, thus avoiding costly mistakes during the design prototyping stage.

The case study above highlights a method of involving industrial stakeholders more actively and practically rather than just at a descriptive and strategic level. This is achieved by engaging them in the project's structure and building entrepreneurial skills, focusing on ethical and commercial considerations. This provides a direct pathway connecting industrial stakeholders as co-developers of the curriculum to meet the needs of current and future industries. It is thus part of the universities' responsibility to flag ethical practice as part of entrepreneurship support.

Case study 2: University of Cape Town, South Africa: engaging the community through service-learning opportunities

Contextual background

This case study examines a curriculum intervention within the Engineering and the Built Environment Faculty at the University of Cape Town that engaged with community stakeholders over more than 14 years (Massyn & Le Jeune, 2007; Le Jeune & Massyn, 2018). This intervention is positioned as a service-learning project incorporating aspects of ethics education that are impossible within the academic environment. It provides additional assessment opportunities relating to ethics education that are inherently distinct from those within the university context.

The initiative is positioned in the first year of the 4-year Construction Management course. Whereas engineering programs in South Africa fall under the remit of the Engineering Council of South Africa (ECSA) to accredit, Construction Management falls under the purview of the Council for the Built Environment (CBE). Both accreditation bodies set out to define and assess learning opportunities that contribute to the development of knowledge, skill, and character attributes required by graduates.

Ethics can be positioned as a key facet of construction management, covering elements such as professional conduct, safety, public interest, and sustainability. Consequently, the focus on a program intervention requiring community engagement provides a helpful case study of how an intervention potentially expands opportunities for ethics education into a practical context.

The first-year community-build project is situated within the South African Council for the Project and Construction Management Professions (SACPCMP) program requirement of service learning (SACPCMP, Criterion 1) to meet "the needs of students and … stakeholders." Service learning is positioned as a "course based, credit bearing educational experience in which students participate in an organized service activity in such a way that meets identified community needs so as to gain further understanding of course content, a broader appreciation of the discipline, and an enhanced sense of civic responsibility" (Bringle & Hatcher, 2009 as quoted by Le Jeune & Massyn, 2018, p. 350). Here community engagement directly contributes to the achievement of program outcomes.

This case study is examined in two ways. First, we evaluate how the service-learning intervention provides students the opportunity to identify a range of factors relating to ethical processes that contribute to effective engagement within the given community. These include factors such as the (co-)determination of objectives, communication with stakeholders, collaborative planning of the different stages of the intervention, and the final evaluation of the success of the intervention. Second, we examine how the intervention provides the faculty with opportunities to assess the ethics education of students as part of the envisaged learning outcomes and the objectives of the broader program.

Description of stakeholder engagement

The process of implementing a service project started with the identification of objectives that could be achieved through community engagement. These original objectives related to the first-year syllabus, which covers the practical construction skills required in the building of a double-story housing unit, including foundations, bricklaying, plastering, and painting, as well as the plumbing and electrical first-fix installations (Massyn & le Jeune, 2007).

The service-learning model was part of the practical training requirement for degree purposes (SACPCMP). This replaced the requirement of partnering with a skills training organization to attain practical skills with an initiative where students were required to contribute to a community-build initiative, incorporating a broader range of potential learning outcomes, combining technical skill with professionalism and social awareness. It is important to note that at the project's initiation, ethics education and professionalism were not an explicit focus of the assessment.

The delivery of the service project changed significantly over the years and required flexibility – from partnering and funding a community-built house for Habitat for Humanity (H4H) to partnering with the Niall Mellon Township Trust (NMTT) company and then through student-activated projects to facilitated community builds. Changes in the organization and management structure of the community-build projects were necessitated when the partnership with NMTT ended. This resulted in senior students taking up the responsibility of initiating their own community-building projects under the guidance of the academic course conveners.

This contrasted the original *facilitated community build* with *student-activated builds*. The overt, practical focus of facilitated community builds was on factors relating to planning, supervision, quality control, and physical toll. In addition, there was scope to develop assessments relating to ethical reflection, skills including communication, and the ability to identify and contrast values such as social or environmental justice as well as to make space for impacting the power balance of actors. Here, community stakeholders were recognized and positioned to have knowledge and resources to contribute rather than be situated as beneficiaries of the project.

The practical focus of student-activated builds has been on skills and funds – as students drive and take responsibility for projects, the level of complexity of the tasks selected needs to be appropriate for the students' current range of skills. As a result, student-activated projects tend to focus on managing more straightforward technical tasks such as painting and the layout and construction of pathways, rather than the complexities of building. In addition, student-activated builds potentially include assessment of professional skills such as project management and teamwork.

Analysis of stakeholder engagement

The intervention has undergone significant changes in design and implementation over time, indicating a shift in its approach. Le Jeune and Massyn (2018) noted differences and similarities between the initial facilitated community build, which involved building according to the external partner's standards and schedule, and student-activated builds, where students approach external parties and identify needs that can be addressed within the project's constraints and the students' skill range.

Changes in implementation were the result of a variety of factors, including efficiency, cost, and risk, the over-supply of labor (50 students working on one house), an associated shortage of production- and learning-related tasks, and the need to weigh up the long-term sustainability of the fund-raising required. Progressively, over 14 years, risk-related matters, such as securing the safety of students on sites in specific residential locations, were recognized as significant threats to the sustainability of the project.

In their reflection on the process of coordinating community builds, Le Jeune and Massyn (2018) unpack the project's contribution in terms of five dimensions relating to skills development (skills variety, task identification, task significance, autonomy, and feedback). It is notable that ethics and professional skills are not explicitly articulated, although they form part of the learning experience. It is important to consider that this is a first-year experience rather than a third- or fourth-year level experience. Nevertheless, it would be possible to set provisional targets regarding the professional skills of learners, such as teamwork, punctuality, and problem-solving.

Whereas the facilitated builds assessed direct outcomes (such as the completion of a house in a specified time period), the student-activated community builds relied on informal and relationship-based responses to assess the impact of the intervention on the community. At completion, the 'handover' has included members of the community singing and expressing appreciation.

This indicates the range of ways the impact of engineering or construction on a community can be measured – through quantitative and technical measures or through qualitative measures such as expressing gratitude. This distinction correlates with differences in assessing ethics education, relating to assessing only technical processes or including assessment of reflection on experience in a way that surfaces human responses and relationships.

In this case, reflective practice is utilized as an assessment tool that directly relates to the objective of explicitly building ethics education into the process of community engagement. This brings reflection to bear on the impact of community engagement in the broader program and on the educational goals relating to ethics. Reflective practice thus contributes to consolidating a deeper understanding of the process.

Because engineering necessarily impacts environments and communities, practical experience developing the skill to reflect on relational processes as part of ethics education is valuable, as this relates to real people in existing communities. This skill of exercising ethical judgment is additional to developing the skill of exercising technical judgment pertaining to scientific and engineering methods. In this case study, the ethical reflection relates to how community interactions and interventions affect the stakeholders and impact future options.

Reflection and recommendations

Evaluating the potential contribution of this community engagement to the assessment of ethics education requires two levels of reflection, firstly regarding the practical choice and organization of activities towards students' educational goals, and secondly the reflection by the students on the significance of the experience itself. This case study draws attention to the importance of evaluating different facets of the service-learning arrangements in terms of skills developed. At the same time, it raises the equally vital requirement to foreground the students' learning relating to decision-making and judgment that connects to practical ethics education beyond the basic level of knowledge.

Learning objectives must be explicitly formulated to address ethical responsibility competencies in community engagement projects aligned to engineering program delivery. These objectives must be translated into explicit learning activities and assessment designs to incorporate and build on students' ability to reflect on their experiences.

Case study 3: Mbarara University of Science and Technology, Uganda: work experience through industrial internship

Contextual background

The Faculty of Applied Sciences and Technology (FAST) Strategy and Implementation Plan at Mbarara University was conceived in 1998 by the vice chancellor, Professor F. I. B. Kayanja, and initially documented by Stephen J. Palmer, a visiting engineer. The faculty's overall vision was to be a center of excellence in the provision of quality training and promotion of research in applied and multidisciplinary science and technology (Mbarara University of Science and Technology, 2023). This case study is based on preliminary research on the contribution of the work experience initiative.

This training involves industrial internship opportunities offered to students during their second and third years of the engineering program. FAST has five undergraduate programs in biomedical, electrical and electronics, civil, mechanical, and petroleum engineering.

The goal of industrial internships is to instill a culture of work and ethical practice in engineering students (FAST, 2021). Industrial internships involve collaboration between the university and industry, where internship placements are offered to students for 2 months as a requirement for fulfilling their undergraduate program at the end of their second and third years in the engineering program.

Description of stakeholder engagement

Stakeholder engagement at FAST, regarding the students' industrial internship program, starts with identifying key stakeholders, namely students, industry partners, university faculty members, and other relevant stakeholders, including regulatory and professional bodies. After stakeholder identification, engagements between faculty and industry regarding the assessment of the industrial internship program in preparing students for ethical engineering practice involve the design of assessment rubrics and gathering feedback at the end of the training period using surveys and in-person meetings.

Regarding assessment requirements, this initially entailed faculty circulating a rubric requiring no direct engagement from industry. Through interaction between the faculty and industry, the lack of clear parameters to the rubric was identified as a shortcoming in assessing the ethical component of the skills developed during the internship.

The faculty realized that the rubric was inadequate, and thus, they decided to collaborate with the industry to improve it. Companies were approached for input based on their size (more than ten placements) and their time associated with the faculty. This engagement resulted in a new rubric. Four cohorts have been assessed using the new rubric design so far.

The assessment process of students involves three areas: the university supervisor's assessment weighted at 45%, the assessment of the students' technical report writing at 30%, and the industry supervisor's assessment at 25%. The evaluation of the students' ethical and professional skills is conducted using the two reports from the university and industry supervisors, which constitute 70% of the total assessment process. This assessment covers communication skills, problem-solving, self-management and development, time management, and teamwork, which, when combined, align well with ethics education. The technical report, constituting 30%, is compiled by the student to reflect on their technical skills learning.

In line with answering the needs of the key stakeholders through the industrial internship program, FAST interacts with industry partners through in-person meetings and surveys to identify the skills, knowledge, and experience related to ethics that industry partners value in interns. This information is later used during engineering curriculum review to tailor the ethics course to industry needs. The same engagement model is used between regulatory and professional bodies and FAST to promote ethical standards of the students throughout the curriculum. It is worth noting that professional bodies such as the Uganda Institute of Professional Engineers (UIPE) and Engineers Registration Board (ERB) are passionate about ethical practice and hold workshops in universities in the form of sensitization.

Analysis of the stakeholder engagement

In line with industrial internship, the current approach demonstrates a commitment to ensuring that the program prepares students for the realities of the engineering profession, including the ethical considerations that are important in the workplace.

The primary areas of stakeholder engagement at FAST involve gathering feedback through surveys and in-person meetings from key stakeholders such as industry partners and regulatory and professional bodies. These surveys and in-person meetings are designed by university staff and administered to the above stakeholders. This information or feedback is currently solely used for curriculum review improvement. It is essential to recognize a variety of areas where this information could be helpful, such as in policy formulations regarding ethical practice in the world of work.

The initial approach at FAST recognized the possibility of bias and limited perspectives in the feedback gathered. The need to incorporate additional perspectives was identified to mitigate this, and industry stakeholders were onboarded to provide input on the rubrics to reflect industry requirements.

Based on the in-person feedback from industry and students' grades, preliminary evidence suggests an improvement in ethical conduct relating to communication skills, self-management, development, and problem-solving using the revised rubric.

Reflections and recommendations

Stakeholder engagement is critical to ensuring the effectiveness of industrial internship programs, and it is crucial to create a space where stakeholders can contribute to the assessment process.

It would be helpful to revise the new rubric and the role of the weighting of the industry supervisor's assessment further, recognizing the potential effect of external evaluation of practical components of ethics education.

The description of stakeholder engagement at FAST highlights the importance of gathering feedback from various stakeholders to ensure that the program answers their respective needs and expectations. This requires critical engagement with the learning objectives of the industrial internship program, thus creating additional opportunities to assess the ethical components that contribute to students' work readiness.

Whereas the current methods for gathering feedback on the level of professional and ethical skills demonstrated by the students are effective, FAST may need to consider expanding the range of stakeholder feedback to include students. This can be done by requiring student reflection as an assignment or course evaluation.

In conclusion, the stakeholder engagement approach at FAST has yielded improved assessment rubrics for the industrial internship and increased student internship placements proffered by industry to the university. This comes despite the competitive nature of these positions and the increasing number of universities offering engineering programs. In an internship program, feedback and reflection from relevant stakeholders are seen to be effective in enhancing the acquisition of ethical and professional skills and knowledge.

Case study 4: University of Cape Town, South Africa: industry engagement as part of the professional practice course

Contextual background

In South Africa, the civil engineering industry is regulated under the Engineering Council of South Africa (ECSA) and the Council of the Built Environment (CBE), where industry bodies, such as the South African Institute of Civil Engineers (SAICE), play an essential role in establishing a culture of ethical compliance and professionalism. Members of the professional industry bodies are expected to mentor graduate engineers into the profession. Standards and objectives for ethical conduct are laid out in the Engineering Council of South Africa's (ECSA) Code of Conduct (ECSA, 2017) and the South African Institute of Civil Engineering's (SAICE) Code of Ethics (SAICE, 2016).

Civil engineering students at South African universities must gain practical experience at a working engineering construction site as part of the undergraduate civil engineering program. Within the Civil Engineering Department of the Faculty of Engineering and the Built Environment at the University of Cape Town, site experience is positioned within a fourth-year Professional Practice course. As a fourth-year capstone course, the course evaluates 5 of the 11 graduate outcomes (ECSA, 2020) built up during the previous years. The five graduate attributes (GAs) that are assessed within the course are:

- GA6 – Professional Communication
- GA7 – Impact of Engineering on the Environment
- GA8 – Individual, Team and Multidisciplinary Working
- GA10 – Engineering Professionalism
- GA11 – Engineering Management

Of these, GA10, Engineering Professionalism, is considered the area where ethics is explicitly assessed, but as will be evidenced, there are areas of alignment across all five of these GAs.

The Professional Practice course includes engineering project management principles and processes; the roles and responsibilities of the different parties, the client, consultant, and contractor; and the elements that form part of an engineering project's life cycle. Students must engage with their professional and ethical responsibilities and duty of care, as outlined in the engineering regulator and the industry body's requirements and codes.

Within the course, learning takes place in terms of a combination of conventional lectures, applying and practicing skills, experiential learning on a construction site, and preparing and submitting several individual and group assignments over the semester (Gwynne-Evans, 2018). The course requires students, working in groups, to visit the construction site and, using the knowledge and insight they have gained on the course, to reflect on their experience in terms of professional practice requirements. This includes but is not limited to the desk study, site investigation, scope of work, feasibility study, health and safety, approvals, and regulatory requirements. This study, experience, and reflection are translated into a group report covering the different aspects. In addition, individual and group assignments require critical and reflective engagement with varying areas of professional practice, explicitly relating to teamwork, communication, professionalism, and ethics. Guest seminars profile representatives of engineering regulatory and industry bodies and representatives from cutting-edge national engineering project sessions. At the end of the course, students produce a group presentation that reflects on areas of the report and relates it to their experience on-site and in interactions with the engineering professionals and site staff.

Description of stakeholder engagement

The stakeholders in this case study include the students, faculty and departmental leadership, regulatory and industry bodies in achieving the graduate attributes, and the industry professionals and contractors involved in the course and on-site.

Students are placed in groups of five or six and allocated to an active engineering site. The group is provided with the contact details of the site liaison person. It is required to make contact, set up a site visit, comply with safety protocol, arrange transport for the group, and communicate professionally with the different parties. Information, documents, and photographic evidence from the site visit must be built into the final report and assignments relating to the individual graduate attributes – subject to individual site authorities' restrictions on documents or photographs.

The industry stakeholders that the students liaise with may be private or public-sector engineers or consultants. The engagement relies on long-term relationship-building, where the university course convener and liaison persons build and maintain respectful and bounded relationships with the engineering firms over time so that student groups will be welcomed and accommodated over multiple years. In the short term, this requires a level of trust in the professionalism and judgment of the students in their interactions on-site. Student groups experience differences in emphases and degrees of support depending on the scope and stage of the engineering project. Students liaise with the engineering companies, drawing on their developing professional communication and teamwork skills and principles and understanding of professionalism and ethics. Students are expected to relate and reflect on their experience of the site and the processes and principles that form the basis of the project report and the assignments.

Currently, there is no explicit requirement for industry stakeholders to contribute to determining the course or assessment objectives or outputs outside of their own internal processes regarding visitor protocol and safety while on the site. There is also no explicit requirement for industry stakeholders to reflect on the value of the experience for their own structures and processes.

Analysis of stakeholder engagement

The engagement between the university and industry stakeholders in the Professional Practice course relies on a respectful and ongoing relationship, extending to the relationship with the engineering contractors and consultants. It requires compliance with industry standards and site protocols, which allows for supported experiential learning and mentorship.

The implementation of processes to involve student groups explicitly reflecting on professional and ethical practice takes place in terms of the five graduate attributes assessed within the capstone course. These graduate attributes form part of the 11 graduate attributes in which the engineering regulator (ECSA) requires students to attain competence and understanding prior to graduation.

Within the course, there are multiple areas where the students are assessed on ethics and professionalism in the context of stakeholder engagement, including:

- The initial group planning around teamwork and professionalism early in the course;
- The reflection on their teamwork experience during and at the end of the course;
- Critical engagement with and reflections on the scope and relevance of the engineering professional code to the engagement with professionals on the site;
- Integrating theoretical knowledge with on-site experience in four reflective essays on the graduate attributes;
- The professional requirements involved in rehearsing and presenting the group presentation to portray the significance of the learning from the site experience; and
- Professional presentation of the draft and final versions of the report – in terms of content and effectiveness in covering areas relating to the scope, roles, and management of the specific project.

Together, these assessments evaluate a wide range of professional skills and knowledge, building into the students' understanding of professionalism and ethics necessary for professional practice based on their experience on the construction site.

Although there are no explicit requirements for industry stakeholders to provide input on the course assignments, the current processes potentially offer a model to stimulate explicit involvement from industry in setting objectives for the interaction. This involvement can potentially feed into the firm's processes and priorities, emphasizing modeling and promoting professionalism and ethics in site interactions – those that relate to the student visitors and the company's processes.

Reflection and recommendations

Student learning relating to ethics and professionalism occurs through formal input within the university and experiential learning on-site. Students learn from their engagement with practicing professionals about professionalism and what protocol is required. Reflection on both formal and informal learning must be articulated through the assessment process and the assignments, requiring reflection and consolidation of the experiences on-site and the interactions with professionals in the field.

Input from industry stakeholders into the professional practice course currently takes place through the formal accreditation process of the engineering program by the industry regulator, in this case, the Engineering Council of South Africa (ECSA), that takes place every 5 years.

There is currently no requirement for industry to provide input or feedback on the interaction with the student groups, and no industry-based report is required. There may be the potential for the engineering firm to provide ongoing, general feedback on the students' professionalism level. This could provide a model that stimulates self-reflection and assessment of the students and the company's own processes. This could potentially emphasize the importance of modeling and promoting professionalism and ethics during on-site interactions, which could support the engagement process through site visits by undergraduate groups.

This analysis and reflection demonstrate the potential of university and industry stakeholder engagements to positively impact the assessment of professionalism and ethics both within the university and within industry engagements. This can contribute to broader industry awareness of the need for and value of professionalism and ethics, potentially impacting the industry stakeholders' processes and values.

Conclusions

The focus on stakeholder engagement in engineering programs recognizes the potential instrumental and normative benefits of that engagement in terms of determining principles of engagement and evaluating the outcomes of the engagement. The four case studies presented in this chapter demonstrate how interactions with stakeholders beyond the academic environment provide opportunities to develop synergies that can potentially strengthen the development and assessment of ethics education for engineering students. Whereas theoretical knowledge tested within the university can assess knowledge about ethics as content, reflection on experience allows for consideration of a broader range of elements related to ethics education, including interpersonal skills and leadership, values, and attitudes. This anticipates that the collaboration with external stakeholders benefits the different stakeholders so that there is mutual learning and mutual contribution. Table 30.1 contrasts different elements of the stakeholder engagement process, indicating a direction to evaluate opportunities for stakeholder engagement as potential sites for assessing ethics education. Our analyses in the case studies suggest it is crucial to foreground a process that enables external stakeholders to explicitly articulate their interests and values alongside those of the university in formulating and assessing educational goals. It is also essential to recognize the university's interdependence with stakeholders in the setting and the achievement of goals relating to ethics education.

Engineering programs in many parts of the world are not explicitly required to assess their engagement or interaction with stakeholders outside the formal accreditation process. The requirement that engineering programs engage stakeholders anticipates that this engagement can provide a normative model of the process and outcomes. This can be potentially valuable to students as they negotiate their own learning, connect to professional responsibility, and positively impact the industry stakeholders' own processes.

Multiple external stakeholders exist in a university engineering program. These include government, accreditation bodies, professional organizations, industry, local authorities, public services, non-governmental organizations (NGOs), and communities external to (or associated with) engineering communities. Each of these stakeholders has particular and potentially overlapping interests in the engineering program and distinct contributions to make.

Opportunities for deepening ethics education are evident across various stakeholder engagement initiatives in the undergraduate engineering curriculum, including areas where the focus is on technical engineering knowledge. The case studies presented in this chapter demonstrate how stakeholder engagement potentially provides opportunities for ethical reflection and learning relating to professional responsibility. These opportunities to experience and reflect on professional practice must be explicitly integrated into program design so they feed into learning objectives and assessment practices.

How stakeholders are engaged and profiled in the engineering program is crucial to how the ethical responsibility of engineers is defined and communicated. This includes how the university positions itself to engage with stakeholders – as an authority or as a partner – and how the engagement process models values and attitudes. Similarly, choices relating to how ethics education is assessed, either individually or as teams, potentially emphasize responsibility as individual or shared.

These engagements provide critical opportunities for students to reflect on how engineering responsibility is connected to and practiced within communities rather than in isolation. Stakeholder engagement thus needs to be recognized as playing a potentially important role in the setting, achievement, and assessment of the objectives of engineering programs explicitly relating to ethics education and engineering responsibility.

References

Balakrishnan, B., Tochinai, F., & Kanemitsu, H. (2019). Engineering ethics education: A comparative study of Japan and Malaysia. *Science and Engineering Ethics, 25*, 1069–1083. https://doi.org/10.1007/s11948-018-0051-3

Basart, J. M., & Serra, M. (2013). Engineering ethics beyond engineers' ethics. *Science and Engineering Ethics 19*, 179–187. https://doi.org/10.1007/s11948-011-9293-z

Bringle, R. G., & Hatcher, J. A. (2009). Implementing service learning in higher education. *Journal of Higher Education, 67*, 221–239. https://doi.org/10.1002/he

Bucciarelli, L. L. (2008). Ethics and engineering education. *European Journal of Engineering Education, 33*(2), 141–149. https://doi.org/10.1080/03043790801979856

Colby, A., & Sullivan, W. M. (2008). Ethics teaching in undergraduate engineering education. *Journal of Engineering Education, 97*(3), 327–338.

Conlon, E., & Zandvoort, H. (2011). Broadening ethics teaching in engineering: Beyond the individualistic approach. *Science and Engineering Ethics, 17*, 217–232. https://doi.org/10.1007/s11948-010-9205-7

Correia, M. G., & Bleicher, R. E. (2008). Making connections to teach reflection. *Michigan Journal of Community Service Learning, Spring 2008, 1*(1), 41–49.

Davis, M., & Feinerman (2012). Assessing graduate student progress in engineering ethics. *Science and Engineering Ethics, 18*, 351–367. https://doi.org/10.1007/s11948-010-9250-2

Donaldson, T., & Preston, L. E. (1995). The stakeholder theory of the corporation: Concepts, evidence, and implications. *The Academy of Management Review*, *20*(1), 65–91. https://www.jstor.org/stable/258887

Engineering Council of South Africa (ECSA). (2017, March 17). Code of conduct. *Engineering Professions Act No 46 of 2000. Government Gazette*. 142 (40691). Government Notice 41. Government Printer. [Online]. https://www.ecsa.co.za/regulation/RegulationDocs/Code_of_Conduct.pdf

Engineering Council of South Africa (ECSA). (2020, September 1). Qualification standard for Bachelor of Science in Engineering (BSc(Eng))/bachelor of Engineering (BEng). NQF Level 08. *Document E-02-PE Revision 6.*

Faculty of Applied Sciences and Technology (FAST), Mbarara University of Science and Technology. (2021). *Bachelor of Engineering in Electrical and Electronic Engineering (BEng. EEE) – Revised Curriculum* (p. 87). Mbarara University: Mbarara (institutional document).

Fore, G. A., & Hess, J. L. (2020). Operationalizing ethical becoming as a theoretical framework for teaching engineering design ethics. *Science and Engineering Ethics, 26*, 1353–1375. https://doi.org/10.1007/s11948-019-00160-w

Freeman, R. E. (1984). *Strategic management: A stakeholder approach.* Pitman.

Gwynne-Evans, A. J. (2018). Student learning at the interface of university and industry relating to engineering professionalism. *Critical Studies in Teaching and Learning, 6*(2), 1–20.

Gwynne-Evans, A. J., Chetty, M., & Junaid, S. (2021). Repositioning ethics at the heart of engineering graduate attributes. *Australasian Journal of Engineering Education, 26*(1), 7–24.

Harris Jr., C. E., Pritchard, M. S., Rabins, M. J., James, R., & Englehardt, E. (2014). *Engineering ethics: Concepts and cases* (5th ed.). International Edition, Cengage Learning.

Hart, M., Bonner, K., Prashar, N., Ri, A., Mwaura, S., Sahasranamam, S., & Levie, J. (2022). *Global entrepreneurship monitor: UK report 2021/22.* Global Entrepreneurship Research Association. https://www.gemconsortium.org/report/global-entrepreneurship-monitor-uk-report-202122

Hess, J. L., & Fore, G. A. (2018). A systematic literature review of US engineering ethics interventions. *Science and Engineering Ethics, 24*(2), 551–553.

Hess, J. L., Kim, D., & Fila, N. D. (2023). Critical incidents in ways of experiencing ethical engineering practice. *Studies in Engineering Education, 3*(2), 1–30.

Kim, D., Jesiek, B. K., Zoltowski, C. B., Loui, M. C., & Brightman, A. O. (2020, Summer). An academic-industry partnership for preparing the next generation of ethical engineers for professional practice. *Advances in Engineering Education.* ASEE, *Summer 2020.*

Le Jeune, K., & Massyn, M. (2018, September 30–October 1). *Reflection of incorporating a service learning activity as part of a construction degree programme* [Paper presentation]. SACQSP 10th International Research Conference, Rosebank, Johannesburg, 2196, South Africa.

Li, J., & Fu, S. (2012). A systematic approach to engineering ethics education. *Science and Engineering Ethics, 18*, 339–349.

Malmqvist, J., Lundqvist, U., Rosén, A., Gupta, R., Leong, H., Cheah, M., Bennedsen, J., Hugo, R., Kamp, A., Leifler, O., Gunnarsson, S., Roslöf, J., & Spooner, D. (2022). *The CDIO Syllabus 3.0 - An updated statement of goals.* http://www.cdio.org/files/CDIO2022_Malmqvist%20et%20al%20Syllabus%203_post_review_v6.pdf

Martin, D. A., & Polmear, M. (2023). The two cultures of engineering education: Looking back and moving forward. In S. Hyldgaard Christensen, A. Buch, E. Conlon, C. Didier, C. Mitcham, & M. Murphy (Eds.), *Engineering, social science, and the humanities: Has their conversation come of age?* (pp. 133–150, Philosophy of Engineering and Technology; Vol. 42). Springer Nature. https://doi.org/10.1007/978-3-031-11601-8_7

Martin, D. A., Conlon, E., & Bowe, B. (2021). Using case studies in engineering ethics education: The case for immersive scenarios through stakeholder engagement and real life data. *Australasian Journal of Engineering Education, 26*(1), 47–63. https://doi.org/10.1080/22054952.2021.1914297

Massyn, M., & Le Jeune, K. (2007, May 14–17), Student's perception of the service learning component of the construction studies programme at the University of Cape Town. In T. C. Haupt & R. Millford (Eds.), *Proceedings of CIB World Building Congress: Construction for Development, Cape Town*, South Africa (pp. 1916–1927).

McGinn, R. E. (2003). "Mind the gaps": An empirical approach to engineering ethics, 1997–2001. *Science and Engineering Ethics, 9*, 517–542. https://doi.org/10.1007/s11948-003-0048-3

Mbarara University of Science and Technology. (2023). *Faculty of applied sciences and technology.* https://www.must.ac.ug/university_unit/faculty-of-applied-sciences-and-technology/

O'Riordan, L., & Fairbrass, J. (2014). Managing CSR stakeholder engagement: A new conceptual framework. *Journal of Business Ethics, 125*, 121–145.

Riley, D. (2022). *Engineering and social justice.* Springer Nature. (Original work published 2008).

South African Council for the Project and Construction Management Professions (SACPCMP). (n/d). *Self-evaluation guidelines for programme accreditation.*SACPCMP.

South African Institute of Civil Engineers (SAICE). (2016). *Code of ethics*. Retrieved November 11, 2023, from https://saice.org.za/wp-content/uploads/2017/09/SAICE-Code-of-Ethics.pdf

The CDIO Standards v 2.0 (with customized rubrics). (2010). Retrieved January 8, 2024, from http://cdio.org/

Worldwide CDIO Initiative. (n.d.). Retrieved January 8, 2024, from http://cdio.org/.

Zhu Q., & Clancy R. (2023). Constructing a role ethics approach to engineering ethics education. *Ethics and Education*, *18*(2), 216–229. https://doi.org/10.1080/17449642.2023.2249740

Zhu, Q., & Jesiek, B. K. (2017). A pragmatic approach to ethical decision-making in engineering practice: Characteristics, evaluation criteria, and implications for instruction and assessment. *Science and Engineering Ethics, 23*, 663–679. https://doi.org/10.1007/s11948-016-9826-6

31

TWO CRITICISMS OF ENGINEERING ETHICS ASSESSMENT

The importance of behaviors and culture

Rockwell Clancy, Xin Luo, Chunping Fan, and Fumihiko Tochinai

Introduction

Ethics is central to engineering. Highlighting the importance of engineering ethics assessment, it is crucial to assess the effectiveness of engineering ethics education, ensuring it has its intended effects. As previous chapters have noted, the intended impact of ethics education has tended to involve moral cognition, including ethical knowledge, awareness, sensitivity, judgments, reasoning, attitudes, values, and so on.[1] Initiatives conducted to date, although useful as first steps, have had three related shortcomings: it is unclear whether (1) measures meant to assess the effectiveness of engineering ethics education measure what they should, namely, *behaviors* and whether, therefore, these measures are *valid*; (2) these measures can be used across different *cultural* groups or are, hence, *reliable*; and (3) such measures adequately incorporate insights and methods from moral and cultural psychology.

This chapter is divided into three parts to address these issues. First, it explains why ethical behaviors should be (but have not been) the goal of engineering ethics education. Next, it outlines why using measures of engineering ethics assessment across different cultural groups would be problematic. Finally, it explores how insights and methods from moral and cultural psychology shed light on and could be used to address these shortcomings, mentioning examples of such work in China and Japan.

Engineering ethics assessment and behaviors[2]

As previous chapters have noted, assessment in engineering ethics education has focused on moral cognition. Here, 'moral cognition' refers to cognitive processes and contents related to morality and ethics, including ethical knowledge, awareness, sensitivity, judgments, reasoning, attitudes, values, and so on. However, adopting these as the goal of engineering ethics education is misguided, and therefore, these measures would be invalid insofar as they fail to measure what they should. Although moral cognition is often conceived as precipitating and resulting in ethical behavior, research into moral psychology calls into question this common-sense understanding of their relations.

DOI: 10.4324/9781003464259-37

This section argues that moral cognition should not be the ultimate goal of engineering ethics education and, therefore, that these measures are invalid, broadly conceived.[3] Rather, the ultimate goal of engineering ethics education should be ethical behaviors. To support this claim and point towards an alternative, this section explains why ethical behaviors should be adopted as the goal of engineering ethics education. We then outline why ethical behaviors have not been adopted as the goal of engineering ethics education. How ethical behaviors might connect to cognitive aspects mentioned before – like values and attitudes – is further discussed near the end of this chapter.

Why should behaviors be the goal of engineering ethics education?

Behaviors should be the goal of engineering ethics education since it is only through behaviors and actions that engineering and engineers affect the world. For example, the ethical knowledge that a mechanical engineer possesses regarding the case of the Ford Pinto counts for nothing if that engineer, nevertheless, designs a compact car with a poorly insulated gas tank. Similarly, a lack of moral awareness among a team of nuclear engineers about the case of Chernobyl would be insignificant if that team still successfully carried out routine tests of a nuclear reactor without incident. Ultimately, it is only through behaviors and actions that engineering and engineers affect the world. This is reflected in professional codes of ethics.

Professional codes of ethics emphasize behaviors and action – for instance, *using* knowledge and skills, *performing* services, *acting* in professional matters, and so on (ASME, 2012). Although engineering ethicists have emphasized the importance of *virtues* – in other words, the importance of not only *behaving* but also of *being* a certain way (Frigo et al., 2021; Harris, 2008) – it is hard to see why one should care about virtues in the absence of the behaviors they produce, for instance, why one would care about *honesty* as a virtue aside from the fact it results in *truth-telling* (Greene, 2014).[4] Presumably, the behaviors and actions of engineers are also what the public cares about. This claim is supported by understandings of 'professions' and 'professionalism' used in engineering ethics and other branches of professional ethics, such as medical, legal, and educational ethics (Davis, 2021).

Professions mediate the public's relationship with engineers. Like professional bodies for doctors, lawyers, and teachers, professional engineering organizations mediate the relationship between individual professionals and society, for example, through the establishment of technical guidelines, professional licensing, and educational accreditation (Luegenbiehl & Clancy, 2017). Although professional organizations and formation vary by country and field of engineering (Didier & Derouet, 2013; Iseda, 2008; Luegenbiehl, 2004), in places with and fields having professions, professions must be responsive to the concerns of citizens in the work they do.[5] In the absence of this concern, the field of engineering would cease to exist, and engineers would be unemployed; if engineers and engineering did not, overall, make the world a better place, then no-one would want the products, processes, and services for which they are responsible (Davis, 2021). Hence, insofar as this concern motivates the engineering profession and professional codes are meant to address these concerns, the behaviors and actions of engineers are what the public ultimately cares about, rather than their ethical knowledge, awareness, and so on.

If the behaviors and actions of engineers are what matter, then why haven't ethical behaviors and actions been adopted as the goal of engineering ethics education and, therefore, assessment? Why have different forms of moral cognition been adopted as the goal of engineering ethics education and assessment?

Why haven't ethical behaviors been adopted as the goal?

There are at least two reasons ethical behaviors and actions have not been adopted as the goal of engineering ethics education and assessment. The first is the assumption that ethical behaviors follow naturally and unproblematically from moral cognition (Fleddermann, 2012) – that ethical behaviors would follow by adopting moral cognition as the goal of engineering ethics education. However, a growing body of work from the psychological and behavioral sciences has called into doubt any simple or straightforward causal relation between antecedent moral cognition and subsequent ethical behaviors.

If moral cognition resulted in more ethical behaviors, then professional ethicists – arguably the most knowledgeable about ethics and capable of ethical reasoning – would behave the most ethically. However, research has consistently failed to find evidence to support this conclusion (Schönegger & Wagner, 2019; Schwitzgebel & Rust, 2014). Not only is ethical reasoning not associated with ethical behaviors (Harding et al., 2007), but it is also associated with more unethical behaviors (Bay & Greenberg, 2001; Ponemon, 1993). Further, considerable research has found that ethical judgments are associated with 'moral intuitions,' closer in nature to emotions than to reasoning (Greene, 2014; Haidt, 2012; Roeser, 2018), and that behaviors are often affected by unconscious, environmental factors (Bazerman & Tenbrunsel, 2012; Doris, 2005). Rather than conscious, rational, and reflective processes, individual behaviors are often driven by implicit expectations regarding what others do and what others think others should do (Bicchieri, 2016; Kahnemann, 2011). Failing to account for these (somewhat counter-intuitive) characteristics of ethical behaviors and moral cognition is one of the main reasons that behaviors have not been adopted as the goal of engineering ethics education and assessment.

The second main reason concerns difficulties associated with adopting ethical behaviors as the goal of engineering ethics education and assessment. These difficulties are both theoretical and practical, having to do with the nature of ethical behaviors and how they would be assessed.

Regarding the first, what it means to 'behave ethically' regarding engineering is ambiguous. People often disagree about what it means to behave ethically, a position known in academic philosophy as descriptive ethical relativism (Rachels, 2001). Further, conceptions of ethics – understood in terms of what should or should not be done and what it means to be good – are affected by social and cultural factors (Henrich, 2020; Nisbett, 2010; Rachels, 2011). This is especially important to engineering ethics since engineering is increasingly cross-cultural and international, with people from different backgrounds working together as never before (Clancy & Zhu, 2022; Wong, 2021). Hence, because of the global environments of engineering and cultural differences, conceptions of ethics are likely to clash. The extent to which there is something fundamental and, therefore, global to ethics in engineering, or whether cultural differences segment conceptions, is an open, ongoing debate between what Clancy and Zhu (2022) have termed 'universalist' and 'particularist' approaches to engineering ethics. In addition to these cultural differences, there are reasons specific to engineering that would make it difficult to specify the nature of ethical behaviors in engineering.

From the perspective of common-sense ethics – in other words, understandings of right and wrong in non-professional, lay terms – what it means to 'behave ethically' in engineering could be counter-intuitive (Stappenbelt, 2013). This counter-intuitive character stems from the nature of engineering, first and foremost, the specific duties and obligations that follow from and are attached to the professional roles engineers occupy. What it means to 'behave ethically' as an engineer could differ from and demand more than what it means to 'behave ethically' as a non-professional layperson. For example, given their tremendous potential impact on public safety, it is essential that engineers only perform within their areas of competence – this is a typical entry within professional engineering codes, even across national and cultural groups (Luegenbiehl, 2010; Luegenbiehl & Clancy, 2017). Failure to do so could have negative impacts on public safety. For similar reasons, engineers must engage in lifelong learning. However, these types of duties are different from and could appear counter-intuitive from the perspective of common-sense ethics. The relation between not harming people, lifelong learning, and only performing within one's area of competence might not be clear. Perhaps as a result, engineering students have ranked these as the least important professional duties (Stappenbelt, 2013). However, they are undoubtedly among the most important from the perspective of professional codes and, therefore, the organizations, practitioners, and public they represent. The same could be true of technical and regulatory guidelines, where very small, seemingly insignificant differences could have vast and dire consequences (Luegenbiehl & Clancy, 2017; Zhu et al., 2022). Similar difficulties stem from the intrinsically novel nature of engineering, the fact that engineering involves technology, and that technology brings into existence situations that did not exist before.

As a result, it is difficult to specify what it would mean to 'behave ethically' in these novel situations, a problem associated with the 'engineering as social experimentation' paradigm (Van de Poel, 2016, 2017). As a result of these difficulties, some have argued it would be impossible to know or, therefore, teach ethical behaviors, which is why ethical behaviors should not be adopted as the goal of engineering ethics education (Baum, 1980; Van de Poel et al., 2001).[6] However, even if the nature of engineering ethical behaviors could be precisely specified, practical difficulties are involved in adopting behaviors as the goal of ethics education.

The most important of these difficulties would be the ability to track long-term ethical behaviors, assessing how they are affected by different kinds of engineering education. Ideally, groups of engineering students could be separated into experimental and control groups during their first year of university. Students in the experimental group would receive ethics education, whereas those in the control group would not. They would then be tracked throughout their university and professional careers, seeing which ones engaged in more (un)ethical behaviors, while controlling for other potentially confounding factors. Obviously, such a procedure would be both unethical and untenable: unethical as it would deny ethics education to one group of students, and untenable as such a procedure would require a tremendous expenditure of resources. In sum, those are additional practical difficulties associated with adopting 'ethical behaviors' as the goal of engineering ethics education.

Engineering ethics assessment and bias

In addition to problems of *validity* discussed above, measures of engineering ethics assessment potentially have problems of *reliability*.[7] This section argues that measures of engineering ethics assessment are unlikely to be reliable because of cultural biases. These biases stem from where and how engineering ethics has developed – the ways it has been conceived and the people by and with whom these measures have been developed. To support this claim and begin to delineate alternatives, this section starts by identifying sources of bias within engineering ethics education, and moves on to locate sources of potential bias in the populations by and with whom measures of engineering ethics assessment have been developed.

Why and how is engineering ethics biased?

As a systematic and reflective study discipline, engineering ethics began in the United States and has evolved (primarily) in the Western world (Clancy & Zhu, 2022; Davis, 1995). This is problematic since there are features of engineering in the United States that are not true elsewhere and do not easily transfer – for example, the professional nature of engineering (Iseda, 2008; Luegenbiehl, 2004) – even within exclusively Western contexts (Didier & Derouet, 2013; Van de Poel & Royakkers, 2011).

Engineering ethics education has tended to be professional and applied, familiarizing students with professional codes of ethics and/or philosophical, ethical theories (Clancy & Zhu, 2022; Harris et al., 1996; Hess & Fore, 2018). Principles contained in these codes and theories are then applied to resolve ethical issues that appear in engineering case studies, typically involving disasters and taking place in the United States (Barry & Herkert, 2015; Harris, 2008; Harris et al., 2018; Van de Poel & Royakkers, 2011). However, this way of thinking about and teaching ethics is particular to a relatively recent Western cultural tradition. By contrast, in Eastern and ancient Western philosophy, the focus has been on what it means to be and *become* good (Ivanhoe & Norden, 2005; Pierre Hadot, 1995). In recent years, the Western professional and applied case-study approach to engineering ethics has begun to change. Engineering ethicists have increasingly emphasized virtue and care ethics and non-Western ethical theories and perspectives (Fleddermann, 2012; Van de Poel & Royakkers, 2011; Zhu, 2010). Additionally, more case studies are now available that focus on incidents and engineering work that has occurred and is occurring outside the United States (Luegenbiehl & Clancy, 2017; Van de Poel & Royakkers, 2011). Such efforts could be understood as attempts to decolonize the engineering curriculum (Fomunyam, 2017, 2019). Additionally, more recent case studies have been 'aspirational' in nature, focusing on engineers doing the right thing (Harris et al., 2018).

How and why are measures of engineering ethics assessment biased?

Due to the biases noted above, measures to assess engineering ethics education could be similarly biased. Measures of ethical knowledge, awareness, and so on have been developed by and with scholars working primarily in the United States. However, there are good reasons for thinking that individuals who belong to this group are poorly representative of global populations and, therefore, that using such measures with non-US groups would be problematic. It is unclear whether measures of engineering ethics would assess the same things with non-US groups or, therefore, whether these measures would be reliable.

The United States is what some have called a 'WEIRD' (Western, educated, industrialized, rich, and democratic) culture. Relative to global populations, samples from WEIRD cultures are consistently outliers on various psychological and social factors, including self-concepts, thought styles, and ethical reasoning (Henrich, 2020). Versus WEIRD cultures, East Asian populations, for instance, tend to think of themselves in interdependent rather than independent terms, reason holistically rather than analytically, and make ethical judgments based on the outcomes of behaviors rather than the intentions of agents (Feinberg et al., 2019; Nisbett, 2010). These general cultural differences provide good reasons for thinking that significant differences regarding engineering ethical knowledge, awareness, and so on would also exist.[8]

Where comparative data is available, it has been found that non-US students perform worse on measures of engineering ethical reasoning than their US counterparts and that they make smaller gains because of ethics education (Borenstein et al., 2010; Canary et al., 2012). However, it has been unclear whether this results from cultural differences, linguistic competence, or some combination of both. Addressing these and related questions is extremely important, given the increasingly cross-cultural and international nature of engineering.

Engineering ethics assessment and improvement

To address the issues of validity and reliability discussed above, those working in engineering ethics should draw from insights and methods from moral and cultural psychology. These fields have resources that could be used to address issues related to ethical behaviors and cultural biases in engineering ethics education and assessment. This third section outlines how these insights and methods could be used to address such issues and areas where they have already been applied and are being applied.

What role can moral and cultural psychology play?

Moral and cultural psychology have resources that could be used to address both theoretical and practical difficulties associated with adopting behaviors as the goal of engineering ethics education. Moral and cultural psychology are empirical, descriptive disciplines concerned with conducting research to describe what and how people think about matters of right and wrong and how these are affected by culture (Doris, 2010; Doris et al., 2017; Heine, 2016). The process of studying and describing what and how people think about right and wrong can be understood in contrast to those of philosophical and applied ethics, which are theoretical and prescriptive, concerned with engaging in reflection to prescribe how and why people *should* think and behave. As such, moral and cultural psychology have resources for addressing the natures of and differences between conceptions of right and wrong behaviors across cultures, addressing both theoretical and practical difficulties explained above (see Chapter 10 for more).

Theoretically, insights and methods from moral and cultural psychology can assess what people think about issues of right and wrong, as well as similarities and differences between cultural and professional groups. This is one of the ways these fields, their insights, and their methods would be relevant to engineering ethics education assessment cross-culturally. For example, large-scale research has found that individuals who identify as politically liberal and are from WEIRD cultures tend to prioritize care and fairness when they think about what it means to be ethical. In contrast, individuals who identify as politically conservative and are from non-WEIRD cultures tend to prioritize not only care and fairness but also loyalty, authority, and sanctity (Graham et al., 2009; Graham et al., 2011; Kim et al., 2012; Talhelm et al., 2015; Zhang & Li, 2015). Although care and fairness appear to be universal features of moral cognition, how people conceive of and practice these could differ between cultural and professional groups. Understanding the effects of professional education/training in general could help to understand and assess the impact of engineering ethics education specifically.

For instance, when confronted with dilemmas involving the allocation of scarce resources to care for patients, hospital administrators are more likely to make outcome-based, 'consequentialist' decisions than doctors and members of the public – sacrificing one individual to allocate more resources among many (Ransohoff, 2011). Similarly, rates of reported cheating are higher among business students than engineering students and higher among engineering students than humanities and social-sciences students (Harding et al., 2012; McCabe et al., 2001). This indicates that professional education and formation can affect ethical judgments like culture can. However, where, how, and why this occurs are unclear; the effects of professional cultures on ethics are understudied. Additional work would be necessary to determine if and how conceptions of ethics among engineers are different from those of the public, as well as how these are affected by national cultures. Such research has been carried out with Chinese engineering students.

China is now a significant engineering country, graduating and employing more science, technology, engineering, and mathematics (STEM) majors than any other country. In recent years, China has invested significantly in engineering ethics education. The Ministry of Education (MOE) now requires the course 'Engineering Ethics' in all engineering Master's programs (Ministry of Education PRC, 2018). Additionally, many engineering universities now offer courses related to engineering ethics at the undergraduate level. Coupled with the increasing influence of China on the world stage, understanding the ethics of Chinese engineering students is essential.

To do so, teams have translated and administered measures of engineering ethical reasoning, such as the Engineering and Sciences Issues Test (ESIT) (Borenstein et al., 2010), and moral intuitions, like the Moral Foundations Questionnaire (MFQ) (Graham et al., 2011), to groups of Chinese engineering students (Clancy, 2020, 2021; Clancy & Hohberger, 2019; Clancy et al., 2022). The results of these studies have been significant.

For example, previous results mentioned above – showing smaller gains in ethical reasoning among non-US students – seem to result from linguistic competence rather than cultural differences. In two recent studies, non-native but high-level English-speaking Chinese students made significant gains in their ethical reasoning abilities, similar to their US counterparts (Clancy, 2020, 2021). These results are supported by the fact that the structure of ethical reasoning among Chinese students is similar to that of their foreign counterparts: one can discern in responses to the ESIT the same structure of pre-conventional (self-based), conventional (rule- and law-based), and post-conventional (universal principles–based) reasoning that one finds in responses to the ESIT from the culturally WEIRD, US participants with which the instrument was developed. This is even though cultural psychologists have called into question the existence of this pre-conventional/conventional/post-conventional taxonomy of ethical reasoning among East Asian populations (Hwang, 2012).

Just as the fields of moral and cultural psychology have resources that could help address theoretical difficulties associated with adopting behaviors as the goal of engineering ethics education, so too do they have resources for addressing practical challenges. Work on moral and cultural psychology could do so by better understanding relation(s) between ethical behaviors and other relevant factors that measurements have been developed to assess, such as ethical knowledge, reasoning, and so on. Such research would be helpful beyond engineering ethics – since engineering ethics is not the only field that neglects the relation between ethical behaviors and other relevant factors.

There is a lack of study regarding the relationship between ethical behaviors and moral cognition in moral psychology, where studies tend to focus on ethical behaviors or moral cognition and less on the relations between them (Ellemers et al., 2019; Villegas de Posada & Vargas-Trujillo, 2015). Although the field of engineering ethics education currently lacks theoretical and empirical resources for addressing this gap, moral psychology has such resources. In recent years, scholars in moral psychology have begun to address the gap by, for example, pairing psychological measures of moral cognition with economic games to precisely quantify and assess the behaviors of individuals (Miranda-Rodríguez et al., 2023). Moreover, simply because ethical knowledge, reasoning, and other facets of moral cognition are insufficient conditions of ethical behaviors does not mean they are unnecessary. Future research should further study these relations, for instance, identifying proxies for or predictors of ethical behaviors. Engineering ethics and education scholars would be well positioned to do so by studying lab spaces and attending workshops in universities.

Labs and workshops in universities can be conceived and studied as microcosms of larger engineering work environments. Although these spaces are smaller and simpler than their corporate or government counterparts, they involve technical work, its practical applications, deadlines and budgets, and social relations and hierarchies that affect technical work. As such, university labs and workshops could be used as touchstones to pilot and assess different kinds of engineering ethics interventions, where (quasi) real-world behaviors could be observed and assessed. In addition to quantitative assessment measures, engineering ethicists and educators could adopt methods from Science and Technology Studies (STS) and anthropology, such as ethnographies of researchers working in engineering labs. Engineering labs and workshops could provide an alternative to manufactured experimental protocols, such as economics games, sometimes used to assess ethical behaviors.

What role can moral and cultural psychology play in making engineering ethics less biased?

Just as moral and cultural psychology have resources to address difficulties with adopting behaviors as the goal of engineering ethics education, so do these fields have resources for mitigating biases associated with measures of engineering ethics assessment.

First, moral and cultural psychology findings support the importance of including larger, more culturally and nationally diverse samples in research on engineering ethics education. Similarities in ethical presuppositions involving knowledge, awareness, and other facets of moral cognition cannot be taken for granted since culture can affect these in unexpected and counter-intuitive ways. Exploring similarities and differences between samples from different cultural and professional groups could help understand if and how this is the case. Again, such work is underway.

In a series of empirical studies, researchers from Malaysia and Japan have explored the ethical perspectives and impact of education on engineering students. These are significant since they use samples of underrepresented engineering students from Malaysia and Japan (Balakrishnan et al., 2018; Balakrishnan et al., 2021, also featured as case studies in Chapter 27). Results from this kind of work can help to mediate debates between 'universalists' who think that professional and disciplinary cultures are more important to the ways people think about ethics than national cultures, and 'particularists' who think that national cultures are more important (Clancy & Zhu, 2022; Davis, 2021). Such debates rest on competing assumptions about the universality of professional standards and ethical understanding between different national and cultural groups and, therefore, if and how engineering ethics education is biased. These methods could also help determine whether psychological or social factors are specific to/distinctive of different professional or disciplinary groups. Such findings could help educators recognize and better respond to these differences.

These differences could be addressed by tailoring curricula in engineering ethics to the cultural backgrounds and professional aspirations of students, for instance, curricula in ethics for civil engineering students from China versus ones for mechanical engineering students from France. Disciplinary specialization is a best practice within education for Responsible Conduct of Research (RCR) (Phillips et al., 2018). Although professional organizations and regulatory bodies might have common ethical expectations of engineers and engineering work, engineers come from diverse backgrounds. Understanding and responding to these backgrounds is necessary to meet common ethical expectations.

Conclusion

Ethics is central to engineering, but ways of assessing engineering ethics education are problematic. These problems concern the *validity* and *reliability* of measures assessing engineering ethics education since these measures do not assess long-term ethical behaviors and are based on biased samples. To address these problems and, thereby, increase the validity and reliability of these measures, the field of engineering ethics education must use insights and methods from moral and cultural psychology to better understand relations between ethical behaviors and moral cognition and how these are affected by culture.

Notes

1 'Ethics,' 'morality,' and their variants are used interchangeably throughout this chapter. For a different understanding of the natures of and relations between these terms regarding engineering and technology, see, for example, (Davis, 2021; Van de Poel & Royakkers, 2011).
2 This section is based on materials included in (Clancy & Gammon, 2021) and expanded on in (Clancy & Zhu, 2023). The interested reader is encouraged to consult that article for a fuller explanation of this argument/line of thought, as well as responses to objections, which could not be addressed here because of restrictions related to space.
3 A reviewer has pointed out that measures are developed according to the objectives of those who define them, such that the measures could be valid even if the objectives defined are questionable. It seems as though the objectives defined in developing measures of engineering ethics education assessment are ultimately ethical behaviors – a point further discussed and justified below – although this might not be the case. Were this not the case, then '(in)valid' and its variants are being used in a looser sense throughout this chapter.
4 One might well wonder whether one could behave or act ethically in the absence of moral cognition. This is a point to which we return below – regarding whether moral cognition might be a *necessary*, if not *sufficient*, condition of ethical behaviors – but, again, the interested reader is encouraged to consult (Clancy & Zhu, 2023) for a fuller consideration of this point.
5 Not all fields are organized as professions, nor is the organization of fields into professions the same across countries. For example, although the field of law has traditionally been organized as a profession, it has fallen into disrepute in recent years, weakening its status as a profession; although engineering seems to be a profession in Canada, for instance, this is less clearly the case elsewhere.
6 It should be pointed out that 'teaching ethical behaviors' is different from 'adopting ethical behaviors as the ultimate goal of engineering ethics education.' One could well think the former is absurd while endorsing the latter. Again, see (Clancy & Zhu, 2023) for more on this point.
7 As with the previous section, the points included here have been dealt with at greater length elsewhere. For instance, see (Clancy & Zhu, 2022) and (Luegenbiehl & Clancy, 2017).
8 However, others disagree about the strength of this evidence – whether *general* cultural differences provide reasons for thinking *specific* differences in engineering ethics would exist. For an extensive, well-reasoned articulation and defense of that view, see (Davis, 2021).

References

ASME. (2012, February 1). Code of ethics of engineers. Retrieved April 26, 2023, from https://www.asme.org/wwwasmeorg/media/resourcefiles/aboutasme/getinvolved/advocacy/policy-publications/p-15-7-ethics.pdf

Balakrishnan, B., Tochinai, F., & Kanemitsu, H. (2018). Engineering ethics education: A comparative study of Japan and Malaysia. *Science and Engineering Ethics*, *23*, 1–15. https://doi.org/10.1007/s11948-018-0051-3

Balakrishnan, B., Tochinai, F., Kanemitsu, H., & Altalbe, A. (2021). Engineering ethics education from the cultural and religious perspectives: A study among Malaysian undergraduates. *European Journal of Engineering Education*, *46*(5), 707–717. https://doi.org/10.1080/03043797.2021.1881449

Barry, B. E., & Herkert, J. R. (2015). Overcoming the challenges of teaching engineering ethics in an international context: A U.S. perspective. In C. Murphy, P. Gardoni, H. Bashir, C. E. Harris, & E. Masad (Eds.), *Engineering ethics for a globalized world* (pp. 167–187). Springer. https://doi.org/10.1007/978-3-319-18260-5

Baum, R. J. (1980). *Ethics and engineering curricula*. The Hastings Center.

Bay, D. D., & Greenberg, R. R. (2001). The relationship of the DIT and behavior: A replication. *Issues in Accounting Education*, *16*(3), 367–380. https://doi.org/10.2308/iace.2001.16.3.367

Bazerman, M. H., & Tenbrunsel, A. (2012). *Blind spots: Why we fail to do what's right and what to do about It*. Princeton University Press.

Bicchieri, C. (2016). *Norms in the wild*. Oxford University Press.

Borenstein, J., Drake, M. J., Kirkman, R., & Swann, J. L. (2010). The Engineering and Science Issues Test (ESIT): A discipline-specific approach to assessing moral judgment. *Science and Engineering Ethics*, *16*(2), 387–407. https://doi.org/10.1007/s11948-009-9148-z

Canary, H. E., Herkert, J. R., Ellison, K., & Wetmore, J. M. (2012). *Microethics and macroethics in graduate education for scientists and engineers: Developing and assessing instructional models*. ASEE Annual Conference and Exposition, Conference Proceedings.

Clancy, R. F. (2020). *Ethical reasoning and moral foundations among engineering students in China*. In Proceedings of the American Society for Engineering Education Annual Conference & Exposition.

Clancy, R. F. (2021). *The relations between ethical reasoning and moral intuitions among engineering students in China*. In Proceedings of the American Society for Engineering Education Annual Conference & Exposition.

Clancy, R. F., & Gammon, A. (2021). *The ultimate goal of ethics education should be more ethical behaviors*. ASEE Annual Conference and Exposition, Conference Proceedings.

Clancy, R. F., & Hohberger, H. (2019). The moral foundations of Chinese engineering students: A preliminary investigation. In *Proceedings of the American Society for Engineering Education Annual Conference & Exposition*.

Clancy, R. F., & Zhu, Q. (2022). Global Engineering Ethics: What? Why? How? and When? *Journal of International Engineering Education*, *4*(1). https://digitalcommons.uri.edu/jiee/vol4/iss1/4?utm_source=digitalcommons.uri.edu%2Fjiee%2Fvol4%2Fiss1%2F4&utm_medium=PDF&utm_campaign=PDFCoverPages

Clancy, R. F., & Zhu, Q. (2023). Why should ethical behaviors be the ultimate goal of engineering ethics education? *Business and Professional Ethics Journal*, *42*(1), 33–53. https://doi.org/10.5840/bpej202346136

Clancy, R. F., Zhu, Q., Streiner, S., Gammon, A., & Thorper, R. (2022). Exploring the relations between ethical reasoning and moral intuitions among first-year engineering students across cultures. In *ASEE Annual Conference and Exposition*.

Davis, M. (1995). An historical preface to engineering ethics. *Science and Engineering Ethics*, *1*(1), 33–48.

Davis, M. (2021). *Engineering as a global profession: Technical and ethical standards*. Rowman & Littlefield Publishers.

Didier, C., & Derouet, A. (2013). Social responsibility in French engineering education: A historical and sociological analysis. *Science and Engineering Ethics*, *19*(4), 1577–1588. https://doi.org/10.1007/s11948-011-9340-9

Doris, J. M. (2005). *Lack of character: Personality and moral behavior*. Cambridge University Press.

Doris, J. M. (Ed.). (2010). *The moral psychology handbook*. Oxford University Press.

Doris, J. M., Stich, S., Phillips, J., & Walmsley, L. (2017). Moral psychology: Empirical approaches. In *Stanford encyclopedia of philosophy*.

Ellemers, N., van der Toorn, J., Paunov, Y., & van Leeuwen, T. (2019). The psychology of morality: A review and analysis of empirical studies published from 1940 through 2017. *Personality and Social Psychology Review*, *23*(4), 332–366. https://doi.org/10.1177/1088868318811759

Feinberg, M., Fang, R., Liu, S., & Peng, K. (2019). A world of blame to go around: Cross-cultural determinants of responsibility and punishment judgments. *Personality and Social Psychology Bulletin*, *45*(4), 634–651. https://doi.org/10.1177/0146167218794631

Fleddermann, C. (2012). *Engineering ethics* (4th ed.). Pearson.

Fomunyam, K. G. (2017). Decolonising teaching and learning in engineering education in a South African university. *International Journal of Applied Engineering Research*, *12*(23), 13349–13358. http://www.ripublication.com

Fomunyam, K. G. (Ed.). (2019). *Decolonising higher education in the era of globalisation and internationalisation*. Sun Press.

Frigo, G., Marthaler, F., Albers, A., Ott, S., & Hillerbrand, R. (2021). Training responsible engineers. Phronesis and the role of virtues in teaching engineering ethics. *Australasian Journal of Engineering Education*, *26*(1), 25–37. https://doi.org/10.1080/22054952.2021.1889086

Graham, J., Haidt, J., & Nosek, B. A. (2009). Liberals and conservatives Rely on different sets of moral foundations. *Journal of Personality and Social Psychology*, *96*(5), 1029–1046. https://doi.org/10.1037/a0015141

Graham, J., Nosek, B. A., Haidt, J., Iyer, R., Koleva, S., & Ditto, P. H. (2011). Mapping the moral domain. *Journal of Personality and Social Psychology*, *101*(2), 366–85. https://doi.org/10.1037/a0021847

Greene, J. D. (2014). *Moral tribes: Emotion, reason, and the gap between us and them*. Penguin Books.
Haidt, J. (2012). *The righteous mind*. Vintage Press.
Harding, T. S., Carpenter, D. D., & Finelli, C. J. (2012). An exploratory investigation of the ethical behavior of engineering undergraduates. *Journal of Engineering Education*, *101*(2), 346–374.
Harding, T. S., Mayhew, M. J., Finelli, C. J., & Carpenter, D. D. (2007). The theory of planned behavior as a model of academic dishonesty in engineering and humanities undergraduates. *Ethics and Behavior*, *17*(3), 255–279. https://doi.org/10.1080/10508420701519239
Harris, C. E. (2008). The good engineer: Giving virtue its due in engineering ethics. *Science and Engineering Ethics*, *14*(2), 153–164. https://doi.org/10.1007/s11948-008-9068-3
Harris, C. E., Davis, M., Pritchard, M. S., & Rabins, M. J. (1996). Engineering ethics: What? Why? How? And when? *Journal of Engineering Education*, *85*(2), 93–96. https://doi.org/10.1002/j.2168-9830.1996.tb00216.x
Harris, C. E., Pritchard, M., Rabins, M., James, R., & Englehardt, E. (2018). *Engineering ethics: Concepts and cases* (6th ed.). Cengage Learning.
Heine, S. J. (2016). *Cultural psychology* (3rd ed.). Norton and Company.
Henrich, J. (2020). *The WEIRDest people in the world: How the west became psychologically peculiar and particularly prosperous*. Farrar, Straus and Giroux.
Hess, J. L., & Fore, G. (2018). A systematic literature review of US engineering ethics interventions. *Science and Engineering Ethics*, *24*(2), 551–583. https://doi.org/10.1007/s11948-017-9910-6
Hwang, K.-K. (2012). *Foundations of Chinese psychology: Confucian social relations* (Vol. 1). Springer New York. https://doi.org/10.1007/978-1-4614-1439-1
Iseda, T. (2008). How should we foster the professional integrity of engineers in Japan? A pride-based approach. *Science and Engineering Ethics*, *14*(2), 165–176. https://doi.org/10.1007/s11948-007-9039-0
Ivanhoe, P. J., & Van Norden, B. W. (Eds.). (2005). *Readings in classical Chinese philosophy* (2nd ed.). Hackett Publishing.
Kahnemann, D. (2011). *Thinking fast and slow*. Macmillan.
Kim, K. R., Kang, J.-S., & Yun, S. (2012). Moral intuitions and political orientation: Similarities and differences between South Korea and the United States. *Psychological Reports*, *111*(1), 173–185. https://doi.org/10.2466/17.09.21.pr0.111.4.173-185
Luegenbiehl, H. C. (2004). Ethical autonomy and engineering in a cross-cultural context. *Techné: Research in Philosophy and Technology*, *8*(1), 57–78. https://doi.org/doi:10.5840/techne20048110
Luegenbiehl, H. C. (2010). Ethical principles for engineers in a global environment. In I. Van de Poel & D. E. Goldberg (Eds.), *Philosophy and engineering: An emerging agenda* (pp. 147–159). Springer.
Luegenbiehl, H. C., & Clancy, R. F. (2017). *Global engineering ethics*. Elsevier.
McCabe, D. L., Treviño, L. K., & Butterfield, K. D. (2001). Cheating in academic institutions: A decade of research. *Ethics and Behavior*, *11*(3), 219–232. https://doi.org/10.1207/S15327019EB1103_2
Ministry of Education PRC. (2018). *Notice on forwarding the "guidance on formulating the training program for engineering masters" and its explanation (in Chinese)*. http://www.moe.gov.cn/s78/A22/tongzhi/201805/t20180511_335692.html
Miranda-Rodríguez, R. A., Leenen, I., Han, H., Palafox-Palafox, G., & García-Rodríguez, G. (2023). Moral reasoning and moral competence as predictors of cooperative behavior in a social dilemma. *Scientific Reports*, *13*(1), 1–11. https://doi.org/10.1038/s41598-023-30314-7
Nisbett, R. E. (2010). *The geography of thought: How Asians and Westerners think differently and why*. Free Press.
Phillips, T., Nestor, F., Beach, G., & Heitman, E. (2018). America COMPETES at 5 years: An analysis of research-intensive universities' RCR training plans. *Science and Engineering Ethics*, *24*(1), 227–249. https://doi.org/10.1007/s11948-017-9883-5
Pierre Hadot. (1995). *Philosophy as a way of life*. Wiley-Blackwell.
Ponemon, L. A. (1993). Can ethics be taught in accounting? *Journal of Accounting Education*, *11*(2), 185–209. https://doi.org/https://doi.org/10.1016/0748-5751(93)90002-Z
Rachels, J. (2001). Subjectivism in ethics. In J. Rachels (Ed.), *The elements of moral philosophy* (pp. 1–14).
Rachels, J. (2011). The challenge of cultural relativism. In J. Rachels (Ed.), *The elements of moral philosophy* (pp. 12–24). McGraw-Hill Education.
Ransohoff, K. J. (2011). *Patients on the trolley track: The moral cognition of medical practitioners and public health professionals*. Harvard University.
Roeser, S. (2018). *Risk, technology, and moral emotions*. Routledge.

Schönegger, P., & Wagner, J. (2019). The moral behavior of ethics professors: A replication-extension in German-speaking countries. *Philosophical Psychology*, *32*(4), 532–559. https://doi.org/10.1080/09515089.2019.1587912

Schwitzgebel, E., & Rust, J. (2014). The moral behavior of ethics professors: Relationships among self-reported behavior, expressed normative attitude, and directly observed behavior. *Philosophical Psychology*, *27*(3), 293–327. https://doi.org/10.1080/09515089.2012.727135

Stappenbelt, B. (2013). Ethics in engineering: Student perceptions and their professional identity development. *Journal of Technology and Science Education*, *3*(1), 86–93. https://doi.org/10.3926/jotse.51

Talhelm, T., Haidt, J., Oishi, S., Zhang, X., Miao, F. F., & Chen, S. (2015). Liberals think more analytically (more "WEIRD") than conservatives. *Personality and Social Psychology Bulletin*, *41*(2), 250–267. https://doi.org/10.1177/0146167214563672

Van de Poel, I. (2016). An ethical framework for evaluating experimental technology. *Science and Engineering Ethics*, *22*(3), 667–686. https://doi.org/10.1007/s11948-015-9724-3

Van de Poel, I. (2017). Society as a laboratory to experiment with new technologies. In D. M. Bowman, E. Stokes, & A. Rip (Eds.), *Embedding new technologies into society: A regulatory, ethical and societal perspective*. Pan Stanford Publishing. https://doi.org/10.1201/9781315379593

Van de Poel, I., & Royakkers, L. (2011). *Ethics, technology, and engineering: An introduction*. Wiley-Blackwell.

Van de Poel, I., Zandvoort, H., & Brumsen, M. (2001). Ethics and engineering courses at Delft University of Technology: Contents, educational setup and experiences. *Science and Engineering Ethics*, *7*(2), 267–282. https://doi.org/10.1007/s11948-001-0048-0

Villegas de Posada, C., & Vargas-Trujillo, E. (2015). Moral reasoning and personal behavior: A meta-analytical review. *Review of General Psychology*, *19*(4), 408–424. https://doi.org/10.1037/gpr0000053

Wong, P.-H. (2021). Global engineering ethics. In D. Michelfelder & N. Doorn (Eds.), *Routledge handbook of philosophy of engineering* (pp. 736–744). Routledge - Taylor & Francis Group.

Zhang, Y., & Li, S. (2015). Two measures for cross-cultural research on morality: Comparison and revision. *Psychological Reports*, *117*(1), 144–166. https://doi.org/10.2466/08.07.PR0.117c15z5

Zhu, Q. (2010). Engineering ethics studies in china: Dialogue between traditionalism and modernism. *Engineering Studies*, *2*(2), 85–107. https://doi.org/10.1080/19378629.2010.490271

Zhu, Q., Martin, M., & Schinzinger, R. (2022). *Ethics in engineering* (5th ed.). McGraw-Hill.

SECTION 6

Accreditation and engineering ethics education

Shannon Chance

Over time, individuals and entities within the engineering education community have developed a multifaceted strategy to synchronize global educational practices and performance metrics. This comprehensive approach involves conducting and disseminating educational research (evident throughout this handbook) and implementing research-informed pedagogies in the classroom (exemplified by the teaching methods section). It also includes establishing and aligning accreditation standards internationally – to guide the content and delivery of engineering education, including ethics. Although this section focuses on the accreditation of engineering ethics education (EEE), understanding the overall accreditation system is essential for grasping the ethics component.

Alignment across culturally and geographically diverse regions and nations has been facilitated by global accords, fostering a shared understanding of expectations in the engineering profession's globalized landscape. Cohesion is vital, as today's engineering students must possess skills to contribute effectively to international teams and projects, impacting environments and lives around the world.

The primary goal of accreditation is to ensure that graduates from engineering programs possess the necessary skills and competencies for effective engineering practice in a globalized world (Chance et al., 2022; Sthapak, 2012). Much of this standardization occurs in English, reflecting its status as the language of global engineering practice and engineering education research (Klassen, 2018), and is based on values first identified and described in the United States (Anwar & Richards, 2013) via organizations like the Accreditation Board for Engineering and Technology (ABET) and the American Society of Civil Engineers (ASCE). Accreditation frameworks for engineering education developed in the United States by ABET and ASCE included components of ethics; their uptake has expanded internationally over time through agreements such as the Washington, Sydney, and Dublin Accords.

Today, these accords are coordinated by the International Engineering Alliance (IEA) (2024), a non-profit alliance with 29 countries and 41 jurisdictions as members. IEA uses seven international agreements to "establish and enforce internationally bench-marked standards for engineering education and expected competence for engineering practice" (IEA, 2024, ¶ 2). Currently, graduates of accredited engineering programs are expected to demonstrate a comprehensive range of abilities, skills, and knowledge informed and aligned by these agreements – and ethics is a component

DOI: 10.4324/9781003464259-38

required in most places, yet its definition is often fuzzy and poorly understood by assessors and engineering educators alike (Gywnne-Evans et al., 2021; Martin, 2020).

This handbook section comprises five chapters explaining how engineering education is regulated through a relatively centralized and top-down approach. It explores crucial aspects of the global drive to accredit and align engineering courses and delves into the *who*, *what*, *when*, *where*, *why*, and *how* of engineering education. The set starts by explaining the systems and how they developed over time but then methodically explores regional differences across engineering ethics education; the influence of the accreditation system on engineering licensure worldwide; what topics, ideas, and voices have been getting left out of the conversations that have defined the system; and strategies for ensuring more diverse and holistic representation in the future of EEE accreditation systems. The contributors to these chapters actively engage with professional accreditation systems and contribute to their ongoing evolution (e.g., the recent conference paper by Chance et al., 2024).

Positionality

The editor who has curated this section of the handbook, Shannon Chance, is a Registered Architect who has been active in institutional and architectural accreditation in the United States, having, for instance, served on and chaired multiple visiting teams for the National Architectural Accrediting Board. Shannon's Ph.D. was in higher education policy, planning, and leadership; these activities informed her interest in professional accreditation (you can read about how these topics overlap with EEE in Chance et al., 2022). Shannon's concern for diversity, inclusivity, and decolonization grew from her experience growing up in Virginia and then teaching at a minority-serving institution for 15 years (where ethics, environment, cultural diversity, and social justice were central themes in her teaching). She also led over a dozen international study programs across Africa and Europe for students from underrepresented groups, deepening her commitment to diversity and localized perspectives. As a result, she felt honored to have helped cultivate this handbook section on EEE accreditation; she and the contributing authors endeavored to include diverse voices and critical perspectives.

Chapter topics

The exploration of engineering education accreditation begins in this section from a historical perspective, tracing its origins and evolution to encompass ethical considerations. In the opening chapter, Chapter 32, titled 'Foundational Perspectives on Ethics in Engineering Accreditation,' authors Brent K. Jesiek, Qin Zhu, and Gouri Vinod delve into the formal integration of ethics outcomes into accreditation criteria for engineering graduates. They provide a comprehensive overview of the historical trajectory of accreditation, focusing on pivotal developments in the United States spanning more than a century. This historical analysis sets the stage for discussing the explicit inclusion of *ethics* in accreditation requirements, which emerged as a more recent trend in the 1970s.

Acknowledging the shift from input-based to output-based accreditation models, the chapter examines alternative quality assurance methods and the widespread global impact of the US-style accreditation approach in engineering ethics education. It explores how ethics and related outcomes have been formally incorporated into accreditation requirements across various global contexts, from the United States to Western/Anglo settings like the United Kingdom and Canada, as well as international agreements such as the Washington Accord and EUR-ACE, and cases in East Asian countries like Japan and China. The authors note a contemporary convergence toward a more consistent set of ideals and target outcomes related to ethics in engineering education. This set increasingly emphasizes the importance of engineering graduates recognizing the societal and environmental impacts of their work.

Overall, Jesiek, Zhu, and Vinod synthesize prior scholarship and conduct new analyses of primary source materials, providing valuable insights into areas of convergence and divergence among accreditation documents worldwide. They highlight the incorporation of equity and diversity in some current documents and probe unique emphases in documents from Japan and China. Their chapter underscores the need for additional comparative research across national and cultural groups to further understand the evolving landscape of engineering ethics education and accreditation.

Following up on this historical perspective, the authors of Chapter 33 on 'Contextual Mapping of Ethics Education and Accreditation Nationally and Internationally,' Sarah Junaid, José Fernando Jiménez Mejía, Kenichi Natsume, Madeline Polmear, and Yann Serreau provide such comparative research. They present an analysis of currently enacted accreditation documents from many different parts of the globe. Their chapter shifts from the historical focus of the prior chapter to focus on the present landscape – aiming to discern the similarities and differences in how ethics are addressed within various nations' EEE accreditation documents. As integral members of a larger research team, the authors monitor how various cultural groups perceive and oversee ethics education within their own specific contexts. They pose critical questions about the global and local perspectives embedded in engineering accreditation documents and introduce a framework for cross-cultural comparative analysis.

Specifically, utilizing the Hofstede model (Hofstede, 2011; Hofstede et al., 2010), the chapter evaluates accreditation documents from four major cultural groups: Latin America, Latin Europe, Confucian Asia, and Anglo countries, represented by case studies from Colombia, France, Japan, and the United Kingdom, respectively. The authors conduct comparative analyses involving seven cultural clusters, shedding light on how engineering education systems conceptualize ethics and articulating previously tacit aspects. They identify trends such as the emphasis on 'application' over 'evaluation' in ethics education and recommend a shift towards higher-order skills development. Both this and the prior chapter endeavor to integrate non-English-language documents. Tackling challenges of language translation, their research aims to comprehensively understand and interpret diverse approaches to ethics education and accreditation practices across different cultural and national contexts.

Overall, the chapter provides insights into the current global landscape of ethics education and accreditation, advocating for continued collaborative efforts to broaden the scope and deepen the understanding of ethical considerations in engineering education.

To probe the far-reaching implications of engineering accreditation, particularly its educational requisites and ethical components, Chapter 34, 'Accreditation and Licensure: Processes and Implications' by Angela R. Bielefeldt, Diana Martin, and Madeline Polmear, examines the interplay between accreditation and licensure across diverse engineering subfields, with a focus on civil engineering. This chapter elucidates the intricate legal frameworks governing the formal recognition of individuals as 'engineers,' highlighting the significant variations in credentialing processes across countries and even within regions. It also navigates the complex terrain of licensure, which has become crucial to ensuring public safety and upholding professional standards.

As the chapter underscores the divergent pathways to engineering licensure worldwide, it emphasizes the role of accredited education in helping ensure the technical proficiency and ethical acumen requisite for competent engineering practice. The linkage between education and licensure is particularly pronounced in civil engineering, which has set the tone for accreditation standards across other engineering disciplines in North America and beyond, influencing ethical considerations in accreditation processes. The chapter probes the multifaceted dynamics between education and professional practice, exemplified by the highly regulated relationships within the United States' civil engineering context, governed by diverse stakeholder perspectives and stringent regulations across culturally diverse states.

In exploring the case studies of engineering education and licensure in the United States and Ireland, the chapter elucidates contrasting approaches to ethics integration within accreditation systems. Whereas the United States exemplifies a robust framework underpinned by non-profit organizations like the American Society of Civil Engineers (ASCE) advocating for ethical considerations, Ireland's accrediting body, Engineers Ireland, has spearheaded transformative change by mandating progressive ethics initiatives. However, research conducted by Martin (2020) reveals past discrepancies between stated objectives and actual practices enacted during accreditation evaluations, underscoring the challenges assessors encounter in evaluating ethical curricular components. The findings accentuate the imperative for ongoing dialogue and collaboration to enhance the assessment of engineering ethics education globally.

Chapter 35, 'A Feminist Critical Standpoint Analysis of Engineering Ethics Education and the Powers at Play in Accreditation, Research, and Practice' by Jillian Seniuk Cicek, Robyn Mae Paul, Diana Martin, and Donna Riley, offers critical perspectives on engineering ethics education. The authors probe the dynamics of engineering ethics education within the realms of accreditation and research, interrogating whose voices are privileged and whose are marginalized in the global discourse surrounding engineering accreditation. They challenge the hegemonic structures that dictate the content of engineering ethics education, shedding light on how these structures perpetuate exclusionary practices and uphold Western-centric perspectives.

The Chapter 35 author team raises a number of key points: the Western-centric nature of accreditation standards, such as those outlined in initiatives like the Washington Accord, often disregards local sensitivities, erasing non-Western perspectives. Engineering's technical epistemology tends to overshadow and marginalize alternative disciplinary perspectives, perpetuating a narrow understanding of ethics. The emphasis on micro-ethics and outcome-based assessment in engineering education separates ethics from broader equity and social justice concepts, limiting its transformative potential. The accreditation process in engineering perpetuates a state of "willful ignorance" (Tuana, 2006, p. 10) regarding its own detrimental effects, hindering meaningful progress. By employing critical feminist analyses, the authors critique the complicity of individuals in reinforcing existing power dynamics through engineering accreditation and encourage more conscientious engagement in the formulation and enactment of accreditation policy.

Overall, the authors contend that engineering educators inadvertently impede transformative change in ethics education by conforming to accreditation standards. They advocate incorporating critical perspectives to challenge and resist the exclusionary status quo, urging the transformation of engineering ethics education to embrace authenticity, significance, and inclusivity. This shift, they argue, is essential for engineers to engage in the profound work of addressing the myriad challenges confronting society and the environment.

The final chapter of this section, Chapter 36 on 'Accreditation Processes and Implications for Ethics Education at the Local Level,' written by Helena Kovacs and Stephanie Hladik, also adopts a critical lens. It explores the disjunction between the implementation of ethics education at grassroots levels and its representation in accreditation documents and formal procedural requirements. Kovacs and Hladik highlight the bureaucratic nature of operationalizing ethics education, which often results in abstract descriptions that fail to capture the nuances of ethical practice in engineering. They note the lack of requirements in accreditation policies for students to demonstrate higher-level cognitive skills, arguing that this can hinder the development of critical thinking skills and practices related to engineering ethics.

Kovacs and Hladik argue that the current approach to accreditation is too impersonal – it limits the local community's ability to shape the learning environment to reflect local interpretations and needs, particularly but not exclusively related to ethics. They describe what this impersonal approach implies at institutional, program, instructor, and student levels. The authors explain that institutions may struggle to integrate ethics into technical coursework effectively when they are not allowed enough room for interpretation or 'personalization.' They note that instructors may rely on historical scenarios that do not address systemic oppressions inherent in engineering design work – the type of oppression so vividly described in the previous chapter. For students, the standalone nature of ethics courses may lead to a perception that ethics is tangential to their core program; they may overlook the complex intersections of culture and decision-making.

In response to these challenges, Kovacs and Hladik propose strategies for bridging the gap between abstract accreditation standards and localized ethical practice. They advocate for collaboration among stakeholders within educational environments to develop scenarios that resonate with students' lived experiences, fostering within students essential skills and behaviors for ethical engineering practice. By empowering students to engage critically with ethics in their local contexts, the authors argue, educators can facilitate meaningful and impactful learning experiences that contribute to societal ethical advancement.

Trends and implications

Bookending the compilation of texts within this handbook, this section on engineering ethics accreditation within this *Routledge International Handbook of Engineering Ethics Education* illuminates the intricate landscape of ethics education and its accreditation practices, tracing a journey from disparate national systems towards a network of accords aimed at fostering global alignment. This final section charts the evolution of accreditation, emphasizing the imperative of equipping graduates with the ethical reasoning skills necessary for today's interconnected and mobile engineering profession.

Throughout the section, a discernible shift over time towards a competency-based approach is evident. This competency-based approach emphasizes technical proficiency and, increasingly, non-technical professional skills, ethical consciousness, and social responsibility. Chapters of this section explore the integration of ethics into accreditation standards from multiple perspectives, acknowledging the diverse interpretations of ethical principles across cultures. There is an overarching theme of reflexivity and criticality – the author teams critically examine power dynamics within education, accreditation, and licensure, highlighting the challenges and, at times, questioning the wisdom of implementing uniform ethical standards in diverse contexts.

A central theme across the chapters is the importance of providing localized, meaningful educational experiences that resonate with societal and environmental needs. By infusing engineering ethics education with local cultural perspectives and personal engagement, engineering educators and engineering education researchers can advocate for a balanced approach that fosters meaningful outcomes and cultural relevance. For them to gain legitimacy and become more mainstream and broadly accepted, the alternative ways of implementing engineering ethics education (i.e., localized, personal, and culturally aware perspectives) need to be translated into policy measures. The definitions of what constitutes 'good' engineering education set by accrediting bodies need to be continually informed by critical reflection, and conscientious objection.

Looking forward, we, the contributors to this section, aspire to amplify diverse voices, expand our understandings, and advance engineering practice and education toward greater ethical awareness and contemplation. We seek to build upon established teaching methods, improve assessment practices, and refine accreditation processes to better articulate the ethical aspirations of the engineering profession.

Ultimately, the message resounding throughout this section is a call to treat each other and our planet with ethical regard and respect, celebrate diversity, and continuously strive for improvement. By embracing localized perspectives and fostering global collaboration, we can navigate the complexities of engineering ethics accreditation, charting pathways for progress toward a more ethically conscious engineering profession.

Conclusions from the editor of this section

Shannon Chance, the lead editor for this section, expresses profound gratitude to all the authors who contributed their expertise, insights, and passion to this project. From the outset, the authors played an integral role in shaping the format and content of the chapters, bringing a diverse range of perspectives and experiences to the table. Despite the challenges of language, distance, and time zones, the authors demonstrated perseverance and remarkable collaboration, working in cross-border teams to produce high-quality analyses and generating rigorous research via enthusiastic and thoughtful engagement. We look forward to pushing these ideas and lines of investigation forward and we welcome others to join us in future collaborations.

References

Anwar, A. A., & Richards, D. J. (2013). Is the USA set to dominate accreditation of engineering education and professional qualifications? Proceedings of the Institution of Civil Engineers, 166(CE1), 42–48.

Chance, S. Martin, D. A., & Deegan, C. (2022). The assessment of ethics: Lessons for planners from engineering education's global strategy. *Educational Planning, 29*(3). 23–40. https://isep.info/wp-content/uploads/2022/09/221259-Journal-29-3_web.pdf#page=22

Chance, S., Seniuk Cicek, J., Bielefeld, A. R., Hladik , S., Martin, D., & Riley, D. M. (2024). Navigating Global Engineering Education Accreditation. Latin American and Caribbean Consortium of Engineering Institutions (LACCEI) conference, 15-19 July, 2024, San Jose, Costa Rica.

Gwynne-Evans, A. J., Chetty, M., & Junaid, S. (2021). Repositioning ethics at the heart of engineering graduate attributes. Australasian Journal of Engineering Education, *26*(1), 7–24.

Hofstede, G. (2011). Dimensionalizing cultures: The Hofstede model in context. *Online readings in psychology and culture, 2*(1), 8.

Hofstede, G., Hofstede, G. J., & Minkov, M. (2005). *Cultures and organizations: Software of the mind* (Vol. 2). New York: McGraw-Hill.

Hofstede, G., Hofstede, G. J., & Minkov, M. (2010). Cultures and organizations: Software of the mind: Intercultural cooperation and its importance for survival. Mcgraw-Hill.

International Engineering Alliance. (2024). Home. https://www.ieagreements.org/

Klassen, M. (2018). The politics of accreditation: A comparison of the engineering profession in five Anglospere countries. Toronto, Canada: University of Toronto.

Martin, D. A. (2020). Towards a sociotechnical reconfiguration of engineering and an education for ethics: A critical realist investigation into the patterns of education and accreditation of ethics in engineering programmes in Ireland. Doctoral Thesis, Technological University Dublin. DOI:10.21427/7M6V-CC71

Sthapak, B. K. (2012). Globalisation of Indian engineering education through the Washington Accord. *Journal of Engineering, Science & Management Education*, *5*(2), 464–466.

Tuana, N. (2006). The speculum of ignorance: The women's health movement and epistemologies of ignorance. *Hypatia, 21*(3), 1–19.

32

FOUNDATIONAL PERSPECTIVES ON ETHICS IN ENGINEERING ACCREDITATION

Brent K. Jesiek, Qin Zhu, and Gouri Vinod

This chapter takes a historical approach to examining and contextualizing the formal incorporation of ethics and related learning outcomes in accreditation criteria for engineering graduates. We begin by examining the origins of modern forms of accreditation in higher education, emphasizing key developments in the United States over more than a century. We also note more recent, widespread moves from inputs- to outputs-based frameworks (i.e., shifting focus from curricula and resources to graduates' capabilities), alternate quality assurance methods in some other contexts, and the continued global influence of American-style accreditation models. We then present a series of specific cases to explore when, where, and how ethics and associated concerns have been formally codified in accreditation requirements for engineering graduates. We begin by examining the United States as a particularly well-documented and influential example, followed by two other Western/Anglo settings (the United Kingdom and Canada). We then turn to two international agreements (the Washington Accord and EUR-ACE) and two East Asian cases (Japan and China).

One main goal of this chapter is to historicize attention to ethics in accreditation policies for higher engineering education. By doing so in a cross-national, comparative manner, we identify broader trends such as increasing attention in accreditation guidelines to an ever-wider range of concerns and considerations linked to engineering ethics, professional responsibility, and associated learning outcomes. Yet our efforts also begin to illustrate how local contextual factors (e.g., cultural, organizational, political) likely inflect accreditation criteria and processes, in turn hinting at reverse salients that counteract global convergence trends.

Our approach to developing this chapter involved synthesizing prior scholarship, including other secondary accounts, and performing new analyses of some primary source materials. We took a broad view of ethics when examining accreditation documents, and our choice of specific cases for this chapter was inflected by the authors' expertise, background, and positionality. All three of us hold undergraduate degrees in engineering and graduate degrees in the humanities and/or social sciences. Our team also includes individuals who are from or have lived in the United States, United Kingdom, and mainland China, and the authors have previously conducted other cross-national comparative studies related to engineering education and practice. While the scope of our inquiry is constrained by limitations such as the availability of source materials and our own expertise (language, etc.), we hope this chapter inspires future research efforts focused on other countries and regions.

DOI: 10.4324/9781003464259-39

Accreditation in higher education: historical origins and US trends

Mechanisms for monitoring the quality and legitimacy of universities can, in part, be traced to the early history of higher education in Europe, from the Middle Ages onward. Historians point to various kinds of oversight, including internal self-governance by student and faculty guilds and external mechanisms such as the formal chartering of institutions by the crown, state, or church (Maassen, 1997; Van Vught & Westerheijden, 1994). Another type of quality assurance emerged much earlier in China, where the Imperial Examination system was used over many millennia to screen candidates for civil-service positions (e.g., Min & Xiuwen, 2001). Yet as Maassen argued, modern accreditation – typically characterized by a focus on quality control mechanisms and the formal recognition of degree programs or entire institutions – "has its roots in American higher education" (1997, p. 124).

Some important early developments in the United States occurred with the establishment of its first colleges. Nine such schools (the 'Colonial Colleges') were operating by the time of the American Revolution, with 'charters' for their establishment usually granted by colonial governors, colonial assemblies, or the British Crown (Stoeckel, 1958). The charters helped establish the legitimacy of these institutions, giving them the formal, legal right to own property and grant degrees. Additional oversight and governance structures started to emerge by the 1780s, such as through the formation of a board of regents in New York to "charter, endow, and control" museums and schools in the state, including colleges (Harcleroad, 1980, p. 15). Nonetheless, historians note that the regulation of US colleges was generally lax into the nineteenth century, even as institutions of widely varying type and quality proliferated (Brittingham, 2019).

In 1847, the first non-profit, voluntary educational association was established in the United States: the National Medical Association, later named the American Medical Association (AMA) (King, 1982). Its founding was partly linked to concerns about the quality of medical education. While initially not very successful in addressing that particular issue, the AMA's efforts to develop a 'Code of Medical Ethics' had lasting impacts (King, 1983). More general calls to regulate US higher education intensified in the latter part of the nineteenth century, especially as new schools proliferated. In response, the late 1800s saw the establishment of regional, non-governmental accreditation bodies, beginning with the New England Association of Schools and Colleges (NEASC) in 1885 (Brittingham, 2009, p. 14). Colleges and universities were members of these voluntary organizations, which were in turn mainly focused on determining "which institutions were legitimately colleges" (Brittingham, 2009, p. 14) and publishing lists of such schools (Harcleroad, 1980, pp. 21–22).

The twentieth century was marked by several trends relevant to this volume. First, new associations focused on specific disciplines and fields multiplied from the 1910s onward. This created a regulatory structure where overall evaluation of universities or colleges was often conducted by regional associations, while discipline-based organizations accredited specific programs. Additionally, key features now associated with the US accreditation model developed during the middle part of the century. Per Brittingham,

> Between 1950 and 1965, the regional accrediting organizations developed and adopted what are considered today's fundamentals in the accreditation process: a mission-based approach, standards, a self-study prepared by the institution, a visit by a team of peers who produced a report, and a decision by a commission overseeing a process of periodic review.
>
> *(2009, pp. 14–15)*

Further, the federal government gradually assumed a larger role in higher education, including through new laws and regulations – many from the post-war period – restricting access to federal funding (and especially student aid) to institutions accredited by recognized non-profit associations.

Another development worth noting involves growing emphasis on *results* and *outcomes* in accreditation processes, particularly in relation to student learning. As Nodine (2016) argued, the basic principles of outcomes-based education (OBE) can be "traced back hundreds of years to craft guilds, apprenticeship training programs, technical training programs (in the military, etc.), and licensure programs (for doctors, lawyers, etc.) where established standards for competence and performance have been identified for specific jobs and roles" (p. 6). He noted the resonance between outcomes-based approaches and the concept of mastery-based learning beginning in the 1920s and a turn toward competency-based education (CBE) from the 1960s onward. Nodine observed three key shifts in this confluence of movements, namely moves toward identifying specific learning outcomes, establishing how to assess or measure those outcomes, and developing more flexible and personalized educational pathways (p. 6).

In summary, the US system of accreditation reflects the country's cultural values and styles of governance, including a 'triad' of federal, state, and non-governmental actors, with the latter especially critical for providing a "self-regulatory, peer review system" for higher education institutions and programs (Brittingham, 2009, p. 10). As Akera et al. summarized,

> the highly decentralized system of educational governance within the U.S., and the great diversity of schools that are both the product and reasons for this ecosystem, have given rise to an extremely heterogeneous system. In the United States, accreditation serves as one of the few central mechanisms for shaping learning; it carries the weight of the state to the extent that it contributes to job and federal loan availability as well as licensure in selected fields.
>
> *(2019, p. 1)*

Such points are salient in relation to other concerns, including questions about the place of learning outcomes related to ethics in degree programs and the diffusion of American-style accreditation models to other countries.

Further, accreditation is one of many kinds of quality assurance (QA), and alternative approaches like "academic audit and inspection" are more prevalent in some settings (Brittingham, 2009, p. 17). Today, accreditation is often associated with defining features like systematic self-assessment, some kind of external review mechanism, and a forward-looking evaluation philosophy (e.g., as reflected in 'continuous improvement' models). Since at least the late twentieth century, rising accountability pressures in higher education in many parts of the world have been accompanied by more widespread implementation of accreditation systems, albeit with notable local variations (El-Khawas, 2007). The number of foreign universities and degree programs directly accredited by US-based or international organizations has also grown considerably, a trend which has, in turn, been critiqued as a new kind of 'academic colonialism' (Altbach, 2003). As Altbach argued, "American accreditation is designed for the realities of American higher education" and exporting that model could pressure foreign institutions to conform to "American patterns of curricular and academic organization" (p. 6) while disregarding local realities.

Accreditation and ethics in engineering education: detailed cases

We now focus on cases focused on specific countries and international agreements. We begin with three Western/Anglo examples (the United States, the United Kingdom, and Canada), followed by two international agreements (the Washington Accord and EUR-ACE) and two East Asian cases (Japan and China). Readers may also want to consult the appendix of this chapter, as it provides verbatim excerpts of ethics-related outcomes/attributes for many of the accreditation frameworks discussed below.

United States

Early efforts to formally evaluate engineering degree programs in the United States were led by the American Institute of Chemical Engineering's (AIChE) Committee on Chemical Engineering Education starting in 1922, followed by the publication of a list of recognized degree programs at 14 schools in 1925 (Prados, 2008, p. 2). Prados claimed that the subsequent Wickenden report on engineering education helped stimulate broader interest in a new national organization with a similar role across engineering fields. As Wickenden declared, "If protection of standards is needed, the accrediting of engineering schools by their own organization and the national professional societies will probably prove to be much more effective than accrediting by educational bodies of a more general character" (1934, p. 1082). An organization of this sort, the Engineers' Council for Professional Development (ECPD), was established in 1932 with seven professional societies as its founding members (Prados, 2008, p. 6). The organization started accrediting engineering degree programs from 1935–1936 onward (Prados, p. 6).

As Stephan documented, the original ECPD accreditation criteria – unchanged from 1933 to 1950 – offered "virtually no specification of minimum standards, except that all accredited programs had to lead to a degree" (2002, p. 11). Yet in 1955, a new set of 'Additional Criteria' mandated more specific curricular requirements in mathematics, basic science, engineering sciences, engineering analysis and design, and humanistic-social studies (Parker, 1961, p. 14). These were specified as the minimum number of years of study (or fraction thereof) in each designated area. The ASEE's *Summary of the Report on Evaluation of Engineering Education* ('Grinter Report'), published in 1955, reflected this period's shift toward a quantitative view of degree requirements: "The consideration of curricula cannot proceed wholly on a philosophical or qualitative basis but must eventually be approached quantitatively in semester hours or at least in terms of fractional percentages of the total program" (CEEE, 1994, p. 85). Yet these new guidelines did not explicitly refer to 'ethics.' The 1955 criteria, for example, noted very generally that a student's humanistic-social studies coursework "should be selected from fields such as history, economics, government, literature, sociology, philosophy, psychology, or fine arts" (Parker, 1961, p. 14). However, the "qualitative" portion of this same document did mention "safety to life and property" as a relevant consideration for engineers doing design work, alongside economic and functional concerns (Parker, p. 14).

By the early 1970s, the ECPD's curricular requirements for accredited engineering degree programs were only a page long. They called for "the equivalent of one-half year to one full year as the minimum content in the area of the humanities and social sciences" (ECPD, 1971, p. 65), but did not explicitly refer to ethics or related themes. Yet, as the length and specificity of the ABET accreditation guidelines steadily increased from the 1970s onward, ethics and associated concerns became more explicit. For example, revised criteria published in 1973 referred to "the extent to which the program develops an ability to apply pertinent knowledge to the practice of engineering in an effective and professional manner," including "development of a sensitivity to the socially related technical problems which confront the profession" (ECPD, 1973, p. 44). The aforementioned humanities and social sciences requirement was also revised to simply specify "one-half year" as the minimum, while clarifying that such coursework was important for "making the young engineer fully aware of his [*sic*] social responsibilities and better able to consider related factors in the decision-making process" (p. 45). In 1974, a new footnote also clarified the meaning of the ECPD's required one-half year of engineering design "in its broadest sense" noting that "sociological, economic, aesthetic, legal, ethical, etc. considerations can be included" (ECPD, 1974, p. 68). In 1975, this same language was moved from a footnote into the body of the guidelines (ECPD, 1975, p. 75). These appear to be the first direct mentions of ethics in ECPD's accreditation guidelines for engineering programs.

These criteria were relatively stable until 1979 when ethics became even more pronounced in the ECPD guidelines. More specifically, the statement "development of an understanding of the characteristics of the engineering profession and the ethics of engineering practice" was added to the overarching preamble statement introducing the general program guidelines (ECPD, 1979, p. 60). This objective was further underscored in a later passage:

> An understanding of the ethical, social, and economic considerations in engineering practice is essential for a successful engineering career. Coursework may be provided for this purpose, but as a minimum it should be the responsibility of the engineering faculty to infuse professional concepts into all engineering coursework.
>
> (p. 61)

As Stephan reported, the latter passage was retained for many years, "substantially unchanged until the issuance of EC 2000" (2002, p. 13).

Ethics and related concerns were also explicit in the general student outcomes presented as part of ABET's Engineering Criteria 2000 (EC2000) framework adopted in 1996 (Lattuca et al., 2006). The new guidelines stipulated that graduates should have "an understanding of professional and ethical responsibility" (Criterion 3.f), "the broad education necessary to understand the impact of engineering solutions in a global and societal context" (3.h), and "a knowledge of contemporary issues" (3.j) (Lattuca et al., pp. 18–19). The most recent version of the Criterion 3 outcomes includes expanded language around graduates having "an ability to apply engineering design to produce solutions that meet specified needs with consideration of public health, safety, and welfare, as well as global, cultural, social, environmental, and economic factors" (ABET, 2018, I.3.2). It also features a multifaceted outcome focused on ethics, namely: "an ability to recognize ethical and professional responsibilities in engineering situations and make informed judgments, which must consider the impact of engineering solutions in global, economic, environmental, and societal contexts" (I.3.4). As reported by Matos et al. (2017), earlier drafts did not mention 'professional responsibility' in the ethics outcome, which generated considerable push-back and led to restoration of the phrase. Some of ABET's field-specific program criteria now also include attention to ethics and related concerns. For example, the criteria for "Civil and Similarly Named Engineering Programs" mandate coverage of "principles of sustainability in design" and the ability to "analyze issues in professional ethics" and "explain the importance of professional licensure" (ABET, 2018, III).

There are at least three key points to take from this brief account. First, explicit attention to ethics and related concepts was included in ABET guidelines earlier than previously reported. Both Pritchard (1990) and Stephan (2002) cited 1985 as the year when "understanding of the ethical characteristics of the engineering practice and profession" first appeared in the guidelines. Yet similar language initially surfaced in 1979, and other relevant statements and concepts appeared even earlier. Second, ABET EC2000 is often framed as a key point of transition where concerns over programmatic 'inputs' were replaced by a focus on 'outcomes' in engineering accreditation processes (Lucena et al., 2008). Yet the preceding account shows how ethics, professional responsibility, and related concerns were framed in outcomes-oriented language as early as the 1970s. This tracks well with other accounts regarding a gradual and more general turn toward outcomes- and competency-based approaches to education and training, especially from the 1960s onward (Hodge, 2007).

Finally, it is worth considering why the aforementioned changes were made. Unfortunately, the official accounts from ECPD offer little explanation. Period reports from the ECPD's ethics committee were primarily focused on a major revision of the ECPD Code of Ethics of Engineers, published in 1974 and then championed for more widespread adoption by other professional societies. Nonetheless, it is reasonable to speculate that the incorporation of ethics-related outcomes in the ECPD guidelines reflected broader movements, such as the efforts of engineer activists in the 1960s to critically interrogate the social and environmental effects of technology (Wisnioski, 2016), as well as the 1970s-era establishment of engineering ethics as a distinct scholarly field (Weil, 1984). More research is needed to establish whether and how these historical trends are connected. And, as Stephan (2002) pointed out, changing language in accreditation documents does not necessarily mean that engineering programs, or even accreditors, have historically treated ethics and related outcomes as key concerns.

United Kingdom

The United Kingdom was an important point of origin for engineering as a modern profession with roots going back to the eighteenth century. Yet fragmentation has been a hallmark of British engineering over this long history, in part reflected in the proliferation of engineering professional societies – and numerous calls to unify the profession (Klassen, 2018, pp. 78–84). As Klassen explained, accreditation of engineering programs in the United Kingdom has historically involved a complex assortment of policies and actors, including individual disciplinary professional societies, and with the Engineering Council providing additional coordination and oversight, especially from the 1980s onward. The United Kingdom's enduring tradition of apprenticeship-based training adds further complexity to this milieu.

Early efforts to unify the profession and improve coordination across the institutes are reflected in the creation of the Joint Council of Engineering Institutions in 1965 (called the Engineering Council since 1981) (Chapman & Levy, 2004). In 1984, the Council's *Standards and Routes to Registration* (SARTOR) established common training pathways and requirements for the three main professional grades recognized in the United Kingdom (Chartered Engineers, Incorporated Engineers, and Engineering Technicians). Second and third editions were published in 1990 and 1997. The latter (SARTOR3) is notable for specifying – like period documents from other countries – five specific outcome areas for each professional grade. One of these areas was specifically dedicated to "Professional Conduct" and declared that qualifying candidates for registration should "Make a personal commitment to live by the appropriate code of professional conduct, recognising obligations to society, the profession and the environment" (UKEC, 1998, p. 3), followed by four precepts that expanded on and clarified aspects of this general statement.

Concerns about the Engineering Council's influence over the accreditation of degree programs – including its efforts in SARTOR3 to raise standards – led to new reforms in the late 1990s and early 2000s. This included the promulgation of a new UK Standard for Professional Engineering Competence and Commitment (UK-SPEC) in 2003 to replace SARTOR. The new UK-SPEC placed greater emphasis on outcomes and eliminated earlier 'input-based' considerations like the quality of students entering degree programs (Temple, 2005). In 2004, the Engineering Council released its Accreditation of Higher Education Programmes (AHEP) policy and stated that it would share the responsibility for regulating engineering education standards with an independent non-profit, namely the Quality Assurance Agency for Higher Education (QAA) (EC, 2004). Additionally, the Engineering Council and Royal Academy of Engineering issued a common "Statement of Ethical Principles" for the engineering profession in 2005 (UKEC, 2017). As the most recent (4th) AHEP document notes, more than 40 engineering institutions are licensed by the Engineering Council to accredit degree programs in their respective fields (UKEC, 2020).

All four versions of the AHEP policy published to date include ethics requirements for engineers seeking registration at the incorporated and chartered levels. The first edition (AHEP1) stated that graduates should have an "Understanding of the need for a high level of professional and ethical conduct in engineering" and "Understanding of appropriate codes of practice and industry standards" and elsewhere repeatedly referred to the importance of health, safety, and risk issues, as well as environmental and sustainability concerns (UKEC, 2004, pp. 11–12). And although the next two editions (AHEP2 in 2013 and AHEP3 in 2014) showed little change in ethics-related outcomes, the most recent AHEP4 (released in 2020 and set to take effect in 2024) includes some notable revisions. First, it featured increasingly nuanced language to distinguish learning outcomes for incorporated and chartered grades, including for three distinct educational pathways associated with each. And while it retains five main outcome categories, it includes an "Engineering and society" category in place of "Economic, legal, social, ethical and environmental context" in AHEP3 (UKEC, 2014) and the even earlier "Economic, social and environmental context" in AHEP1 and AHEP2 (UKEC, 2004; UKEC, 2013). This category of outcomes also featured a revised preamble stating:

> Engineering activity can have a significant societal impact and engineers must operate in a responsible and ethical manner, recognise the importance of diversity, and help ensure that the benefits of innovation and progress are shared equitably and do not compromise the natural environment or deplete natural resources to the detriment of future generations.
>
> *(UKEC, 2020, p. 30)*

As this statement suggests, the new standard incorporates wide-ranging outcomes that refer to ethical conduct, risk management, sensitivity to the broader impacts of engineered solutions, and attention to diversity and equity concerns. Indeed, among educators interviewed by Xavier et al. (2023), "AHEP4 was believed to constitute a step change that encouraged the inclusion of [the] 'social'" (p. 4) in engineering programs.

As a final development worth noting, the Engineering Professors' Council (EPC) and Royal Academy of Engineering released an Engineering Ethics toolkit in 2021 "to help engineering educators integrate ethics content into their teaching" (EPC, 2022). As background, they note "growing advocacy for bringing engineering ethics to the fore in engineering programmes – alongside technical skills," including as reflected in current AHEP and UK-SPEC standards.

Canada

Since the early decades of the twentieth century, engineering has been legally regulated as a profession in Canada, mainly at the provincial/territorial level but with national co-ordination (Klassen, 2018, pp. 33–34). Oversight of engineering degree programs originated with establishment of the Canadian Accreditation Board (CAB) in 1965 as a standing committee of the Canadian Council of Professional Engineers (or Engineers Canada from 2007 onward), with the first assessments of undergraduate degree programs occurring in 1969 (CAB, 1975, p. 4). By 1975, the CAB's accreditation criteria specified required program content in five main areas, including "a minimum of one-half year of appropriate humanities and social sciences" (CAB, 1975, p. 15). While this document did not explicitly mention ethics, it did note the need for students to develop "social consciousness" and receive a "sufficient liberal education" (p. 12).

In 1976, a revised set of "specific objectives" included a section (B-1.7) stating that "Students must be made aware of the vital role of the professional engineer in society and the interaction of engineering work with the economic, social and human goals of the nation" (CAB, 1976, p. 10). The document went on to explain that students in accredited programs must understand:

a) the quality of the natural and human environment and the impact of technology;
b) the function and activities of our society, business and government in shaping our society and its values;
c) the legal responsibilities and ethical guidelines and constraints applied to the profession.

(CAB, 1976, p. 10)

As the report emphasized, "Every opportunity should be seized to weave into the fabric of engineering education an awareness of such matters through course material and through liaison with practicing engineers and other groups outside of the educational establishments" (p. 10). Another stipulation regarding a "minimum one half year of appropriate humanities, social sciences and administrative studies" clarified that the aim of such coursework was to "develop a social awareness as related to the philosophy of section B-1.7" (p. 14).

Similar language was retained until 1986 (under the renamed Canadian Engineering Accreditation Board, or CEAB), when a streamlined version of the accreditation criteria removed any direct mention of ethics. A new section of the guidelines (2.1.4) instead simply stated: "The criteria are intended to ensure that students are made aware of the role of the professional engineer in society and the impact that engineering in all its forms makes on the environmental, economic, social and cultural aspirations of society" (CEAB, 1987, p. 14). In the CEAB's 1989–1990 annual report, this statement was revised to refer to the "role *and responsibilities* [emphasis added] of the professional engineer" (CEAB, 1990, p. 14). Requirements published in 1993 also added language in the "Engineering Design" area to acknowledge "constraints which may be governed by standards or legislation to varying degrees depending upon the discipline. These constraints may relate to economic, health, safety, environmental, social or other pertinent factors" (CEAB, 1993, p. 17). In 1996, a new criterion was added (2.2.8) stipulating that "Each program must ensure that students are made aware of the role and responsibilities of the professional engineer in society. Appropriate exposure to ethics, equity, public and worker safety and health considerations and concepts of sustainable development and environmental stewardship must be an integral component of the engineering curriculum" (CEAB, 1996, p. 14).

The preceding language was retained verbatim until 2008 when it was replaced by a new set of 12 "Graduate Attributes" (CEAB, 2008, pp. 12–13). Four of the attributes refer to ethics or related concerns, namely (1) design, (2) professionalism, (3) impact of engineering on society and the environment, and (4) ethics and equity. This same document also retained quantitative requirements for curricular coverage in specific areas, including a stated expectation that all programs include studies of "The impact of technology on society," "Health and safety," "Professional ethics, equity and law," and "Sustainable development and environmental stewardship" (CEAB, 2008, p. 18). These requirements were subjected to only minor editorial changes in more recent versions of the CEAB guidelines. As this overview suggests, the current CEAB framework includes a fairly comprehensive set of ethics-related attributes similar to what can be found in policy documents promulgated in many other contexts.

Washington Accord

Western nations have had deep and lasting impacts on engineering education and professional practice around the world, both through colonial legacies and other influences. As a more specific example, the US-based ABET describes how it engages globally through four mechanisms: "1) accreditation of academic programs; 2) mutual recognition of accreditation organizations; 3) Memoranda of Understanding with accreditation/quality assurance organizations; and 4) engagement in global STEM education organizations" (ABET, n.d.). The third mechanism (regarding MOUs) includes specific cross-national agreements (e.g., between the United States and Canada concerning the accreditation of engineering degree programs, first signed in 1979) and more general agreements like the Washington Accord.

The latter is a multilateral framework that sets standards for mutual recognition of engineering degree programs and professional mobility among signatories, including six countries when initially signed in 1989 (Australia, Canada, Ireland, New Zealand, the United States, and the United Kingdom). Founded in 2007, the associated International Engineering Alliance (IEA) is a global non-profit organization that manages seven such agreements among members representing 41 jurisdictions in 29 countries (IEA, 2015). The IEA also maintains a set of "Graduate Attributes and Competency Profiles" developed from 2001 to 2005 by signatories of the Washington Accord (the preceding six countries, plus Hong Kong and South Africa). "Ethics" was one of 13 attributes in Version 1.1 of this framework ("Understand and commit to professional ethics, responsibilities, and norms of engineering practice"), along with other relevant concerns listed under "The Engineer and Society" and "Environment and Sustainability" (ABET, 2006). Similar categories and language were retained in later revisions (e.g., see IEA, 2013). Today, the Washington Accord has 23 full signatories and seven provisional ones (IEA, n.d.). As this overview suggests, a relatively small group of actors – primarily representing Western, anglophone nations or former colonies thereof – have spearheaded the development of global standards for accrediting engineering programs using outcomes-based approaches. As discussed in more detail below, Japan and China are contrasting examples of Washington Accord adoption, each likely inflected by distinct cultural and ideological factors.

European Accredited Engineer (EUR-ACE)

Beginning in the late 1990s, the intergovernmental initiative known as the 'Bologna Declaration' stimulated efforts to harmonize higher education programs across Europe (Augusti, 2007). Field-specific initiatives like the EUR-ACE (European Accredited Engineer) standard grew out of this larger trend. They became linked to a desire to increase the global mobility of engineering graduates, establish minimum quality standards for engineering degree programs, and encourage quality improvements (Augusti, 2007; Sánchez-Chaparro et al., 2022). EUR-ACE is a comprehensive standard with multifaceted attention to physical facilities; staff qualifications; program management; teaching, learning, and assessment practices; and so on (ENAEE, 2021).

Like other contemporary frameworks, EUR-ACE, from the beginning, also emphasized programmatic aims and student learning outcomes. Regarding the initial development of EUR-ACE, Augusti (2010) noted that a study of engineering accreditation systems across Europe "revealed striking similarities behind different façades" which in turn made "compilation of a set of shared accreditation standards and procedures comparatively easy" (p. 2). The resulting outcomes for EUR-ACE were organized around six core dimensions, with the sixth ("Transferable Skills") stressing the importance of graduates committing to "professional ethics, responsibilities and norms of engineering practice" (Augusti, Birch, & Payzin, 2011). The framework also underscored the importance of societal, environmental, ethical, and other "non-technical" considerations in three other outcome areas. Similar language and outcomes have been retained in more recent versions of the EUR-ACE standard (e.g., ENAEE, 2021).

From its 2006 inception to the present, the EUR-ACE designation has been granted to more than 4,000 degree programs at more than 700 higher education institutions in 46 countries, in Europe and beyond (ENAEE, n.d.). As EUR-ACE continues to spread, commentators have pointed out that the complexity and diversity of European higher education institutions and policy bodies introduce both benefits and challenges for cross-border quality assurance and accreditation efforts. For example, Sánchez-Chaparro et al. noted "difficulties in interpretation and consistency" of the European standards, while at the same time opening up "learning opportunities" as accreditation agencies work to adopt common standards while respecting cross-national contextual differences (2022, p. 322). How ethics is specifically treated in such processes is beyond the scope of this chapter, but worthy of further exploration.

Japan

Engineering as a modern field of practice originated in Japan in the late nineteenth and early twentieth centuries. Over time, engineers were primarily identified as members of corporate 'households' aligned with broader national goals for economic and technological development (Downey et al., 2007). Thus, the Western concept of autonomous professionalism is relatively new for Japanese engineers, and engineering societies in Japan have historically not operated like their Western counterparts, instead mainly focusing on creating standards for education and industrial practices. Indeed, most have had little historical engagement with codes of ethics or accreditation-related activities. Downey et al. reported that the Japanese Society of Civil Engineers has had a statement of "Beliefs and Principles of Practice for Civil Engineers" since at least 1938 but argued that it was "of relatively little consequence" (2007, p. 480). Another notable exception is the Japan Consulting Engineer Association's (JCEA) first ethics codes (published in 1951 and 1961), which reflected influences from counterpart American societies (Kenichi, 2021). Kenichi additionally reports that Kimura Hisao, Chair of the IEEE Computer Society's Japan Chapter, advocated for developing ethics codes among Japanese engineering societies in the 1960s and 1970s. Yet others expressed reluctance, arguing that (1) codes of ethics might encourage engineers to demand their own rights to the detriment of their social responsibilities and (2) it was unnecessary to develop codes of ethics for individual fields when there should be a code of ethics for *all* professional societies (Kenichi, 2021). Some IEEE Japan board members were also worried that establishing a code of ethics for a particular association (e.g., IEEE Japan) might be a selfish act, disturbing the harmony of the scientific community in Japan (Kenichi, 2021).

In the 1990s, the Japanese government undertook initiatives to internationalize engineering education programs and qualifications with the goal of making their engineers more globally competitive, in turn setting in motion a burgeoning professionalization movement. Engineering societies also started to establish their own ethics codes (Kenichi, 2021), and in 2000 the Japanese *diet* (legislature) passed an updated Professional Engineers Law, which explicitly referred to the ethical duties of engineers (Downey et al., 2007). Another key development involved the 1999 founding of the Japan Accreditation Board of Engineering Education (JABEE), which created an accreditation system similar to the US model. And in 2002, an ethics outcome was added to Japan's accreditation criteria, stipulating that graduates of accredited programs should demonstrate "understanding of … engineers' social responsibilities (engineering ethics)" (Downey et al., 2007). Yet early efforts to develop and roll out accreditation criteria came with growing recognition that there was a lack of ethics education in Japanese engineering education and uncertainty about how it should be taught (Iseda, 2008; Kanemitsu, 2021). Nonetheless, Sato and Harada (2005) found that 76.1% of surveyed institutions were soon thereafter offering courses in engineering ethics.

During this same period, the Japanese Society for Engineering Education (JSEE) established a committee to study the syllabi of engineering ethics courses in Japan and found that they incorporated some core ideas and key concepts from Western engineering ethics, such as the analogy between ethical problem-solving and design thinking and specific tools for ethical decision-making (Kobayashi & Fudano, 2004). The JSEE's Engineering Ethics Research Committee also assumed an instrumental role in providing nationwide guidance and resources related to engineering ethics education. Since 2012, this committee has developed three versions of the "Learning and Educational Objectives of Engineering Ethics Education." The most recent version features four learning objectives: (1) understanding the relationship between science, technology, society, and the environment (cognitive domain); (2) understanding the role, responsibilities, and duties of engineers (cognitive domain); (3) ethical judgment abilities and problem-solving abilities (cognitive domain); and (4) attitudes and shared values as professional engineers (affective domain) (Kobayashi & Fudano, 2016). These are in general alignment with current JABEE requirements, with one of the nine learning criteria focused specifically on "understanding of effects and impact of professional activities on society and nature, and of professionals' social responsibility." This criterion is in turn elaborated with a series of more specific statements:

- "Understanding of impact of technology of related engineering fields on public welfare"
- "Understanding of implication of technology of related engineering fields on environmental safety and sustainable development of society"
- "Understanding of engineering ethics"
- "An ability to take action based on the understanding mentioned above" (JABEE, 2016, p. 4).

Additionally, a dedicated design criterion specifies that graduates should be able to "specify constraints from public welfare, environmental safety, and economy" (JABEE, 2016). Such statements reflect a fairly typical range of concerns found in many accreditation frameworks. (For more on Japan, see Chapter 33.)

China (mainland)

Contemporary concerns about quality assurance in Chinese higher education must be situated against a much longer historical legacy and backdrop, including the civil-service examination system in Imperial China, which serves as one of the very first examples of a standardized test system (O'Sullivan & Cheng, 2022). This system ensured that students met the criteria (or 'learning outcomes,' in a modern sense) for professional politicians and bureaucrats serving the Imperial government – some of whom later became what we would now call engineers (Dodgen, 2001). The state employed various efforts and tactics to indoctrinate examinees, including through government-issued textbooks and the contents of the exam itself (Lin, 2021).

In more contemporary terms, developing countries such as China have often taken a pragmatic approach to developing professional standards and accreditation systems. This can take the form of borrowing from the West, as reflected in a series of ethics codes published from 1933 onward by the Chinese Institute of Engineers (CIE) (Zhang & Davis, 2018). As Zhang and Davis (2018) argue, the adaptation and evolution of these early codes seemed to reflect practical realities and national development objectives rather than Confucian cultural values. They and others (e.g., Cao,

2015) have additionally noted a lack of formal ethical codes for engineers in mainland China from the Communist Revolution (which ended in 1949) to the present. Yet this is not surprising given China's ideological context, that is, where Western ideas of autonomous, independent professionalism stand in tension with Communist party authority and values.

Nonetheless, ethics and related concepts have recently surfaced in engineering education, particularly against the backdrop of a pragmatic approach to accreditation policy-making. Given the lack of a pre-existing accreditation model, the Washington Accord was used as an actionable 'startup template' in China, but without fully acknowledging or challenging its fundamental ideas, concepts, and assumptions (Zhu, Jesiek, & Yuan, 2014). Chinese policy-makers made adjustments to the ABET accreditation process to ensure that the resulting policies were better aligned with China's unique cultural and political context (e.g., by seeing ethics and ideological education as related or interchangeable). Accreditation expert and former university administrator Li (2017) observed that the adoption of the Washington Accord accreditation criteria in the early development of China's engineering accreditation system served the pragmatic goal of ensuring that the professional qualifications of Chinese students who graduate from accredited programs would be recognized by other Washington Accord signatories – thus enabling global mobility of Chinese talent.

In 2013, China became a provisional member of the Washington Accord, and in 2016 a full signatory member. Scholars have argued that a major motivation for establishing an accreditation system for engineering education was in part linked to concerns over academic quality and administration (Wang, Zhao & Lei, 2014). Wang et al. also pointed out that, in contrast to other countries, China's accreditation system exhibited more 'top-down' characteristics. Rather than primarily relying on representatives from industry to shape the standards for accreditation, the central government spearheaded coordination and policy-making, including organizing expert panels for formulating learning outcomes for engineering programs.

Current Chinese accreditation standards include ethics-related statements in four different outcome categories, namely (c) Design/Development Solutions, (f) Engineering and Society, (g) Environment and Sustainable Development, and (h) Professional Ethics. Notably, the only direct mention of social responsibility in engineering is in outcome (h), which states that students who graduate from accredited programs should "possess literacy in humanities and social sciences and social responsibility," be able to "understand and comply with professional morality and norms in engineering practice," and "exercise [their] responsibilities" (CEEAA, 2022a, section 4.3). Like many other accreditation policies, the other learning outcomes noted here (i.e., (c), (f), and (g)) variously indicate that engineering practice – including design, analysis, and problem-solving activities – should include attention to social, environmental, health, legal, and cultural, and other impacts.

Nevertheless, Li (2017) reminded engineering educators in China that accreditation criteria should not be considered equivalent to engineering program quality standards. In other words, the accreditation standards are a minimum benchmark, and the ethics-related learning outcomes may not wholly satisfy the government's expectations regarding graduate engineers' ethical and political qualities. For instance, some moral and ideological educational goals set by the central government – such as cultivating the builders and successors of Socialism with comprehensive development in morality, intelligence, physical fitness, and aesthetic appreciation – are not explicit in the accreditation policies but are nonetheless central to the training of Chinese engineers.

Given the top-down governance structure of China's policy-making, China has also employed multiple tactics to ensure that engineering programs and accreditation experts accurately interpret the accreditation criteria set by the central government and incorporate them into educational reforms and program evaluations. To begin, the government implemented several 'innovative' organizational structures to purportedly guarantee the 'autonomy' of accreditation activities while also maintaining the central government's influence in accreditation practices. It designated the Chinese Association for Science and Technology (CAST) as the official agency responsible for representing China's membership within the Washington Accord. The major accreditation body, the Chinese Engineering Education Accreditation Association (CEEAA), then became a corporate member of CAST, despite the fact that CEEAA was initiated by and located in the Chinese Ministry of Education. As "the largest national non-governmental organization of scientific and technological workers in China," CAST oversees other engineering societies such as the China Civil Engineering Society. Additionally, these societies were granted the authority to offer expert guidance and direction concerning engineering accreditation within their respective fields of expertise. Therefore, one notable aspect of engineering ethics education in China is that engineering societies organize nationwide professional development activities that train faculty in their specific engineering fields to teach discipline-based engineering ethics (e.g., civil engineering ethics, safety engineering ethics, etc.).

From as early as 2016, the central government has also regularly published guidelines on how to interpret and implement the accreditation criteria appropriately. The Chinese Engineering Accreditation Association (CEEAA) published the two most recent guidelines in 2020 and 2022. These guidelines provide details on how each learning outcome should be evaluated and how to understand certain key terms such as 'ethics' and 'social responsibility' in students' learning outcomes. In the most recent revision, one of the six major guiding principles is related to the cultivation of responsible engineers:

> To further clarify the requirements for implementing the fundamental task of "cultivating moral character and nurturing talented individuals," it is demanded that the educational objectives of professional training reflect the education policy of fostering socialist constructors and successors who possess comprehensive development in morality, intelligence, physical fitness, aesthetics, and labor. The graduation requirements should also incorporate relevant content regarding socialist core values.
>
> *(CEEAA, 2022b, pp. 5–6)*

As this statement suggests, the pragmatic adaptation of Western professional standards and processes in the Chinese context reflects a core concern with positioning ideological allegiance to the Chinese Communist Party (CCP) above other types of ethical commitments and values.

Discussion and conclusion

As documented in this chapter, modern forms of accreditation in higher education have strong historical roots in the US system of higher education. The first formal mechanisms to accredit engineering degree programs also originated in the United States, evolving considerably over a century-long period and ultimately having a marked global influence. However, explicit attention to ethics and related concerns in accreditation requirements is a more recent trend. For the countries examined in this chapter, such statements first appeared in 1970s-era policies in the United States and Canada. These same guidelines additionally reflected the early presence and influence of outcomes-based educational philosophies, albeit in tandem with period expectations for content and curricula as 'inputs' for engineering degree programs seeking accreditation. A more widespread transition to outcomes-based standards for engineering education occurred from the 1990s onward, accompanied by growing attention to ethics and related concerns.

Further, the preceding account suggests considerable convergence toward a common, core set of ethics-related outcomes in accreditation frameworks in many different countries and regions. Such documents most often refer to (1) professional/ethical responsibilities in general; (2) ethics as an 'upstream' constraint or consideration in problem solving, design, and so on; and (3) the 'downstream' impacts of engineered solutions on society. Further, most accreditation policies now mention environmental and/or sustainability concerns, in some cases as dedicated learning outcomes. Interestingly, the scope of ethics-related outcomes in the two global policies introduced above (Washington Accord and EUR-ACE) essentially cover this outcome space.

It is worth pondering whether and how a kind of global 'standard' for accreditation has been developed and advanced in recent decades, in part linked to broader processes of globalization. Yet our analysis suggests notable points of difference and divergence. For example, we observe the somewhat recent appearance of diversity and equity considerations in some accreditation criteria, such as Canadian policies that jointly refer to 'ethics and equity.'

It remains to be seen whether similar statements start to appear in other accreditation frameworks. We also find that explicit mention of ethical codes of conduct or practice only appears in general accreditation guidelines from the United Kingdom, even though such codes are well established in many other countries discussed above. Further, our analysis suggests important contextual nuances in two East Asian settings. The overarching storyline in Japan seems most significantly inflected by local cultural values (e.g., collectivistic ways of being, promoting social harmony) and particular understandings of how Japanese engineers contribute to national progress. The Chinese case is likely also shaped by similar cultural values, but with political and ideological forces at the forefront, especially in terms of ensuring that the ethical and social responsibilities of engineers align with party values and priorities.

There were, of course, practical limits to the breadth and depth of analysis we were able to present here, and we acknowledge a growing body of scholarship exploring engineering accreditation histories and trends in other contexts, developed and developing countries alike. We hope our efforts inspire more cross-national comparative research, and indeed highlight emerging opportunities for bringing more ethics-related outcomes into accreditation guidelines worldwide.

Acknowledgments

This material is based upon work supported by the National Science Foundation under Grant Nos. 2024301, 2027519, 2124984, and 2316634. The authors also extend thanks to a number of informants who kindly shared information and resources with us as we developed the chapter and two anonymous peer reviewers who provided helpful feedback on an earlier draft.

Appendix. Ethics-related outcomes/attributes from select accreditation frameworks

ABET - EC2000 (Lattuca et al., 2006)	**ABET - Current** (ABET, 2018)	**UK - SARTOR3** (UKEC, 1998, p. 3)	**AHEP1** (UKEC, 2004, pp. 11–12) (same for IEng and CEng programmes)
Outcome 3.f; "an understanding of professional and ethical responsibility" Outcome 3.h; "the broad education necessary to understand the impact of engineering solutions in a global and societal context" Outcome 3.j; "a knowledge of contemporary issues"	Outcome 3.2; "an ability to apply engineering design to produce solutions that meet specified needs with consideration of public health, safety, and welfare, as well as global, cultural, social, environmental, and economic factors." Outcome 3.4; "an ability to recognize ethical and professional responsibilities in engineering situations and make informed judgments, which must consider the impact of engineering solutions in global, economic, environmental, and societal contexts."	"E. Make a personal commitment to live by the appropriate code of professional conduct, recognising obligations to society, the they must: E.1 comply with the Codes and Rules of Conduct; E.2 manage and apply safe systems of work; E.3 undertake their engineering work in compliance with the Codes of Practice on Risk and the Environment; E.4 carry out the continuing professional development necessary to ensure competence in their areas of future intended practice."	*Design;* "[I]dentify constraints including environmental and sustainability limitations, health and safety and risk assessment issues" *Economic, social and environmental context;* "Understanding of the requirement for engineering activities to promote sustainable development"; "Awareness of the framework of relevant legal requirements governing engineering activities, including personnel, health, safety, and risk (including environmental risk) issues"; "Understanding of the need for a high level of professional and ethical conduct in engineering" *Engineering Practice;* "Understanding of appropriate codes of practice and industry standards"

AHEP4 - Current *(UKEC, 2020)*	**Canadian Accreditation Board (CAB) - 1976** *(CAB, 1976, p. 10)*	**Canadian Engineering Accreditation Board (CEAB) - 2008** *(CAB, 2008, p. 13)*	**Washington Accord - Version 1.1** *(ABET, Inc., 2006, p. 13)*
Design and Innovation: B5. "Design solutions for [IEng: broadly-defined; CEng: complex] problems that meet a combination of societal, user, business and customer needs as appropriate. This will involve consideration of applicable health and safety, diversity, inclusion, cultural, societal, environmental and commercial matters, codes of practice and industry standards." *The engineer and society:* B7. "Evaluate the environmental and societal impact of solutions to [IEng: broadly-defined; CEng: complex] problems."; B8. "Identify and analyse ethical concerns and make reasoned ethical choices informed by professional codes of conduct."; B9. "Use a risk management process to identify, evaluate and mitigate risks (the effects of uncertainty) associated with a particular project or activity." (IEng); "B11. Recognise the responsibilities, benefits and importance of supporting equality, diversity and inclusion." (CEng); "C11. Adopt an inclusive approach to engineering practice and recognise the responsibilities, benefits and importance of supporting equality, diversity and inclusion."	B-1.7. "An understanding must be acquired of: a) the quality of the natural and human environment and the impact of technology; b) the function and activities of our society, business and government in shaping our society and its values; c) the legal responsibilities and ethical guidelines and constraints applied to the profession."	3.1.4. "Design: An ability to design solutions for complex, open-ended engineering problems and to design systems, components or processes that meet specified needs with appropriate attention to health and safety risks, applicable standards, economic, environmental, cultural and societal considerations." 3.1.8. "Professionalism: An understanding of the roles and responsibilities of the professional engineer in society, especially the primary role of protection of the public and the public interest." 3.1.9. "Impact of engineering on society and the environment: An ability to analyse social and environmental aspects of engineering activities. Such abilities include an understanding of the interactions that engineering has with the economic, social, health, safety, legal, and cultural aspects of society; the uncertainties in the prediction of such interactions; and the concepts of sustainable design and development and environmental stewardship." 3.1.10. "Ethics and equity: An ability to apply professional ethics, accountability, and equity."	Attribute 9. *The Engineer and Society*, "Demonstrate understanding of the societal, health, safety, legal, and cultural issues and the consequential responsibilities relevant to engineering practice." Attribute 10. *Ethics*, "Understand and commit to professional ethics, responsibilities, and norms of engineering practice." Attribute 11. *Environment and Sustainability*, "Understand the impact of engineering solutions within a societal context, and demonstrate knowledge of and need for sustainable development."

EUR-ACE - 2008 *(Augusti, Birch, & Payzin, 2011, pp. 7–9)*	**Japan Accreditation Board of Engineering Education - Current** *(JABEE, 2016)*	**China Engineering Education Accreditation Criteria (CEEAA) - Current** *(CEEAA, 2022a)*
Outcome 2. Engineering Analysis, "Graduates should be able to … recognise the importance of societal, health and safety, environmental and commercial constraints." Outcome 3. Engineering Design, "[A]n awareness of societal, health and safety, environmental and commercial considerations." Outcome 5. Engineering Practice, "They should also recognise the wider, non-technical implications of engineering practice, ethical, environmental, commercial and industrial." Outcome 6. Transferable Skills, "[D]emonstrate awareness of the health, safety and legal issues and responsibilities of engineering practice, the impact of engineering solutions in a societal and environmental context, and commit to professional ethics, responsibilities and norms of engineering practice."	Criterion 1(2)(b), "An ability of understanding of effects and impact of professional activities on society and nature, and of professionals' social responsibility" • "Understanding of impact of technology of related engineering fields on public welfare" • "Understanding of implication of technology of related engineering fields on environmental safety and sustainable development of society" • "Understanding of engineering ethics" • "An ability to take action based on the understanding mentioned above" Criterion 1(2)(e), "Design ability to respond to requirements of the society by utilizing various sciences, technologies and information" (including) "An ability to specify constraints from public welfare, environmental safety, and economy to be taking account of"	c) Design/Development Solutions: "Able to design solutions for complex engineering problems, design systems, units (components), or process flows that meet specific requirements, and demonstrate innovative thinking during the design phase while considering social, health, safety, legal, cultural, and environmental factors." f) Engineering and Society: "Able to conduct rational analysis based on engineering-related background knowledge, evaluate the impact of professional engineering practices and solutions to complex engineering problems on society, health, safety, law, and culture, and understand the responsibilities to be assumed." g) Environment and Sustainable Development: "Able to understand and evaluate the impact of engineering practices for complex engineering problems on the environment and social sustainable development." h) Professional Ethics: "Possessing humanistic and social science literacy, a sense of social responsibility, and the ability to understand and abide by engineering professional ethics and standards, fulfilling responsibilities in engineering practice."

References

ABET. (2006). *Graduate attributes and professional competency profiles for the engineer, Engineering technologist, and engineering technician*. ABET, Inc.

ABET. (2018). *Criteria for accrediting engineering programs, 2019–2020*. ABET, Inc. https://www.abet.org/accreditation/accreditation-criteria/criteria-for-accrediting-engineering-programs-2019-2020/#GC3

ABET. (n.d.). *ABET's role in global accreditation*. ABET, Inc. https://www.abet.org/accreditation/get-accredited/accreditation-outside-the-u-s/

Akera, A., Appelhans, S., Cheville, A., De Pree, T., Fatehiboroujeni, S., Karlin, J., & Riley, D. M. (2019). ABET & engineering accreditation - History, theory, practice: Initial findings from a national study on the governance of engineering education. *Proceedings of the 2019 ASEE Annual Conference & Exposition*. https://peer.asee.org/32020

Altbach, P. G. (2003). Academic colonialism in action: American accreditation of foreign universities. *International Higher Education, 32*, 5–7. https://ejournals.bc.edu/index.php/ihe/article/view/7373

Augusti, G. (2007). Accreditation of engineering programmes: European perspectives and challenges in a global context. *European Journal of Engineering Education, 32*(3), 273–283.

Augusti, G. (2010). EUR-ACE: A common European quality label for accredited engineering programs. *Proceedings of the 2010 International Conference on Engineering Education (ICEE-2010)*.

Augusti, G., Birch, J., & Payzin, A. E. (2011). *EUR-ACE: A system of accreditation of engineering programmes allowing national variants.* https://www.mudek.org.tr/doc/sun/20110405%28Augusti-Birch-Payzin-INQAAHE2011-paper%29.pdf

Brittingham, B. (2009). Accreditation in the United States: How did we get to where we are? *New Directions for Higher Education, 145*, 7–27. https://doi.org/10.1002/he.331

Canadian Accreditation Board (CAB), The. (1975). *1st annual report*. The Canadian Council of Professional Engineers.

Canadian Accreditation Board (CAB), The. (1976). *Annual report - June 1976*. The Canadian Council of Professional Engineers.

Canadian Engineering Accreditation Board (CEAB), The. (1987). *Canadian engineering accreditation board - 1986/1987 report.* The Canadian Council of Professional Engineers.

Canadian Engineering Accreditation Board (CEAB), The. (1990). *Canadian engineering accreditation board - 1989/1990 annual report.* The Canadian Council of Professional Engineers.

Canadian Engineering Accreditation Board (CEAB), The. (1993). *Canadian engineering accreditation board - 1993 annual report.* The Canadian Council of Professional Engineers.

Canadian Engineering Accreditation Board (CEAB), The. (1996). *1996 accreditation criteria and procedures*. The Canadian Council of Professional Engineers.

Canadian Engineering Accreditation Board (CEAB), The. (2008). *Accreditation criteria and procedures - 2008*. The Canadian Council of Professional Engineers.

Cao, G.H. (2015). Comparison of China-US engineering ethics educations in sino-western philosophies of technology. *Sci Eng Ethics, 21*, 1609–1635. https://doi.org/10.1007/s11948-014-9611-3

Chapman, C. R., & Levy, J. (2004). An engine for change: A chronicle of the engineering council. Engineering Council UK.

China Engineering Education Accreditation Association (CEEAA). (2022a). 工程教育认证标准[Engineering education accreditation criteria]. https://www.ceeaa.org.cn/gcjyzyrzxh/rzcxjbz/gcjyrzbz/tybz/630662/index.html

China Engineering Education Accreditation Association (CEEAA). (2022b). 工程教育认证通用标准解读及使用指南(2022版) [Interpretation and user guide of universal standards for engineering education accreditation (2022 ed.)]. http://jxyzyrz.swu.edu.cn/info/1005/1380.htm

Committee on Evaluation of Engineering Education (CEEE). (1994). Report on evaluation of engineering education (reprint of the 1955 report). *Journal of Engineering Education, 93*(1), 74–94.

Dodgen, R. A. (2001). *Controlling the dragon: Confucian engineers and the Yellow River in late Imperial China*. University of Hawai'i Press.

Downey, G. L., Lucena, J. C., & Mitcham, C. (2007). Engineering ethics and identity: Emerging initiatives in comparative perspective. *Science and Engineering Ethics, 13*(4), 463–487.

El-Khawas, E. (2007). Accountability and quality assurance: New issues for academic inquiry. In J. F. James & P. G. Altbach (Eds.), *International handbook of higher education* (pp. 23–37). Springer.

Engineers' Council for Professional Development (ECPD). (1971). *39th annual report year ending Sept. 30, 1971*. Engineers' Council for Professional Development.

Engineers' Council for Professional Development (ECPD). (1973). *41st annual report year ending Sept. 30, 1973*. Engineers' Council for Professional Development.
Engineers' Council for Professional Development (ECPD). (1974). *42nd annual report year ending Sept. 30, 1974*. Engineers' Council for Professional Development.
Engineers' Council for Professional Development (ECPD). (1975). *43rd annual report year ending Sept. 30, 1975*. Engineers' Council for Professional Development.
Engineers' Council for Professional Development (ECPD). (1979). *47th annual report year ending Sept. 30, 1979*. Engineers' Council for Professional Development.
European Network for Accreditation of Engineering Education (ENAEE). (2021). *EUR-ACE framework standards and guidelines (EAFSG)*. https://www.enaee.eu/wp-content/uploads/2022/03/EAFSG-04112021-English-1-1.pdf
European Network for Accreditation of Engineering Education (ENAEE). (n.d.). *Database of EUR-ACE labelled programmes*. https://eurace.enaee.eu/node/163
Engineering Professors' Council (EPC). (2022). Welcome to the engineering ethics toolkit. https://epc.ac.uk/article/welcome-to-the-engineering-ethics-toolkit/
Harcleroad, F. F. (1980). *Accreditation: History, process, and problems* (AHE-ERIC/Higher education research report No. 6). National Institute of Education.
Hodge, S. (2007). The origins of competency-based training. *Australian Journal of Adult Learning*, *47*(2), 179–209.
International Engineering Alliance (IEA). (2013). *Graduate attributes and competency profiles*. https://www.ieagreements.org/assets/Uploads/Documents/Policy/Graduate-Attributes-and-Professional-Competencies.pdf
International Engineering Alliance (IEA). (2015). *A history of the International Engineering Alliance and its constituent agreements: Toward global engineering education and professional competence standards*. https://www.ieagreements.org/assets/Uploads/Documents/History/IEA-History-1.1-Final.pdf
International Engineering Alliance (IEA). (n.d.). Washington Accord – Signatories. https://www.ieagreements.org/accords/washington/signatories/
Iseda T. (2008). How should we foster the professional integrity of engineers in Japan? A pride-based approach. *Science and Engineering Ethics*, *14*(2), 165–176.
Japan Accreditation Board of Engineering Education (JABEE). (2016). *JABEE category-dependent criteria for accreditation of professional education programs*. JABEE. https://jabee.org/doc/12334.pdf
Kanemitsu, H. (2021). *How accreditation leads to Japanese-specific teaching materials for Japanese-specific engineering ethics*. SEFI. https://www.sefi.be/2021/02/24/how-accreditation-leads-to-japanese-specific-teaching-materials-for-japanese-specific-engineering-ethics/
Kenichi, N. (2021). *Japan's engineering ethics and Western culture: Social status, democracy, and economic globalization*. Lexington Books.
King, L. S. (1982). IV. The founding of the American Medical Association. *JAMA*, *248*(14), 1749–1752. https://doi.org/10.1001/jama.1982.03330140059036
King, L. S. (1983). IX. The AMA gets a new code of ethics. *JAMA, 249*(10), 1338–1342. https://doi.org/10.1001/jama.1983.03330340072038
Klassen, M. (2018). *The politics of accreditation: A comparison of the engineering profession in five anglosphere countries* (Unpublished master's thesis). University of Toronto, Toronto, Ontario, Canada.
Kobayashi, Y., & Fudano, J. (2016). 「技術者倫理教育における学習・教育目標2016」および「モジュール型モデル・シラバス」解説 ["Learning and educational objectives in technical ethics education 2016" and the explanation of the "module-based model syllabus"]. *Journal of Engineering Education/工学教育*, *64*(5), 141–159.
Lattuca, L., Terenzini, P. T., & Wolkwein, J. F. (2006). *Engineering change: A study of the impact of EC2000*. ABET, Inc.
Li, Z. (2017). 对我国工程教育专业认证十年的回顾与反思之二 [A review and reflection on ten years of engineering education professional accreditation in our country: Part two]. http://meea.cmes.org/article?id=188
Lin, H. (2021). Examination essays, paratext, and Confucian orthodoxy: Negotiating the public and private in knowledge authority in early seventeenth-century China. In M. Green, L C. Norgaard & M. B. Bruun (Eds.), *Early modern privacy* (pp. 297–314). Brill.
Lucena, J., Downey, G. L., Jesiek, B. K., & Ruff, S. (2008). Competencies beyond countries: The re-organization of engineering education in the United States, Europe, and Latin America. *Journal of Engineering Education*, *97*(4), 433–447. https://doi.org/10.1002/j.2168-9830.2008.tb00991.x

Maassen, P. A. M. (1997). Quality in European higher education: Recent trends and their historical roots. *European Journal of Education*, *32*(2), 111–127. http://www.jstor.org/stable/1503543
Matos, S. M., Riley, D. M., & Akera, A. (2017). WannABET? Historical and organizational perspectives on governance in engineering education. *Proceedings of the 2017 ASEE Annual Conference & Exposition*. https://peer.asee.org/29110
Min, H., & Xiuwen, Y. (2001). Educational assessment in China: Lessons from history and future prospects. *Assessment in Education: Principles, Policy & Practice*, *8*(1), 5–10. https://doi.org/10.1080/09695940120033216
Nodine, T. R. (2016). How did we get here? A brief history of competency-based higher education in the United States. *Competency-based Education*, *1*(1), 5–11. https://doi.org/10.1002/cbe2.1004
O'Sullivian, B., & Cheng, L. (2022). Lessons from the Chinese imperial examination system. *Language Testing in China*, *12*, Article 52.
Parker, J. M., III. (1961). Geological engineering curricula. *Journal of Geological Education*, *9*(1), 13–18.
Prados, J. (2008). Accreditation of undergraduate curricula. In *AIChE Centennial 1908–2008: A century of achievements* (Chapter 19). American Institute of Chemical Engineering (AIChE).
Pritchard, M. (1990). Beyond disaster ethics. *The Centennial Review*, *34*(2), 295–318.
Sánchez-Chaparro, T., Remaud, B., Gómez-Frías, V., Duykaerts C., & Jolly, A.-M. (2022). Benefits and challenges of cross-border quality assurance in higher education. A case study in engineering education in Europe. *Quality in Higher Education*, *28*(3), 308–325.
Sato, Y., & Harada, S. (2005). JABEEに関するアンケート集計結果 [Survey Results on JABEE]. *Journal of Engineering Education*/工学教育, *53*(3), 101–112.
Stephan, K. D. (2002). All this and accreditation too: A history of accreditation requirements. *IEEE Technology and Society Magazine*, Fall 2002, 8–15.
Stoeckel, A. L. (1958). *Politics and administration in the early colonial colleges* (Unpublished doctoral thesis). University of Illinois, Urbana, Illinois.
UK Engineering Council (UKEC). (1998). 2.1.1 - Roles and responsibilities of chartered engineers. *SARTOR 3rd Edition Part 2 Document*. https://web.archive.org/web/20000903215442/http://www.engc.org.uk/sartor/
UK Engineering Council (UKEC). (2004). *The Accreditation of Higher Education Programmes (AHEP1) first edition*.
UK Engineering Council (UKEC). (2013). *The Accreditation of Higher Education Programmes (AHEP2) second edition*.
UK Engineering Council (UKEC). (2014). *The Accreditation of Higher Education Programmes (AHEP3) third edition*. https://www.engc.org.uk/EngCDocuments/Internet/Website/Accreditation%20of%20Higher%20Education%20Programmes%20third%20edition%20(1).pdf
UK Engineering Council (UKEC). (2017). *Ethical statement*. https://www.engc.org.uk/media/2334/ethical-statement-2017.pdf
UK Engineering Council (UKEC). (2020). Accreditation of Higher Education Programmes (AHEP4) fourth edition. https://www.engc.org.uk/media/3410/ahep-fourth-edition.pdf
Van Vught, F. A., & Westerheijden, D. F. (1994). Towards a general model of quality assessment in higher education. *Higher Education, 28*, 355–371. https://doi.org/10.1007/BF01383722
Wang, S., Zhao, Z., & Lei, H. (2014). 中国工程教育认证制度的构建与完善 [Construction and improvement of China's engineering education accreditation system]. *Research in Higher Engineering Education*/高等工程教育研究, *5*, 23–34.
Wickenden, W. E. (1934). Final report of the director. In *Report of the Investigation of Engineering Education 1923-1929, Volume II* (pp. 1041–1116). Society for the Promotion of Engineering Education.
Weil, V. (1984). The rise of engineering ethics. *Technology in Society*, *6*(4), 341–345. https://doi.org/10.1016/0160-791X(84)90028-9
Wisnioski, M. (2016). *Engineers for change: Competing visions of technology in 1960s America*. The MIT Press.
Xavier, P., Wint, N., & Orbaek White, G. (2023). A snapshot of how 'social' considerations are currently being interpreted and addressed within engineering education and accreditation. In *Engineering, social sciences, and the humanities: Have their conversations come of age?* (pp. 65–92). Cham: Springer International Publishing.
Zhang, H., & Davis, M. (2018). Engineering ethics in China: A century of discussion, organization, and codes. *Business & Professional Ethics Journal*, *37*(1), 105–135. http://www.jstor.org/stable/45149312
Zhu, Q., Jesiek, B., & Yuan, J. (2014). Engineering education policymaking in cross-national context: A critical analysis of engineering education accreditation in China. *Proceedings of the 2014 ASEE Annual Conference & Exposition*. https://peer.asee.org/20388

33

CONTEXTUAL MAPPING OF ETHICS EDUCATION AND ACCREDITATION NATIONALLY AND INTERNATIONALLY

Sarah Junaid, José Fernando Jiménez Mejía, Kenichi Natsume, Madeline Polmear, and Yann Serreau

Introduction

The previous chapter in this handbook outlined the development of accreditation practices and documents over time. This chapter picks up where Chapter 32 ends, describing the 'here and now' and assessing the state of the multicultural context at a point when people are connected trans-nationally more than ever before. This chapter probes the words used in current accreditation policy documents as these words influence curriculum design. We consider how terms used in various countries' documents compare and look for values-related patterns using an established cultural framework. In summary, this chapter aims to initiate discussion and probe the following questions:

1) *How is engineering ethics described in accreditation documents?*
2) *What commonalities and differences are evident trans-nationally?*
3) *What patterns can be observed in the way learning outcomes or competencies are written that can provide insight into how ethics might be taught to engineering students?*

The chapter is presented in two parts: (part 1) a global analysis that addresses the first two research questions and (part 2) four case studies to address the third research question. We begin by describing our positionalities to illustrate why we care about this topic.

Acknowledgments

It is important to acknowledge all contributors who played a significant role in building the premise for the work presented here. All the authors on the papers Junaid et al. (2021) and Junaid et al. (2022) were key contributors to the work reported in this chapter. In alphabetical order by surname, they are Alison Gwynn-Evans (AGE, South Africa), Sarah Junaid (SJ, United Kingdom), Helena Kovacs (HK, Switzerland), Johanna Lönngren (JL, Sweden), Diana Adela Martin (DM, Ireland), José Fernando Jiménez Mejía (JFJM, Colombia), Kenichi Natsume (KN, Japan), Madeline Polmear (MP, the United States and Canada), Yann Serreau (YS, France), Corrinne Shaw (CS,

DOI: 10.4324/9781003464259-40

South Africa), Mircea Toboşaru (MT, Romania), and Fumihiko Tochinai (FT, Japan). This chapter would not have been possible without the tremendous efforts – in translating texts, extracting data, reviewing and critically evaluating national accreditation documents – presented in our previous studies by our colleagues.

The idea of critically analyzing ethics education through the lens of accreditation was sparked by an organic discussion at the European Society of Engineering Education (SEFI) special interest group on ethics (SIG-Ethics) at the SEFI 2020 conference, held virtually. It was from this discussion that the policy subgroup of SIG-Ethics was born with the aim of critically analyzing the portrayal of ethics in engineering accreditation documents across countries in order to observe regional and global trends and differences. The group started with four members (SJ, HK, DM, YS), where the quantitative framework and qualitative analysis was first developed and published with four countries analyzed in a European-focused study (Junaid et al., 2021). The group expanded with colleagues from Africa, Asia, Europe, North America, and South America joining the policy group (AGE, JL, JFJM, KN, MP, CS, MT, FT) to carry out a global analysis, with a global team and diverse voices. This resulted in a publication in 2022, further developing our quantitative approach and introducing a new qualitative cultural framework (Junaid et al., 2022). This chapter builds on the work and brings more focus on values (through case study narratives) and a new statistical approach to pattern observation through principal component analysis (PCA). This is an ongoing project by the SEFI Ethics SIG, and any researcher or practitioner working on accreditation is welcome to join by contacting Sarah Junaid (the project lead) or Diana Adela Martin (the SIG-Ethics co-chair).

Authors' positionalities

Wishing to compare the place of ethics in engineering education in different cultures, this chapter presents the findings and personal narratives derived from the words used in accreditation policy documents. The discussion of values is core to this chapter, and as such, it is important to describe the background of the authoring team that shapes our perspectives and informs our contribution to this handbook. The five authors are engineers and experimental physicists by training, with expertise in various fields (biomedical, civil, electrical, environmental, and general engineering). We are all active in higher education at the levels of practical teaching, curriculum development, and influencing national frameworks for engineering ethics education. We vary culturally and we identify across four different cultural clusters (more about cultural clusters is presented below). We seek increasingly global representation in our group of collaborators. However, this chapter has a heavy representation of policy documents available in English or documents that our team could translate into English. We recognize we have yet to capture the complete landscape, and the chapter is biased toward Western values. However, we have tried to bring diverse voices into this discussion through our affiliations, global networks, and cultural identities.

The premise of the chapter is based on the policy work carried out by the European Society of Engineering Education (SEFI) special interest group Ethics (SIG-Ethics).

Together, we believe that, in educational practice today, ethics and moral decision-making are usually taught peripherally to technical subjects; they feature in curricula in limited ways. We assert that ethics must be integrated more into higher education engineering programs. Although some practitioners advocate for change – and communities of practice in engineering ethics are starting to gain critical mass – such grassroots efforts need top-down enablers to facilitate wider adoption. Accreditation is an essential mechanism for precipitating change. Thus, we evaluate *how much* and *in what ways* ethics is articulated in the documents of various nations and accrediting bodies, representing as much of the globe as we have been able to access to date.

Part one: analysis of how engineering ethics is portrayed in accreditation documents within various clusters

Background

This study constitutes a step towards a global comparative analysis of policy documents that inform the design of engineering curricula. It aims to identify differences in ethics learning outcomes and competencies required for engineering courses trans-nationally. Research indicates that social and cultural dimensions are critical in normative ethics and moral reasoning (Alas, 2006). As such, we believe that countries' policy documents will reflect cultural norms and societal values that will consequently influence graduates' ethical awareness and engineers' ethical practices. Honest reflection and decision-making are integral parts of our social construct where a social collective creates and accepts ideas and concepts. Yet the demarcation and presumed separation of moral reflection and logical thinking have been noted at engineering, governmental, and policy levels (Bacchi, 2007). Bacchi posits that a broader engagement of ethical reflection in policy-making is needed; she recommends reducing the dependence on ethicists shaping policy – more engineers and experts from diverse fields must get involved. Indeed, higher-education policy is influenced by the political and economic framework it sits and operates within (Ball, 2015a), and these voices should chime in on policy formation. Policy, by its nature, is not neutral; it reflects the collective value system. A well-rounded constituency can help shape policies that more effectively support the collective.

Bardi and Goodwin (2011) assert that values – the ideals people perceive as necessary – drive how they think, perceive, act, and behave. Although values can be considered universal, cultures vary in the hierarchy they allot and the importance they place on various aspects, and these vary on individual and collective bases. Collective values can be regarded as part of the cultural identity of a group; however, Bardi and Goodwin note that they are subject to change due to time, critical events, personal choices, and environmental factors.

Understanding 'policy as text' means assessing how policy is written, whereas understanding 'policy as discourse' requires considering policy implementation (Ball, 2015b). Both can provide insight into underlying values expressed in policy. The power of policy language can express neoliberal structures and economic values embedded in our social structure (Beasley & Bacchi, 2007). This has implications for our higher education institutions (HEIs) and the engineers trained to work within these structures.

Policies play a significant role in the interpretation and design of engineering curricula. They can, therefore, be used to shift values by recommending or requiring specific ways of thinking, perceiving, acting, and behaving. The implications at this level are systemic, and an analysis of how ethics is conveyed (policy as text) can provide insight into how it (policy as discourse) could be used. Viewing national engineering accreditation documents through the lens of cultural structures can help us understand such implications regarding the degree to which ethics education is embedded in engineering locally, nationally, and globally. The caveat to this work is an awareness that there is no universally agreed definition of 'ethics' and, as such, ethics has been explored by identifying associated terms – ones that appear implicitly in the context of the engineering profession.

A conceptual framework for the study

We endeavor to produce a trans-national review of accreditation documents, which presents some challenges, and the work reported in this chapter is part of our ongoing efforts to do so. In revealing the political dynamics in engineering accreditation systems in five anglosphere countries, Klassen (2018) highlighted the importance of analyzing relevant literature and policy documents of the engineering curricula in political, organizational, and historical contexts, using pluralist political theory. Our study builds on Klassen's groundwork, acknowledging the importance of the dimensions he noted. So, realizing that not every country can be included in exploratory analyses of the type we are conducting, we sought to establish a framework and research methodology to help us address the challenges of representativeness, sampling bias, and language interpretation. We considered how we would categorize countries and what meaningful classification tools we could use to analyze ethics at the policy level.

We adopted a model based on cultural identities in the workplace pioneered by Geert Hofstede et al. (2010, 2011). The seminal 'Hofstede model' is not specific to engineering education; it was drawn from comparative research of corporate environments and organizational cultures across more than fifty countries. According to Hofstede, "Culture is the collective programming of the mind that distinguishes the members of one group or category of people from others" (Hofstede, 2011, p.3). Therefore, culture is a collective phenomenon that groups of people construct.

Hofstede's model was built using IBM's database of 100,000 surveys collected via the company's network involving 50 countries. It was later expanded to other corporations but was made without input from non-corporate organizations, such as rural or not-for-profit entities.

Despite its limitations, the Hofstede model provides an empirically grounded way to classify various cultural contexts regarding six factors. Any cultural group can be located somewhere along a continuum for each of the six factors that Hofstede presented as dichotomous or 'bi-polar' dimensions, with extremes labeled at either end. The six dimensions of culture involve power distance, uncertainty avoidance, masculinity versus femininity, individualism versus collectivism, long-term versus short-term orientation, and indulgence versus restraint. Power distance entails acceptance of those without power and the level of equity in power distribution within the group, society, or culture. Uncertainty avoidance depicts the degree of desire for predictable outcomes and avoiding an uncertain future. Masculinity versus femininity is the societal preference towards achievements and ambition versus altruistic motivations, traditionally associated with the distribution of roles between men and women. Individualism versus collectivism regards how people relate to each other and make decisions based on individual needs versus group needs. Long-term versus short-term orientation entails the preference of efforts towards the future, such as pragmatic problem solving, forward-looking, and adaptability of traditions (balancing the present and past, related to steadfastness and preservation of traditions). Indulgence versus restraint relates to relative control in allowing gratification of human needs and desires versus social norms and codes that may regulate them.

Some criticisms were voiced against Hofstede's work (Kirkman et al., 2006), precipitating refinement; the Hofstede model was built upon by the Global Leadership and Organizational Behavior Effectiveness (GLOBE) initiative to cover 150 countries. It expanded from six dimensions to nine: performance orientation, future orientation, gender egalitarianism, assertiveness, institutional collectivism, in-group collectivism, power distance, humane orientation, and uncertainty avoidance. Like the Hofstede model, the GLOBE model is also based on business interests.

In this chapter, we reference 11 cultural clusters. In this study, we used the GLOBE cultural clusters and the dimensional factors formulated by Hofstede to cluster and analyze policy documents related to engineering ethics education. Despite the emergence of the more elaborate GLOBE model, we also drew from the Hofstede model due to its usefulness and simplicity for mapping features to distinguish different cultures. We used GLOBE as a primary guide for clustering and the Hofstede dimensions to compare various aspects of the clusters' accreditation documents.[1] Based on these models, we identified 11 global cultural clusters: Anglo, Arab, Confucian Asia, East Europe, Germanic Europe, Latin America, Latin Europe, Middle East, Nordic Europe, South Asia, and Sub-Saharan Africa (GLOBE Foundation, n.d.; Hadwick, 2011; House et al., 2002; Ronen & Senkar, 2013). Although these models were developed in the business sector, they hold applicability for engineering, which has corporate, business, and management dimensions. From a philosophical perspective, however, there is a distinction between engineering practice and engineering business. The roles of engineering in civic life, social entrepreneurship, and policymaking are not restricted to engineering within corporate practices. These should be explored further to challenge assumptions regarding the role of engineering and engineers in society. The results could help engineering downplay its corporate identity and provide increased focus on the needs of future societies. Despite their corporate emphasis, these two models offer a starting point for assessing historic and present structures; they can help us contextualize the different roles of engineering. The Hofstede and GLOBE models provide a valuable lens for clustering cultures in meaningful ways based on explicit dimensional factors. We recognize that, by applying this conceptual framework, we neglected other temporal and regional factors that influence curriculum development. For example, competencies identified by recent document of France's Commission des Titres d'Ingénieur (CTI) 2022 include sustainability-related changes. The changes address current and imminent socio-environmental needs that have led to legislative changes within national and European Union laws.

Collection of accreditation documents

The authors of this chapter are part of an international team of colleagues who have previously analyzed accreditation documents across 12 countries within five continents to identify trends and differences (Junaid et al., 2021; Junaid et al., 2022). These studies used a mixed-methods approach, which involved quantitative analyses of key ethical terms used both explicitly and implicitly and qualitative analyses of learning outcomes stated in accreditation documents based on the cognitive level (degree of thinking versus doing) they require of students. We used the same accreditation documents collected earlier and, for this chapter, conducted additional analysis using the culture-based conceptual framework presented above. By including 12 countries, our analyses have represented parts of seven different cultural clusters, identified using the abovementioned GLOBE clustering and Hofstede dimensions.

Thus, accreditation documents for both studies came from the following 12 countries. The cultural cluster each represents is indicated first, and the name of the country or countries involved is shown (in parentheses): Africa (South Africa); Anglo (Canada, Ireland, South Africa, the United Kingdom, and the United States); Confucian Asia (Japan); East Europe (Romania); Latin America (Colombia); Latin Europe (Belgium, France, French-speaking Switzerland); and Nordic Europe (Sweden). Note that South Africa was considered in two clusters, Anglo and African, due to the disparate cultures within the country. A total of ten accreditation documents were analyzed in the previous study. In most cases, we used the country's accreditation documents. However, the CTI French accreditation has been sought by Swiss and Belgian institutions on a voluntary basis. Therefore, we consider this CTI French accreditation document as representative of Latin Europe cluster. For this chapter, the author team analyzed previously collected accreditation documents to answer new research questions and conducted additional exploratory analyses using statistical procedures to yield a deeper understanding.

Cultural analysis

We carried out the cultural analysis for this chapter in two phases: (1) quantitative analyses to investigate learning outcomes explicit and implicit to ethics using the list of terms derived from the studies (Junaid et al., 2021; Junaid et al., 2022) and (2) qualitative analyses to investigate verb usage associated with the relevant learning outcomes. Quantitatively, in the previous studies, we counted the frequency of explicit terms (where the word 'ethics' or 'ethical' appeared) and terms that implied ethics (using a list of keywords). To develop the list, we extracted a range of keywords from the contents page of five engineering ethics textbooks (Junaid et al., 2021; Junaid et al., 2022). Then, we collated the words and achieved consensus across the authors regarding the refined list for use in coding. The 14 key terms we identified that implicitly reference ethics are 'global view,' 'values,' 'profession,' 'responsibility,' 'charters and codes,' 'critical reasoning,' 'organization,' 'safety and risks,' 'sustainability,' 'international context,' 'integrity,' 'technologies,' 'justice,' and 'society.' Variations of these terms, for example, 'professional' and 'societal,' were also included in the word search.

In the previous study, our subsequent qualitative analyses of verb usage investigated the context of the explicit terms concerning learning outcomes and competencies. We incorporated advice from 'A pragmatic master list of action verbs for Bloom's Taxonomy' published by Newton et al. (2020). Verbs appearing in accreditation documents regarding clear requirements for 'ethics' or 'ethical' indicated what students should be able to do. We cross-examined the verbs we found using Bloom's learning taxonomies. Our reason for excluding learning outcomes that did not explicitly mention ethics was due to the broad interpretation that people using the documents can adopt; for example, sustainability in curriculum design may focus wholly on technical solutions to the problem rather than also addressing the ethical implications of the solutions and of how the solutions (such as products) are created or manufactured. Equality, diversity, and inclusion were also not considered in our analyses for three reasons:

1) These topics have cultural and political contexts, which are highly sensitive to regional and national variability.
2) These subjects are often managed at the organizational level and may not necessarily be captured at the policy level. This may wrongly skew any interpretation of these subjects, possibly introducing new prejudices that will be counter-productive to the aim of this work.
3) These topics require more data and need to be addressed in further depth than the study presented here. This chapter reports a solid start and offers a way forward.

We did not explore the nuances in linguistics or local differences in word interpretations. To avoid introducing differences in linguistic translations, the previous study was carried out in English. This included using official English translations of accreditation documents where possible. Exceptions to this were Colombia, Romania, and France because authors on our team are native speakers and were able to interpret applicable documents (Junaid et al., 2022).

Analysis of explicit and implicit ethical terms

The results of earlier work, presented in full in Junaid et al. (2022), indicate that the word 'ethics' and 'ethical' appear from two to ten times across the ten documents (average 4.7 +/- 2.6 across cultural clusters) across the learning outcomes and competencies. This contrasted with the 14 implicit reference terms, which were found more frequently than the explicit terms. The implicit terms occurred between 6 and 187 times (average 88.7 +/- 62.4 across clusters). The documents under review varied in length from approximately 500 words (Sweden) to over 100 pages (Canada), and this is one reason for the wide variation in frequency found.

Moreover, some implicit terms were more heavily emphasized than others. Most notably, 'profession,' 'society,' 'charters and codes,' 'international context,' and 'responsibility' represented around 70% of occurrences of implicit terms across all ten documents. The remaining words such as 'values' and 'integrity' accounted for less than 10% of term usage across groups. The word 'justice' did not appear in documents of any culture (except the root word 'just,' which we will discuss in the case study of Colombia), and this supports the claim that engineers have generally assumed an apolitical and asocial stance. This omission could be seen as suppressive due to the principal role justice plays in respectful current theories about the origin and aims of morality, particularly in the development of moral philosophy to support moral decision-making. From a philosophical viewpoint, empathy and justice are keywords used to describe morality. Therefore, we found it surprising that neither term was present despite the increasing trends toward considering ethics in policy documents and curricula. More central to this, justice is recognized as one of the fundamental pillars of personal ethics and morality, which are values developed during a person's infancy. Studies have also shown human species developing moral capabilities for over two million years (Tomasello, 2016).

Comparing cultural clusters, our analyses revealed clear differences in emphasis of implicit ethical words. These may be explained by considering the different historical, cultural, and legal frameworks for accreditation in each country, but, due to the complexity of the problem and the scarcity of information on the matter, a complete comparison is beyond the scope of this chapter. To provide some examples, however, within documents from the Anglo, Latin European, and African clusters, the words 'profession' and 'security and risks' occur most frequently. We believe that certain similarities apparent among these three cultural clusters could relate to (a) the geographical and cultural proximity between Latin Europe and some Anglo countries (Ireland and the United Kingdom) and (b) the presence of the British Empire in South Africa during the nineteenth and twentieth centuries. These clusters' focus on safety and risk is perhaps a by-product of industrial revolutions from the eighteenth century onwards, where manufacturing, mass production, and rapid economic growth in places like the United Kingdom also brought in an increasing regard for human safety, the need for standardization, and the introduction of professional institutions. The differences may also highlight the social and historical contexts at play. For example, the Latin European cluster emphasizes 'sustainability' more than the African cluster, which more often refers to 'security and risk,' perhaps reflecting areas of concern nationally. Our analyses suggest additional variations and commonalities among clusters that warrant future investigation using historical, economic, political, and/or social lenses.

For this chapter, we conducted PCA (principal component analysis) to explore the occurrence of the implicit terms across the cultural clusters (including a total of ten countries) to help us understand the results of the prior work reported by Junaid et al. (2022). The PCA projected vectors in a two-dimensional space using (a) the seven cultural clusters as dependent variables and (b) the 14 implicit terms as dependent variables (Figure 33.1). By setting these as 'dependent variables,' we can identify patterns in variance between clusters and between implicit terms used in the accreditation documents. It is important to note that although South Africa is identified as belonging to both Anglo and African clusters, the country was included in the African group and not included in the Anglo group to avoid data duplication.

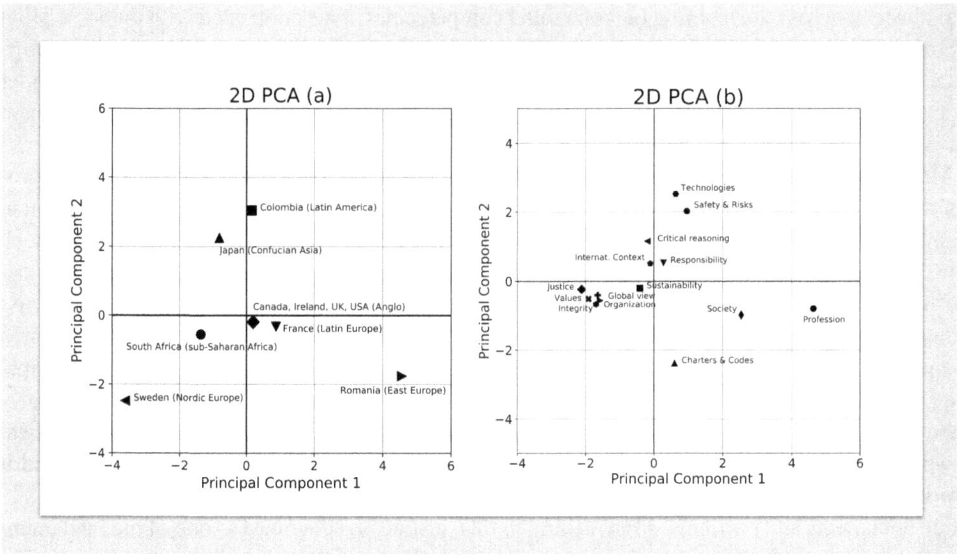

Figure 33.1 PCA diagrams in two dimensions using (a) the seven cultural clusters as dependent variables and (b) the 14 implicit terms as dependent variables. NOTE: the United Kingdom and United States are shown as UK and USA respectively.

Looking first at Figure 33.1a, the first two principal components add up to 67% of the cumulative explained variance, and the first three, together, explain up to 84% (note: the cumulative explained variance is not shown in the figure). It's important to note that the matrix contains a relatively small amount of non-homogeneous data – among the documents consulted, some refer exclusively to accreditation processes in engineering and others to a broader set of professions – and the documents are very dissimilar in their structure. As a result, this analysis represents an exploratory first step to help us better understand the situation. Due to the small sample group, we graph and report our findings with reference to the individual countries and their respective clusters, and we use the terms 'country' and 'cluster' interchangeably in our reports.

Some points we observed in the Figure 33.1a statistical matrix were:

1) The Anglo cluster (Canada, Ireland, the United Kingdom, and the United States) published the most inclusive range of terms, and thus sits closest to the center of null variance. In addition, at least in the two-dimensional analysis, the cluster with the second lowest variance is France (representing Latin Europe), closely followed by South Africa (of the African cluster).
2) The data for Colombia (Latin America), Romania (Eastern Europe), Sweden (Nordic), and Japan (Confucian) are peripheral compared to the central core of the rest of the countries in the figure. The proximity between Japan and Colombia does not imply significant cultural similarities between both countries. This pattern suggests that statistically speaking, South Africa, France, and the Anglo cluster have similarities across their accreditation documents regarding ethics. In contrast, the distances between the other countries indicate that their ethics accreditation words differ substantially in these two principal components that are, as of yet, unknown.

3) In the direction of the first principal component (PC1), Sweden is markedly opposite to Romania. This indicates differences in the data corresponding to the categories 'charters and codes' and 'profession,' which are the most important for Romania but do not even appear in the Swedish documents. The Hofstede index relating to PC1 is possibly most influenced by power distance (PDI) due to the differences across countries for this index. Power distance entails acceptance of those without power and the level of equity in power distribution within the group, society, or culture. Among these countries, Romania has the highest PDI at 90, while Sweden has the lowest at 31. Listed from highest to lowest PDI: Romania (90), France (68), Colombia (67), Japan (54), South Africa (49), Anglo (36), and Sweden (31) (Hofstede et al., 2010, pp. 57–59). Although PC1 is not correlated with PDI in all countries, it is reasonable to interpret that 'profession,' which is oriented towards higher social status, and 'charters and codes,' which are the requirements for the status, are correlated with PDI.
4) In the direction of the second principal component (PC2), The category most influencing it is 'society,' and the Hofstede index relating to it is individualism (IDV). Collectivist countries with low IDV tend to have higher PC2: from highest, Colombia (13), Romania (30), Japan (46), South Africa (65), France (71), Sweden (71), and Anglo (83) (Hofstede et al., 2010, pp. 95–97). Here, IDV and PC2 do not precisely correlate, as Romania, with an IDV of 30, which is lower than Japan and can be interpreted as collectivist, has a lower value in PC2. However, it is reasonable to interpret that society orientation and collectivism are correlated. At least, the tendency towards individualism or collectivism should have a significant influence on their engineering ethics.

Shifting now to Figure 33.1b, where the implicit ethical terms were defined as dependent variables, the first two principal components added up to 46% of the cumulative explained variance, and the first three added up to 68%. For this case, the following patterns were observed:

1) In the direction of the first principal component, the terms with the most variance were 'profession' and 'society,' which, together with 'charters and codes,' were those with the highest percentage in frequency of occurrence in the global analysis. These three terms were used heavily in the referenced documents, in contrast to the other terms.
2) In the direction of the second component, the greatest distance, and thus most considerable variance, corresponded to 'technologies' and 'charters and codes,' but at opposite ends of the scale. At the same time, the term 'safety and risks' appeared close to 'technologies.'
3) Several terms grouped tightly together: 'global views,' 'integrity,' 'organization,' 'values,' and 'justice.' Moreover, 'sustainability,' 'responsibility', 'international context,' and even 'critical reasoning' were not far removed from this tight cluster. However, several of these terms ('justice,' 'values,' 'integrity,' and 'global views') were consistently underrepresented across the accreditation documents, which explains the lack of variance seen. On the other side, the terms 'society,' 'profession,' and 'charters and codes' stand almost in opposition to the tight cluster and the rest of the terms. These peripheral terms indicate the most variance between accreditation documents.

Interpretation of the cultural analysis

Our analyses of implicit ethical terms found that words such as 'justice,' 'integrity,' and 'values' are sorely missing. These may be hidden or assumed to be covered under other umbrella terms like 'ethics'; however, ethics is an ambiguous term, and accreditation documents typically define terms with a greater level of clarity to reduce confusion and help ensure reliable results across assessment teams, for instance. Values, like ethics, can be complex and can change due to time, environment, events, and personal reflection. Impermanent meanings and shifting interpretations may be reasons why these words have sometimes been avoided in writing accreditation documents (Beasley & Bacchi, 2007). We found considerable differences among cultural clusters. For example, the term 'charters and codes' was highly emphasized in Latin America and East Europe, whereas the term was not mentioned in the Confucian Asia and Nordic Europe clusters. This may be due to the historical, political, social, or religious contexts or a combination of the four. It might also be due to mandating that engineers register with professional bodies to work in the profession. For example, graduates in Colombia (Latin America) from an unaccredited degree will not have their qualification recognized as a higher education engineering degree. It is necessary for graduates from Colombian universities to register with a professional body to work as engineers. Therefore, it is unusual that the term 'profession' showed a stark drop in emphasis in Latin America compared to the other clusters. There may be social context that can explain this and would need to be explored further.

One way of examining the place given to ethics in engineering training curricula is to analyze the verb types related to the way ethics-related learning objectives are described. To this end, the taxonomies initiated by Bloom provide categories of verbs used to define learning objectives. Most variations of Bloom's original taxonomy include six levels, from the most elementary to the most complex. The six levels can be summarized as follows: remembering, understanding, applying, analyzing, evaluating, and creating (Anderson et al., 2001; Krathwohl, 2002; Mallalieu, 2023). For the verbs in the repositories we studied, the 'apply' level was the most represented when relating directly to ethics learning (39%). The 'evaluate' level was the least represented.

Nevertheless, if we consider that ethics will be an essential component of the role of engineers in the coming years, we might wish to move the level of objectives toward the highest level, 'create.' This would mean that the passive 'apply' level would no longer be in the first rank – it would have to cede this top rank to a higher level that requires higher-order thinking.

The universal emphasis on 'apply' is understandable, with engineering requiring technical skill-based competencies. However, the low use of 'evaluate' verbs within subjects that link to ethical practice serves as an interesting area for further study. The more limited mention of 'ethics' and 'ethical' learning outcomes that we found skewed towards more cognitive-based learning, that is, ethics education rather than ethical practice. A general analysis can hide national nuances that could help in understanding how ethical practice is influenced by accreditation-level learning outcomes. For example, France's accreditation process is competency-based and, therefore, requires demonstrable practice of the competencies. This is reflected in the emphasis, in French accreditation documents, on 'applying,' 'analyzing,' and 'synthesizing' (Bloom's original term) or 'creating' (a modification made in later adaptations of Bloom's taxonomy).

The analyses of verbs we present in this chapter have several limitations. Firstly, we acknowledge the limitations of inferences derived from one (or only a few) representative countries within a cultural cluster; we do not intend to extrapolate the values from one country and act as if they represent the complete set. Rather, we use the Hofstede and GLOBE models to help us work toward wider inclusion of diverse cultures in our overall effort to understand ethics-related accreditation characteristics and trends. For example, Japan, the only country in the Confucian Asia cluster, has unique cultural and historical structures that can be quite different to other countries within the same cluster (for more on this, see Chapter 32). A second limitation is that an ethical model for drawing the quantitative analysis of terms was not used; rather, the research team collated a list of terms (Junaid et al., 2022). A potential benefit to this approach was reducing biases embedded in an existing model. Nevertheless, this study has a clear Anglo bias since the terms were collated in English, and few non-English terms were considered. Thirdly, the Hofstede model is limited to six bi-polar dimensions and focuses on organizational cultures; it may not consider the cultural identities that define other value systems beyond the corporate realm. Fourth, we have presented only a general overview of the data and data patterns due to the small data set. Observations from this exploratory study must be viewed cautiously; it is impossible to infer causation. Finally, the master list of action verbs to define learning that we derived using Newton et al. (2020) is limited to Anglo papers and therefore presents an Anglo and British bias of cognitive learning. Despite these limitations, this study provides early pilot data and has helped highlight nuances in engineering ethics education trans-nationally and trans-culturally to explore more extensively for further research.

Part two: case studies of four countries' accreditation documents and their cultural context

Part two of this chapter discusses four different regional contexts, identifying similar and dissimilar qualities of how ethics is framed in four case-study countries drawn from the overall set of ten countries analyzed above and reported previously by Junaid et al. (2022). The countries investigated in depth below, with regard to engineering ethics accreditation documents, are Colombia (Latin America), France (Latin Europe), Japan (Confucian Asia), and the United Kingdom (Anglo). These four case studies were selected from distinctly different cultural clusters to give readers a broad global overview. The four also represent the authors' home countries, allowing our team to highlight nuances.

The following four cases contextualize commonalities and differences, suggesting a pathway for understanding diversity and inclusion globally. The case studies help compare and contrast various scenarios related to engineering education to increase our understanding of what various countries value. They can help us and our readers build cultural awareness and develop stronger global interpersonal skills.

Latin America case study: Colombia

In Latin America, most accreditation processes are voluntary and regulated by state entities. This condition does not prevent the application of a varied set of quality accreditation models and proposals for higher education institutions and university programs (UNESCO et al., 2018).

For this chapter, the Colombian case study focuses on Agreement 02 of 2020 (CESU, 2020), interpreted from Spanish, as there were no official English translations. The analyses yielded the results summarized below using the methodological approach from Junaid et al. (2022).

A list of key terms explicitly defined in the Colombian legal framework are 'accreditation' (CESU, 2014, art.12, p. 30), 'competence' (CESU, 2020, art.2, p. 8), 'graduate attributes' (Colombia, 2019, numeral 2.5.3.2.3.2.3, p. 12), 'learning outcomes' (Colombia, 2019; CESU, 2020, p. 8), and 'responsibilities of engineering practice' (Colombia, 2003, art. 33, p. 16). The precise definition of each term has value in cross-culture analyses (Junaid et al., 2022). The set of definitions provided in the Colombian legal framework can facilitate nuanced understanding – regarding how terms are used, the meaning behind their use, and how they may be interpreted differently in other places. For instance, most defined terms in the Colombia document correspond to legal acts approved in the last decade. In the case of 'accreditation,' the Colombian document defines the term as "the act by which the State adopts and makes public the recognition that academic peers make of the quality of a program or institution based on a previous evaluation process in which the institution, the academic communities, and the Council participate" (CESU, 2014, art. 12, p. 30). Thus, although accreditation is voluntary, the Colombian state is the agent that evaluates and recognizes the quality of engineering programs, making the accreditation process essentially public. In comparison, other countries like the United Kingdom administer their accreditation process through professional institutions, devolving that responsibility to the collective community of professionals in engineering.

The count of the implicit ethical terms – the order of recurrence in parentheses – is as follows: 'society' (59), 'charters and codes' (32), 'international context' (26), 'profession' (21), 'critical reasoning' (19), 'global view' (11), 'responsibility' (7), 'technologies' (6), 'integrity' (3), 'values' (2) and 'sustainability' (1). It is worth noting that the previous analyses by Junaid et al. (2022) did not consider 'inclusion' and 'diversity,' but these are mentioned within the Colombian document in the following sentence: "A declared commitment to the comprehensive training of people to face, with ethical, social, and environmental responsibility, the endogenous development challenges and to participate in the construction of a more just and inclusive society that recognizes and promotes diversity" (CESU, 2020, p. 20). This sentence references a more 'just' society and, by using the root of the word 'justice,' it indicates an affinity with the term. In this case, an explicit intention is to preserve the national ecosystems, peoples, and ethnicities – this constitutes a critical focus for the professions and a reflection of historical and political contexts. Changes found in France's CTI 2022 document (when comparing it with the earlier CTI 2018 that it supersedes) indicate emerging emphasis on sustainability goals; likewise, this Colombian case demonstrates how social debates are expressed through legal and political forms on accreditation processes.

In the Colombian document, among the set of verbs describing learning outcomes relevant to ethics, we found that about 24 were action verbs (e.g., 'apply,' 'demonstrate,' 'participate,' 'transform'), whereas 30 prioritized cognition (e.g., 'analyze,' 'define,' 'know,' 'understand'). Additionally, 32 blended the realms of action and cognition (e.g., 'create,' 'inquire,' 'research,' 'think'). This finding contrasts with the broader analyses by Junaid et al. (2022), in which cognitive verbs predominated widely over action verbs across the sample of ten countries. Nevertheless, there is a need for both verb types in education curricula. On one hand, cognition is the process of thinking that includes self-awareness, reflection, and consciousness about the world as it is; metacognitive development is essential to develop in the engineering profession (Cervin-Ellqvist et al., 2021). On the other hand, there is the need for action, which necessitates developing skills and translating practical abilities through ethical decision-making that experienced engineers have developed into educational frameworks in engineering.

Based on the master list of action verbs suggested by Newton et al. (2020), according to the original Bloom's categories, it is possible to compare the number of verbs related to learning ethics reported in the accreditation documents analyzed by Junaid et al. (2022) with the equivalent verbs of Colombia's Agreement 02/2020. The Colombian document uses the verbs 'apply,' 'analyze,' 'evaluate,' and 'create' (levels 3–6, the higher levels of Bloom's taxonomy), but the first two categories of remembering and understanding are missing. This bias may initially appear to be a positive shift toward applying knowledge. However, if we follow the premise behind Bloom's taxonomy, the lower learning levels should provide scaffolding to loftier levels of cognitive learning; in this sense, it is assumed that middle and primary education provides these learning fundamentals in the national education system. Whether this aim is achieved or not is essential for fulfilling professional training.

Latin Europe case study: France

In France, 200 schools, 51 of which are private, are accredited to deliver at least one engineering degree course. Engineering degrees are issued at the school level, which is not the case in other professions in France such as medicine, where the qualifications are issued at the national level (Grelon, 2021, p. 68). These engineering schools are accredited by the Commission des titres d'ingénieur (CTI), created in 1934 as an autonomous joint body (CTI, 2022). The French engineering degree corresponds to a master's degree, level 7 of the European Qualifications Framework (CTI, 2022). Since French engineers are not constituted as a professional order, the practice of engineering is not governed by such a professional order nor is the training of engineers linked to it. Instead, the training of engineers in France is situated within the European framework for higher education and the Bologna process (CTI, 2022; Djurovic & Lubarda, 2014; European Education Area, n.d.). Thus, the CTI promotes the quality assurance of engineering education, and it delivers the European quality label for engineering education EUR-ACE® of European Network for Accreditation of Engineering Education (ENAEE) (Augusti, 2009; Augusti, 2013).

Engineering schools are required to apply ethics initiatives and define a strategy for social and environmental responsibility, with objectives that are monitored. The school must also ensure "compliance with the requirements of scientific integrity, deontology and ethics. It conducts awareness-raising activities among students on these subjects" (CTI, 2022, p. 8). This starkly contrasts the Anglo cluster, which does not mention integrity. CTI describes the engineer as someone who identifies "professional, societal and environmental, ethical and deontological problems created by technological innovations" (CTI, 2022, p. 19). It makes CTI (France) a document that emphasizes ethics, among the documents analyzed, explicitly as a piece of a framework for engineering decision-making.

A set of competencies proposed by schools is associated with each engineering curriculum. Among the set of competencies required by the CTI, training in social and environmental responsibility constitutes a major criterion for accreditation (CTI, 2022). This includes societal issues, basic teaching of environmental and societal responsibilities, life-cycle analysis and design, *et cetera*, highlighting the ecological and climatic imperatives currently at play in Europe and globally.

Among the more generic competencies required by the CTI, themes that can be closely linked to ethics include ethical and professional responsibilities, issues of life at work (relations at work, health and safety, and diversity), transition, ecological and climatic imperatives, and needs of society.

More directly, the in-depth discussion of "concepts of ethics, deontology and occupational health and safety" (CTI, 2022, p. 27) is explicit. The document stipulates that a part of the teaching must be allocated to ethics, health and safety at work, social relations, sustainable development, and the ecological transition. With such imperative and structured guidance, one would expect a clear link to what is taught and/or delivered to students regarding ethics.

Our textual analyses showed that the engineer's postures associated with the ethical themes were defined by specific verbs in the French documents: 'consider,' 'report,' 'integrate,' and 'accompany' (CTI, 2022, p. 21). The implicit wordlists used in our earlier analyses that are highly represented in this framework are 'profession,' 'international context,' 'responsibility,' and 'sustainability.' Considering the all-encompassing term 'ethics,' there is no universally agreed definition, and as such, the implicit terms are invaluable in manifesting what 'ethics' means in the context of the engineering profession.

Compared to other clusters, what delimitates this cluster, according to our method, is the major part taken by sustainable development, international context and global view. To add to this, the lack of references to rules and codes as emphasized in clusters such as Latin America or East Europe is interesting. This could reflect civic rights over authoritative power, drawing on the legacy of French enlightenment and the constitutional right of liberty, equality, and fraternity.

Through this lens, this constitutional right has filtered into how professions as structures of authority are required to behave and operate for the good of society.

Confucian Asia case study: Japan

Some East, Southeast, and South Asian countries have accreditation bodies for engineering education as part of national engineering councils, while others have them as independent organizations. In either case, they have prepared their programs since the late 1990s due to the growing need for global alliances regarding education and licensing, including the Washington Accord and APEC Engineer, the Asia-Pacific Economic Cooperation as part of the International Engineering Alliance (IEA). It is an interesting challenge to reveal the earlier relationship between their education policies and ethics in the context of each country prior to that time. However, such research requires in-depth historical analysis of each country. Furthermore, small countries such as Vietnam do not have their own accreditation bodies but are accredited by Western programs such as ABET of the United States. This is not ideal as it does not account for embedding the value system of a country. Therefore there is a real danger of transplanting Western value systems that may not reflect the nuanced differences in the region. Supporting smaller countries to develop their own accreditation systems can allow an authentic reflection of what society needs from engineers and suggest how that training could be developed regionally.

Adopting others' standards is a reasonable decision given the burden of launching their own programs when accreditation is emphasized in the context of globalization, but local nuances are particularly critical within the actual practices of engineering ethics education. Cultural context can often play a significant role. However, such research requires a great deal of effort for this cultural cluster. Therefore, this section will focus on the current criteria of Asian countries that have their own accreditation bodies and original criteria, which are available in English, with particular attention to the case of Japan (for more on Japan and China, see Chapter 32).

In Japan, in the broader sense, the Ministry of Education (MOE) has accredited educational programs. A non-governmental, United States–style accreditation system was introduced in 1947 when the Allied Forces led the establishment of the Japan University Accreditation Association (JUAA). However, the MOE neglected the system after the restoration of sovereignty in 1952. Later, in the 1990s, the need for global quality assurance in education led to the establishment of Japan Accreditation Board for Engineering Education (JABEE), and the JUAA also regained its presence over the same period. JABEE offers a rigorous accreditation process running every 6 years with a 3-year review that individual programs undergo, which can be cumbersome for higher education institutions (HEIs). In contrast, JUAA accredits institutions rather than programs and therefore covers all degrees across the HEI. The MOE has also encouraged the autonomous development of each university by relaxing the standards and introducing an individual voluntary assessment system, resulting in three different accreditation bodies. However, the vision of an accreditation system relevant to Japanese society has not been achieved yet.

The engineering education reform in the 1990s thus aimed to conform to global standards, modeled on the United States system. From this perspective, JABEE attracted a great deal of attention at first. However, it could not resolve the incongruity with the predominant cultural style. The number of JABEE accredited programs has been declining since the late 2000s because of the system's unclear effectiveness for graduates and the cumbersome preparations for the accreditation.

The characteristics of the JABEE accreditation criteria are derived from the following historical background. They begin with Criterion 1 as follows:

1.1) Profile of Autonomous Profession (establishment, disclosure, and dissemination of the image of an autonomous engineer).
1.2) An ability of multi-dimensional thinking with knowledge from a global perspective.

In criterion 1.1, the English translation of the document uses the word *autonomous*, whereas the Japanese document uses another word that has the meaning of *independence*. Both words are pronounced 'jiritsu' in Japan. Partly because of the same pronunciation, the two words are sometimes used interchangeably in relation to individualism. The difference of the two meanings is not so clear for many Japanese people; however, when written in Kanji (Chinese characters used in Japan), the difference is evident: 'jiritsu' as autonomous is written as '自律' and independence as '自立.' In both words, the first character '自,' which is pronounced 'ji,' means 'self.' The second characters of both of the words have the same pronunciation, 'ritsu.' The fact that they have different meanings is very obvious for Japanese people: '律' means 'rule,' 'law,' 'code'; '立' means 'standing up.' Therefore, even if we do not know the definition of these words, we can guess that '自律' implies something about autonomy (rule for oneself to act) and '自立' about independence (standing up by oneself).

The importance of autonomy in ethics can be said to be the definition, but this is not obvious in Japan, where harmony with the organization and not disturbing it have been emphasized. In the 1990s, the emphasis on individualism and autonomy as opposed to collectivism became a major social issue in Japan. It was an important philosophy in the establishment of JABEE that clashes with Japanese norms.

In criterion 1.2, globalization was another issue in the 1990s. 'Multi-dimensional thinking' from a global perspective is related to relativism in post-war Japanese education, as well as consideration for the global economy. In Japan, consideration for other countries is inevitably linked to memories of the Pacific War. The year 1995 marked the fiftieth anniversary of the end of the War, and thus review of post-war values became a major social concern.

This multi-dimensional global perspective is like that of Latin America. Unlike Latin American countries, however, Japan's accreditation documents do not emphasize the need to overcome anthropocentrism with respect to the planet's biological diversity. Latin America's attitude toward environmental issues may be related to the region's evident deterioration of strategic ecosystems and the neoliberal economic development model that has plunged much of the population into severe inequity and violence. The historical and environmental interests of each country may influence this difference.

Furthermore, equity for people is not stated in Japan. While it may be implied in the accreditation requirements, Japan's focus is on something else: the development of independent and autonomous leaders who can respond to the globalized society that became more evident in the 1990s. The accreditation guideline states the following:

> This item indicates education and intellect required for the independent globally active individuals who take leading roles to structure sustainable and changing society emphasized [*sic*] on spiritual value shifting from the materialized society.
>
> (*JABEE, n.d.*)

As demonstrated above, Japan focuses on cultural diversity in a global society. This context is emphasized because Japan has adopted accreditation for engineering education as a Western system that is indispensable for economic globalization.

Design and communication were also important ideas of the 1990s in Japan. Japanese engineering faculties were training engineering scientists rather than engineering professionals. The engineering scientist conducts research and development at a university or corporate laboratory, while the engineering professional engages in engineering practice in a company or independently. American engineering design education was an innovative idea for Japan. These basic ideas are also important when understanding Japanese engineering ethics and the influence of the United States in teaching engineering ethics.

In comparison to the Latin European case, 'dialogue structure' is a French approach that emphasizes philosophical dialogue in education. In Japan, the similar competency is described as "(f) Communication skills including logical writing, presentation and debating," but it can be read as a prerequisite for communication that values harmony within a group as much as, or more than, critical discussion. Furthermore, the older generation promoting the accreditation system in Japan often complains about the lack of communication skills of the younger generation.

In Japan's first constitution established at the beginning of the seventh century, the first article emphasized respect for harmony, '以和為貴,' based on the Confucian *Analects*. The *Analects* (13.23) also states: "子曰, 君子和而不同, 小人同而不和" [the Master said, "the superior man is affable, but not adulatory; the mean man is adulatory, but not affable"] (Legge, 1861, p.137). The word '和' translated here as 'affable' is the word translated as 'harmonious' or 'peaceful' in general. Harmony is not inherently favorable, but the two are often confused in collectivism. It has been a major cause of corporate misconduct in Japan.

Anglo case study: United Kingdom

Degree accreditation for engineering programs in higher education institutions (HEIs) in the United Kingdom is not legally required. However, accreditation is an essential component to validating engineering programs and ensuring they are fit for purpose. The Engineering Council is an umbrella organization that sets and unifies the professional competencies for all engineering disciplines and their corresponding institutions. Thirty-nine licensed specialized engineering institutions use these competencies as authorized bodies to accredit degrees within their respective disciplines in HEIs such as the Institute of Mechanical Engineers (IMechE) and the Institute of Civil Engineers (ICE). The following case study focuses on the IMechE. Like the Japanese accreditation requirements, the UK process is rigorous and requires several review stages and visits. The advantage of accreditation is its alignment with other internationally recognized teaching quality benchmarks for engineering education, including the EUR-ACE Accord, the Washington Accord, and the Sydney Accord. This gives UK graduates the advantage of having a degree that is internationally and nationally recognized and that satisfies the educational requirements on the pathway to professional chartered engineering status in their disciplines.

Upon finding satisfactory evidence of the program meeting the requirements, the accrediting body awards the HEI with accreditation for the program for 4 years, which remains valid on the conditions that (a) annual reports and assessment samples are sent for review and (b) any changes to the program, including learning outcomes required within modules, are ratified by the accrediting body. Renewal for accreditation at the end of the 4 years requires a complete review of the program with a site visit. In addition to industry-specific accreditation, all HEIs must satisfy their responsibilities to students according to the Higher Education and Research Act 2017 (HERA), which led to the government establishing the Office for Students (OfS) as a public body under the Department of Education to oversee and regulate Higher Education in England and hold HEIs accountable. (For more on the UK system, see Chapter 32.)

Earlier multi-country analyses conducted by the authors (Junaid et al, 2022) revealed three key findings from the UK perspective. Firstly, the definition of terms from the competencies guidelines of the United Kingdom's AHEP-4 (Association of Higher Education Professionals) includes the third most comprehensive list of terms defined (of all ten documents analyzed). This UK document included eight target terms: 'accreditation,' 'competence,' 'delivery,' 'graduate outcomes,' 'higher education,' 'learning outcomes,' 'module,' and 'program.' From the previous AHEP-3 permutation, three definition terms had been removed for AHEP-4: 'awareness,' 'knowledge,' and 'know-how' – and interestingly, all three of these verbs are ones that Newton's taxonomy analysis recommends avoiding when defining learning outcomes (Newton et al., 2020). Secondly, the number of implicit ethical terms (shown in parentheses here) heavily emphasized 'profession' (24), 'safety and risks' (21), and 'society' (20). These constituted 62% of the terms found. Combined with the terms 'charters and codes' (13), 'technologies' (11), and 'responsibility' (9), 93% of all terms identified for the study were covered. There appears to be a greater emphasis on safety and risk in the UK documents compared to the multi-country average, and this reflects the United Kingdom's reputation for high safety standards in the workplace and the influence of the legally binding Health and Safety Executive (HSE). The HSE was established after the Health and Safety at Work Act was passed in 1974. It set a precedent in criminal and civil law by assigning responsibility for protecting their employees to the highest senior levels in organizations. Furthermore, the UK engineering industries' contribution to the industrial revolutions also necessitated the focus on health and safety, charters, and codes. However, our analysis found no terms for 'global view,' 'organization,' 'international context,' 'integrity,' or 'justice.' These are unusual omissions, considering the first industrial revolution put UK engineering on the global map and onto the international stage. These exclusions will inevitably be reflected in curricular designs lacking both international outlook and impetus to address inequalities more widely, even though these competencies are required through being signatories of international accords. Membership in these accords may allow graduates to work as engineers in cross-national teams; however, our study indicates that more emphasis is needed on these qualities ('global view,' 'organization,' 'international context,' 'integrity,' and 'justice') in the learning outcomes to prepare students to navigate these roles on a global stage.

Observations and discussion

Accreditation documents can help bring ethics to the fore of engineering programs. However, this chapter does not explore the translation of policy into curriculum design. The results of integrating these terms into accreditation documents may not go far enough in challenging (future) engineers to take active roles in protecting and nurturing society and the natural environment (see Chapters 6, 9, and 35 for more on these topics). Engineering solves human-conceived problems, which in turn creates new problems to solve.

Our research into differences and commonalities identified through comparative, trans-national study is driven by the belief that nuances embedded in policy documents drive engineering curriculum development and, hence, influence how and what our engineers may be taught. Dialogue that considers the historical, socio-political, socio-economic, and environmental influences can bring new insights regarding what engineering curricula are doing (and how they are doing it) to develop competent engineers from nation to nation, region to region, and from one cultural cluster to another. It is essential to explore how these aspects drive curriculum delivery and expose students to value-driven contextual nuances in ethical awareness, ethical decision-making, and ethical practice to prepare them for working in global teams. The scope of this work presented here represents an initial step toward realizing these ambitions.

Realizing our goal of comprehensive, global cross-cultural analysis will require a larger study, to include more countries – so that we can more fully understand culturally nuanced differences in engineering ethics education. This will help us understand how ethics is conveyed in accreditation documents globally, to support more purposeful curriculum design and bring new insight regarding the ethical competencies that engineers need to work in locations around the globe. How ethics is seen and contextualized in parts of the world we haven't yet covered may help us understand ethics more fully. Therefore, to extend what we have achieved in this chapter, we will need to collect and analyze more data. We therefore put forth a call and an invitation to readers – those with interest in supporting or collaborating in the work – to join us in the work that still needs to be done.

The limitations of the analyses conducted to date serve as areas for further research. For example, more can be done via linguistic and discursive analysis: analyzing power through language, the uneven influence and dominance between languages and cultures, and how these play into accreditation processes globally (for more on this, see Chapters 35 and 36). The interplay of different fields influencing ethics education needs experts in disparate fields (see Chapters 14–18) to work together to synthesize new insights from these analyses. We need linguists, engineers, policy-makers, sociologists, political scientists, and philosophers (see Chapters 1–13). Finally, exploring the role of engineering in the context of corporate and non-corporate social structures (see Chapters 9 and 11) also challenges our institutions and graduates to consider the different roles engineers can play as, for instance, the civic engineer, the entrepreneurial engineer, the policy-making engineer, and so on.

Conclusion

Analyzing the rhetoric and discursive linguistics in accreditation documents (beyond the granular analysis presented above) is necessary. Such analysis can help develop insight into how these policy documents shape program design and impact the pedagogical structures we observe in our own institutions and, consequently, in the engineers who graduate and work in society. All clusters in our study used action-oriented learning levels: 'apply,' 'analyze,' 'evaluate,' and 'create.' We can, therefore, assume that most reference systems in the context of ethics are designed to inspire action. Nevertheless, if we consider that ethics will be an important component of the role of engineers in the coming years, we might wish to move the level of objectives towards the highest learning level, 'create,' which is more representative of responsibility. This would mean that simply 'remembering' (the passive level) would no longer be in the first rank. It would have to cede this current rank to a higher level to 'apply.'

This chapter has touched on the social, political, and environmental realms that engineers can influence. The authors posit that the future engineer should be actively involved in these spheres, even more than before, because of the power and risk that emerging technologies have on our societies.

Note

1 The GLOBE system was used for clustering the countries included in the accreditation analysis. The GLOBE clustering was more comprehensive and included more countries. The Hofstede dimensions were used to analyze and compare patterns between accreditation documents.

References

Alas, R. (2006). Ethics in countries with different cultural dimensions. *Journal of Business Ethics*, *69*, 237–247. https://doi.org/10.1007/s10551-006-9088-3

Anderson, L. W., Krathwohl D. R., Airasian P. W., Cruikshank K. A., Mayer R. E., Pintrich P. R., Raths J., & Wittrock M. C. (2001). A revision of Bloom's Taxonomy of educational objectives. https://www.uky.edu/~rsand1/china2018/texts/Anderson-Krathwohl%20-%20A%20taxonomy%20for%20learning%20teaching%20and%20assessing.pdf

Augusti, G. (2009). Format for proceedings papers. EUR-ACE: The European accreditation system of engineering education and its global context. In A. Patil & P. Gray (Eds.), *Proceedings of the International Conference Global Cooperation in Engineering Education*. Engineering Education Quality Assurance. https://doi.org/10.1007/978-1-4419-0555-0_3

Augusti, G. (2013). Origins, present status and perspectives of the European EUR-ACE engineering accreditation system. *Engineering Education, 8*. https://aeer.ru/filesen/io/m12/art_3.pdf

Bacchi, C. (2007). The ethics of problem representation: Widening the scope of ethical debate. *Policy and Society, 26*, 5–20. http://doi.org/10.1016/S1449-4035(07)70112-1

Ball, S. J. (2015a). What is policy? 21 years later: Reflections on the possibilities of policy research. *Discourse: Studies in the Cultural Politics of Education, 36*, 306–313. https://doi.org/10.1080/01596306.2015.1015279

Ball, S. J. (2015b). Living the neo-liberal University. *Journal of Education, Research, Development and Policy, Special Issue: Education and Social Transformation, 50*, 258–261. https://doi.org/10.1111/ejed.12132

Bardi, A., & Goodwin, R. (2011). Journal of cross-cultural psychology the dual route to value change: Individual processes and cultural moderators. *Journal of Cross-Cultural Psychology, 42*, 271–287. https://doi.org/10.1177/0022022110396916

Beasley, C., & Bacchi, C. (2007). Envisaging a new politics for an ethical future. *Feminist Theory, 8*, 279–298. https://doi.org/10.1177/1464700107082366open_in_new

Cervin-Ellqvist, M., Larsson, D., Adawi, T. Stöhr, C., & Negretti, R. (2021). Metacognitive illusion or self-regulated learning? Assessing engineering students' learning strategies against the backdrop of recent advances in cognitive science. *Higher Education, 82*, 477–498. https://doi.org/10.1007/s10734-020-00635-x

CESU. (2014). Acuerdo 03 de 2014, por el cual se aprueban los Lineamientos para la Acreditación Institucional. https://www.cna.gov.co/1779/

CESU. (2020). Acuerdo 02 de 2020, por el cual se actualiza el modelo de acreditación de alta calidad. https://www.mineducacion.gov.co/portal/Educacion-superior/CESU/399567:Acuerdo-02-del-1-de-julio-de-2020

Colombia, Congreso de la República. (2003). Ley 842 de 2003, por la cual se modifica la reglamentación del ejercicio de la ingeniería, de sus profesiones afines y de sus profesiones auxiliares, se adopta el Código de Ética Profesional y se dictan otras disposiciones. https://www.mineducacion.gov.co/1621/articles-105031_archivo_pdf.pdf

Colombia, Presidencia de la República. (2019). Decreto 1330 de 2019, por el cual se sustituye el Capítulo 2 y se suprime el Capítulo 7 del Título 3 de la Parte 5 del Libro 2 del Decreto 1075 de 2015 -Único Reglamentario del Sector Educación. https://www.mineducacion.gov.co/portal/normativa/Decretos/387348:Decreto-1330-de-julio-25-de-2019

Commission des Titres d'Ingénieurs (CTI). (2022). *Références et orientations*. Accreditation Criteria, Guidelines and Procedures. Retreived April, 2022, from https://www.cti-commission.fr/wp-content/uploads/2022/03/RO_Referentiel_2022_VF_2022-03-15.pdf

Djurovic, M., & Lubarda, V. (2014). *Engineering education and the Bologna process*. http://maeresearch.ucsd.edu/~vlubarda/research/pdfpapers/Beirut_Confer.pdf

GLOBE Foundation. (n.d.). Www.globeproject.com. https://www.globeproject.com/results/clusters/anglo?menu=list#list

Grelon, André. « L'organisation de la formation des ingénieurs en France ». Artefact. *Techniques, histoire et sciences humaines*, n° 13 (7 janvier 2021), 4975. https://doi.org/10.4000/artefact.6252

Hadwick, R. (2011). *Should I use GLOBE or Hofstede? Some insights that can assist cross-cultural scholars, and others, choose the right study to support their work*. https://www.anzam.org/wp-content/uploads/pdf-manager/574_ANZAM2011-335.PDF

Hofstede, G. (2011). Dimensionalizing cultures: The Hofstede model in context. *Online Readings in Psychology and Culture, 2*, 1–16. https://doi.org/10.9707/2307-0919.10145

Hofstede, G., Hofstede, G. J., & Minkov, M. (2010). *Cultures and organizations: Software of the mind: Intercultural cooperation and its importance for survival*. Mcgraw-Hill.

House, R., Javidan, M., Hanges, P., & Dorfman, P. (2002). Understanding cultures and implicit leadership theories across the globe: An introduction to project GLOBE. *Journal of World Business, 37*, 3–10. https://doi.org/10.1016/s1090-9516(01)00069-4

JABEE. (n.d.). *JABEE criteria guide for accreditation of engineering education programs at bachelor level applicable in the year 2019 and later*. Retrieved June 30, 2023, from https://jabee.org/doc/Criteria_Guide_ENB_2019-.pdf

Junaid, S., Gwynne-Evans, A., Kovacs, H., Lönngren, J., Mejia, J.F.J., Natsume, K., Polmear, M., Serreau, Y., Shaw, C., Toboşaru, M. , & Martin, D. A. (2022). What is the role of ethics in accreditation documentation from a global view? In H.-M. Jarvinen, S. Silvestre, A. Llorens, & B. V. Nagy (Eds.), *SEFI 2022 - 50th Annual Conference of the European Society for Engineering Education, Proceedings* (pp. 369–378). European Society for Engineering Education (SEFI). https://doi.org/10.5821/conference-9788412322262.1336

Junaid, S., Kovac, H., Martin, D. A., & Serreau, Y. (2021). What is the role of ethics in accreditation guidelines for engineering programmes in Europe? In H.-U. Heiß, H.-M. Järvinen, A. Mayer, & A. Schulz (Eds.), *Proceedings SEFI 49th Annual Conference: Blended Learning in Engineering Education: challenging, enlightening – and lasting?* (pp. 274–282). European Society for Engineering Education (SEFI). https://www.sefi.be/wp-content/uploads/2021/12/SEFI49th-Proceedings-final.pdf

Kirkman, B., Lowe, K., & Gibson, C. (2006). A quarter century of culture's consequences: A review of empirical research incorporating Hofstede's cultural values framework. *Journal of International Business Studies, 37*, 285–320. https://doi.org/10.1057/palgrave.jibs.8400202

Klassen, M. (2018). *The politics of accreditation: A comparison of the engineering profession in five anglosphere countries*. University of Toronto (Canada).

Krathwohl, D. R. (2002). A revision of Bloom's taxonomy: An overview. *Theory into Practice, 41*, 212–218. https://doi.org/10.1207/s15430421tip4104_2

Legge, J. (1861). *The Chinese classics (Vol. 1): Confucian Analects, the great learning, and the Doctrine of the Mean*. N. Trübner.

Mallalieu, S. (2023). Appendix: Bloom's taxonomy of educational objectives. Retrieved June 30, 2023, from https://ocw.mit.edu/courses/16-540-internal-flows-in-turbomachines-spring-2006/351f515bb32e0ea5fe661f8317169220_syllbloomtaxon.pdf

Newton, P. M., Da Silva, A., & Peters, L. G. (2020). A pragmatic master list of action verbs for Bloom's taxonomy. *Frontiers in Education, 5*. https://doi.org/10.3389/feduc.2020.00107

Ronen, S., & Shenkar, O. (2013). Mapping world cultures: Cluster formation, sources and implications. *Journal of International Business Studies, 44*, 867–897. https://doi.org/10.1057/jibs.2013.42.

The Bologna Process and the European Higher Education Area | European Education Area. (n.d.). Education.ec.europa.eu. Retrieved September 8, 2022, from https://education.ec.europa.eu/node/1522

Tomasello, M. (2016). *A natural history of human morality*. Harvard University Press. http://www.jstor.org/stable/j.ctv1g13jjn

UNESCO – IESALC y Universidad Nacional de Córdoba. (2018). *Tendencias de la educación superior en América Latina y el Caribe 2018*. Coordinado por Pedro Henríquez Guajardo. Córdoba, Argentina. https://unesdoc.unesco.org/ark:/48223/pf0000372645/PDF/372645spa.pdf.multi

34
ACCREDITATION AND LICENSURE
Processes and implications

Angela R. Bielefeldt, Diana Adela Martin, and Madeline Polmear

Introduction

Striving toward the ethical and competent practice of engineers in the workforce motivates linkages between the individual professional licensure of engineers and accreditation to control the quality of the educational preparation of engineers. Significant differences exist globally, regionally, and even among engineering subdisciplines in the requirement and/or importance of engineering licensure for employability. The requisites for engineering licensure and the processes for setting these rules also vary widely. Further, there are complex and differing relationships between the accreditation of engineering degree programs and the licensure or certification of engineers globally. It is beyond our scope to present an extensive range of global examples of these conditions. Instead, the ethics of these requirements and processes from a few examples will be examined in this chapter, including issues of power dynamics, inclusion, and transparency. After exploring these topics at a high level, the authors leverage their personal experience and empirical work to reveal nuances not typically evident via two in-depth case studies set in the context of two original signatories of the Washington Accord. The first case examines civil engineering in the United States, probing the ethics of licensure requirements and the processes for setting educational accreditation requirements. It reveals the complex interactions of multiple organizations, including state governments, multiple non-profit groups, and a professional society. A second case study in Ireland examines the consequences of licensure and accreditation policies on engineering ethics education. Here, there is more direct government control at multiple levels, but it manifests differently through engineering education at different higher education institutions. These examples provide a grounding that others can use when considering their locally relevant specifics.

Licensure

Licensure is intended to help ensure professional competence and responsibility, such that an individual engineer can fulfill their primary ethical requirement to protect human health, safety, and welfare. Licensure can occur at the level of 'engineering,' at the discipline level (e.g., civil engineering, mechanical engineering), or at the subdiscipline level (e.g., structural engineering). Given the heterogeneity of licensure requirements and processes globally, this chapter provides

DOI: 10.4324/9781003464259-41

examples from different countries. The examples are primarily drawn from the United States, the United Kingdom, Canada, and Ireland. The rationale for this focus is multifaceted, including the context and expertise of the authors (further explained under Author Positionality), the availability of English resources and documents related to licensure, and the cultural and structural focus on licensure in these countries. For example, in the United States, an individual can lose their license to practice engineering due to ethical violations.

The extent to which engineering should be viewed as a profession and demand licensure for individuals to call themselves engineers and conduct engineering work as their job and career is contested and varies by geographical context. Each government individually determines licensure requirements, and these vary substantially. Engineering licensing occurs within individual states in the United States and provinces in Canada – a practice that has been critiqued as overly restrictive compared to licensure at the country level, which is more common (e.g., Cleary, 2018). Licensure typically requires a combination of educational preparation (judged of sufficient quality) and relevant on-the-job work experience under the mentoring of a qualified engineer, with some jurisdictions, such as the United States, additionally requiring examinations to prove competence. Sometimes, this work experience must be within a particular geographic jurisdiction (e.g., in Canada or the United States). Geddie (2002) found that the local work experience requirement was the "most significant obstacle noted by foreign-trained engineers" (p. 129) in becoming licensed to practice in the province of British Columbia, Canada.

In addition to country-level licensure, efforts are being made across countries to standardize and recognize certification. For example, the International Professional Engineers Agreement (IPEA) has 16 countries as full members and 3 countries as provisional members, and the Asia Pacific Economic Cooperation (APEC) Agreement offers substantial equivalence of professional competence requirements across 14 countries and 2 provisional members; the IPEA and APEC countries have significant overlap. Within Europe, the EUR ING certificate under Engineers Europe (formerly FEANI) applies across 33 countries.

The laws and policies concerning non-licensed individuals working as engineers differ among countries. The Netherlands doesn't require licensure or registration (Davis, 2015), and in France, engineering is "both a job *and* a title" (Didier, 1999, p. 474). Within the United States, many mechanical, electrical, and chemical engineers working 'in-house' for a manufacturing or other business firm function without a license under the industrial exemption, which has been characterized as a threat to the profession (Spinden, 2015; Swenty & Swenty, 2017). Most US states have significant exemptions to engineering licensure laws, with an average of 14 different exemptions per state (Swenty & Swenty, 2023). Similar industrial exemption to engineering licensure occurs in the United Kingdom. In contrast, this industrial exemption does not exist within Canada, except for the province of Ontario. These examples speak to the heterogeneity of licensure practices across countries and cultures, with differences even between industries and regions.

Walesh (2022) proposes that many engineering disasters could have been avoided by requiring that professionally licensed engineers direct projects:

> All the engineering organizations behind these failures were exempt from placing licensed engineers in charge. Engineering did not need to be conducted under the direction of competent and accountable engineers whose paramount ethical and legal responsibility was public protection. Instead, the "engineering" was primarily driven by bottom-line-oriented managers and executives.
>
> *(p. 1)*

However, the significance of licensure and its role in determining ethical behavior and quality in engineering remain contentious topics, as exemplified by the comments in opposition to licensure posted in response to Walesh's article.

Engineering licensure requirements rest on the argument that engineering is a profession and, thus, should be licensed similarly to disciplines like medicine. However, "Marxists (Braverman, 1998), Foucauldians (Nettleton, 1992) and others [have] used power lenses to question and challenge the control and authority vested in professionals due to their esoteric knowledge base and supposedly superior ethics" (Klassen, 2018, p. 13). Professions more broadly have been critiqued as "sites of substantial inequity and marginalization" (Klassen, 2018, p. 13).

Accreditation role in licensure

Engineering licensure is commonly linked to receiving education from accredited engineering programs. Program accreditation is assumed to ensure that the quality of educational preparation is sufficient to ensure engineering competence and ethical behavior. In many countries, governmental entities control both engineering licensure and educational accreditation. For example, in the United Kingdom, the Engineering Council controls both engineering licensure and accreditation; in Ireland, Engineers Ireland (EI) also has a dual function. Klassen (2018) states (italics added here for emphasis):

> There is a widespread assumption in Anglo-American contexts that accreditation exists to align the focuses of professional education in universities with the needs of professional bodies and ultimately employers, where professionals go to work. [This] perspective provides an underpinning assumption for legislation whereby the state delegates regulatory power to the professional body *in return for a commitment to serving the public good and upholding high standards of ethics*. This assumes a very clear definition of the scope of practice being regulated, and proactive steps taken by the professional body to intervene and discipline their members if they malpractice or operate without a license. Interestingly, neither of these assumptions appear to hold well in the case of the engineering profession.
>
> *(p. 14)*

Countries typically tie their licensure requirements to accredited degrees within their own country. This creates barriers for individuals who have earned degrees that are not accredited. It also creates mobility problems for individuals possessing engineering degrees from outside the country. Various international groups are trying to address global mobility issues by determining substantial equivalency of engineering accreditation standards. The first significant effort to establish accreditation equivalency across countries was the Washington Accord (see more information in Chapter 32). Countries participate in the Washington Accord through representation by governmental or private entities; for example, the Accreditation Board for Engineering and Technology (ABET) represents the United States. However, there continue to be barriers for individuals receiving engineering degrees from countries not signatories to the Washington Accord. For example, Geddie (2002) found significant financial and time barriers associated with the examinations and interviews used to evaluate the competence of foreign-educated individuals to be licensed in the province of British Columbia, Canada.

Klassen (2018) argued that the accreditation process generally fails to acknowledge that a high percentage of students who graduate with degrees in engineering pursue careers outside of engineering. In the United States, 65% of degreed engineers were working in occupations not considered engineering, and 18% of those working in engineering occupations did not have a degree in engineering (NAE, 2018). Fortunately, there is a significant overlap between the knowledge and skills embedded in engineering accreditation requirements and the skills needed for careers at large (OECD, 2021).

In addition to the accreditation requirements that apply uniformly to all engineering disciplines, there may be additional requirements for specific engineering disciplines. Within ABET, these 'program criteria' are largely set by the professional societies that relate to each discipline. Each professional society uses different processes to modify these criteria. Other examples vary by country and discipline. For example, in the United Kingdom, the Joint Board of Moderators, comprised of five different professional groups, accredits civil engineering and related programs; the Institution of Mechanical Engineers accredits mechanical engineering degrees; the Institution of Engineering and Technology accredits electrical and electronic engineering degrees.

Processes to determine accreditation standards

How accreditation processes are structured and who determines and controls these structures have ethical implications. The professional groups and regulatory bodies involved in accreditation have self-interest and therefore may "act to maintain their own privileged and powerful position as a controlling body," which might confound their commitment to the public interest for high quality and ethical engineering (Harvey, 2004, p. 212). Further, "goals and decisions emerge from bargaining and negotiation among competing stakeholders jockeying for their own interests" (Bolman & Deal, 2013, p. 194-195). Goals differ among countries with respect to global competitiveness (e.g., intellectual property), between for-profit companies and public agencies (profit vs. wise stewardship of resources), among disciplines (differences in salary and prestige), and privileged versus less privileged groups (e.g., particular nations over others; and in the United States, white men vs. minoritized groups). The extent to which accreditation processes prioritize true public good versus other interests merits consideration.

While the processes for engineering programs at higher education institutions to become accredited have been well documented, uncovering the processes used to set these rules is more challenging. From the outside, there might seem to be a broad consensus on accreditation requirements and procedures. But this is far from the case. There is typically a fairly small number of people who develop accreditation policies and procedures. The extent to which these individuals develop criteria that match their personal opinions versus the broader views of diverse stakeholders is generally not apparent. The process by which individuals are selected to serve on these committees and their qualifications, expertise, representation of diverse stakeholders, and true level of engagement in the process is also unclear. Who has a seat at the table and is included or excluded has embedded ethical considerations. For example, Case (2017) contrasted the 'shop culture' of working engineers versus the 'school culture' of engineering academics, which differ in the value placed on particular knowledge and skills. In Ireland, accreditation bodies strive to include two academics and an industry practitioner on each accreditation panel, which helps assess whether the program under evaluation is substantially equivalent to the programs that the academics deliver in their own institutions and aligns with the requirements of industry (Murphy et al., 2019).

The International Engineering Alliance [IEA] (2021) has established *Graduate Attributes and Professional Competencies*, which are closely related to the Washington Accord and the International Professional Engineers Agreement. The Washington Accord has 23 signatories and seven provisional signatories as of 2023. Signatories have committed to mutual recognition of substantial equivalency of accreditation standards and processes (Hanrahan, 2013). If mutual recognition is truly the case, analyzing the process of any one signatory would accurately represent all. Underlying cultural differences may shape the "individual accreditation processes and variations in accreditation criteria as well as different documentation requirements and reporting processes" (Patil & Gray, 2009, p. 20). Thus, no one set of processes should be deemed optimal or the most ethical. The basic process for accreditation of engineering programs consists of a repeatable cycle of review (3–6 years being most common); documentation of self-assessment that the program meets the accreditation requirements (typically a combination of student learning outcomes and/or curriculum structure and content, qualifications and number of faculty members, processes for student admissions, and verification of fulfillment of graduation requirements); program review by individuals typically including an on-site visit (number, qualifications, and training of the reviewers are specified); and specific outcomes/decisions of the accreditation process. Some governments have a single set of requirements for all engineering degrees; others have varying requirements for different engineering disciplines (as is the case under ABET and in the United Kingdom).

In the United States, some highly respected universities have opted not to accredit some of their engineering degrees, viewing ABET accreditation as unnecessary, burdensome, and/or restrictive. These highly ranked programs at research-intensive universities do not believe they need traditional accreditation to vouch for their quality (Klassen, 2023). Examples of universities and programs opting out of ABET accreditation include Stanford University (Electrical Engineering 2013, Environmental Systems Engineering 2015, Chemical Engineering 2020), the California Institute of Technology (Caltech, Chemical and Electrical Engineering in 2018), the University of California Berkeley (Electrical Engineering 2017), and Tufts University (Biomedical Engineering 2022). In its announcement that it would not re-accredit its environmental engineering degree, Stanford University stated: "The accreditation process ... isn't quite at the cutting edge of the field" (Stanford, 2015). Another university discontinuing ABET accreditation stated (Caltech, 2017):

> The undergraduate program in Chemical Engineering at Caltech is widely regarded as one of the most rigorous in the world. In our efforts to maintain that rigor in light of the rapid pace of change in this discipline, Caltech's Chemical Engineering faculty have concluded that the process of engineering accreditation by ABET limits our ability to offer the best possible education.

The letter cited limitations to flexibility, specifically that "the restrictions and requirements imposed by ABET criteria and examiners have led to an excessively structured curriculum," and concerns with the vagaries of individual program evaluators (PEVs). Despite the lack of requirements to meet ABET accreditation outcomes, engineering ethics content remains embedded within required courses in the Caltech chemical engineering curriculum, including a senior chemical engineering lab course that embeds ethics within team projects and the analysis of case studies (Caltech, 2022). Alternatively, some programs have opted to accredit under the general criteria rather than the appropriate program criteria, which impose additional restrictions (e.g., Massachusetts Institute of Technology (MIT) Civil Engineering).

The following sections provide case studies of accreditation and licensure in the United States and Ireland to complement the broader overview of the preceding sections. The US case study focuses on civil engineering, exploring ethical issues within accreditation and licensing processes. These overall processes are drivers for the ethics requirements in engineering education and licensure. The second case study, in Ireland, reveals the consequences of the licensure and accreditation requirements on the ethics education in engineering.

Author positionality

The cases were selected based on the authors' expertise. The first author has led accreditation efforts in the civil engineering program at her institution for 15 years, served on the American Society of Civil Engineers (ASCE) Body of Knowledge 3 Task Committee and Program Criteria Task Committee, and is a licensed professional engineer in Colorado, United States. Her experiences offer insight into the development and implementation of civil engineering accreditation in the United States. The second author provides a case study in Ireland based on her doctoral dissertation. During her dissertation research, the second author interviewed instructors and evaluators to understand the accreditation process in Ireland and the role of ethics within it, and observed several accreditation events. The second and third authors are involved in an international study on the role of ethics in engineering accreditation (Junaid et al., 2022) and use this work as well as their engineering education research experience in different countries (Ireland, the Netherlands, Belgium, and the United Kingdom) to inform their perspectives.

Case study: United States, ABET, civil engineering

In the United States, state governments control engineering licensure. The licensing process typically involves the shortest path when the individual has graduated with a Bachelor's degree from an ABET-accredited engineering program, passed two national examinations, and has 4 years of qualifying practice vouched for by professionally licensed engineers. An individual graduating without an ABET-accredited degree may be eligible for licensure after an additional 1–4 years of qualifying experience. Some states issue a general Professional Engineer (PE) license with the expectation to practice in one's area of competence; other states license PEs in specific disciplines (e.g., Professional Civil Engineer). In addition, structural engineering (SE) is separately licensed in many states. Most states allow individuals licensed in one state to easily become licensed in another through the process of comity.

The two licensing examinations are common nationally and controlled by the National Council of Examiners for Engineering and Surveying (NCEES), a nonprofit organization. The Fundamentals of Engineering (FE) exam is largely multiple-choice, is proctored, online, and has versions for different engineering disciplines. The exam is commonly taken by senior-level (i.e., final-year) undergraduate engineering students, and some engineering degree programs require their students to take the FE exam before graduation. The FE exam can be taken at any time and repeated if not passed. The FE exam includes a few questions on engineering ethics and professional practice or societal impacts (3–8 out of 110 total). The quality of these questions and the ability to actually evaluate ethical reasoning abilities have been critiqued (French, 2006). The NCEES is not transparent about who writes the exam questions, including those related to ethics, stating only that "NCEES exams are developed by licensed engineers and surveyors who volunteer to write and evaluate exam questions in conjunction with NCEES procedures and accepted psychometric standards" (NCEES, 2021, p. 4). The second exam, Principles and Practice of Engineering (PE), was historically 8 hours and tested higher-order engineering design skills; professional, licensed engineers scored it. However, the PE exam has also moved to a multiple-choice format. The PE exam includes no content on ethics.

Some states have additional licensure provisions that target ethics, such as a further examination that covers ethics and/or local legal issues (e.g., Texas, California, Nevada, and South Dakota). Most of these exams appear to be in 'take home' format with simple multiple choice or true/false questions. Some states require that an individual seeking professional licensure obtain letters that attest to personal and/or professional character and/or integrity (e.g., Texas, Oregon, Mississippi, Rhode Island, Montana). For example:

> The Texas Engineering Practice Act states that a person seeking to obtain a license to practice professional engineering shall provide evidence of good professional character and reputation which, in the judgment of the Board, is sufficient to ensure that the individual can consistently act in the best interest of clients and the public in any practice setting. Such evidence shall establish that the person is able to distinguish right from wrong, is able to think and act rationally, is able to keep promises and honor obligations, and is accountable for his/her own behavior.
>
> *(Texas, 2022, p. 2)*

Most states require continuing professional development via education to retain one's license, ranging from 8 to 15 hours per year (which may be documented on an annual, biennial, or triennial basis), called professional development hours (PDH) or continuing education units (CEU) (E1 Education, 2020). Nine states have no requirements for continuing education documentation. Fifteen states have some minimum requirements for continuing education hours related to ethics education (ranging from 0.33 to 1.5 hours per year). Professional development hours are often earned by attending professional conferences or online education sessions. The quality of this education is uncertain and, therefore, has been critiqued in some cases as simply a money-making business for groups and professional societies that exercise their power to create requirements for PDH and then also offer those hours (Nevada, 2020).

Overall, the importance and extent of licensing in the United States varies significantly among disciplines, appearing the highest in civil engineering, where nearly all civil engineering graduates take the FE exam and three-fourths go on to attempt professional licensure, compared to less than half even starting the professional licensure path by taking the FE exam among mechanical, electrical, and chemical graduates (based on author calculations from ASEE (2020) and NCEES (2019) data).

ABET is a non-profit organization that sets accreditation standards in the United States. ABET relies heavily upon volunteers to lead the development of the accreditation requirements and to implement the requirements by reviewing programs. ABET recently reported the age, gender, race/ethnicity, and job-sector demographics of its volunteers, but these were not disaggregated among roles (ABET, 2022). Akera et al. (2019) note that "many ABET volunteers are older, retired, and tend to have more conventional views about their discipline" (p. 13). The ABET Engineering Accreditation Commission (EAC) has been critiqued for lacking transparency and open feedback processes. Within the EAC, professional societies are primarily responsible for establishing the program-level criteria associated with specific disciplines and selecting and training the program evaluators. Civil engineering has been one of the most transparent, publishing widely on its processes. In 2023, there were 365 ABET-accredited civil engineering programs across 268 US institutions and 90 institutions outside the United States (representing 20 countries).

The American Society of Civil Engineers (ASCE) is the lead society associated with civil engineering within ABET. The ASCE Code of Ethics has explicitly included sustainability since 1997 and added diversity and equity provisions in 2017. The civil engineering program criteria (CEPC) include the requirement that "faculty teaching courses that are primarily design in content are qualified to teach the subject matter by virtue of professional licensure, or by education and design experience" (ABET, 2023, p. 23). There are curriculum requirements in the CEPC related to professional licensure dating back to 2002, and sustainability and professional ethics since 2016. The

ASCE publishes a commentary document to explain the rationale and expectations associated with the criteria to guide civil engineering PEVs and faculty. The commentary (ASCE, 2019a) states:

> Graduates should be able to explain the unique nature of civil engineers' responsibility to the general public and the consequent emphasis on professional licensure in civil engineering professional practice.
>
> *(p. 24)*

> The program Criteria … reflects an expectation for a higher level of achievement in professional ethics than required by General Criterion … requiring a curriculum to include an opportunity for students to go beyond a simple understanding of ethical responsibility and have students analyze issues.
>
> *(p. 22)*

> The Civil Engineering Code of Ethics includes as one of the Fundamental Cannons that "Engineers shall strive to comply with the principles of sustainable development …" … The criterion simply requires coverage of sustainability in the curriculum be sufficient so graduates can include key concepts of sustainability in an engineering design.
>
> *(p. 18)*

By comparison, ethics is lacking from other ABET EAC program criteria with the exception of construction (where ASCE is also the lead society) and cybersecurity.

The ABET CEPC are derived from the Civil Engineering Body of Knowledge (CEBOK). The CEBOK "defines the knowledge, skills, and attitudes necessary for entry into the practice of civil engineering at the professional level" (ASCE, 2019b, p. vii). The 2019 edition (CEBOK3) specifies the expected cognitive level of achievement of 21 outcomes using Bloom's taxonomy verbs and recommends pathways to meet these requirements, which include undergraduate education (the lower levels of all 21 outcomes), postgraduate education (2 outcomes), and mentored experience on-the-job (14 outcomes). The CEBOK3 also includes seven affective outcomes. The ethical responsibilities outcome in the CEBOK3 was cross-linked with the outcomes of design, professional responsibilities, professional attitudes, sustainability, and lifelong learning.

The ASCE has established a repeatable cycle every 8 years whereby it reviews the CEBOK, then determines the extent to which the ABET criteria are aligned with the CEBOK and if changes to the CEPC are warranted (Ressler & Lynch, 2011). The ethics outcome in the 2019 CEBOK compared to the 2008 CEBOK had a lowered level of achievement from undergraduate education (to Bloom's level 2 'explain' from 4 'analyze') and entry to professional practice (to Blooms level 5 'develop' from 6 'justify'). In addition, the 2019 CEBOK3 added the affective domain expectation to "advocate for ethical behavior in the practice of civil engineering" (level 5), achieving level 2 as part of undergraduate education ("comply with applicable ethical codes") (ASCE, 2019b, p. 61).

The development of both the CEBOK and CEPC included numerous cycles of soliciting and responding to stakeholder feedback. This occurred via specific committees in ASCE, discussion boards, and open surveys that were broadly distributed. Nevertheless, limited outside participation in these forums occurred. The development and results of feedback were carefully documented and distributed via peer-reviewed, open-access papers (e.g., Bielefeldt et al., 2019; Nolen et al., 2022). The CEBOK and CEPC represent a compromise as consensus on these topics was not reached.

The ASCE tries to comprise committees that are broadly representative, including individuals from academia (faculty members) and practicing engineers, individuals representing multiple subdiscipline areas in civil engineering (e.g., structures, geotechnical, construction, transportation, water resources, environmental), and a variety of personal demographics (age, gender, etc.). Despite these efforts, the committees have recently been predominated by academics with a low representation of traditionally underrepresented groups (e.g., women of color). The recent CEBOK3 committee included four individuals with ethics expertise, and the current CEPC committee includes individuals with expertise in ethics, sustainability, licensure, and diversity, equity, and inclusion (DEI). The CEBOK and CEPC groups opened their meetings to corresponding members not serving on the committee but wishing to provide input. Thus, while the committees have recently included 10–18 members, there were also 20–70 corresponding members who provided additional perspectives.

The CEBOK2 review resulted in the addition of ethics to the ABET CEPC in 2015. When reviewing the CEPC with respect to the CEBOK3 in 2020–2022, the ethics outcome was revised to state that "the curriculum must include application of: an engineering code of ethics to ethical dilemmas; principles of sustainability, risk, resilience, diversity, equity, and inclusion to civil engineering problems" and "explanation of professional attitudes and responsibilities of a civil engineer, including licensure and safety" (ABET, 2023, p. 22). There was extensive discussion around the ethics outcome, with an early proposal of "apply the ASCE Code of Ethics to an ethical dilemma." This reflected the fact that the ASCE Code of Ethics (2020) uses a hierarchical stakeholder model that embeds sustainability and DEI elements. However, stakeholder feedback on the practicality of this suggestion noted that many programs co-educate civil engineering students alongside other engineering majors with respect to professional ethics and, therefore, requiring the specific civil ethics code would be problematic.

ASCE also sets the requirements for, approves, and trains ABET program evaluators (PEVs) for civil engineering programs. The qualifications to be a civil engineering PEV include registration as a PE, at least 10 years of experience in the practice of engineering, and membership in the ASCE at the Member or Fellow grade. From ABET, all PEV candidates complete about 20 hours of online training and a 1.5-day experiential workshop simulating an ABET accreditation visit; this PEV training may qualify as PDH for licensure. PEVs also agree to a code of conduct policy that includes confidentiality and conflict-of-interest issues. Akera et al. (2021) noted that some of the individuals they interviewed "spoke about consistency, PEV training and variation" (p. 5) as part of their frustrations with the ABET review process.

A recent paper by Ressler and Lenox (2020) explored the potential for ASCE to withdraw from ABET, identifying the benefits and costs of ABET membership from the perspective of ASCE. These authors have deep engagement, leadership, and service with both ABET and ASCE. They recognize "ASCE's ability to establish, promulgate, and enforce educational standards through ABET accreditation represents a powerful tool for advancing the Society's strategic interests" (p. 7). "However, these benefits are not being fully realized, because ASCE's perspectives and interests so often diverge from those of ABET and many of its Member Societies" (p. 10).

This case illustrates the complexity and interconnected nature of groups that influence engineering ethics education in civil engineering in the United States through specification and affect licensure and accreditation processes. Compared to other disciplines, civil engineering appears to be at the forefront in the United States regarding concern for ethics in the education and practice of engineers. Yet the practical implications of these regulations on ethics education are unclear, given the strong role of the engineering culture (which preferences technical expertise and business or profit motives in the workforce) and the significant level of control of individual teachers in their classrooms. The next case study illustrates the consequences of licensing and accreditation on ethics education in engineering programs in Ireland.

Case study: licensure and accreditation in Ireland

Moving from process to practice, this case study presents empirical data rooted in the Irish context of engineering education to illustrate the impact of accreditation on curriculum development and the link between accreditation and educational change. Engineers Ireland (EI) has formally accredited engineering programs in the Republic of Ireland since 1982. Graduates of accredited programs may achieve one of the professional titles of Chartered Engineer, Associate Engineer, or Engineering Technician. From 2013, in order to apply to become a Chartered Engineer, candidates need to hold a Master's Degree from an engineering program accredited by EI. Under the terms of the Washington Accord, EI also recognizes qualifications obtained outside Ireland that meet a similar educational level. Ethics is part of the licensure process for becoming a Chartered Engineer via a dedicated requirement, which requires professionals to provide examples of ethical practice in their written application. This is understood to comprise evidence that the candidate has "complied with appropriate codes and rules of conduct" (competence 5.1), "managed and applied safe systems of work" (competence 5.2), "ensured that their engineering work complies with the code of practice on risk and the environment" (competence 5.3), and "ensured their continuing professional development to maintain the currency of their professional engineering knowledge and skills" (competence 5.4). Evidence pertaining to ethics can also be provided for different competencies, which require the candidate to show that they understood and applied advanced knowledge of the widely applied engineering principles underpinning good practice (competence 1.2). The written application is followed by an interview, comprising a presentation and a discussion with the panel where candidates are further asked about how they meet the five competencies.

EI was one of the six original signatories of the Washington Accord in 1989, which targeted the mutual equivalence of Bachelors of Engineering degrees (International Engineering Alliance, 2015). The Washington Accord included a focus on ethical responsibilities and the societal role of the engineering profession, including sustainability (see Chapter 32) (International Engineering Alliance, 2014). The emphasis of global accords on ethical and societal considerations in the practice of engineering is considered to have led to the establishment of engineering ethics education as a mandatory accreditation requirement in signatory countries (Coates, 2000).

In Ireland, ethics first appeared in 2007 in the accreditation criteria; they were revised and extended in 2014 and 2021 (Engineers Ireland, 2014, 2021). In addition to a program outcome dedicated to professional and ethical responsibilities, the most recent formulation of the accreditation criteria includes an outcome on sustainability (Engineers Ireland, 2021).

The accreditation process in Ireland

EI, like ABET, accredits an individual program rather than an entire college or institution. Each program offered by an engineering college or faculty undergoes a separate accreditation process, for which it prepares its own set of documents based on guidance and objectives set by EI. This is a quality review process occurring approximately every 5 years. It encompasses three steps that are quite similar to the ABET process: (1) internal self-study documentation, (2) a site visit over 2 days, and (3) an external evaluation report submitted by the representative of EI and the accreditation panel.

The EI Registrar is responsible for managing the evaluation process, selecting the accreditation panel members, and preparing the agenda for the accreditation visit. Each program has an internal team preparing the documents and overseeing the organizational aspects of the visit, including the guided tour of facilities and separate sessions with students, alumni, and employers. The accreditation panel responsible for evaluating the program is comprised of two external academics and one industry representative.

Ethics in the context of accreditation

Martin's (2020) doctoral study examined the evaluation of ethics for accreditation in Ireland. The study used internal documentation prepared by the programs, accreditation reports, interviews with evaluators and instructors, and observation of accreditation events. It identified how ethics was being considered and evaluated at the three stages of the accreditation process mentioned above in 23 engineering programs offered by six institutions across Ireland.

Internal self-study documentation

Within the qualitative descriptions in the self-study documents prepared by participant programs, ethics was often described as 'complementary' or 'ancillary' to four 'technical' core program outcomes, that is, (A) technical and scientific knowledge, (B) problem-solving, (C) design, and (D) conducting experiments. One self-study noted: "Programme outcomes E [ethics], F [communication], and G [teamwork] are associated with developing a *complementary skill set* in graduates and are generic to most branches of engineering" (Martin, 2020, p. 250). Ethics was described in similar terms by a program that had the objective of "equipping students with 'advanced technical, design, research and complementary skills to be of direct benefit to the profession in particular and society in general'" (p. 250–251). Another program at the same institution mentioned a similar distinction between two types of outcomes. According to the documents submitted by one of the participant university's programs:

> while the first four outcomes relate to the acquisition of a sound technical and analytic base and a mastery of the necessary discipline-specific knowledge, the last three outcomes relate to the practice of engineering in a work and professional context.
>
> *(p. 251)*

The internal self-study documentation highlighted that the programs had a stronger focus on attaining scientific and technical outcomes distributed throughout the 4 years of study, while ethics was integrated into just a few courses and course units. Notably, the outcomes purporting to technical and scientific knowledge and problem-solving were described as core technical outcomes. Ethics had an ancillary role in several programs, which was reinforced by descriptions that the program aspired to produce "graduates with the necessary theoretical foundations, domain-specific technical knowledge and practical and ancillary skills" (p. 251–252).

The description of the 'complementary' status of ethics within the engineering curricula was reflected in the numerical self-assessment of how programs deemed each of their courses to have met the ethics outcome. For this numerical self-assessment, Engineers Ireland (2015) recommends using a five-point scale for the programs to indicate how the learning outcomes set for each of their modules meet the seven program outcomes set by the accrediting body. These scores range from 0 (module does not contribute to the outcome) to 4 (strongly contributes). Engineers Ireland (2015, p. 4) states this is "the most important section of the accreditation document."

The analysis showed that ethics was the program outcome with the lowest weight in the curriculum of the engineering programs that participated in Martin's (2020) study.[1] The average for the ethics program outcome considering all courses offered by 17 of the 23 programs participating in Martin's (2020) study was 1.56/4.00, less than half the average for the outcome purporting to technical and scientific knowledge (3.18/4.00) and problem-solving (3.12/4.00).[2] Considering disciplinary differences, the lowest averages registered in the numerical self-assessment of ethics were encountered in the programs of Electric and Electronic Engineering (average 0.81/4.00) and Electronic and Computer Science (average 0.88/4.00) (Martin et al., 2019).

The curricular weight given to each of the seven program outcomes together with the explanation provided about the implementation of these outcomes seem to place nontechnical skills on a different par than technical skills. Engineering programs tend to emphasize the attainment of technical, scientific, experimental, and design outcomes throughout the four years of study, viewing them as 'fundamental,' 'core,' and 'discipline specific' skills. Ethics, alongside the learning outcomes of communication and teamwork, have their place in a smaller number of courses and are described as providing a 'complementary' or 'ancillary' skill set. Engineering programs are thus seen to explicitly cultivate the dichotomy between what traditionally have been called 'hard' skills and 'soft' skills. This distinction and language are problematic for privileging technical skills, sending the message to students that professional skills are optional, and marginalizing educators and engineers who practice and promote professional skills (Berdanier, 2022).

Accreditation visit

The accreditation events observed during the doctoral research study (Martin, 2020) revealed different strategies for approaching evidence that distinguished between technical and professional outcomes. As such, during accreditation, discussions related to the analysis of evidence led to an agreement among evaluators to distribute their responsibilities such that the panel was "split into a hard and soft outcome" for each evaluator (Martin, 2020, p. 254). Reflecting on the approach to evidence, ethics "tends to be not singled out," such that the program outcomes purporting to technical and scientific knowledge, problem-solving, and design were discussed as a group, while the program outcomes purporting to ethics, communications, and teamwork were discussed as another group (Martin, 2020, p. 254). During the peer assessment process, less time was dedicated to discussing how programs met the non-technical outcomes compared to the time allocated for discussing the technical outcomes.

Some evaluators expressed their belief that ethics did not need to have the same emphasis as technical outcomes (Martin, 2020, p. 244–246). This view seems to have been shared by the instructors of the programs evaluated. The final plenary sessions of the accreditation events observed also reflected a lower threshold for what was deemed an acceptable provision of ethics education. Discussions between evaluators and the internal program team focused mostly on how technical outcomes had been met. As long as the evaluators' comments about technical outcomes were positive, the seemingly weaker curricular presence of ethics outcomes (based on the quantitative rubric scores or evidence) was deemed acceptable, and the programs were recommended for accreditation.

Overall, the participant programs' low focus on ethics outcomes (rendered via both low self-assessment scores and internal evidence) was perceived as a common state of affairs. Evaluators noted that there was "mostly low scoring" in ethics or that ethics outcomes "are hit lightly" (Martin, 2020, p. 245). During one accreditation event, two evaluators were in agreement that "this is mostly the case everywhere." Evaluators noted it was common for programs to give a low priority to the implementation of ethics. During interviews following the accreditation events observed by Martin (2020, p. 258), an evaluator stated that "sometimes it might appear like it is tagged on a bit at the end … not quite an afterthought, but it is probably not given as much importance." Another evaluator shared a similar opinion, considering that "ethics is way down the priority list" and is "mainly there just to cover the requirements of Engineers Ireland, … but the amount of module[3] content dedicated to it would be minimal" (p. 258).

Martin's (2020) study found that evaluators regard ethics as not having a fundamental role in the engineering curriculum. Reflecting on his experience as an evaluator, one participant stated that "programs do not see it as important. They probably prioritize having the core skills as an engineer or as a technician as being the primary skills requirement coming from the course" (p. 256). While technical outcomes are implemented in a systematic manner in the curriculum of engineering programs, ethics does not receive the same treatment according to another evaluator, who claimed that:

> if you take technical subjects, like structures or signal processing, the academics will make sure that the design of the program incorporates these, and in a logical and coherent way. But they do not take the same approach about the ethical material.
>
> *(p. 258)*

The lack of comprehensive implementation of ethics, often incorporated via an individual champion's efforts, is also reflected in an evaluator's remark that "programs were all relying on this person to show that ethics has been integrated into the program" (p. 258). The outcome is a normalization of the lower presence of ethics in the engineering curriculum. As such, teaching ethics in one or two courses will "hit the target sufficiently to avoid being a problematic issue" and for the program to avoid ending up in "a condition territory" when it comes to receiving the accreditation (p. 246).

Members of accreditation panels expressed difficulties in evaluating the ethics program outcome. According to an evaluator, the ethics outcome is the most challenging to evaluate because "we are not specialists in ethics. … there is this part of us that believes that we are not really qualified to evaluate that … because we are not trained to do that. So first of all, this is something new. Second of all, a lot of us, and especially people teaching highly technical tools, never thought about it and they never asked that question" (Martin, 2020, p. 263). One evaluator even stated that "I just do not like the ethics" outcome, and considers the technical outcomes are "the easiest" to assess (Martin, 2020, p. 263).

External evaluation reports

The evaluation reports contained little information to guide programs in strengthening or increasing the presence of ethics. In the recommendations related to ethics, there was a notable absence of suggestions for specific content or increasing the curricular presence of ethics. In contrast, there often was the recommendation to "strengthen ties with industry" or "introducing employers in the advisory board" (Martin et al., 2021, p. 369).

Summary

More attention was given to the procedural aspects of how programs prepared and displayed their evidence of the ethics outcome than to ensuring sufficient weight was given to ethics in the curriculum or exploring the broadness or societal relevance of its treatment. Although evaluators noticed the lower weight given to ethics in the curricula of engineering programs, it tended to be considered a common state of affairs, with the evaluators reasoning that ethics does not need the same emphasis in the engineering curriculum compared to technically oriented outcomes. Martin's (2020) study found that the existence of an accreditation criterion dedicated to ethics does not necessarily lead to a curriculum that addresses the social and political dimensions of engineering practice broadly (Conlon, 2013; Murphy et al., 2019). The study thus points to a lower threshold for the ethics outcome, compared to technical outcomes, of what was judged to be a satisfactory education. It also suggests that while accreditation can offer an impetus for including specific content in the engineering curriculum, it is not a guarantee that programs offer the best education to meet the requirements set by accrediting bodies. The impact of accreditation on curriculum development and educational change can be limited by the programs' resistance and sometimes self-limiting beliefs as to what engineering education is about. This adds to the current debates (mentioned in the previous section) of whether accreditation is indeed necessary for offering high-quality education (Caltech, 2017).

Closure

There is a complex interplay between accreditation and licensure that connects engineering education and practice while regulating the competencies and expectations of engineers. Important in grounding these linkages is the notion that engineering is a profession. Licensing and accreditation are key considerations in the conversation around engineering ethics education. These processes and the standards they set guide curriculum development, including the role of ethics; highlight what is valued and required in engineering practice, such as ethical responsibilities; and inform professional conduct, such as behaving ethically and the implications of not doing so. Establishing ethical processes for setting and enforcing engineering licensure and educational requirements, such as attending to inclusion and transparency, is critical. Accreditation and licensure continue to grow in importance with the globalization of the engineering workforce and the need for cross-cultural understanding of ethics (Chung, 2015). International efforts such as the Washington Accord and the European Network for Accreditation of Engineering Education (ENAEE, 2021) provide a level of global alignment among the processes and criteria for undergraduate engineering programs to be accredited. Although many accreditation documents include a marginal consideration of ethics compared to other outcomes, there is still variety in how ethics is implicitly and explicitly treated and defined (Junaid et al., 2022). This heterogeneity is greater for licensure, where processes and standards, including whether they are mandated, vary across disciplines, industries, and regions within the same countries. Disciplinary variation in licensure and education is notable and appropriate. However, given the increasingly interdisciplinary roles and implications of engineering work and continually evolving challenges, lifelong learning concerning ethics is critical for all working engineers. Globalization and mobility in the engineering workforce call into question the applicability of standardizing ethics (Clancy & Zhu, 2021; Zhu & Jesiek, 2020). The role of accreditation and licensure thus have implications for engineering ethics education not only via curricular and professional expectations but also via the cultural and power dynamics through which they are developed and implemented.

Notes

1 Note that the analysis follows Engineers Ireland's (2014) formulation of program outcomes. These were subject to redesign following the study, with a revised set of criteria being published in 2021.
2 The other six programs could not be included in the analysis of the self-assessment scores due to using a different self-assessment scale that could not be converted to the scale recommended by Engineers Ireland (2015).
3 In the Irish higher education system, a 'module' is the typical designation for a 'course' in the US.

References

ABET. (2022). *Ahead of the curve: Annual impact report 2021*. ABET.

ABET. (2023). *Criteria for accrediting engineering programs, effective for reviews during the 2024–2025 accreditation cycle*. ABET.

Akera, A., Appelhans, S., Cheville, A., De Pree, T., Fatehiboroujeni, S., Karlin, J., & Riley, D. M. (2019). ABET & engineering accreditation – History, theory, practice: Initial findings from a national study on the governance of engineering education. *American Society for Engineering Education (ASEE) Annual Conference & Exposition Proceedings,* 19 pp. https://peer.asee.org/32020

Akera, A., Appelhans, S., Cheville, A., De Pree, T., Fatehiboroujeni, S., Karlin, J., & Riley, D. M. (2021). ABET's maverick evaluators and the limits of accreditation as a mode of governance in engineering education. *American Society for Engineering Education (ASEE) Annual Conference & Exposition Proceedings*. 22 pp. https://peer.asee.org/36632

American Society for Engineering Education (ASEE). (2020). *Engineering & engineering technology by the numbers. 2019 edition*. ASEE.

ASCE. (2019a). *Commentary on the ABET program criteria for civil and similarly named programs. Effective for 2019 2020 accreditation review cycle*. ASCE. https://www.asce.org/-/media/asce-images-and-files/career-and-growth/educators/civil-engineering-program-commentary-eac.pdf

ASCE. (2019b). *Civil engineering body of knowledge: Preparing the future civil engineer* (3rd ed., 172 pp). ASCE.

ASCE. (2020, October 26). *Code of ethics*. https://www.asce.org/career-growth/ethics/code-of-ethics

Berdanier, C. G. P. (2022). A hard stop to the term "soft skills". *Journal of Engineering Education, 111*, 14–18. https://doi.org/10.1002/jee.20442

Bielefeldt, A. R., Barry, B. E., Fridley, K. J., Nolen, L., & Hains, D. B. (2019). Constituent input in the process of developing the third edition of the Civil Engineering Body of Knowledge (CEBOK3). *American Society for Engineering Education (ASEE) Annual Conference & Exposition Proceedings*. 22 pp. https://doi.org/10.18260/1-2-32540

Bolman, L. G., & Deal, T. E. (2013). *Reframing organizations: Artistry, choice, and leadership* (3rd ed.). Jossey-Bass.

Braverman, H. (1998). *Labor and monopoly capital: The degradation of work in the twentieth century*. NYU Press.

Caltech. (2017, September 27). *Chemical engineering at Caltech will no longer pursue ABET accreditation*. https://cce.caltech.edu/graduate/graduate-chemical-engineering/abet-accreditation

Caltech. (2022). *Division of chemistry and chemical engineering*. Filtered CHCHE Courses (2022–23). https://cce.caltech.edu/academics/courses?q=&level=undergrad&index_id=59831&departments=12

Case, J. (2017). The historical evolution of engineering degrees: Competing stakeholders, contestation over ideas, and coherence across national borders. *European Journal of Engineering Education, 42*(6), 974–986. https://doi.org/10.1080/03043797.2016.1238446

Chung, C. (2015). Comparison of cross culture engineering ethics training using the simulator for engineering ethics education. *Science and Engineering Ethics, 21*(2), 471–478. https://doi.org/10.1007/s11948-014-9542-z

Clancy, R. F., & Zhu, Q. (2021). Global engineering ethics: What? Why? How? And when? *American Society for Engineering Education (ASEE) Annual Conference & Exposition Proceedings*. https://peer.asee.org/37227

Cleary, J. (2018). Multistate licensure: Engineers need to figure this out! *Journal of Professional Issues in Engineering Education Practi*ce, *144*(4), 02518005, 2 pp. https://doi.org/10.1061/(ASCE)EI.1943-5541.0000383

Coates, G. (2000). Developing a values-based code of engineering ethics. *IPENZ Transactions, 27*(1), 11–16.

Conlon, E. (2013). Broadening engineering education: Bringing the community in. *Science and Engineering Ethics, 19*, 1589–1594.

Davis, M. (2015). Engineering as profession: Some methodological problems in Its study. In S. Christensen, C. Didier, A. Jamison, M. Meganck, C. Mitcham, & B. Newberry (Eds.), *Engineering identities, epistemologies and values. Philosophy of engineering and technology* (Vol. 21). Springer. https://doi.org/10.1007/978-3-319-16172-3_4

Didier, C. (1999). Engineering ethics in France: A historical perspective. *Technology in Society*, *21*, 471–486.

E1 Education. (2020, October 7). PDH License Renewal Requirements for professional engineers PEs. https://www.e1education.com/p/pe-pdh-requirements

Engineers Ireland. (2014). *Accreditation criteria for engineering education programmes.* https://www.engineersireland.ie/listings/resource/198

Engineers Ireland. (2015). *Procedure for accreditation of engineering education programmes.* https://www.engineersireland.ie/Resources/Documents/resource/199

Engineers Ireland. (2021). *Accreditation criteria for engineering education programmes.* https://www.engineersireland.ie/listings/resource/519

ENAEE - European Network for Accreditation of Engineering Education. (2021). *EUR-ACE® framework standards and guidelines.* https://www.cti-commission.fr/wp-content/uploads/2022/12/EAFSG_ENAEE_2021_EN.pdf

French, K. W. (2006). *Ethics in and of the fundamentals of engineering exam*. Proceedings of the 6th Christian Engineering Education Conference, June 21–23, Olivet Nazarene University.

Geddie, K. P. (2002). *License to labour: A socio-institutional analysis of employment obstacles facing Vancouver's foreign-trained engineers* [Master's thesis]. University of British Columbia.

Hanrahan, H. (2013). *Toward global recognition of engineering qualifications accredited in different systems.* ENAEE Conference, Leuven, Belgium. https://www.enaee.eu/wp-content/uploads/2018/11/HANRAHAN-Paper-130820.pdf

Harvey, L. (2004). The power of accreditation: Views of academics. *Journal of Higher Education Policy and Management*, *26*(2), 207–223.

International Engineering Alliance. (2014). *25 years of the Washington Accord.* https://www.ieagreements.org/assets/Uploads/Documents/History/25YearsWashingtonAccord-A5booklet-FINAL.pdf

International Engineering Alliance. (2015). *A history of the international engineering alliance and its constituent agreements: Toward global engineering education and professional competence standards. Version* 1. Wellington NZ.

International Engineering Alliance. (2021, June 21). *Graduate attributes & professional competences, version 2021.1.* Approved version 4.1.

Junaid, S., Gwynne-Evans, A., Kovacs, H., Lönngren, J., Mejia, J. F. J., Natsume, K., Polmear, M., Serreau, Y., Shaw, C., Toboaru, M., & Martin, D. A. (2022). What is the role of ethics in accreditation documentation from a global view? In H.-M. Jarvinen, S. Silvestre, A. Llorens, & B. V. Nagy (Eds.), *SEFI 2022 - 50th Annual Conference of the European Society for Engineering Education, Proceedings* (pp. 369–378). European Society for Engineering Education (SEFI). https://doi.org/10.5821/conference-9788412322262.1336

Klassen, M. (2018). *The politics of accreditation: A comparison of the engineering profession in five Anglosphere countries* [Master's thesis]. University of Toronto.

Klassen, M. (2023). *Curriculum governance in the professions: A comparative and sociological analysis of engineering accreditation* [Doctoral thesis]. University of Toronto.

Martin, D. A. (2020). *Towards a sociotechnical reconfiguration of engineering and an education for ethics: A critical realist investigation into the patterns of education and accreditation of ethics in engineering programs in Ireland* [Doctoral dissertation]. Technological University Dublin. https://doi.org/10.21427/7M6V-CC71

Martin, D. A., Bombaerts, G., & Johri, A. (2021). Ethics is a disempowered subject in the engineering curriculum. In H. U. Heiss, H. M. Jarvinen, A. Mayer, & A. Schulz (Eds.), *Proceedings - SEFI 49th Annual Conference: Blended learning in engineering education: Challenging, enlightening – and lasting?* (pp. 357–365). European Society for Engineering Education (SEFI). https://www.sefi.be/wp-content/uploads/2021/12/SEFI49th-Proceedings-final.pdf

Martin, D. A., Conlon, E., & Bowe, B. (2019). Engineering education, split between two cultures: An examination into patterns of implementation of ethics education in engineering programs in Ireland. In V. N. Balazs, H. M. Järvinen, M. Murphy, & A. Kálmán (Eds.), *47th SEFI Annual Conference 2019: Varietas delectat, complexity is the new normality* (pp. 1742–1752). European Society for Engineering Education (SEFI).

Murphy, M., O'Donnell, P., & Jameson, J. (2019). Business in engineering education: Issues, identities, hybrids, and limits. In S. H. Christensen, B. Delahousse, C., Didier, M. Meganck, & M. Murphy (Eds.), *The engineering-business nexus.* Philosophy of Engineering and Technology Series, 32. Springer.

National Academy of Engineering (NAE). (2018). *Understanding the educational and career pathways of engineers*. The National Academies Press. https://doi.org/10.17226/25284

NCEES. (2019). *Squared.* https://ncees.org/about/publications/past-annual-reports-squared/
NCEES. (2021). *Squared.* https://ncees.org/wp-content/uploads/Squared-2021_web.pdf
Nettleton, S. (1992). *Power, pain and dentistry.* Open University Press.
Nevada. (2020). Nevada Board of Professional Engineers and Land Surveyors Legislative Committee Continuing Education Survey -Q5 Any comments related to answer to Question 4. https://nvbpels.org/wp-content/uploads/2020/12/All-Respondents-Q4-Q6-comments.pdf
Nolen, L., Puckett, J., Dzombak, D., & Bergstrom, W. (2022). Preparing the future civil engineer: ASCE's proposed revision of the ABET civil engineering program criteria. In *American Society for Engineering Education (ASEE) Annual Conference & Exposition Proceedings* (p. 21). https://peer.asee.org/41592
OECD. (2021). *OECD skills outlook 2021: Learning for Life*. OECD Publishing. https://doi.org/10.1787/0ae365b4-en
Patil, A. S., & Gray, P. (Eds.). (2009). *Engineering education quality assurance: A global perspective.* Springer.
Ressler, S. J., & Lynch, D. R. (2011). The civil engineering body of knowledge and accreditation criteria: A plan for long-term management of change. *American Society for Engineering Education (ASEE) Annual Conference & Exposition Proceedings,* 18 pp. https://peer.asee.org/18392
Ressler, S. J., & Lenox, T. A. (2020). Is it time for ASCE to withdraw from ABET? *American Society for Engineering Education (ASEE) Annual Conference & Exposition Proceedings,* 18 pp. https://peer.asee.org/34886
Spinden, P. M. (2015). The enigma of engineer's industrial exemption to licensure: The exception that swallowed a profession. *UMKC Law Review, 83*(3), 637–686. http://digitalcommons.liberty.edu/lusol_fac_pubs/72
Stanford University. (2015, May 11). *New environmental systems engineering major allows more freedom in course selection.* https://cee.stanford.edu/news/new-environmental-systems-engineering-major-allows-more-freedom-course-selection
Swenty, M., & Swenty, B. J. (2017). Professional licensure: The core of the civil engineering body of knowledge. *American Society for Engineering Education (ASEE) Annual Conference & Exposition Proceedings.* Paper ID #19425. 14 pp. https://peer.asee.org/28762
Swenty, M. K., & Swenty, B. J. (2023). A student of EAC-ABET civil ngineering accreditation curriculum requirements and exemption provisions of state licensure laws and rules. *American Society for Engineering Education (ASEE) Annual Conference & Exposition Proceedings.* Paper ID #37663. 18 pp. https://peer.asee.org/42507
Texas. (2022). *Texas Board of professional engineers and land surveyors, professional engineer reference statement.* https://pels.texas.gov/downloads/ref.pdf
Walesh, S. (2022, February 9). Engineering licensing-exemptions put the public at unnecessary risk. *MachingDesign.* https://www.machinedesign.com/community/guest-commentary/article/21216453/engineering-licensingexemptions-put-the-public-at-unnecessary-risk
Zhu, Q., & Jesiek, B. K. (2020). Practicing engineering ethics in global context: A comparative study of expert and novice approaches to cross-cultural ethical situations. *Science and Engineering Ethics, 26*(4), 2097–2120. https://doi.org/10.1007/s11948-019-00154-8

35

A FEMINIST CRITICAL STANDPOINT ANALYSIS OF ENGINEERING ETHICS EDUCATION AND THE POWERS AT PLAY IN ACCREDITATION, RESEARCH, AND PRACTICE

Jillian Seniuk Cicek, Robyn Mae Paul,
Diana Adela Martin, and Donna Riley

Introduction

Ethics is essential to the engineering profession, embroiled within engineering regulations, codes, canons, and decision-making. As such, it's also part of engineering education, mandated and monitored in engineering programs accredited by national and international regulatory bodies (for more on this, see Chapters 19 and 22). However, ethics is complex, and even if engineering educators were to develop methods to teach to this complexity (and we argue that to date, they largely don't), it would still be difficult to assess under accreditation. Ethics requires practice; it cannot (just) be leveled and measured.

This practice must be muddy and messy, and prioritized over the achievement of reductionist individual learning outcomes (Woolston, 2008). Although we shouldn't teach engineering ethics for the sake of accreditation, it's impossible to separate our pedagogical decisions and approaches from accreditation regimes. The reduction of ethics to serve accreditation decouples it from its complexity and connections to broader concepts that are inextricably linked, including equity, diversity, inclusion, and social and environmental justice.

Here, we hesitate to define ethics. Wilson (2008) draws on Tafoya's (1995) Theory of Uncertainty to explain how you can't know both context and definition simultaneously; we're more concerned with understanding engineering ethics education in the context of accreditation and research. Therefore, to begin this journey, we offer Walker's *Moral Understandings: Alternative 'Epistemology' for a Feminist Ethics*. She writes about the necessity of ethics and moral legitimacy as a fight for moral justice to "end male domination, or perhaps to end domination generally" (Walker, 1989, p. 15). In our experience, this is not an approach commonly taken in engineering. In fact, in this chapter, we argue that current engineering ethics education and accreditation, if anything, promote and maintain demographic and social injustices.

DOI: 10.4324/9781003464259-42

Engineering ethics content and pedagogy are often rooted in objectivity and reductionism, (unconsciously) promoting Western, educated, industrialized, rich, democratic (WEIRD) ideals (Martin et al., 2023) that call to mind colonial practices encountered in engineering practice (Davies, 2021; see Chapter 9 for more on this topic). There is less attention paid to critical perspectives on what engineering ethics is and who decides; on how engineering ethics interacts with professional codes, societies, and licensure organizations; and on the power influences that exist within engineering ethics structures and approaches.

Thus, how engineering ethics education and accreditation are emphasized – or not – in the literature and the power dynamics influencing them need to be questioned. We examine how three parallel goals of engineering ethics education come together and into conflict with each other: teaching future engineers, obtaining and maintaining accreditation, and conducting pedagogical research. *How are we, as engineering educators, motivated to engage in engineering ethics teaching and learning? How are these efforts self-limiting within the epistemological and ontological frames that comprise engineering ethics education, accreditation, and practice? What roles do accreditation artifacts play in thought leadership, in enacting accountability, in perpetuating limiting epistemologies, and in driving or resisting change?*

This chapter puts forth a critical feminist standpoint analysis of engineering ethics education in the context of accreditation, research, and practice using a narrative methodology. We challenge engineering ethics education and accreditation as Western/Global North concepts exclusionary to other perspectives. We explore the epistemic power relations within engineering accreditation and its reductionism and assess-ability. We examine how boundaries are drawn. We conceptualize engineering ethics education accreditation as a rhetorical justification, performative discourse artifact, and feckless change strategy. We address the silences in the literature and then close by arguing that, inadvertently, we are puppeteers of accreditation, perpetuating inauthentic change and limiting transformative engineering ethics education. Throughout the chapter, we leverage scholars who have engaged in these critical discussions and consider how to turn these philosophical discussions into action.

Positionalities

Each author has engaged with engineering ethics education and accreditation processes in diverse contexts and systems. These experiences come from our roles as accreditors, researchers, faculty members, administrators, and board members through a range of engineering education positions affording us different freedoms and power and forms of constraint. We are (at the time of writing this chapter) an assistant professor, doctoral candidate, senior researcher, and engineering dean. We engage in critical and sociotechnical engineering education research, and combined, have lengthy careers as theorists in social justice, decolonialization, and critical theory, where we aim to challenge the status quo in engineering cultures and identities. We are all women of various European ethnicities, working in, studying, and promoting justice-oriented perspectives in engineering education through research, practice, and community service. We come from different backgrounds, including diverse socio-economic statuses, sexual orientations, nationalities, religious or spiritual beliefs, political structures, ages, educational journeys, (dis)abilities, and family structures. We have each struggled to undo habits of behaviors and thoughts, including implicitly engrained WEIRD traditions for three of us raised as settlers in North America, and for one of us born during communism in Romania, the afterlife of Marxist philosophy. These are not easy tasks. Due to language, awareness, and our choice to lean into our authority of experience in this chapter, our literature review comes from WEIRD-centric journals, conferences, and experiences that cannot and should not be generalized for all possible experiences. We acknowledge the limitations of our perspectives and experiences, rooted in our demographic standpoint. We offer this critical feminist standpoint analysis as a culmination of our collective experiences, understandings, acts of resistance, criticisms, and hopes for advancing more critical scholarship and research of WEIRD engineering ethics education in the context of accreditation, as it hardly exists.

Approach

For this work, we ask: *How is engineering ethics education represented in the literature and in our experiences? What stories are told, and what is silenced? What are the power dynamics at play, the implicit gaps in research, and the silences in the texts on accreditation and engineering ethics education?*

To explore these questions, we combine a review of the literature with critical feminist standpoint analysis, leaning on authority of experience and reading the silences in texts (Olsen, 1978; Scott, 1999). It is important to note that "standpoint theory has been criticized for its tendency to universalize white, western, middle class women" (Beddoes et al. 2011, p. 286). We thus use this analysis cautiously, deploying it with the intention of uncovering the silences and bringing it into conversation with critiques of Western normativity.

As a methodology, the authority of experience is closely aligned with feminist standpoint theory, which asserts that women's experiences are essential for knowledge-building, particularly within our WEIRD and patriarchal society, and are "legitimate sites of knowledge" (Beddoes & Borrego, 2011, p. 286). Feminist standpoint theory is connected to the sciences (Haraway, 1988), "maintain[ing] that scientific knowledge in a 'gender-stratified society' has marginalized women's experiences and has therefore produced knowledge biased by male interests and perspectives" (Beddoes & Borrego, 2011, p. 286). It acknowledges that the personal insights gained from women's experiences are "distinctive resources" (p. 286) unable to be perceived by dominant groups and thus essential to be considered alongside scientific understandings to produce knowledge with "theoretically richer explanations" (Harding, 2001, p. 145, as quoted in Beddoes & Borrego, 2011, p. 284). Feminist standpoint theory involves "studying … interpretation and intention" (p. 292), with its value found in how "the findings then are used to challenge existing power relations and guide future research" (p. 292).

We intentionally embody standpoint theory in this work, and explicitly name and claim it as feminist work (Riley, 2013). We identify as women, a group outside the dominant demographic of engineering; we hold that engineering and its cultures and driving forces, such as accreditation, exist in this patriarchal world, rooted in and dominated by patriarchal "ideologies, values, and institutions" (Beddoes & Borrego, 2011, p. 285); and we consciously choose to "listen to women's voices" (p. 292) – our voices – and be guided by our own experiences of accreditation. In this way, we are conscious "participants" in this work, aiming to produce "new and less coerced information" by harnessing our viewpoints (McLoughlin, 2005, p. 374). We employ a narrative style to augment our findings with our personal experiences and anecdotes rather than tracing causality by inferring from empirical data. We shared stories as part of our research approach, exchanging anecdotes about our experiences with and knowledge of accreditation in our contexts to kindle this work and forge our pathways into the literature.

Story-telling as a research methodology has been advocated for by Indigenous (Wilson, 2008) and critical race scholars as a valid research method and also a necessary method (Datta, 2018). Through story-telling, researchers are given agency and power, leading to more "connected, collaborative, and comprehensive" research (p. 42). Story-telling as a methodology deconstructs our positions as researchers and provides us with a space to apply critical self-consciousness and self-realization through our interactions and relationships with the research.

In this way, story-telling, as part of our approach, supports our feminist standpoint position. "For a position to count as a standpoint," Harding (2001, p. 147, quoted in Beddoes & Borrego, 2011, p. 292) has written, "we must insist on an objective location – women's lives – as the place from which feminist research should begin." Through this, we're able to model and exemplify the importance of adjusting research methodologies to unlearn, disrupt, and resist normative tendencies. Therefore, unlike systematic literature reviews, our review was intentionally not comprehensive. In this chapter, our discussion is influenced by the literature within the spheres of our experiences, and the gaps rather than the continuities woven into our stories.

As such, we critically reviewed and reflected on journal articles, textbooks, conference proceedings, and reports from these sources:

- Journals (*Australasian Journal of Engineering Education*; *Engineering Studies*; *European Journal of Engineering Education*; *IEEE Transactions on Education*; *International Journal of Engineering Pedagogy*; *Journal of Women and Minorities in Science and Engineering*; *Journal of Engineering Education Transformations; Science and Engineering Ethics; and International Journal of Engineering, Social Justice and Peace*)
- Textbooks (*Springer POET, Philosophy of Engineering and Technology* book series)
- Conference proceedings (*American Society for Engineering Education*; *Canadian Engineering Education Association*; *European Society for Engineering Education*; *Frontiers in Education*; *Global Engineering Education Conference*; *World Engineering Education Forum*)
- Reports we've encountered (often confidential, so we include only our observations and reflections)

Western normativity

Engineering ethics education and accreditation are dominated by a Western/Global North system that perpetuates a neoliberal and colonial worldview. The introduction of ethics as an outcomes-based accreditation criterion is often credited for the increase in engineering ethics classes (Martin et al., 2021) and "potentially elevat[ing] the prominence of instruction in engineering ethics and the societal context of engineering" (Herkert, 2000, p. 303). However, these assumptions place too much power in accreditation while ignoring a rich history of engineering ethics education that was multifaceted and interdisciplinary.

Engineering ethics education was required well before accreditation and outcomes-based assessment (Wacker, 1990). For example, focusing on the United States: As early as 1968, Sterling Olmsted developed *Liberal Learning for the Engineer*, which prompted experimentation with teaching interventions addressing technology's societal implications (Wisnioski, 2012, p. 165). In 1970, the Punderson Conference brought together 35 engineering, humanities, and social-sciences instructors to discuss greater coherence between the technical and liberal components of the engineering curricula and increase their societal relevance (Gravander, 2004). In 1977, a report commissioned by the Hastings Center mapped the status and prospects of engineering ethics and addressed for the first time the aims and content of engineering ethics education, instructor qualifications, and available teaching materials (Baum, 1980; Mitcham & Englehardt, 2019). The late 1970s and 1980s saw the publication of the first textbooks on engineering ethics (Mitcham, 2009; Weil, 1984).

Nevertheless, the literature and public discourse (mostly from the Global North) often credit outcomes-based accreditation processes as a prompt for introducing and developing ethics in engineering programs. These changes did increase research on designing curricula to measure engineering competencies, an increase seen between 2000 and 2010 across the United States, Canada (Brennan 2018), and Europe in parallel with the timing of outcomes-based processes.

The expansive adoption of the Washington Accord, which has strong status and power in the international community, led to the alignment of accreditation systems in countries worldwide (Patil & Codner, 2007). The Washington Accord, signed in 1989, has six original members, representing engineering education systems from WEIRD and Anglo-Saxon countries. Since its inception over 30 years ago, it's grown in scope and power (Klassen, 2018). Currently, it includes 23 countries with full rights and eight provisional signatories, an expansion explained as the outcome of globalization (Sthapak, 2012). The current provisional signatories (i.e., those on track to receive approval in the future) are all from countries in the Global South. This process of assimilating into Western norms is fraught with power issues, where we are implicitly using the mainstream definition of ethics and seeing ourselves as 'ethically' colonizing the Global South.

The influence of the Washington Accord has spread to signatory and non-signatory countries alike, where accreditation requirements, although not completely overlapping (Patil & Gray, 2009), nevertheless have a similar focus (Hanrahan, 2008; Paul et al., 2015). Engineering accreditation competencies often aim to broaden engineers' scope, and "inform them about their ethical, social, and professional responsibilities" (Sethy, 2017, p. 987). However, by focusing on competencies, accreditation processes overly endorse outcomes-based education, which leads to reductionist approaches to engineering ethics education. By proximity, engineering ethics education becomes steeped in serving neoliberal interests, where competencies and their assessment prioritize graduating engineers who can contribute to the economy rather than engineers who could serve society's needs (Handford et al., 2019; Leyva, 2009; Riley, 2012). In this sense, *globalization* in the engineering ethics education landscape is more a domination exercise by Western domains than an attempt to create a global engineering ethics education encompassing a broad range of cultural perspectives and ethical frameworks (Anwar & Richards, 2013; Gray et al., 2009; Haug, 2003).

This homogenizing of engineering ethics education creates a singular conceptualization that can be harmful when mismatched with local populations. With singular conceptualizations of accreditation requirements 'spreading' worldwide, we see a diffusion of the considerations as to what engineering ethics is and how ethical decision-making is taught. Diverse views sit apart from this: Confucian philosophy (Zhu, 2020), anti-colonial, feminist, African values, and others are not represented in canonical engineering ethics. Typically, the major ethical theories in engineering ethics education are consequentialism or deontology frameworks, which developed in the cultural space of Europe and Ancient Greece, "a specific product of the Western philosophical tradition" (Luegenbiehl, 2009, p. 149). This domination and the perpetuation of uniform ideals can create a dissonance for 'non-WEIRD' student populations (Clancy et al., 2022), which is a significant portion of engineering students worldwide given that WEIRD demographics are "the least representative populations one could find for generalizing about humans" (Henrich et al., 2010, p. 61).

Literature acknowledging the power dynamics in the development of professional codes of ethics and accreditation metrics is nascent (Klassen, 2023; Seron & Silbey, 2009). Ethics literature often focuses on the pedagogies used to teach ethics (Hess & Fore, 2017); however, these pedagogies rarely take a critical perspective on what engineering ethics is, how engineering ethics interacts with professional codes, and how power influences engineering ethics structures and approaches (Martin et al., 2021). The critical literature that exists is cited heavily throughout this chapter; however, it is typically disconnected from having any power in the process. Thus, the educational ideal set in the UNESCO World Declaration on Higher Education for the 21st Century to "understand, interpret, preserve, enhance, promote and disseminate national and regional, international and historic cultures, in a context of cultural pluralism and diversity" (UNESCO, 1998, p. 3) is lost in the articulation of accreditation ethics requirements.

We argue, based on our personal beliefs and first-hand experiences, that ABET and its efforts at globalization is a colonial neoliberal project intentionally planned 'to take over the world'. In 2009, George Peterson, the executive director of ABET from 1993 to 2009, who led the organization through EC 2000, wrote about the importance of ABET's venture into non-domestic accreditation, emphasizing that it would produce engineering graduates qualified to work "in any country on earth" (Peterson, 2009, p. 82). ABET "oozes zealotry, bewildering vocabulary, unexamined tenets, reliance on imperatives rather than indicatives, irrefutable claims, and support from administrators and politicians, not practitioners" (Woolston, 2008, p. 4). Accreditation seems not to be about education and learning but about power, with the goal of maintaining and expanding a particular cultural role and status in engineering (Slaton, 2012). Accreditation is big (financial) business that supports "corporate instrumentalism" and "corporatism in engineering," reducing the "professional independence of engineers" and the importance of public interest while increasing capitalist market expansion (Handford et al., 2019, p. 171). Yet, non-Western locales, even if conscious of this domination, still adopt these framings, as the power of being ABET-accredited outweighs fighting Western domination (Balakrishnan et al., 2021). To change the system, a double anti-colonial push is recommended: first, the formulation of requirements must be opened to include diverse non-Western values and theoretical perspectives and authentic reckoning of the harms resulting from WEIRD ethics shaping engineering practice; and second, research must be consolidated to form more inclusive and broader engineering ethics education and accreditation processes that confront, resist, and avoid defaulting to WEIRD samples, research methodologies, and theoretical lenses.

Epistemic power relations

In addition to WEIRD dominance, the biases and devaluations of engineering ethics education are deeply embedded in the cultural epistemological power hierarchy of knowledge. This has multiple threads, including (1) *othering*: the perpetuation of the superiority of technical knowledge and skills over the 'complementary' engineering competencies; (2) *false authority*: those with objective, technical knowledge have independent authority to make decisions, including ethical decisions; (3) *unquestionable rigor*: assuming that the positivistic, structured approaches of ethics accreditation are thorough and indisputable; and (4) *engineering elitism*: undervaluing and/or discounting non-expert and non-engineering perspectives during accreditation.

Firstly, engineering ethics is often seen as *other* than the 'core' knowledge required for engineering students (Cech, 2014; Monteiro et al., 2017; Stevens et al., 2007). Although recognized within most accreditation processes, it is often lumped with *professionalism* and *social context* and described using language such as 'soft,' 'complementary,' or 'ancillary' (Parker et al., 2019). It has historically been described as a "complement to the technical content" (Wacker, 1990, p. 97), positioning ethics as a less important, non-essential topic in engineering education. Whether through direct comments or indirect actions, we've observed accreditation visitors perpetuate this othering hierarchy in their attention and efforts to evaluate the 'more essential' (and thus superior) technical knowledge and skills and gloss over the 'other' 'professional skills.'

To counteract these discourses, engineering ethics education is often packaged to be more like technical topics, such as by emphasizing quantitative methodologies like decision trees and risk factors (Harris et al., 1997). This also devalues it. As Newberry (2004) writes, "we should resist the tendency to engineer-ize ethics" by viewing it as another rational-scientific problem to be solved. Commandeering ethics as a calculative skill perpetuates the bias that "soft skills are 'easy', and perhaps don't require formal education, while hard skills are 'difficult' and must be continually reinforced in multiple classes" (Bauschpies et al., 2018, p. 2). This reflects "the profession's tendency to marginalize, ignore, silence, and/or atrophy the … central elements of ethical engineering practice" (Riley & Lambrinidou, 2015, p. 2), which include the non-technical dimensions of engineering, local knowledge, agency of all persons, and the public as the profession's primary client.

Often engineering ethics is conflated with the social, economic, environmental, and global impacts of engineering solutions (e.g., within ABET student outcomes 1–7). It's not to argue these macro topics aren't also critical concepts to ethical engineering, but rather that these blended and ancillary approaches demote essential competencies into "diluted … everything-but-the-kitchen-sink outcome[s]" (Riley, 2016) and 'othered' topics, which devalue and limit the attention on the teaching and assessing of ethics in engineering education. We don't skip calculus courses and blend calculus learning outcomes into thermodynamics courses; all these topics (ethics, social, economic, environmental, etc.) are important to teach independently, as well as in integration.

The second epistemological tension is the perpetuation of a *false sense of authority* and how this relates to engineering ethics education. There is a dichotomy: although ethics is othered and devalued in engineering education, in a parallel discourse it is frequently emphasized as essential to engineering identity. Ethics is synonymous with engineering as a profession and has been acknowledged as core to professional engineering conduct for over 50 years (Rottmann & Reeve, 2020). Students acknowledge this importance in how they define engineering (Doré et al., 2021); however, it is combined with the belief that positivist approaches are superior, and engineering ethics education tends to "favor sober deliberation and reason over passion and sentiment" (Fernandez, 2021, p. 3). A belief in objectivity and the ability to rationally solve problems, including ethical problems, gives students a false sense of power and authority over engineering decisions. Gary Downey (2012) argued how this bias for 'normative holism,' embedded in engineering's rhetoric of improving the welfare of humanity, creates a false logic in which engineers assume that when they engineer, they must be doing 'good.' This implies that an engineer doesn't need ethics education and can consider themselves ethical as long as they graduate from an accredited program, behave professionally, and abide by the law (Bauschpies et al., 2018).

Building on this is the "flaw of the awe" (Bauschpies et al., 2018, p. 1), a culture that positions engineers as purely objective, the ultimate authority, and even as the public 'savior.' Engineers are perceived as 'heroic', which perpetuates a greater sense of obligation to the public. Although 'awe' of nature (waterfalls, sunsets, etc.) is typically humbling, 'awe' of engineers is very individualistic and leads to awe of oneself, and to the assumption that engineers have all the power and greatness needed to make (ethical) decisions (Fernandez, 2021). This sense of superiority and feeling of being "technological guardians of the public good" furthers engineers' belief in their elite (though arguably unethical) ability to speak on behalf of communities without engagement, "often delegitimizing or discounting local knowledge, agency, and voice" (Bauschpies et al., 2018, p. 1).

Third, given that engineering education tends to perpetuate and attract systematic and analytical approaches to knowledge building, there is an underlying *unquestionable rigor* to said processes, and it becomes difficult to challenge approaches such as outcomes-based assessment. These analytical, systematic, and structured processes inherently hold power over engineers, engineering educators, and engineering accreditors because they *appear* to be robust (Woolston, 2008). Heywood (2016) confirms: "It seems that if an objective (outcome) is stated in terms of what a person is able to do that there can be no question about its validity" (p. 3). Not surprisingly, this appearance of robustness creates inconsistent and inauthentic approaches to teaching engineering ethics. This competency lives within a knowledge paradigm that actually cannot be systematically and objectively analyzed. For example, we observed accreditors being unconcerned by low-scoring and failing ethics accreditation evaluations, remarking, 'That's normal.' The belief in the positivistic, systematic approach is so strong that it is maintained even when there is methodological evidence it is not working. These notions are cradled within the false notions of engineering as objective (Lord et al., 2019), and in the pursuit of "rigor" (Riley, 2017) and validity.

The final thread of epistemic power we address is *engineering elitism* within engineering ethics accreditation. This is the idea that only experts can teach engineering ethics – similar to the accreditation argument that only engineers can teach design (Hladik et al., 2023). What constitutes an expert is curious and contradictory, however. On the one hand, we have the inexplicable continuing assumption that engineering instructors with accredited engineering degrees are unprepared and unqualified to teach engineering ethics. When engineers abdicate their responsibility to teach ethics and send students across campus to philosophers, this emphasizes the devalued 'other' status of ethics. On the other hand, within the elitism of engineering, having engineers teach engineering ethics is believed to lend "credibility to the course in the students' perception" (Wacker, 1990, p. 4). In fact, to improve ethics education, we've observed accreditors recommend strengthening ties with industry and engaging more professional engineers as guest speakers. Although professional experience and practice can be essential in making decisions (Klein, 1998; Walther et al., 2007), this assumption is problematic. Experience helps to recognize patterns in decision-making, but given the underlying *unquestionable rigor*, these 'patterns' in a rational engineering culture are often rooted in assumptions of superiority and correctness, and not necessarily in ethics (Perlman & Varma, 2002).

Each of these epistemic tensions (techno-superiority, false authority, unquestionable rigour, and engineering elitism) creates a null curriculum for students: They get the message that ethics isn't really part of the central body of engineering knowledge, yet they still have the authority and power to make ethical decisions. This cycle of power is further perpetuated by the reductionist approach we use to teach and assess engineering ethics.

Reductionism and assess-ability

Accreditation drives outcomes-based assessment, where the goal becomes teaching engineering competencies ('graduate attributes' or 'learning outcomes') so they can be assessed (Shuman et al., 2005). As such, 'ethics' is condensed into content (rather than behavior) that can be (easily) taught and measured.

In engineering education, teaching and assessing ethics often mimic other technical skills and courses, using reductionist, calculative, and objective approaches. This takes a Newtonian, determinist, and mechanist view of the universe, a belief that the human and more-than-human world can be broken into its mechanical parts and described by mathematical equations (Bauschpies et al., 2018). Rottmann and Reeve (2020) dub this reduction as the "rules and codes approach," which "remains a baseline feature in engineering ethics education" (p. 148). Although more accessible (and assessable) for students, this approach "may unintentionally omit ethical principles that have not been codified, implicitly treating ethical codes as uncontested statements of moral good rather than historically contextualized settlements negotiated by professional regulators, their constituents and the public" (p. 148–149).

This reductionist approach overemphasizes teaching ethics through case studies (Martin et al., 2020; Polmear et al., 2019) that focus on the micro ethics of engineering failures, often through analysis of mistakes in calculations or simple processes (Perlman & Varma, 2002). Inclusions of engineering ethics in the curriculum often cite engineering accidents or disasters, such as the Turkish Airlines cargo door failure, the chemical release at Bhopal, the Ford Pinto issue, or the Challenger shuttle explosion (Didier, 2000). We teach ethics – or 'failure' – hoping engineers will *act* more responsibly in the future. This reductionism is also maintained by calculative approaches to ethics, where ethical problems use measurement tools, "such as 'line drawing' to weigh options, creating flow charts, and using cost-benefit analyses," suggesting that the "right" decision can be found by applying "a set of heuristics" (Bauschpies et al., 2018, p. 3). Ethics is reduced to concrete concepts such as "harm," "safety," "disclosure," "honesty," or "fairness" (p. 42). These are important elements of ethics; however emphasis on these micro-ethics implies they are the primary and most important ethical considerations within engineering.

One justification for reductionism is that one cannot measure ethical *action*, only student *understanding* of ethical proscriptions or principles (Davis, 1991). As such, educators often do not hold students accountable to a high enough standard of responsible action and behavior, and rest rather on teaching and assessing rote 'understanding.' Engineering ethics is often assessed through multiple choice exams that imply ethical action is fulfilling a duty to do *one* right action, when ethical decisions often have significant complexity (Swan et al., 2019).

Heywood (2016) argued that reductionist assessment tools influence what students learn about ethics (content) and how they learn to think about ethics (strategies). Woolston (2008) argued similarly, specifically incriminating *outcomes-based assessment*, which "simply shifts the focus to assessment, specifically to the results of whatever educational process is at hand, not how the learning itself takes place" (p. 1). As such, students are trained to focus on the 'answer' being sought rather than on the processes of critical thinking and behaviors required to engage in ethical considerations for authentic real-world ethical practice (Perlman & Varma, 2002). These tactics arguably perpetuate "the disinterest shown by students for the lack of their emotional engagement with ethics issues" (Barros-Castro et al., 2022, p. 2) and result in engineering ethics education (and accreditation) that is disconnected, "unsystematic," and unimpactful (p. 1).

In our experience, many accreditation visitors don't hold programs to more than a standard of minimum student understanding, either. Ironically, most ABET assessors don't have assessment backgrounds (Akera et al., 2021). Accreditation processes are fraught with anecdotes, side comments, and direct declarations from accreditors not clearly knowing how to assess ethics as an outcome. This irony extends as the accreditors who do the assessments perpetuate the belief that professional 'soft' skills (such as ethics) can't be assessed.

While many engineering education researchers conclude that we need more interactive pedagogies that position students as agents (Whitbeck, 1998), there is a deeper consideration. Engineers' behavior has a broader set of influences and power dynamics within disasters, ones that education can affect in limited ways, at best, and that accreditation influences even more indirectly (Vaughan, 2004). The reduction of ethics to serve accreditation decouples it from its complexity and removes critical connections to broader concepts of equity, and social and environmental justice. Further, outcomes-based assessment eliminates variation in student learning. Thus, rather than supporting and increasing student diversity, a 'wicked' problem engineering has been wrestling to solve for decades, it rather 'de-diversifies' students, drawing boundaries and reducing them to a 'variable' to be controlled for (Woolston, 2008).

Drawing boundaries

Mitcham and Englehardt (2019) argue that the boundaries between the internal and external STEM communities perpetuate the boundaries between ethics as code (inside the community) and ethics as social justice (outside the community). We argue this is enacted in engineering ethics education and propagated by accreditation. This insider–outsider boundary also fosters secrecy, which Perlman and Varma (2002) contend is counter to engineering ethics, "which requires transparency" (p. 47).

In Canada, *The Ritual of the Calling of an Engineer*, a 100-year-old secretive engineering graduate ceremony, is an enactment of this power boundary. A recent analysis demonstrated this secretive ritual upholds and maintains boundaries of engineering ethics that are reductionist and positivistic (Paul et al., 2023). The ritual and ceremony, written by known imperialist Rudyard Kipling (see www.retoolthering.ca for more), is an engineering tradition and example of safeguarding the problematic boundaries of ethics discussed in this chapter.

As discussed, engineering codes are frequently used in teaching and assessing ethics to legitimize engineering ethics education as being within the *boundaries* of engineering. Colby and Sullivan (2008) describe how engineering codes are a "valuable framework for thinking about the goals of educating for engineering ethics" (p. 327). However, they clarify that educators must look below the surface to consider the skills, behaviors, and mindsets required to achieve the codes and avoid losing the broader implications of engineering. Codes provide an accessible approach to ethics education, but engineering educators may omit ethical principles *outside* the codes, treat the codes as indisputable statements of morality, and deny the historical framing and boundaries embedded *within* the codes (Mitcham & Englehardt, 2019; Rottmann & Reeve, 2020; Tang & Nieusma, 2017; Vesilind, 1995).

Riley, Slaton, and Herkert (2015) further this insider–outsider boundary by drawing on Slaton's (2001) earlier work on reinforced concrete standards that served to shift expertise from artisans (who aren't subject to the standards) to engineers. They argue that engineering ethics codes create a class of workers capable of "proper technical conduct" (p. 4), set apart from non-engineers who are not subject to the code. Superficially, ethics codes reflect a common set of values for the profession or lay out essential capacities for engineers to develop in their professional formation. Yet in practice, as argued, ethics is relegated to the fringes of most curricula and given little emphasis by accreditors. Codes are inherently political: created within past histories, attempting to predict future transgressions, and interpreted through the present (Mitcham & Englehardt, 2019). By maintaining a vagueness in writing codes, power interests are met while holding fast the boundaries and controlling who has access. Building on Pfatteicher's work (2005), Slaton (2012) explains: "codes of ethics that historically have urged engineers to practice only within the limits of their own competence have rarely defined those limits clearly," which makes engineering standards and codes "virtually impossible for non-experts to apply" (p. 100). These examples emphasize how codes produce professional identity and continue to maintain (*police?*) the profession's boundaries.

Engineering ethics education via accreditation also maintains a boundary between the technical and social (Friedensen et al., 2020; Martin & Polmear, 2023), the danger of which has been discussed for decades (Didier, 2000). McAuliffe (2006) argues that our education system teaches "the technician worldview, with its ideological tunnel vision and disinterest in stepping outside of professional standards," making professionals "less attuned to the situational contextual dynamics" (p. 493). The boundary between technical and social reflects Cech's (2014) work on the social disengagement culture that "seems to be inherent to engineering education" (Mitcham & Englehardt, 2019, p. 1756). The irony is that this is counter-intuitive to engineering, which "is the interaction of people and 'things,' or people and any engineered technology, that brings ethics to the fore" (Ballentine, 2008, p. 332).

The authority wielded through professional codes upholds a power boundary, giving a profession elite status. However, any attempts at changes to codes (e.g., sustainability, gender and sexual diversity, equity, inclusion, decolonization, the iron ring) cause significant stirs (Paul et al., 2023;

Riley et al., 2015) because change requires shifting a stiff boundary upheld by people with significant power. Through codes and standards, these boundaries are passed down and perpetuated within engineering ethics education, accreditation, and literature.

Rhetorical justification, performative artifact, and change strategy

Ethics, *accreditation*, and *ethics accreditation* as a rhetorical justification, performative discourse artifact, or change strategy were also prevalent across the literature.

Although our review method intentionally did not seek to describe quantitative findings, in a significant portion of articles, the terms 'engineering ethics education' and 'accreditation' were mentioned only briefly. These mentions provided context and justification and were found mainly in the introductions and/or conclusions, at times in the abstracts or background sections, and generally nowhere in the body of the literature. This was especially common in conference papers, where outcomes-based accreditation requirements were used as a rhetorical justification to validate the 'why' of pedagogical innovations.

Others positioned the addition of outcomes-based accreditation as a change strategy for engineering ethics, and ethics education as a 'response' to the call for change to engineering education (e.g., Rottmann & Reeve, 2020). Typically, this worked from an implicit or explicit assumption that accreditation could drive change in engineering education; at times, this too was merely performative, and at other times more substantive and sincere, if ignorant of accreditation's power dynamics and self-limiting nature.

Although rhetorical justification, performative discourse, and change strategy are three different narratives, we argue that the power dynamics within each are interconnected. There is an underlying harmful assumption that accreditation is beneficial to engineering ethics education and, more broadly, to engineering education. Accreditation further drives the rhetorical justification of outcomes-based assessment as a change strategy as it gives power to the data as 'evidence' that 'ethics' is included. However, the inclusion is rarely authentic and is perhaps actively encouraged to be inauthentic: "The implementation of this component was typically characterized by a 'laissez-faire' approach … strongly suggest[ing] that the engineering academic community was not motivated to achieve the spirit of the accreditation objectives" (Wacker, 1990, p. 98). There is a lack of clarity on what ethics is (rightly so, given its muddiness) and the purpose of engineering ethics education; accreditation tries to answer these by wordsmithing the perfect outcome. But there are invalid assumptions: (1) that measurement and assessment will lead to improved education; (2) that improved education will lead to improved learning; and (3) that improved learning will lead to improved ethical actions (Biesta, 2007). One could argue, as Woolston (2008) does, that accreditation and outcomes-based assessment as a pedagogical change strategy is feckless; it really only changed accreditation and outcomes-based assessment, rather than actually improving students' learning.

Silences

The power we've exposed in this chapter is also wielded in the silences. Engineering ethics, as developed by instructors and policed by accreditors, advances segregation. These are the 'holes' in disciplinary knowledge created by the exclusionary boundaries (Pawley, 2012). The WEIRD biases of engineering ethics mean we're missing feminist ethics frames (Walker, 1989) around care (Pantazidou & Nair, 1999) and justice (Jaggar, 1995), and those that call out the military-industrial-prison complex (Nieusma, 2013; Philip et al., 2018) (see Chapter 15). We are biased toward thinking about ethics rather than feeling it (Hess & Fore, 2017) or embodying it (Bombaerts et al., 2023). Winner (1990) shows how our reduction of ethics instruction to case studies and our unwillingness to challenge engineering's militarism create some WEIRD (and just plain weird) cases. The outcomes-framing of accreditation ignores and silences the active pedagogies and learning processes that can produce ethical practice when instructors and learners work as partners. As Bauschpies et al. (2018) offer:

> a determination that justice has not been served is easier to make than a determination that justice has been served. This approach would also avoid repeating the colonialist mistake. Rather than coming in as the ethics expert who knows all about justice and simply expects students to conform, the instructor engages students in listening for signals of injustice – a skill that, along with humility, should stand them in good stead when it comes to community engagement.
>
> *(p. 4)*

For over 20 years, there have been calls to consider not just professional values, but also public values in engineering ethics and social, environmental, and energy justice (Chance et al., 2021; Riley, 2023), with a need "for an extension of traditional ethical frameworks to incorporate treatment of questions of social responsibility, including the issue of sustainability" (Johnston et al., 2000, p. 315). In recent years, there have been attempts to build DEI (diversity, equity, and inclusion) considerations into engineering ethics (e.g. Chance et al., 2021; Hess & Fore, 2017; Rottmann & Reeve, 2020), founded in a 'do no harm' approach (Harris et al., 1997). The current focus is trying to shift to understanding ethics as 'global responsibility' – for example, Pawley (2019) noted the exclusion of the climate crisis and engineers' responsibility to urgently respond – and projecting a "holistic sense of ethics, sustainability, and obligation" (Chance et al., 2022, p. 164).

Nevertheless, although we are seeing engineering ethics being integrated with other important topics in the literature such as global competency, leadership, policy, stewardship, and sustainability, the "impact" or macro-framing of ethics and the social antecedents of technology remain underrepresented. It has been decades since Herkert (2000) pointed out the bias toward micro-ethics problems and away from macro-problems, focusing on individual decision-making and excluding the collective decisions of organizations, institutions, and governments. Activist work (e.g. *Science for the People*, *Engineering Social Justice and Peace*, and many other local and national community groups) that addresses engineering in society or consumer rights is excluded. Optional extra-curricular service-learning is typically ignored for accreditation, which is geared toward measuring the outcome of *every* student in *every* course *within* the boundaries of accredited engineering programs.

Engineering lacks a self-reflective mechanism to see the system in which we work (Foucault, 1990). We do not understand that accreditation regimes are ultimately about the regimes themselves, and how power and knowledge are enacted within. Accreditors remind us that engineering educators are free to go above the minimally low bar they set for ethics, never recognizing the rhetorical constraint and the shaping of the standard that occurs in those exchanges. If something is 'above' or 'beyond,' it is not *in* – it is extra (Chance et al., 2021). It is Other.

Closing: we are puppeteers

Assessment is the goal for accreditation, reducing engineering ethics to objective, measurable packages to teach and assess. This reductionism is influenced by a Western dominant culture, and rational approaches common in teaching engineering technical competencies. It results in an over-simplification of ethics, a perpetuation of linear rather than critical thinking, minimal rote understanding rather than critically reflective praxis, and a disregard of ethics' integral links to diversity, equity, inclusion, and social, environmental, and energy justice. Our collective inability to recognize the power dynamics of accreditation regimes results in us giving power over to accreditation, effectively abolishing much of what is valuable in engineering ethics education. Accreditation fosters feckless, inauthentic, and performative change, not transformation of our classrooms, our students, and ourselves. Our participation in this system makes us puppeteers of accreditation. We are caught in 6-year cycles (at best) of incremental improvements, buried in and busy with measures that themselves comprise the establishment. As such, we argue that accreditation does not support authentic, significant, and inclusive engineering ethics education and, ultimately, leads to a 'de-diversification' of engineering students.

We make four critical arguments in this chapter:

1) Accreditation is Western/Global North-centered, and when non-Western countries (and/or countries in the Global South) join initiatives like the Washington Accord, they must adopt Western/Northern standards, and local sensitivities vanish (Western normativity).
2) Technical epistemology outbalances and marginalizes other disciplinary perspectives (epistemic power relations).
3) The emphasis on micro-ethics and outcome-based assessment in ethics teaching decouples engineering ethics education from moral action and broader concepts of equity and social justice (reductionism and assess-ability and drawing boundaries).
4) The accreditation process produces "willful ignorance" (Tuana, 2006, p. 10) of its own undesirable effects (rhetorical justification, performative artifacts, and change strategy).

These four conditions work together to homogenize engineering ethics education, rooted in and puppeteered by patriarchal ideologies, values, and institutions.

Biesta (2007) offers a way out: to stop focusing on 'what works' or 'what is required' (e.g., for accreditation) and focus instead on what students ought to learn. When we bring in different ways of knowing, when we take critical perspectives, we can begin to consider why, and challenge or resist the power dynamics maintaining the status quo. We wonder: *What would engineering ethics education look like if truly accountable, if it was rooted in a diversity of experiences, realities, intersectionalities (e.g., race, class, ethnicity, sexual orientation, culture, and nationality), and ways of knowing* (Beddoes & Borrego, 2011; Riley & Lambrinidou, 2015)? Accreditation might come from the various publics we serve, rather than a self-policing that reflects existing patriarchal hierarchies of knowledge. *What would accreditation look like if it were locally situated, with accountability to the land and waters and their stewards?* Moving away from outcomes-based assessments toward those that are relational, process-driven, reflective, participatory, and liberatory may be far more appropriate for engineering ethics education.

Relying on accreditation systems to ensure ethical engineers is self-limiting at best. It distracts us from the profound work we have yet to do in reckoning with what trauma engineers have wrought on this planet and in our communities, and exploring imaginaries knit from below the surface, from the silences, and from the interstices of our knowledges.

Acknowledgments

This chapter was created through a collaborative wisdom process. All authors share first authorship, and we are very grateful for the opportunity to work and celebrate in this journey together.

Thank you to our editor and reviewers for their careful and insightful feedback, which added to the collective knowledge in this chapter. Finally, in story-telling traditions, we acknowledge the role and contributions of the reader, and thank them for engaging in this work.

References

Akera, A., Appelhans, S., Cheville, A., De Pree, T., Fatehiboroujeni, S., Karlin, J., & Riley, D. (2021). ABET's Maverick evaluators and the limits of accreditation as a mode of governance in engineering education. *2021 ASEE Virtual Annual Conference Content Access Proceedings*, 36632. https://doi.org/10.18260/1-2--36632

Anwar, A. A., & Richards, D. J. (2013). Is the USA set to dominate accreditation of engineering education and professional qualifications? *Proceedings of the Institution of Civil Engineers, 166*, 42–48.

Balakrishnan, B., Tochinai, F., Kanemitsu, H., & Altalbe, A. (2021). Engineering ethics education from the cultural and religious perspectives: A study among Malaysian undergraduates. *European Journal of Engineering Education, 46*(5), 707–717. https://doi.org/10.1080/03043797.2021.1881449

Ballentine, B. (2008). Professional communication and a 'whole new mind': Engaging with ethics, intellectual property, design, and globalization. *IEEE Transactions on Professional Communication, 51*(3), 328–340. https://doi.org/10.1109/TPC.2008.2001251

Barros-Castro, R. A., Suarez-Medina, G., & Barrero, L. H. (2022). Ethics and professional role to the society: A proposal to promote ethical attitudes in engineering students. *2022 IEEE IFEES World Engineering Education Forum - Global Engineering Deans Council (WEEF-GEDC)*, 1–5. https://doi.org/10.1109/WEEF-GEDC54384.2022.9996256

Baum, R. J. (1980). *Ethics and engineering curricula*. The Hastings Center, Institute of Society, Ethics, and the Life Sciences.

Bauschpies, W., Holbrook, J. B., Douglas, E. P., Lambrinidou, Y., & Lewis, E. Y. (2018). Reimagining ethics education for peace engineering. *2018 World Engineering Education Forum - Global Engineering Deans Council (WEEF-GEDC)*, 1–4. https://doi.org/10.1109/WEEF-GEDC.2018.8629655

Beddoes, K., & Borrego, M. (2011). Feminist Theory in three engineering education journals: 1995–2008. *Journal of Engineering Education, 100*(2), 281–303. https://doi.org/10.1002/j.2168-9830.2011.tb00014.x

Biesta, G. (2007). Why 'what works' won't work: Evidence-based practice and the democratic deficit in educational research. *Educational Theory, 57*(1), 1–22. https://doi.org/10.1111/j.1741-5446.2006.00241.x

Bombaerts, G., Anderson, J., Dennis, M., Gerola, A., Frank, L., Hannes, T., Hopster, J., Marin, L., & Spahn, A. (2023). Attention as practice: Buddhist ethics responses to persuasive technologies. *Global Philosophy, 33*(2), 25. https://doi.org/10.1007/s10516-023-09680-4

Brennan, R. (2018). A systematic review of Canadian engineering education research 2004–2017. *Proceedings of the Canadian Engineering Education Association (CEEA)*. https://doi.org/10.24908/pceea.v0i0.13079

Cech, E. A. (2014). Culture of disengagement in engineering education? *Science, Technology, & Human Values, 39*(1), 42–72. https://doi.org/10.1177/0162243913504305

Chance, S., Direito, I., & Mitchell, J. (2022). Opportunities and barriers faced by early-career civil engineers enacting global responsibility. *European Journal of Engineering Education, 47*(1), 164–192. https://doi.org/10.1080/03043797.2021.1990863

Chance, S., Lawlor, R., Direito, I., & Mitchell, J. (2021). Above and beyond: Ethics and responsibility in civil engineering. *Australasian Journal of Engineering Education, 26*(1), 93–116. https://doi.org/10.1080/22054952.2021.1942767

Clancy, R. F., Ge, Y., & An, L. (2022). Investigating factors related to ethical expectations and motivations among Chinese engineering students. *European Journal of Engineering Education, 47*(5), 762–773. https://doi.org/10.1080/03043797.2022.2066509

Colby, A., & Sullivan, W. M. (2008). Ethics teaching in undergraduate engineering education. *Journal of Engineering Education, 97*(3), 327 338. https://doi.org/10.1002/j.2168-9830.2008.tb00982.x

Datta, R. (2018). Traditional storytelling: An effective Indigenous research methodology and its implications for environmental research. *AlterNative: An International Journal of Indigenous Peoples, 14*(1), 35–44. https://doi.org/10.1177/1177180117741351

Davies, A. (2021). The coloniality of infrastructure: Engineering, landscape and modernity in Recife. *Environment and Planning D: Society and Space, 39*(4), 740–757. https://doi.org/10.1177/02637758211018706

Davis, M. (1991). Thinking like an engineer: The place of a code of ethics in the practice of a profession. *Philosophy & Public Affairs*, *20*(2), 150–167.
Didier, C. (2000). Engineering ethics at the Catholic University of Lille (France): Research and teaching in a European context. *European Journal of Engineering Education*, *25*(4), 325–335.
Doré, S., Seniuk Cicek, J., Jamieson, M. V., Terriault, P., Belleau, C., & Bezerra Rodrigues, R. (2021). What is an engineer: Study description and codeboook development. *Proceedings of the Canadian Engineering Education Association (CEEA)*. https://doi.org/10.24908/pceea.vi0.14850
Downey, G. L. (2012). The local engineer: Normative holism in engineering formation. In S. H. Christensen, C. Mitcham, B. Li, & Y. An (Eds.), *Engineering, development and philosophy* (Vol. 11, pp. 233–251). Springer Netherlands. https://doi.org/10.1007/978-94-007-5282-5_14
Fernandez, L. (2021). The role of the emotions in teaching engineering ethics. *2021 IEEE International Symposium on Ethics in Engineering, Science and Technology (ETHICS)*, 1–4. https://doi.org/10.1109/ETHICS53270.2021.9632673
Foucault, M. (1990). *The history of sexuality: An introduction*. Vintage.
Friedensen, R. E., Rodriguez, S., & Doran, E. (2020). The making of 'ideal' electrical and computer engineers: A departmental document analysis. *Engineering Studies*, *12*(2), 104–126. https://doi.org/10.1080/19378629.2020.1795182
Gravander, J. (2004). Toward the "integrated liberal arts:" Reconceptualizing the role of the liberal arts in engineering education. *Humanities and Technology Review*, *23*, 1–18.
Gray, P. J., Patil, A., & Codner, G. (2009). The background of quality assurance in higher education and engineering education. In A. Patil & P. Gray (Eds.), *Engineering education quality assurance* (pp. 3–25). Springer US. https://doi.org/10.1007/978-1-4419-0555-0_1
Handford, M., Van Maele, J., Matous, P., & Maemura, Y. (2019). Which "culture"? A critical analysis of intercultural communication in engineering education. *Journal of Engineering Education*, *108*(2), 161–177. https://doi.org/10.1002/jee.20254
Hanrahan, H. (2008). The Washington Accord: History, development, status and trajectory. *Proceedings of the 7th ASEE Global Colloquium on Engineering Education*, 19–23.
Haraway, D. (1988). Situated knowledges: The science question in feminism and the privilege of partial perspective. *Feminist Studies, 14*(3), 575–599.
Harding, S. (2001). Feminist standpoint epistemology. In M. Lederman & I. Bartsch (Eds.), *The gender and science reader* (pp. 145–154). Routledge.
Harris, C. E., Pritchard, M. S., & Rabins, M. J. (1997). *Practicing engineering ethics*. Institute of Electrical and Electronics Engineers.
Haug, G. (2003). Quality assurance/accreditation in the emerging European higher education area: A possible scenario for the future. *European Journal of Education*, *38*(3), 229–241. https://doi.org/10.1111/1467-3435.00143
Henrich, J., Heine, S. J., & Norenzayan, A. (2010). The weirdest people in the world? *Behavioral and Brain Sciences*, *33*(2–3), 61–83. https://doi.org/10.1017/S0140525X0999152X
Herkert, J. R. (2000). Engineering ethics education in the USA: Content, pedagogy and curriculum. *European Journal of Engineering Education*, *25*(4), 303–313. https://doi.org/10.1080/03043790050200340
Hess, J. L., & Fore, G. (2017). A systematic literature review of US engineering ethics interventions. *Science and Engineering Ethics*. https://doi.org/10.1007/s11948-017-9910-6
Heywood, J. (2016). More by luck than good judgement: Moral purpose in engineering education policy making for change. *2016 IEEE Frontiers in Education Conference (FIE)*, 1–6. https://doi.org/10.1109/FIE.2016.7757552
Hladik, S., Zacharias, K., & Seniuk Cicek, J. (2023). *Experiences of engineering education research faculty members in Canada: A collaborative autoethnography*. Canadian Engineering Education Association Annual Conference.
Jaggar, A. M. (1995). Caring as a feminist practice of moral reason [1995]. In V. Held (Ed.), *Justice and care* (1st ed., pp. 179–202). Routledge. https://doi.org/10.4324/9780429499463-16
Johnston, S., McGregor, H., & Taylor, E. (2000). Practice-focused ethics in Australian engineering education. *European Journal of Engineering Education*, *25*(4), 315–324. https://doi.org/10.1080/03043790050200359
Klassen, M. (2018). *The politics of accreditation: A comparison of the engineering profession in five anglosphere countries*. University of Toronto.
Klassen, M. (2023). *Curriculum governance in the professions: A comparative and sociological analysis of engineering accreditation* [Doctoral]. University of Toronto.

Klein, G. (1998). *Sources of power: How people make decisions*. MIT Press.
Leyva, R. (2009). No child left behind: A neoliberal repackaging of social darwinism. *Journal for Critical Education Policy Studies*. http://www.jceps.com/archives/606
Lord, S. M., Mejia, J. A., Hoople, G. D., & Chen, D. A. (2019). *Special session: Starting a dialogue on decolonization in engineering education*. IEEE Frontiers in Education Conference.
Luegenbiehl, H. C. (2009). Ethical principles for engineers in a global environment. In I. Poel & D. Goldberg (Eds.), *Philosophy and engineering* (Vol. 2, pp. 147–159). Springer Netherlands. https://doi.org/10.1007/978-90-481-2804-4_13
Martin, D. A., Clancy, R. F., Zhu, Q., & Bombaerts, G. (2023). Why do we need norm sensitive design? A WEIRD critique of value sensitive approaches to design. *Journal of Global Philosophy*, *33*(40). https://doi.org/10.1007/s10516-023-09689-9
Martin, D. A., Conlon, E., & Bowe, B. (2020). Exploring the curricular content of engineering ethics education in Ireland. *2020 IFEES World Engineering Education Forum - Global Engineering Deans Council (WEEF-GEDC)*, 1–5. https://doi.org/10.1109/WEEF-GEDC49885.2020.9293664
Martin, D. A., Conlon, E., & Bowe, B. (2021). A multi-level review of engineering ethics education: Towards a socio-technical orientation of engineering education for ethics. *Science and Engineering Ethics*, *27*(5), 60. https://doi.org/10.1007/s11948-021-00333-6
Martin, D. A., & Polmear, M. (2023). The two cultures of engineering education: Looking back and moving forward. In S. H. Christensen, A. Buch, E. Conlon, C. Didier, C. Mitcham, & M. Murphy (Eds.), *Engineering, social science, and the humanities: Has their conversation come of age?* (Vol. 42, pp. 133–150). Springer International Publishing. https://doi.org/10.1007/978-3-031-11601-8_7.
McAuliffe, G. (2006). The evolution of professional competence. In C. Hoare (Ed.), *Handbook of adult learning and development*. Oxford University Press.
McLoughlin, L. A. (2005). Spotlighting: Emergent gender bias in undergraduate engineering education. *Journal of Engineering Education*, *94*(4), 373–381.
Mitcham, C. (2009). A historico-ethical perspective on engineering education: From use and convenience to policy engagement. *Engineering Studies*, *1*(1), 35–53. https://doi.org/10.1080/19378620902725166
Mitcham, C., & Englehardt, E. E. (2019). Ethics across the curriculum: Prospects for broader (and Deeper) teaching and learning in research and engineering ethics. *Science and Engineering Ethics*, *25*(6), 1735–1762. https://doi.org/10.1007/s11948-016-9797-7
Monteiro, F., Leite, C., & Rocha, C. (2017). The influence of engineers' training models on ethics and civic education component in engineering courses in Portugal. *European Journal of Engineering Education*, *42*(2), 156–170. https://doi.org/10.1080/03043797.2016.1267716
Newberry, B. (2004). The dilemma of ethics in engineering education. *Science and Engineering Ethics*, *10*(2), 343–351. https://doi.org/10.1007/s11948-004-0030-8
Nieusma, D. (2013). Engineering, social justice, and peace: Strategies for educational and professional reform. In J. C. Lucena (Ed.), Engineering education for social justice (pp. 19–40). Springer.
Olsen, T. (1978). *Silences*. Delacorte Press.
Pantazidou, M., & Nair, I. (1999). Ethic of care: Guiding principles for engineering teaching & practice. *Journal of Engineering Education*, *88*(2), 205–212. https://doi.org/10.1002/j.2168-9830.1999.tb00436.x
Parker, A. M., Watson, E., Ivey, M., & Carey, J. P. (2019). Approaches to graduate attributes and continual improvement processes in faculties of engineering across Canada: A narrative review of the literature. *Proceedings of the Canadian Engineering Education Association (CEEA)*. https://doi.org/10.24908/pceea.vi0.13736
Patil, A., & Codner, G. (2007). Accreditation of engineering education: Review, observations and proposal for global accreditation. *European Journal of Engineering Education*, *32*(6), 639–651. https://doi.org/10.1080/03043790701520594
Patil, A., & Gray, P. (Eds.). (2009). *Engineering education quality assurance*. Springer US. https://doi.org/10.1007/978-1-4419-0555-0
Paul, R., Hugo, R., & Falls, L. C. (2015). *International expectations of engineering graduate attributes*. 11th International CDIO Conference.
Paul, R. M., Zacharias, K., Nolan, E. M., Monkman, K., & Thomsen, V. (2023). Stubborn boundaries: The iron ring ritual as a case of mapping, resisting, and transforming Canadian engineering ethics. *Frontiers in Education*, *8*, 1177035. https://doi.org/10.3389/feduc.2023.1177035
Pawley, A. L. (2012). Engineering faculty drawing the line: A taxonomy of boundary work in academic engineering. *Engineering Studies*, *4*(2), 145–169. https://doi.org/10.1080/19378629.2012.687000

Pawley, A. L. (2019). "Asking questions, we walk": How should engineering education address equity, the climate crisis, and its own moral infrastructure? *Journal of Engineering Education*, *108*(4), 447–452. https://doi.org/10.1002/jee.20295

Perlman, B., & Varma, R. (2002). Improving ethical engineering practice. *IEEE Technology and Society Magazine*, *21*(1), 40–47. https://doi.org/10.1109/44.993600

Peterson, G. D. (2009). Quality assurance in the preparation of technical professionals: The ABET perspective. In A. Patil & P. Gray (Eds.), *Engineering education quality assurance* (pp. 73–83). Springer US. https://doi.org/10.1007/978-1-4419-0555-0_5

Pfatteicher, S. (2005). Anticipating engineering's ethical challenges in 2020. *IEEE Technology and Society Magazine*, *24*(4), 4–43. https://doi.org/10.1109/MTAS.2005.1563495

Philip, T. M., Gupta, A., Elby, A., & Turpen, C. (2018). Why ideology matters for learning: A case of ideological convergence in an engineering ethics classroom discussion on Drone Warfare. *Journal of the Learning Sciences*, *27*(2), 183–223. https://doi.org/10.1080/10508406.2017.1381964

Polmear, M., Bielefeldt, A. R., Knight, D., Canney, N., & Swan, C. (2019). Analysis of macroethics teaching practices and perceptions in engineering: A cultural comparison. *European Journal of Engineering Education*, *44*(6), 866–881. https://doi.org/10.1080/03043797.2019.1593323

Riley, D. (2012). Aiding and ABETing: The bankruptcy of outcomes-based education as a change strategy. *2012 ASEE Annual Conference & Exposition Proceedings*, 25.141.1–25.141.13. https://doi.org/10.18260/1-2--20901

Riley, D. (2013). Hidden in plain view: Feminists doing engineering ethics, engineers doing feminist ethics. *Science Engineering Ethics, 19*(1), 189–206. https://doi.org/10.1007/s11948-011-9320-0. Epub 2011 Oct 28. PMID: 22033855.

Riley, D. (2017). Rigor/Us: Building boundaries and disciplining diversity with standards of merit. *Engineering Studies*, *9*(3), 249–265. https://doi.org/10.1080/19378629.2017.1408631

Riley, D. (2016). Don't lower the bar: Remarks at NAE workshop on ABET criteria changes. *Against ABET: Defending the Broad Education of Engineers*. https://aabet.org/

Riley, D. (2023). Engineering principles: Restoring public values in professional Life. In A. Fritzsche & A. Santa-María (Eds.), *Rethinking technology and engineering* (Vol. 45, pp. 13–23). Springer International Publishing. https://doi.org/10.1007/978-3-031-25233-4_2

Riley, D., & Lambrinidou, Y. (2015). Canons against Cannons? Social justice and the engineering ethics imaginary. *2015 ASEE Annual Conference and Exposition Proceedings*, 26.322.1-26.322.19. https://doi.org/10.18260/p.23661

Riley, D., Slaton, A., & Herkert, J. (2015). What is gained by articulating non-canonical engineering ethics Canons? *2015 ASEE Annual Conference and Exposition Proceedings*, 26.1723.1-26.1723.16. https://doi.org/10.18260/p.25059

Rottmann, C., & Reeve, D. (2020). Equity as Rebar: Bridging the micro/macro divide in engineering ethics education. *Canadian Journal of Science, Mathematics and Technology Education*, *20*(1), 146–165. https://doi.org/10.1007/s42330-019-00073-7

Scott, J. W. (1999). *Gender and the politics of history*. Columbia University Press.

Seron, C., & Silbey, S. S. (2009). The dialectic between expert knowledge and professional discretion: Accreditation, social control and the limits of instrumental logic. *Engineering Studies*, *1*(2), 101–127. https://doi.org/10.1080/19378620902902351

Sethy, S. S. (2017). Undergraduate engineering students' attitudes and perceptions towards 'professional ethics' course: A case study of India. *European Journal of Engineering Education*, *42*(6), 987–999. https://doi.org/10.1080/03043797.2016.1243656

Shuman, L. J., Besterfield-Sacre, M., & McGourty, J. (2005). The ABET "professional skills" - Can they be taught? Can they be assessed? *Journal of Engineering Education*, *94*(1), 41–55. https://doi.org/10.1002/j.2168-9830.2005.tb00828.x

Slaton, A. (2001). *Reinforced concrete and the modernization of American building, 1900–1930*. Johns Hopkins University Press.

Slaton, A. E. (2012). Engineering improvement: Social and historical perspectives on the NAE's "grand challenges". *International Journal of Engineering, Social Justice, and Peace*, *1*(2). https://doi.org/10.24908/ijesjp.v1i2.4305

Stevens, R., Amos, D., Jocuns, A., & Garrison, L. (2007). *Engineering as lifestyle and a meritocracy of difficulty: Two pervasive beliefs among engineering students and their possible effects*. American Society of Engineering Education (ASEE) Annual Conference & Exposition.

Sthapak, B. K. (2012). Globalisation of Indian engineering education through the Washington Accord. *Journal of Engineering, Science & Management Education*, *5*(2), 464–466.

Swan, C., Kulich, A., & Wallace, R. (2019). A review of ethics cases: Gaps in the engineering curriculum. *2019 ASEE Annual Conference & Exposition Proceedings*, 31988. https://doi.org/10.18260/1-2--31988

Tafoya, T. (1995). 2-Finding harmony: Balancing traditional values with western science in therapy. *Canadian Journal of Native Education*, 21.

Tang, X., & Nieusma, D. (2017). Contextualizing the code: Ethical support and professional interests in the creation and institutionalization of the 1974 IEEE code of ethics. *Engineering Studies*, *9*(3), 166–194. https://doi.org/10.1080/19378629.2017.1401630

Tuana, N. (2006). The speculum of ignorance: The women's health movement and epistemologies of ignorance. *Hypatia*, *21*(3), 1–19.

UNESCO. (1998). *World declaration on higher education for the twenty-first century*. https://unesdoc.unesco.org/ark:/48223/pf0000141952

Vaughan, D. (2004). Theorizing disaster: Analogy, historical ethnography, and the challenger accident. *Ethnography*, *5*(3), 315–347. https://doi.org/10.1177/1466138104045659

Vesilind, P. A. (1995). Evolution of the American society of civil engineers code of ethics. *Journal of Professional Issues in Engineering Education and Practice*, *121*(1), 4–10. https://doi.org/10.1061/(ASCE)1052-3928(1995)121:1(4)

Wacker, G. (1989). Teaching impacts of technology/professional practice. *Proceedings of Delicate balance: Technics, culture and consequences* (pp. 97–103). https://doi.org/10.1109/TCAC.1989.697051

Walker, M. U. (1989). Moral understandings: Alternative "epistemology" for a feminist ethics. *Hypatia*, *4*(2), 15–28. https://doi.org/10.1111/j.1527-2001.1989.tb00570.x

Walther, J., Radcliffe, D., & Mann, L. (2007). *Analysis of the use of an accidental competency discourse as a reflective tool for professional placement students*. Frontiers in Education Conference.

Weil, V. (1984). The rise of engineering ethics. *Technology in Society*, *6*(4), 341–345. https://doi.org/10.1016/0160-791X(84)90028-9

Whitbeck, C. (1998). *Ethics in engineering practice and research*. Cambridge University Press.

Wilson, S. (2008). *Research is ceremony: Indigenous research methods*. Fernwood Publishing.

Winner, L. (1990). Engineering ethics and political imagination. In P. T. Durbin (Ed.), *Broad and narrow interpretations of philosophy of technology* (pp. 53–64). Springer Netherlands. https://doi.org/10.1007/978-94-009-0557-3_6

Wisnioski, M. H. (2012). *Engineers for change: Competing visions of technology in 1960s America*. MIT Press.

Woolston, D. C. (2008). Outcomes-based assessment in engineering education: A critique of its foundations and practice. *2008 38th Annual Frontiers in Education Conference*, S4G-1-S4G-5. https://doi.org/10.1109/FIE.2008.4720266

Zhu, Q. (2020). Ethics, society, and technology: A Confucian role ethics perspective. *Technology in Society*, *63*, 101424. https://doi.org/10.1016/j.techsoc.2020.101424

36

ACCREDITATION PROCESSES AND IMPLICATIONS FOR ETHICS EDUCATION AT THE LOCAL LEVEL

Helena Kovacs and Stephanie Hladik

Introduction

Engineering ethics is a critical component of engineering professional practice and must be included in professional engineering education (Harris Jr. et al., 1996). As engineers design new technologies that introduce new ethical dilemmas, regarding, for instance, autonomous vehicles (Martinho et al., 2021) and algorithms (Benjamin, 2019; Noble, 2018), postsecondary institutions must ensure that their engineering students are graduating with the knowledge and skills to meet emerging challenges. Accreditation documents and processes have been designed as guidelines for how engineering ethics should be taught and assessed to standardize these goals and metrics for different institutions across the nation (e.g., see the Canadian Engineering Accreditation Board (CEAB) in Canada and Commission des Titres d'Ingénieur (CTI) in Switzerland). However, questions emerge about how the broad definition of ethics put forward by accreditation bodies is understood and enacted within the local contexts of specific institutions and classrooms, each with its own heterogeneous group of instructors and students.

In this chapter, we discuss the implications accreditation has on engineering ethics education at the local level. More specifically, the chapter debates the potential gap between how ethics is articulated in accreditation documents and processes and what ethics in engineering education means locally for institutions, instructors, and students (i.e., by 'gap,' we mean its impersonality). In essence, this chapter aims to problematize the potential impersonality of ethics hidden in the documents that are bureaucratically operationalized at the level of educational programs and typically do not consider the different histories, demographics, and needs of local engineering communities. Accreditation of engineering education programs includes varied definitions and requirements, often broad, ill-defined, and implicit. This is particularly true concerning engineering ethics, where words and phrases such as 'profession,' 'society,' 'responsibility,' and 'integrity' are often used in place of 'ethics' in learning and program outcomes (Junaid et al., 2022).

Impersonality in accreditation documents and processes is in tension with how inherently personal engineering ethics is in local contexts; it must respond to the needs and experiences of institutions, instructors, students, and industry partners hiring program graduates. Lack of synchronization across different levels and stakeholders and the fact that these levels are not homogeneous can complicate the impact at the local level (Martin et al., 2021). Different political landscapes,

DOI: 10.4324/9781003464259-43

local industry demands, institution-community partnerships, and faculty and student body diversity contribute to heterogeneity, making it difficult to argue that one shared understanding of engineering ethics is desirable or possible. Levels, accreditation standards, and local implementation are complex and contextual. Therefore, it is necessary to more deeply theorize and investigate the potential gap between the accreditation of engineering ethics and what engineering ethics means at the local level.

Setting this as an opening for the chapter, we genuinely want to question: *How do the broad, impersonal conceptions of engineering ethics contained in accreditation documents and processes impact how they can be translated, interpreted, or implemented in local and personal contexts?* As such, in the first part of this chapter, we emphasize the importance of understanding the history and the character of accreditation in engineering education, as well as the messages and ideas of the peculiar language the accreditation documents suggest. We will do this by drawing connections to the previous chapters in the theme. We will also briefly address the accreditation process and interrogate how it may contribute to the gap between the broad, high-level accreditation standards and local contexts. Our work regards existing accreditation documents and processes as broad and impersonal. The second part of the chapter will address educational questions related to how this gap is related to institutional positionings, curriculum design across different engineering disciplines, teacher agency and relations to ethics in engineering education, and the student as the focal point in education. In addition, we will highlight the heterogeneity in each of these levels that influences the degree of disconnect between the local contexts and the accreditation of engineering ethics. We will also graphically illustrate the very complex and contextualized picture of engineering ethics education at the local level. Finally, we will offer suggestions at both the accreditation and local levels to bridge the gap between them. The value of this chapter is very much in its potential to offer a way in which different and diverse positionalities of ethics at the local level can be considered when examining accreditation and vice versa.

Regarding the positionality of the authors of this chapter, Helena is a social scientist with a Ph.D. in Teacher Education obtained as a Marie Curie Fellow at two different institutions and countries in Europe. Her Masters was in Lifelong Learning: Policy and Management, and her Bachelor's in Community Youth Work and Non-formal Education. All educational degrees were obtained in different countries and followed different educational provisions, allowing Helena to tap into various systems and practices. Currently, she works on translational research that serves institutional changes and practice development. Her engagement with engineering ethics education comes from working on many transversal skills and exploring the difficulties in teaching and learning. Further interest in accreditation and engineering ethics education came from her previous work on policy and curricular matters and her recent research on teaching transversal skills. The fact that she works in a French-speaking context makes her close and familiar with CTI.

What would accreditation look like if it were locally situated, with accountability to the land and waters and their stewards? and this interdisciplinary educational background provides a solid grounding in the traditions of engineering education and critical approaches to education that challenge systemic inequities such as racism and sexism. An even more profound understanding of the different impacts that technologies can have on various groups arose from her postdoctoral work in an Information Science department. In her current role as an assistant professor of engineering education, she uses this critical lens to teach an undergraduate course focused on the impact of engineering on society and engineering ethics. She hopes that her students will be able not only to discuss and understand the broad ethical issues arising from the development of new technologies (which are often intertwined with systemic issues such as racism) but also apply these lessons to ask critical questions in every phase of the engineering design process – from forming a design team to considering what happens at the end of a product's life cycle. Part of Stephanie's role is to ensure that the course content and assessments align with CEAB guidelines (CEAB is used in Canada and is similar to the Accreditation Board of Engineering and Technology (ABET), which is used in many English-speaking countries) and to collect data indicating how well students are meeting those guidelines. As an authorial team, we both draw from our social-sciences backgrounds, which treated ethics as a nuanced, contextual, and complex topic. We both believe that this complex view of ethics is critical in a field such as engineering, as the new technologies and infrastructures that engineers design can considerably impact society – impacts that are not the same for different locations and groups of people.

The impersonality of accreditation

Accreditation is a process by which an external organization evaluates an institution's or program's quality and standards. Accreditation aims to ensure that specific standards are met and that quality education and services are provided to its stakeholders (Adreani et al., 2020). As stated in previous chapters of this section, accreditation also involves documents and processes that convey a very vague, broad perspective of ethics. Hence, the essence of engineering ethics is detached from the local context in many ways. While the emphasis on standards and quality is often enumerated through accreditation, the essential aspects of ethics may be so broadly stated that they are decontextualized, creating an impersonal relation between accreditation and the local contexts of the institutions, instructors, and students. In this section, we discuss three dimensions of accreditation – quality, values, and language – using a critical lens to look for how they impact perspectives on engineering ethics in ways that are enacted at a local level.

Quality

As a tool for ensuring quality, accreditation has become a crucial component in higher education. National and international accreditation systems "ascertain the existence of qualitative requirements through an evaluation process" to ensure a given service complies with the quality standards and encourage ongoing improvement (Adreani et al., 2020, p. 691). In engineering education, "accreditation programmes identify specific areas of knowledge and skills that need to be addressed in order for students to qualify as engineering graduates" (Junaid et al., 2022, p. 371). Thus, institutions use accreditation tags to argue that their programs are of a certain quality and that their graduates uphold specific knowledge, skills, and values.

In that way, accreditation creates a specific guarantee for the institutions that their students will be seen and considered worthy of the professional standards within a particular discipline and employable in a specific context. In addition, accreditation can be regional, national, or international. Regional and national accreditation is important to ensure the running and funding of the programs; international accreditation facilitates recognition across borders and is an important aspect of graduates' mobility.

In light of this, Cardoso et al. (2016) questioned whether quality assurance agencies can ensure quality, especially if quality is perceived as culture, compliance, and consistency. While those authors point to quality as negotiated and assured mainly within the institutional setting, the same is observed when the quality assurance process is externalized. Quality is often non-negotiated and fails to recognize the full range of essential factors and processes within an academic institution. Obstacles to ensuring quality, including lack of incentives and overvaluing research against other activities such as teaching, drive a restrictive agenda in quality assurance. Additionally, in discussing the role of accreditation, particularly in engineering, the relevance of quality assurance is more than often related and defined according to programs' perceived relevance to the labor market (Bendixen & Jacobsen, 2020) and often not sensitive to societal and environmental aspects that are particularly important to ethics in the engineering professions.

Furthermore, standardization of ethics in accreditation can be perceived as a double-edged sword. While standardization is important to achieve coherence in quality assurance across contexts, it disconnects accreditation from specific contexts. It perhaps undermines, misrepresents, or erases the local aspects of quality and locally important aspects of ethics. For instance, two large accreditation systems in engineering education are ABET and CTI, both used in various countries. As a result, the standards that CTI, a French accreditation body, places in France need to correspond to standards in other countries such as Canada, Belgium, Switzerland, and China.

On the other hand, each accrediting body or agency has its own set of criteria for evaluating institutions, which can create confusion and inconsistency across engineering education programs (Wysocka et al., 2022). Different demands, often very vague, can lead to disparities in the quality of engineering ethics education in different institutions. However, over-standardizing or over-generalizing an aspect such as ethics can 'wash off' the much-needed local aspects of ethics.

Values

The question of quality relates directly to the question of values. Although accreditation constitutes an important way to ensure that engineering programs provide their students with the knowledge and skills necessary to enter the profession and protect the safety and welfare of the public, often what is considered a professional standard in engineering is set up and maintained by one dominant social group. This can lead to issues where engineering designs may be viewed as successful, even if they do not work for – or worse yet, actively cause harm to – people from communities underrepresented in engineering (e.g., through misidentification of people of color by facial recognition algorithms; Buolamwini & Gebru, 2018). Attention to diversity issues in engineering education has been growing, and the need to ensure that accreditation standards are inclusive and equitable comes along the same lines. Several organizations, including ABET and the National Society of Black Engineers (NSBE) in the United States, have developed initiatives and resources to promote diversity and inclusion in engineering education and accreditation (Zoltowski et al., 2020).

Andreani et al. (2020) noted that accreditation serves neo-institutional behaviors through which quality assurance mechanisms present a delegation of power, sourcing it from institutional management to external accreditation body and that these processes support the New Public Management paradigm, focusing on "the benefits of trade in terms of efficiency and consumer freedom" (p. 694). Power relations among institutions, industry, and professional bodies have been shown to play a role in the accreditation of engineering programs, with specific impacts on engineering ethics education (Martin et al., 2021). Industry professionals often serve on accreditation boards, give feedback on accreditation documents, and participate in accreditation visits, influencing which ideas about engineering ethics are most important to new graduates working in their industries. Similarly, professional organizations may strive to ensure alignment between engineering ethics education and their own definitions and codes of ethics. While this alignment can promote greater attention to the local ethical needs of industry and professional organizations, there is the potential for bias in terms of what definitions of ethics are considered good (or good enough) and should be included or not included in accreditation.

In many cases, the values that guide accreditation documents and processes may be implicit or invisible to the institutions, instructors, and students. To that end, we wish to amplify the call for more systematic and comprehensive policy documents (e.g., Davies et al., 2010), including accreditation documents, to clarify underlying values. Knowing exactly which values are being reified in these accreditation documents can help us to identify instances in which those values may be aligned or misaligned with those upheld by institutions, teachers, and students, therefore highlighting potential gaps to be bridged.

Language

A global analysis of accreditation documents conducted by Junaid and her colleagues (2021, 2022) has shown that language in accreditation documents is very implicit and that a gap exists in strongly supporting ethics as part of an engineering degree. Furthermore, the study shows a large degree of misalignment on ethics in different accreditation documents worldwide and much space for individual interpretations. This brings us to question the perceived value of ethics in engineering education and practice, being both informed and partially constructed by accreditation bodies.

Language is crucial as it constructs a discourse, and it is through words that action (or inaction) is created. For instance, language carries implicit and explicit values. As such, policy documents, including those related to accreditation, can create an imagery of what is essential in, what should be taught in, and what is not considered part of engineering ethics. Junaid and colleagues (2022) analyzed the language of different accreditation documents. They found that words used to describe expected learning outcomes around ethics were mainly lower-level verbs from Bloom's taxonomy (e.g., 'know,' 'define,' 'be aware of'). The absence of higher-level verbs such as 'compare' or 'justify' can lead to a lack of critical thinking around engineering ethics in the curriculum. Furthermore, the lack of words such as 'global' or 'justice' gives a perception that these are not part of ethics education in engineering (Junaid et al., 2022).

Regarding the place of ethics, Junaid et al. (2022) analyzed the content of accreditation documents worldwide. They found that definitions vary to a great extent, and this leaves room for interpretation of what engineering ethics is. Very few terms were comparable across contexts, even though engineers are highly mobile professionals. For instance, in South Africa, the definitions are focused more on technical than educational terms (Gwynne-Evans et al., 2021). Observing ethics as a technical aspect of engineering assumes that ethics is a technical checkpoint, not a multi-perspective issue that must be carefully discussed. Additionally, some accreditation documents are used internationally, such as ABET (adopted in 41 countries) or CTI, using identical processes for all contexts. Having this in mind, "accreditation bodies could bring into focus the terms they use and what they mean as part of their ethical due diligence in the construction of engineering programmes across a range of countries where the context of the intended meaning of the learning outcomes as state might otherwise be missed" (Junaid et al., 2022, p. 375).

Local context(s), cultures, and positionalities

Interrogating engineering ethics education locally requires considering how 'the local' is defined. In educational research, the local can be a geographic descriptor. However, beyond this and particularly in the realm of higher education, local can be considered as an institution or department, as a discipline or a subject, as a singular program or a course, as a teaching practice, and quite certainly as a student learning experience in classes and afterward, as graduate engineers. Further, considering the complexity of 'the local' also highlights that these layers are not self-standing. Hence they are tied by power practices and are heavily relational. These layers of the local are also rarely homogeneous, meaning the interactions and impacts are different in different institutions, classrooms, and communities and, therefore, must be addressed through a comprehensive lens that includes socio-economic-political dimensions of educational practice. Beyond this, in their careers as ethical practitioners, engineering graduates are tasked with 'wicked' complex problems that concern diverse societies and the environment, some of which are critical to the people and communities around them (one sense of local), and others that have global considerations, potentially bringing another dimension into the discussion of 'the local.' Understanding the nuances of ethics at this local level is critical for their participation in society as *people*, not just as engineers.

In the following sections, we break down how high-level accreditation documents and processes are translated (or not) within three local levels: the institution and program level, the classroom and instructor level, and the student level. We use the word 'translation' (as opposed to implementation) as it encompasses the idea of a higher-level concept or framework undergoing some change in order to suit the local context to which it is being applied (Völker et al., 2023). We highlight how these local contexts matter in our interpretation of engineering ethics due to how the levels interact through power structures (e.g., instructors following a curriculum set by their institutions), as well as the heterogeneity that is inherent within a single institution or classroom as well as collectively across a nation. We wish to be clear that many of the challenges we discuss here are not caused *solely* by a mistranslation of accreditation documents or processes and may have other contributing factors, including institutional context and instructor and student values. However, by pointing out potential mismatches between accreditation definitions of engineering ethics and those at these local levels, we hope to call attention to more nuanced ways of approaching engineering ethics education that may require more specificity than currently afforded by current accreditation documents and processes.

Institutional and program level

Unlike pre-tertiary education, higher education is known to be relatively autonomous in preparing its programs and course offerings. While institutions must manage the needs and desires of various stakeholders – such as national or provincial/state professional organizations, industry members, faculty, and students – accreditation is a critical hurdle that institutions and programs desire to 'pass,' as it strengthens claims regarding the quality of degrees to their students (future and current) and to the employers who will hire those students. With the emergence of engineering ethics as an explicit criterion in many accreditation documents over the past few decades, institutions have made some progress in incorporating ethics into their programs. However, it is rare to have a systemic institutional push towards comprehensively incorporating ethics into engineering academic programs. Instead, institutions often meet accreditation standards by creating one or two standalone courses related to engineering ethics (Hamad et al., 2013). This raises some issues in how engineering ethics is integrated and perceived locally.

When all discussion of ethics is relegated to a standalone course rather than being carefully incorporated into technical courses, students may need help to connect the broad rules and lessons of engineering ethics with their day-to-day technical design work. Ethics is, therefore, decontextualized and invisible in their local work contexts; it is discussed in terms of historical case studies or professional codes of conduct, and ethics only emerges as relevant in its breakdown (i.e., an engineering disaster with ethical implications). These issues are compounded when students or even faculty do not consider these standalone courses as 'core' engineering courses (Martin & Polmear, 2023). Their positioning in the program (e.g., as an option that can be taken in any year) and the possibility that this is one of the few courses that students will encounter being taught by a 'non-engineer' can devalue these courses in students' minds. *How can ethics be core to their daily prac-*

tice when it is not core to their educational program? Suppose accreditation standards insisted on the systematic incorporation of ethics across multiple courses. In that case, it may become more accessible to locate ethics in the daily practices and discussions of engineering work and apply the broad ideas in meaningful ways in engineers' communities. Asking, in various courses, *What are the ethical issues here?* allows students to understand ethics more deeply and in ways that can be applied not just to their professional contexts, but also to situations and experiences outside of engineering. Integrating engineering ethics across the curriculum would also go a long way toward perpetuating consistency and a culture of engineering ethics in the institution rather than simply focusing on compliance as an indicator of the program's quality.

A systemic incorporation of ethics across a program would require a robust and convincing narrative of engineering ethics – something not present in current accreditation standards. The language currently used in accreditation documents is broad, unspecified, and open to interpretation, which weakens its ability to convey a cohesive narrative of engineering ethics (Junaid et al., 2022). Creating a robust and program-level narrative of engineering ethics responsive to the local context may be challenging. For example, *Who defines ethics at the institutional level, and what values do they hold?* Careful consideration is necessary regarding what ethical commitments are important for a particular institution. Local geographies, histories, and politics impact these different ethical commitments. For example, an institution may view ethical partnership and reconciliation with Indigenous communities as a key commitment to its research, teaching, and service. In this case, a local narrative of engineering ethics may prioritize ethical ways of partnering with Indigenous communities harmed by past engineering projects, such as hydroelectric dams (Martin & Hoffman, 2008). Students can then use these understandings of ethics in other professional and personal connections with Indigenous communities. However, challenges may arise if those local definitions of engineering ethics and their implicit and explicit values do not match the accreditation expectations.

In some cases, the definition of engineering ethics in accreditation documents may focus on personal accountability or ethical principles that do not recognize ethical partnerships and reconciliation with Indigenous communities as meaningful indicators of engineering ethics. On the other hand, if institutions have the power to define their own indicators, they can fine-tune their indicators to respond to the ethical histories and situations that are relevant in their local contexts. The responsibility for ensuring that engineering ethics at the program level responds to local contexts falls to the institution to define and revise its own indicators over time.

A final aspect of program-level narratives of engineering ethics that may be disconnected from local contexts and needs emerges when we consider the different ways that engineering ethics is conceptualized in different engineering disciplines. The heterogeneity in engineering programs and practices contributes to this complexity. In civil engineering, for example, ethics may be discussed in the context of the safety of roads, bridges, and buildings, showing popular case studies (e.g., the Tacoma Narrows Bridge collapse in 1940) that highlight how faulty design can lead to the loss of human life. Biomedical engineering may put a slightly different spin on ethics, drawing from codes of ethics of medical fields to highlight the importance of understanding the impacts of new biomedical devices on patients, including how to design ethical research and testing protocols. Students in software engineering may struggle to connect these types of case studies and principles to their work – after all, they are not building large physical structures and may not be writing software that will be used in the medical field. These students require yet another approach to ethics that highlights the dangers of black-boxed algorithms that perpetuate systemic inequities in often invisible ways (Benjamin, 2019; Buolamwini, 2022; Noble, 2018). While only three different engineering disciplines have been mentioned here, it is already easy to see how broad accreditation standards and indicators have very different local interpretations within the different engineering disciplines.

Once again, the broad sketch of engineering ethics in accreditation may be helpful because it allows for multiple interpretations of ethics, including all of the different ethical considerations that are more or less important in various engineering disciplines. At the same time, without explicit guidance regarding what ethics can look like in different engineering disciplines, engineering ethics education may be limited to the more visible and historical scenarios (e.g., ensuring your bridge does not fall and injure or kill anyone), while complex ethical issues that are increasingly important in today's digital world (e.g., algorithmic biases) may not be explicitly identified as essential or necessary for engineering students to learn.

To summarize this section, we see that the broad standards and descriptions of engineering equity in accreditation documents are often several levels removed from what engineering ethics means locally for institutions and engineering programs. While this may seem like a good thing in that it can allow for multiple interpretations of engineering ethics that can meet local needs, we must keep in mind that this is extra work that the institution and program must do, which may not be explicitly recognized, valued, or accepted during the accreditation process. It may be easier and cheaper for institutions to continue to design standalone ethics courses that technically meet the standards for accreditation without actually preparing students for the specific ethical challenges in their engineering disciplines and broader communities.

Course and instructor level

The local level of individual courses and instructors is also crucial to consider. Related to the ideas discussed at the institutional and program levels, the perceived value of engineering ethics comes from the messages and directives instructors receive from more senior faculty members and administrators. If internal stakeholders perceive that ethics is only being taught in order to meet accreditation standards rather than because it is a critical topic for their students to know and apply in their future work, they may simply 'teach to accreditation' (similar to 'teaching to the test') and include the bare minimum of ethics education to meet accreditation standards satisfactorily (Martin et al., 2021). If ethics coverage is met through a single standalone ethics course, other instructors may believe they do not need to cover ethics in their technical and design courses. This can create the impression that ethics is not critical to engineers' daily technical and design work. It also results in a missed opportunity for instructors to touch upon the ethical issues that arise within their local technical and design contexts, such as the ethics of user testing a new iPhone app, the ethics of sourcing raw materials, or the ethics involved in working with clients in an engineering capstone course. These examples may also be relevant to students' lives outside of their engineering work, such as purchasing new technologies or engaging in professional relationships. In essence, we can find local instances of engineering ethics in most, if not all, engineering courses, but the accreditation process can stifle the desire to think critically about these local definitions of ethics and teach them to students.

In considering how engineering ethics is personalized in courses, we must also consider each individual instructor's relationships and experiences with engineering ethics. Instructors may need help determining what definitions and examples of engineering ethics are appropriate for their students due to the broadness of the topic in both the literature and the accreditation documents. Other instructors may be grounded in the tradition/assumption of engineering as apolitical and objective (Cech, 2013). While those instructors may be comfortable teaching traditional engineering ethics case studies such as the Challenger disaster or the Tacoma Narrows Bridge, they may be less likely to include issues that appear to align more with social justice discourses, such as systemic racism in algorithms (Benjamin, 2019; Noble, 2018) or recent movements to modernize and address harm in engineering graduation rituals (Retool the Ritual, 2023). This is perhaps not surprising, given that the definitions of engineering ethics in accreditation documents, as explored by Junaid et al. (2022), did not reference terms such as "global," "value," and "justice." To attend to local definitions of justice, we must explicitly call out engineering values (including historical values of objectivism) and consider how larger systemic injustices can and should be addressed within and beyond engineering practice.

On top of the *content* of engineering ethics, the accreditation documents need to address issues of *pedagogy*. Junaid et al. (2022) note that many accreditation documents use lower-level verbs from Bloom's taxonomy (e.g., know, define, be aware of) to describe students' understandings of ethics than technical items required (e.g., analyze, synthesize, evaluate, design, or create). *Precisely how are engineering students expected to come to know engineering ethics, and what contexts (local and global) are they expected to be aware of? Is there an expectation that instructors should find ways to connect the classroom with the local community, and if so, how should they be supported in doing so?* These questions are not answered in the accreditation documents. Instructors who try to move beyond memorized definitions and historical case studies may struggle due to a lack of training and support in pedagogical approaches, including small group discussions, debates, or other activities with high student-to-student interaction (Martin et al., 2021). As students begin to engage deeply in discussions of ethics, locating engineering ethics within their personal histories and professional experiences, conflict may arise as different viewpoints clash – something that is especially likely if the course is diving into ideas of engineering ethics that may be viewed as more controversial and political. Instructors must not only deal with the content of ethics within their classroom but also with the idea of ethics as it relates to their pedagogical choices: *How can they support all students and minimize harm while still ensuring that students are prepared to address these real-world applications of engineering ethics they will face in their future lives?* Once again, the broad and impersonal definitions of ethics in accreditation documents may cause instructors to shy away from such challenging topics, classroom dynamics, or pedagogical approaches. Even if instructors find ways to create a classroom environment that allows for multiple personal definitions and understandings of engineering ethics to come together, they must still figure out how to assess students' understandings of engineering ethics, which has its challenges.

Finally, the accreditation process is chore-like and can place a heavy load on already burdened instructors. In some cases, instructors must carefully document low, average, and high-quality student work for every assignment and exam while carefully articulating which accreditation standards are being addressed by each question or section of a rubric. This work is done in addition to the regular lecturing, planning, prepping, and grading in each of their courses. In the end, these piles of documentation are sent to other faculty and staff members on a committee to be compiled and presented to those carrying out the accreditation evaluation. While these instructors may eventually hear, months later, that their program has passed accreditation, they are unlikely to receive meaningful and actionable feedback that they can implement in their courses. Therefore, accreditation risks becoming a bureaucratic procedure without local, actionable impact, leading to feelings of disempowerment and disengagement from instructors.

Student level

We can also understand the local level of engineering ethics as the experience of students – who are both learners and graduates headed out into the workforce. While accreditation standards and processes may guide the courses they are expected to take throughout their degree, students may be unaware of this, as they are often not directly part of the accreditation process other than to provide permission for instructors to use their coursework as examples of student learning outcomes. Therefore, their notion of engineering ethics is built through their classroom experiences. As we have already mentioned, a standalone ethics course (among many more technical courses) may send students the message that ethics is not at the core of their program and duties as future engineers (Martin & Polmear, 2023). This message is reinforced by the perception that an ethics course that relies predominantly upon content and pedagogies from the social sciences (such as group discussions or written work) should automatically be more accessible than their technical courses. Therefore it is viewed as less critical and often pushed to the bottom of students' lengthy to-do lists as they complete their semester's work.

Students may also need help locating themselves within traditional ethics case studies. At times, these traditional cases may be viewed as irrelevant when they deal with outdated technologies or ethical codes that are not part of the students' everyday lives and will not be part of their future professional practice. In addition, when complex techno-socio-political situations, such as that of the Challenger disaster, are reduced to a 1–2-page case description, it is easy for students to read the facts and say, 'Of course, I would never do that!' Students argue that they would do proper testing and speak up against management, even though they are missing critical context that is not easily summed up in the case. Sociologist Diane Vaughan has pointed out that the Challenger disaster was not the result of individual bad decisions but rather of years of practices and norms at NASA that created a corporate culture that normalized deviance and missed routine signals of impending disaster (Vaughan, 1996). She argues that the NASA managers conformed to requirements and did not break any rules during the launch. In this way, we can see how students' understanding of engineering ethics at the local level can be limited to individual decisions versus the cultures and norms that heavily influence those decisions in ways that are either invisible or considered unimportant until disaster strikes. Students do not realize that, in their local contexts, they can unintentionally reify those cultures and norms. Further, students' hyperfocus on individual actions may cause them to ignore or be unable to address the role that engineers play in complex systemic issues, including climate change, racism, and other biases in technology design and reconciliation efforts with Indigenous communities.

Finally, we must consider the impact of heterogeneity in the student population on the translation of engineering ethics at the local level. Students in engineering programs have different histories, backgrounds, and identities, and all of these factors can influence what engineering ethics means to them (Castagno et al., 2022). For example, a racialized woman in engineering may argue that professional codes of ethics do not adequately respond to equity, diversity, and inclusion goals in the profession. A student from a country that has been dealing with war and violence may question the idea of engineers holding paramount the safety, health, and welfare of the public (Engineers Geoscientists Manitoba, 2018, p. 1): *Which public is being protected through the use of drones and automated weaponry?* Some may argue that a professional code of ethics can be interpreted in ways that address emerging ethical dilemmas in engineering and can respond to personal codes of ethics; others may turn that argument on itself, arguing that allowing for room for interpretation means that decisions can be made that do not fully address the ethical issue or that lead to contradictions (e.g., be faithful to your employer but also hold the safety of the public paramount – *But what if your employer is doing something harmful?*).

In short, students will respond to the broad ideas of engineering ethics that are handed down to them through the curriculum. However, they will also internalize and personalize (or fail to do so) these ideas in accordance with their own histories and identities.

Summary

This section has highlighted how engineering ethics is grounded in local contexts, cultures, and positionalities. Accreditation documents and processes have implications for how engineering ethics is taught and understood by institutions, instructors, and students. Without institutional or instructor drive to connect engineering ethics as articulated in engineering accreditation documents and processes to local contexts, students can be left with fragmented, vague, and surface-level understandings of engineering ethics that will not address the ethical challenges they will face in their future professional work and as members of society. We have created a visual (see Figure 36.1) that graphically illustrates engineering ethics at these various local levels that can act as a tool for stimulating discussion and reflection.

Although we have mainly discussed the impersonality of engineering ethics accreditation documents and processes as a challenge that should be overcome, it is also possible to understand this impersonality as something that allows space for flexibility and adaptation. As mentioned in our discussion of institutions and programs, vagueness requires institutions to define their own indicators for engineering ethics, which can, in turn, respond to local contexts and priorities. Additionally, instructors may feel a greater sense of freedom in choosing the content and classroom activities that resonate most strongly with them, their students, and their communities if they are not burdened by a detailed list of ethical concepts and considerations in the accreditation documents. Our concern arises from the fact that although some institutions, programs, and instructors may thrive in this ill-defined space, others may require more guidance, which they cannot get from the accreditation documents and processes. There may also be those who continually fall back upon traditional, depersonalized notions of engineering ethics (due to a lack of experience, lack of desire or resources to change, or general resistance to the idea of engineering as politicized) who would need more of an external push to contextualize engineering ethics in their local contexts. As accreditation documents and processes are used to both shape and assess engineering programs worldwide, they are well-placed to instigate change in this way.

Conclusion

Our goal for this chapter was to problematize the idea of engineering ethics education at the local level as it relates to engineering accreditation documents and processes. We have discussed how vague language, assumptions of standardized values, and the assumption of quality in accreditation can lead to disconnects between how engineering ethics is defined and assessed in accreditation and how engineering ethics lives within institutions, classrooms, students, and their communities. We would encourage everyone involved in engineering ethics education – accreditation bodies, assessors, institution leadership, instructors, and students – to engage in deeper reflection on what engineering ethics means in diverse and heterogeneous local contexts.

We encourage institutions, programs, educators, and students to ground engineering ethics within their local contexts. This may mean connecting to institutional and community values, reaching out to industry and non-governmental partners, including ethics in more technical engineering courses, and incorporating classroom discussion that encourages students to share their definitions of engineering ethics and its importance to them. At the accreditation level, we call upon accreditation boards to explicitly acknowledge and value the heterogeneity in local engineering ethics within the accreditation documents and processes. We hope that by calling attention to what engineering ethics looks like at the local level, we can create curricula and classroom environments that prepare engineering students for the complex ethical challenges they will encounter in their workplaces and communities.

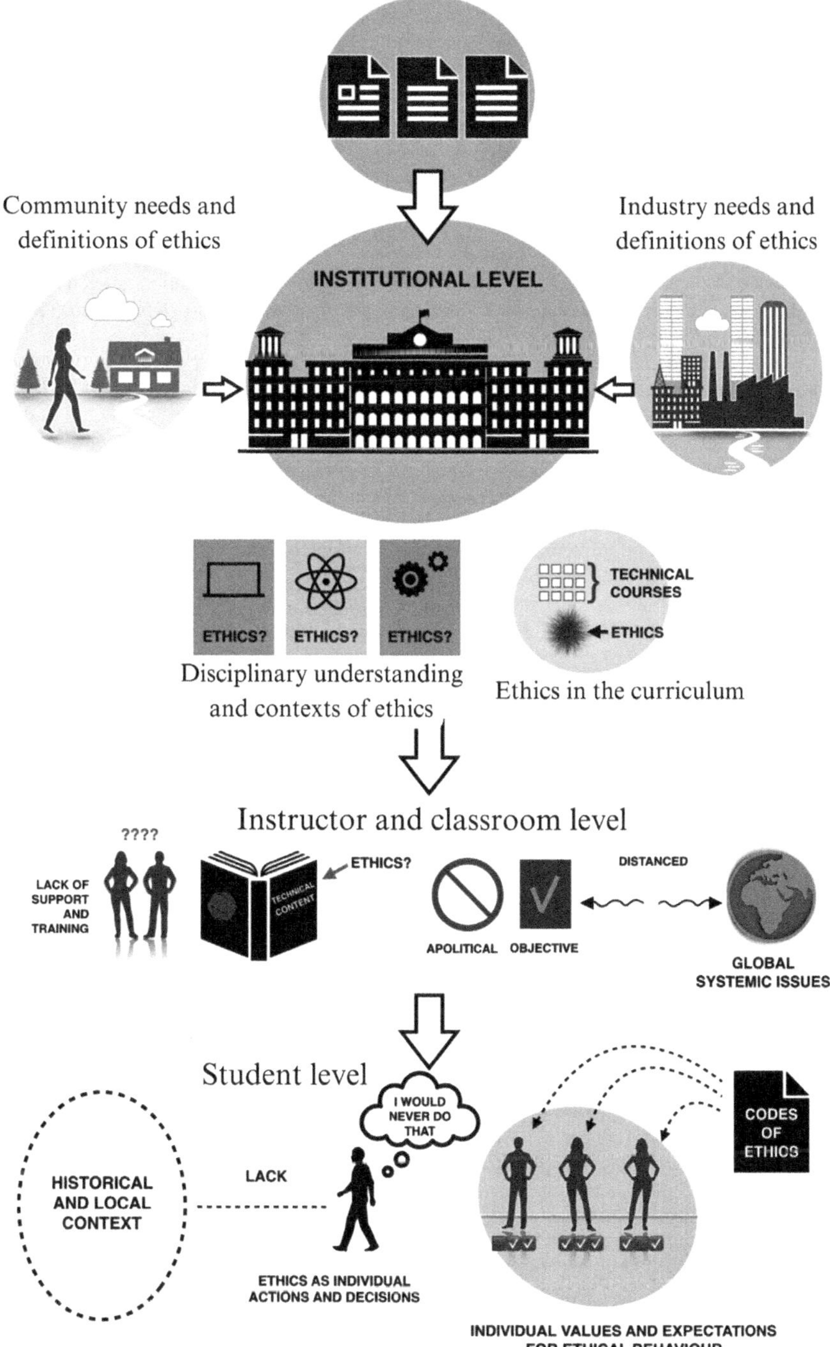

Figure 36.1 Graphical representation of potential mismatches between accreditation documents/processes for engineering ethics and implications at the institutional, instructor, and student levels.

References

Andreani, M., Russo, D., Salini, S., & Turri, M. (2020). Shadows over accreditation in higher education: Some qualitative evidence. *Higher Education, 79*, 691–709. https://doi.org/10.1007/s10734-019-00432-1

Bendixen, C., & Jacobsen, J. C. (2020). Accreditation of higher education in Denmark and European Union: From system to substance? *Quality in Higher Education, 26*(1), 66–79. https://doi.org/10.1080/13538322.2020.1729310

Benjamin, R. (2019). *Race after technology*. Polity.

Buolamwini, J. (2022). *Facing the coded gaze with evocative audits and algorithmic audits* [Doctoral thesis, Massachusetts Institute of Technology]. MIT DSpace. https://dspace.mit.edu/handle/1721.1/143396

Buolamwini, J., & Gebru, T. (2018). Gender shades: Intersectional accuracy disparities in commercial gender classification. *Proceedings of Machine Learning Research, 81,* 1–15. 2018 Conference on Fairness, Accountability, and Transparency.

Cardoso, S., Rosa, M. J., & Stensaker, B. (2016). Why is quality in higher education not achieved? The view of academics. *Assessment & Evaluation in Higher Education, 41*(6), 950–965. https://doi.org/10.1080/02602938.2015.1052775

Castagno, A. E., Ingram, J, C., Camplain, R., & Blackhorse, D. (2022). "We constantly have to navigate": Indigenous students' and professionals' strategies for navigating ethical conflicts in STEMM. *Cultural Studies of Science Education, 17,* 683–700.

Cech, E. A. (2013). The (mis) framing of social justice: why ideologies of depoliticization and meritocracy hinder engineers' ability to think about social injustices. In J. Lucena (Ed.), *Engineering education for social justice: Critical explorations and opportunities* (pp. 67–84).

Davies, P., Walker, A.E. & Grimshaw, J.M. (2010). A systematic review of the use of theory in the design of guideline dissemination and implementation strategies and interpretation of the results of rigorous evaluations. *Implementation Science* **5**, 1–6. https://doi.org/10.1186/1748-5908-5-14

Engineers Geoscientists Manitoba. (2018). *Code of ethics.* https://www.enggeomb.ca/pdf/CodeOfEthics.pdf

Gwynne-Evans, A. J., Chetty, M., & Junaid, S. (2021). Repositioning ethics at the heart of engineering graduate attributes. *Australasian Journal of Engineering Education, 26*(1), 7–12. https://doi.org/10.1080/22054952.2021.1913882

Hamad, J. A., Hasanain, M., Abdulwahed, M., & Al-Ammari, R. (2013, October). Ethics in engineering education: A literature review. In *2013 IEEE Frontiers in Education Conference (FIE)* (pp. 1554–1560). IEEE.

Junaid, S., Gwynne-Evans, A., Kovacs, H., Lönngren, J., Jiménez Mejía, J. F., Natsume, K., Polmear, M., Serreau, Y., Shaw, C., Toboşaru, M., and Martin, D. A. (2022). What is the role of ethics in accreditation documentation from a global view? *Proceedings - SEFI 50th Annual Conference* (pp. 369-378). European Society for Engineering Education (SEFI). https://upcommons.upc.edu/bitstream/handle/2117/383000/p369-p378.pdf?sequence=1&isAllowed=y

Harris, C. E. Jr., Davis, M., Pritchard, M. S., & Rabins, M. J. (1996). Engineering ethics: What? Why? How? And when? *Journal of Engineering Education, 85,* 93–96. https://doi-org.uml.idm.oclc.org/10.1002/j.2168-9830.1996.tb00216.x

Martin, D. A., Bombaerts, G., & Johri, A. (2021). Ethics is a disempowered subject in the engineering curriculum. In H.-U. Heiss, H.-M. Jarvinen, A. Mayer, & A. Schulz (Eds.), *Proceedings - SEFI 49th Annual Conference: Blended Learning in Engineering Education: challenging, enlightening – and lasting?* (pp. 357–365). European Society for Engineering Education (SEFI). https://www.sefi.be/wp-content/uploads/2021/12/SEFI49thProceedings-final.pdf

Martin, D. A., & Polmear, M. (2023). The two cultures of engineering education: Looking back and moving forward. In S. Hyldgaard Christensen, A. Buch, E. Conlon, C. Didier, C. Mitcham, & M. Murphy (Eds.), *Engineering, social science, and the humanities: Has their conversation come of age?* (pp. 133–150). (Philosophy of Engineering and Technology; Vol. 42). Springer Nature. https://doi.org/10.1007/978-3-031-11601-8_7

Martin, T., & Hoffman, S. M. (Eds.). (2008). *Power struggles: Hydroelectric development and First Nations in Manitoba and Quebec.* University of Manitoba Press.

Martinho, A., Herber, N., Kroesen, M., & Chorus, C. (2021). Ethical issues in focus by the autonomous vehicles industry. *Transport Reviews, 41*(5), 556–577. https://doi.org/10.1080/01441647.2020.1862355

Noble, S. U. (2018). *Algorithms of oppression: How search engines reinforce racism.* NYU Press.

Retool The Ritual. (2023). *The iron ring ritual.* https://www.retoolthering.ca/iron-ring-ritual

Vaughan, D. (1996). *The challenger launch decision: Risky technology, culture, and deviance at NASA*. The University of Chicago Press.

Völker, T., Strand, R., & Slaattelid, R. (2023). Transformative translations? Challenges and tensions in territorial innovation governance. NOvation - Critical Studies of Innovation, 5, 56–85. https://doi.org/10.5380/nocsi.v0i5.93603https://www.ssoar.info/ssoar/handle/document/92598

Wysocka, K., Jungnickel, C., & Szelągowska-Rudzka, K. (2022). Internationalization and quality assurance in higher education. *Management, 26*(1), 204–230. https://doi.org/10.2478/manment-2019-0091

Zoltowski, C., Brightman, A., Buzzanell, P., Eddinton, S., Corple, D., Matters, M., & Booth-Womack, V. (2020). *Using design to understand diversity and inclusion within the context of professional formation of engineers*. Paper presented the American Society for Engineering Education Virtual Conference, Virtual, June 22–26, 2020.

INDEX

For Product Safety Concerns and Information please contact our EU representative GPSR@taylorandfrancis.com Taylor & Francis Verlag GmbH, Kaufingerstraße 24, 80331 München, Germany

Batch number: 10399613

Printed by Printforce, the Netherlands